Conversion Factors
U.S.–British Units to SI Units

To convert from	To	Multiply by
(Area)		
ft^2	m^2	9.2903×10^{-2}
in.^{2}	m^2	6.4516×10^{-4}
(Density)		
lb/in.$^{3.}$	kg/m^3	2.7680×10^4
lb/ft^3	kg/m^3	16.018
(Energy)		
Btu	J	1.0544×10^3
Btu/lb	J/kg	2.3244×10^3
Btu/lb · °F	J/kg · °K	4.1840×10^3
(Force)		
lbf	N	4.4482
(Length)		
ft	m	0.30480
in.	m	2.5400×10^{-2}
(Mass)		
lb	kg	0.45359
(Power)		
Hp (550 ft · lbf/sec)	J/sec (W)	7.4570×10^2
(Pressure)		
in. of water (60°F)	Pa	2.4884×10^2
atm	Pa	1.0133×10^5
psi	Pa	6.8948×10^3
torr (mmHg, 0°C)	Pa	1.3332×10^2
(Temperature)		
°C	°K	°C + 273.15
°F	°K	(°F + 459.67)/1.8
°R	°K	°R/1.8
(Velocity)		
ft/min	m/sec	5.0800×10^{-3}
ft/sec	m/sec	0.30480
in./min	m/sec	4.2333×10^{-4}
in./sec	m/sec	2.5400×10^{-2}
(Volume)		
ft^3	m^3	2.8317×10^{-2}
in.^{3}	m^3	1.6387×10^{-5}
gal	m^3	3.7854×10^{-3}
bbl (42 gal)	m^3	0.15899

Up She Goes!

A 350-ton deisobutanizer distillation column, 212 feet high, was raised into position in one piece at the El Segundo refinery of Standard Oil Co. of California, Western Operations, Inc. The lift was one of the heaviest ever accomplished in the U.S. with a load of this type. Macco Refinery and Chemical Division, California, was the prime contractor for construction. [*Petroleum Refiner*, **37**, No. 2, 184 (1958)]. Column shown was designed by one of the authors.

Equilibrium-Stage Separation Operations in Chemical Engineering

Ernest J. Henley

Professor of Chemical Engineering
University of Houston

and

J. D. Seader

Professor of Chemical Engineering
University of Utah

JOHN WILEY & SONS

New York · Chichester · Brisbane · Toronto · Singapore

Library of Congress Cataloging in Publication Data

Henley, Ernest J
    Equilibrium-stage separation operations in
chemical engineering.

    Includes index.
    1. Separation (Technology)   2. Chemical
equilibrium.   I. Seader, J. D., joint author.
II. Title.
TP156.S45H43   660.2'842   80-13293
ISBN 0-471-37108-4

Printed in the United States of America

20 19 18 17

The literature abounds with information on all phases of distillation calculations and design. There has been such a bewildering flow of information, dealing especially with the principles of stage calculations, that the engineer who is not a distillation expert finds himself at a loss as to how to select the best procedures for solving his distillation problems.

James R. Fair and
William L. Bolles, 1968

PREFACE

No other area of chemical engineering has changed so dramatically in the past decade as that of design procedures for separation operations based on the equilibrium-stage concept. Ten years ago, design of fractionators, absorbers, strippers, and extractors was often done by approximate calculation procedures; and reboiled absorbers and extractive distillation columns were often "guesstimated" from experience and pilot plant data. Today, accurate thermodynamics packages coupled with sufficiently rigorous computational algorithms enable engineers to solve rapidly on time-shared computer terminals, without leaving their desks, what were once considered perversely difficult problems. Commercially available computer programs for stagewise computations are now so robust and reliable that one can say of them, as was once said of the army, that they were organized by geniuses to be run by idiots.

One of the premises of this book is that what was once good for the army is not necessarily good for the engineering profession. The availability of commercial process simulation computing systems such as CONCEPT, DESIGN/2000, FLOWTRAN, GPS-II, and PROCESS has, in many instances, reduced the engineer to the status of an army private. Most often, his undergraduate training did not cover the modern algorithms used in these systems, the User's Manual contains only vague or unobtainable references to the exact computational techniques employed, and the Systems Manual may be proprietary, so the design exercise degenerates into what is often a "black-box" operation, the user being left in the dark.

The aim of this book is to bring a little light into the darkness. We made a careful study of all major publicly available computing systems, ran a fairly large number of industrially significant problems, and then used these problems as vehicles to bring the reader to an in-depth understanding of modern calculation procedures. This approach enabled us to trim the book by eliminating those techniques that are not widely used in practice or have little instructional value.

Instead, we include topics dealing with sophisticated and realistic design problems. Indeed, we hope the terms "realistic" and "industrially significant" are the adjectives that reviewers will use to characterize this book. We did our best to "tell it like it is."

The material in the book deals with topics that are generally presented in undergraduate and graduate courses in equilibrium-stage processes, stagewise separation processes, mass transfer operations, separations processes, and/or distillation. Chapter 1 includes a survey of mass transfer separation operations and introduces the equilibrium-stage concept and its use in countercurrent multistage equipment. The student gains some appreciation of the mechanical details of staged and packed separations equipment in Chapter 2.

The thermodynamics of fluid-phase equilibria, which has such a great influence on equilibrium-stage calculations, is presented in Chapter 3 from an elementary and largely graphical point of view and further developed in Chapters 4 and 5, where useful algebraic multicomponent methods are developed for predicting fluid-phase densities, enthalpies, fugacities, and activity coefficients. Included are substantial treatments of Redlich–Kwong-type equations of state and the local composition concept of Wilson.

The presentation of equilibrium-stage calculation techniques is preceded by Chapter 6, which covers an analysis of the variables and the equations that relate the variables so that a correct problem specification can be made. Single-stage calculation techniques are then developed in Chapter 7, with emphasis on so-called isothermal and adiabatic flashes and their natural extension to multistage cascades.

The graphical calculation procedures based on the use of the McCabe–Thiele diagram, the Ponchon–Savarit diagram, and the triangular diagram for binary or ternary systems are treated in Chapters 8 to 11, including a brief chapter on batch distillation. When applicable, graphical methods are still widely employed and enable the student to readily visualize the manner in which compositional changes occur in multistage separators. The most useful approximate methods for multicomponent distillation, absorption, stripping, and extraction are covered in detail in Chapter 12. Methods like those of Fenske–Underwood–Gilliland, Kremser, and Edmister are of value in early considerations of design and simulation and, when sufficiently understood, can permit one to rapidly extend to other design specifications the results of the rigorous computer methods presented in Chapter 15.

Determination of separator flow capacity and stage efficiency is best done by proprietary data, methods, and experience. However, Chapter 13 briefly presents acceptable methods for rapid preliminary estimates of separator diameter and overall stage efficiency. An introduction to synthesis, the determination of optimal arrangements of separators for processes that must separate a mixture into more than two products, is presented in Chapter 14.

The rigorous multicomponent, multistage methods developed in Chapter 15 are based largely on treating the material balance, energy balance, and phase equilibrium relationships as a set of nonlinear algebraic equations, as first shown by Amundson and Pontinen for distillation and subsequently for a variety of separation operations by Friday and Smith. Emphasis is placed on developing an understanding of the widely used tridiagonal and block-tridiagonal matrix algorithms.

Chapter 16 presents an introduction to the subject of mass transfer as it applies to separation operations with particular emphasis on the design and operation of packed columns. The concluding chapter deals with the important topic of energy conservation in distillation and includes a method for computing thermodynamic efficiency so that alternative separation processes can be compared.

The book is designed to be used by students and practicing engineers in a variety of ways. Some possible chapter combinations are as follows.

1. Short introductory undergraduate course:
   Chapters 1, 2, 3, 8, 10, 11, 16

2. Moderate-length introductory undergraduate course:
   Chapters 1, 2, 3, 6, 7, 8, 9, 10, 11, 12.

3. Comprehensive advanced undergraduate course:
   Chapters 1, 4, 5, 6, 7, 8, 9, 10, 11, 12, 15, 17.

4. Graduate course and self-study by practicing engineers:
   Chapters 4, 5, 6, 7, 12, 13, 14, 15, 17.

Almost every topic in the book is illustrated by a detailed example. A liberal number of problems, arranged in the order of topical coverage, are given at the end of each chapter. Because we are in a transition period, both U.S.–British and SI units are used, often side by side. Useful conversion factors are listed on the inside back cover.

Two appendices are included. In the first, 16 physical property constants and coefficients from the Monsanto FLOWTRAN data bank are given for each of 176 chemicals. Sources of FORTRAN computer programs for performing most of the tedious and, sometimes, complex calculation procedures discussed in the book are listed in Appendix II.

Ernest J. Henley
J. D. Seader

ACKNOWLEDGMENTS

This book started in 1970 as a second edition of *Stagewise Process Design* by E. J. Henley and H. K. Staffin. However, Dr. Staffin, who is now President of Procedyne Corp., New Brunswick, N.J., was unable to continue as a coauthor. Nevertheless, his influence on the present book is greatly appreciated.

The original outline for this book was developed with the assistance of Professor E. C. Roche, Jr., and was reviewed by Professor D. B. Marsland of the North Carolina State University at Raleigh. We are grateful to both of them for their help in the initial planning of the book.

Stimulating conversations, which helped to determine the topical coverage, were held with Professors R. R. Hughes of the University of Wisconsin—Madison, C. J. King of the University of California, Berkeley, R. S. H. Mah of Northwestern University, R. L. Motard of Washington University, D. L. Salt of the University of Utah, W. D. Seider of the University of Pennsylvania, and A. W. Westerberg of Carnegie–Mellon University; and with J. R. Fair and A. C. Pauls of Monsanto Company.

We are indebted to Professor J. R. F. Alonso of the Swinburn College of Technology, Melbourne, Australia, who contributed to Chapters 8 and 16 and who helped reorganize Chapters 1 and 2.

Professors Buford D. Smith of Washington University and Vincent Van Brunt of the University of South Carolina carefully read the draft of the manuscript and offered many invaluable suggestions for improvement. We were fortunate to have had the benefit of their experience and guidance.

A number of students at the University of Houston and the University of Utah, including A. Gallegos, A. Gomez, S. Grover, F. R. Rodrigo, N. Sakakibara, J. Sonntag, L. Subieta, and R. Vazquez, provided valuable assistance in the preparation of examples and problems.

The draft of the manuscript was typed by Vickie Jones, Marilyn Roberts, and Trace Faulkner.

If, in the seemingly endless years that we have been working on this book, we have forgotten major and minor contributors whose assistance is not ac- knowledged here, please forgive us—our strength is ebbing, and our memories are dimming.

We are grateful to John Wiley's patient editorial staff, including Thurman R. Poston, Andy Ford, and Carol Beasley, for their kind forebearance through 10 years of changes of outlines, authorship, and contents and for their polite acceptance of countless apologies and mañanas. Our thanks also to Ann Kearns and Deborah Herbert of the John Wiley production staff for their fine work.

E.J.H.
J.D.S.

CONTENTS

Notation   xv

1. Separation Processes   1
2. Equipment for Multiphase Contacting   47
3. Thermodynamic Equilibrium Diagrams   88
4. Phase Equilibria from Equations of State   140
5. Equilibrium Properties from Activity Coefficient Correlations   183
6. Specification of Design Variables   239
7. Equilibrium Flash Vaporization and Partial Condensation   270
8. Graphical Multistage Calculations by the McCabe–Thiele Method   306
9. Batch Distillation   361
10. Graphical Multistage Calculations by the Ponchon-Savarit Method   372
11. Extraction Calculations by Triangular Diagrams   410
12. Approximate Methods for Multicomponent, Multistage Separations   427
13. Stage Capacity and Efficiency   501
14. Synthesis of Separation Sequences   527
15. Rigorous Methods for Multicomponent, Multistage Separations   556
16. Continuous Differential Contacting Operations: Gas Absorption   624
17. Energy Conservation and Thermodynamic Efficiency   677

Appendices

I. Physical Property Constants and Coefficients   714
II. Sources of Computer Programs   725

Author Index   729

Subject Index   742

xiii

NOTATION

‡Latin Capital and Lowercase Letters

A	absorption factor of a component as defined by (12-48); total cross-sectional area of tray
A, B	parameters in Redlich–Kwong equation as defined in (4-42) to (4-45); parameters in Soave–Redlich–Kwong equation as defined in (4-108) to (4-111)
$A_0, A_1, \ldots, A_{14}$	constants in the Chao–Seader equations (5-12) and (5-13)
*A_1, A_2, A_3	constants in Antoine vapor pressure equation (4-69)
A_a	active cross-sectional area of tray
A_d	cross-sectional area of tray downcomer
A_h	hole area of tray
A_i, B_i, C_i	constants in empirical vapor enthalpy equation given in Example 12.8
A_j, B_j, C_j, D_j	material balance parameters defined by (15-8) to (15-11)
A_{ij}	constant in Soave–Redlich–Kwong equation as defined by (4-113); binary interaction parameter in van Laar equation, (5-26)
A'_{ij}	binary interaction constant in van Laar equation as defined by (5-31)
a	activity of a component in a mixture as defined by (4-18); interfacial area per unit volume
a	parameter defined in (4-103)

* Denotes that constant or coefficient is tabulated for 176 species in Appendix I.
‡ Boldface letters are vectors; boldface letters with overbars are matrices.

a, b	constants in van der Waals equation (4-35); constants in Redlich–Kwong equation, (4-38); constants in Soave–Redlich–Kwong equation, (4-102)
*$a_1, a_2, \ldots, a_5$	constants in (4-59) for the ideal gas heat capacity
a_i, b_i, c_i	constants in empirical liquid enthalpy equation given in Example 12.8
a_{ij}	binary group interaction parameter in UNIFAC equation
B	bottoms product flow rate; flow rate of solvent-free raffinate product; availability function defined by (17-21)
B'	flow rate of raffinate product
b	component flow rate in bottoms product
C	number of components in a mixture; Souders and Brown capacity parameter defined by (13-3); molar concentration
C_1, C_2	integration constants in (4-10) and (4-11), respectively
C_D	drag coefficient in (13-2)
C_F	capacity factor in (13-5) as given by Fig. 13.3; mass transfer parameter in (16-43)
$C_{P_V}^o$	ideal gas heat capacity
D'	distillate flow rate; flow rate of solvent-free extract product
D	flow rate of extract product, density on page 48; distillate flow rate
D_T	column or vessel diameter
D_{AB}	mass diffusivity of A in B
d	component flow rate in distillate
d_p	droplet diameter
E	mass flow rate of extract phase; extraction factor defined by (12-80); phase equilibria function defined by (15-2), (15-59), and (15-74)
E_o	overall plate (stage) efficiency defined by (13-8)
E_L	Murphree plate efficiency based on the liquid phase
E_V	Murphree plate efficiency based on the vapor phase
e	entrainment flow rate
F	feed flow rate; force; general function; packing factor given in Table 16.6
F_b	buoyant force
F_d	drag force

F_F	foaming factor in (13-5)
F_g	gravitational force
F_{HA}	hole area factor in (13-5)
F_{LV}	kinetic energy ratio defined in Fig. 13.3
F_{ST}	surface tension factor in (13-5)
$\mathscr{F}$	variance (degrees of freedom) in Gibbs phase rule
f	fugacity defined by (4-11); function; component flow rate in the feed
f'	derivative of a function
f^o	fugacity of a pure species
f_1, f_2, f_3	property corrections for mass transfer given by (16-47) to (16-49)
G	Gibbs free energy; number of subgroups given by (14-2); gas flow rate
G'	flow rate of inert (carrier) gas
G''	gas mass velocity
G_{ij}	binary interaction parameter in the NRTL equation as defined by (5-59)
g	Gibbs free energy per mole; acceleration due to gravity
$\bar{g}$	partial molal Gibbs free energy
g_c	force–mass conversion factor
g^E	excess Gibbs free energy per mole defined by (5-1)
$\bar{g}^E$	partial molal excess Gibbs free energy defined by (5-2)
g_{ij}	energy of interaction in the NRTL equation as given in (5-60) and (5-61)
H	enthalpy per mole; vapor enthalpy per mole in Ponchon–Savarit method of Chapter 10; vessel height; energy balance function defined by (15-5) and (15-60)
$\bar{H}$	partial molal enthalpy
H'	Henry's law constant $= y_i/x_i$
HETP, HETS	height of packing equivalent to a theoretical (equilibrium) plate (stage)
HTU, H	height of a mass transfer unit defined in Table 16.4
H^E	excess enthalpy per mole
$\bar{H}^E$	partial molal excess enthalpy defined by (5-3)

H_V°	ideal gas enthalpy per mole
h	height of liquid; parameter in Redlich–Kwong equation as defined by (4-46); liquid enthalpy per mole in Ponchon–Savarit method of Chapter 10
J	molar flux relative to stream average velocity
K	vapor–liquid equilibrium ratio (K-value) defined by (1-3); overall mass transfer coefficient
K'	overall mass transfer coefficient for unimolecular diffusion
K_D	liquid–liquid equilibrium ratio (distribution coefficient) defined by (1-6)
K_D'	modified liquid–liquid equilibrium ratio defined by (1-9)
$K_G a$	overall volumetric mass transfer coefficient based on the gas phase
$K_L a$	overall volumetric mass transfer coefficient based on the liquid phase
k	mass transfer coefficient
k'	mass transfer coefficient for unimolecular diffusion
k_h	Henry's law constant defined in Section 3.14
k_{ij}	binary interaction parameter in (4-113)
L	liquid flow rate; liquid flow rate in rectifying section; flow rate of underflow or raffinate phase in extraction
$\bar{L}$	liquid flow rate in stripping section
L'	flow rate of inert (carrier) liquid; liquid flow rate in intermediate section
L''	liquid mass velocity
LW	thermodynamic lost work in (17-22)
L_R	reflux flow rate
L_V	length of vessel
ℓ	constant in UNIQUAC and UNIFAC
ℓ_T	height of packing; component flow rate in liquid stream
$*M$	molecular weight; material balance function as defined by (15-1) and (15-58)
m	parameter in Soave–Redlich–Kwong equation as given below (4-103)
$\mathcal{M}$	number of moles

N	number of equilibrium stages; molar flux relative to a stationary observer
NTU, N	number of transfer units defined in Table 16.4
N_A	number of additional variables
N_a	actual number of trays
N_D	number of independent design variables (degrees of freedom or variance) as given by (6-1)
N_E	number of independent equations or relationships
N_R	number of redundant variables
N_V	number of variables
n	number of moles; number of components
P	pressure; difference point defined by (11-3); number of products
P', P''	difference points in Ponchon–Savarit method of Chapter 10; difference points in triangular-diagram method for extraction in Chapter 11 when both stripping and enriching sections are present
$*P_c$	critical pressure of a species
P_r	reduced pressure $= P/P_c$
P_s	vapor pressure (saturation pressure of a pure species)
$\mathscr{P}$	number of phases present
p	partial pressure [given by (3-2) when Dalton's law applies]; function defined by (15-14)
Q	heat transfer rate
Q_k	area parameter for group k in UNIFAC equation
q	relative surface area of a molecule as used in the UNIQUAC and UNIFAC equations; parameter in McCabe–Thiele method as defined by (8-29); heat transferred per unit flow; function defined by (15-15)
R	reflux ratio $= L/D$; mass flow rate of raffinate phase; number of components; universal gas constant, 1.987 cal/gmole $\cdot$ °K or Btu/lbmole $\cdot$ °R, 8314 J/kgmole $\cdot$ °K or Pa $\cdot$ m³/kgmole $\cdot$ °K, 82.05 atm $\cdot$ cm³/gmole $\cdot$ °K, 0.7302 atm $\cdot$ ft³/lbmole $\cdot$ °R, 10.73 psia $\cdot$ ft³/lbmole $\cdot$ °R
R_k	volume parameter for group k in UNIFAC equation
r	relative number of segments per molecule as used in the UNIQUAC and UNIFAC equations; function defined by (15-18)

S	solids flow rate; sidestream flow rate; entropy; flow rate of solvent in extraction; stripping factor defined by (12-61); number of separation sequences; cross-sectional area
Sc	Schmidt number defined by (16-45)
S_j	bubble-point function defined by (15-21); dimensionless vapor side-stream flow rate $= W_j/V_j$
S_x, S_y	mole fraction summation functions defined by (15-3) and (15-4)
s	entropy per mole
s^E	excess entropy per mole
$\bar{s}^E$	partial molal excess entropy defined by (5-3)
s_j	dimensionless liquid side-stream flow rate $= U_j/L_j$
T	temperature; number of separation methods
T'	temperature in °F
$*T_c$	critical temperature of a species
T_o	datum temperature for enthalpy in (4-60)
T_r	reduced temperature $= T/T_c$
T_{ij}	binary interaction parameter as defined by (5-71) for the UNIQUAC equation and by (5-80) for the UNIFAC equation
t	time; scalar attenuation factor in (15-49), (15-67), and (15-78)
U	superficial velocity; average velocity; reciprocal of extraction factor as defined by (12-81); number of unique splits given by (14-3); liquid side-stream flow rate
U_c	average superficial velocity of the continuous phase in the downward direction in an extractor
U_d	average superficial velocity of the discontinuous (droplet) phase in the upward direction in an extraction
U_f	flooding velocity
u_o	characteristic rise velocity for a single droplet as given by (13-19)
$\bar{u}_c$	average actual velocity of the continuous phase as defined by (13-13)
$\bar{u}_d$	average actual velocity of the discontinuous phase as defined by (13-12)
$\bar{u}_r$	average droplet rise velocity relative to the continuous phase in an extractor
u_{ij}	energy of interaction in the UNIQUAC equation as given in (5-71)

V	vapor flow rate; volume; vapor flow rate in rectifying section; flow rate of overflow or extract phase in extraction; velocity on p. 81
$\bar{V}$	vapor flow rate in stripping section
V_V	vessel volume
v	volume per mole; component flow rate in vapor stream
$\bar{v}$	partial molal volume
W	liquid remaining in still; vapor side-stream flow rate; rate of work
W_s	shaft work
X	mass ratio of components in liquid phase or in raffinate phase; parameter in (12-40); general output variable; group mole fraction in (5-79)
x	mole fraction in liquid phase; mass fraction in liquid phase or in raffinate (underflow) phase
Y	mass ratio of components in vapor or extract phase; parameter in (12-40)
y	mole fraction in vapor phase; mass fraction in vapor phase or in extract (overflow) phase
y^*	vapor mole fraction in equilibrium with liquid composition leaving stage
Z	compressibility factor defined by (4-33); elevation; distance
$\bar{Z}$	lattice coordination number in UNIQUAC and UNIFAC equations
$^*Z_c$	compressibility factor at the critical point
z	mole fraction

Greek Letters

$\alpha_i, \beta_i, \gamma_i, \delta_i$	constants in empirical K-value equation given in Example 12.8
$\alpha_j, \beta_j, \gamma_j,$	energy balance parameters defined by (15-24) to (15-26)
α_{ij}	relative volatility of component i with respect to component j as defined by (1-7); constant in the NRTL equation, (5-29)
β_{ij}	relative selectivity of component i with respect to component j as defined by (1-8)
Γ_k	residual activity coefficient of group k in the actual mixture as given by (5-77)
$\Gamma_k^{(i)}$	residual activity coefficient of group k in a reference mixture containing only molecules of type i

γ	activity coefficient of a component in a mixture as defined by (4-19) and (4-20)
Δ	difference operator; net component flow in (8-3)
$*\delta$	solubility parameter defined by (5-6)
ϵ	convergence tolerance defined by (15-31)
ϵ_1	convergence tolerance for (15-51)
ϵ_2	convergence tolerance for (15-53)
ϵ_3	convergence tolerance for (15-76)
ζ	constant in Winn equation for minimum equilibrium stages as defined by (12-14)
η	constant in (5-52); Murphree tray efficiency defined by (15-73); thermodynamic efficiency defined by (17-24) and (17-25)
Θ	parameter in (12-118)
θ	area fraction defined by (5-70); root of the Underwood equation, (12-34)
θ_m	area fraction of group m defined by (5-78)
Λ_{ij}	binary interaction parameter in the Wilson equation, (5-28)
λ	latent heat of vaporization of a liquid per mole
λ_{ij}	energy of interaction in the Wilson equation
μ	chemical potential defined by (4-4); viscosity
ν^o	pure species fugacity coefficient defined by (4-13)
$\nu^{(o)}$	fugacity coefficient of a simple pure fluid ($\omega = 0$) in (5-11)
$\nu^{(1)}$	correction to the fugacity coefficient in (5-11) to account for the departure of a real pure fluid ($\omega \neq 0$) from a simple fluid ($\omega = 0$)
$\nu_k^{(i)}$	number of groups of kind k in molecule i
ξ	liquid volume constant in (4-79)
ρ	density
σ	surface tension
τ	sum of squares of differences defined by (15-32)
τ_1	sum of differences in (15-51)
τ_2	sum of squares of normalized differences in (15-53)
τ_3	sum of squares of discrepancy functions in (15-75)
τ_{ij}	binary interaction parameter in the NRTL equation, (5-29)

Φ	volume fraction defined by (5-5); parameters in Underwood equations (12-30) and (12-31)
$\bar{\Phi}$	local volume fraction defined by (5-38)
ϕ	fugacity coefficient of a component in a mixture as defined by (4-16) and (4-17); mass transfer parameter in (16-43)
$\phi_A, \phi_{AE}, \phi_{AX}$	fraction of a species not absorbed as given by (12-56) and (12-58) where E denotes enricher and X denotes exhauster in Fig. 12.24
ϕ_E	fraction of species not extracted
$\phi_S, \phi_{SE}, \phi_{SX}$	fraction of species not stripped as given by (12-60) where E denotes enricher and X denotes exhauster in Fig. 12.24
ϕ_U	fraction of species not transferred to the raffinate
ϕ_d	average fractional volumetric holdup of dispersed phase in an extractor
φ	constant in Winn equation for minimum equilibrium stages as defined by (12-14)
Ψ	segment fraction defined by (5-69); association parameter in (16-4)
ψ	mole ratio of vapor to feed; mass transfer parameter in (16-46)
$^*\omega$	acentric factor of Pitzer as defined by (4-68)

Subscripts

B	bottoms; reboiler, bottom stage
BA	bottom stage in absorber
BE	bottom stage in enricher
BX	bottom stage in exhauster
C	Cth component; condenser
D	distillate; condenser; discontinuous phase
F	feed
G	gas
HK	heavy key
HHK	heavier than heavy key
K	key component
L	liquid phase

LK	light key
LLK	lighter than light key
m	mean
R	reboiler; rectifying section
S	side stream; solvent in extract phase; stripping section; top stage in stripper
T	top stage, total
TE	top stage in enricher
TP	triple point
TX	top stage in exhauster
V	vapor phase
W	inerts in raffinate phase; liquid in still
avg	average
b	normal boiling point
c	continuous phase
d	discontinuous phase
e	element; average effective value
eq	equivalent
f	flooding condition
i	particular component; at phase interface
i, j	particular component *i* in a stream leaving stage *j*
irr	irreversible
m	mean (average); two-phase mixture
min	minimum
o	initial condition; infinite reservoir condition
r	reference component
s	saturation condition; shaft
t	total
∞	pinch zone at minimum reflux conditions

Superscripts

C	combinatorial contribution
E	extract phase; excess

I, II	particular liquid phase
R	raffinate phase; residual contribution
T	transpose of a vector
(k)	kth phase; iteration index
o	pure species
∞	at infinite dilution
$'$	pertaining to the stripping section
$*$	at equilibrium

Abbreviations

atm	atmosphere
bbl	barrel
Btu	British thermal unit
°C	degrees Celsius, °K − 273.15
cal	calorie
cfs	cubic feet per second
cm	centimeter
cp	centipoise
cw	cooling water
ESA	energy separating agent
°F	degrees Fahrenheit, °R − 459.67
ft	foot
g	gram
gal	gallon
gmole	gram-mole
gpm	gallons per minute
Hp	horsepower
hr	hour
in.	inch
J	joule
°K	degrees Kelvin
kg	kilogram

kgmole	kilogram-mole
lb	pound of mass
lbf	pound of force
lbmole	pound-mole
m	meter
min	minute
mm	millimeter
MSA	mass separating agent
N	Newton
psi	pounds force per square inch
psia	pounds force per square inch absolute
°R	degrees Rankine
sec	seconds
Stm	steam

Prefixes

M	mega (10^6)
k	kilo (10^3)
m	milli (10^{-3})
μ	micro (10^{-6})

Mathematical Conventions

exp	exponential function
Lim	limiting value
ln	natural logarithm
log	logarithm to the base 10
π	pi (3.1416)
Π	product
Σ	summation
$\leftarrow$	is replaced by
{ }	braces enclose arguments of a function; e.g., $f\{x, y\}$

1
Separation Processes

The countercurrent separation process may be defined as a process in which two phases flow in countercurrent contact to produce a concentration gradient between inlet and outlet. The dominant feature is the concentration gradient; the purpose of the process is to achieve a change in concentration of one or more components of the feed.

Mott Souders, Jr., 1964

The separation of mixtures into essentially pure components is of central importance in the manufacture of chemicals. Most of the equipment in the average chemical plant has the purpose of purifying raw materials, intermediates, and products by the multiphase mass transfer operations described qualitatively in this chapter.

Separation operations are interphase mass transfer processes because they involve the creation, by the addition of heat as in distillation or of a mass separation agent as in absorption or extraction, of a second phase, and the subsequent selective separation of chemical components in what was originally a one-phase mixture by mass transfer to the newly created phase. The thermodynamic basis for the design of equilibrium staged equipment such as distillation and extraction columns are introduced in this chapter. Various flow arrangements for multiphase, staged contactors are considered.

Included also in this chapter is a qualitative description of separations based on intraphase mass transfer (dialysis, permeation, electrodialysis, etc.) and discussions of the physical property criteria on which the choice of separation operations rests, the economic factors pertinent to equipment design, and an introduction to the synthesis of process flowsheets.

1.1   Industrial Chemical Processes

Industrial chemical processes manufacture products that differ in chemical content from process feeds, which are naturally occurring raw materials, plant or animal matter, intermediates, chemicals of commerce, or wastes. Great Canadian Oil Sands, Ltd. (GCOS), in a process shown in Fig. 1.1, produces naphtha, kerosene, gas oil, fuel gas, plant fuel, oil, coke, and sulfur from Canadian Athabasca tar sands, a naturally occurring mixture of sand grains, fine clay, water, and a crude hydrocarbon called *bitumen*.[1] This is one of a growing number of processes designed to produce oil products from feedstocks other than petroleum.

A chemical plant involves different types of operations conducted in either a *batchwise* or a *continuous* manner. These operations may be classified as:

Key operations.
   Chemical reaction.
   Separation of chemicals.
   Separation of phases.

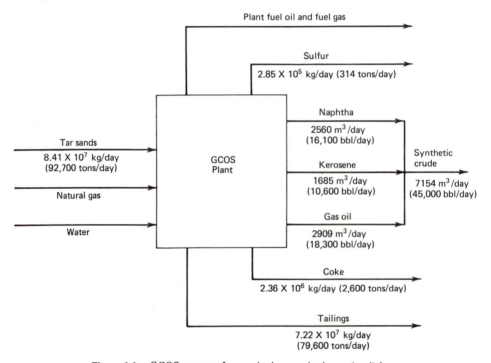

Figure 1.1. GCOS process for producing synthetic crude oil from Canadian Athabasca tar sands.

or

Auxiliary operations.
   Heat addition or removal (to change temperature or phase condition).
   Work addition or removal (to change pressure).
   Mixing or dividing.
   Solids agglomeration by size reduction.
   Solids separation by size.

   Block flow diagrams are frequently used to represent chemical processes. These diagrams indicate by blocks the key processing steps of chemical reaction and separation. Considerably more detail is shown in *process flow diagrams*, which also include the auxiliary operations and utilize symbols that depict the type of equipment employed.

   The block flow diagram of a continuous process for manufacturing anhydrous hydrogen chloride gas from evaporated chlorine and electrolytic hydrogen[2] is shown in Fig. 1.2. The heart of the process is a chemical reactor where the high-temperature combustion reaction $H_2 + Cl_2 \rightarrow 2HCl$ occurs. The only other equipment required consists of pumps and compressors to deliver the feeds to the reactor and the product to storage, and a heat exchanger to cool the product. For this process, no feed purification is necessary, complete conversion of chlorine occurs in the reactor with a slight excess of hydrogen, and the

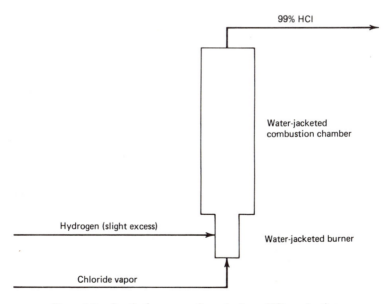

Figure 1.2. Synthetic process for anhydrous HCl production.

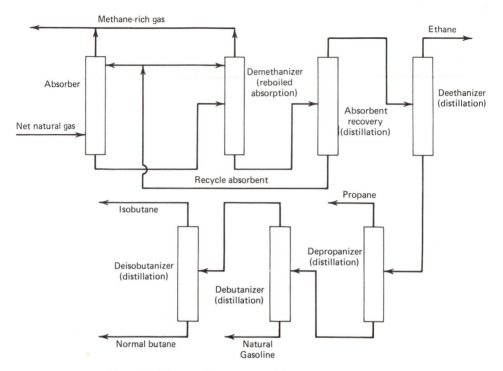

Figure 1.3. Process for recovery of light hydrocarbons from casinghead gas.

product, consisting of 99% hydrogen chloride with small amounts of H_2, N_2, H_2O, CO, and CO_2, requires no purification. Such simple commercial processes that require no equipment for separation of chemical species are very rare.

Some industrial chemical processes involve no chemical reactions but only operations for separating chemicals and phases together with auxiliary equipment. A typical process is shown in Fig. 1.3, where wet natural gas is continuously separated into light paraffin hydrocarbons by a train of separators including an absorber, a reboiled absorber,* and five distillation columns.[3] Although not shown, additional separation operations may be required to dehydrate and sweeten the gas. Also, it is possible to remove nitrogen and helium, if desired.

Most industrial chemical processes involve at least one chemical reactor accompanied by a number of chemical separators. An example is the continuous direct hydration of ethylene to ethyl alcohol.[4] The heart of the process is a

*See Table 1.1.

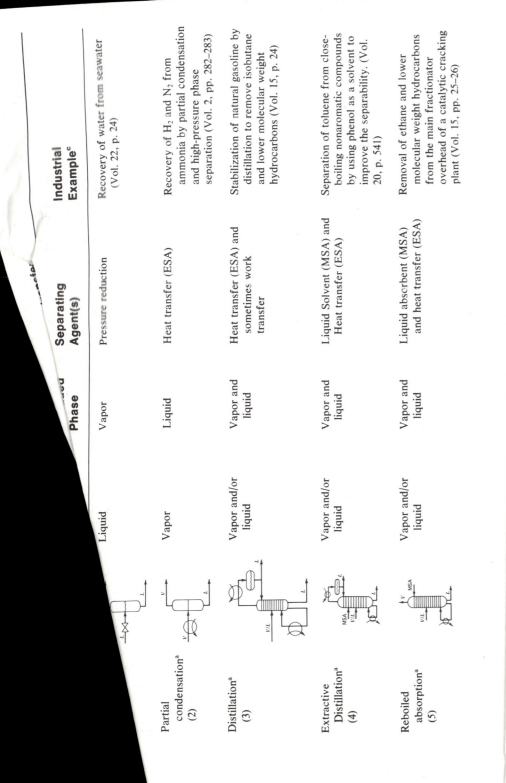

Operation	Created or Added Phase	Separating Agent(s)	Industrial Example[c]
	Vapor / Liquid	Pressure reduction	Recovery of water from seawater (Vol. 22, p. 24)
Partial condensation[a] (2)	Liquid / Vapor	Heat transfer (ESA)	Recovery of H₂ and N₂ from ammonia by partial condensation and high-pressure phase separation (Vol. 2, pp. 282–283)
Distillation[a] (3)	Vapor and liquid / Vapor and/or liquid	Heat transfer (ESA) and sometimes work transfer	Stabilization of natural gasoline by distillation to remove isobutane and lower molecular weight hydrocarbons (Vol. 15, p. 24)
Extractive Distillation[a] (4)	Vapor and liquid / Vapor and/or liquid	Liquid Solvent (MSA) and Heat transfer (ESA)	Separation of toluene from close-boiling nonaromatic compounds by using phenol as a solvent to improve the separability. (Vol. 20, p. 541)
Reboiled absorption[a] (5)	Vapor and liquid / Vapor and/or liquid	Liquid absorbent (MSA) and heat transfer (ESA)	Removal of ethane and lower molecular weight hydrocarbons from the main fractionator overhead of a catalytic cracking plant (Vol. 15, pp. 25–26)

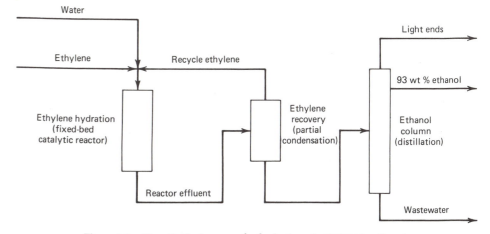

Figure 1.4. Hypothetical process for hydration of ethylene to ethanol.

fixed-bed catalytic reactor operating at 299°C and 6.72 MPa (570°F and 975 psia) in which the reaction $C_2H_4 + H_2O \rightarrow C_2H_5OH$ takes place. Because of thermodynamic equilibrium limitations, the conversion of ethylene is only 5% per pass through the reactor. Accordingly, a large recycle ratio is required to obtain essentially complete overall conversion of the ethylene fed to the process. If pure ethylene were available as a feedstock and no side reactions occurred, the relatively simple process in Fig. 1.4 could be constructed. This process uses a reactor, a partial condenser for ethylene recovery, and distillation to produce aqueous ethyl alcohol of near-azeotropic composition. Unfortunately, as is the prevalent situation in industry, a number of factors combine to greatly increase the complexity of the process, particularly with respect to separation requirements. These factors include impurities in the ethylene feed and side reactions involving both ethylene and feed impurities such as propylene. Consequently, the separation system must also handle diethyl ether, isopropyl alcohol, acetaldehyde, and other products. The resulting industrial process is shown in Fig. 1.5. After the hydration reaction, a partial condenser and water absorber, operating at high pressure, recover ethylene for recycle. Vapor from the low-pressure flash is scrubbed with water to prevent alcohol loss. Crude concentrated ethanol containing diethyl ether and acetaldehyde is distilled overhead in the crude distillation column and catalytically hydrogenated in the vapor phase to convert acetaldehyde to ethanol. Diethyl ether is removed by distillation in the light-ends tower and scrubbed with water. The final product is prepared by distillation in the final purification tower, where 93% aqueous ethanol product is withdrawn several trays below the top tray, light ends are concentrated in the tray section above the product withdrawal tray and recycled to the catalytic hydrogenation

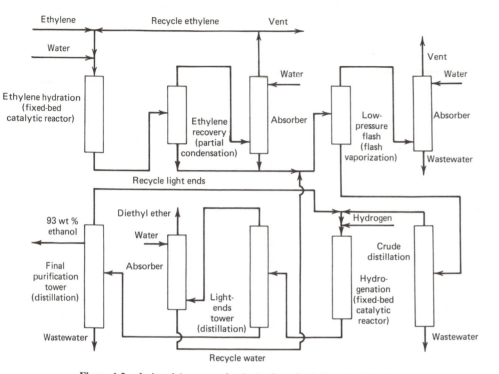

Figure 1.5. Industrial process for hydration of ethylene to ethanol.

reactor, and wastewater is removed from the bottom of the tower. Besides the separators shown, additional separators may be necessary to concentrate the ethylene feed to the process and remove potential catalyst poisons.

The above examples serve to illustrate the great importance of separation operations in the majority of industrial chemical processes. Such operations are employed not only to separate feed mixtures into their constituents, to recover solvents for recycle, and to remove wastes, but also, when used in conjunction with chemical reactors, to purify reactor feed, recover reactant(s) from reactor effluent for recycle, recover by-product(s), and recover and purify product(s) to meet certain product specifications. Sometimes a separation operation, such as SO_2 absorption in limestone slurry, may be accompanied simultaneously by a chemical reaction that serves to facilitate the separation.

1.2 Interphase Mass Transfer Separation Operations

If the mixture to be separated is a homogeneous single-phase solution (gas, liquid, or solid), a second phase must generally be developed before separation of

chemical species can be achieved economically. This second phase can be created by an *energy separating agent* (ESA) or by a *mass separating agent* (MSA) such as a solvent or absorbent. In some separations, both types of agents may be employed.

Application of an ESA involves heat transfer and/or work transfer to or from the mixture to be separated. Alternatively, a second phase may be created by reducing the pressure.

An MSA may be partially immiscible with one or more of the species in the mixture. In this case, the MSA is frequently the constituent of highest concentration in the second phase. Alternatively, the MSA may be completely miscible with the mixture but may selectively alter species volatilities to facilitate a more complete separation between certain species when used in conjunction with an ESA, as in extractive distillation.

In order to achieve a separation of chemical species, a potential must exist for the different species to partition between the two phases to different degrees. This potential is governed by equilibrium thermodynamics, and the rate of approach to the equilibrium composition is controlled by interphase mass transfer. By intimately mixing the two phases, we enhance mass transfer and the maximum degree of partitioning is more quickly approached. After sufficient phase contact, the separation operation is completed by using gravity and/or a mechanical technique to disengage the two phases.

Table 1.1 is a list of the commonly used continuous separation operations based on interphase mass transfer. Symbols for the operations suitable for process flow diagrams are included in the table. Entering vapor, liquid and/or solid phases are designated by V, L, and S. Design procedures have become fairly well standardized for those operations marked by the superscript letter a in Table 1.1. These are batch distillation, respectively, and they are treated in considerable detail in subsequent chapters of this book. Batchwise versions of these operations are commonly used.

When the mixture to be separated includes species that differ widely in their tendency to vaporize and condense, *flash vaporization* and *partial condensation* operations, (1) and (2) in Table 1.1, may be adequate to achieve the desired separation. In the former operation, liquid feed is partially vaporized by reducing the pressure (e.g., with a valve), while in the latter operation, vapor is partially condensed by removing heat. In both operations, after sufficient phase contact, interphase mass transfer has occurred, the resulting vapor phase is enriched with respect to the species that are most volatile, while the liquid phase is enriched with respect to the least volatile species. After sufficient phase contact, the phases, which are of different density, are separated.

Often, the degree of species separation achieved by a single flash or partial condensation is inadequate because the volatility differences are not sufficiently large. In that case, it may

Separation operation	Symbol	Initial or feed phase	Developed or added phase	Separating agent(s)	Industrial example
Absorption[a] (6)		Vapor	Liquid	Liquid absorbent (MSA)	Separation of carbon dioxide from combustion products by absorption with aqueous solutions of an ethanolamine (Vol. 4, p. 358, 362)
Stripping[a] (7)		Liquid	Vapor	Stripping vapor (MSA)	Stream stripping of naphtha, kerosene, and gas oil side cuts from a crude distillation unit to remove light ends (Vol. 15, pp. 17–18)
Refluxed stripping (steam distillation)[a] (8)		Vapor and/or liquid	Vapor and liquid	Stripping vapor (MSA) and heat transfer (ESA)	Distillation of reduced crude oil under vacuum using steam as a stripping agent (Vol. 15, p. 55)
Reboiled stripping[a] (9)		Liquid	Vapor	Heat transfer (ESA)	Removal of light ends from a naphtha cut (Vol. 15, p. 19)
Azeotropic distillation (10)		Vapor and/or liquid	Vapor and liquid	Liquid entrainer (MSA); heat transfer (ESA)	Separation of acetic acid from water using n-butyl acetate as an entrainer to form an azeotrope with water (Vol. 2, p. 851)
Liquid–liquid extraction[a] (11)		Liquid	Liquid	Liquid solvent (MSA)	Use of propane as a solvent to deasphalt a reduced crude oil (Vol. 2, p. 770)

Table 1.1 (Cont.)

Unit Operation	Symbol[b]	Initial or Feed Phase	Developed or Added Phase	Separating Agent(s)	Industrial Example[c]
Liquid-liquid extraction (two-solvent) (12)		Liquid	Liquid	Two liquid solvents (MSA$_1$ and MSA$_2$)	Use of propane and cresylic acid as solvents to separate paraffins from aromatics and naphthenes (Vol. 15, pp. 57–58)
Drying (13)		Liquid and often solid	Vapor	Gas (MSA) and/or heat transfer (ESA)	Removal of water from polyvinyl-chloride with hot air in a rotary dryer (Vol. 21, pp. 375–376)
Evaporation (14)		Liquid	Vapor	Heat transfer (ESA)	Evaporation of water from a solution of urea and water (Vol. 21, p. 51)
Crystallization (15)		Liquid	Solid (and vapor)	Heat transfer (ESA)	Crystallization of p-xylene from a mixture with m-xylene (Vol. 22, pp. 487–492)

					Industrial example[c]
Desublimation (16)		Vapor	Solid	Heat Transfer (ESA)	Recovery of phthalic anhydride from gas containing N_2, O_2, CO_2, CO, H_2O, and other organic compounds by condensation to the solid state (Vol. 15, p. 451)
Leaching (17)		Solid	Liquid	Liquid solvent (MSA)	Aqueous leaching of slime to recover copper sulfate (Vol. 6, p. 167)
Adsorption (18)		Vapor or liquid	Solid	Solid adsorbent (MSA)	Removal of water from air by adsorption on activated alumina (Vol. 1, p. 460)

[a] Design procedures are fairly well standardized.
[b] Trays are shown for columns, but alternatively packing can be used. Multiple feeds and side streams are often used and may be added to the symbol (see example in Fig. 1.7).
[c] Citations refer to volume and page(s) of *Kirk-Othmer Encyclopedia of Chemical Technology* 2nd ed., John Wiley and Sons, New York, 1963-1969.

chemical separation without introducing an MSA, by employing *distillation* (3), the most widely utilized industrial separation method. Distillation involves multiple contacts between liquid and vapor phases. Each contact consists of mixing the two phases for partitioning of species, followed by a phase separation. The contacts are often made on horizontal trays (usually referred to as *stages*) arranged in a vertical column as shown schematically in the symbol for distillation in Table 1.1.* Vapor, while proceeding to the top of the column, is increasingly enriched with respect to the more volatile species. Correspondingly, liquid, while flowing to the bottom of the column, is increasingly enriched with respect to the less volatile species. Feed to the distillation column enters on a tray somewhere between the top tray and the bottom tray; the portion of the column above the feed is the *enriching section* and that below is the *stripping section.* Feed vapor passes up the column; feed liquid passes down. Liquid is required for making contacts with vapor above the feed tray and vapor is required for making contacts with liquid below the feed tray. Often, vapor from the top of the column is condensed to provide contacting liquid, called *reflux.* Similarly, liquid at the bottom of the column passes through a reboiler to provide contacting vapor, called *boilup.*

When volatility differences between species to be separated are so small as to necessitate very large numbers of trays in a distillation operation, *extractive distillation* (4) may be considered. Here, an MSA is used to increase volatility differences between selected species of the feed and, thereby, reduce the number of required trays to a reasonable value. Generally, the MSA is less volatile than any species in the feed mixture and is introduced near the top of the column. Reflux to the top tray is also utilized to minimize MSA content in the top product.

If condensation of vapor leaving the top of a distillation column is not readily accomplished, a liquid MSA called an *absorbent* may be introduced to the top tray in place of reflux. The resulting operation is called *reboiled absorption* (or fractionating absorption) (5). If the feed is all vapor and the stripping section of the column is not needed to achieve the desired separation, the operation is referred to as *absorption* (6). This procedure may not require an ESA and is frequently conducted at ambient temperature and high pressure. Constituents of the vapor feed dissolve in the absorbent to varying extents depending on their solubilities. Vaporization of a small fraction of the absorbent also generally occurs.

The inverse of absorption is *stripping* (7). Here, a liquid mixture is separated, generally at elevated temperature and ambient pressure, by contacting liquid feed with an MSA called a *stripping vapor.* The MSA eliminates the need

---

* The internal construction of distillation, absorption, and extraction equipment is described in Chapter 2.

to reboil the liquid at the bottom of the column, which is important if the liquid is not thermally stable. If contacting trays are also needed above the feed tray in order to achieve the desired separation, a *refluxed stripper* (8) may be employed. If the bottoms product from a stripper is thermally stable, it may be reboiled without using an MSA. In that case, the column is called a *reboiled stripper* (9).

The formation of minimum-boiling mixtures makes *azeotropic distillation* a useful tool in those cases where separation by fractional distillation is not feasible. In the example cited for separation operation (10) in Table 1.1, *n*-butyl acetate, which forms a heterogeneous minimum-boiling azeotrope with water, is used to facilitate the separation of acetic acid from water. The azeotrope is taken overhead, the acetate and water layers are decanted, and the MSA is recirculated.

Liquid–Liquid Extraction (11) and (12) using one or two solvents is a widely used separation technique and takes so many different forms in industrial practice that its description will be covered in detail in later Chapters.

Since many chemicals are processed wet and sold dry, one of the more common manufacturing steps is a *drying* operation (13) which involves removal of a liquid from a solid by vaporization of the liquid. Although the only basic requirement in drying is that the vapor pressure of the liquid to be evaporated be higher than its partial pressure in the gas stream, the design and operation of dryers represents a complex problem in heat transfer, fluid flow, and mass transfer. In addition to the effect of such external conditions as temperature, humidity, air flow, and state of subdivision on drying rate, the effect of internal conditions of liquid diffusion, capillary flow, equilibrium moisture content, and heat sensitivity must be considered.

Although drying is a multiphase mass transfer process, equipment design procedures differ from those of any of the other processes discussed in this chapter because the thermodynamic concepts of equilibrium are difficult to apply to typical drying situations, where concentration of vapor in the gas is so far from saturation and concentration gradients in the solid are such that mass transfer driving forces are undefined. Also, heat transfer rather than mass transfer may well be the limiting rate process. The typical dryer design procedure is for the process engineer to send a few tons of representative, wet, sample material for pilot plant tests by one or two reliable dryer manufacturers and to purchase the equipment that produces a satisfactorily dried product at the lowest cost.

Evaporation (14) is generally defined as the transfer of a liquid into a gas by volatilization caused by heat transfer. Humidification and evaporation are synonymous in the scientific sense; however, usage of the word *humidification* or *dehumidification* implies that one is intentionally adding or removing vapor to or from a gas.

The major application of evaporation is *humidification*, the conditioning of air and cooling of water. Annual sales of water cooling towers alone exceed $200 million. Design procedures similar to those used in absorption and distillation can be applied.

Crystallization (15) is a unit operation carried out in many organic and almost all inorganic chemical manufacturing plants where the product is sold as a finely divided solid. Since crystallization is essentially a purification step, the conditions in the crystallizer must be such that impurities remain in solution while the desired product precipitates. There is a great deal of art in adjusting the temperature and level of agitation in a crystallizer in such a way that proper particle sizes and purities are achieved.

Sublimation is the transfer of a substance from the solid to the gaseous state without formation of an intermediate liquid phase, usually at a relatively high vacuum. Major applications have been in the removal of a volatile component from an essentially nonvolatile one: separation of sulfur from impurities, purification of benzoic acid, and freeze drying of foods, for example. The reverse process, *desublimation* (16), is also practiced, for example in the recovery of phthalic anhydride from reactor effluent. The most common application of sublimation in everyday life is the use of dry ice as a refrigerant for storing ice cream, vegetables and other perishables. The sublimed gas, unlike water, does not puddle and spoil the frozen materials.

Solid–liquid extraction is widely used in the metallurgical, natural product, and food industries. *Leaching* (17) is done under batch, semibatch, or continuous operating conditions in stagewise or continuous-contact equipment. The major problem in leaching is to promote diffusion of the solute out of the solid and into the liquid. The most effective way of doing this is to reduce the solid to the smallest size feasible. For large-scale applications, in the metallurgical industries in particular, large, open tanks are used in countercurrent operation. The major difference between solid–liquid and liquid–liquid systems centers about the difficulty of transporting the solid, or the solid slurry, from stage to stage. For this reason, the solid is often left in the same tank and only the liquid is transferred from tank to tank. In the pharmaceutical, food, and natural product industries, countercurrent solid transport is often provided by fairly complicated mechanical devices. Pictures and descriptions of commercial machinery can be found in Perry's handbook.[5]

Until very recently, the use of *adsorption* systems (18) was generally limited to the removal of components present only in low concentrations. Recent progress in materials and engineering techniques has greatly extended the applications, as attested by Table 1.2, which lists only those applications that have been commercialized. Adsorbents used in effecting these separations are activated carbon, aluminum oxide, silica gel, and synthetic sodium or calcium aluminosilicate zeolite adsorbents (molecular sieves). The sieves differ from the

Table 1.2   Important commercial adsorptive separations

Dehydration Processes		Miscellaneous Separations and Purifications	
Gases	**Liquids**	**Material Adsorbed**	**Material Treated**
Acetylene	Acetone	Acetylene	Liquid oxygen
Air	Acetonitrile	Ammonia	Cracked ammonia
Argon	Acrylonitrile	Ammonia	Reformer hydrogen
Carbon dioxide	Allyl chloride	2-Butene	Isoprene
Chlorine	Benzene	Carbon dioxide	Ethylene
Cracked gas	Butadiene	Carbon dioxide	Air
Ethylene	n-Butane	Carbon dioxide	Inert gases
Helium	Butene	Carbon monoxide, methane	Hydrogen
Hydrogen	Butyl acetate		
Hydrogen chloride	Carbon tetrachloride	Compressor oil	Many kinds of gases
Hydrogen sulfide	Cyclohexane	Cyclic hydrocarbons	Naphthenes and paraffins
Natural gas	Dichloroethylene	Ethanol	Diethyl ether
Nitrogen	Dimethyl sulfoxide	Gasoline components	Natural gas
Oxygen	Ethanol	Hydrogen sulfide	Liquefied petroleum gas
Reformer hydrogen	Ethylene dibromide	Hydrogen sulfide	Natural gas
Sulfur hexafluoride	Ethylene dichloride	Hydrogen sulfide	Reformer hydrogen
	No. 2 fuel oil	Krypton	Hydrogen
	n-Heptane	Mercaptans	Propane
	n-Hexane	Methanol	Diethyl ether
	Isoprene	Methylene chloride	Refrigerant 114
	Isopropanol	Nitrogen	Hydrogen
	Jet fuel	NO, NO_2, N_2O	Nitrogen
	Liquefied petroleum gas	Oil vapor	Compressed gases
	Methyl chloride	Oxygen	Argon
	Mixed ethyl ketone	Unsaturates	Diethyl ether
	Others	Color, odor, and taste formers	Vegetable and animal oils, sugar syrups, water, and so on
		Vitamins	Fermentation mixes
		Turbidity formers	Beer, wines

Source. E. J. Henley and H. K. Staffin, *Stagewise Process Design*, John Wiley & Sons, Inc., New York, 1963, 50.

other adsorbents in that they are crystalline and have pore openings of fixed dimensions.

Adsorption units range from the very simple to the very complex. A simple device consists of little more than a cylindrical vessel packed with adsorbent through which the gas or liquid flows. Regeneration is accomplished by passing a hot gas through the adsorbent, usually in the opposite direction. Normally two or more vessels are used, one vessel desorbing while the other(s) adsorb(s). If the vessel is arranged vertically, it is usually advantageous to employ downward flow to prevent bed lift, which causes particle attrition and a resulting increase in pressure drop and loss of material. However, for liquid flow, better distribution is achieved by upward flow. Although regeneration is usually accomplished by thermal cycle, other methods such as pressure cycles (desorption by decompression), purge-gas cycles (desorption by partial pressure lowering), and displacement cycles (addition of a third component) are also used.

In contrast to most separation operations, which predate recorded history, the principles of *ion exchange* were not known until the 1800s. Today ion exchange is a major industrial operation, largely because of its wide-scale use in water softening. Numerous other ion-exchange processes are also in use. A few of these are listed in Table 1.3.

Ion exchange resembles gas adsorption and liquid–liquid extraction in that, in all these processes, an inert carrier is employed and the reagent used to remove a component selectively must be regenerated. In a typical ion-exchange application, water softening, an organic or inorganic polymer in its sodium form removes calcium ions by exchanging calcium for sodium. After prolonged use

Table 1.3   Applications of ion exchange

Process	Material Exchanged	Purpose
Water treatment	Calcium ions	Removal
Water dealkylization	Dicarbonate	Removal
Aluminum anodization bath	Aluminum	Removal
Plating baths	Metals	Recovery
Rayon wastes	Copper	Recovery
Glycerine	Sodium chloride	Removal
Wood pulping	Sulfate liquor	Recovery
Formaldehyde manufacture	Formic acid	Recovery
Ethylene glycol (from oxide)	Glycol	Catalysis
Sugar solution	Ash	Removal
Grapefruit processing	Pectin	Recovery
Decontamination	Isotopes	Removal

Source. E. J. Henley and H. K. Staffin. *Stagewise Process Design*, John Wiley & Sons, Inc., New York, 1963, 59.

the (spent) polymer, which is now saturated with calcium, is regenerated by contact with a concentrated brine, the law of mass action governing the degree of regeneration. Among the many factors entering into the design of industrial exchangers are the problems of:

1. Channeling. The problem of nonuniform flow distribution and subsequent bypass is generic to all flow operations.

2. Loss of resins. Ultimately, the exchange capacity of the resin will diminish to the point where it is no longer effective. In a system where resin is recirculated, loss by attrition is superimposed on the other losses, which also include cracking of the resin by osmotic pressure.

3. Resin utilization. This is the ratio of the quantity of ions removed during treatment to the total capacity of the resin; it must be maximized.

4. Pressure drop. Because ion exchange is rapid, the limiting rate step is often diffusion into the resin. To overcome this diffusional resistance, resin size must be reduced and liquid flow rate increased. Both of these measures result in an increased bed pressure drop and increased pumping costs.

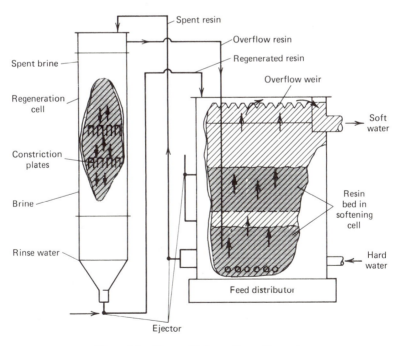

Figure 1.6. Dorrco Hydro-softener for water.

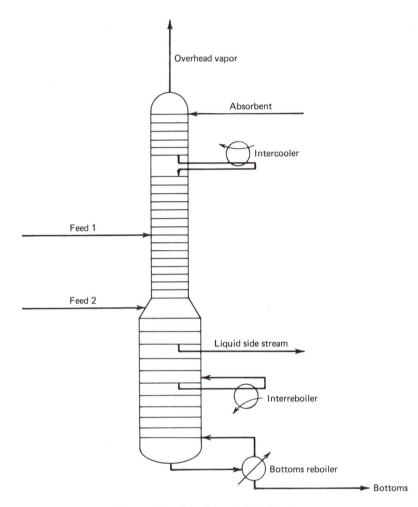

Figure 1.7. Complex reboiled absorber.

Methods of operation used in ion exchange reflect efforts to overcome design problems. Ion-exchange units are built to operate batchwise, where a fixed amount of resin and liquid are mixed together, as fixed beds, where the solution is continuously pumped through a bed of resin, or as continuous countercurrent contactors. In general, fixed beds are preferred where high purities and recoveries are desired; batch processes are advantageous where very favorable equilibrium exists or slurries must be handled; continuous countercurrent operation offers more effective utilization of regeneration chemicals and geometric compactness.

One of the more interesting methods for continuous countercurrent ion exchange is the use of fluidized bed techniques for continuous circulation of the resin. Figure 1.6 shows the Dorrco Hydro-softener. In the fluidized bed, a solid phase is suspended in a liquid or gas. Consequently, the solid behaves like a fluid and can be pumped, gravity fed, and handled very much like a liquid. The fluidized resin moves down through the softener on the right and is then picked up by a brine-carrier fluid and transferred to the regenerator on the left.

Each equipment symbol shown in Table 1.1 corresponds to the simplest configuration for the represented operation. More complex versions are possible and frequently desirable. For example, a more complex version of the reboiled absorber, item (5) in Table 1.1, is shown in Fig. 1.7. This reboiled absorber has two feeds, an intercooler, a side stream, and both an interreboiler and a bottoms reboiler. Acceptable design procedures must handle such complex situations.

1.3 Intraphase Mass Transfer Separation Operations

Changing environmental and energy constraints, a source of despair and frustration to those who are most comfortable with the status quo, are an opportunity and a challenge to chemical engineers, who, by nature of their training and orientation, are accustomed to technological change. Distillation and extraction are highly energy-intensive operations, the latter also requiring elaborate solvent recovery or cleanup and disposal procedures where environmental standards are enforced. New technologies, responsive to changing social needs and economic conditions, are emerging in the chemical industry and elsewhere. The appliance industry, for example, is turning to dry, electrostatic coating methods to avoid paint solvent recovery and pollution problems; the chlorine-caustic industry is developing electrolytic membranes to solve its mercury disposal woes; fresh-water-from-seawater processes based on membrane and freezing principles rather than evaporation suddenly appear to be considerably more competitive. Similarly, liquid-phase separation of aromatic from paraffinic hydrocarbons by adsorption as an alternative to extraction or extractive distillation; or gas-phase separation of low-molecular-weight hydrocarbons by adsorption as an alternative to low-temperature distillation; or alcohol dehydration by membrane permeation instead of distillation—are all processes where technical feasibility has been demonstrated, and large-scale adoption awaits favorable economics.

The separation operations described so far involve the creation or removal of a phase by the introduction of an ESA or MSA. The emphasis on new, less energy- or material-intensive processes is spurring research on new processes to effect separations of chemical species contained in a single fluid phase without the energy-intensive step of creating or introducing a new phase. Methods of accomplishing these separations are based on the application of barriers or fields

Table 1.4 Separation operations based on intraphase mass transfer

Unit Operation	Phase State	Method of Separation	Separating Agent	Industrial Example[a]
Pressure diffusion (1)	Gas	Pressure gradient induced by centrifugal force	Centrifugal force	Separation of isotopic mixtures, Vol. 7, p. 149
Gaseous diffusion (2)	Gas	Forced flow through porous barriers	Porous barrier	Separation of isotopes of uranium, Vol. 7, p. 110
Reverse osmosis (3)	Liquid	Pressure gradient to overcome osmotic pressure	Membrane	Desalination of water, Vol. 14, p. 347
Permeation (4)	Gas or liquid	Forced flow through semipermeable membrane	Membrane	Dehydration of Isopropanol, Vol. 9, p. 284
Dialysis (5)	Liquid	Difference in diffusion rate across membrane	Membrane	Recovery of purified caustic soda from rayon process liquid, Vol. 7, p. 14
Foam fractionation (6)	Liquid	Selective concentration of species at interface	Foam interface	Enzyme and dye separations, Vol. 9, p. 896
Chromatographic separations (7)	Gas or liquid	Selective concentration in and on solids	Solids	Mixed vapor solvent recovery, Vol. 5, p. 413
Zone melting (8)	Solid	Liquid zone travels through metal ingot	Temperature gradient	Germanium purification, Vol. 22, p. 680
Thermal diffusion (9)	Gas or liquid	Temperature-induced concentration gradient	Temperature gradient	Separation of gaseous isotopic mixtures, Vol. 7, p. 138
Electrolysis (10)	Liquid	Electric field plus membranes	Electric field and membranes	Separation of hydrogen and deuterium, Vol.6, p. 895
Electrodialysis (11)	Liquid	Electric field plus charged membranes	Membrane and electric field	Desalination of water, Vol. 7, p. 857

[a] Citations refer to H. F. Mark, J. J. McKetta, and D. F. Othmer, Eds., *Kirk-Othmer Encyclopedia of Chemical Technology*, 2nd ed., John Wiley & Sons, New York, 1969.

to cause species to diffuse at different velocities. Table 1.4 summarizes a number of these operations.

An important example of *pressure diffusion* (1) is the current worldwide competition to perfect a low-cost gas centrifuge capable of industrial-scale separation of the ^{235}U, and ^{238}U, gaseous hexafluoride isotopes. To date, atomic bomb and atomic power self-sufficiency has been limited to major powers since the underdeveloped countries have neither the money nor an industrial base to build the huge multi–billion dollar *gaseous diffusion* (2) plants currently required for uranium enrichment. In these plants, typified by the U.S. Oak Ridge Operation, ^{235}UF$_6$ and ^{238}UF$_6$ are separated by forcing a gaseous mixture of the two species to diffuse through massive banks of porous fluorocarbon barriers across which a pressure gradient is established.

Membrane separations involve the selective solubility in a thin polymeric membrane of a component in a mixture and/or the selective diffusion of that component through the membrane. In *reverse osmosis* (3) applications, which entail recovery of a solvent from dissolved solutes such as in desalination of brackish or polluted water, pressures sufficient to overcome both osmotic pressure and pressure drop through the membrane must be applied. In *permeation* (4), osmotic pressure effects are negligible and the upstream side of the membrane can be a gas or liquid mixture. Sometimes a phase transition is involved as in the process for dehydration of isopropanol shown in Fig. 1.8. In addition, polymeric liquid surfactant and immobilized-solvent membranes have been used.

Dialysis (5) as a unit operation considerably antedates gas and liquid permeation. Membrane dialysis was used by Graham in 1861 to separate colloids from crystalloids. The first large industrial dialyzers, for the recovery of caustic from rayon steep liquor, were installed in the United States in the 1930s. Industrial dialysis units for recovery of spent acid from metallurgical liquors have been widely used since 1958. In dialysis, bulk flow of solvent is prevented by balancing the osmotic pressure, and low-molecular-weight solutes are recovered by preferential diffusion across thin membranes having pores of the order of 10^{-6} cm. Frequently diffusion is enhanced by application of electric fields.

In adsorptive bubble separation methods, surface active material collects at solution interfaces and, thus, a concentration gradient between a solute in the bulk and in the surface layer is established. If the (very thin) surface layer can be collected, partial solute removal from the solution will have been achieved. The major application of this phenomenon is in ore flotation processes where solid particles migrate to and attach themselves to rising gas bubbles and literally float out of the solution. This is essentially a three-phase system.

Foam fractionation (6), a two-phase adsorptive bubble separation method, is a process where natural or chelate-induced surface activity causes a solute to migrate to rising bubbles and thus be removed as a foam. Two government-

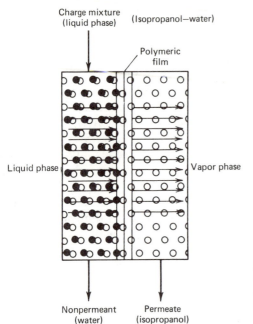

Charge mixture
(liquid phase) (Isopropanol—water)

Polymeric
film

Liquid phase

Vapor phase

Nonpermeant
(water)

Permeate
(isopropanol)

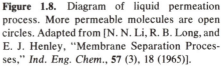

Figure 1.8. Diagram of liquid permeation process. More permeable molecules are open circles. Adapted from [N. N. Li, R. B. Long, and E. J. Henley, "Membrane Separation Processes," *Ind. Eng. Chem.*, **57** (3), 18 (1965)].

funded pilot plants have been constructed: one for removing surface-active chelated radioactive metals from solution; the other for removing detergents from sewage. The enrichment is small in concentrated solutions, and in dilute solutions it is difficult to maintain the foam. A schematic of the process is shown in Fig. 1.9.

When *chromatographic separations* (7) are operated in a batch mode, a portion of the mixture to be separated is introduced at the column inlet. A solute-free carrier fluid is then fed continually through the column, the solutes separating into bands or zones. Some industrial operations such as mixed-vapor solvent recovery and sorption of the less volatile hydrocarbons in natural gas or natural gasoline plants are being carried out on pilot plant and semiworks scales. Continuous countercurrent systems designed along the basic principles of distillation columns have been constructed.

Zone melting (8) relies on selective distribution of impurity solutes between a liquid and solid phase to achieve a separation. Literally hundreds of metals have been refined by this technique, which, in its simplest form, involves nothing more than moving a molten zone slowly through an ingot by moving the heater or drawing the input past the heater, as in Fig. 1.10.

If a temperature gradient is applied to a homogeneous solution, concentration gradients can be established and *thermal diffusion* (9) is induced. Al-

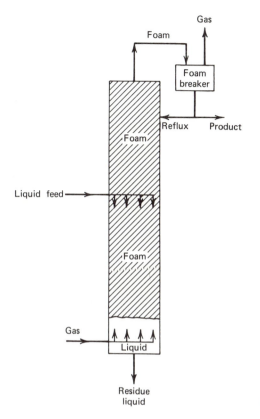

Figure 1.9.   Foam fractionation column.

though there are no large-scale commercial applications of this technique, it has been used to enhance the separation in uranium isotope gaseous diffusion cascades.

Natural water contains 0.000149 atom fraction of deuterium. When water is decomposed by *electrolysis* (10)* into hydrogen and oxygen, the deuterium concentration in the hydrogen produced at the cathode is lower than that in the water remaining in the air. Until 1953 when the Savannah River Plant was built, this process was the only commercial source of heavy water.

The principle of operation of a multicompartmented *electrodialysis* unit (11) is shown in Fig. 1.11. The cation and anion permeable membranes carry a fixed charge; thus they prevent the migration of species of like charge. In a commercial version of Fig. 1.11, there would be several hundred rather than three compartments, multicompartmentalization being required to achieve electric power economics, since electrochemical reactions take place at the electrodes.

---

* Excluded from this discussion are electrolysis and chemical reactions at the anode and cathode.

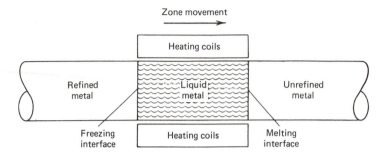

Figure 1.10. Diagram of zone refining.

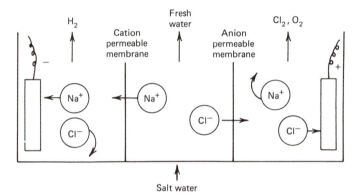

Figure 1.11. Principle of electrodialysis.

1.4 The Equilibrium-Stage Concept

The intraphase mass transfer operations of Table 1.4 are inherently nonequilibrium operations. Thus the maximum attainable degree of separation cannot be predicted from thermodynamic properties of the species. For the interphase operations in Table 1.1, however, the phases are brought into contact in *stages*. If sufficient stage contact time is allowed, the chemical species become distributed among the phases in accordance with thermodynamic equilibrium considerations. Upon subsequent separation of the phases, a single *equilibrium contact* is said to have been achieved.

Industrial equipment does not always consist of stages (e.g., trays in a column) that represent *equilibrium stages*. Often only a fraction of the change from initial conditions to the equilibrium state is achieved in one contact.

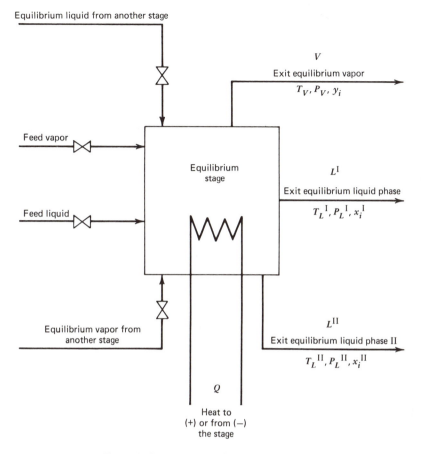

Figure 1.12.   Representative equilibrium stage.

Nevertheless, the concept of the equilibrium stage has proved to be extremely useful and is widely applied in design procedures which calculate the number of *equilibrium* (or so-called *theoretical*) *stages* required for a desired separation. When coupled with a *stage efficiency* based on mass transfer rates, the number of equilibrium stages can be used to determine the number of *actual stages* required.

A representative equilibrium stage is shown schematically in Fig. 1.12. Only four incoming streams and three outgoing equilibrium streams are shown, but the following treatment is readily extended to any number of incoming or outgoing streams. Any number of chemical species may exist in the incoming streams, but no chemical reactions occur. Heat may be transferred to or from the stage to regulate the stage temperature, and the incoming streams may be throttled by valves to regulate stage pressure.

All exit phases are assumed to be at thermal and mechanical equilibrium; that is

$$T_V = T_L^I = T_L^{II} \tag{1-1}$$

$$P_V = P_L^I = P_L^{II} \tag{1-2}$$

where T = temperature, P = pressure, and V and L refer, respectively, to the vapor phase and the liquid phase, the two liquid phases being denoted by superscripts I and II. The exit phase compositions are in equilibrium and thus related through thermodynamic equilibrium constants.

For vapor–liquid equilibrium, a so-called K-value (or vapor–liquid equilibrium ratio) is defined for each species i by

$$K_i \equiv \frac{y_i}{x_i} \tag{1-3}$$

where y = mole fraction in the vapor phase and x = mole fraction in the liquid phase. In Fig. 1.12, two K-values exist, one for each liquid phase in equilibrium with the single vapor phase. Thus

$$K_i^I = \frac{y_i}{x_i^I} \tag{1-4}$$

$$K_i^{II} = \frac{y_i}{x_i^{II}} \tag{1-5}$$

For liquid–liquid equilibrium, a *distribution coefficient* (or liquid–liquid equilibrium ratio), K_D is defined for each species by

$$K_{D_i} \equiv \frac{x_i^I}{x_i^{II}} \tag{1-6}$$

As discussed in Chapters 4 and 5, these equilibrium ratios are functions of temperature, pressure, and phase compositions.

When only a small number of species is transferred between phases, equilibrium phase compositions can be represented graphically on two dimensional diagrams. Design procedures can be applied directly to such diagrams by graphical constructions without using equilibrium ratios. Otherwise, design procedures are formulated in terms of equilibrium ratios, which are commonly expressed by analytical equations suitable for digital computers.

For vapor–liquid separation operations, an index of the relative separability of two chemical species i and j is given by the *relative volatility* α defined as the ratio of their K-values

$$\alpha_{ij} \equiv \frac{K_i}{K_j} \tag{1-7}$$

The number of theoretical stages required to separate two species to a desired degree is strongly dependent on the value of this index. The greater the departure of the relative volatility from a value of one, the fewer the equilibrium stages required for a desired degree of separation.

For liquid–liquid separation operations, a similar index called the *relative selectivity* β may be defined as the ratio of distribution coefficients

$$\beta_{ij} \equiv \frac{K_{D_i}}{K_{D_j}} \tag{1-8}$$

Although both α_{ij} and β_{ij} vary with temperature, pressure, and phase compositions, many useful approximate design procedures assume these indices to be constant for a given separation problem.

1.5 Multiple-Stage Arrangements

When more than one stage is required for a desired separation of species, different arrangements of multiple stages are possible. If only two phases are involved, the *countercurrent flow* arrangement symbolized for many of the operations in Table 1.1 is generally preferred over *cocurrent flow*, *crosscurrent flow*, or other arrangements, because countercurrent flow usually results in a higher degree of separation efficiency for a given number of stages. This is illustrated by the following example involving liquid–liquid extraction.

Example 1.1. Ethylene glycol can be catalytically dehydrated to p-dioxane (a cyclic diether) by the reaction $2HOCH_2CH_2HO \rightarrow H_2CCH_2OCH_2CH_2O + 2H_2O$. Water and p-dioxane have boiling points of 100°C and 101.1°C, respectively, at 1 atm and cannot be separated by distillation. However, liquid–liquid extraction at 25°C (298.15°K) using benzene as a solvent is reasonably effective. Assume that 4,536 kg/hr (10,000 lb/hr) of a 25% solution of p-dioxane in water is to be separated continuously by using 6,804 kg/hr (15,000 lb/hr) of pure benzene. Assuming benzene and water are mutually insoluble, determine the effect of the number and arrangement of stages on the percent extraction of p-dioxane. The flowsheet is shown in Fig. 1.13.

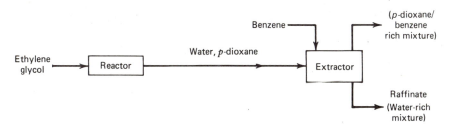

Figure 1.13. Flowsheet for Example 1.1.

Solution. Because water and benzene are mutually insoluble, it is more convenient to define the distribution coefficient for *p*-dioxane in terms of mass ratios, X_i^I = mass dioxane/mass benzene and X_i^{II} = mass dioxane/mass water, instead of using mole fraction ratios as in (1-6).

$$K'_{D_i} \equiv X_i^I/X_i^{II} \tag{1-9}$$

Since *p*-dioxane is the only species transferring between the two phases, the subscript *i* will be dropped. Also, let I = *B* for the benzene phase and let II = *W* for the water phase. From the equilibrium data of Berndt and Lynch,[6] K'_D varies from approximately 1.0 to 1.4 over the concentration range of interest. For the purposes of this example, we assume a constant value of 1.2.

(a) **Single equilibrium stage.** For a single equilibrium stage, as shown in Fig. 1.14, a mass balance on dioxane gives

$$WX_o^W + BX_o^B = WX_1^W + BX_1^B \tag{1-10}$$

where *W* and *B* are, respectively, the mass flow rates of benzene and water. Assuming the exit streams are at equilibrium

$$K'_D = X_1^B/X_1^W \tag{1-11}$$

Combining (1-10) and (1-11) to eliminate X_1^B and solving for X_1^W, we find that the mass ratio of *p*-dioxane to water in the exit water phase is given by

$$X_1^W = \frac{X_o^W + (B/W)X_o^B}{1 + E} \tag{1-12}$$

where *E* is the extraction factor BK'_D/W. The percent extraction of *p*-dioxane is $100(X_o^W - X_1^W)/X_o^W$.

For this example, $X_o^W = 2{,}500/7{,}500 = 1/3$, $X_o^B = 0$, and $E = (15{,}000)(1.2)/7{,}500 = 2.4$. From (1-12), $X_1^W = 0.0980$ kg *p*-dioxane/kg water, and the percent extraction is 70.60. From (1-11), $X_1^B = 0.1176$.

To what extent can the extraction of *p*-dioxane be increased by adding additional stages in cocurrent, crosscurrent, and countercurrent arrangements?

Two-stage cases are shown in Fig. 1.14. With suitable notation, (1-12) can be applied to any one of the stages. In general

$$X_{out}^W = \frac{X_{in}^W + (S/W)X_{in}^B}{1 + (SK'_D/W)} \tag{1-13}$$

where *S* is the weight of benzene *B* entering the stage about which the mass balance is written.

(b) **Cocurrent arrangement.** If a second equilibrium stage is added in a cocurrent arrangement, the computation of the first stage remains as in Part (a). Referring to Fig. 1.14, we find that computation of the second stage is based on $X_1^B = 0.1176$, $X_1^W = 0.0980$, $S/W = B/W = 2$, and $SK'_D/W = BK'_D/W = 2.4$. By (1-13), $X_2^B = 0.1176$. But this value is identical to X_1^B. Thus, no additional extraction of *p*-dioxane occurs in the second stage. Furthermore, regardless of the number of cocurrent equilibrium stages, the percent extraction of *p*-dioxane remains at 70.60%, the value for a single equilibrium stage.

(c) **Crosscurrent arrangement.** Equal Amounts of Solvent to Each Stage. In the crosscurrent flow arrangement, the entire water phase progresses through the stages. The total benzene feed, however, is divided, with equal portions sent to each stage

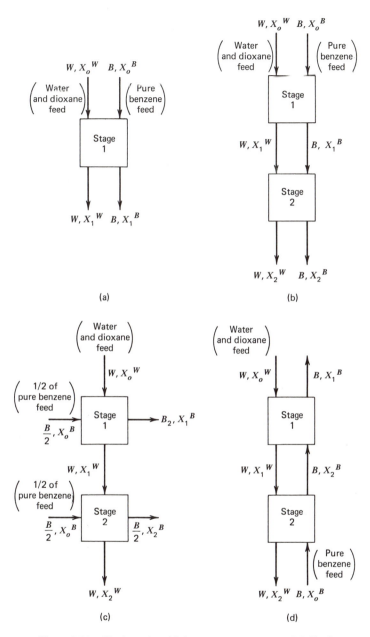

Figure 1.14.   Single and multiple-stage arrangements. (*a*) Single-stage arrangement. (*b*)-(*d*) Two-stage arrangements. (*b*) Cocurrent. (*c*) Crosscurrent. (*d*) Countercurrent.

as shown in Fig. 1.14. Thus, for each stage, $S = B/2 = 7{,}500$ in (1-13). For the first stage, with $X_o^W = 1/3$, $X_o^B = 0$, $S/W = 1$, and $SK_D'/W = 1.2$, (1-13) gives $X_1^W = 0.1515$. For the second stage, X_2^W is computed to be 0.0689. Thus, the overall percent extraction of p-dioxane is 79.34. In general, for N crosscurrent stages with the total solvent feed equally divided among the stages, successive combinations of (1-13) with all $X_{in}^B = 0$ lead to the equation

$$X_N^W = \frac{X_o^W}{(1 + E/N)^N} \qquad (1\text{-}14)$$

where E/N is the effective extraction factor for each crosscurrent stage. The overall percent extraction of p-dioxane is $100(W_o^W - X_N^W)/X_o^W$. Values of the percent extraction

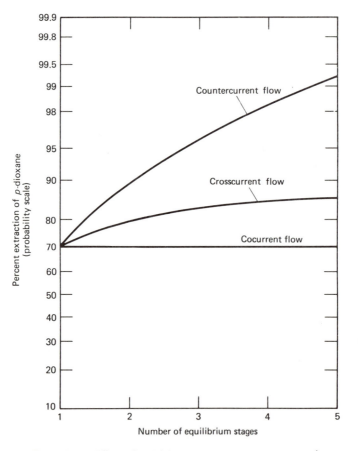

Figure 1.15. Effect of multiple-stage arrangement on extraction efficiency.

for values of X_N^W computed from (1-14) are plotted in Fig. 1.15 for up to 5 equilibrium stages. The limit of (1-14) as $N \to \infty$ is

$$X_\infty^W = \frac{X_o^W}{\exp(E)} \tag{1-15}$$

With $E = 2.4$, X_∞^W is computed to be 0.0302, which corresponds to an overall 90.93% extraction of p-dioxane from water to benzene. A higher degree of extraction than this could be achieved only by increasing the benzene flow rate.

(d) **Countercurrent arrangement.** In the countercurrent flow arrangement shown in Fig. 1.14, dioxane–water feed enters the first stage, while the entire benzene solvent enters the final stage. The phases pass from stage to stage counter to each other. With this arrangement, it is not possible to apply (1-13) directly to one stage at a time. For example, to calculate X_1^W, we require the value for X_2^B, but X_2^B is not initially known. This difficulty is circumvented by combining the following stage equations to eliminate X_1^W and X_2^B.

$$X_1^W = \frac{X_o^W + (B/W)X_2^B}{1 + E} \tag{1-16}$$

$$X_2^W = \frac{X_1^W + (B/W)X_0^B}{1 + E} \tag{1-17}$$

$$X_2^B = X_2^W K_D' \tag{1-18}$$

Solving for X_2^W gives the following relation for the two countercurrent equilibrium stages with $X_o^B = 0$:

$$X_2^W = \frac{X_o^W}{1 + E + E^2} \tag{1-19}$$

With $X_o^W = 1/3$ and $E = 2.4$, (1-19) gives $X_2^W = 0.0364$, which corresponds to 89.08% overall extraction of p-dioxane. In general, for N countercurrent equilibrium stages, similar combinations of stage equations with $X_o^B = 0$ lead to the equation

$$X_N^W = \frac{X_o^W}{\sum_{n=0}^{N} E^n} \tag{1-20}$$

Values of overall percent extraction are plotted in Fig. 1.15 for up to five equilibrium stages, at which better than 99% extraction of p-dioxane is achieved. The limit of (1-20) as $N \to \infty$ depends on the value of E as follows.

$$X_\infty^W = 0, \ 1 \le E \le \infty \tag{1-21}$$

$$X_\infty^W = (1 - E)X_o^W, \ E \le 1 \tag{1-22}$$

For this example, with $E = 2.4$, an infinite number of equilibrium stages in a countercurrent flow arrangement can achieve complete removal of p-dioxane. This can not be done in a crosscurrent flow arrangement. For very small values of E, (1-15) gives values of X_∞^W close to those computed from (1-22). In this case, countercurrent flow may not be significantly more efficient than crosscurrent flow. However, in practical applications, solvents and solvent rates are ordinarily selected to give an extraction factor greater than 1. Then, as summarized in Fig. 1.15, the utilization of a countercurrent flow arrangement is distinctly advantageous and can result in a high degree of separation.

1.6   Physical Property Criteria for Separator Selection

An industrial *separation problem* may be defined in terms of a process feed and specifications for the desired products. An example adapted from Hendry and Hughes,[7] based on a separation process for a butadiene processing plant, is given in Fig. 1.16.

For separators that produce two products, the minimum number of separators required is equal to one less than the number of products. However, additional separators may be required if mass separating agents are introduced and subsequently removed and/or multicomponent products are formed by blending.

In making a preliminary selection of feasible separator types, we find our experience to indicate that those operations marked by an a in Table 1.1 should be given initial priority unless other separation operations are known to be more attractive. To compare the preferred operations, one will find certain physical properties tabulated in handbooks and other references useful.[8, 9, 10, 11, 12, 13] These properties include those of the pure species—normal boiling point, critical point, liquid density, melting point, and vapor pressure—as well as those involving the species and a solvent or other MSA—liquid diffusivity, gas solubility, and liquid solubility. In addition, data on thermal stability are important if elevated temperatures are anticipated.

As an example, Table 1.5 lists certain physical properties for the compounds in Fig. 1.16. The species are listed in the order of increasing normal boiling point. Because these compounds are essentially nonpolar and are similar in size and

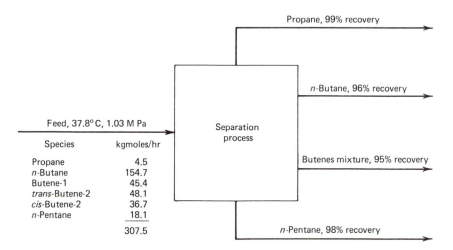

Figure 1.16. Typical separation problem. [Adapted from J. E. Hendry and R. R. Hughes, *Chem. Eng. Progr.*, **68** (6), 71–76 (1972).]

Table 1.5   Certain physical properties of some light hydrocarbons

Species	Normal boiling point, °C	Critical temperature, °C	Critical pressure, MPa	Approximate relative volatility at 0.101 MPa (1 atm)
Propane	−42.1	96.7	4.17	4.4
Butene-1	−6.3	146.4	3.94	1.25
n-Butane	−0.5	152.0	3.73	1.055
trans-Butene-2	0.9	155.4	4.12	1.11
cis-Butene-2	3.7	161.4	4.02	3.2
n-Pentane	36.1	196.3	3.31	

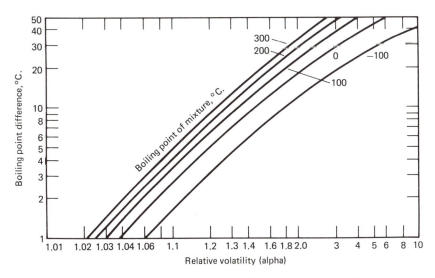

Figure 1.17.   Approximate relative volatility of binary hydrocarbon mixtures at one atmosphere. [Reproduced by permission from F. W. Melpolder and C. E. Headington, *Ind. Eng. Chem.*, **39**, 763–766 (1947).]

shape, the normal boiling points are reasonable indices of differences in volatility. In this case, Fig. 1.17 from Melpolder and Headington[14] can be applied to obtain approximate relative volatilities of the adjacent species in Table 1.5. These values are included in Table 1.5. They indicate that, while propane and *n*-pentane are quite easy to remove from the mixture by ordinary distillation, the separation of *n*-butane from butenes would be extremely difficult by this means. As a further example, the industrial separation of ethyl benzene from *p*-xylene, with a normal boiling-point difference of 2.1°C and a relative volatility of approximately 1.06, is conducted by distillation; but 350 trays contained in three columns are required.[15]

1.7 Other Factors in Separator Selection

When physical property criteria indicate that ordinary distillation will be difficult, other means of separation must be considered. The ultimate choice will be dictated by factors such as:

1. Engineering design and development. If the operation is an established one, equipment of standard mechanical design can be purchased. The difference in cost between a conventional separator using standard auxiliary equipment and an unconventional separator that requires development and testing can be excessive.

2. Fixed investment. Included in fixed investment are fabricated equipment and installation costs. Fabrication costs are highly dependent on geometric complexity, materials of construction, and required operating conditions. The latter factor favors separators operating at ambient conditions. When making comparisons among several separator types, one must include cost of auxiliary equipment (e.g., pumps, compressors, heat exchangers, etc.). An additional separator and its auxiliary equipment will usually be needed to recover an MSA.

3. Operating expense. Different types of separators have different utility, labor, maintenance, depreciation, and quality control costs. If an MSA is required, 100% recovery of this substance will not be possible, and a makeup cost will be incurred. Raw material costs are often the major item in operating expense. Therefore, it is imperative that the separator be capable of operating at the design efficiency.

4. Operability. Although no distinct criteria exist for what constitutes operable or non-operable separator design, the experience and judgment of plant engineers and operators must be given careful consideration. Understandably, resistance is usually great to unusual mechanical design, high-speed rotating equipment, fragile construction, and equipment that may be difficult to maintain. In addition, handling of gaseous and liquid phases is generally favored over solids and slurries.

5. Safety. It is becoming increasingly common to conduct quantitative assessments of process risks by failure modes and effects, fault tree, or other analytical alternatives. Thus, the probability of an accident times the corresponding potential loss is a cost factor which, although probabilistic,

should be considered. Vacuum distillation of highly combustible mixtures, for example, involves a hazard to which a dollar figure should be attached. Such an operation should be avoided if it involves a major risk.

6. Environmental and social factors. Farsighted engineering involves not only meeting current standards but also anticipating new ones. Plants designed for cooling systems based on well water, in areas of high land subsidence, are invitations to economic and social disaster not only for the particular plant but for the fabric of our free enterprise system. The cheapest is not necessarily the best in the long run.

1.8 Synthesis of Separation Sequences

Consider the separation problem of Fig. 1.18, which is adapted from Heaven.[16] Three essentially pure products and one binary product (pentanes) are to be recovered. Table 1.6 is a list of the five species ranked according to increasing normal boiling point. Corresponding approximate relative volatilities at 1 atm between species of adjacent boiling points, as determined from Fig. 1.17, are also included. At least three two-product separators are required to produce the four products. Because none of the relative volatilities are close to one, ordinary distillation is probably the most economical method of making the separations. As shown in the block flow diagrams of Fig. 1.19, five different sequences of three distillation columns each are possible, from which one must be selected.

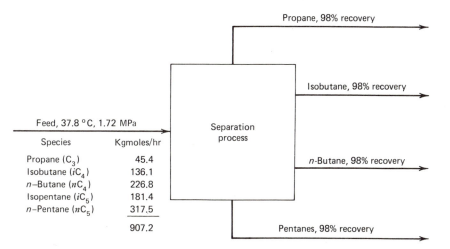

Figure 1.18. Paraffin separation problem. [Adapted from D. L. Heaven, M.S. Thesis (1969).]

Table 1.6 Certain physical properties of some paraffin hydrocarbons

Species	Normal boiling point,°C	Approximate relative volatility at 1 atm (0.103 MPa)
Propane	−42.1	
Isobutane	−11.7	3.6
n-Butane	−0.5	1.5
Isopentane	27.8	2.8
n-Pentane	36.1	

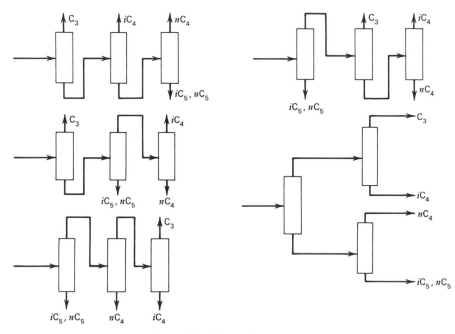

Figure 1.19. Separation sequences.

For the more difficult separation problem of Fig. 1.16, suppose that propane and n-pentane can be recovered by ordinary distillation, but both ordinary distillation and extractive distillation with an MSA of furfural and water must be considered for the other separations. Then, 227 different sequences of separators are possible assuming that, when utilized, an MSA will be immediately recovered by ordinary distillation after the separator in which the MSA is introduced. An industrial sequence for the problem in Fig. 1.16 is shown in Fig. 1.20.

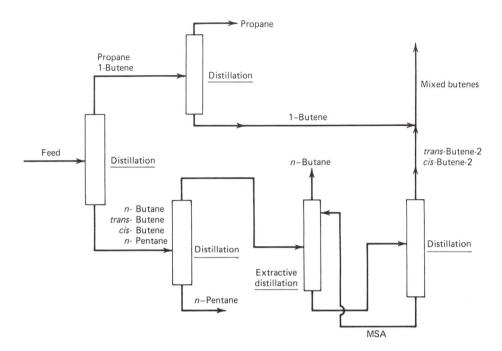

Figure 1.20.   Industrial separation sequence.

It is one of five sequences, generated by a computer program, that are relatively close in cost.

Systematic procedures for synthesis of the most economical separation sequences are available. These do not require the detailed design of all possible configurations and the theoretical framework for these procedures is discussed in Chapter 14.

1.9   Separators Based on Continuous Contacting of Phases

Large industrial chemical separators are mainly collections of trays or discrete stages. In small-size equipment, however, a common arrangement is a vertical column containing a fixed packing that is wetted by a down-flowing liquid phase that continuously contacts the other, up-flowing, phase as shown in Fig. 1.21. As described in Chapter 2, packings have been developed that provide large interfacial areas for efficient contact of two phases.

Equipment utilizing continuous contacting of phases cannot be represented strictly as a collection of equilibrium stages. Instead, design procedures are generally based on mass transfer rates that are integrated over the height of the

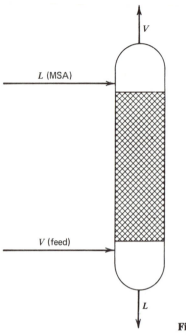

L (MSA)

V (feed)

Figure 1.21. Packed column absorber.

region of phase contacting. Two key assumptions that simplify the design computations are (1) plug flow of each phase (i.e., absence of radial gradients of velocity, temperature, and composition for the bulk flows) and (2) negligible axial diffusion of thermal energy or mass (i.e., predominance of bulk flow as the transport mechanism in the axial direction). In some cases, it is convenient to determine the height equivalent to a theoretical stage. Detailed design procedures for continuous contacting equipment are presented in Chapter 16.

References

1. Fear, J. V. D., and E. D. Innes, "Canada's First Commercial Tar Sand Development," *Proceedings 7th World Petroleum Congress*, Elsevier Publishing Co., Amsterdam, 1967, **3**, 633–650.

2. Maude, A. H., "Anhydrous Hydrogen Chloride Gas," *Trans. AIChE*, 1942, **38**, 865–882.

3. Considine, D. M., Ed., *Chemical and Process Technology Encyclopedia*,

McGraw–Hill Book Co., New York, 1974, 760–763.

4. Carle, T. C., and D. M. Stewart, "Synthetic Ethanol Production," *Chem. Ind.* (London), 830–839 (May 12, 1962).

5. Perry, R. H., and C. H. Chilton, Eds., *Chemical Engineers Handbook*, 5th ed., McGraw–Hill Book Co., New York 1973, Section 19.

6. Berndt, R. J., and C. C. Lynch, "The Ternary System: Dioxane-Benzene-Water," *J. Amer. Chem. Soc.*, 1944, **66**, 282–284.

7. Hendry, J. E., and R. R. Hughes, "Generating Separation Processs Flowsheets," *Chem. Eng. Progr.*, 1972, **68** (6), 71–76.

8. Timmermans, J., *Physical-Chemical Constants of Pure Organic Compounds*, Elsevier Publishing Co., Amsterdam, 1950.

9. *International Critical Tables*, McGraw–Hill Book Co., New York, 1926–1933.

10. *Technical Data Book—Petroleum Refining*, American Petroleum Institute, New York, 1966.

11. Perry, R. H., and C. H. Chilton, Eds., *Chemical Engineers' Handbook*, 5th ed., McGraw–Hill Book Co., New York, 1973, Section 3.

12. Weast, R. C., Ed., *Handbook of Chemistry and Physics*, 54th ed., CRC Press, Cleveland, Ohio, 1973.

13. Reid, R. C., J. M. Prausnitz, and T. K. Sherwood, *The Properties of Gases and Liquids*, 3rd ed., McGraw–Hill Book Co., New York, 1977.

14. Melpolder, F. W., and C. E. Headington, "Calculation of Relative Volatility from Boiling Points," *Ind. Eng. Chem.*, 1947, **39**, 763–766.

15. Chilton, C. H., "Polystyrene via Natural Ethyl Benzene," *Chem. Eng.*, (N.Y.) **65** (24), 98–101 (December 1, 1958).

16. Heaven, D. L., *Optimum Sequencing of Distillation Columns in Multicomponent Fractionation*, M.S. Thesis in Chemical Engineering, University of California, Berkeley, 1969.

Problems

1.1 In *Hydrocarbon Processing*, **54** (11), 97–222 (1975), process flow diagrams and descriptions are given for a large number of petrochemical processes. For each of the following processes, list the separation operations in Table 1.1 that are used.

(a) Acrolein.
(b) Acrylic acid.
(c) Acrylonitrile (Sohio process).
(d) Ammonia (M. W. Kellogg Co.).
(e) Chloromethanes.
(f) Cresol.
(g) Cyclohexane.
(h) Ethanolamines (Scientific Design Co.).
(i) Ethylbenzene (Alkar).
(j) Ethylene (C-E Lummus).
(k) Isoprene.
(l) Polystyrene (Cosden).
(m) Styrene (Union Carbide–Cosden–Badger).
(n) Terephthalic acid purification.
(o) Vinyl acetate (Bayer AG).
(p) *p*-Xylene.

1.2 Select a separation operation from Table 1.1 and study the Industrial Example in the reference cited. Then describe an alternative separation method based on a different operation in Table 1.1 that could be used in place of the one selected and compare it with the one selected on a basis of

(a) Equipment cost.
(b) Energy requirement.
(c) Pollution potential.

(With reference to row 6 in Table 1.1, for example, describe a separation process based on an operation other than absorption in ethanolamine to separate carbon dioxide from combustion products.)

1.3 In the manufacture of synthetic rubber, a low-molecular-weight, waxlike fraction is obtained as a by-product that is formed in solution in the reaction solvent, normal heptane. The by-product has a negligible volatility. Indicate which of the following separation operations would be practical for the recovery of the solvent and why. Indicate why the others are unsuitable.

(a) Distillation.
(b) Evaporation.
(c) Filtration.

1.4 Consider the separation of A from a mixture with B. In Table 1.1, liquid–liquid extraction (11) is illustrated with only one solvent, which might preferentially dissolve B. Under what conditions should the use of two different solvents (12) be considered?

1.5 Figure 1.7 shows a complex reboiled absorber. Give possible reasons in a concise manner for the use of:

(a) Absorbent instead of reflux.
(b) Feed locations at two different stages.
(c) An interreboiler.
(d) An intercooler.

1.6 Discuss the similarities and differences among the following operations listed in Table 1.1: flash vaporization, partial condensation, and evaporation.

1.7 Discuss the similarities and differences among the following operations listed in Table 1.1: distillation, extractive distillation, reboiled absorption, refluxed stripping, and azeotropic distillation.

1.8 Compare the advantages and disadvantages of making separations using an ESA, an MSA, combined ESA and MSA, and pressure reduction.

1.9 An aqueous acetic acid solution containing 6.0 gmoles of acid per liter is to be extracted with chloroform at 25°C to recover the acid from chloroform-insoluble impurities present in the water. The water and chloroform are essentially immiscible:

If 10 liters of solution are to be extracted at 25°C, calculate the percent recovery of acid obtained with 10 liters of chloroform under the following conditions.

(a) Using the entire quantity of solvent in a single batch extraction.
(b) Using three batch extractions with one third of the total solvent used in each batch.
(c) Using three batch extractions with 5 liters of solvent in the first, 3 liters in the second, and 2 liters in the third batch.

Equilibrium data for the system at 25°C are given below in terms of a distribution coefficient K''_D, where $K''_D = A_B/A_C$; A_C = concentration of acetic acid in water, gmoles/liter; and A_B = concentration of acetic acid in chloroform, gmoles/liter.

A_B	K''_D
3.0	2.40
2.5	2.60
2.0	2.80
1.5	3.20
1.0	3.80
0.7	4.45

1.10 (a) When rinsing clothes with a given amount of water, would one find it more efficient to divide the water and rinse several times; or should one use all the water in one rinse? Explain.

(b) Devise a clothes-washing machine that gives the most efficient rinse cycle for a fixed amount of water.

1.11 In the four-vessel CCD process shown below, 100 mg of A and 100 mg of B initially dissolved in 100 ml of aqueous solution are added to 100 ml of organic solvent in Vessel 1; a series of phase equilibrium and transfer steps follows to separate A from B.

(a) What are the equilibrium distribution coefficients for A and B if the organic phase is taken as phase I?

(b) What is the relative selectivity for A with respect to B?

(c) Compare the separation achieved with that obtainable in a single batch equilibrium step.

(d) Why is the process shown not a truly countercurrent operation? Suggest how this process can be made countercurrent. What would be the advantage? [O. Post and L. C. Craig, *Anal. Chem.*, **35**, 641 (1963).]

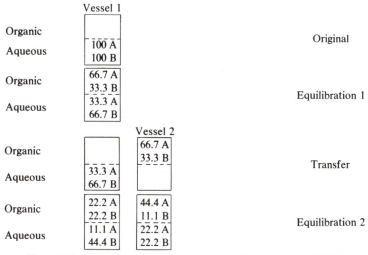

Figure 1.22. Four-vessel countercurrent distribution process (CCD) for separating substances A and B.

Vessel 3

Organic		22.2 A / 22.2 B	44.4 A / 11.1 B	
Aqueous	11.1 A / 44.4 B	22.2 A / 22.2 B		

Transfer

Organic	7.4 A / 14.8 B	29.6 A / 14.8 B	29.6 A / 3.7 B	
Aqueous	3.7 A / 29.6 B	14.8 A / 29.6 B	14.8 A / 7.4 B	

Equilibration 3

Vessel 4

Organic		7.4 A / 14.8 B	29.6 A / 14.8 B	29.6 A / 3.7 B
Aqueous	3.7 A / 29.6 B	14.8 A / 29.6 B	14.8 A / 7.4 B	

Transfer

Organic	2.5 A / 9.9 B	14.8 A / 14.8 B	29.6 A / 7.4 B	19.7 A / 1.2 B
Aqueous	1.2 A / 19.7 B	7.4 A / 29.6 B	14.8 A / 14.8 B	9.9 A / 2.5 B

Equilibration 4

Figure 1.22. Continued.

1.12 A 20 wt % solution of uranyl nitrate (UN) in water is to be treated with TBP to remove 90% of the uranyl nitrate. All operations are to be batchwise equilibrium contacts. Assuming that water and TBP are mutually insoluble, how much TBP is required for 100 g of solution if at equilibrium (g UN/g TBP) = 5.5(g UN/g H_2O) and

(a) All the TBP is used at once?
(b) Half is used in each of two consecutive stages?
(c) Two countercurrent stages are used?
(d) An infinite number of crosscurrent stages is used?
(e) An infinite number of countercurrent stages is used?

1.13 The uranyl nitrate (UN) in 2 kg of a 20 wt % aqueous solution is to be extracted with 500 g of tributyl phosphate. Using the equilibrium data in Problem 1.12, calculate and compare the percentage recoveries for the following alternative procedures.

(a) A single-stage batch extraction.
(b) Three batch extractions with one third of the total solvent used in each batch (the solvent is withdrawn after contacting the entire UN phase).
(c) A two-stage cocurrent extraction.
(d) A three-stage countercurrent extraction.
(e) An infinite-stage countercurrent extraction.
(f) An infinite-stage crosscurrent extraction.

1.14 One thousand kilograms of a 30 wt % dioxane in water solution is to be treated with benzene at 25°C to remove 95% of the dioxane. The benzene is dioxane free, and the equilibrium data of Example 1.1 can be used. Calculate the solvent requirements for:

(a) A single batch extraction.
(b) Two crosscurrent stages using equal amounts of benzene.

(c) Two countercurrent stages.
(d) An infinite number of crosscurrent stages.
(e) An infinite number of countercurrent stages.

1.15 Chloroform is to be used to extract benzoic acid from wastewater effluent. The benzoic acid is present at a concentration of 0.05 gmoles/liter in the effluent, which is discharged at a rate of 1,000 liters/hr. The distribution coefficient for benzoic acid at process conditions is given by

$$C^{I} = K_{D}^{"} C^{II}, \quad \text{where} \quad K_{D}^{"} = 4.2$$

and

C^{I} = molar concentration of solute in solvent

C^{II} = molar concentration of solute in water

Chloroform and water may be assumed immiscible.

If 500 liters/hr of chloroform is to be used, compare the fraction benzoic acid removed in

(a) A single equilibrium contact.
(b) Three crosscurrent contacts with equal portions of chloroform.
(c) Three countercurrent contacts.

1.16 The distribution coefficient of ethanol (E) between water (W) and ester (S) is roughly $2 = (\text{mole}\% \ E \ \text{in} \ S)/(\text{mole}\% \ E \ \text{in} \ W) = x^{I}/x^{II}$ at 20°C. A 10 mole% solution of E in W is to be extracted with S to recover the ethanol. Compare the separations to be obtained in countercurrent, cocurrent, and crosscurrent (with equal amounts of solvent to each stage) contacting arrangements with feed ratios of S to W of 0.5, 5, and 50 for one, two, three, and infinite stages. Assume the water and ester are immiscible.

Repeat the calculations, assuming the equilibrium data are represented by the equation $x^{I} = (2x^{II})/(1 + x^{II})$.

1.17 Prior to liquefaction, air is dried by contacting it with dry silica gel adsorbent. The air entering the dryer with 0.003 kg water/kg dry air must be dried to a minimum water content of 0.0005 kg/kg dry air. Using the equilibrium data below, calculate the kg gel per kg dry air required for the following.

(a) A single-stage batch contactor.
(b) A two-stage countercurrent system.
(c) A two-stage crossflow contactor with equally divided adsorbent flows.

$\dfrac{\text{Kg H}_2\text{O}}{\text{Kg dry air}}$	0.00016	0.0005	0.001	0.0015	0.002	0.0025	0.003
$\dfrac{\text{Kg H}_2\text{O}}{\text{Kg gel}}$	0.012	0.029	0.044	0.060	0.074	0.086	0.092

Data from L.C. Eagleton and H. Bliss, *Chem. Eng. Prog.*, **49**, 543 (1953).

1.18 A new EPA regulation limits H_2S in stack gases to 3.5 g/1000 m^3 at 101.3 kPa and 20°C. A water scrubber is to be designed to treat 1000 m^3 of air per year containing 350 g of H_2S prior to discharge. The equilibrium ratio for H_2S between air and water at 20°C and 101.3 kPa is approximated by $y = 500x$.

Assuming negligible vaporization of water and negligible solubility of air in the water, how many kg of water are required if the scrubber (gas absorber) to be used

(a) Has one equilibrium stage?
(b) Has two countercurrent stages?
(c) Has infinite countercurrent stages?

   If there is also an EPA regulation that prohibits discharge water containing more than 100 ppm of H_2S, how would this impact your design and choice of process?

1.19 Repeat Example 1.1 for a solvent for which $E = 0.90$. Display your results with a plot like Fig. 1.15. Does countercurrent flow still have a marked advantage over crosscurrent flow? It is desirable to choose the solvent and solvent rate so that $E > 1$? Explain.

1.20 Derive Equations 1-14, 1-15, 1-19, and 1-20.

1.21 Using Fig. 1.17 with data from Appendix I, plot $\log \alpha$ (relative volatility) of C_5 to C_{12} normal paraffins as referred to n-C_5 against the carbon number (5 to 12) at a reasonable temperature and pressure. On the same plot show $\log \alpha$ for C_5 to C_{10} aromatics (benzene, toluene, ethylbenzene, etc.), also with reference to n-C_5. What separations are possible by ordinary distillation, assuming a mixture of normal paraffins and aromatic compounds? (E. D. Oliver, *Diffusional Separation Processes: Theory, Design & Evaluation*, J. Wiley & Sons, New York, 1966, Chapter 13.

1.22 Discuss the possible methods of separating mixtures of o-Xylene, m-Xylene, p-Xylene, and ethylbenzene in light of the following physical properties.

	o-Xylene	m-Xylene	p-Xylene	Ethylbenzene
Normal boiling point, °F	291.2	282.7	281.3	277.2
Freezing point, °F	−13.4	54.2	55.9	−138.8
Change in boiling point with change in pressure, °F/mmHg	0.0894	0.0883	0.0885	0.0882
Dipole moment, 10^{-18} esu	0.62	0.36	0.0	
Dielectric constant	2.26	2.24	2.23	2.24
Density at 20°C, g/cm^3	0.8802	0.8642	0.8610	0.8670

1.23 Using the physical property constants from Appendix I and various handbooks, discuss what operations might be used to separate mixtures of

(a) Propane and methane.
(b) Acetylene and ethylene.
(c) Hydrogen and deuterium.
(d) o-Xylene and p-xylene.
(e) Asphalt and nonasphaltic oil.
(f) Calcium and Strontium 90.
(g) Aromatic and Nonaromatic hydrocarbons.
(h) Polystyrene (MW = 10,000) and polystyrene (MW = 100,000).
(i) Water and NaCl.

1.24 It is required to separate the indicated feed into the indicated products. Draw a simple block diagram showing a practical sequence of operations to accomplish the required results. State briefly the reasons for your choice.

Feed	Products
45% Alcohol.	1. 98% Alcohol, 2% B.
45% Organic solvent B.	2. 98% Organic solvent B containing 2% alcohol.
10% Soluble, nonvolatile wax.	3. Soluble wax in solvent B.

Properties

Normal boiling pt. alcohol = 120°C.
Normal boiling pt. organic solvent B = 250°C.
Viscosity of soluble wax at 10% concentration is similar to water.
Viscosity of soluble wax at 50% concentration is similar to heavy motor oil.
The alcohol is significantly soluble in water.
The organic solvent B and water are essentially completely immiscible.
The nonvolatile wax is insoluble in water.

1.25 Apply the Clapeyron vapor pressure equation to derive an algebraic expression for the type of relationship given in Fig. 1.17. Use your expression in conjunction with Trouton's rule to check Fig. 1.17 for a boiling-point difference of 20°C and a mixture boiling point of 100°C.

1.26 Consider the separation of a mixture of propylene, propane, and propadiene. Show by diagrams like those in Fig. 1.19

(a) Two sequences of ordinary distillation columns.
(b) One sequence based on the use of extractive distillation with a polar solvent in addition to ordinary distillation.

1.27 A mixture of 70% benzene, 10% toluene, and 20% ortho-xylene is to be separated into pure components by a sequence of two ordinary distillation columns. Based on each of the following heuristics (rules), make diagrams like those of Fig. 1.19 of the preferred sequence.

(a) Separate the more plentiful components early, if possible.
(b) The most difficult separation is best saved for last.

1.28 Develop a scheme for separating a mixture consisting of 50, 10, 10, and 30 mole % of methane, benzene, toluene, and ortho-xylene, respectively, by distillation using the two heuristics given in Problem 1.27.

1.29 Suggest a flowsheet to separate the reactor effluents from acrylonitrile manufacture $(2C_3H_6 + 3O_2 + 2NH_3 \rightarrow 2C_3H_3N + 6H_2O)$.

The reactor effluent is 40% inert gases, 40% propylene, 8% propane, 6% acrylonitrile, 5% water, and 1% heavy by-product impurities. Design requirements are to

1. Vent inert gases without excessive loss of acrylonitrile or propylene.
2. Recycle the propylene to the reactor without recycling the acrylonitrile.
3. Purge the propane to prevent propane buildup in the reactor.

4. Recover the acrylonitrile for final purification. (D. F. Rudd, G. J. Powers, and J. J. Siirola, *Process Synthesis*, Prentice–Hall Book Co., 1973, p. 176.

1.30 Show as many separation schemes as you can for the problem of Fig. 1.16, assuming that, in the presence of furfural as an MSA, *n*-butane is more volatile than 1-butene and that ordinary distillation cannot be used to separate *n*-butane from *trans*-butene-2. Furfural is less volatile than *n*-pentane. One such scheme is shown in Fig. 1.20.

2

Equipment for Multiphase Contacting

Because distillation is a very important industrial process for separating chemical components by physical means, much effort has been expended to increase the performance of existing distillation equipment and to develop new types of vapor-liquid contacting devices which more closely attain equilibrium (the condition of 100% efficiency) between the distilling phases.

William G. Todd and Matthew Van Winkle, 1972

Given thermodynamic data, efficient algorithms for accurately predicting the stages required for a specific separation are readily available. However, when it comes to the mechanical design of the equipment, to quote D. B. McLaren and J. C. Upchurch:

The alchemist is still with us, although concealed by mountains of computer output. The decisions and actions that lead to trouble free operation depend on the "art" of the practitioner. One dictionary defines art as a "personal, unanalyzable creative power." It is this power, plus experience, that makes the difference between success or failure of process equipment.[1]

Chemical companies usually do not attempt the mechanical design of their own process equipment. Their engineers perform the calculations and conduct the experiments required to fill out a vendor's design data sheet such as shown in Fig. 2.1 for a vapor–liquid contactor. The vendor, who very frequently is not told what chemical is to be processed, submits a quotation and a recommended mechanical design. This must then be evaluated by the chemical companies' engineers because, even though there may be performance guarantees, plant shutdowns for modification and replacement are costly.

Process Design Data Sheet

Item No. or Service				
Tower diameter, I.D.				
Tray spacing, inches				
Total trays in section				
Max. ΔP, mm Hg				
Conditions at Tray No.				
Vapor to tray, °F				
Pressure				
Compressibility				
°Density, lb./cu. ft.				
°Rate, lb./hr.				
cu. ft./sec. (cfs)				
cfs$\sqrt{D_V/(D_L - D_V)}$				
Liquid from tray, °F				
Surface tension				
Viscosity, cp				
°Density, lb./cu. ft.				
°Rate, lb./hr.				
GPM hot liquid				
Foaming tendency	None_____	Moderate_____	High_____	Severe_____

°These values are required in this form for direct computer input.

NOTES:

1. This form may be used for several sections of trays in one tower, for several towers, or for various loading cases. Use additional sheets if necessary.
2. Is maximum capacity at constant vapor-liquid ratio desired?_____
3. Minimum rate as % of design rate:_____%
4. Allowable downcomer velocity (if specified):_____ft/sec
5. Number of flow paths or passes:_____Glitsch Choice;_____
   Bottom tray downcomer: Total draw_____;Other _____
6. Trays numbered: top to bottom_____; bottom to top _____
7. Enclose tray and tower drawings for existing columns.
8. Manhole size, I.D.,_____inches.
9. Manways removable: top_____; bottom_____; top & bottom_____
10. Corrosion allowance: c.s._____; other _____
11. Adjustable weirs required: yes_____no _____
12. Packing material if required_____; not required _____
13. Tray material and thickness _____
14. Valve material _____
15. Ultimate user _____
16. Plant location_____
17. Other_____

Figure 2.1. Typical vendor data sheet for vapor-liquid contactor with trays or packing. (Courtesy of F. Glitsch and Sons.)

Prior to the 1950s, all but the very large chemical and petroleum companies that had research staffs and large test facilities relied on equipment vendors and experience for comparative performance data on phase-contacting equipment. This reliance was considerably reduced by establishment, in 1953, of a nonprofit, cooperative testing and research company, Fractionation Research Inc., which collects and disseminates to member companies performance data on all fractionation devices submitted for testing. These data have taken much of the "art" out of equipment design. Nevertheless, much of the evaluative literature on process equipment performance is fragmentary and highly subjective.

This chapter includes a qualitative description of the equipment used in the most common separation operations—that is, distillation, absorption, and extraction—and discusses design parameters and operating characteristics. More quantitative aspects of designs are given in Chapter 13.

2.1 Design Parameters for Multiphase Mass Transfer Devices

Section 1.4 introduced the concept of an equilibrium stage or tray, wherein *light* and *heavy* phases are contacted, mixed, and then disengaged. The light and heavy phases may be gas and liquid as in distillation and gas absorption, two immiscible liquids in liquid–liquid extraction, or a solid and liquid in leaching. Generally, the phases are in countercurrent flow; and, as was done in Example 1.1, the number of stages, temperature and pressure regimes, and feed flow rate and composition are firmly established prior to the attempt at a mechanical design. The design problem then consists of constructing a device that has the required number of stages and is economical, reliable, safe, and easy to operate. The two classes of mass transfer devices of overwhelming commercial importance are staged contactors and continuous contactors. Figure 2.2a depicts a staged liquid-vapor contactor, where each tray is a distinguishable stage, and the phases are disengaged after each contact. Liquid–vapor mixing on the trays is promoted by dispersing the gas into the liquid, which is the continuous phase. The most commonly used trays have bubble caps or valves to hold the liquid on the tray and to direct the gas flow, or they are sieve trays where the gas is dispersed through small openings, the liquid being held on the tray only by gas pressure drop.

In the continuous contactors of Fig. 2.2b and c, there are no distinct stages, contact is continuous, and phase disengagement is at the terminals of the apparatus. Usually the gas phase is continuous, the purpose of the packing being to promote turbulence by providing a tortuous gas flow path through the liquid, which flows in and around the packing. The packing may be ceramic, metal, or plastic rings or saddles dumped into the tower randomly, or carefully stacked metal meshes, grids, or coils. Packed columns and tray columns can be used also

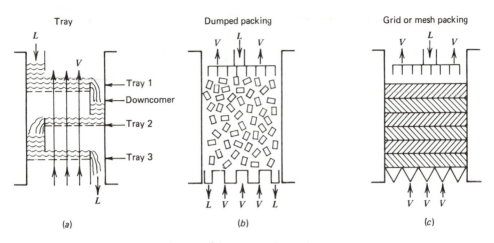

Figure 2.2. Commercial mass transfer devices. (*a*) Tray. (*b*) Dumped packing. (*c*) Grid or mesh packing.

for liquid–liquid separations, but may require additional agitation to be effective.

In the following discussion of design parameters, it should be evident that not all parameters are of equal importance in all operations. Pressure drop, for example, is of central importance in vacuum crude stills but of little import in the liquid–liquid extraction of penicillin from fermentation mashes; stage efficiency is not important in the design of a whiskey distilling column, which requires few stages, but it can be critical for a deisobutanizer, which may require over 100 stages.

Capacity

That a certain number of stages is required to achieve a given separation is dictated by equilibrium relationships. This number is required whether 1 kg or 100 million kg per year of material is to be processed. Although the number of stages would not vary, physical dimensions (particularly the diameter) of the equipment vary directly with throughput, these dimensions being a function of the hydrodynamics and the required contact time in the stage. Assume, for example, that hydrodynamic considerations dictate that the average superficial vapor velocity should be one m/sec. If the vapor density is 2 kg/m³, a stage can handle a mass flux of (1 m/sec) (2 kg/m³) = 2 kg/sec · m², and we have established the diameter of the column as a function of throughput; if 5 kg/sec of vapor is to be processed, the column cross-sectional area needs to be (5 kg/sec)/(2 kg/sec · m²) = 2.5 m².

It is important that a column be designed to handle as wide a range of compositions and vapor and liquid loadings as possible. In the 1950s and 1960s some petroleum companies built refinery units that could operate efficiently only over narrow ranges of throughputs. Some expensive lessons were learned during the Arab oil embargo when lack of Mideastern crudes and feedstock necessitated production cutbacks.

Structural parameters such as the slenderness ratio frequently govern what can and cannot be done. An engineer would be hard pressed to find a reputable contractor willing to build a 40-m-long column, half a meter in diameter.

Pressure Drop

Temperature lability and the possibility of chemical reactions such as poly- merization and oxidation frequently necessitate high-vacuum operation. Then pressure drop through the column becomes a critical parameter. Pressure drop is also important in systems that tend to foam; a high ΔP accentuates this tendency (although high absolute pressures reduce it).

Cost

The column shell plus auxiliary pumps, heat exchangers, boilers, and reflux drum cost anywhere from three to six times as much as the trays or packing in a typical installation that is outdoors and operating under pressure or vacuum. Specific cases can vary by orders of magnitude; Sawistowski and Smith[2] present a cost analysis for a column where the shell plus auxiliary equipment costs less than the plates.

Highly important also are the utilities (electricity, steam, cooling water, etc.) since operating costs are usually from two to six times yearly equipment depreciation. Given the very high cost ratio of structure and utilities to column internals, it is apparent that an engineer who specifies anything other than the most efficient and versatile column packing or trays is the modern equivalent of "pence-wise, pound-foolish."

Operability

Here one quickly gets into the realm of human factors. In many large companies, new designs are developed by a central engineering department in cooperation with vendors and are subject to approval by the plant manager in whose plant the equipment is to be installed.

A typical conversation between the engineer from Central Engineering, the sales engineer from the equipment vendor, the plant manager, and his main- tenance engineer could sound approximately like this.

Plant Manager   How about some coffee? We've got the best damn coffee in Louisiana right here. Grind it ourselves. (The next hour's conversation, which relates to hunt- ing, fishing, and the weather, has been deleted.)

Central Engineer	Isn't it getting near time to put out the order on the glycol tower? Lead time on equipment purchases is running 12 to 15 months. Did you get a chance to look at our test data on the pilot plant runs?
Plant Manager	You know how busy we are down here. What did you fellows have in mind?
Sales Engineer	We ran three tons of raw glycol through our test column, and we think our new 240-7 packing gets you a more efficient column at a lower price.
Plant Manager	Sounds good. What's this packing made of?
Sales Engineer	It's a proprietary plastic developed for us by Spillips Petroleum. It has a heat distortion point of 350°F and a tensile—
Maintenance Engineer	Didn't we once try a plastic packing in our laboratory column? Took Pete four days to drill it out.
Central Engineer	You must have lost vacuum and had a temperature excursion. Do you have a report on your tests?
Sales Engineer	There have been a lot of improvements in plastics in the last five years. Our 240-Z packing has stood up under much more severe operating conditions.
Plant Manager	Seems I heard old Jim Steele say they tried mesh packing in a glycol column in Baytown. It flooded, and when they opened her up the packing was squashed flatter than a pancake.
Sales Engineer	Must have been one of our competitor's products. Not that we never make mistakes, but—
Plant Manager	Can you give me a few names and phone numbers of people using your packing in glycol columns?
Sales Engineer	This will be our first glycol column, but I can give you the names of some other customers who've had good luck with our 240-Z.
Maintenance Engineer	Those old bubble caps we got in column 11 work real good.
Central Engineer	I seem to remember your having a few problems with liquid oscillations and dumping a few years back... etc.

The postscript to this story is that at least one of the large new glycol distillation columns built in the last two years has (supposedly obsolete) bubble caps. The plant manager's reluctance to experiment with new designs is understandable. There are a large number of potentially unpleasant operating problems that can make life difficult for him, and, at one time or another, he has probably seen them all. Packed column vapor–liquid contactors can:

Flood. This condition occurs at high vapor and/or liquid rates when the gas pressure drop is higher than the net gravity head of the liquid, which then backs up through the column.

Channel (bypass). The function of the packing is to promote fluid turbulence and mass transfer by dispersing the liquid, which, ideally, flows as a film over the surface of the packing and as droplets between and inside the packing. At low liquid and/or vapor flows, or if the liquid feed is not distributed evenly over the packing, it will tend to flow down the wall, bypassing the vapor flowing up the middle. At very low flow rates, there may be insufficient liquid to wet the surface of the packing.

Flooding and channeling restrict the range of permissible liquid and vapor flows in packed columns. In Fig. 2.3 the operable column limits, in terms of gas and vapor flow rates, are shown schematically for a typical distillation application. Although the maximum operability range is seen to be dictated by flooding and bypassing, practical considerations limit the range to between a minimal allowable efficiency and a maximum allowable pressure drop.

Although tray columns are generally operable over wider ranges of gas and liquid loadings than packed columns, they have their own intriguing problems.

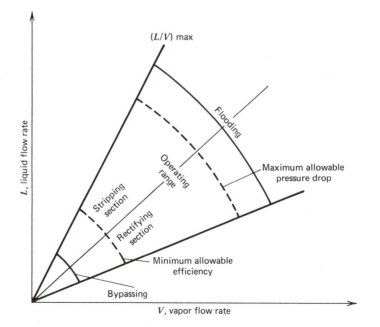

Figure 2.3. Flooding and by-passing in packed columns.

These tray malfunctions depend, to some extent on whether valve, sieve, or bubble cap trays are in service. Common malfunctions leading to inefficiencies and inoperability are:

Foaming. This problem, which is especially likely to appear in extractive distillation and absorption service, is aggravated by impurities (frequently present during start-up), low pressures, and high gas velocities. In a moderately high foam regime, liquid will be carried up by the gas into the next stage and separation efficiencies will drop by as much as 50%. Alternatively, foam can carry vapor down to the tray below. In extreme cases the downcomers (or downspouts), which direct the liquid flow between stages, fill with foam and flooding occurs, much like in a packed column. Indeed, plate columns can flood even without foam at pressure drops or liquid flow rates large enough so that the liquid level exceeds the tray spacing, causing liquid backup in the downcomers.

Entrainment. In a properly functioning column, much of the mass transfer takes place in a turbulent, high-interfacial-area froth layer which develops above the liquid on the plate. Inadequate disengagement of the liquid and vapor in the froth results in the backmixing of froth with the liquid from the tray above, and a lowered efficiency. Entrainment is frequently due to inadequate downcomer size or tray spacing.

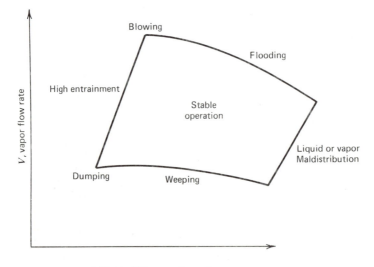

L/V, Liquid-to-vapor flow rate

Figure 2.4. Tray malfunctions as a function of loading.

Liquid maldistribution. Weirs, positioned at the entrance to the down-comers, are used to control the height of liquid on the tray. In improperly designed or very large trays, the height of liquid across the tray will vary, causing a substantial hydraulic gradient. This can result in nonuniform gas flows and, in extreme cases, unstable, cyclic oscillations in liquid and gas flows (alternate *blowing* and *dumping*). Common preventive measures include use of multiple downcomers or passes and split trays, and directing of the vapor flow to propel the liquid across the tray.

Weeping. Perforated plate and other types of trays that rely solely on gas pressure to hold liquid on the tray will, at the *weep point*, start leaking liquid through the gas orifices. Extreme weeping or *showering* is called *dumping.*

Tray malfunctions as a function of liquid and vapor flow are shown schematically in Fig. 2.4. Some of the malfunctions will not occur with certain types of trays.

Stage Efficiency

Column packings used in continuous contactors are characterized by the height of packing equivalent to a theoretical stage HETP. Two related quantities are: (a) the height of a transfer unit HTU, which is roughly proportional to the HETP and usually somewhat smaller, and (b) the mass transfer (capacity) coefficients K_Ga or K_La, which are inversely proportional to the HTU. Packing efficiency is inversely proportional to the HETP, which can be as low as 10 cm for high-performance mesh packing or as high as 1 m for large ring packings.

Plate contactors such as valve trays are evaluated in terms of a plate efficiency, which is inversely proportional to how closely the composition of the streams leaving a stage approach the predicted compositions at thermodynamic equilibrium. To obtain the number of stages required for any given separation, it is necessary to divide the computed number of theoretical stages by an average empirical fractional plate efficiency.

Plate efficiencies and HETP values are complex functions of measurable physical properties: temperature, pressure, composition, density, viscosity, diffusivity, and surface tension; measurable hydrodynamic factors: pressure drop and liquid and vapor flow rates; plus factors that cannot be predicted or measured accurately: foaming tendency, liquid and gas turbulence, bubble and droplet sizes, flow oscillations, emulsification, contact time, froth formation, and others. Values for plate efficiency, HETP, or HTU, particularly those that purport to compare various devices, are usually taken over a limited range of concentration and liquid-to-vapor ratios. The crossovers in Fig. 2.5 and the rather strange behavior of the ethyl alcohol–water system, Fig. 2.6, demonstrate the critical need for test data under expected operating conditions.[3]

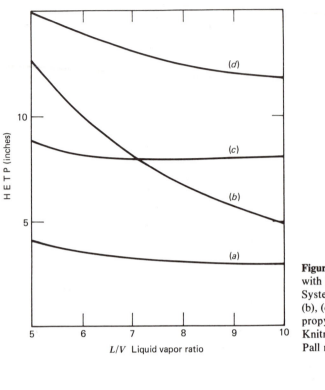

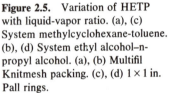

Figure 2.5. Variation of HETP with liquid-vapor ratio. (a), (c) System methylcyclohexane-toluene. (b), (d) System ethyl alcohol–n-propyl alcohol. (a), (b) Multifil Knitmesh packing. (c), (d) 1 × 1 in. Pall rings.

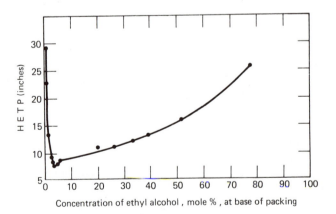

Figure 2.6. Variation of HETP with composition for the system ethyl alcohol—water. [S. R. M. Ellis and A. P. Boyes, *Trans. Inst. Chem. Eng.*, **52**, 202 (1974) with permission.]

2.2 Column Packings

So-called random or dumped tower packings are fabricated in shapes such that they fit together with small voids without covering each other. Prior to 1915, packed towers were filled with coke or randomly shaped broken glass or pottery; no two towers would perform alike.

The development of the Raschig Ring, shown in Fig. 2.7, in 1915 by Fredrick Raschig introduced a degree of standardization into the industry. Raschig Rings, along with Berl Saddles, were, up to 1965, the most widely used packing materials. By 1970, however, these were largely supplanted by Pall Rings and more exotically shaped saddles such as Norton's Intalox® Saddle, Koch's Flexisaddle®, Glitsch's Ballast Saddle®, and so on. The most widely used packings at present are: (a) modified Pall Rings that have outside ribs for greater

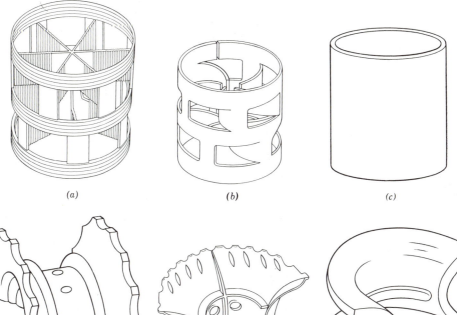

(a)   (b)   (c)

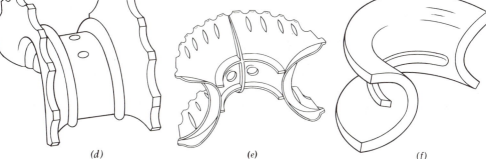

(d)   (e)   (f)

Figure 2.7. Tower packings. (*a*) Plastic pall ring®. (*b*) Metal pall ring®. (*c*) Raschig ring. (*d*) Super Intalox® saddle. (*e*) Plastic Intalox® saddle. (*f*) Intalox® saddle. (Courtesy of the Norton Co.).

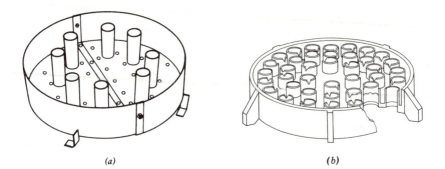

(a)   *(b)*

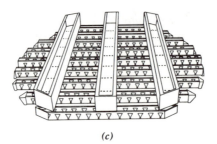

(c)

Figure 2.8. Feed-liquid distributors (*a*) Orifice type. (*b*) Weir type. (*c*) Weir-trough type. (Courtesy of the Norton Co.).

strength and many ribbonlike protrusions on the inside to promote turbulence and provide more liquid transfer points and (b) saddles with scalloped edges, holes, and protrusions. Of the two, the saddles are more widely used, partly because they are available in ceramic materials and the rings are not. Figure 2.7 is taken from the Norton Company catalog; vendors such as Glitsch, Koch, and Hydronyl have similar products.

2.3 Packed Tower Internals

Feed–Liquid Distributors

Packing will not, of itself, distribute liquid feed adequately. An ideal distributor (Norton Company Bulletin TA-80) has the following attributes.

1. Uniform liquid distribution.

2. Resistance to plugging and fouling.

3. High turndown ratio (maximum allowable throughput to minimum allowable throughput).

4. High free area for gas flow.

5. Adaptability for fabrication from many materials of construction.

6. Sectional construction for installation through manways.[4]

The two most widely used distributors are the orifice and weir types, shown in Fig. 2.8. In the weir type, cylindrical risers with V-shaped weirs are used as downcomers for the liquid, thus permitting greater flow as head increases. The orifice type, where the liquid flows down through the holes and the gas up through the risers, is restricted to relatively clean liquids and narrow liquid flow ranges. Weir–trough distributors are more expensive but more versatile. Liquid is fed proportionately to one or more parting boxes and thence to troughs. Spray nozzles and perforated ring distributors are also fairly widely used.

Liquid Redistributors

These are required at every 3 to 6 m of packing to collect liquid that is running down the wall or has coalesced in some area of the column and then to redistribute it to establish a uniform pattern of irrigation. Design criteria are similar to those for a feed-liquid distributor.

Fig. 2.9a shows a Rosette-type, wall-wiper redistributor that must be sealed to the tower wall. Figure 2.9b is a redistributor that achieves total collection of the liquid prior to redistribution. It is designed to be used under a gas-injector support plate.

Gas Injection Support Plates

In addition to supporting the weight of the packing, the support plate by its design must allow relatively unrestricted liquid and gas flow. With the types of plates shown in Fig. 2.10a and 2.10b, liquid flows through the openings at the bottom and gas flows through the top area.

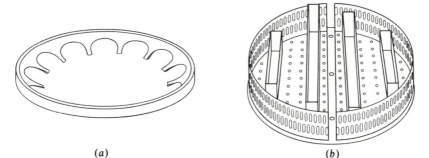

(a)   (b)

Figure 2.9.   Liquid redistributors (a) Rosette type. (b) Metal type. (Courtesy of the Norton Co.).

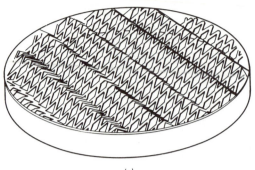

(a)

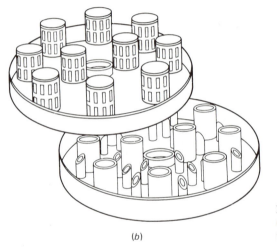

(b)

Figure 2.10. Gas-injector support plates. (*a*) Light-duty type. (*b*) Perforated riser type. (Courtesy of Hydronyl Ltd.).

Hold-Down Plates (Bed Limiters)

Hold-down plates are placed at the top of the packing to prevent shifting, breaking, and expanding of the bed during high pressure drops or surges. They are used primarily with ceramic packing, which is subject to breakage, and plastic packing, which may float out of the bed. Various types are shown in Fig. 2.11. Frequently mesh pads (demisters) are used above the packing in conjunction with or in addition to hold-down plates to prevent liquid entrainment in exit vapor.

Liquid–Liquid Disperser Support Plate

Liquid–liquid disperser support plates are used in packed towers in liquid–liquid extraction service. At the base of the tower, they function as a support and as a

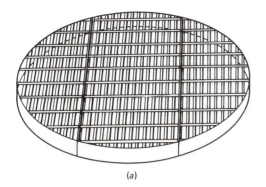

(a)

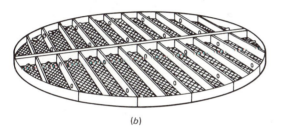

(b)

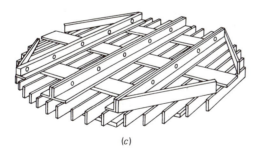

(c)

Figure 2.11. Hold-down and retainer plates (*a*) Retaining plate. (*b*) Hold-down plate. (*c*) Hold-down plate. (Courtesy of Koch Engineering Co.).

disperser for the light phase. They are also placed 6 to 12 ft apart in the bed as resupports and redispersers for the dispersed phase, which tends to coalesce. When placed at the top of the tower, they can be used to disperse the heavy phase when it is desired to make the light phase continuous. In general, the dispersed phase enters through the orifices, and the heavy phase through the risers. A typical plate is shown in Fig. 2.12.

Of considerably less commercial importance than the random or dumped packings are the oriented grids, meshes, wires, or coils. These range from inexpensive open-lattice metal stampings that stack 45° or 90° to each other and

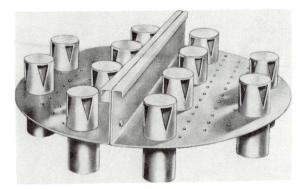

Figure 2.12. Liquid-liquid disperser support plates. (Courtesy of the Norton Co.).

Figure 2.13. Koch Sulzer packing, 7-ft diameter, installed in DMT purification vacuum tower. (Courtesy of Koch Engineering Co.).

look somewhat like the gas distributors in Fig. 2.10*b* to the very expensive regular (vertical) arrangements of corrugated, woven metal or glass gauze in Fig. 2.13. As opposed to trays and dumped packing where the vapor–liquid interface is created by a combination of surface penetration, bubbling, spray, and froth effects, the interface in packings such as the Koch Sulzer is stationary and depends largely on surface wetting and capillarity. One would therefore expect good performance even at low liquid rates. Recent literature also describes some interesting vertically stacked helical coils in contact, with the liquid rapidly spinning about each coil and then mixing and redistributing at points of contact between the coils.[5] Very low HETP values and pressure drops are claimed.

2.4 Characterization and Comparison of Packings

Test data involving comparisons between packings are not universally meaningful. The Glitsch Company for example, markets a highly and irregularly perforated stamped-metal, stacked packing for vacuum crude stills that has 97% free space and looks like a by-product from one of their tray manufacturing operations. If one were to compare this packing with the Koch Sulzer packing in Fig. 2.13 using a standard system such as ethanol–propanol, it would undoubtedly be a poor second best in terms of HETP. The comparison, however, is meaningless because the Sulzer packing could probably not operate at all in the viscous, high-vacuum environment of a crude still. Another major factor is the liquid-to-gas mass flow ratio, which in absorption and stripping can be much larger than 4, but in distillation much smaller than 4. Thus, the anticipated hydrodynamic regime must be factored into a comparison.

Further questions that make comparative test data on "model" systems such as air–water of less than universal importance are: (a) Will the liquid wet the packing? (b) Are there heat or chemical effects? (c) Do we want to generate gas-phase or liquid-phase turbulence? That is, is most of the mass transfer resistance in the gas or liquid phase? It is only after the nature of the service and the relative importance of the mass transfer factors are established that meaningful evaluations of packing characteristics and performance can be made.

Available technical data on packings generally pertain to the physical

Table 2.1 Representative F factors

Type of Packing	Material	Nominal Packing Size, In.				
		$\frac{1}{2}$	1	$1\frac{1}{2}$	2	3
Raschig rings	Metal	350	120	80	60	32
	Ceramic	600	150	95	65	37
Saddles (1965)	Plastic	—	—	—	—	—
	Ceramic	380	110	65	45	—
Saddles (1975)	Plastic	—	33	—	21	16
	Ceramic	240	60	—	30	22
Pall rings (1965)	Metal	—	48	28	20	—
	Plastic	—	52	32	25	—
Pall rings (1975)	Metal	70	45	28	20	16
	Plastic	95	50	40	25	16

characteristics (surface area, free area, tensile strength, and temperature and chemical stability), hydrodynamic characteristics (pressure drop and permissible flow rates), and efficiency (HETP, HTU, and $K_G a$ or $K_L a$).

Modern correlations of permissible flow rates with fluid properties ΔP and packing geometry have evolved from the pioneering research of T. Sherwood, W. Lobo, M. Leva, J. Eckert, and collaborators. The packing factor F, which is an experimentally determined constant related to the surface area of the packing divided by the cube of the bed voidage, is used to predict pressure drop and flooding at given flow rates and fluid properties. Alternatively, permissible flow

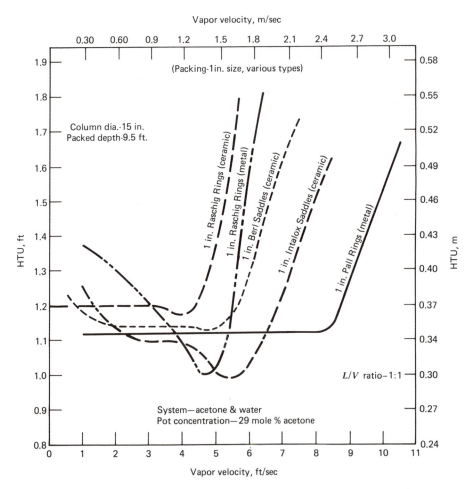

Figure 2.14. HTU variation for various packings. [J. S. Eckert, *Chem. Eng. Progr.*, **59** (5), 78 (1963) with permission.]

rates can be calculated given allowable pressure drops. In assessing the representative values of F in Table 2.1, it should be borne in mind that as F increases ΔP increases at a given flow rate and that column capacity is inversely proportional to $\sqrt{F}$. The improvement in packing performance between 1965 and 1975 should be noted. Also, more sizes and materials of construction are now available. High-performance rings are made only of plastics or metal; saddles are available in plastic or ceramic only.

A crude comparison by Eckert of the efficiency of the various packings (ca 1963) is given in Fig. 2.14 for a particular system and column.[6] The superiority of Pall Rings and Intalox Saddles over the older Raschig Rings and Berl Saddles is clearly evident.

2.5 Plate Columns for Vapor–Liquid Contacting

Before 1950, the bubble cap tray was the only plate device in popular service for vapor–liquid contacting. The early 1950s marked the emergence of a number of competitors including sieve, Ripple®, Turbogrid®, Kittle®, Venturi®, Uniflux®, Montz®, Benturi®, and a number of different valve trays. Of these, only sieve trays achieved immediate popularity, and they quickly captured the bulk of the market. However, improvements in valve tray design, particularly in the realm of pressure drop, cost, and valve reliability, led to its increasing use until, today, the valve tray dominates the market. This is not to say, however, that sieve and particularly bubble cap trays are passé. Recently, large nitric acid and glycol plants using bubble cap columns were constructed, and there are other applications where low tray-to-tray leakage and high liquid residence times are critical, so that bubble caps are preferred. Sieve trays will continue to be used because they are inexpensive, easily fabricated, and have performed well in many applications.

Valve Trays

Typical of the valves used are the Koch type K-8, A, and T and the Glitsch A-1 and V-1 in Fig. 2.15. According to the Koch Engineering Company Inc., the type-T valve provides the best liquid seal, the type-A valve, which has the guide legs and lift studs integral, is more economical, and the K-8 has the lowest pressure drop because it utilizes streamline flow of a venturi orifice to lower the inlet and outlet friction losses. The two Glitsch Ballast valves are mounted on decks, the vapor flowing into the valve through flat or extruded orifices. Each A-1 ballast unit consists of an orifice or vapor port, an orifice seat, orifice cover plate, ballast plate, and travel stop. The V-1 unit sits on three tabs when closed, and the lip under the slot edge is shaped to give a *vena contracta* at the position where the vapor enters the liquid. This increases the turbulence and vapor–liquid interfacial area. The additional orifice plate in the A–1 valve is useful when no

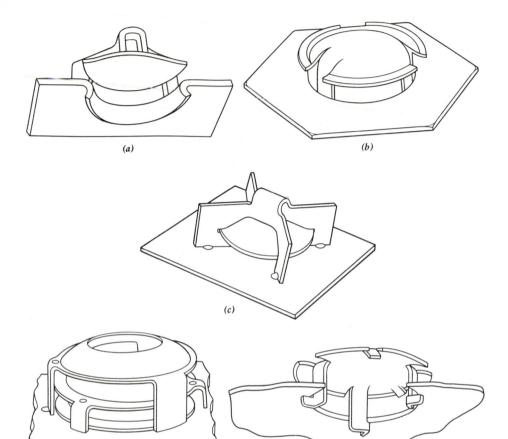

Figure 2.15.   Representative valves. (*a*) Koch type K-8. (*b*) Koch
type A. (*c*) Koch type T. (*d*) Glitsch type A-1. (*e*) Glitsch type V-1.
(Courtesy of Koch Engineering Co., Inc., and Glitsch, Inc.).

liquid leakage is permitted, even with interrupted flow. The turndown ratio of the
V-1 units is said to be as high as nine.

A single-pass ballast tray is shown in Fig. 2.16. Larger trays have multipass
split flow or cascade arrangements to reduce the detrimental effects of hydraulic
gradients. Figure 2.17 shows some of the possible arrangements. The recommended
design for ballast trays is to hold liquid rates between 0.02 and 0.05 m³/sec per meter
of flow rate width (the active tray area divided by flow path length) by means of
increasing the number of passes.

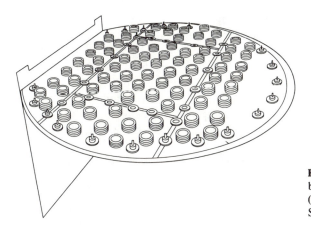

Figure 2.16. Single-pass Glitsch ballast tray with A-1 valves. (Courtesy of F. L. Glitsch and Sons.).

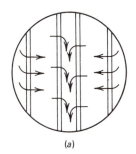

(a)

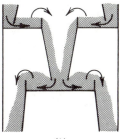

(b)

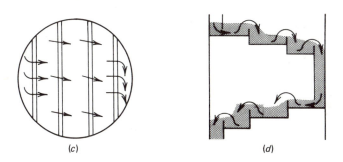

(c)   (d)

Figure 2.17. Two-pass split flow and cascade trays. (a) Split flow (top view). (b) Split flow (side view). (c) Cascade cross flow (top view). (d) Cascade cross flow (side view).

Sieve Trays

By far the most widely used sieve trays have perforated plates with the liquid flow directed across the tray. However, counterflow "shower" trays, without downcomers, where the liquid and vapor flow through the same openings, also are used. One version, the Turbogrid® tray of the Shell Development Company, is a flat grid of parallel slots; the other, a Ripple® tray, is a corrugated tray with small perforations. Devices are available that are hybrids of sieves and valves, thus combining the low-pressure-drop and low-cost advantages of sieve trays with the extended operating range of valve trays.

Contact action on a sieve tray, as in a valve tray, is between vapor rising through orifices and the mass of liquid moving across the tray. With reference to Fig. 2.18, one sees that the liquid descends through the downcomer onto the tray at point A. No inlet weir is shown, but it is used in many applications to seal off upward vapor flow through the downcomer. Clear liquid of height h_{li} is shown from A to B because there are usually no holes in this part of the tray.

From B to C is the so-called active, high-aeration portion of froth height h_f. The liquid height h_l in the manometer on the right may be thought of as the settled head of clear liquid of density ρ_l. Collapse of the froth begins at C, there

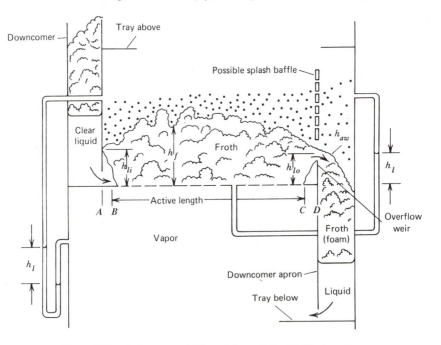

Figure 2.18. A sieve tray. (Adapted from B. Smith, *Design of Equilibrium Stage Processes*, McGraw-Hill Book Co., New York, © 1963, 542.)

being no perforation from C to D. The outlet liquid height is h_{lo}, and $(h_{li} - h_{lo})$ is the hydraulic gradient (essentially zero here).

The design of sieve trays has many features in common with that of valve trays, differences being primarily in the substitution of holes for valves as gas inlet ports. Hole diameters are generally from 0.3 to 1.3 cm in diameter, the larger holes being preferable where potential fouling exists. A high hole area contributes to weeping; a low hole area increases plate stability but it also increases the possibility of entrainment and flooding along with pressure drop. Frequently, hole sizes and hole spacings are different in various sections of a column to accommodate to flow variations. Another common practice is to blank out some of the holes to provide flexibility in terms of possible future increases in vapor load.

Bubble Cap Trays

Bubble caps have an ancient and noble history, dating back to the early 1800s. As one might expect, they come in a large variety of sizes and shapes, as shown in Fig. 2.19. A bubble cap consists of a riser bolted, welded, riveted, or wedged to the plate and a cap bolted to the riser or plate. Study of the caps shows that, although most of them have slots (from 0.30 to 0.95 cm wide and 1.3 to 3.81 cm long), some, such as the second from the right in the second row of Fig. 2.19, have no slots, the vapor leaving the cap from under the edge of the skirt, which may be as much as 3.81 cm from the plate. Commercial caps range in size from 2.54 to 15 cm in diameter. They are commonly arranged on the plate at the corners of equilateral triangles with rows oriented normal to the direction of flow.

Figure 2.19. Some typical bubble caps. (Courtesy of F. W. Glitsch and Sons.).

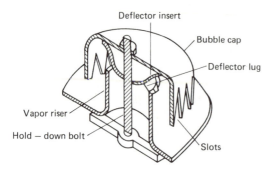

Figure 2.20. Bubble cap design for streamlined vapor flow. [Adapted from *Chem. Eng.*, p. 238 (January, 1955).]

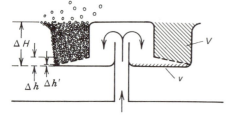

Figure 2.21. Lateral screen skirt cap. [Adapted from F. A. Zenz, *Petrol. Refiner*, p. 103 (June, 1950).]

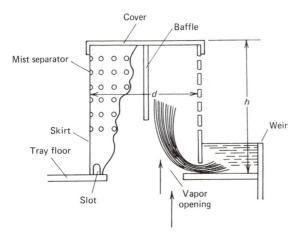

Figure 2.22. A VST cap. [S. Tanigawa, *Chem. Economy Eng. Rev.*, **5** (2), 22 (1973) with permission.]

Many modified designs have been proposed; usually they are based on some hydrodynamic principle. Figure 2.20 shows a cap that includes an insert to give streamline vapor flow. Another experimental type, shown in Fig. 2.21, is provided with fins for better vapor dispersal. Lateral screened skirt caps have also been proposed, the most recent entry being the VST cap developed by the

Mitsui Shipbuilding and Engineering Company Ltd.[7] Figure 2.22 shows the basic principle. Liquid enters the base of the cap through the slot, is entrained in the rising gas, and discharged through holes in the cap. Some extravagant claims are made for this cap, which appears to be similar to the "siphon cap" developed by L. Cantiveri and tested at Stevens Institute of Technology in 1956.

The unique advantages of bubble cap trays are that: (1) if properly sealed to the tower, they permit no leakage, and (2) there is a wealth of published material and user experience. The disadvantages are rather apparent:

1. The flow reversals and multiplicity of expansions and contractions lead to high pressure drops.

2. Stage efficiencies are 10 to 20% below that of sieve or valve trays.

3. They are, on a tray-for-tray basis, 25 to 50% more expensive than sieve trays and 10 to 30% more expensive than valve trays.

Design procedures for sizing columns, which are considered in detail in Chapter 13, generally start with an estimation of the tower diameter and tray spacing. Capacity, pressure drop, and operating range at that diameter are then compared to the process specifications. The diameter, downcomer dimensions, cap spacing, or tray spacing can then be made to meet specifications, and to obtain a minimum cost design, or a design optimized to capacity, efficiency, or operating cost.

The lifeblood of engineering fabricators is their proprietary design manual, which contains formulas and graphs for calculating column parameters such as flooding, capacity, downcomer velocities, vapor capacity, tray diameters, column diameters, downcomer area, pressure drop, flow path width, tray layout, and cap, valve, or perforation size. Of key importance also are some of the other column internals such as:

Antijump baffles. These are sometimes used with the splitflow downcomers shown in Fig. 2.17 to keep liquid from jumping over downcomers into an adjacent section of the same tray.

Picket fence splash baffles. These are placed on top of downcomers or weirs to break up foam entrainment or froth.

Inlet weirs. These are used to insure a positive downcomer seal at high vapor rates and low liquid rates as shown in Fig. 2.23*a*.

Inlet and drawoff sumps and seals. These are used to provide positive seals under all conditions as shown in Fig. 2.23*b* and 2.23*c*.

Splash panels. These are used to prevent splashing and promote flow uniformity as shown in Fig. 2.23*d*.

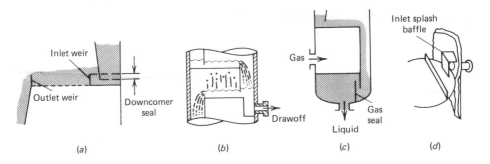

Figure 2.23. Column internals. (*a*) Inlet weir. (*b*) Drawoff sump. (*c*) Gas seal. (*d*) Splash panels. [F. A. Zenz, *Chem. Eng.*, **79**(25), 120 (1972) with permission.]

Demister pads. On large columns, these are sometimes placed between stages, as well as at the top of the columns, to promote liquid–vapor disengagement.

Tower manholes. Manhole diameter is a major factor in tray design. It also affects the number of pieces to be installed and plate layout.

Trusses, rings, supports. In large-diameter towers, trays are supported on channels, or I-beams. The method used in bolting and clamping trays to the shell requires experience and careful planning. Trays must be level to insure uniform distribution of flows.

2.6 Packed Columns Versus Plate Columns

The difference in cost between plate and packed columns is not too great, although packing is more expensive than plates.[8] Also, the difference in column height is not usually significant if the flow rates are such that efficiencies are near maximum. Table 2.2 shows that 2-in. Pall Rings are equivalent to valve trays on 24-in. spacings. As a rule of thumb, plates are always used in columns of large diameter and towers that have more than 20 to 30 stages. The efficiency of packed towers decreases with diameter while plate tower efficiency increases. Packed columns find their greatest application in gas absorption (scrubbing) service where corrosive chemical reactions frequently occur, laboratory and pilot plant equipment, and high-vacuum duty. Other guidelines are:
Conditions favoring packed columns.

Small-diameter columns (less than 0.6 m).
Corrosive service.
Critical vacuum distillations where low pressure drops are mandatory.

Table 2.2 Comparison between plates and rings

Pall Ring	Equivalent Valve Tray
Size, In.	Spacing, In.
1	18
2	24

Low liquid holdups (if the material is thermally unstable).
Foamy liquids (because there is less liquid agitation in packed columns).

Conditions favoring plate columns.

Variable liquid and/or vapor loads.
Exotherms requiring cooling coils inside the column.
Pressures greater than atmospheric.
Low liquid rates.
Large number of stages and/or diameter.
High liquid residence times.
Dirty service (plate columns are easier to clean).
Thermal or mechanical stresses (which might lead to cracked packing).

2.7 Modern Tray Technology—a Case Study

In many respects, the case history presented by D. W. Jones and J. B. Jones of the DuPont Company is typical.[9] They reported studies conducted on a pair of columns in use at Dana, Indiana, and Savannah River, Georgia, for a heavy-water process using dual temperature exchange of deuterium between water and hydrogen sulfide at elevated pressures.

Constructed in the early 1950s, the columns were originally equipped with bubble cap trays. Corroded trays at Dana were replaced, beginning in 1957, by sieve trays, which were found to have a lower pressure drop and to be more efficient and cheaper. At Dana, liquid pumping capacity limited tower F-factor (vapor velocity times the square root of the vapor density) based on tray bubbling area to 1.6 with sieve trays, versus 1.43 with bubble cap trays, where flooding was limiting. Later tests at Savannah River in a 1.98-m-diameter column showed sieve plate flooding at F-factors of 1.88. However, at high water turbidity, foaming reduced the allowable vapor rates to F-factors of 1.25 (which could be raised somewhat by antifoam addition).

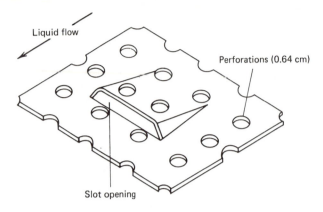

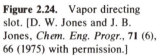

Figure 2.24. Vapor directing slot. [D. W. Jones and J. B. Jones, *Chem. Eng. Progr.*, **71** (6), 66 (1975) with permission.]

By 1972 the seventy 3.35-m-diameter bubble cap plates in the column at Savannah River were corroded and subject to retraying, so tests were conducted on a proprietary sieve tray designed by the Linde Division of Union Carbide, and two nonproprietary trays designed by Glitsch, Inc. These trays, designated as A, B, and C, all had 0.64-cm holes and about an 11% hole area. Type A had some unusual patented features, including vapor directing slots, shown in Fig. 2.24 which are claimed to help reduce hydraulic gradients and increase mass transfer.

Tests conducted with the three trays showed very comparable pressure drops and weep points at an F-factor of 1.0. A typical curve is shown in Fig. 2.25. The ΔP at the preferred operating F-factor of 1.7 to 1.8 is 30% higher for the bubble cap trays. However, small deviations in feed-water quality or antifoam concentration had previously led to foaming and resulting bubble cap flooding at an F-factor as low as 1.55. Flooding did not occur with type-A or -C trays even without an antifoam agent at an F of 1.8. The type-B tray did not flood with loss of antifoams but did flood at poor water quality (an example of the "art" factor). Tray efficiencies for all trays were roughly comparable.

In many respects, the DuPont experience is typical; in other ways it is not. Heavy water is not a normal chemical of commerce and is not subject to the vagaries of the market place. In the mid-1970s the combination of a recession, feedstock shortages, a three- to fourfold energy cost increase, plus the fact that the process industry had swung heavily to large, single-train plants, forced the industry to adopt new distillation strategies. Heat-intensive, high-reflux operation of fractionating columns became uneconomical, and ways to operate plants at much less than design capacity had to be found. The latter factor accelerated the trend toward valve trays, since these can operate over a greater range of liquid and vapor rates than sieve plates. The lowering of reflux ratios to save heat can be accomplished in an existing column only by increasing the number of stages. To get more stages into an existing column, some companies have

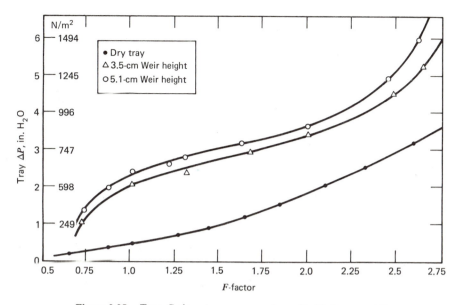

Figure 2.25. Type C sieve tray pressure drop. [D. W. Jones and J. B. Jones, *Chem. Eng. Progr.*, **71** (6), 66 (1975) with permission.]

replaced stages with high-performance packing, despite the additional expense.

As the art becomes more of a science, designers become more confident and innovative. Thus we are now seeing applications involving mixed-mode equipment, that is, trays that have valves plus sieve holes, and columns that have alternate sections of grids and packing, or mesh and trays. These mixed-mode devices are particularly useful in the not-uncommon situation where liquid and vapor loads vary appreciably over the length of the apparatus.

2.8 Less Commonly Used Liquid–Vapor Contactors

Spray Columns

In gas absorption applications such as the absorption of SiF_4 by water, the solvent has such great affinity for the gas that very few stages are required. In this case, one may bubble a gas through an agitated liquid, or use a spray column. The simplest spray absorption column consists of nothing more than an empty chamber in which liquid is sprayed downward and gas enters at the bottom. In more sophisticated devices, both phases may be dispersed through relatively complicated atomization nozzles, pressure nozzles, venturi atomizers, or jets. This dispersion, however, entails high pumping cost.

Spray units have the advantage of low gas pressure drop; they will not plug

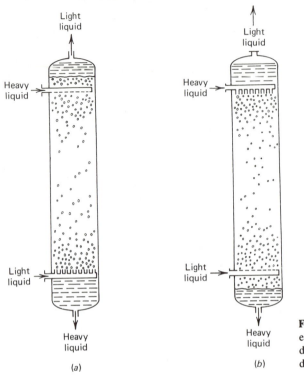

Figure 2.26. Spray tower for extractions. (*a*) Light liquid dispersed. (*b*) Heavy liquid dispersed.

should solids form, and they never flood. (They may also be used for extraction service, as shown in Fig. 2.26.)

Baffle Towers and Shower Trays

Baffle columns and shower tray columns, shown in Fig. 2.27, are characterized by relatively low liquid dispersion and very low pressure drops. The major application of this type of flow regime is in cooling towers, where the water flows across wooden slats and very large volumes of gas are handled. Here economics dictate that fans rather than compressors be used. Some gas absorption and vacuum distillation columns employ baffle or shower trays.

2.9 Liquid–Liquid Extraction Equipment

The petroleum industry represents the largest volume and longest standing application for liquid–liquid extraction; over 100,000 m³ per day of feedstocks are processed using physically selective solvents.[10] Extraction processes are well

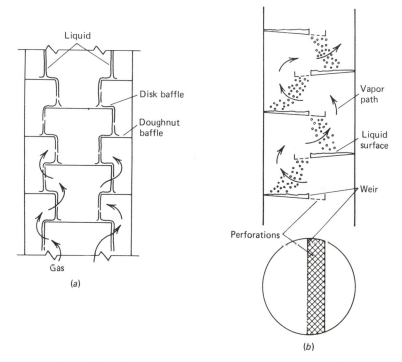

Figure 2.27. Baffle tray columns. (*a*) Disk and doughnut baffle column. (*b*) Shower tray column.

suited for the petroleum industry where heat-sensitive feeds are separated as to type (such as aliphatic or aromatic) rather than molecular weight. Table 2.3 shows some of the petrochemical industries' proposed and existing extraction processes. Other major applications lie in the biochemical industry (here emphasis is on the separation of antibiotics and protein recovery from natural substrates); in the recovery of metals (copper from ammoniacal leach liquors) and in separations involving rare metals and radioactive isotopes from spent fuel elements; and in the inorganic chemical industry where high-boiling constituents such as phosphoric acid, boric acid, and sodium hydroxide need be recovered from aqueous solutions.

In general, extraction is preferred to distillation for the following applications:

1. In the case of dissolved or complexed inorganic substances in organic or aqueous solutions, liquid–liquid extraction, flotation, adsorption, ion exchange, membrane separations, or chemical precipitation become the processes of choice.

Table 2.3 Developments in extraction solvents for petroleum products and petrochemicals[a]

Single solvent Systems[b]			Mixed Solvent Systems[d]		
Solvent	Feedstock	Main Products	Solvent	Feedstock	Main Products
Dimethyl formamide (DMF)	Catalytic cycle oil (204 to 316°C boiling point)	Alkyl naphthalenes aliphatic-rich stream	Phenol and ethyl alcohol	Petroleum residues	Lube stocks
Dimethyl formamide	C₄ hydrocarbons	Butadiene	Diglycol amine and NMP, mono-ethanol amine and NMP, glycerol and NMP	Mixed hydrocarbons	C₆–C₈ aromatics
β-Methoxy proprionitrile (β-MPON)	C₄ hydrocarbons	Butadiene			
Dimethyl formamide (diethyl formamide)	Urea adducts (2.6% aromatics impurities)	Paraffins (240 to 360°C boiling point)	Furfural, furfural alcohol, water	Cycle oil	Heavy aromatics
Nitromethane	Catalytic naphtha	Diesel fuel and aromatics (94% pure)	Furfural and C₆–C₁₀ mono-ketones	Lube oil feed	Lube oil
Sulphur dioxide	Straight run distillates	Burning kerosenes	Furfural and alcohols	Lube oil feed	Lube oil
Furfural	Process gas oils	Carbon black feed, and catalogue cracking feed	N-Alkyl pyrrolidine, urea (or thio urea), water	Hydrocarbon mixture	Aromatics
	FC cycle gas oil	Dinuclear aromatics			
	Coker distillates	Carbon black feed	Hydrotropic salt solutions (e.g. sodium aryl sulfonates)	Hydrocarbon mixtures	Paraffins and aromatics
	Vacuum distillates	lube oil blend stocks polycyclic aromatics			
Phenol	Distillates	Lube oil blend stocks			
Sulfolane	Vacuum distillate	Polycyclic aromatics	E-Caprolactam and water, alkyl carbamates and water	Hydrocarbon mixtures	Paraffins and C₆–C₇ aromatics
	Catalytic naphthas	C₆–C₈ aromatics			
	Light distillates	Special solvents			

Solvent	Feedstock	Special solvents	Dual Solvent System		
Dimethyl sulphoxide	Naphthas	Olefinic rich extracts aromatics; paraffinic raffinates			
Ammonia	*Dearomatized naphthas, process naphthas, heavy distillates*	$C_6 - C_8$ aromatics	*Carbamate, thiocarbamate esters, and water*	Steam-cracked naphtha	$C_6 - C_7$ aromatics
Methyl Carbamate (Carmex)	Hydrogenated process naphthas	Aromatics	*Hydrogen fluoride and borontrifluoride* (complexing)	*Mixed xylenes* $C_6 - C_9$	*p* and *o*-Xylene, *m*-xylene (extract)
Substituted phospholanes	Hydrocarbon mixtures	Aromatics			
N-Hydroxyethyl propylene-diamine	Hydrocarbon mixtures	$C_6 - C_8$ aromatics and *n*-paraffins			
Glycols {dipropylene, diethylene, triethylene, tetraethylene}	Heavy naphtha, catalytic reformate, urea adducts, cracked gasoline	$C_6 - C_8$ aromatics *e.g.* 5 tert-butyl isophthalic acid			
Mixed xylenes[c]	Carboxylic acid, salt mixtures	Lube oil stocks	*Dimethyl formamide* (or amide)+glycerol (hydroxy corpound) for reextraction of extract	Hydrocarbon oil	Paraffinic and aromatic oil
N-methyl-2-pyrrolidone (NMP)	Heavy distillate, naphtha	High-purity aromatics (C_6, C_7, C_8)			
W-Methoxy-alkyl-pyrrolidine	C_4 streams	Butadiene			
Fluorinated hydrocarbons and alkanols	—	Aromatics			
Hydrogen fluoride	Test hydrocarbon mixtures	—			
1,3-Dicyanobutane (methyl glutaronitrile)	Petroleum tars / Naphtha	Naphthalene / Aromatics, unsaturates			

Note:

[a] Solvents known to be used commercially are italicized.

[b] This definition does not exclude use of minor proportions of water as "antisolvent".

[c] Dissociative Extraction.

[d] Term excludes displacement solvents.

79

2. For the removal of a component present in small concentrations, such as a color former in tallow, or hormones in animal oil, extraction is preferred.

3. When a high-boiling component is present in relatively small quantities in a waste stream, as in the recovery of acetic acid from cellulose acetate, extraction becomes competitive with distillation.

4. In the recovery of heat-sensitive materials, extraction is well suited.

The key to an effective process lies with the discovery of a suitable solvent. In addition to being nontoxic, inexpensive, and easily recoverable, a good solvent should be relatively immiscible with feed components(s) other than the solute and have a different density. It must have a very high affinity for the solute, from which it should be easily separated by distillation, crystallization, or other means.

If the solvent is a good one, the distribution coefficient for the solute between the phases will be at least 5, and perhaps as much as 50. Under these circumstances, an extraction column will not require many stages, and this indeed is usually the case.

Given the wide diversity of applications, one would expect a correspondingly large variety of liquid–liquid extraction devices. Most of the equipment as well as the design procedures, however, is similar to those used in absorption and distillation. Given the process requirement and thermodynamic data, the necessary number of stages are computed. Then the height of the tower for a continuous countercurrent process is obtained from experimental HETP or mass transfer performance data that are characteristic of a particular piece of equipment. (In extraction, some authors use HETS, height equivalent to a theoretical stage, rather than HETP.)

Some of the different types of equipment available include:

Mixer-Settlers. This class of device can range from a simple tank with an agitator in which the phases are mixed and then allowed to settle prior to pump-out, to a large, compartmented, horizontal or vertical structure. In general, settling must be carried out in tanks, unless centrifuges are used. Mixing, however, can be carried out by impingement in a jet mixer; by shearing action, if both phases are fed simultaneously into a centrifugal pump or in-line mixing device; by injectors where the flow of one liquid is induced by another; or in orifices or mixing nozzles.

A major problem in settlers is emulsification, which occurs if the dispersed droplet size falls below 1 to 1.5 micro meters (μm). When this happens coalescers, separator membranes, meshes, electrostatic forces, ultrasound, chemical treatment, or other ploys are required to speed the settling action.

Spray Columns. As in gas absorption, axial dispersion (backmixing) in the continuous phase limits these devices to applications where only one or two stages are required. They are rarely used, despite their very low cost. Typical configurations were shown in Fig. 2.26.

Packed Columns. The same types of packings used in distillation and absorption service are employed for liquid–liquid extraction. The choice of packing material, however, is somewhat more critical. A material preferentially wetted by the continuous phase is preferred. Figure 2.28 shows some performance data for Intalox Saddles in extraction.[11] As in distillation, packed extractors are used in applications where the height and/or diameter need not be very large. Backmixing is a problem in packed columns and the HETP is generally larger than for staged devices.

Plate Columns. The much preferred plate is the sieve tray. Columns have been built successfully in diameters larger than 4.5 m. Holes from 0.64 to 0.32 cm in diameter and 1.25 to 1.91 cm apart are commonly used. Tray spacings are much closer than in distillation—10 to 15 cm in most applications involving low-interfacial-tension liquids. Plates are usually built without outlet weirs on the downspouts. A variation of the simple sieve column is the Koch Kascade Tower®, where perforated plates are set in vertical arrays of moderately complex designs.

If operated in the proper hydrodynamic flow regime, extraction rates in sieve plate columns are high because the dispersed phase droplets coalesce and

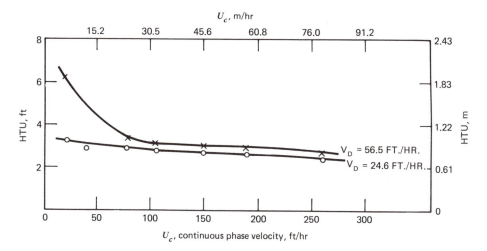

Figure 2.28. Efficiency of 1-in. Intalox saddles in a column 60 in. high with MEK–water–kerosene. [R. R. Neumatis, J. S. Eckert, E. H. Foote, and L. R. Rollinson, *Chem. Eng. Progr.*, **67** (1), 60 (1971) with permission.]

reform on each stage. This helps destroy concentration gradients, which can develop if a droplet passes through the entire column without disturbance. Sieve plate columns in extraction service are subject to the same limitations as distillation columns: flooding, entrainment, and, to a lesser extent, weeping. Additional problems such as scum formation due to small amounts of impurities are frequently encountered.

Mechanically Assisted Gravity Devices. If the surface tension is high and or density differences between the two liquid phases are low, gravity forces will be inadequate for proper phase dispersal and the creation of turbulence. In that case, rotating agitators driven by a shaft that extends axially through the column are used to create shear mixing zones alternating with settling zones in the column. Differences between the various columns lie primarily in the mixers and settling chambers used. Three of the more popular arrangements are shown in Fig. 2.29. Alternatively, agitation can be induced by moving the plates back and forth in a reciprocating motion.

The RDC, rotating disc contactor, has been used in sizes up to 12 m tall and 2.4 m in diameter for petroleum deasphalting, as well as for furfural extraction of lubricating oil, desulfurization of gasoline, and phenol recovery from wastewater. Rapidly rotating discs provide the power required for mixing, and the annular rings serve the purpose of guiding the flow and preventing backmixing.

In the original Scheibel (York–Scheibel) device, mixing was by unbaffled turbine blades; in later versions baffles were added. The wire mesh packing between the turbines promotes settling and coalescence. The Oldshue-Rushton (Lightnin CM Contactor®), which is no longer widely used, has deep compartments with turbine agitators and no separate settling zones.

Other devices in commercial use include a mixer-settler cascade in column form invented by R. Treybal,[12] and pulsed sieve or plate columns with a

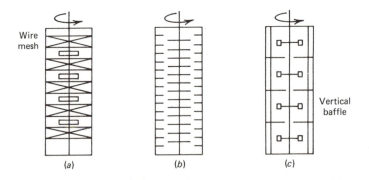

Figure 2.29.   Mechanically assisted gravity devices. (*a*) Scheibel. (*b*) RDC. (*c*) Oldshue-Rushton.

reciprocating plunger or piston pump to promote turbulence and improve efficiency. Although a great deal of government-sponsored research has gone into development of pulsed columns, their only use has been in nuclear materials processing on a pilot scale.

Centrifugal Extractors. Centrifugal forces, which can be thousands of times larger than gravity, can greatly facilitate separations where emulsification is a problem, where low-density differences exist, or where very low residence times are required because of rapid product deterioration, as in the antibiotic industry. Usually, centrifugal extractors have only one or two stages; however four-stage units have been built.

Traditionally, the chemical industry has eschewed high-velocity rotational equipment because of maintenance difficulties and poor performance in continuous duty. Advances in equipment design have overcome some of the unreliability problems, and they are becoming more popular despite their high initial cost and power requirements.

Table 2.4   Advantages and disadvantages of different extraction equipment

Class of equipment	Advantages	Disadvantages
Mixer-settlers	Good contacting Handles wide flow ratio Low headroom High efficiency Many stages available Reliable scale-up	Large holdup High power costs High investment Large floor space Interstage pumping may be required
Continuous counterflow contactors (no mechanical drive)	Low initial cost Low operating cost Simplest construction	Limited throughput with small density difference Cannot handle high flow ratio High headroom Sometimes low efficiency Difficult scale-up
Continuous counterflow (mechanical agitation)	Good dispersion Reasonable cost Many stages possible Relatively easy scale-up	Limited throughput with small density difference Cannot handle emulsifying systems Cannot handle high flow ratio
Centrifugal extractors	Handles low density difference between phases Low holdup volume Short holdup time Low space requirements Small inventory of solvent	High initial costs High operating cost High maintenance cost Limited number of stages in single unit

Source. R. B. Akell, *Chem. Eng. Prog.,* 62, *No.* (9), 50–55, (1966).

2.10 Comparison of Extraction Equipment

A summary of the advantages and disadvantages of the contacting devices used in extraction, as well as a preference ordering, is given in Tables 2.4 and 2.5, respectively.

Table 2.5 Order of preference for extraction contacting devices

Factor or Condition	Preferred Device(s)	Exceptions
1. Very low power input desired:		
(a) One equilibrium stage	Spray column	
(b) Few equilibrium stages	Baffle column	
(c) Many equilibrium stages	1. Perforated plate column	
	2. Packed column	
2. Low-to-moderate power input desired, three or more stages:		
(a) General and fouling service		Strongly emulsifying systems
(b) Nonfouling service requiring low residence time or small space	1. Centrifugal extractors	
	2. Columns with rotating stirrers or reciprocating plates	
3. High power input	Centrifugal extractors	Use mixer-settlers for one to two stages
4. High phase ratio	1. Perforated plate column	
	2. Mixer-settler	
5. Emulsifying conditions	Centrifugal extractors	Fouling systems
6. No design data on mass transfer rates for system being considered	Mixer-settlers	
7. Radioactive systems	Pulsed extractors	

Adapted from E. D. Oliver, *Diffusional Separation Processes: Theory, Design, and Evaluation*, John Wiley & Sons, New York, 1966.

References

1. McLaren, D. B., and J. C. Upchurch, *Chem. Eng.* (N.Y.), **77** (12), 139-152 (1970).

2. Sawistowski, H., and W. Smith, *Mass Transfer Process Calculations*, Interscience Publishing Co., New York, 1963, 99.

3. Ellis, S. R. M., and A. P. Boyes, *Trans. Inst. Chem. Eng.*, **52**, 202-210 (1974).

4. Bulletin TA-80, Norton Co., 1974.

5. Ellis, S. R. M., *Chem. Eng.*, (London), **259**, 115-119 (1972).

6. Eckert, J. S., *Chem. Eng. Progr.*, **59** (5), 76–82 (1963).

7. Tanigawa, S., *Chem. Economy and Engr. Rev.*, **5** (2), 22-27 (1973).

8. Fair, J., *Chem. Eng. Progr.*, **66** (3), 45-49 (1970).

9. Jones, D. W., and J. B. Jones, *Chem. Eng. Progr.*, **71** (6), 65-72 (1975).

10. Bailes, P. J., and A. Winward, *Trans. Inst. Chem. Eng.*, **50**, 240-258 (1972).

11. Neumaites, R. R., J. S. Eckert, E. H. Foote, and L. R. Rollinson, *Chem. Eng. Progr.*, **67** (11), 60-67 (1971).

12. Treybal, R. E., *Chem. Eng. Progr.*, **60** (5), 77–82 (1964).

Problems

2.1 In distillation, how is the number of stages related qualitatively to:
(a) The difficulty of the separation (relative volatility)?
(b) The tower height?
(c) The tower diameter?
(d) The liquid and vapor flow rates?

2.2 A gas absorption column to handle 3630 kg/hr of a gas is being designed. Based on pressure drop, entrainment, and foaming consideration, the maximum vapor velocity must not exceed 0.61 m/sec. If the density of the vapor is 0.801 kg/m^3, what is the column diameter?
Discuss, qualitatively, the factors governing the height of the column.

2.3 The overall distillation plate efficiency, E_o, may be correlated in terms of the following variables: liquid density, vapor density, liquid viscosity, vapor viscosity, liquid diffusivity, surface tension, pressure, temperature, pressure drop, liquid flow rate, vapor flow rate, bubble size, contact time.
Discuss what range of values for the exponents $a, \ldots, m$ in an expression such as

$$E_o = f\{(A^a)(B^b)(C^c)(D^d)\ldots(M^m)\}$$

might reasonably be expected.

2.4 In *Chemical Engineering Progress*, **74** (4), 2, 61–65 (1978), Eastham and co-workers describe a new packing called the Cascade Mini-Ring®.
(a) List the advantages and disadvantages of this new packing compared to Raschig Rings, Berl Saddles, and Pall Rings.
(b) Under what conditions have Cascade Mini-Rings been used to replace trays successfully?

2.5 The bubble caps shown in Figs. 2.20 and 2.21 have never achieved popularity. Can you suggest some reasons why?

2.6 Shown below is a novel bubble cap called the siphon cap that was developed at the Stevens Institute of Technology by L. Cantiveri. Discuss the operation of this cap and compare it to the VST cap shown in Fig. 2.22.

2.7 Absorption of sulfur oxides from coal-fired power plant flue gases by limestone slurries is the current method of choice of the EPA. List some of the more important problems you would anticipate in the design of operable equipment, and draw a sketch and describe the internal column flow arrangement of a limestone scrubber.

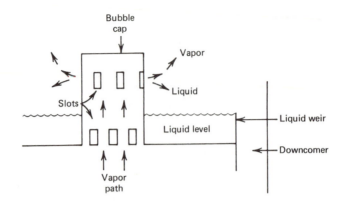

Problem 2.6. The siphon cap.

2.8   The following appeared in the New Product and Services section of *Chemical Engineering*, **82** (2), 62 (January 20, 1975).

> Buoyed by high grades in a recent commercial application, the Angle Tray strives to graduate to wider CPI roles. Entirely different from other tray plates, it combines low manufacturing costs, high rigidity, and simple construction with low pressure drop, good efficiency and high capacity.
>
>   The tray's free area is set by merely arranging the carbon steel angle members at the design distance; it requires no plate work or other modification. According to the manufacturer, the unique design eliminates support members, uses less materials than conventional trays, and boasts low tray flexure—making it ideal for large distillation towers.
>
>   Vapor and liquid contact in openings between the angles. The angles' sides rectify rising vapors so that vapors do not strike the tray plate at right angles—as is the case with sieve trays. Lower pressure drop reportedly occurs.
>
>   **Scaled-Up Tests.** Independent tests performed at the Fractionation Research Incorporation (Los Angeles, Calif.), showed that the Angle Tray has a 102% maximum efficiency at a 1.32 superficial F-factor (Fs). Efficiency remained above 80% at Fs ranging from 0.8-1.6. Compared to conventional sieve trays, the Angle Tray boasts better high-load efficiency, while sieve trays excel in low-load treatment.
>
>   The trays, installed in an 8-ft-dia., 6-stage distillation tower, were spaced 18 in apart. Each tray contained 69 angle elements (0.787 × 0.787 × 0.118 in. apiece) at 0.158 in. spacings. Slot area was 4.38 ft²/tray. Distilling a mixture of cyclohexane and n-hexane with total reflux at 24 psi showed the Angle Tray's capacity factor is approximately 20% larger than sieve trays, while its pressure drop is less than half. Also, as vapor load increases, pressure drop rises slowly—which the developer considers another good result.

Commercial Use. Applied to a 5-stage, 4-ft-dia. commercial water-diethylene glycol distillation tower, the Angle Tray reportedly was designed, constructed and installed in just 15 days.

In this system 5 theoretical plates (operating with reflux) were needed to hold the diethylene glycol concentration to below several hundred ppm in the wastewater. Using five Angle Trays, the concentration was actually lower than the design point; therefore, the device's efficiency was above 100%. Good efficiency was maintained even with reduced operational pressure, and low pressure drop also resulted. (*Ishikawajima-Harima Heavy Industries Co., Ltd., Tokyo, Japan.*)

Discuss the possible merits of this new angle tray.

2.9 When only one or a few equilibrium stages are required for liquid–liquid extraction, mixer-settler units are often used. Determine the main distinguishing features of the different mixer-settler units described by Bailes and coworkers in *Chemical Engineering*, **83** (2), 96-98 (January 19, 1976).

2.10 What tower packing and/or type of tray would you recommend for each of the following applications.
(a) Distillation of a very viscous crude oil.
(b) Distillation of a highly heat-sensitive and -reactive monomer.
(c) Distillation under near-cryogenic conditions.
(d) Gas absorption accompanied by a highly exothermic reaction (such as absorption of nitric oxide in water).
(e) Absorption of a noxious component present in parts-per-million quantities in a very hot gas stream (500°C).
(f) Absorption of a very corrosive gas, such as HF, where absorption is highly exothermic.
(g) Extraction of a labeled organic compound with a half-life of 1 min.

2.11 As a major chemical process equipment manufacturer, your company wishes to determine how a projected fourfold increase in energy costs would impact the sale of their present equipment lines and whether or not new marketing opportunities will be created.

With respect to presently manufactured products, they would like to know the effect on the relative sales of:
(a) Sieve, valve, and bubble cap distillation columns.
(b) Ring, saddle, and mesh packings for gas absorption columns.
(c) Packed columns and plate columns for distillation.
(d) RDC extraction columns.

They also wish to establish long-term trends in the equipment market for:
(a) Adsorption units.
(b) Membrane permeation processes.
(c) Ion-exchange resins.
(d) Chromatographic columns.

3

Thermodynamic Equilibrium Diagrams

When a gas is brought into contact with the surface of a liquid, some of the molecules of the gas striking the liquid surface will dissolve. These dissolved molecules will continue in motion in the dissolved state, some returning to the surface and re-entering the gaseous state. The dissolution of gas in the liquid will continue until the rate at which gas molecules leave the liquid is equal to the rate at which they enter. Thus a state of dynamic equilibrium is established, and no further changes will take place in the concentration of gas molecules in either the gaseous or liquid phases.

Olaf A. Hougen and Kenneth M. Watson, 1943

Stagewise calculations require the simultaneous solution of material and energy balances with equilibrium relationships. It was demonstrated in Example 1.1 that the design of a simple extraction system reduces to the solution of linear algebraic equations if (1) no energy balances are needed and (2) the equilibrium relationship is linear.

In cases involving complex equilibrium functions and/or energy balances, solutions to large sets of nonlinear simultaneous equations are required. If many stages are involved, the system of equations becomes so large that rigorous analytical solutions cannot be easily obtained by manual means, and computers are required. However, when separation problems involve only two or three components or when only approximate solutions are needed, graphical techniques provide a convenient alternative to the computer. Furthermore, graphical methods provide a lucid visual display of the stage-to-stage extent of component separation. Some of the more commonly used thermodynamic equilibrium diagrams for distillation, absorption, and extraction and their application to simple material and energy balance problems are described in this introductory chapter. Such diagrams can be constructed from experimental measurements of equilibrium compositions, or from compositions computed by analytical thermodynamic equations described in Chapter 4.

The first phase-equilibrium diagrams discussed are for two-component liquid–vapor systems. Next, three-component diagrams used in extraction, absorption, leaching, and ion exchange are developed. Finally, enthalpy-composition diagrams, which include energy effects, are constructed.

3.1 Homogeneous and Heterogeneous Equilibrium

If a mixture consisting of one or more components possesses uniform physical and chemical properties throughout, it is said to be a *single-phase, homogeneous* system. If, however, a system consists of one or more parts that have different properties and are set apart from each other by bounding surfaces, so that the phases are mechanically separable, the system is *heterogeneous*. When equilibrium exists between the distinct parts of the system, this condition is known as *heterogeneous equilibrium*.

3.2 The Phase Rule

The phase rule of J. Willard Gibbs relates the variance (degrees of freedom) $\mathscr{F}$ for a nonchemically reactive system at heterogeneous equilibrium to the number of coexisting phases $\mathscr{P}$ and the number of components (chemical species), C present.

$$\mathscr{F} = C - \mathscr{P} + 2$$

The variance designates the number of intensive properties that must be specified to completely fix the state of the system. For the systems to be treated here only the intensive properties T, P, and concentration are considered.

For a gas having n components, $C = n$, so $\mathscr{F}$ is $n + 1$, and the specification of the temperature, pressure, and $n - 1$ concentration variables completely defines the state of the system.

Figure 3.1 is a schematic one-component, three-phase equilibrium diagram. The three different phase regions are separated by lines $D\text{-}TP$ (solid vapor pressure, or sublimation curve), $F\text{-}TP$ (melting point curve), and $TP\text{-}C$ (liquid vapor pressure or boiling-point curve). Point C is the critical point where the vapor and liquid phases become indistinguishable and TP is the triple point where solid, liquid, and vapor phases can coexist. There are only two in-

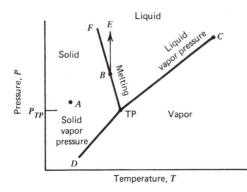

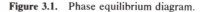

Figure 3.1. Phase equilibrium diagram.

dependent variables, T and P. Applying the phase rule, we note that, at point A, $\mathscr{P} = 1$; hence $\mathscr{F} = 2$. There are two independent variables, T and P, which we can change by small amounts without creating a new phase. At B, which is on the solid–liquid equilibrium line, there are two phases in equilibrium; hence $\mathscr{F} = 1$. If we raise the pressure to E, the temperature, which is now a dependent variable, must be lowered if we are to continue to have two phases in equilibrium. We note that, at TP, $\mathscr{P} = 3$ and $\mathscr{F} = 0$. There are no independent variables, and any changes in temperature or pressure will immediately result in the disappearance of one of the phases. It is thus impossible to make an equilibrium mixture of solid, liquid, and vapor by cooling water vapor at a constant pressure other than P_{TP}, which for water is 610 Pa.

3.3 Binary Vapor–Liquid Mixtures

For vapor–liquid mixtures of component A and B, $\mathscr{F} = 2$. The two independent variables can be selected from T, P, and, since both liquid and vapor phases are present, the concentration of one of the components in the vapor y_A and in the liquid x_A. The concentrations of B, y_B, and x_B are not independent variables, since $y_A + y_B = 1$ and $x_A + x_B = 1$. If the pressure is specified, only one independent variable remains (T, y_A, or x_A).

Four isobaric (constant pressure) phase equilibrium diagrams involving the variables T, x, and y can be constructed: T-y, T-x, combined T-x-y, and x-y diagrams. Figures 3.2a and 3.2b are schematic T-x-y and x-y diagrams for a two-component vapor–liquid system. In Fig. 3.2a the temperatures T_A and T_B are the boiling points of the pure components A and B at a given pressure. The lower curve connecting T_A to T_B is the isobaric *bubble-point temperature* (saturated liquid) curve. The upper curve connecting T_A and T_B is the *dew-point temperature* (saturated vapor) curve. A "subcooled" liquid of composition x_A at

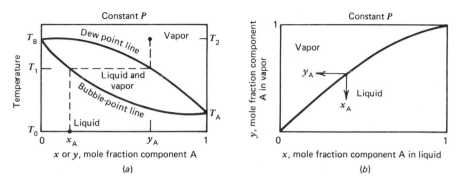

Figure 3.2.   Vapor–liquid phase equilibrium. (E. J. Henley and E. M. Rosen, *Material and Energy Balance Computations*, John Wiley & Sons, New York, © 1969.)

T_0, when heated to T_1, will produce the first bubble of equilibrium vapor of composition y_A. Conversely, a superheated vapor of composition y_A at T_2, when cooled to T_1, will condense, the first drop of liquid being of composition x_A. Figure 3.2a also shows that, in general, complete vaporization or condensation of a binary mixture occurs over a range of temperatures, rather than at a single temperature as with a pure substance.

It is important to note that, since $\mathscr{F} = 2$ and the pressure is fixed, the specification of only one additional thermodynamic variable completely defines a binary vapor–liquid mixture. If the composition of the liquid is x_A, both the vapor–phase composition y_A and the bubble-point temperature T_1 are uniquely fixed.

Figure 3.2b, an x-y vapor–liquid equilibrium diagram, is an alternative way of presenting some of the information in Fig. 3.2a. Here each point on the x-y equilibrium curve is at a different but undesignated temperature. Figure 3.2b is widely used in calculating equilibrium-stage requirements even though it contains less information than Fig. 3.2a.

3.4   Use of Physical Properties to Predict Phase Equilibrium Composition

Chapter 4 describes the generation of vapor–liquid and liquid–liquid equilibria data using analytical correlations based on physical properties. It will be seen that correlations based solely on properties of pure components are successful only for homologous systems when molecular size differences are small and interactions among like molecules are similar to the interactions among unlike molecules. For mixtures of liquid *n*-hexane and *n*-octane, for example, we might correctly predict that the components will be miscible, and bubble- and dew-point temperatures will be between the boiling points of the pure components and closer to that of the component present in higher concentrations. In addition, we might predict that, to a first approximation, no heat will be released upon mixing and that the total solution volume will equal the sum of the volumes of the pure components. For such mixtures, which are termed *ideal solutions*, it is possible to predict the distribution of components between phases at equilibrium from the molecular properties of the pure components. A rigorous thermodynamic definition of ideal solutions will be given in Chapter 4.

3.5 Raoult's Law for Vapor–Liquid Equilibrium of Ideal Solutions

If two or more liquid species form an ideal liquid solution with an equilibrium vapor mixture, the partial pressure p_i of each component in the vapor is proportional to its mole fraction in the liquid x_i. The proportionality constant is the vapor pressure P_i^s of the pure species at the system temperature, and the relationship is named *Raoult's law* in honor of the French scientist who developed it.

$$p_i = P_i^s x_i \qquad (3\text{-}1)$$

Furthermore, at low pressure, Dalton's law applies to the vapor phase, and

$$p_i = P y_i \qquad (3\text{-}2)$$

where P is the total pressure and y_i the vapor-phase mole fraction. Combining (3-1) and (3-2), we have

$$y_i = (P_i^s/P) x_i \qquad (3\text{-}3)$$

With this equation, values for the vapor pressures of the pure components suffice to establish the vapor–liquid equilibrium relationship.

Departures from Raoult's law occur for systems in which there are differing interactions between the constituents in the liquid phase. Sometimes the interaction takes the form of a strong repulsion, such as exists between hydrocarbons and water. In a liquid binary system of components A and B, if these repulsions lead to essentially complete immiscibility, the total pressure P over the two liquid phases is the sum of the vapor pressures of the individual components, and

$$P = P_A^s + P_B^s \qquad (3\text{-}4)$$

Example 3.1. Vapor pressures for *n*-hexane, H, and *n*-octane, O, are given in Table 3.1.
(a) Assuming that Raoult's and Dalton's laws apply, construct $T\text{-}x\text{-}y$ and $x\text{-}y$ plots for this system at 101 kPa (1 atm).

Table 3.1 Vapor pressures for *n*-hexane and *n*-octane

Temperature		Vapor Pressure, kPa	
°F	°C	*n*-Hexane	*n*-Octane
155.7	68.7	101	16
175	79.4	137	23
200	93.3	197	37
225	107.2	284	58
250	121.1	400	87
258.2	125.7	456	101

Source. J. B. Maxwell, *Data Book on Hydrocarbons*, D. Van Nostrand and Co., Inc., New York, 1950, 32, 34.

(b) When a liquid containing 30 mole % H is heated, what is the composition of the initial vapor formed at the bubble-point temperature?

(c) Let the initial (differential amount) of vapor formed in (b) be condensed to its bubble point and separated from the liquid producing it. If this liquid is revaporized, what is the composition of the initial vapor formed? Show the sequential processes (b) and (c) on the *x-y* and *T-x-y* diagrams.

Solution. (a) According to Raoult's law (3-1)

$$p_H = P_H^s x_H \qquad \text{and} \qquad p_O = P_O^s x_O$$

By Dalton's law (3-2)

$$p_H = P y_H \qquad \text{and} \qquad p_O = P y_O$$

Also

$$p_H + p_O = P \qquad x_H + x_O = 1 \qquad \text{and} \qquad y_H + y_O = 1$$

From (3-3)

$$y_H = \frac{P_H^s x_H}{P} \qquad y_O = \frac{P_O^s x_O}{P} \tag{3-5}$$

Combining the expressions for y_H and y_O, we have

$$x_H = \frac{P - P_O^s}{P_H^s - P_O^s} \tag{3-6}$$

Equations (3-5) and (3-6) permit calculation of y_H and x_H at a specified temperature. Using the vapor pressures from Table 3.1, 79.4°C, for example, we find that

$$x_H = \frac{101 - 23}{137 - 23} = 0.684$$

$$y_H = \frac{137}{101}(0.684) = 0.928$$

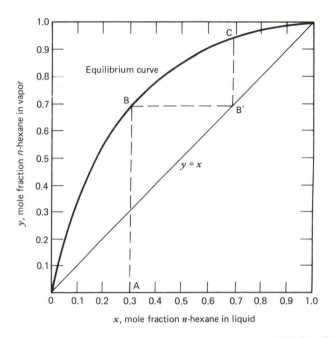

Figure 3.3. The x-y diagram for n-hexane–n-octane, at 101 kPa. (E. J. Henley and E. M. Rosen, *Material and Energy Balance Computations*, John Wiley & Sons, New York, © 1969.)

It should be noted that, if alternatively one assumes a value for either x_H or y_H, the result is a trial-and-error calculation since $P_i^s = P_i^s\{T\}$ and T is not easily expressed as $T = T\{P_i^s\}$. The results of the calculations are shown in Figs. 3.3 and 3.4 as solid lines. The 45° line $y = x$ also shown in Fig. 3.3 is a useful reference line. In totally condensing a vapor, we move horizontally from the vapor–liquid equilibrium line to the $y = x$ line, since the newly formed liquid must have the same composition as the (now condensed) vapor.

(b) The generation of an infinitesimal amount of vapor such that x_H remains at 0.30 is shown by line $\overline{AB}$ in Fig. 3.3 and by line A_OA in Fig. 3.4. The paths $A \rightarrow B$ and $A \rightarrow A_O$ represent isobaric heating of the liquid, $x_H = 0.3$. From Fig. 3.4 we see that boiling takes place at 210°F (98.9°C), the vapor formed (B) having the composition $y_H = 0.7$. Although Fig. 3.3 does not show temperatures, it does show a saturated liquid of $x_H = 0.3$ in equilibrium with a saturated vapor of $y_H = 0.7$ at point B.

(c) When the vapor at B is totally condensed (B $\rightarrow$ B′) and then brought to the bubble point (B′ $\rightarrow$ C) the concentration of hexane in the vapor is 0.93. Thus, starting with a liquid containing only 30% hexane, one produces a vapor containing 93% hexane. However, only a differential amount of this vapor is produced. Practical techniques for producing finite amounts of pure products will be discussed in subsequent chapters.

□

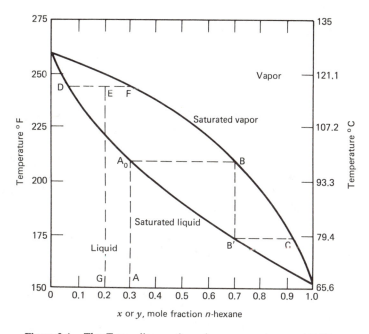

Figure 3.4. The T-x-y diagram for n-hexane–n-octane, at 101 kPa.

Example 3.2. A liquid mixture of 25 kgmoles of benzene (B), 25 kgmoles of toluene (T), and 50 kgmoles of water (W) is at equilibrium with its vapor at 50°C. Assuming that benzene and toluene follow Raoult's law, but that neither are miscible with water, calculate:

(a) The total pressure above the mixture.

(b) The composition of the vapor assuming that Dalton's law applies.

Solution. Vapor pressures of the three components at 50°C are

$$P^s_W = 12.3 \text{ kPa} \qquad P^s_B = 40.0 \text{ kPa} \qquad P^s_T = 11.3 \text{ kPa}$$

(a) Mole fractions in the hydrocarbon liquid phase are

$$x_B = \frac{25}{50} = 0.5 \qquad x_T = \frac{25}{50} = 0.5$$

From (3-1)

$$p_B = (40.0)(0.5) = 20.0 \text{ kPa}$$

$$p_T = (11.3)(0.5) = 5.65 \text{ kPa}$$

The separate liquid–water phase exerts a partial pressure equal to its pure vapor pressure. Thus, $p_W = P^s_W = 12.3$ kPa. Extending (3–4), the total pressure is

$$P = p_B + p_T + p_W = 20.0 + 5.65 + 12.3 = 37.95 \text{ kPa}$$

(b) From (3-2), $y_i = p_i/P$. Thus,

$$y_B = \frac{20.0}{37.95} = 0.527 \qquad y_T = \frac{5.65}{37.95} = 0.149 \qquad y_W = \frac{12.3}{37.95} = 0.324$$

☐

3.6 Vapor–Liquid Material Balances Using Phase Equilibrium Diagrams

Graphical solutions to material balance problems involving equilibrium relationships offer the advantages of speed and convenience. Fundamental to all graphical methods is the so-called inverse lever rule, which is derived in Example 3.3 and applied in Example 3.4.

Example 3.3. Prove that the ratio of the moles of liquid to vapor in the two-phase mixture E (at 240°F, 115.6°C) shown in Fig. 3.4 is in the ratio of the line segments $\overline{FE}/\overline{ED}$.

Solution. Letting $\mathcal{M}_E$, $\mathcal{M}_D$ and $\mathcal{M}_F$ represent the moles of total mixture, liquid, and vapor, respectively, and z_E, x_D, and y_F the corresponding mole fractions of hexane, a material balance for hexane yields

$$\mathcal{M}_E z_E = (\mathcal{M}_D + \mathcal{M}_F) z_E = \mathcal{M}_D x_D + \mathcal{M}_F y_F$$

Solving for the mole ratio of liquid L to vapor V, we have

$$\frac{\mathcal{M}_D}{\mathcal{M}_F} = \frac{L}{V} = \frac{y_F - z_E}{z_E - x_D} = \frac{\overline{FE}}{\overline{ED}}$$

Similarly

$$\frac{\mathcal{M}_D}{\mathcal{M}_E} = \frac{L}{L+V} = \frac{y_F - z_E}{y_F - x_D} = \frac{\overline{FE}}{\overline{FD}}$$

and

$$\frac{\mathcal{M}_F}{\mathcal{M}_E} = \frac{V}{L+V} = \frac{z_E - x_D}{y_F - x_D} = \frac{\overline{ED}}{\overline{FD}}$$

☐

Example 3.4. A solution F containing 20 mole % n-hexane and 80 mole % n-octane is subject to an equilibrium vaporization at 1 atm such that 60 mole % of the liquid is vaporized. What will be the composition of the remaining liquid?

Solution. This process can be shown directly on Fig. 3.4, the T-x-y diagram for hexane–octane. We move along the path G→E until, by trial and error, we locate the isotherm $\overline{DEF}$ such that it is divided by the $x = 0.2$ vertical line into two segments of such lengths that the ratio of liquid to vapor $L/V = 0.4/0.6 = \overline{FE}/\overline{ED}$. The liquid remaining D has the composition $x = 0.07$; it is in equilibrium with a vapor $y = 0.29$. This method of solving the problem is essentially a graphical trial-and-error process and is equivalent to solving the hexane material balance equation

$$Fx_F = yV + xL = (1)(0.2)$$

or

$$y(0.6) + x(0.4) = 0.2$$

where y and x are related by the equilibrium curve of Fig. 3.3. We thus have two equations in two unknowns, the equilibrium relation and the material balance.

□

3.7 Binary Vapor–Liquid Equilibrium Curves Based on Constant Relative Volatility

For systems where the liquid phase is an ideal solution that follows Raoult's law and where the gas phase follows the ideal gas laws, it is possible to formulate relative volatilities that are functions only of temperature. For component i of a mixture, in accordance with (1-4) and (3-3)

$$K_i = \frac{y_i}{x_i} = \frac{P_i^s}{P} \tag{3-7}$$

If the mixture also contains component j, the relative volatility of i to j can be expressed as a ratio of the K-values of the two components

$$\alpha_{ij} = \frac{K_i}{K_j} = \frac{P_i^s}{P_j^s} = \frac{y_i/x_i}{y_j/x_j} \tag{3-8}$$

In a two-component mixture, where $y_j = (1 - y_i)$, and $x_j = (1 - x_i)$, (3-8) becomes

$$y_i = \frac{\alpha_{ij} x_i}{1 + x_i(\alpha_{ij} - 1)} \tag{3-9}$$

It is possible to generate x-y equilibrium curves such as Fig. 3.3 using (3-9) by assuming that the relative volatility is a constant independent of temperature. This is convenient for close-boiling mixtures forming ideal solutions, but can lead to erroneous results for mixtures of components with widely different boiling points because it assumes that both P_i^s and P_j^s are identical functions of T. For example, inspection of the vapor pressure data for the hexane–octane system, Table 3.1, reveals that α varies from $101/16 = 6.3$ at $68.7°C$ to $456/101 = 4.5$ at $125.7°C$. Calculation of relative volatilities by more accurate methods will be considered in Chapter 4.

3.8 Azeotropic Systems

Departures from Raoult's law frequently manifest themselves in the formation of *azeotropes*, particularly for mixtures of close-boiling species of different chemical types. Azeotropes are liquid mixtures exhibiting maximum or minimum

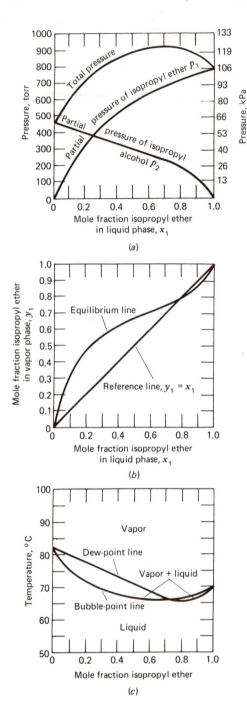

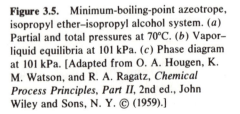

Figure 3.5. Minimum-boiling-point azeotrope, isopropyl ether–isopropyl alcohol system. (*a*) Partial and total pressures at 70°C. (*b*) Vapor–liquid equilibria at 101 kPa. (*c*) Phase diagram at 101 kPa. [Adapted from O. A. Hougen, K. M. Watson, and R. A. Ragatz, *Chemical Process Principles*, *Part II*, 2nd ed., John Wiley and Sons, N. Y. © (1959).]

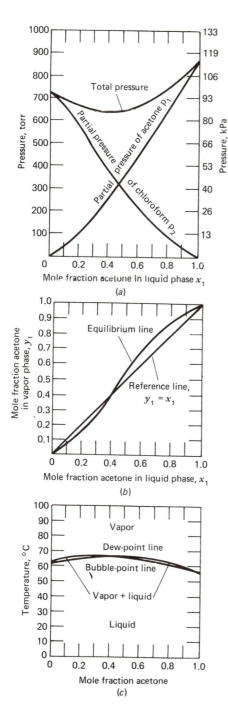

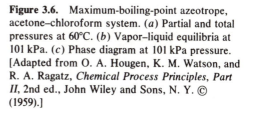

Figure 3.6. Maximum-boiling-point azeotrope, acetone–chloroform system. (*a*) Partial and total pressures at 60°C. (*b*) Vapor–liquid equilibria at 101 kPa. (*c*) Phase diagram at 101 kPa pressure. [Adapted from O. A. Hougen, K. M. Watson, and R. A. Ragatz, *Chemical Process Principles, Part II*, 2nd ed., John Wiley and Sons, N. Y. © (1959).]

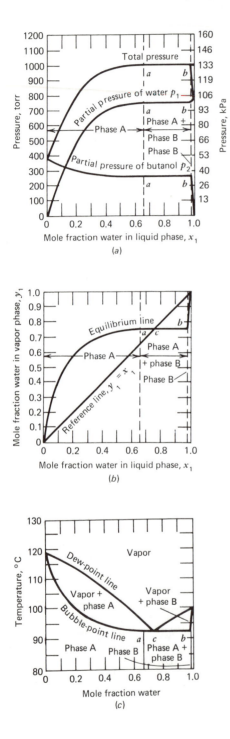

Figure 3.7. Minimum-boiling-point (two liquid phases) water–n-butanol system. (a) Partial and total pressures at 100°C. (b) Vapor–liquid equilibria at 101 kPa. (c) Phase diagram at 101 kPa pressure. [Adapted from O. A. Hougen, K. M. Watson, and R. A. Ragatz, *Chemical Process Principles*, *Part II*, 2nd ed., John Wiley and Sons, N.Y. © (1959).]

boiling points that represent, respectively, negative or positive deviations from Raoult's law. Vapor and liquid compositions are identical for azeotropes.

If only one liquid phase exists, the mixture is said to form a *homogeneous* azeotrope; if more than one liquid phase is present, the azeotrope is said to be *heterogeneous*. In accordance with the Phase Rule, at constant pressure in a two-component system the vapor can coexist with no more than two liquid phases, while in a ternary mixture up to three dense phases can coexist with the vapor.

Figures 3.5, 3.6, and 3.7 show three types of azeotropes commonly encountered in two-component mixtures.

For the *minimum-boiling* isopropyl ether–isopropyl alcohol mixture in Fig. 3.5a, the maximum total pressure is greater than the vapor pressure of either component. Thus, in distillation, the azeotropic mixture would be the overhead product. The y-x diagram in Fig. 3.5b shows that at the azeotropic mixture the liquid and vapor have the same composition. Figure 3.5c is an isobaric diagram at 101 kPa, where the azeotrope, at 78 mole % ether, boils at 66°C. In Fig. 3.5a, which displays isothermal (70°C) data, the azeotrope, at 123 kPa, is 72 mole % ether.

For the *maximum-boiling azeotropic* acetone–chloroform system in Fig. 3.6a, the minimum total pressure is below the vapor pressures of the pure components, and the azeotrope would concentrate in the bottoms in a distillation operation. *Heterogeneous* azeotropes are always minimum-boiling mixtures. The region a-b in Fig. 3.7a is a two-phase region where total and partial pressures remain constant as the relative amounts of the two phases change. The y-x diagram in Fig. 3.7b shows a horizontal line over the immiscible region and the phase diagram of Fig. 3.7c shows a minimum constant temperature.

Azeotropes limit the separation that can be achieved by ordinary distillation techniques. It is possible, in some cases, to shift the equilibrium by changing the pressure sufficiently to "break" the azeotrope, or move it away from the region where the required separation must be made. Ternary azeotropes also occur, and these offer the same barrier to complete separation as binaries.

Azeotrope formation in general, and heterogeneous azeotropes in particular, can be usefully employed to achieve difficult separations. As discussed in Chapter 1, in azeotropic distillation an *entrainer* is added (frequently near the bottom of the column) for the purpose of removing a component that will combine with the agent to form a minimum boiling azeotrope, which is then recovered as the distillate.

Figure 3.8 shows the Keyes process[1,2,3] for making pure ethyl alcohol by *heterogeneous azeotropic distillation*. Water and ethyl alcohol form a binary minimum-boiling azeotrope containing 95.6% by weight alcohol and boiling at 78.15°C at 101 kPa. Thus it is impossible to obtain absolute alcohol (bp 78.40°C) by ordinary distillation. The addition of benzene to alcohol–water results in the formation of a minimum-boiling heterogeneous ternary azeotrope containing by

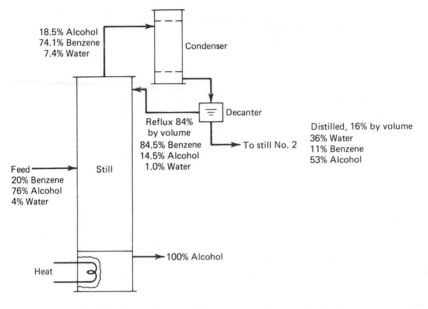

Figure 3.8. The Keyes process for absolute alcohol. All compositions weight percent.

weight, 18.5% alcohol, 74.1% benzene, and 7.4% water and boiling at 64.85°C. Upon condensation, the ternary azeotrope separates into two liquid layers: a top layer containing 14.5% alcohol, 84.5% benzene, and 1% water, and a bottoms layer of 53% alcohol, 11% benzene, and 36% water, all by weight. The benzene-rich layer is returned as reflux. The other layer is processed further by distillation for recovery and recycle of alcohol and benzene. Absolute alcohol, which has a boiling point above that of the ternary azeotrope, is removed at the bottom of the column. A graphical method for obtaining a material balance for this process is given later in this chapter as Example 3.7.

In *extractive distillation*, as discussed in Chapter 1, a solvent is added, usually near the top of column, for the purpose of increasing the relative volatility between the two species to be separated. The solvent is generally a relatively polar, high-boiling constituent such as phenol, aniline, or furfural, which concentrates at the bottom of the column.

3.9 Vapor–Liquid Equilibria in Complex Systems

Petroleum and coal extracts are examples of mixtures that are so complex that it is not feasible to identify the pure components. Vaporization properties of these substances are conventionally characterized by standard ASTM (American

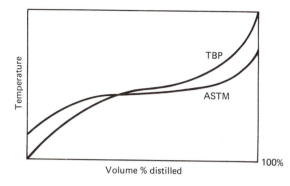

Figure 3.9. Typical distillation curve.

Society for Testing and Materials) boiling-point curves obtained by batch fractionation tests, (e.g., ASTM D86, D158, D1160). Figure 3.9 is a representative curve. Alternatively, data are obtained from more elaborate tests, including equilibrium flash vaporization (EFV), or true-boiling batch distillation (TBP) involving a large number of stages and a high reflux ratio. If either a TBP, EFV, or ASTM curve is available, the other two can be predicted.[4,5] Techniques for processing TBP curves to characterize complex mixtures, for relating them to pseudocomponents, and for obtaining K-values to design fractionators to produce jet fuel, diesel fuel base stock, light naphtha, etc. are available.[6]

3.10 Liquid–Liquid Systems, Extraction

A convenient notation for classifying mixtures employed in liquid–liquid extraction is $C/\mathcal{N}$, where C is the number of components and $\mathcal{N}$ the number of partially miscible pairs. Mixtures 3/1, 3/2, and 3/3 are called "Type I, Type II, and Type III" by some authors. A typical 3/1 three-component mixture with only one partially miscible pair is furfural–ethylene glycol–water, as shown in Fig. 3.10, where the partially miscible pair is furfural–water. In practice, furfural is used as a solvent to remove the solute, ethylene glycol, from water; the furfural-rich phase is called the *extract*, and the water-rich phase the *raffinate*. Nomenclature for extraction, leaching, absorption, and adsorption always poses a problem because, unlike distillation, concentrations are expressed in many different ways: mole, volume, or mass fractions; mass or mole ratios; and special "solvent-free" designations. In this chapter, we will use V to represent the extract phase and L the raffinate phase, and y and x to represent solute concentration in these phases, respectively. The use of V and L does not imply that the extract phase in extraction is conceptually analogous to the vapor phase in distillation; indeed the reverse is more correct for many purposes.

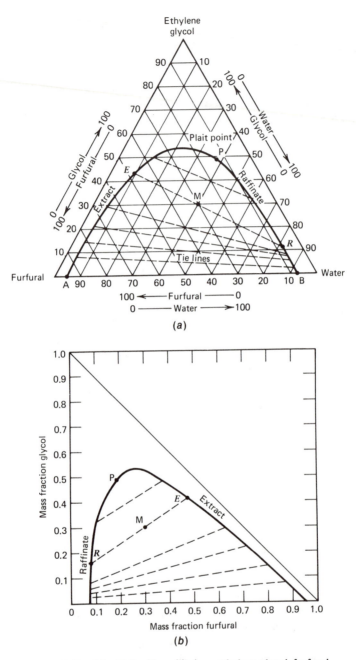

Figure 3.10. Liquid–liquid equilibrium, ethylene glycol–furfural–water, 25°C, 10 kPa.

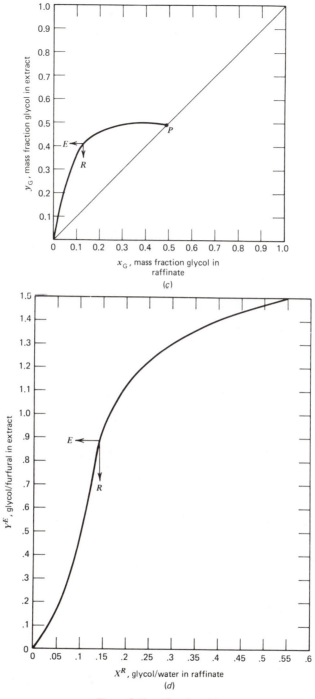

Figure 3.10. (Continued.)

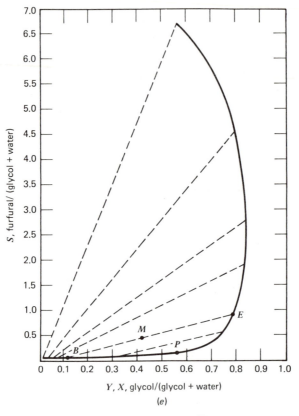

Figure 3.10. (Continued.)

Figure 3.10*a* is the most common way that ternary liquid–liquid equilibrium data are displayed in the chemical literature. Such an equilateral triangular diagram has the property that the sum of the lengths of the perpendiculars drawn from any interior point to the sides equals the altitude. Thus, if each of the altitudes is scaled from 0 to 100, the percent of, say, furfural, at any point such as M, is simply the length of the line perpendicular to the base opposite the pure furfural apex, which represents 100% furfural.

The miscibility limits for the furfural-water binary are at A and B. The miscibility boundary (saturation curve) $\overline{AEPRB}$ is obtained experimentally by a *cloud-point titration*; water, for example, is added to a (clear) 50 wt% solution of furfural and glycol, and it is noted that the onset of cloudiness due to the formation of a second phase occurs when the mixture is 10% water, 45% furfural, 45% glycol by weight. Further miscibility data are given in Table 3.2.

Table 3.2   Equilibrium miscibility data in weight percent: furfural, ethylene glycol, water, 25°C, 101 kPa

Furfural%	Ethylene Glycol%	Water%
94.8	0.0	5.2
84.4	11.4	4.1
63.1	29.7	7.2
49.4	41.6	9.0
40.6	47.5	11.9
33.8	50.1	16.1
23.2	52.9	23.9
20.1	50.6	29.4
10.2	32.2	57.6
9.2	28.1	62.2
7.9	0.0	92.1

To obtain data to construct *tie lines*, such as ER, it is necessary to make a mixture such as M (30% glycol, 40% water, 30% furfural), equilibrate it, and then chemically analyze the resulting extract and raffinate phases E and R (41.8% glycol, 10% water, 48.2% furfural and 11.5% glycol, 81.5% water, 7% furfural, respectively). At point P, called the *plait point*, the two liquid phases have identical compositions. Therefore, the tie lines converge to a point and the two phases become one phase. Tie-line data are given in Table 3.3 for this system.

When there is mutual solubility between two phases, the thermodynamic variables necessary to define the system are temperature, pressure, and the concentrations of the components in each phase. According to the phase rule, for a three-component, two-liquid-phase system, there are three degrees of freedom. At constant temperature and pressure, specification of the concen-

Table 3.3   Mutual equilibrium (tie-line) data for furfural–ethylene glycol–water at 25°C, 101 kPa

Glycol in Water Layer, wt%	Glycol in Furfural Layer, wt%
49.1	49.1
32.1	48.8
11.5	41.8
7.7	28.9
6.1	21.9
4.8	14.3
2.3	7.3

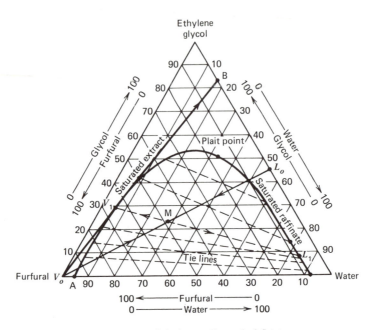

Figure 3.11. Solution to Example 3.5 (a).

.ration of one component in either of the phases suffices to completely define the state of the system.

Figure 3.10*b* is a representation of the same system on a right-triangular diagram. Here the concentration of any two of the three components is shown (normally the solute and solvent are used), the concentration of the third being obtained by difference. Diagrams like this are easier to read than equilateral triangles, and the scales are more readily distended to display regions of interest.

Figures 3.10*c* and 3.10*d* are representations of the same ternary system in terms of weight fraction and weight ratios of the solute. In Fig. 3.10*d* the ratio of coordinates for each point on the curve is a distribution coefficient $K'_{D_i} = Y_i^E/X_i^R$. If K'_D were a constant, independent of concentration, the curve would be a straight line. In addition to their other uses, *x-y* or *X-Y* curves can be used to obtain interpolate tie lines, since only a limited number of tie lines can be shown on triangular graphs. Because of this, *x-y* or *X-Y* diagrams are often referred to as *distribution diagrams*. Numerous other methods for correlating tie-line data for interpolation and extrapolation purposes exist.

In 1906 Janecke[7] suggested the equilibrium data display shown as Fig. 3.10*e*. Here, the mass of solvent per unit mass of solvent-free material, $S =$ furfural/(water + glycol), is plotted as the ordinate versus the concentration, on a solvent-free basis, of glycol/(water + glycol) as abscissa. Weight or mole ratios can be used also.

Any of the five diagrams in Fig. 3.10 (or others) can be used for solving problems involving material balances subject to equilibrium constraints, as is demonstrated in the next three examples.

Example 3.5. Calculate the composition of the equilibrium phases produced when a 45% by weight glycol (G)–55% water (W) solution is contacted with its own weight of pure furfural (F) at 25°C and 101 kPa. Use each of the five diagrams in Fig. 3.10 if possible. What is the composition of the water–glycol mixture obtained by removing all of the furfural from the extract?

Solution. Assume a basis of 100 g of 45% glycol–water feed (see process sketch below).

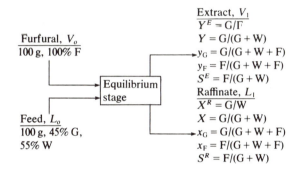

(a) **By equilateral triangular diagram, Fig. 3.11.**

Step 1. Locate the two feed streams at points L_o (55% W/45% G) and (100% F) on Fig. (3.11).

Step 2. Define M, the mixing point, $M = L_o + V_o = L_1 + V_1$.

Step 3. Apply the lever rule to the equilateral triangular phase equilibrium diagram. Letting x_{ij} be the mass fraction of species i in the raffinate stream j and y_{ij} the mass fraction of species i in extract stream j.

Solvent balance: $M x_{F,M} = (L_o + V_o) x_{F,M} = L_o x_{F,L_o} + V_o y_{F,V_o}$

$$\frac{L_o}{V_o} = \frac{y_{F,V_o} - x_{F,M}}{x_{F,M} - x_{F,L}} \tag{1}$$

Thus, points V_o, M, and L_o lie on a straight line, and, by the lever rule, $L_o/V_o = \overline{V_o M}/\overline{M L_o}$.

Step 4. Since M lies in the two-phase region, the mixture must separate into the extract phase V_1 (27.9% G, 6.5% W, 65.6% F) and the raffinate L_1 (8% G, 84% W, 8% F).

Step 5. The lever rule applies to V_1, M, and L_1, so $V_1 = M(\overline{L_1 M}/\overline{L_1 V_1}) = 145.8$ g and $L_1 = 200 - 145.8 = 54.2$ g.

Step 6. The solvent-free extract composition is obtained by extending the line through $\overline{V_o V_1}$ to the point B (81.1% G, 18.9% W, 0% F), since this line is the locus of all possible mixtures that can be obtained by adding or subtracting pure solvent to or from V_1.

(b) **By right-triangular diagram, Fig. 3.12.**

Step 1. Locate the two feed streams L_o, V_o.

Step 2. Define the mixing point $M = L_o + V_o$.

Step 3. The lever rule, as will be proven in Example 3.8, applies to right-triangular diagrams so $\overline{M V_o}/\overline{M L_o} = 1$, and the point M is located.

Step 4. L_1 and V_1 are on the ends of the tie line passing through M.

Step 5. Then B, the furfural stripped extract, is found by extending the line $\overline{V_1 V_o}$ to the zero furfural axis.

The results of Part (b) are identical to those of Part (a).

(c) **By *x-y* diagram, Fig. 3.10c.**

The glycol material balance

$$L_o x_{G,L_o} + V_o y_{G,V_o} = 45 = L_1 x_{G,L_1} + V_1 y_{G,V_1} \tag{2}$$

must be solved simultaneously with the equilibrium relationship. It is not possible to do this graphically using Fig. 3.10c in any straightforward manner. The outlet stream composition can, however, be found by a trial-and-error algorithm.

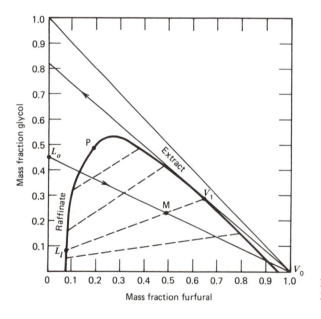

Figure 3.12. Solution to Example 3.5 (b).

Step 1. Pick a value of y_G, x_G from Fig. 3.10c.

Step 2. Substitute this into the equation obtained by combining (2) with the overall balance, $L_1 + V_1 = 200$. Solve for L_1 or V_1.

Step 3. Check to see if the furfural (or water) balance is satisfied using the equilibrium data from Fig. 3.10a, 3.10b, or 3.10e. If it is not, repeat Steps 1 to 3. This procedure leads to the same results obtained in Parts (a) and (b).

(d) **By Y^E-X^R diagram, Fig. 3.10d.** This plot suffers from the same limitations as Fig. 3.10c in that a solution must be achieved by trial and error. If, however, the solvent and carrier are completely immiscible, Y^E-X^R diagrams may be conveniently used. Example 3.6 will demonstrate the methodology.

(e) **By Janecke diagram, Fig. 3.13.**

Step 1. The feed mixture is shown at $X_o = 0.45$. With the addition of 100 g of furfural, $M = L_o + V_o$ is located at $X_M = 0.45$, $S_M = 100/100 = 1$.

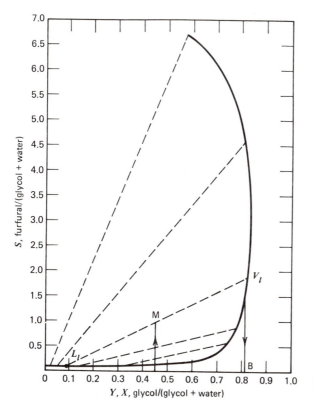

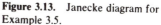

Figure 3.13. Janecke diagram for Example 3.5.

Step 2. This mixture separates into the two streams V_1, L_1 with the coordinates ($S^E = 1.91$, $Y = 0.81$; $S^R = 0.087$, $X = 0.087$).

Step 3. Let $Z^E = (W + G)$ in the extract and $Z^R = (W + G)$ in the raffinate. Then the following balances apply.

$$\text{Solvent: } 1.91 \, Z^E + 0.087 \, Z^R = 100$$

$$\text{Glycol: } 0.81 \, Z^E + 0.087 \, Z^R = 45$$

Solving, $Z^E \doteq 50.00$, $Z^R = 51.72$.

Thus, the furfural in the extract $= (1.91)(50.00) = 95.5$ g, the glycol in the extract $= (0.81)(50.00) = 40.5$ g, and the water in the extract $= 50 - 40.5 = 9.5$ g.

The total extract is $(95.5 + 40.5 + 9.5) = 145.5$ g, which is almost identical to the results obtained in Part (a). The raffinate composition and amount can be obtained just as readily.

It should be noted that on the Janecke diagram $\overline{MV_1}/\overline{ML_1}$ does *NOT* equal L_1/V_1; it equals the ratio of L_1/V_1 on a solvent free basis.

Step 4. Point B, the furfural-free extract composition, is obtained by extrapolating the vertical line through V_1 to $S = 0$. The furfural-free extract mixture is 81.1% glycol and 18.9% water by weight.

□

Example 3.6. As was shown in Example 1.1, *p*-dioxane (D) can be separated from water (W) by using benzene (B) as a liquid–liquid extraction solvent. Assume as before a distribution coefficient $K'_D = (\text{mass D}/\text{mass B})/(\text{mass D}/\text{mass W}) = 1.2$, independent of

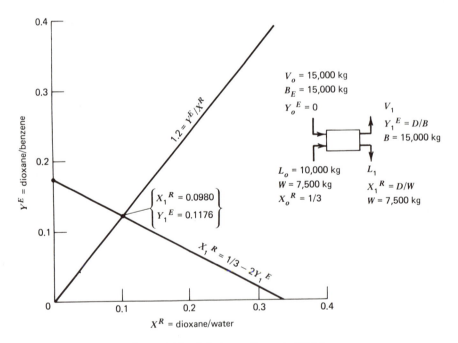

Figure 3.14. Solution to Example 3.6 (a).

composition at 25°C and 101 kPa, the process conditions. Water and benzene may be assumed to be completely immiscible.

For this example, 10,000 kg/hr of a 25 wt% solution of D in water is to be contacted with 15,000 kg of B. What percent extraction is achieved in (a) one single stage and (b) in two cross-flow stages with the solvent split as in Figure 1.14c?

Solution

(a) **Single equilibrium stage.** The constant distribution coefficient plots as a straight line on a Y^E-X^R phase equilibrium diagram, Fig. 3.14. A mass balance for p-dioxane, where B and W are mass flow rates of benzene and water, is

$$WX_o^R + BY_o^E = WX_1^R + BY_1^E \tag{1}$$

Substituting $W = 7500$, $B = 15,000$, $X_o^R = 0.333$, and $Y_o^E = 0$ and solving for X_1^R, we find

$$X_1^R = \frac{1}{3} - 2Y_1^E \tag{2}$$

The intersection of (2), the *material balance operating line,* with the equilibrium line

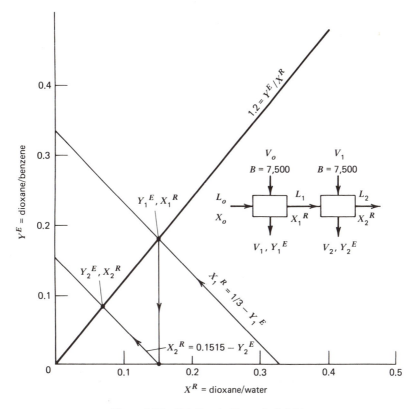

Figure 3.15. Solution to Example 3.6 (b).

marks the composition of the streams V_1, L_1. The fraction p-dioxane left in the raffinate is

$$\frac{WX_1^R}{WX_0^R} = \frac{(1/3 - 2Y_1^E)}{X_o^R} = \frac{0.333 - 2(0.1176)}{0.333} = 0.294$$

Therefore, the percent extracted is 70.6, as in Example 1.1(a).

(b) **Two-stage crossflow.** Equation (1) again applies to the first stage with $W = 7500$, $B = 7500$, $X_o^R = 0.333$, and $Y_o^E = 0$. Thus, $X_1^R = 1/3 - Y_1^E$, which intersects the equilibrium curve at $Y_1^E = 0.1818$ and $X_1^R = 0.1515$, in Fig. 3.15.

For the second stage, the subscripts in (1) are updated, and, since the equation is recursive, $X_2^R = 0.1515 - Y_2^E$; X_1^R replacing X_o^R. Thus, $X_2^R = 0.0689$ and $Y_2^E = 0.0826$. The percent extracted is $(0.333 - 0.0689)/0.333 = 79.34\%$.

☐

Example 3.7. In the Keyes process (Fig. 3.8) for making absolute alcohol from alcohol containing 5 wt% water, a third component, benzene, is added to the alcohol feed. The benzene lowers the volatility of the alcohol and takes the water overhead in a constant minimum boiling azeotropic mixture of 18.5% alcohol, 7.4% H_2O, and 74.1% C_6H_6 (by weight). It is required to produce 100 m³/day of absolute alcohol by this process, as shown in Fig. 3.16. Calculate the volume of benzene that should be fed to the column. Liquid specific gravities are 0.785 and 0.872 for pure ethanol and benzene, respectively. The ternary phase diagram at process conditions is given as Fig. 3.17.

Solution. The starting mixture lies on the line $\overline{DB}$, since this is the locus of all mixtures obtainable by adding benzene to a solution containing 95% alcohol and 5% water. Likewise the line $\overline{CE}$ is the locus of all points representing the addition of absolute alcohol bottom product to the overhead product mixture E (18.5% alcohol, 7.4% water, 74.1% benzene). The intersection of lines $\overline{CE}$ and $\overline{DB}$, point G, represents the combined feedstream composition, which is (approximately) 34% benzene, 63% alcohol, and 3% water. The alcohol balance, assuming 1000 g of absolute alcohol, is

$$0.63 \, F = W + 0.185 \, D \qquad (1)$$

The overall material balance is

$$F = D + W \qquad (2)$$

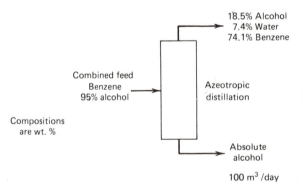

18.5% Alcohol
7.4% Water
74.1% Benzene

Combined feed
Benzene
95% alcohol

Azeotropic
distillation

Absolute
alcohol

100 m³/day

Compositions
are wt. %

Figure 3.16. Flowsheet, Example 3.7. Compositions are weight percent. (Adapted from E. J. Henley and H. Beiber, *Chemical Engineering Calculations*, McGraw-Hill Book Co., New York, © 1959.]

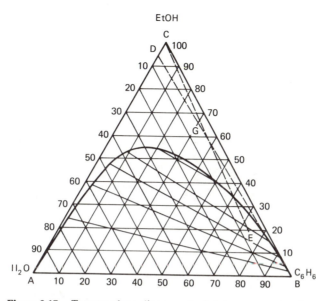

Figure 3.17. Ternary phase diagram, alcohol–benzene–water, in weight percent, 101 kPa, $T = 25°C$. (Data from E. J. Henley and H. Bieber, *Chemical Engineering Calculations*, McGraw-Hill Book Company, New York, © 1959.)

Combining (1) and (2) to solve for F, with W = 1000 g

$$F = \frac{1000 - 0.185(1000)}{0.630 - 0.185} = 1831 \text{ g}$$

The mass of benzene in the combined feed is

$$1831(0.34) = 623 \text{ g}$$

The actual benzene feed rate is $(100)(0.785/0.872)(623/1000) = 56.1 \text{ m}^3/\text{day}$.

□

3.11 Other Liquid–Liquid Diagrams

Some of the cases that arise with 3/2 systems are as shown in Fig. 3.18. Examples of mixtures that produce these configurations are given by Francis[8] and Findlay[9]. In Fig. 3.18*a*, two separate two-phase regions are formed, while in Fig. 3.18*c*, in addition to the two-phase regions, a three-phase region RST is formed. In Fig. 3.18*b*, the two-phase regions merge. For a ternary mixture, as temperature is reduced, phase behavior may progress from Fig. 3.18*a* to 3.18*b* to 3.18*c*. In Fig. 3.18*a*, 3.18*b*, and 3.18*c* all tie lines slope in the same direction. In some systems of importance, *solutropy*, a reversal of tie-line slopes, occurs.

Quaternary mixtures are encountered in some extraction processes, parti-

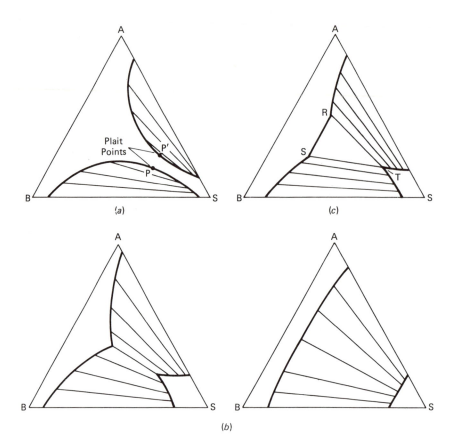

Figure 3.18. Equilibria for 3/2 systems. At (*a*), the miscibility boundaries are separate; at (*b*), the miscibility boundaries and tie-line equilibria merge; at (*c*), the tie lines do not merge and the three-phase region RST is formed.

cularly when two solvents are used for fractional liquid–liquid extraction. In general, multicomponent equilibria are very complex and there is no compact graphical way of representing the data. The effect of temperature on equilibria is very acute; usually elevating the temperature narrows the range of immiscibility.

3.12 Liquid–Solid Systems, Leaching

From a phase rule standpoint, there is no difference between a liquid–liquid or a solid–liquid system. Phase equilibrium data for a three-component mixture of solute, solid, and solvent at constant temperature and pressure can therefore be represented on equilateral or right-triangular, *x-y*, or mass ratio diagrams.

However, there are major differences between liquid–liquid and solid–liquid contacting because, in the latter case, diffusion in the solid is so slow that true equilibrium is rarely achieved in practice. Also, drainage is frequently slow, so complete phase separations are seldom realized in mixer–settlers, the most common type of leaching equipment employed. It is necessary, therefore, to take a rather pragmatic approach to equipment design. Instead of using thermodynamic equilibrium data to calculate stage requirements, one uses pilot plant or bench-scale data taken in prototype equipment where residence times, particle size, drainage conditions, and level of agitation are such that the data can be extrapolated to plant-size leaching equipment. Stage efficiencies are therefore inherently included in the so-called equilibrium diagrams. Also, instead of having equilibrium phases, we have an *overflow* solution in equilibrium with solution adhering to an *underflow* of solids and solution.

If (1) the carrier solid is completely inert and is not dissolved or entrained in the solvent, (2) the solute is infinitely soluble in the solvent, and (3) sufficient contact time for the solvent to penetrate the solute completely is permitted, ideal leaching conditions exist and the phase equilibrium diagrams will be as shown in Fig. 3.19a. Here the following nomenclature is employed.

X_s = solute/(solvent + solute) in the overflow effluent

Y_s = solute/(solvent + solute) in the underflow solid or slurry

Y_I = inerts/(solute + solvent)

y = mass fraction solvent

x = mass fraction solute

In Fig. 3.19a $Y_s = X_s$ since the equilibrium solutions in both underflow and overflow have the same composition. Also $Y_I = 0$ in the overflow when there is complete drainage and the carrier is not soluble in the solvent. In the y-x diagram, the underflow line $\overline{AB}$ is parallel to the overflow line $\overline{FD}$, and the extrapolated tie lines (e.g., $\overline{FE}$) pass through the origin (100% inerts).

In nonideal leaching, Fig. 3.19b, the tie lines slant to the right, indicating that the solute is more highly concentrated in the underflow, either because of equilibrium solubility or because of incomplete leaching (the latter is more likely when the solute and solvent are completely miscible). Also, curve $\overline{CD}$ does not coincide with the $Y_I = 0$ axis, indicating a partially miscible carrier or incomplete settling. In the right-triangular diagram, the tie line $\overline{FE}$ does not extrapolate to $y = 0$.

If the solubility of solute in the solvent is limited, the underflow curve $\overline{AEB}$ would dip down and intersect the abscissa before $X_s = 1$, and the Y_s-X_s curve would be vertical at that point.

Construction of material balance lines on solid–liquid diagrams depends

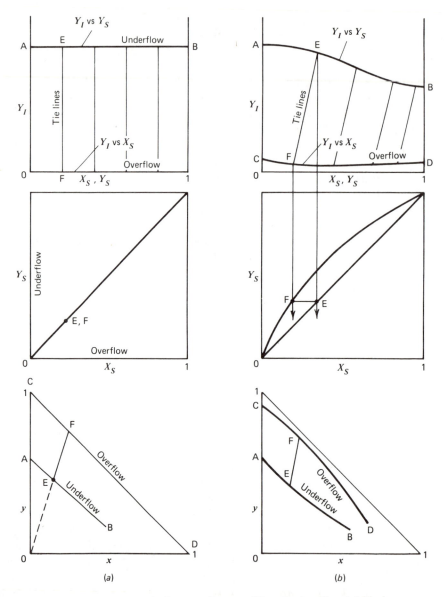

Figure 3.19. Underflow-overflow conditions for leaching. (*a*) Ideal leaching conditions. (*b*) Nonideal leaching conditions.

critically on the coordinates used to represent the experimental data. In the following example, data are given as y-x mass fractions on a right-triangular diagram, with no solubility of the inerts in the overflow and with constant underflow, such that diagrams of the type shown in Fig. 3.19a apply. The method of solution, however, would be identical if the diagrams resembled Fig. 3.19b.

Example 3.8. Given the experimental data for the extraction of oil from soybeans by benzene in Fig. 3.20, calculate effluent compositions if one kilogram of pure benzene is mixed with one kilogram of meal containing 50% by weight of oil. What are the amounts

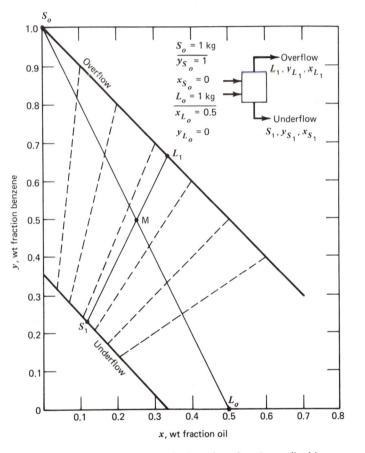

Figure 3.20. Experimental data for leaching of soybean oil with benzene. (Modified from W. L. Badger and J. T. Banchero, *Introduction to Chem. Engr.* McGraw-Hill Book Co., New York, © 1955 p. 347.)

of underflow and overflow leaving the extractor? What percentage of oil is recovered in the benzene overflow?

Solution

Step 1. Locate the two feeds L_o at $y_{L_o} = 0$, $x_{L_o} = 0.5$ and S_o at $y_{S_o} = 1$, $x_S = 0$.

Step 2. Define the mixing point $M = L_o + S_o$, where quantities are in kilograms.

Step 3. The lever rule can be applied, making an oil balance

$$Mx_M = (L_o + S_o)x_M = L_o x_{L_o} + S_o x_{S_o}$$

$$\frac{L_o}{S_o} = \frac{x_{S_o} - x_M}{x_M - x_{L_o}} \tag{1}$$

or a solvent balance

$$My_M = (L_o + S_o)y_M = L_o y_{L_o} + S_o y_{S_o}$$

$$\frac{L_o}{S_o} = \frac{y_{S_o} - y_M}{y_M - y_{L_o}} \tag{2}$$

The point M must lie on a straight line connecting S_o and L_o. Then by (1) or (2) $L_o/S_o = S_o M/ML_o$.

Step 4. The mixture M is in the two-phase region and must split into the two equilibrium streams L_1, at $y = 0.667$, $x = 0.333$, and S_1, at $y = 0.222$, $x = 0.111$.

Step 5. Since $L_1 + S_1 = M$, the ratio of $L_1/M = \overline{MS_1}/\overline{L_1 S_1} = 0.625$, so $L_1 = 1.25$ kg and $S_1 = 2.00 - 1.25 = 0.75$ kg. The underflow consists of 0.50 kg of solid and 0.25 kg of solution adhering to the solid.

Step 6. Confirm the results by a solvent balance

$$L_1 y_{L_1} + S_1 y_{S_1} = L_o y_{L_o} + S_o y_{S_o}$$

$$1.25(0.667) + 0.75(0.222) = 1.00 \text{ kg benzene}$$

Step 7. Percent recovery of oil $= (L_1 x_{L_1}/L_o x_{L_o})100 = [(1.25)(0.333)/(1.0)(0.5)](100) = 83.25\%$

□

3.13 Adsorption and Ion Exchange

Calculation procedures for adsorption and ion exchange differ only in detail from liquid–liquid extraction since an ion-exchange resin or adsorbent is analogous to the solvent in extraction. All coordinate systems used to represent solvent–solute or liquid–vapor equilibria may be used to display three-component solid–liquid, or solid–gas phase equilibria states. For the case of gas adsorption, equilibria are usually a function of pressure and temperature, and so isobaric and isothermal displays such as Fig. 3.21, which represents the propane–propylene–silica gel system, are convenient.

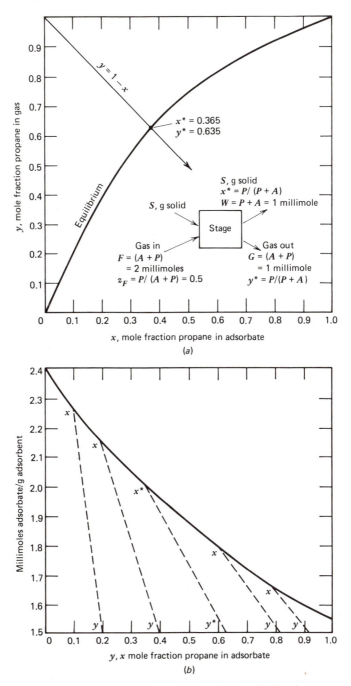

Figure 3.21. Adsorption equilibrium at 25°C and 101 kPa of propane and propylene on silica gel. [Adapted from W. K. Lewis, E. R. Gilliland, B. Chertow, and W. H. Hoffman, *J. Amer. Chem. Soc.*, **72**, 1153 (1950).]

Equilateral and right-triangular equilibrium diagrams can also be constructed, but they are generally not as useful because the weight percent of adsorbed gas is frequently so small that tie lines at the adsorbent axis become very crowded and difficult to read. Since Fig. 3.21 represents a ternary, two-phase system, only three variables need be specified to satisfy the phase rule. If the mole fraction of propane in a binary gaseous mixture in equilibrium with adsorbed gas at 101 kPa and 25°C is 0.635, then all other quantities describing the equilibrium state are fixed. From Fig. 3.21, the mole fraction propane in the adsorbate must be 0.365 (the fact that these two numbers add to one is coincidental). The concentration of adsorbate on silica adsorbent is also a dependent variable and can be obtained from Fig. 3.21b as 1.99. The ratio 0.365/0.635 can be viewed as a separation factor similar to the K-value in distillation or distribution coefficient in extraction. A separation factor analogous to the relative volatility can also be defined for adsorption of propylene relative to propane. For this example it is $(1 - 0.365)(0.635)/[(1 - 0.635)/(0.365)]$ or 3.03, which is much larger than the relative volatility for distillation. Nevertheless the separation of propylene from propane by adsorption is not widely practiced.

Just as a T-x-y diagram contains more information than an x-y diagram, Fig. 3.21b displays an additional parameter not present in Fig. 3.21a. Example 3.9 demonstrates how Fig. 3.21a can be used to make an equilibrium-stage calculation, Fig. 3.21b being used to obtain auxiliary information. Alternative solutions involving only Fig. 3.21b are possible. Also, diagrams based on mole or weight ratios could be used, with only slight adjustments in the material balance formulations.

Liquid–solid adsorption, and ion-exchange equilibrium data and material balances, are handled in a manner completely analogous to gas–solid systems. An example of a liquid–solid ion-exchange design calculation is included in Chapter 8.

Example 3.9. Propylene (A) and propane (P), which are difficult to separate by distillation, have been separated on an industrial scale by preferential adsorption of propylene on silica gel (S), the equilibrium data at 25°C and 101 kPa being as shown in Fig. 3.21.

Two millimoles of a gas containing 50 mole% propane is equilibrated with silica gel at 25°C and 101 kPa. Manometric measurements show that 1 millimole of gas was adsorbed. What is the concentration of propane in the gas and adsorbate, and how many grams of silica gel were used?

Solution. A pictorial representation of the process is included in Fig. 3.21, where W = millimoles of adsorbate, G = millimoles of gas leaving, and z_F = mole fraction of propane in the feed.
The propane mole balance is

$$Fz_F = Wx^* + Gy^* \tag{1}$$

With $F = 2$, $z_F = 0.5$, $W = 1$, and $G = 1$, (1) becomes $1 = x^* + y^*$.

The operating (material balance) line $y^* = 1 - x^*$, the locus of all solutions of the material balance equations, is shown on Fig. 3.21a. It intersects the equilibrium curve at $x^* = 0.365$, $y^* = 0.635$. From Fig. 3.21b at the point x^*, there must be 1.99 millimoles adsorbate/g adsorbent; therefore there were $1.0/1.99 = 0.5025$ g of silica gel in the system.

☐

3.14 Gas–Liquid Systems, Absorption, and Henry's Law

When a liquid S is used to absorb gas A from a gaseous mixture of $A + B$, the thermodynamic variables for a single equilibrium stage are P, T, x_A, x_B, x_S, y_A, y_B, and y_S. There are three degrees of freedom; hence, if three variables P, T, and y_A are specified, all other variables are determined and phase equilibrium diagrams such as Fig. 3.22a and 3.22b can be constructed. Should the solvent S have negligible vapor pressure and the carrier gas B be insoluble in S, then the only variables remaining are P, T, x_A, and y_A, then Fig. 3.22a is of no value.

When the amount of gas that dissolves in a liquid is relatively small, a linear equilibrium relationship may often be assumed with reasonable accuracy. Henry's law, $p_A = k_h x_A$, where p_A is the partial pressure of gas A above the solution, x_A is the mole fraction of A in solution, and k_h is a constant, is such a linear expression. Figure 3.23 gives Henry's law constants as a function of temperature for a number of gases dissolved in water. The following two examples demonstrate calculation procedures when Henry's law applies and when it does not.

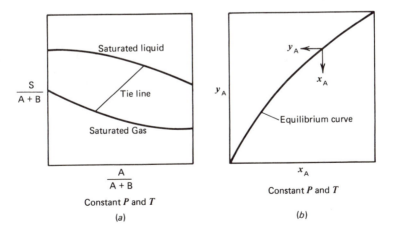

Figure 3.22. Gas–liquid phase equilibrium diagram.

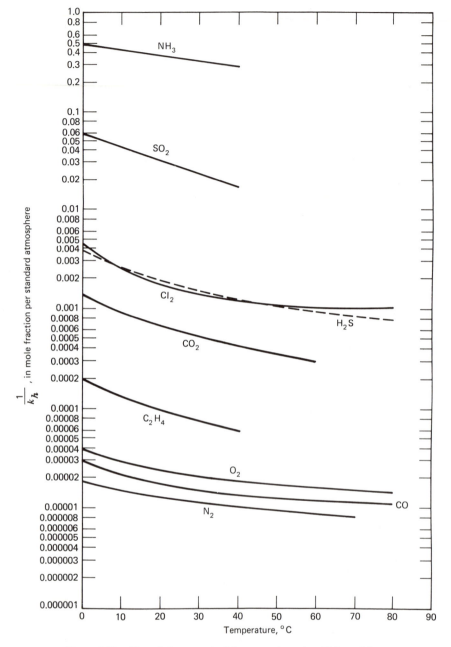

Figure 3.23. Henry's law constant for gases in water. [Adapted from A. X. Schmidt, and H. L. List, *Material and Energy Balances*, Prentice-Hall, Englewood Cliffs, N.J. ©, 1962.]

Example 3.10. The DuPont Company's Nitro West Virginia Ammonia Plant, which is located at the base of a 300-ft (91.44-m) mountain, employed a unique adsorption system for disposing of by-product CO_2. The CO_2 was absorbed in water at a CO_2 partial pressure of 10 psi (68.8 kPa) above that required to lift water to the top of the mountain. The CO_2 was then vented to the atmosphere at the top of the mountain, the water being recirculated as shown in Fig. 3.24. At 25°C, calculate the amount of water required to dispose of 1000 ft³ (28.31 m³) (STP) of CO_2.

Solution. Basis: 1000 ft³ (28.31 m³) of CO_2 at 0°C and 1 atm (STP). From Fig. 3.23 the reciprocal of Henry's law constant for CO_2 at 25°C is 6×10^{-4} mole fraction/atm. The CO_2 pressure in the absorber (at the foot of the mountain) is

$$p_{CO_2} = \frac{10}{14.7} + \frac{300 \text{ ft } H_2O}{34 \text{ ft } H_2O/\text{atm}} = 9.50 \text{ atm} = 960 \text{ kPa}$$

At this partial pressure, the equilibrium concentration of CO_2 in the water is

$$x_{CO_2} = 9.50(6 \times 10^{-4}) = 5.7 \times 10^{-3} \text{ mole fraction } CO_2$$

The corresponding ratio of dissolved CO_2 to water is

$$\frac{5.7 \times 10^{-3}}{1 - 5.7 \times 10^{-3}} = 5.73 \times 10^{-3} \text{ mole } CO_2/\text{mole } H_2O$$

The total moles of gas to be absorbed are

$$\frac{1000 \text{ ft}^3}{359 \text{ ft}^3/\text{lbmole (at STP)}} = \frac{1000}{359} = 2.79 \text{ lbmole}$$

or

$$(2.79)(44)(0.454) = 55.73 \text{ kg}$$

Assuming all the absorbed CO_2 is vented at the mountain top, the moles of water required

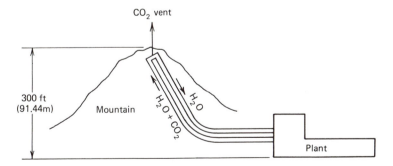

Figure 3.24. Flowsheet, Example 3.10.

are

$$2.79/(5.73 \times 10^{-3}) = 485 \text{ lbmole} = 8730 \text{ lb} = 3963 \text{ kg}$$

If one corrects for the fact that the pressure on top of the mountain is 101 kPa, so that not all of the CO_2 is vented, 4446 kg (9810 lb) of water are required.
□

Example 3.11.   The partial pressure of ammonia in air–ammonia mixtures in equilibrium with their aqueous solutions at 20°C is given in Table 3.4. Using these data, and neglecting the vapor pressure of water and the solubility of air in water, construct an equilibrium diagram at 101 kPa using mole ratios Y_A = moles NH_3/mole air, X_A = moles NH_3/mole H_2O as coordinates. Henceforth, the subscript A for ammonia will be dropped.
(a) If 10 moles of gas, of composition $Y = 0.3$, are contacted with 10 moles of a solution of composition $X = 0.1$, what will be the composition of the resulting phases at equilibrium? The process is isothermal and at atmospheric pressure.

Solution.   The equilibrium data given in Table 3.4 are recalculated, in terms of mole

Table 3.4   Partial pressure of ammonia over ammonia–water solutions at 20°C

NH₃ Partial Pressure, kPa	g NH₃/g H₂O
4.23	0.05
9.28	0.10
15.2	0.15
22.1	0.20
30.3	0.25

Source. Data from *Chemical Engineers Handbook,* 4th ed., R. H. Perry, C. H. Chilton, and S. D. Kirkpatrick, Eds., McGraw–Hill Book Co., New York, 1963, p. 14–4.

Table 3.5   Y-X data for ammonia–water, 20°C

Y, Moles NH₃/Mole Air	X, Moles NH₃/Mole H₂O
0.044	0.053
0.101	0.106
0.176	0.159
0.279	0.212
0.426	0.265

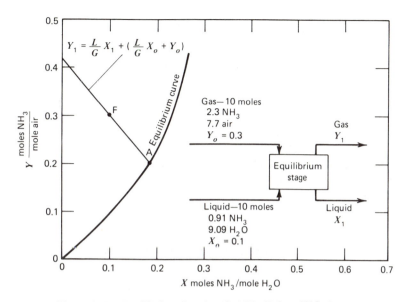

Figure 3.25. Equilibrium data for air–NH$_3$–H$_2$O at 20°C, 1 atm.

ratios in Table 3.5, and plotted in Fig. 3.25.

$$\text{Moles NH}_3 \text{ in entering gas} = 10(Y/(1+Y)) = 10(0.3/1.3) = 2.3$$

$$\text{Moles NH}_3 \text{ in entering liquid} = 10(X/(1+X)) = 10(0.1/1.1) = 0.91$$

A molal material balance for ammonia about the equilibrium stage is

$$GY_o + LX_o = GY_1 + LX_1 \tag{1}$$

where G = moles of air and L = moles of H$_2$O. Then $G = 10 - 2.3 = 7.7$ moles and $L = 10 - 0.91 = 9.09$ moles.
Solving for Y_1 from (1), we have

$$Y_1 = -\frac{L}{G}X_1 + \left(\frac{L}{G}X_o + Y_o\right) \tag{2}$$

This material balance relationship is an equation of a straight line of slope $L/G = 9.09/7.7 = 1.19$, with an intercept of $(L/G)(X_o) + Y_o = 0.42$.

The intersection of this material balance line with the equilibrium curve, as shown in Fig. 3.25, gives the composition of the gas and liquid phases leaving the stage $Y_1 = 0.195$, $X_1 = 0.19$. This result can be checked by an NH$_3$ balance, since the amount of NH$_3$ leaving is $(0.195)(7.70) + (0.19)(9.09) = 3.21$, which equals the total amount of NH$_3$ entering.

It is of importance to recognize that (2), the material balance line, is the locus of all passing stream pairs; thus, X_o, Y_o (Point F) also lies on this operating line.

□

3.15   Variables Other Than Concentration

The phase equilibrium diagrams shown thus far have been in terms of T, P, and concentrations; however, other thermodynamic functions can be used in place of these. In graphical methods for distillation, for instance, it is sometimes convenient to work in terms of enthalpy rather than temperature, because the diagram can then be used to show heat addition or removal as well as composition changes.

Figure 3.26 is a composition enthalpy diagram for the n-hexane/n-octane system at constant pressure. An example demonstrating the construction and utility of this enthalpy-composition diagram follows.

Example 3.12.
(a) Using the thermodynamic data in Table 3.6 and vapor–liquid equilibrium data developed in Example 3.1, construct an enthalpy-composition diagram (H-y, h-x) for the system n-hexane/n-octane at 101 kPa total pressure, where H_V is vapor enthalpy and H_L is liquid enthalpy.

(b) Solve Part (b) of Example 3.1, assuming that the liquid is initially at 100°F (37.8°C). Calculate the amount of energy added per mole in each case.

(c) Calculate the energy required for 60 mole% vaporization at 101 kPa of a mixture initially at 100°F (37.8°C) and containing 0.2 mole fraction n-hexane.

Solution.   Basis: 1 lbmole (0.454 kgmole) of hexane–octane.
(a)   From Example 3.1, vapor–liquid equilibrium data at 101 kPa are listed in Table 3.7. Corresponding saturated liquid and vapor phase enthalpies in Table 3.7 are obtained after converting the enthalpy data in Table 3.6 to Btu/lbmole using molecular weights of 86.2 and 114.2.

For example, the data for 200°F (93.3°C) are calculated as follows, assuming no heats of mixing (ideal solution).

$$H_L = (0.377)(117)(86.2) + (1.0 - 0.377)(109)(114.2)$$
$$= 11,557 \text{ Btu/lbmole } (2.686 \times 10^7 \text{ J/kgmole})$$

$$H_V = (0.773)(253)(86.2) + (1.0 - 0.773)(248)(114.2)$$
$$= 23,287 \text{ Btu/lbmole } (5.413 \times 10^7 \text{ J/kgmole})$$

Enthalpies for subcooled liquid and superheated vapor are obtained in a similar manner. For example, enthalpy of a subcooled liquid equimolar mixture at 100°F (37.8°C) is computed as follows using data from Table 3.6.

$$H_L = (0.5)(55.5)(86.2) + (0.5)(52)(114.2)$$
$$= 5361 \text{ Btu/lbmole } (1.246 \times 10^7 \text{ J/kgmole})$$

From enthalpy calculations of subcooled liquid, saturated liquid, saturated vapor, and superheated vapor, Fig. 3.26 is constructed.

(b)   The path $\overline{AB}$, Fig. 3.26 denotes heating of 1 mole of liquid with 0.3 mole fraction of hexane until the bubble point is reached at 210°F (98.9°C). Heat added = $12,413 - 5991 = 6422$ Btu/lbmole (1.493×10^7 J/kgmole).

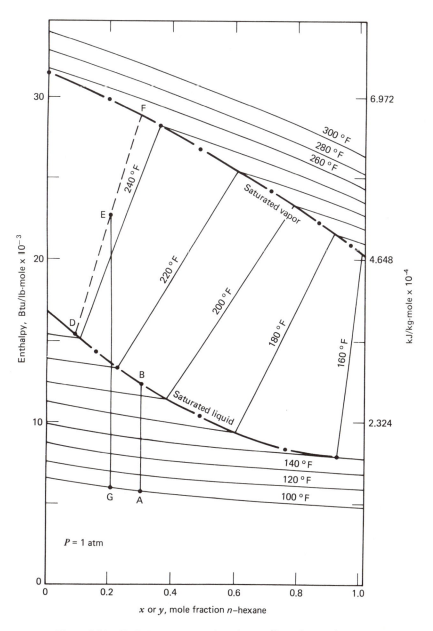

Figure 3.26. Enthalpy concentration diagram for n-hexane/n-octane. Solution to Example 3.12.

Table 3.6 Enthalpy data for *n*-hexane/*n*-octane at 101 kPa. Enthalpy datum: $H_L = 0$ @ 0°F

T, °F	n-Hexane H_L	n-Hexane H_V	n-Octane H_L	n-Octane H_V
100	55.5	210	57	203
125	70.5	220	65	214
150	85	230.5	80	225
175	100.5	241	95	237
200	117	253	109	248
225	133	266	125	260
250	150	278	140	272.5
275	167	290.5	157.5	285
300	185	303	173	298

Source. Data of J. B. Maxwell, *Data Book on Hydrocarbons*, D. Van Nostrand Co., New York, 1950, pp. 103, 105.

Table 3.7 Tabulated *H-y*, *H-x* data for *n*-hexane/*n*-octane at 101 kPa

T, °F	n-Hexane Mole Fractions x	n-Hexane Mole Fractions y	Mixture Enthalpy, Btu/lbmole H_L	Mixture Enthalpy, Btu/lbmole H_V
155.7	1.0	1.0	7,586	20,085
160	0.917	0.986	8,030	20,309
170	0.743	0.947	8,198	20,940
180	0.600	0.900	9,794	21,642
19C	0.481	0.842	10,665	22,433
200	0.377	0.773	11,557	23,287
210	0.295	0.693	12,413	24,286
220	0.215	0.592	13,309	25,432
230	0.151	0.476	14,193	26,709
240	0.099	0.342	15,064	28,140
250	0.045	0.178	15,850	29,845
258.2	0.0	0.0	16,559	31,405

(c)   The path $\overline{GE}$ in Fig. 3.26 denotes heating of 1 lbmole (0.454 kgmole) of liquid until 60 mole% has been vaporized. Terminals of tie line $\overline{DEF}$ are equilibrium vapor and liquid mole fractions. Heat added = 22,900 − 6164 = 16,736 Btu/lbmole (3.890 × 10⁷ J/kgmole).

□

References

1. Keyes, D. B., *Ind. Eng. Chem.*, **21**, 998–1001 (1929).

2. Keyes, D. B., U.S. Pat. 1,676,735, June 10, 1928.

3. Schreve, N., *Chemical Process Industries*, McGraw–Hill Book Co., New York, 1945, 659.

4. Edmister, W. C., and K. K. Okamoto, *Petroleum Refiner*, **38**, (8), 117–129 (1959).

5. Edmister, W. C., and K. K. Okamoto,

Petroleum Refiner, **38**, (4), 271–280 (1959).

6. Taylor, D. L., and W. C. Edmister, *AIChE J.*, **17**, 1324–1329 (1971).

7. Janecke, E., *Z. Anorg. Allg. Chem.*, **51**, 132–157 (1906).

8. Francis, A. W., *Liquid–Liquid Equilibriums*, Interscience Publishing Co., New York, 1963.

9. Findlay, A., *Phase Rule*, Dover Publications, New York, 1951.

Problems

3.1 A liquid mixture containing 25 mole% benzene and 75 mole% ethyl alcohol, which components are miscible in all proportions, is heated at a constant pressure of 1 atm (101.3 kPa, 760 torr) from a temperature of 60°C to 90°C.

(a) At what temperature does vaporization begin?

(b) What is the composition of the first bubble of equilibrium vapor formed?

(c) What is the composition of the residual liquid when 25 mole% has evaporated? Assume that all vapor formed is retained within the apparatus and that it is completely mixed and in equilibrium with the residual liquid.

(d) Repeat Part (c) for 90 mole% vaporized.

(e) Repeat Part (d) if, after 25 mole% vaporized as in Part (c), the vapor formed is removed and an additional 35 mole% is vaporized by the same technique used in Part (c).

(f) Plot the temperature versus the percent vaporized for Parts (d) and (e).

  Use the vapor pressure data below in conjunction with Raoult's and Dalton's laws to construct a T-x-y diagram, and compare it and the answers obtained in Parts (a) and (f) with those obtained using the experimental T-x-y data given below. What do you conclude?

Vapor pressure data

Vapor pressure, torr	20	40	60	100	200	400	760
Ethanol, °C	8	19.0	26.0	34.9	48.4	63.5	78.4
Benzene, °C	−2.6	7.6	15.4	26.1	42.2	60.6	80.1

Experimental T-x-y data for benzene–ethyl alcohol at 1 atm

Temperature, °C	78.4	77.5	75	72.5	70	68.5	67.7	68.5	72.5	75	77.5	80.1
Mole% benzene in vapor	0	7.5	28	42	54	60	68	73	82	88	95	100
Mole% benzene in liquid	0	1.5	5	12	22	31	68	81	91	95	98	100

3.2 Repeat Example 3.2 for the following liquid mixtures at 50°C.
(a) 50 mole% benzene and 50 mole% water.
(b) 50 mole% toluene and 50 mole% water.
(c) 40 mole% benzene, 40 mole% toluene, and 20 mole% water.

3.3 A gaseous mixture of 75 mole% water and 25 mole% n-octane at a pressure of 133.3 kPa (1000 torr) is cooled under equilibrium conditions at constant pressure from 136°C.
(a) What is the composition of the first drop to condense?
(b) What is the composition and temperature of the last part of the vapor to condense?
Assume water and n-octane are immiscible liquids.

3.4 Stearic acid is to be steam distilled at 200°C in a direct-fired still, heat jacketed to prevent condensation. Steam is introduced into the molten acid in small bubbles, and the acid in the vapor leaving the still has a partial pressure equal to 70% of the vapor pressure of pure stearic acid at 200°C. Plot the kilograms of acid distilled per kilogram of steam added as a function of total pressure from 101.3 kPa down to 3.3 kPa at 200°C. The vapor pressure of stearic acid at 200°C is 0.40 kPa.

3.5 The relative volatility, α, of benzene to toluene at 1 atm is 2.5. Construct an x-y diagram for this system at 1 atm. Repeat the construction using vapor pressure data for benzene from Problem 3.1 and for toluene from the table below in conjunction with Raoult's and Dalton's laws. Also construct a T-x-y diagram.
(a) A liquid containing 70 mole% benzene and 30 mole% toluene is heated in a container at 1 atm until 25 mole% of the original liquid is evaporated. Determine the temperature. The phases are then separated mechanically, and the vapors condensed. Determine the composition of the condensed vapor and the liquid residue.
(b) Calculate and plot the K-values as a function of temperature at 1 atm.

Vapor pressure of toluene

Vapor pressure, torr	20	40	60	100	200	400	760	1520
Temperature, °C	18.4	31.8	40.3	51.9	69.5	89.5	110.6	136

3.6 The vapor pressures of toluene and n-heptane are given in the accompanying tables.

Vapor pressure of n-heptane

Vapor pressure, torr	20	40	60	100	200	400	760	1520
Temperature, °C	9.5	22.3	30.6	41.8	58.7	78.0	98.4	124

(a) Plot an x-y equilibrium diagram for this system at 1 atm by using Raoult's and Dalton's laws.
(b) Plot the T-x bubble-point curve at 1 atm.
(c) Plot α and K-values versus temperature.
(d) Repeat Part (a) using the arithmetic average value of α, calculated from the two extreme values.
(e) Compare your x-y and T-x-y diagrams with the following experimental data of Steinhauser and White [*Ind. Eng. Chem.*, **41**, 2912 (1949)].

Vapor–liquid equilibrium data for n-heptane/toluene at 1 atm

$x_{n\text{-heptane}}$	$y_{n\text{-heptane}}$	$T,\ °C$
0.025	0.048	110.75
0.129	0.205	106.80
0.250	0.349	104.50
0.354	0.454	102.95
0.497	0.577	101.35
0.692	0.742	99.73
0.843	0.864	98.90
0.940	0.948	98.50
0.994	0.993	98.35

3.7 Saturated liquid feed, at $F = 40$, containing 50 mole% A in B is supplied continuously to the apparatus below. The condensate from the condenser is split so that half of it is returned to the still pot.
(a) If heat is supplied at such a rate that $W = 30$ and $\alpha = 2$, as defined below, what will be the composition of the overhead and the bottom product?
(b) If the operation is changed so that no condensate is returned to the still pot and $W = 3D$ as before, what will be the composition of the products?

$$\alpha = \text{relative volatility} = \frac{P_A^s}{P_B^s} = \frac{y_A x_B}{x_A y_B}$$

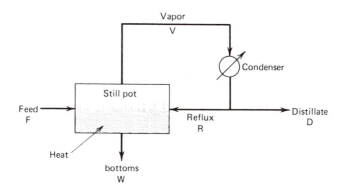

3.8 Vapor–liquid equilibrium data for mixtures of water and isopropanol at 1 atm (101.3 kPa, 760 torr) are given below.
(a) Prepare T-x-y and x-y diagrams.
(b) When a solution containing 40 mole% isopropanol is slowly vaporized, what will be the composition of the initial vapor formed?
(c) If this same 40% mixture is heated under equilibrium conditions until 75 mole% has been vaporized, what will be the compositions of the vapor and liquid produced?

Vapor–liquid equilibrium for isopropanol and water

T, °C	x	y
93.00	1.18	21.95
89.75	3.22	32.41
84.02	8.41	46.20
83.85	9.10	47.06
82.12	19.78	52.42
81.64	28.68	53.44
81.25	34.96	55.16
80.62	45.25	59.26
80.32	60.30	64.22
80.16	67.94	68.21
80.21	68.10	68.26
80.28	76.93	74.21
80.66	85.67	82.70
81.51	94.42	91.60

Notes: All compositions in mole% isopropanol.
Composition of the azeotrope: $x = y = 68.54\%$.
Boiling point of azeotrope: 80.22°C.
Boiling point of pure isopropanol: 82.5°C.

(d) Calculate K-values and α's at 80°C and 89°C.
(e) Compare your answers in Parts (a), (b), and (c) to those obtained from T-x-y and x-y diagrams based on the following vapor pressure data and Raoult's and Dalton's laws.

Vapor pressures of isopropanol and water

Vapor pressure, Torr	200	400	760
Isopropanol, °C	66.8	82	97.8
Water, °C	66.5	83	100

3.9 Forty-five kilograms of a solution containing 0.3 wt fraction ethylene glycol in water is to be extracted with furfural. Using Fig. 3.10a and 3.10e, calculate:
(a) Minimum quantity of solvent.
(b) Maximum quantity of solvent.
(c) The weights of solvent-free extract and raffinate for 45 kg solvent and the percent glycol extracted.
(d) The maximum possible purity of glycol in the finished extract and the maximum purity of water in the raffinate for one equilibrium stage.

3.10 Prove that in a triangular diagram, where each vertex represents a pure component, the composition of the system at any point inside the triangle is proportional to the length of the respective perpendicular drawn from the point to the side of the triangle opposite the vertex in question. It is not necessary to assume a special case (i.e., a right or equilateral triangle) to prove the above.

3.11 A mixture of chloroform (CHCl₃) and acetic acid at 18°C and 1 atm (101.3 kPa) is to be extracted with water to recover the acid.

(a) Forty-five kilograms of a mixture containing 35 wt% CHCl₃ and 65 wt% acid is treated with 22.75 kg of water at 18°C in a simple one-stage batch extraction. What are the compositions and weights of the raffinate and extract layers produced?

(b) If the raffinate layer from the above treatment is extracted again with one half its weight of water, what will be the compositions and weights of the new layers?

(c) If all the water is removed from this final raffinate layer, what will its composition be?

Solve this problem using the following equilibrium data to construct one or more of the types of diagrams in Fig. 3.10.

Liquid–liquid equilibrium data for CHCl₃–H₂O–CH₃COOH at 18°C and 1 atm

Heavy Phase (wt%)			Light Phase (wt%)		
CHCl₃	H₂O	CH₃COOH	CHCl₃	H₂O	CH₃COOH
99.01	0.99	0.00	0.84	99.16	0.00
91.85	1.38	6.77	1.21	73.69	25.10
80.00	2.28	17.72	7.30	48.58	44.12
70.13	4.12	25.75	15.11	34.71	50.18
67.15	5.20	27.65	18.33	31.11	50.56
59.99	7.93	32.08	25.20	25.39	49.41
55.81	9.58	34.61	28.85	23.28	47.87

3.12 Isopropyl ether (E) is used to separate acetic acid (A) from water (W). The liquid–liquid equilibrium data at 25°C and 1 atm (101.3 kPa) are given below.

(a) One hundred kilograms of a 30 wt% A-W solution is contacted with 120 kg of ether in an equilibrium stage. What are the compositions and weights of the

Liquid–liquid equilibrium data for acetic acid (A), water (W), and isopropanol ether (E) at 25°C and 1 atm

Water-Rich Layer			Ether-Rich Layer		
Wt% A	Wt% W	Wt% E	Wt% A	Wt% W	Wt% E
1.41	97.1	1.49	0.37	0.73	98.9
2.89	95.5	1.61	0.79	0.81	98.4
6.42	91.7	1.88	1.93	0.97	97.1
13.30	84.4	2.3	4.82	1.88	93.3
25.50	71.1	3.4	11.4	3.9	84.7
36.70	58.9	4.4	21.6	6.9	71.5
45.30	45.1	9.6	31.1	10.8	58.1
46.40	37.1	16.5	36.2	15.1	48.7

resulting extract and raffinate? What would be the concentration of acid in the (ether-rich) extract if all the ether were removed?

(b) A mixture containing 52 kg A and 48 kg W is contacted with 40 kg of E in each of 3 cross-flow stages. What are the raffinate compositions and quantities?

3.13 In its natural state, zirconium, which is an important material of construction for nuclear reactors, is associated with hafnium, which has an abnormally high neutron-absorption cross section and must be removed before the zirconium can be used. Refer to the accompanying flowsheet for a proposed liquid/liquid extraction process wherein tributyl phosphate (TBP) is used as a solvent for the separation of hafnium from zirconium. [R. P. Cox, H. C. Peterson, and C. H. Beyer, *Ind. Eng. Chem.*, **50** (2), 141 (1958).]

One liter per hour of 5.10 N HNO_3 containing 127 g of dissolved Hf and Zr oxides/liter is fed to stage 5 of a 14-stage extraction unit. The feed contains 22,000 ppm Hf. Fresh TBP enters stage 14 while scrub water is fed to stage 1. Raffinate is removed in stage 14, while the organic extract phase which is removed at stage 1 goes to a stripping unit. The stripping operation consists of a single contact between fresh water and the organic phase. The table below gives the experimental data obtained by Cox and coworkers. (a) Use these data to fashion a complete material balance for the process. (b) Check the data for consistency in as many ways as you can. (c) What is the advantage of running the extractor as shown? Would you recommend that all the stages be used?

Stagewise analyses of mixer-settler run

Stage	Organic Phase			Aqueous Phase		
	g oxide/liter	N HNO$_3$	$\frac{Hf}{Zr}(100)$	g oxide/liter	N HNO$_3$	$\frac{Hf}{Zr}(100)$
1	22.2	1.95	<0.010	17.5	5.21	<0.010
2	29.3	2.02	<0.010	27.5	5.30	<0.010
3	31.4	2.03	<0.010	33.5	5.46	<0.010
4	31.8	2.03	0.043	34.9	5.46	0.24
5	32.2	2.03	0.11	52.8	5.15	3.6
6	21.1	1.99	0.60	30.8	5.15	6.8
7	13.7	1.93	0.27	19.9	5.05	9.8
8	7.66	1.89	1.9	11.6	4.97	20
9	4.14	1.86	4.8	8.06	4.97	36
10	1.98	1.83	10	5.32	4.75	67
11	1.03	1.77	23	3.71	4.52	110
12	0.66	1.68	32	3.14	4.12	140
13	0.46	1.50	42	2.99	3.49	130
14	0.29	1.18	28	3.54	2.56	72
Stripper	.....	0.65	....	76.4	3.96	<0.01

[From R. P. Cox, H. C. Peterson, and C. H. Beyer, *Ind. Eng. Chem.*, **50** (2), 141 (1958).]

(Problem 3.13 is adapted from E. J. Henley and H. Bieber, *Chemical Engineering Calculations*, McGraw-Hill Book Co., p. 298, 1959).

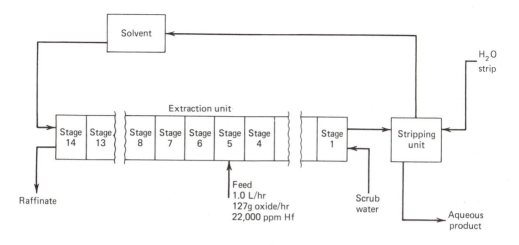

3.14 Repeat Example 3.8 for each of the following changes.

(a) Two kilograms of pure benzene is mixed with 1 kg of meal containing 50 wt% oil.

(b) One kilogram of pure benzene is mixed with 1 kg of meal containing 25 wt% oil.

3.15 At 25°C and 101 kPa, 2 gmoles of a gas containing 35 mole% propylene in propane is equilibrated with 0.1 kg of silica gel adsorbent. Using the equilibrium data of Fig. 3.21, calculate the gram-moles and composition of the gas adsorbed and the equilibrium composition of the gas not adsorbed.

3.16 Vapor–liquid equilibrium data for the system acetone–air–water at 1 atm (101.3 kPa) are given as:

y, mole fraction acetone in air	0.004	0.008	0.014	0.017	0.019	0.020
x, mole fraction acetone in water	0.002	0.004	0.006	0.008	0.010	0.012

(a) Plot the data as (1) a graph of moles acetone/mole air versus moles acetone/mole water, (2) partial pressure of acetone versus g acetone/g water, (3) y versus x.

(b) If 20 moles of gas containing 0.015 mole fraction acetone is brought into contact with 15 moles of water in an equilibrium stage, what would be the composition of the discharge streams? Solve graphically.

(Problem 3.16 is adapted from E. J. Henley and H. K. Staffin, *Stagewise Process Design*, J. Wiley & Sons, New York, 1963.)

3.17 It has been proposed that oxygen be separated from nitrogen by absorbing and desorbing air in water. Pressures from 101.3 to 10,130 kPa and temperatures between 0 and 100°C are to be used.

(a) Devise a workable scheme for doing the separation assuming the air is 79 mole% N_2 and 21 mole% O_2.

(b) Henry's law constants for O_2 and N_2 are given in Fig. 3.23. How many batch absorption steps would be necessary to make 90 mole% pure oxygen? What yield of oxygen (based on total amount of oxygen feed) would be obtained?

3.18 A vapor mixture having equal volumes of NH_3 and N_2 is to be contacted at 20°C and 1 atm (760 torr) with water to absorb a portion of the NH_3. If 14 m³ of this mixture is brought into contact with 10 m³ of water and if equilibrium is attained, calculate the percent of the ammonia originally in the gas that will be absorbed. Both temperature and total pressure will be maintained constant during the absorption. The partial pressure of NH_3 over water at 20°C is:

Partial Pressure of NH_3 in air, torr	Grams of Dissolved $NH_3/100$ g of H_2O
470	40
298	30
227	25
166	20
114	15
69.6	10
50.0	7.5
31.7	5.0
24.9	4.0
18.2	3.0
15.0	2.5
12.0	2.0

3.19 Using the y-x and T-x-y diagrams constructed in Problem 3.5 and the enthalpy data provided below,
(a) Construct an H-x-y diagram for the benzene–toluene system at 1 atm (101.3 kPa). Make any assumptions necessary.

	Saturated Enthalpy, kJ/kg			
T, °C	**Benzene**		**Toluene**	
	H_L	H_V	H_L	H_V
60	79	487	77	471
80	116	511	114	495
100	153	537	151	521

(b) Calculate the energy required for 50 mole% vaporization of a 30 mole% liquid solution of benzene and toluene initially at saturation temperature. If the vapor is then condensed, what is the heat load on the condenser if the condensate is saturated and if it is subcooled by 10°C?

3.20 It is required to design a fractionation tower to operate at 101.3 kPa to obtain a distillate consisting of 95 mole% acetone (A) and 5 mole% water, and a residue containing 1 mole% A. The feed liquid is at 125°C and 687 kPa and contains 57 mole% A. The feed is introduced to the column through an expansion valve so that it enters the column partially vaporized at 60°C. Construct an H-x-y diagram and determine the molar ratio of liquid to vapor in the partially vaporized feed. Enthalpy and equilibrium data are as follows.

Molar latent heat of A = 29,750 kJ/kgmole (assume constant)

Molar latent heat of H_2O = 42,430 kJmole (assume constant)
Molar specific heat of liquid A = 134 kJ/kgmole · °K
Molar specific heat of liquid H_2O = 75.3 kJ/kgmole · °K
Enthalpy of high-pressure, hot feed before adiabatic expansion = 0
Enthalpies of feed phases after expansion: H_V = 27,200 kJ/kgmole
H_L = −5270 kJ/kgmole

Vapor–liquid equilibrium data for acetone–H_2O at 101.3 kPa

	T, °C						
	56.7	57.1	60.0	61.0	63.0	71.7	100
Mole% A in liquid	100	92.0	50.0	33.0	17.6	6.8	0
Mole% A in vapor	100	94.4	85.0	83.7	80.5	69.2	0

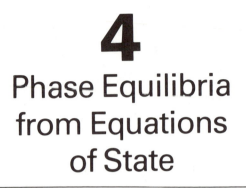

4

Phase Equilibria from Equations of State

Almost all small, nonpolar molecules satisfy the theorem of corresponding states; their P-V-T relation is quite well represented by a two-parameter equation proposed in 1949. A third individual parameter is known to be required for long chains and polar molecules.

Otto Redlich, 1975

For multicomponent mixtures, graphical representations of properties, as presented in Chapter 3, cannot be used to determine equilibrium-stage requirements. Analytical computational procedures must be applied with thermodynamic properties represented preferably by algebraic equations. Because mixture properties depend on temperature, pressure, and phase composition(s), these equations tend to be complex. Nevertheless the equations presented in this chapter are widely used for computing phase equilibrium ratios (K-values and distribution coefficients), enthalpies, and densities of mixtures over wide ranges of conditions. These equations require various pure species constants. These are tabulated for 176 compounds in Appendix I. By necessity, the thermodynamic treatment presented here is condensed. The reader can refer to Perry and Chilton[1] as well as to other indicated sources for fundamental classical thermodynamic background not included here.

The importance of accurate thermodynamic property correlations to design of operable and economic equipment cannot be overemphasized. For example, Gully[2] showed for a deethanizer that reboiler vapor rate varied by approximately 20% depending upon which of four enthalpy correlations was used. Stocking, Erbar, and Maddox[3] found even more serious differences for a hypothetical depropanizer. Using six K-value and seven enthalpy correlations, they found that reboiler duties varied from 657 to 1111 MJ/hr (623,000 to 1,054,000 Btu/hr) and condenser duties varied from 479 to 653 MJ/hr (454,000 to 619,000 Btu/hr).

In another study, Grayson[4] examined the effect of K-values on bubble-point, dew-point, equilibrium flash, distillation, and tray efficiency calculations. He noted a wide range of sensitivity of design calculations to variations in K-values.

4.1 Fugacity—a Basis for Phase Equilibria

For each phase in a multiphase, multicomponent system, the Gibbs free energy is given functionally as

$$G = G\{T, P, n_1, n_2, \ldots, n_c\}$$

where n = moles and subscripts refer to species. The total differential of G is

$$dG = \left(\frac{\partial G}{\partial T}\right)_{P,n_i} dT + \left(\frac{\partial G}{\partial P}\right)_{T,n_i} dP + \sum_{i=1}^{C} \left(\frac{\partial G}{\partial n_i}\right)_{P,T,n_j} dn_i \tag{4-1}$$

where $j \neq i$. From classical thermodynamics

$$\left(\frac{\partial G}{\partial T}\right)_{P,n_i} = -S \tag{4-2}$$

and

$$\left(\frac{\partial G}{\partial P}\right)_{T,n_i} = V \tag{4-3}$$

where S = entropy and V = volume. Defining the *chemical potential*, μ, of species i as

$$\mu_i = \left(\frac{\partial G}{\partial n_i}\right)_{P,T,n_j} \tag{4-4}$$

and substituting into (4-1), we have

$$dG = -S\,dT + V\,dP + \sum_{i=1}^{C} \mu_i\,dn_i \tag{4-5}$$

When (4-5) is applied to a closed system consisting of two or more phases in equilibrium at uniform temperature and pressure, where each phase is an open system capable of mass transfer with another phase[5]

$$dG_{\text{system}} = \sum_{k=1}^{\wp} \left[\sum_{i=1}^{c} \mu_i^{(k)}\,dn_i^{(k)}\right] = 0 \tag{4-6}$$

where the superscript (k) refers to each of $\wp$ phases. Conservation of moles of each species requires that

$$dn_i^{(1)} = -\sum_{k=2}^{\wp} dn_i^{(k)}$$

which, upon substitution into (4-6), gives

$$\sum_{k=2}^{p}\left[\sum_{i=1}^{C} (\mu_i^{(k)} - \mu_i^{(1)})\, dn_i^{(k)}\right] = 0 \tag{4-7}$$

With $dn_i^{(1)}$ eliminated in (4-7), each $dn_i^{(k)}$ term can be varied independently of any other $dn_i^{(k)}$ term. But this requires that each coefficient of $dn_i^{(k)}$ in (4-7) be zero. Therefore,

$$\mu_i^{(1)} = \mu_i^{(2)} = \mu_i^{(3)} = \ldots \mu_i^{(p)} \tag{4-8}$$

Thus, the chemical potentials of any species in a multicomponent system are identical in all phases at physical equilibrium.

Chemical potential cannot be expressed as an absolute quantity, and the numerical values of chemical potential are difficult to relate to more easily understood physical quantities. Furthermore, the chemical potential approaches an infinite negative value as pressure approaches zero. For these reasons, the chemical potential is not directly useful for phase equilibria calculations. Instead, fugacity, as defined below, is employed as a surrogate.

Equation (4-3) restated in terms of chemical potential is

$$\left(\frac{\partial \mu_i}{\partial P}\right)_T = \bar{v}_i \tag{4-9}$$

where $\bar{v}_i$ = partial molal volume. For a pure substance that behaves as an ideal gas, $\bar{v}_i = RT/P$, and (4-9) can be integrated to give

$$\mu_i = RT \ln P + C_1\{T\} \tag{4-10}$$

where C_1 depends on T.

Unfortunately, (4-10) does not describe real multicomponent gas or liquid behavior. However, (4-10) was rescued by G. N. Lewis, who in 1901 invented the fugacity f, a pseudopressure, which, when used in place of pressure in (4-10), preserves the functional form of the equation. Thus, for a component in a mixture

$$f_i = C_2\{T\} \exp{(\mu_i/RT)} \tag{4-11}$$

where C_2 is related to C_1.

Regardless of the value of C_1, it is shown by Prausnitz[6] that, at physical equilibrium, (4-8) can be replaced with

$$f_i^{(1)} = f_i^{(2)} = \ldots \tag{4-12}$$

For a pure, ideal gas, fugacity is equal to the pressure and, for a component in an ideal gas mixture, it is equal to its partial pressure, $p_i = y_i P$.

4.2 Definitions of Other Useful Thermodynamic Quantities

Because of the close relationship between fugacity and pressure, it is convenient to define their ratio for a pure substance as

$$\nu_i^o \equiv f_i^o/P \tag{4-13}$$

where ν_i^o is the pure species fugacity coefficient and f_i^o is the pure species fugacity. The fugacity concept was extended to mixtures by Lewis and Randall and used to formulate the *ideal solution* rule

$$f_{iV} = y_i f_{iV}^o \tag{4-14}$$

$$f_{iL} = x_i f_{iL}^o \tag{4-15}$$

where subscripts V and L refer, respectively, to the vapor and liquid phases. Ideal liquid solutions occur when molecular diameters are equal, chemical interactions are absent, and intermolecular forces between like and unlike molecules are equal. These same requirements apply to the gas phase, where at low pressures molecules are not in close proximity and an ideal gas solution is closely approximated.

It is convenient to represent the departure from both types of ideality (ideal gas law and ideal gas solution) by defining the following *mixture fugacity coefficients*.

$$\phi_{iV} \equiv \frac{f_{iV}}{y_i P} \tag{4-16}$$

$$\phi_{iL} \equiv \frac{f_{iL}}{x_i P} \tag{4-17}$$

In the limit, as ideal gas behavior is approached, $f_{iV}^o \to P$; and in the vapor, $\nu_{iV}^o = 1.0$. Similarly, $f_{iV} \to p_i$, and $\phi_{iV} = 1.0$. However, as ideal solution behavior is approached in the liquid, $f_{iL}^o \to P_i^s$ and, as shown below, $\nu_{iL}^o = P_i^s/P$. Similarly, $f_{iL} \to x_i P_i^s$ and $\phi_{iL} = P_i^s/P$ where $P_i^s =$ vapor pressure.

At a given temperature, the ratio of the fugacity of a component in a mixture to its fugacity in some standard state is termed the *activity*. If the standard state is selected as the pure species at the same pressure and phase condition as the mixture, then

$$a_i \equiv \frac{f_i}{f_i^o}. \tag{4-18}$$

For an ideal solution, substitution of (4-14) and (4-15) into (4-18) shows that $a_{iV} = y_i$ and $a_{iL} = x_i$.

To represent departure of activity from mole fraction when solutions are

Table 4.1 Thermodynamic quantities for phase equilibria

Thermodynamic Quantity	Definition	Physical Significance	Limiting Value for Ideal Gas and Ideal Solution Conditions
Chemical potential	$\mu_i = \left(\dfrac{\partial G}{\partial n_i}\right)_{P,T,n_j}$	Partial molal free energy, $\bar{g}_i$	$\mu_i = \bar{g}_i$
Fugacity	$f_i = C_2\{T\}\exp(\mu_i/RT)$	Thermodynamic pressure	$f_{iV} = y_i P$ $f_{iL} = x_i P_i^s$
Fugacity coefficient for pure species	$\nu_i^o \equiv f_i^o/P$	Deviation to fugacity due to pressure	$\nu_{iV}^o = 1.0$ $\nu_{iL}^o = P_i^s/P$
Fugacity coefficient for species in a mixture	$\phi_{iV} \equiv f_{iV}/y_i P$ $\phi_{iL} \equiv f_{iL}/x_i P$	Deviations to fugacity due to pressure and composition	$\phi_{iV} = 1.0$ $\phi_{iL} = P_i^s/P$
Activity	$a_i \equiv f_i/f_i^o$	Relative thermodynamic pressure	$a_{iV} = y_i$ $a_{iL} = x_i$
Activity coefficient	$\gamma_{iV} = a_{iV}/y_i$ $\gamma_{iL} = a_{iL}/x_i$	Deviation to fugacity due to composition	$\gamma_{iV} = 1.0$ $\gamma_{iL} = 1.0$

nonideal, *activity coefficients* based on concentrations in mole fractions are commonly used

$$\gamma_{iV} \equiv \frac{a_{iV}}{y_i} \tag{4-19}$$

$$\gamma_{iL} \equiv \frac{a_{iL}}{x_i} \tag{4-20}$$

For ideal solutions, $\gamma_{iV} = 1.0$ and $\gamma_{iL} = 1.0$.

For convenient reference, thermodynamic quantities useful in phase equilibria calculations are summarized in Table 4.1.

4.3 Phase Equilibrium Ratios

It is convenient to define an *equilibrium ratio* as the ratio of mole fractions of a species in two phases in equilibrium. For the vapor–liquid case, the constant is referred to as the *K-value* or vapor–liquid equilibrium ratio as defined by (1-3) as $K_i \equiv y_i/x_i$. For the liquid–liquid case, the constant is frequently referred to as the *distribution coefficient* or liquid–liquid equilibrium ratio as defined by (1-6) as $K_{D_i} \equiv x_i^I/x_i^{II}$.

For equilibrium-stage calculations involving the separation of two or more components, separability factors are defined by forming ratios of equilibrium ratios. For the vapor–liquid case, *relative volatility* is defined by (1-7) as $\alpha_{ij} \equiv K_i/K_j$. For the liquid–liquid case, the *relative selectivity* is defined by (1-8) as $\beta_{ij} \equiv K_{D_i}/K_{D_j}$.

Equilibrium ratios can be expressed in terms of a variety of formulations starting from (4-12).

Vapor–Liquid Equilibrium

For vapor–liquid equilibrium, (4-12) becomes

$$f_{iV} \equiv f_{iL} \tag{4-21}$$

To form an equilibrium ratio, fugacities are replaced by equivalent expressions involving mole fractions. Many replacements are possible. Two common pairs derived from (4-16) through (4-20) are

Pair 1:

$$f_{iV} = \gamma_{iV} y_i f_{iV}^o \tag{4-22}$$

and

$$f_{iL} = \gamma_{iL} x_i f_{iL}^o \tag{4-23}$$

Pair 2:

$$f_{iV} = \phi_{iV} y_i P \qquad (4\text{-}24)$$

and

$$f_{iL} = \phi_{iL} x_i P \qquad (4\text{-}25)$$

These equations represent two symmetrical and two unsymmetrical formulations for K-values. The symmetrical ones are:

$$K_i = \left(\frac{\gamma_{iL}}{\gamma_{iV}}\right)\left(\frac{f_{iL}^o}{f_{iV}^o}\right) \qquad (4\text{-}26)$$

$$K_i = \left(\frac{\phi_{iL}}{\phi_{iV}}\right) \qquad (4\text{-}27)$$

Equation (4-26) was developed by Hougen and Watson.[7] More recently, Mehra, Brown, and Thodos[8] utilized it to determine K-values for binary hydrocarbon systems up to and including the true mixture critical point.

Equation (4-27) has received considerable attention. Applications of importance are given by Benedict, Webb, and Rubin;[9] Starling and Han[10,11]; and Soave.[12]

Two unsymmetrical formulations are

$$K_i = \frac{\phi_{iL} P}{\gamma_{iV} f_{iV}^o} = \frac{\phi_{iL}}{\gamma_{iV} \nu_{iV}^o} \qquad (4\text{-}28)$$

$$K_i = \frac{\gamma_{iL} f_{iL}^o}{\phi_{iV} P} = \frac{\gamma_{iL} \nu_{iL}^o}{\phi_{iV}} \qquad (4\text{-}29)$$

Equation (4-28) has been ignored; but, since 1960, (4-29) has received considerable attention. Applications of (4-29) to important industrial systems are presented by Chao and Seader;[13] Grayson and Streed;[14] Prausnitz et al.;[15] Lee, Erbar, and Edmister;[16] and Robinson and Chao.[17] An important modification of (4-29) not presented here is developed by Prausnitz and Chueh.[18]

Liquid–Liquid Equilibrium

For liquid–liquid equilibrium, (4-12) is

$$f_{iL}^{I} = f_{iL}^{II} \qquad (4\text{-}30)$$

where superscripts I and II refer to immiscible liquid phases. A distribution coefficient is formed by incorporating (4-23) to yield the symmetrical formulation

$$K_{D_i} = \frac{x_i^{I}}{x_i^{II}} = \left(\frac{\gamma_{iL}^{II}}{\gamma_{iL}^{I}}\right)\left(\frac{f_{iL}^{oII}}{f_{iL}^{oI}}\right) = \frac{\gamma_{iL}^{II}}{\gamma_{iL}^{I}} \qquad (4\text{-}31)$$

Regardless of which thermodynamic formulation is used for predicting

K-values or distribution coefficients, the accuracy depends upon the veracity of the particular correlations used for the various thermodynamic quantities required. For practical applications, choice of K-value formulation is a compromise among considerations of accuracy, complexity, and convenience. The more important formulations are (4-27), (4-29), and (4-31). They all require correlations for fugacity coefficients and activity coefficients. The application of (4-27) based on fugacity coefficients obtained from equations of state is presented in this chapter. Equations (4-29) and (4-31) require activity coefficient correlations, and are discussed in Chapter 5.

4.4 Equations of State

Equipment design procedures for separation operations require phase enthalpies and densities in addition to phase equilibrium ratios. Classical thermodynamics provides a means for obtaining all these quantities in a consistent manner from P-v-T relationships, which are usually referred to as *equations of state*. Although a large number of P-v-T equations have been proposed, relatively few are suitable for practical design calculations. Table 4.2 lists some of these. All the equations in Table 4.2 involve the universal gas constant R and, in all cases except two, other constants that are unique to a particular species. All equations of state can be applied to mixtures by means of *mixing rules* for combining pure species constants.

Equation (4-32) in Table 4.2, the ideal gas equation, is widely applied to pure gases and gas mixtures. This equation neglects molecular size and potential energy of molecular interactions. When each species in a mixture, as well as the mixture, obeys the ideal gas law, both Dalton's law of additive partial pressures and Amagat's law of additive pure species volumes apply. The mixture equation in terms of molal density ρ/M is

$$\frac{\rho}{M} = \frac{1}{v} = \frac{P}{RT} = \frac{\sum_{i=1}^{C} n_i}{V} \qquad (4\text{-}39)$$

The ideal gas law is generally accurate for pressures up to one atmosphere. At 50 psia (344.74 kPa), (4-39) can exhibit deviations from experimental data as large as 10%.

No corresponding simple equation of state exists for the liquid phase other than one based on the use of a known pure species liquid density and the assumptions of incompressibility and additive volumes. When a vapor is not an ideal gas, formulation of an accurate equation of state becomes difficult because of the necessity to account for molecular interactions.

The principle of corresponding states,[19] which is based on similitude of

Table 4.2 Equations of state

Name	Equation and Equation Number		Evaluation of Species Constants or Functions
Ideal gas law	$P = \dfrac{RT}{v}$	(4-32)	None
Generalized	$P = \dfrac{ZRT}{v}$	(4-33)	$Z = Z\{P_r, T_r, Z_c \text{ or } \omega\}$ as derived from experimental data
Virial (Onnes)	$P = \dfrac{RT}{v}\left(1 + \dfrac{B}{v} + \dfrac{C}{v^2} + \ldots\right)$	(4-34)	From volumetric data, generalized correlations, or statistical mechanics, where B, C, and so on depend on temperature
van der Waals	$P = \dfrac{RT}{v - b} - \dfrac{a}{v^2}$	(4-35)	$a = \dfrac{27}{64}\dfrac{R^2 T_c^2}{P_c}, b = \dfrac{RT_c}{8P_c}$
Beattie–Bridgeman	$P = \dfrac{RT}{v^2}\left(1 - \dfrac{c}{vT^3}\right)\left[v + B_o\left(1 - \dfrac{b}{v}\right)\right] - \dfrac{A_o}{v^2}\left(1 - \dfrac{a}{v}\right)$	(4-36)	Constants, a, A_o, b, B_o, and c from experimental data
Benedict–Webb–Rubin (B-W-R)	$P = \dfrac{RT}{v} + \left[B_o RT - A_o - \dfrac{C_o}{T^2}\right]\dfrac{1}{v^2} + (bRT - a)\dfrac{1}{v^3}$ $+ \dfrac{a\alpha}{v^6} + \dfrac{c\left(1 + \dfrac{\gamma}{v^2}\right)\exp\left(-\dfrac{\gamma}{v^2}\right)}{v^3 T^2}$	(4-37)	Constants a, A_o, b, B_o, c, C_o, α, γ from experimental data or correlations
Redlich–Kwong (R-K)	$P = \dfrac{RT}{v - b} - \dfrac{a}{T^{0.5}v(v + b)}$	(4-38)	$a = 0.4278\dfrac{R^2 T_c^{2.5}}{P_c}, b = 0.0867\dfrac{RT_c}{P_c}$

molecular behavior particularly at the critical point, can be used to derive generalized graphical or tabular correlations of the vapor or liquid compressibility factor Z in (4-33) as a function of reduced (absolute) temperature $T_r = T/T_c$, reduced (absolute) pressure $P_r = P/P_c$, and a suitable third parameter. Generalized equations of state are often fitted to empirical equations in T_r and P_r for computerized design methods.

The virial equation of state in Table 4.2 provides a sound theoretical basis for computing P-v-T relationships of polar as well as nonpolar pure species and mixtures in the vapor phase. Virial coefficients B, C, and higher can, in principle, be determined from statistical mechanics. However, the present state of development is such that most often (4-34) is truncated at B, the second virial coefficient, which is estimated from a generalized correlation.[20,21] In this form, the virial equation is accurate to densities as high as approximately one half of the critical. Application of the virial equation of state to phase equilibria is discussed and developed in detail by Prausnitz et al,[15] and is not considered further here.

The five-constant equation of Beattie and Bridgeman,[22] the eight-constant equation of Benedict, Webb, and Rubin (B-W-R),[23] and the two-constant equation of Redlich and Kwong (R-K),[24] are empirical relationships applicable over a wide range of pressure. The R-K equation is particularly attractive because it contains only two constants and these can be determined directly from the critical temperature T_c and critical pressure P_c. Furthermore, the R-K equation has an accuracy that compares quite favorably with more complex equations of state;[25] and it has the ability to approximate the liquid region, as is illustrated in the following example. The two-constant van der Waals equation can fail badly in this respect.

Example 4.1. Thermodynamic properties of isobutane were measured at subcritical temperatures from 70°F (294.29°K) to 250°F (394.26°K) over a pressure range of 10 psia (68.95 kPa) to 3000 psia (20.68 MPa) by Sage and Lacey.[26] Figure 4.1 is a log–log graph of pressure (psia) versus molal volume (ft³/lbmole) of the experimental two-phase envelope (saturated liquid and saturated vapor) using the tabulated critical conditions from Appendix I to close the curve. Shown also is an experimental isotherm for 190°F (360.93°K). Calculate and plot 190°F isotherms for the R-K equation of state and for the ideal gas law and compare them to the experimental data.

Solution. From (4-38), pressure can be calculated as a function of molal volume with $T = 649.67°R$ (190°F) and $R = 10.731$ psia · ft³/lbmole · °R. Redlich–Kwong constants in Table 4.2 are

$$a = 0.4278 \left[\frac{(10.731)^2(734.7)^{2.5}}{529.1} \right] = 1.362 \times 10^6 \frac{\text{psia} \cdot °\text{R}^{0.5} \cdot \text{ft}^6}{\text{lbmole}^2}$$

$$b = 0.0867 \left[\frac{(10.731)(734.7)}{529.1} \right] = 1.2919 \frac{\text{ft}^3}{\text{lbmole}}$$

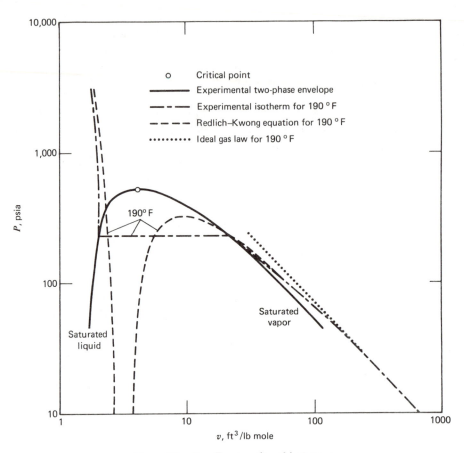

Figure 4.1. P-v-T properties of isobutane.

The calculated values of pressure from the R-K equation are shown as a dashed line in Fig. 4.1. For example, for $v = 100 \text{ ft}^3/\text{lbmole}$, substitution into (4-38) gives

$$P = \frac{(10.731)(649.67)}{(100 - 1.2919)} - \frac{1.362 \times 10^6}{(649.58)^{0.5}(100)(100 + 1.2919)} = 65.34 \text{ psia (450.5 kPa)}$$

The dotted line represents calculations from (4-32) for the ideal gas law. For example, for $v = 100 \text{ ft}^3/\text{lbmole}$

$$P = \frac{(10.731)(649.67)}{100} = 69.71 \text{ psia (480.6 kPa)}$$

At 190°F, for pressures up to 30 psia, both the R-K equation and the ideal gas law are in very good agreement with experimental data. For 30 psia up to the saturation pressure, the R-K equation continues to agree with experimental data, but the ideal gas law shows increasing deviations. A quantitative comparison is shown in the following table for the vapor region at 190°F.

| P, psia | Percent Deviation of Calculated Vapor Molal Volume | |
	R-K Equation	Ideal Gas Law
10	0.2	1.2
30	0.6	3.5
60	1.3	7.5
100	1.9	13.1
150	3.0	21.9
229.3 (saturation)	4.1	41.6

At 190°F, the R-K equation consistently predicts liquid molal volumes larger than those measured. The following table provides a quantitative comparison.

P, psia	Percent Deviation of Calculated Liquid Molal Volume from R-K Equation
229.3 (saturation)	18.8
500	14.5
1000	10.8
2000	7.5
3000	5.8

The deviations are much larger than for the vapor region, but not grossly in error.

Within the two-phase envelope, the R-K equation has continuity, but it fails badly to predict an isobaric condition. This has no serious practical implications. If the molal volume for a two-phase mixture is required, it can be computed directly from individual phase volumes.

□

Either (4-38) or the following equivalent compressibility factor forms are used when the R-K equation is applied to mixtures.

$$Z = \frac{1}{1-h} - \frac{A^2 h}{B(1+h)} \tag{4-40}$$

$$Z^3 - Z^2 + BP\left(\frac{A^2}{B} - BP - 1\right)Z - \frac{A^2}{B}(BP)^2 = 0 \tag{4-41}$$

where the *mixing rules* are

$$A = \sum_{i=1}^{C} A_i y_i \quad or \quad \sum_{i=1}^{C} A_i x_i \tag{4-42}$$

$$B = \sum_{i=1}^{C} B_i y_i \quad or \quad \sum_{i=1}^{C} B_i x_i \tag{4-43}$$

with

$$A_i = \left(\frac{a_i}{R^2 T^{2.5}} \right)^{1/2} = \left(\frac{0.4278}{P_c T_r^{2.5}} \right)^{1/2} \tag{4-44}$$

$$B_i = \frac{b_i}{RT} = \frac{0.0867}{P_c T_r} \tag{4-45}$$

$$h = \frac{BP}{Z} \tag{4-46}$$

If (4-38) is used directly, the equivalent mixing rules for the vapor phase are

$$a = \sum_{i=1}^{C} \left(\sum_{j=1}^{C} y_i y_j a_{ij} \right) \tag{4-47}$$

where

$$a_{ij} = (a_i a_j)^{\frac{1}{2}} \tag{4-48}$$

and

$$b = \sum_{i=1}^{c} y_i b_i \tag{4-49}$$

    Equations (4-40) and (4-41) are cubic equations in Z. As shown by Edmister[27] and as observed in Fig. 4.1, only one positive real root is obtained at supercritical temperatures where only a single phase exists. Otherwise, three real roots are obtained, the largest value of Z corresponding to a vapor phase, and the smallest value of Z to a liquid phase.

Example 4.2.   Glanville, Sage, and Lacey[28] measured specific volumes of vapor and liquid mixtures of propane and benzene over wide ranges of temperature and pressure. Use the R-K equation to calculate specific volume of a vapour mixture containing 26.92 weight % propane at 400°F (477.59°K) and a saturation pressure of 410.3 psia (2.829 MPa). Compare the computed value to the measured quantity.

    Solution.   Let propane be denoted by P and benzene by B. The mole fractions are

$$y_P = \frac{\dfrac{0.2692}{44.097}}{\dfrac{0.2692}{44.097} + \dfrac{0.7308}{78.114}} = 0.3949$$

$$y_B = 1 - 0.3949 = 0.6051$$

Using the critical constants from Appendix I with (4-44) and (4-45), we have

$$A_P = \left[\frac{0.4278}{(617.4)\left(\frac{859.67}{665.9}\right)^{2.5}}\right]^{\frac{1}{2}} = 0.01913 \text{ psia}^{-\frac{1}{2}}$$

Similarly, $A_B - 0.03004 \text{ psia}^{-\frac{1}{2}}$, $B_P = 1.088 \times 10^{-4} \text{ psia}^{-1}$, and $B_B = 1.430 \times 10^{-4} \text{ psia}^{-1}$. Applying the mixing rules, (4-42) and (4-43), we have

$A = 0.3949(0.01913) + 0.6051(0.03004) = 0.0257 \text{ psia}^{-1/2}$

$B = 0.3949(1.088 \times 10^{-4}) + 0.6051(1.43 \times 10^{-4}) = 1.295 \times 10^{-4} \text{ psia}^{-1}$

$BP = 1.295 \times 10^{-4}(410.3) = 0.0531$

Substituting these quantities into (4-41) gives

$$Z^3 - Z^2 + 0.2149 Z - 0.01438 = 0$$

for which there is only one real root, which is the vapor compressibility factor $Z = 0.7339$. Molal volume is computed from (4-33).

$$v = \frac{ZRT}{P} = \frac{(0.7339)(10.731)(859.67)}{(410.3)} = 16.50 \frac{\text{ft}^3}{\text{lbmole}} (1.03 \frac{\text{m}^3}{\text{kgmole}})$$

The average molecular weight of the mixture is

$$M = \sum_{i=1}^{C} y_i M_i = 0.3949(44.097) + 0.6051(78.114) = 64.68 \frac{\text{lb}}{\text{lbmole}}$$

The specific volume is

$$\frac{v}{M} = \frac{16.50}{64.68} = 0.2551 \text{ ft}^3/\text{lb} (0.01593 \text{ m}^3/\text{kg})$$

This value is 2.95% higher than the measured value of 0.2478 ft³/lb corresponding to a compressibility factor of 0.7128. The density of the vapor is the reciprocal of the specific volume

$$\rho = \frac{1}{v/M} = \frac{1}{0.2551} = 3.92 \text{ lb/ft}^3 (62.8 \text{ kg/m}^3)$$

The molal density is the reciprocal of the molal volume

$$\frac{\rho}{M} = \frac{1}{v} = \frac{1}{16.50} = 0.0606 \text{ lbmole/ft}^3 (0.971 \text{ kgmole/m}^3)$$

☐

4.5 Derived Thermodynamic Properties of Liquid and Vapor Phases

If ideal gas (zero pressure) specific heat or enthalpy equations for pure species are available, as well as an equation of state, thermodynamic properties can be derived in a consistent manner by applying the equations of classical thermo-

Table 4.3 Useful equations of classical thermodynamics

Integral equations

$$(H - H_V^o) = Pv - RT - \int_{\infty}^{v} \left[P - T\left(\frac{\partial P}{\partial T}\right)_v \right] dv \tag{4-50}$$

$$\nu^o = \exp\left[\frac{1}{RT} \int_o^P \left(v - \frac{RT}{P}\right) dP \right] = \exp\left[\frac{1}{RT} \int_v^{\infty} \left(P - \frac{RT}{v}\right) dv - \ln Z + (Z - 1) \right] \tag{4-51}$$

$$\phi_i = \exp\left\{ \frac{1}{RT} \int_V^{\infty} \left[\left(\frac{\partial P}{\partial n_i}\right)_{T,V,n_j} - \frac{RT}{V}\right] dV - \ln Z \right\} \tag{4-52}$$

where V is the total volume equal to $v \sum n_i$.

Differential equations

$$(H_{iV}^o - H_i) = RT^2 \left(\frac{\partial \ln f_i}{\partial T}\right)_P \tag{4-53}$$

$$v_i = RT \left(\frac{\partial \ln f_i}{\partial P}\right)_T \tag{4-54}$$

$$(H_V^o - H) = RT^2 \left(\frac{\partial \ln \nu^o}{\partial T}\right)_P \tag{4-55}.$$

$$v = RT \left[\left(\frac{\partial \ln \nu^o}{\partial P}\right)_T + \frac{1}{P}\right] \tag{4-56}$$

$$(\overline{H}_{iL} - H_{iL}) = \overline{H}_{iL}^E = -RT^2 \left(\frac{\partial \ln \gamma_{iL}}{\partial T}\right)_{P,x_i} \tag{4-57}$$

$$(\bar{v}_{iL} - v_{iL}) = \bar{v}_{iL}^E = RT \left(\frac{\partial \ln \gamma_{iL}}{\partial P}\right)_{T,x_i} \tag{4-58}$$

---

dynamics compiled in Table 4.3. Equations (4-50) through (4-56) are applicable to vapor or liquid phases, where the superscript ^{o} refers to the ideal gas.

Enthalpy

The molal specific heat of gases is conventionally given as a polynomial in temperature.

$$c_{P_V}^o = a_1 + a_2 T + a_3 T^2 + a_4 T^3 + a_5 T^4 \tag{4-59}$$

Integration of (4-59) provides an equation for the ideal gas molal enthalpy at temperature T referred to a datum temperature T_o.

$$H_V^o = \int_{T_o}^T C_{P_V}^o \, dT = \sum_{k=1}^5 \frac{a_k(T^k - T_o^k)}{k} \tag{4-60}$$

Values of the five constants a_1 through a_5, with T in °F and $T_o = 0$°F are given in Appendix I for 176 compounds.

When the ideal gas law assumption is not valid, (4-50) is used to correct the enthalpy for pressure. For a pure species or mixtures at temperature T and pressure P, the vapor enthalpy is

$$H_V = \left[\sum_{i=1}^{C} (y_i H_{iV}^o)\right] + (H_V - H_V^o) \tag{4-61}$$

Equations (4-61) and (4-50) are particularly suitable for use with equations of state that are explicit in pressure (e.g., those shown in Table 4.2). The same two equations can be used to determine the liquid-phase enthalpy. Application is facilitated if the equation of state is a continuous function in passing between vapor and liquid regions, as in Figure 4.1. Thus

$$H_L = \left[\sum_{i=1}^{C} (x_i H_{iV}^o)\right] + (H_L - H_V^o) \tag{4-62}$$

For pure species at temperatures below critical, (4-62) can be divided into the separate contributions shown graphically in Fig. 4.2 and analytically by (4-63), where the subscript s refers to saturation pressure conditions.

$$H_L = \underbrace{H_V^o + (Pv)_{V_s} - RT}_{\substack{(1)\ \text{vapor at} \\ \text{zero pressure}}} - \underbrace{\int_{\infty}^{v_{V_s}} \left[P - T\left(\frac{\partial P}{\partial T}\right)_v\right] dv}_{\substack{(2)\ \text{pressure correction for vapor} \\ \text{to saturation pressure}}}$$

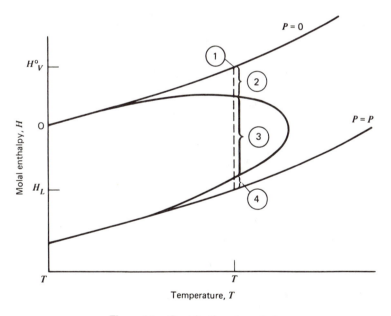

Figure 4.2.   Contributions to enthalpy.

$$-T\left(\frac{\partial P}{\partial T}\right)_s (v_{V_s} - v_{L_s}) + (Pv)_L - (Pv)_{L_s} - \int_{v_{L_s}}^{v_L} \left[P - T\left(\frac{\partial P}{\partial T}\right)_v\right] dv \qquad (4\text{-}63)$$

$$\underbrace{}_{\substack{\text{(3) latent heat of} \\ \text{vaporization}}} \quad \underbrace{}_{\substack{\text{(4) correction to liquid for pressure} \\ \text{in excess of saturation pressure}}}$$

Equation (4-62) is the preferred form, especially if the equation of state is a continuous function.

If the R-K equation of state is substituted into (4-50), the required integration is performed as shown in detail by Edmister,[29] and the result is substituted into (4-61) and (4-62), the equations for mixture molal enthalpy become

$$H_V = \sum_{i=1}^{C} (y_i H_{iV}^o) + RT\left[Z_V - 1 - \frac{3A^2}{2B}\ln\left(1 + \frac{BP}{Z_V}\right)\right] \qquad (4\text{-}64)$$

$$H_L = \sum_{i=1}^{C} (x_i H_{iV}^o) + RT\left[Z_L - 1 - \frac{3A^2}{2B}\ln\left(1 + \frac{BP}{Z_L}\right)\right] \qquad (4\text{-}65)$$

Example 4.3. Use the R-K equation of state to obtain the change in enthalpy for isobutane vapor during isothermal compression from 10 psia (68.95 kPa) to 229.3 psia (1.581 MPa) at 190°F (360.93°K). Compare the estimate to the measured value reported by Sage and Lacey.[26]

Solution. By the procedure of Example 4.2, $A = 0.03316 \text{ psia}^{-\frac{1}{2}}$, $B = 1.853 \times 10^{-4} \text{ psia}^{-1}$, Z at 10 psia = 0.991, and Z at 229.3 psia = 0.734. Let the 10-psia condition be denoted by 1 and the 229.3-psia condition by 2. From (4-64)

$$H_{V_2} - H_{V_1} = RT\left\{Z_{V_2} - Z_{V_1} - \frac{3A^2}{2B}\left[\ln\left(1 + \frac{BP_2}{Z_{V_2}}\right) - \ln\left(1 + \frac{BP_1}{Z_{V_1}}\right)\right]\right\}$$

$$= 1.987(649.58)\left\{0.734 - 0.991 - \frac{3(0.03316)^2}{2(1.853 \times 10^{-4}}\right.$$

$$\left[\ln\left(1 + \frac{(1.853 \times 10^{-4})(229.3)}{0.734}\right) - \ln\left(1 + \frac{(1.853 \times 10^{-4})(10)}{0.991}\right)\right]\right\}$$

$$= -957 \text{ Btu/lbmole } (-2.22 \text{ MJ/kgmole})$$

The change in specific enthalpy is

$$\frac{H_{V_2} - H_{V_1}}{M} = \frac{-957}{58.124} = -16.78 \text{ Btu/lb } (-39.00 \text{ kJ/kg})$$

This is 4.7% less than the measured change of $(184.1 - 201.7) = -17.60$ Btu/lb reported by Glanville, Sage, and Lacey.[28]

□

Example 4.4. Estimate the enthalpy of a liquid mixture of 25.2 mole % propane in benzene at 400°F (477.59°K) and 750 psia (5.171 MPa) relative to zero-pressure vapor at the same temperature using the R-K equation of state. Compare the result to the measurement by Yarborough and Edmister.[30] The liquid saturation pressure for the mixture at 400°F is estimated to be 640 psia (4.41 MPa) from the data of Glanville, Sage, and Lacey.[28]

Solution.   From (4-42) and (4-43), using the pure species R-K constants computed in Example 4.2, we find

$$A = 0.252(0.01913) + 0.748(0.03004) = 0.0273 \text{ psia}^{-\frac{1}{2}}$$

$$B = 0.252(1.088 \times 10^{-4}) + 0.748(1.430 \times 10^{-4}) = 1.344 \times 10^{-4} \text{ psia}^{-1}$$

$$BP = 1.344 \times 10^{-4}(750) = 0.1008$$

Substituting into (4-41) gives

$$Z^3 - Z^2 + 0.4480\,Z - 0.05634 = 0.$$

Solution of this cubic equation yields only the single real root 0.1926. This corresponds to Z_L and compares to an interpolated value of 0.17 from measurements by Glanville, Sage, and Lacey.[28]

From (4-65)

$$H_L - H_V^o = 1.987(859.67)\left[0.1926 - 1 - \frac{3(0.0273)^2}{2(1.344 \times 10^{-4})} \ln\left(1 + \frac{0.1008}{0.1926}\right)\right]$$

$$= -7360 \text{ Btu/lbmole of mixture } (-17.1 \text{ MJ/kgmole})$$

The molecular weight of the mixture is

$$M = 0.252(44.097) + 0.748(78.114) = 69.54 \text{ lb/lbmole}$$

The specific enthalpy difference is

$$\frac{H_L - H_V^o}{M} = \frac{-7360}{69.54} = -105.84 \text{ Btu/lb of mixture } (-246.02 \text{ kJ/kg})$$

This corresponds to a 16.9% deviation from the value of -127.38 Btu/lb measured by Yarborough and Edmister.[30] If the experimental value of 0.17 for Z_L is substituted into (4-65), a specific enthalpy difference of -115.5 Btu/lb is computed. This is approximately midway between the measured value and that computed using Z_L from the R-K equation of state.

□

Pure Species Fugacity Coefficient

The fugacity coefficient v^o of a pure species at temperature T and pressure P can be determined directly from an equation of state by means of (4-51). If $P < P_i^s$, v^o is the vapor fugacity coefficient. For $P > P_i^s$, v^o is the fugacity coefficient of the liquid. Saturation pressure corresponds to the condition $v_L^o = v_V^o$. Integration of (4-51) with the R-K equation of state gives

$$v_V^o = \exp\left[Z_V - 1 - \ln(Z_V - BP) - \frac{A^2}{B}\ln\left(1 + \frac{BP}{Z_V}\right)\right] \tag{4-66}$$

$$v_L^o = \exp\left[Z_L - 1 - \ln(Z_L - BP) - \frac{A^2}{B}\ln\left(1 + \frac{BP}{Z_L}\right)\right] \tag{4-67}$$

Vapor Pressure

At temperature $T < T_c$, the saturation pressure (vapor pressure) P_i^s can be estimated from the R-K equation of state by setting (4-66) equal to (4-67) and solving for P by an iterative procedure. The results, as given by Edmister,[27] are plotted in reduced form in Fig. 4.3. The R-K vapor pressure curve does not satisfactorily represent data for a wide range of molecular shapes as witnessed by the experimental curves for methane, toluene, n-decane, and ethyl alcohol on the same plot. This failure represents one of the major shortcomings of the R-K equation. Apparently, critical constants T_c and P_c alone are insufficient to generalize thermodynamic behavior. However, generalization is substantially improved by incorporating into the equation a third constant that represents the generic differences in the reduced vapor pressure curves.

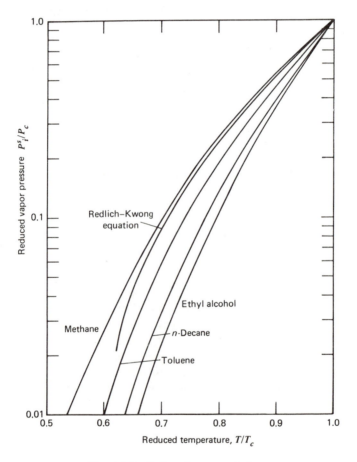

Figure 4.3.   Reduced vapor pressure.

A very effective third constant is the *acentric factor* introduced by Pitzer et al.[31] It is widely used in thermodynamic correlations based on the theorem of corresponding states. The acentric factor accounts for differences in molecular shape and is defined by the vapor pressure curve as

$$\omega_i \equiv \left[-\log \left(\frac{P_i^s}{P_c} \right)_{T_r=0.7} \right] - 1.000 \qquad (4\text{-}68)$$

This definition gives values for ω of essentially zero for spherically symmetric molecules (e.g., the noble gases). Values of ω for 176 compounds are tabulated in Appendix I. For the species in Fig. 4.3, values of ω are 0.00, 0.2415, 0.4869, and 0.6341, respectively, for methane, toluene, n-decane, and ethyl alcohol. Modification of the R-K equation by the addition of ω as a third constant greatly improves its ability to predict vapor pressures and other liquid-phase thermodynamic properties.

Alternatively, vapor pressure may be estimated from any of a large number of empirical correlations. Although not the most accurate over a wide range of temperature, the Antoine equation is convenient and widely used.[32] The reduced pressure form of the Antoine equation is

$$\ln \left(\frac{P_i^s}{P_c} \right) = A_1 - \frac{A_2}{T' + A_3} \qquad (4\text{-}69)$$

Values of A_1, A_2, and A_3, with T' in °F, are given in Appendix I. Because (4-69) is in reduced pressure form, the constant A_1 is not the same as that commonly given in tabulations of Antoine constants found in other sources.

When an equation of state is suitable for the vapor phase but unsuitable for the liquid phase, an alternative to (4-67) can be used to compute v_L^o. From (4-51), by integrating from 0 to P_i^s and P_i^s to P,

$$\ln v_L^o = \ln v_{V_s}^o - \ln \left(\frac{P}{P_i^s} \right) + \frac{1}{RT} \int_{P_i^s}^P v_L \, dP \qquad (4\text{-}70)$$

The last term is the *Poynting correction*. If the liquid is incompressible, (4-70) reduces to

$$v_L^o = \left(\frac{P_i^s}{P} \right) v_{V_s}^o \exp \left[\frac{v_L(P - P_i^s)}{RT} \right] \qquad (4\text{-}71)$$

At very high pressures, such as exist in petroleum reservoirs, the Poynting correction for light gases is appreciable. At low pressure, $v_{V_s}^o$ equals one, and the argument of the exponential Poynting term approaches zero. Then v_L^o approaches the ratio of vapor pressure to total pressure as shown in Table 4.1.

Example 4.5.   Estimate the fugacity of liquid isobutane at 190°F (360.93°K) and 500 psia (3.447 MPa) from
(a)   The R-K equation of state (4-67).

(b) Equation (4-71) using the Antoine equation for vapor pressure, the experimental liquid specific volume at saturation of 0.03506 ft³/lb (0.002189 m³/kg) from Sage and Lacey,[26] and the R-K equation for vapor fugacity coefficient at saturation. Compare the estimates with the experimental value of Sage and Lacey.[26]

Solution. (a) From Examples 4.1 and 4.3, $Z_L = 0.1633$, $A = 0.03316$ psia$^{-\frac{1}{2}}$, $B = 1.853 \times 10^{-4}$ psia^{-1}. Thus, $BP = 0.09265$ and $A^2/B = 5.934$. Substitution into (4-67) gives

$$v_L^o = \exp\left[0.1633 - 1 - \ln(0.1633 - 0.09265) - 5.934 \ln\left(1 + \frac{0.09265}{0.1633}\right)\right] = 0.4260$$

From (4-13)

$$f_L^o = v_L^o P = 0.4260(500) = 213 \text{ psia } (1.47 \text{ MPa})$$

This is 12.4% greater than the experimental value of 189.5 psia (1.307 MPa).

(b) From (4-69) using Antoine vapor pressure constants and the critical pressure from Appendix I

$$P_i^s = 529.056 \exp\left[5.611805 - \frac{3870.419}{190 + 409.949}\right] = 228.5 \text{ psia } (1.575 \text{ MPa})$$

This is only 0.35% less than the experimental value of 229.3 psia.

By the procedure of Example 4.2, real roots of 0.7394, 0.1813, and 0.07939 are found for Z from (4-41) at saturation pressure. Thus, $Z_{V_s} = 0.7394$ and $Z_{L_s} = 0.07939$. From (4-66)

$$v_{V_s} = \exp\left[0.7394 - 1 - \ln(0.7394 - 0.04234) - 5.934 \ln\left(1 + \frac{0.04234}{0.7394}\right)\right] = 0.7373$$

From (4-13) at $P_i^s = 228.5$ psia

$$f_{V_s}^o = v_{V_s}^o P_i^s = 0.7373(228.5) = 168.5 \text{ psia } (1.162 \text{ MPa})$$

This is 3.82% less than the experimental value of 175.2 psia. From (4-71)

$$v_L^o = \left(\frac{228.5}{500}\right)(0.7373) \exp\left[\frac{0.03506(58.12)(500 - 228.5)}{(10.731)(649.67)}\right] = 0.3648$$

From (4-13) at $P = 500$ psia

$$f_L^o = v_L^o P = 0.3648(500) = 182.4 \text{ psia } (1.258 \text{ MPa})$$

This is only 3.75% lower than the experimental value.[26] The Poynting correction is a factor of approximately 1.083. Without it, $f_L^o = 168.4$ psia.

☐

Mixture Fugacity Coefficients

Fugacity coefficients of species in liquid or vapor mixtures can be obtained from (4-52). If the R-K equation of state is applied, a rather tedious procedure, as given by Redlich and Kwong,[24] leads to the following working equations.

$$\phi_{iV} = \exp\left[(Z_V - 1)\frac{B_i}{B} - \ln(Z_V - BP) - \frac{A^2}{B}\left(\frac{2A_i}{A} - \frac{B_i}{B}\right)\ln\left(1 + \frac{BP}{Z_V}\right)\right] \quad (4\text{-}72)$$

$$\phi_{iL} = \exp\left[(Z_L - 1)\frac{B_i}{B} - \ln(Z_L - BP) - \frac{A^2}{B}\left(\frac{2A_i}{A} - \frac{B_i}{B}\right)\ln\left(1 + \frac{BP}{Z_L}\right)\right] \quad (4\text{-}73)$$

Example 4.6. Estimate the fugacity coefficient of benzene in the vapor mixture of Example 4.2 using the R-K equation of state.

Solution. Substituting the quantities computed in Example 4.2 into (4-72)

$$
\phi_{iV} = \exp \left\{ (0.7339 - 1) \frac{(1.430 \times 10^{-4})}{(1.295 \times 10^{-4})} - \ln (0.7339 - 0.0531) \right.
$$

$$
\left. - \frac{(0.0257)^2}{(1.295 \times 10^{-4})} \left[\frac{2(0.03004)}{0.0257} - \frac{1.430 \times 10^{-4}}{1.295 \times 10^{-4}} \right] \ln \left[1 + \frac{0.0531}{0.7339} \right] \right\} = 0.7055
$$

This is substantially less than the value of one for an ideal gas and an ideal solution of gases.

□

4.6 Thermodynamic Properties for Ideal Solutions at Low Pressures

Separation operations are frequently conducted under vacuum or at atmospheric pressure where the ideal gas law is valid. If, in addition, the species in the mixture have essentially identical molecular size and intermolecular forces, ideal solutions are formed, and a very simple expression can be derived for estimating K-values. From (4-26) with $\gamma_{iL} = \gamma_{iV} = 1$

$$
K_{\text{Ideal}_i} = \frac{f_{iL}^o}{f_{iV}^o} = \frac{\nu_{iL}^o}{\nu_{iV}^o} \tag{4-74}
$$

When the ideal gas law is assumed, (4-51) reduces to $\nu_{iV}^o = 1$. Similarly if system pressure is close to species vapor pressure so that the Poynting correction is negligible, (4-71) becomes $\nu_{iL}^o = P_i^s/P$. Substituting these expressions for ν_{iV}^o and ν_{iL}^o in (4-74) gives the so-called *Raoult's law K-value* expression

$$
K_{\text{Raoult}_i} = \frac{P_i^s}{P} \tag{4-75}
$$

An alternate derivation was given in Chapter 3. The Raoult's law K-value varies inversely with pressure, varies exponentially with temperature, and is independent of phase compositions.

The relative volatility defined by (1-7) is the ratio of vapor pressures and, thus, depends only on temperature

$$
\alpha_{\text{Raoult}_{ij}} = \frac{P_i^s}{P_j^s} \tag{4-76}
$$

A convenient experimental test for the validity of the Raoult's law K-value is to measure total pressure over a liquid solution of known composition at a given temperature. The experimental value is then compared to that calculated from the summation of a variation of (4-75)

$$P = \sum_{i=1}^{C} y_i P = \sum_{i=1}^{C} x_i P_i^s \qquad (4\text{-}77)$$

By this technique, Redlich and Kister[33] showed that isomeric binary mixtures of isopentane/n-pentane, o-xylene/m-xylene, m-xylene/p-xylene, and o-xylene/p-xylene were represented by Raoult's law to within approximately 0.4% over the entire composition range at 1-atm. Binary mixtures of each of the three xylene isomers with ethylbenzene showed a maximum deviation of 0.7%. However, if values of relative volatility are close to one, such deviations may be significant when designing separation equipment.

For ideal solutions at low pressure, the vapor-phase molal density may be obtained from (4-39). An estimate of the liquid-phase molal density is obtained by applying the following equation, which is based on additive liquid molal volumes

$$\frac{\rho_L}{M} = \frac{1}{\displaystyle\sum_{i=1}^{C} x_i v_{iL}} \qquad (4\text{-}78)$$

At low to moderate pressures, assuming liquid incompressibility, the pure species molal volumes in (4-78) can be estimated by the method of Cavett[34] using the empirical equation

$$v_{iL} = \xi_i(5.7 + 3.0 T_{r_i}) \qquad (4\text{-}79)$$

where the liquid volume constant ξ_i is back-calculated from the measured liquid molal volume at a known temperature. Values of the liquid molal volume are tabulated in Appendix I.

From (4-61), with the assumptions of the ideal gas law and an ideal gas solution, vapor enthalpy is simply

$$H_V = \sum_{i=1}^{C} (y_i H_{iV}^o) \qquad (4\text{-}80)$$

Liquid enthalpy for ideal solutions is obtained from (4-63), which simplifies to

$$H_L = \sum_{i=1}^{C} x_i [H_{iV}^o - \lambda_i] \qquad (4\text{-}81)$$

where λ = molal heat of vaporization, which can be obtained from (4-53). For ideal solutions at low pressures where $f_{iL} = P_i^s$, (4-53) becomes

$$\lambda_i = RT^2 \left(\frac{d \ln P_i^s}{dT}\right) \qquad (4\text{-}82)$$

From (4-69) for the Antoine vapor pressure

$$\ln P_i^s = A_1 - \frac{A_2}{T' + A_3} + \ln P_c \tag{4-83}$$

Differentiation of (4-83) followed by substitution of the result into (4-82) gives an expression that is valid at low pressures

$$\lambda = \frac{A_2 R T^2}{(T' + A_3)^2} \tag{4-84}$$

where $T' = °F$ and $T = $ absolute temperature.

Example 4.7. Styrene is manufactured by catalytic dehydrogenation of ethylbenzene. The separation sequence includes vacuum distillation to separate styrene from unreacted ethylbenzene.[35] Estimate the relative volatility between these two compounds based on Raoult's law K-values at typical distillation operating conditions of 80°C (176°F) and 100 torr (13.3 kPa) (the low temperature is employed to prevent styrene polymerization). Compare the computed relative volatility to the experimental value of Chaiyavech and Van Winkle.[36] Also calculate the heat of vaporization of ethylbenzene at 80°C.

Solution. Let $E = $ ethylbenzene and $S = $ styrene. From (4-69) with Antoine constants and critical pressures given in Appendix I, vapor pressures at 176°F are computed as in Example 4.5 to be $P_E^s = 125.9$ torr (16.79 kPa) and $P_S^s = 90.3$ torr (12.04 kPa). From (4-75), Raoult's law K-values are

$$K_E = \frac{125.9}{100} = 1.259$$

$$K_S = \frac{90.3}{100} = 0.903$$

From (4-76), relative volatility from Raoult's law is

$$\alpha_{E,S} = \frac{1.259}{0.903} = 1.394$$

By interpolation of smoothed experimental data of Chiyavech and Van Winkle

$$K_E = 1.280 \,(1.64\% \text{ higher than estimated})$$

$$K_S = 0.917 \,(1.53\% \text{ higher than estimated})$$

$$\alpha_{E,S} = 1.396 \,(\text{only } 0.14\% \text{ higher than estimated}).$$

This close agreement indicates that ethylbenzene and styrene obey Raoult's law quite closely at vacuum distillation conditions.

Because the pressure is low, the heat of vaporization can be estimated from (4-84)

$$\lambda = \frac{(5862.905)(1.987)(635.67)^2}{(176 + 349.8527)^2} = 17023 \text{ Btu/lbmole (39.6 MJ/kgmole)}$$

and

$$\frac{\lambda}{M} = \frac{17023}{106.168} = 160.3 \text{ Btu/lb (372.8 kJ/kg)}$$

The API *Technical Data Book—Petroleum Refining* (1966) gives a value of 159 Btu/lb.
□

4.7 Thermodynamic Properties for Ideal Solutions at Low to Moderate Pressures

For ideal solutions where $\gamma_{iL} = \gamma_{iV} = 1$, when the pressure is superatmospheric and the ideal gas law is invalid, (4-74) as

$$K_{\text{Ideal}_i} = \frac{\nu_{iL}^o}{\nu_{iV}^o} \tag{4-85}$$

applies. The *ideal K-value*, defined by (4-85) was developed by G. G. Brown, W. K. Lewis, and others.[37] Like the Raoult's law K-value, the ideal K-value is independent of phase compositions.

Despite widespread use of the ideal K-value concept in industrial calculations, particularly during years prior to digital computers, a sound thermodynamic basis does not exist for calculation of the fugacity coefficients for pure species as required by (4-85). Mehra, Brown, and Thodos[8] discuss the fact that, for vapor–liquid equilibrium at given system temperature and pressure, at least one component of the mixture cannot exist as a pure vapor and at least one other component cannot exist as a pure liquid. For example, in Fig. 4.3, at a reduced pressure of 0.5 and a reduced temperature of 0.9, methane can exist only as a vapor and toluene can exist only as a liquid. It is possible to compute ν_L^o or ν_V^o for each species but not both, unless $\nu_L^o = \nu_V^o$, which corresponds to saturation conditions. An even more serious problem is posed by species whose critical temperatures are below the system temperature. Attempts to overcome these difficulties via development of pure species fugacity correlations for hypothetical states by extrapolation procedures are discussed by Prausnitz.[38]

The ideal solution assumptions are applicable to mixtures of isomers or homologs that have relatively close boiling points. Here, only limited extrapolations of pure-component fugacities into hypothetical regions are required. A frequently used extrapolation technique is as follows.[39] Whether hypothetical or not, (4-71), which assumes incompressible liquid, is utilized to determine ν_{iL}^o. Substitution of (4-71) into (4-85) gives

$$K_{\text{Ideal}_i} \cong \left(\frac{P_i^s}{P}\right)\left(\frac{\nu_{iV_s}^o}{\nu_{iV}^o}\right) \exp\left[\frac{\nu_{iL}(P - P_i^s)}{RT}\right] \tag{4-86}$$

To determine pure vapor fugacity coefficients (4-51) is rewritten as

$$\nu^o_{iV} = \exp\left(\int_0^P \frac{(Z-1)}{P}\, dP\right) \tag{4-87}$$

At low to moderate pressures, the integrand of (4-87) is approximately constant for a given temperature. The integration is thus readily performed, with the following result in reduced form

$$\nu^o_{iV} = \exp\left[P_{r_i}\left(\frac{Z-1}{P_{r_i}}\right)_{P_{r_i}=0}\right] \tag{4-88}$$

The term $[(Z-1)/P_{r_i}]$ at $P_{r_i} = 0$ is determined from an appropriate equation of state. If the R-K equation is used, a reduced form is convenient. It is derived by combining (4-38), (4-40), (4-44), (4-45), and (4-46) to give

$$Z = \frac{1}{1 - (0.0867 P_r/T_r Z)} - \frac{0.4278 P_r}{T_r^{2.5} Z(1 + (0.0867\, P_r/T_r Z))} \tag{4-89}$$

In the limit

$$\left(\frac{Z-1}{P_r}\right)_{P_r=0} = \frac{0.0867}{T_r} - \frac{0.4278}{T_r^{2.5}} \tag{4-90}$$

Substituting (4-90) into (4-88) gives

$$\nu^o_{iV} = \exp\left(\frac{0.0867\, P_{r_i}}{T_{r_i}} - \frac{0.4278\, P_{r_i}}{T_{r_i}^{2.5}}\right) \tag{4-91}$$

A working equation for the ideal K-value applicable for low to moderate pressures is obtained by combining (4-91) with (4-86)

$$K_{\text{Ideal}_i} = \left(\frac{P_i^s}{P}\right)\exp\left[\left(\frac{0.0867}{T_{r_i} P_{c_i}} - \frac{0.4278}{T_{r_i}^{2.5} P_{c_i}} - \frac{\nu_{iL}}{RT}\right)(P_i^s - P)\right] \tag{4-92}$$

The exponential term is a correction to the Raoult's law K-value and cannot be neglected at moderate pressures as will be shown in the next example.

Example 4.8. The propylene–1–butene system at moderate pressures might be expected to obey the ideal solution laws. Use (4-92) to compute K-values at 100°F (310.93°K) over a pressure range of approximately 60 psia (413.69 kPa) to 200 psia (1.379 MPa). Compare the results to Raoult's law K-values and to experimental data of Goff, Farrington, and Sage.[40]

Solution. Using constants from Appendix I in (4-69), vapor pressures of 229.7 psia (1.584 MPa) and 63.0 psia (434 kPa), respectively, for propylene and 1-butene are computed at 100°F in the same manner as in Example 4.5. The values compare to measured values of 227.3 psia and 62.5 psia, respectively. Raoult's law K-values follow directly from (4-75) and are plotted in Fig. 4.4. Ideal K-values are determined from (4-92) using liquid molal volumes computed from (4-79). For instance, from Appendix I for 1-butene at 25°C (536.67°R), the liquid molal volume is 95.6 cm³/gmole (1.532 ft³/lbmole). From (4-79)

$$\xi = \frac{1.532}{[5.7 + 3(536.67/755.3)]} = 0.1956 \text{ ft}^3/\text{lbmole}$$

At 100°F (559.67°R), from (4-79) with this value of ξ

$$v_L = 0.1956\left[5.7 + 3\left(\frac{559.67}{755.3}\right)\right] = 1.55 \text{ ft}^3/\text{lbmole}$$

At 150 psia (1.034 MPa) and 100°F, the ideal K-value for 1-butene from (4-92) is

$$K_{\text{Ideal}} = \left(\frac{63.0}{150}\right) \exp\left\{\left[\frac{0.0867}{(559.67/755.3)(583)} - \frac{0.4278}{(559.67/755.3)^{2.5}(583)} - \frac{1.55}{10.73(559.67)}\right](63\text{-}150)\right\} = 0.483$$

Ideal K-values are also plotted in Fig. 4.4 together with experimental data points. In general, agreement between experiment and computed ideal K-values is excellent. Raoult's law K-values tend to be high for propylene relative to 1-butene. At 125 psia (862 kPa), relative volatilities from (1-7) for propylene relative to 1-butene are

$$\alpha_{\text{Raoult}} = \frac{1.838}{0.504} = 3.65$$

$$\alpha_{\text{Ideal}} = \frac{1.627}{0.557} = 2.92$$

$$\alpha_{\text{Expt}} = \frac{1.625}{0.560} = 2.90$$

Agreement is almost exact between experiment and the ideal relative volatility; but the Raoult's law relative volatility shows a significant positive deviation of 26%.
□

The problem of hypothetical states also causes difficulties in formulating thermodynamic equations for the density and enthalpy of ideal solutions. However, for low to moderate pressures, the fugacity expressions used in formulating the ideal K-value given by (4-92) can also be used to derive consistent expressions for the other thermodynamic quantities. Equation (4-54), for example, provides a way to obtain pure vapor molal volumes

$$v_{iV} = RT\left(\frac{\partial \ln f_{iV}^o}{\partial P}\right)_T \tag{4-93}$$

From (4-91) with $f_{iV}^o = v_{iV}^o P$

$$\ln f_{iV}^o = \ln P - \left(\frac{0.4278 P_{r_i}}{T_{r_i}^{2.5}}\right) \tag{4-94}$$

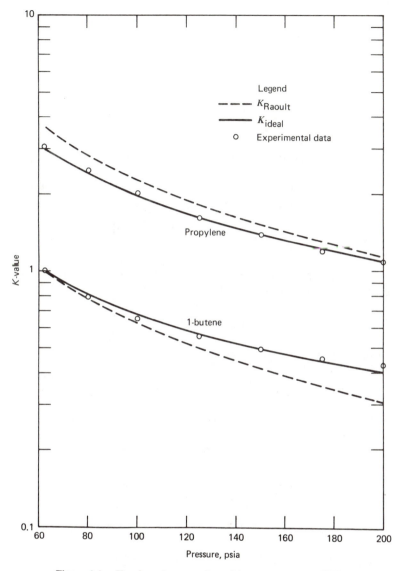

Figure 4.4. K-values for propylene–1-butene system at 100°F.

Differentiating (4-94) and substituting into (4-93) gives

$$v_{iV} = \frac{RT}{P} - \frac{0.4278\, RT_{c_i}}{P_{c_i} T_{r_i}^{1.5}} \tag{4-95}$$

For ideal solutions, species molal volumes are summed to obtain the molal density for the vapor mixture.

$$\frac{\rho}{M} = \frac{1}{\sum\limits_{i=1}^{C} y_i v_{iv}} \tag{4-96}$$

Equation (4-53) gives the effect of pressure on vapor enthalpy of a pure component as

$$H_{iv}^o - H_{iv} = RT^2 \left(\frac{\partial \ln f_{iv}^o}{\partial T} \right)_P \tag{4-97}$$

Differentiating (4-94), substituting the result into (4-97), and summing yields a working equation for vapor mixture enthalpy.

$$H_V = \sum_{i=1}^{C} y_i \left[H_{iv}^o - \frac{1.0695 \, RTP_{r_i}}{T_{r_i}^{2.5}} \right] \tag{4-98}$$

Liquid mixture density can be approximated by (4-78) and (4-79). A relation for liquid-phase enthalpy of ideal solutions is derived by applying (4-53).

$$H_{iv}^o - H_{iL} = RT^2 \left(\frac{\partial \ln f_{iL}^o}{\partial T} \right)_P \tag{4-99}$$

The liquid-phase fugacity is obtained by combining (4-13) and (4-71)

$$\ln f_{iL}^o = \ln v_{iL}^o + \ln P = \ln P_i^s + \ln v_{iv_s}^o + \frac{v_{iL}(P - P_i^s)}{RT} \tag{4-100}$$

Replacing P_i^s by (4-83)—the Antoine vapor pressure relation, $v_{iv_s}^o$ by (4-91) with P_{r_i} evaluated at saturation, and v_{iL} by (4-79); differentiating the resulting equation for $\ln f_{iL}^o$; solving for the species liquid enthalpy; and summing, yields the following expression for the ideal liquid mixture enthalpy

$$H_L = \sum_{i=1}^{C} x_i \left\{ H_{iv}^o - \frac{1.0695 RTP_i^s}{P_{c_i} T_{r_i}^{2.5}} - \frac{P_i^s TA_{2_i}}{(T' + A_3)_i^2} \left[\frac{RT}{P_i^s} - \frac{0.4278 RT_{c_i}}{P_{c_i} T_{r_i}^{1.5}} - v_{iL} \right] \right.$$
$$\left. + 5.7 \, \xi_i (P - P_i^s) \right\} \tag{4-101}$$

The four terms on the right-hand side of (4-101) correspond to the four contributions on the right-hand side of (4-63).

Example 4.9. At 100°F (310.93°K) and 150 psia (1.034 MPa), Goff et al.[40] measured experimental equilibrium phase compositions for the propylene–1–butene system. The liquid-phase composition was 56 mole% propylene. Assuming an ideal solution, estimate the liquid-phase enthalpy relative to vapor at 0°F (−17.8°C) and 0 psia.

Solution. From (4-60), using vapor enthalpy constants from Appendix I for propylene, we have

$$H_V^\circ = 13.63267(100) + \frac{0.2106998 \times 10^{-1}(100)^2}{2} + \frac{0.249845 \times 10^{-5}(100)^3}{3}$$

$$\frac{0.1146863 \times 10^{-7}(100)^4}{4} + \frac{0.5247386 \times 10^{-11}(100)^5}{5}$$

$$= 1469.17 \text{ Btu/lbmole (3.415 MJ/kgmole)}$$

From (4-101) using results from Example 4.8, for the propylene component only with $\xi = 0.1552$ ft^3/lbmole and v_{iL} from (4-79)

$$H_L = 1469.17 - \frac{1.0695(1.987)(559.67)(229.7/667.0)}{(559.67/657.2)^{2.5}} - \frac{229.7(559.67)(3375.447)}{(100 + 418.4319)^2}$$

$$\times \left\{ \frac{1.987(559.67)}{229.7} - \frac{0.4278(1.987)(657.2)}{667.0(559.67/657.2)^{1.5}} - 0.1552 \left[5.7 + 3 \left(\frac{559.67}{657.2} \right) \right] \left(\frac{1.987}{10.73} \right) \right\}$$

$$+ \frac{5.7(0.1552)(150 - 229.7)1.987}{(10.73)}$$

$$= 1469.17 - 612.02 - 5712.72 - 13.06$$

$$= -4868.63 \text{ Btu/lbmole propylene } (-11.32 \text{ MJ/kgmole})$$

A similar calculation for 1-butene using a vapor pressure of 63.0 psia (434.4 kPa) from Example 4.8 gives

$$H_L = -6862.10 \text{ Btu/lbmole } (-15.95 \text{ MJ/kgmole}) \text{ 1-butene}$$

Enthalpy for the ideal liquid mixture is computed from

$$H_L = \sum_{i=1}^{2} (x_i H_{iL}) = 0.56(-4863.63) + 0.44(-6862.10)$$

$$= -5745.76 \text{ Btu/lbmole } (-13.36 \text{ MJ/kgmole}) \text{ mixture}$$

☐

4.8 Soave–Redlich–Kwong Equation of State

Ideal solution thermodynamics is most frequently applied to mixtures of non-polar compounds, particularly hydrocarbons such as paraffins and olefins. Figure 4.5 shows experimental K-value curves for a light hydrocarbon, ethane, in various binary mixtures with other less volatile hydrocarbons at 100°F (310.93°K) at pressures from 100 psia (689.5 kPa) to *convergence pressures* between 720 and 780 psia (4.964 MPa to 5.378 MPa). At the convergence pressure, K-values of all species in a mixture become equal to a value of one, making separation by operations involving vapor–liquid equilibrium impossible. The temperature of 100°F is close to the critical temperature of 550.0°R (305.56°K) for ethane. Figure 4.5 shows that ethane does not form ideal solutions with all the other components because the K-values depend on the other component,

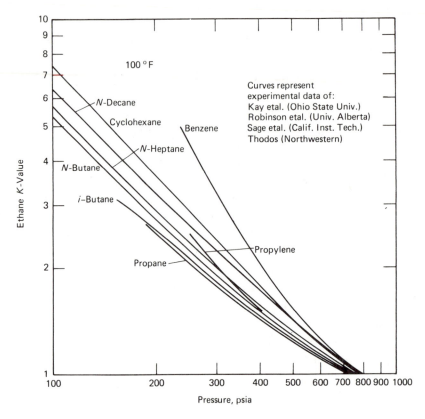

Figure 4.5. *K*-values of ethane in binary hydrocarbon mixtures at 100°F.

even for paraffin homologs. At 300 psia, the *K*-value of ethane in benzene is 80% greater than that in propane.

Thermodynamic properties of nonideal hydrocarbon mixtures can be predicted by a single equation of state if it is valid for both the vapor and liquid phases. Although the Benedict–Webb–Rubin (B-W-R) equation of state has received the most attention, numerous attempts have been made to improve the much simpler R-K equation of state so that it will predict liquid-phase properties with an accuracy comparable to that for the vapor phase. The major difficulty with the original R-K equation is its failure to predict vapor pressure accurately, as was exhibited in Fig. 4.3. Following the success of earlier work by Wilson,[41] Soave[12] added a third parameter, the Pitzer acentric factor, to the R-K equation and obtained almost exact agreement with pure hydrocarbon vapor pressure data. The limit of accuracy of such three-parameter equations of state is discussed by Redlich.[42]

The Soave modification of the R-K equation, referred to here as the S-R-K equation, is

$$P = \frac{RT}{v-b} - \frac{a_a}{v(v+b)} \tag{4-102}$$

where a, as given by (4-103), is temperature dependent and replaces the $1/T^{0.5}$ term in (4-38). From (4-51) and (4-103), expressions can be derived for v_{iL}^o and v_{iV}^o. Soave back-calculated values of a in (4-102) for various hydrocarbon species over a range of reduced temperatures using vapor pressure data and the saturation condition $v_{iL}^o = v_{iV}^o$ to obtain the following correlation for a

$$a = [1 + m(1 - T_r^{0.5})]^2 \tag{4-103}$$

where $m = 0.480 + 1.574\,\omega - 0.176\,\omega^2$.

Working equations for computing thermodynamic properties of interest are derived from (4-102) and (4-103) and the equations of Table 4.3, in the same manner as for the original R-K equation. The resulting expressions are applicable to either the liquid or vapor phases provided that the appropriate phase composition and compressibility factor are used.

$$Z^3 - Z^2 + Z(A - B - B^2) - AB = 0 \tag{4-104}$$

$$v_i^o = \exp\left[Z - 1 - \ln(Z - B_i) - \frac{A_i}{B_i}\ln\left(\frac{Z + B_i}{Z}\right)\right] \tag{4-105}$$

$$\phi_i = \exp\left\{(Z - 1)\frac{B_i}{B} - \ln(Z - B) - \frac{A}{B}\left[\frac{2A_i^{0.5}}{A^{0.5}} - \frac{B_i}{B}\right]\ln\left(\frac{Z + B}{Z}\right)\right\} \tag{4-106}$$

$$H = H_V^o + RT\left\{Z - 1 - \frac{1}{B}\ln\left(\frac{Z + B}{Z}\right)\sum_{i=1}^{C}\sum_{j=1}^{C} y_i y_j A_{ij}\left[1 - \frac{m_i T_{r_i}^{0.5}}{2a_i^{0.5}} - \frac{m_j T_{r_j}^{0.5}}{2a_j^{0.5}}\right]\right\} \tag{4-107}$$

In (4-105), Z is the compressibility factor for the pure species.

Constants A_i, B_i, A, and B depend on T_{r_i}, P_{r_i}, and ω_i. For pure species

$$*A_i = 0.42747\, a_i \frac{P_{r_i}}{T_{r_i}^2} \tag{4-108}$$

$$*B_i = 0.08664 \frac{P_{r_i}}{T_{r_i}} \tag{4-109}$$

Mixing rules for nonpolar species are those of the original R-K equation. For

---

* Equations (4-108) and (4-109) differ from (4-44) and (4-45).

vapor mixtures, for example

$$A = \sum_{i=1}^{C} \sum_{j=1}^{C} y_i y_j A_{ij} \qquad (4\text{-}110)$$

where,

$$A_{ij} = (A_i A_j)^{0.5} \qquad (4\text{-}111)$$

$$B = \sum_{i=1}^{C} y_i B_i \qquad (4\text{-}112)$$

Once the appropriate roots of Z are obtained from (4-104), other thermodynamic properties are readily computed including K-values from (4-27). A FORTRAN computer program for performing the calculations is available.[43]

Except for hydrogen, the S-R-K equations can also be applied to light gases such as nitrogen, carbon monoxide, carbon dioxide, and hydrogen sulfide if binary interaction parameters k_{ij} are incorporated into a revision of (4-111), given as

$$A_{ij} = (1 - k_{ij})(A_i A_j)^{1/2} \qquad (4\text{-}113)$$

Values of k_{ij} for many binary pairs have been back-calculated from experimental data.[32,43]

West and Erbar[44] used extensive experimental data for hydrocarbon mixtures to evaluate the S-R-K equations with results summarized in Table 4.4. Also

Table 4.4 Summary of evaluation of thermodynamic correlations for hydrocarbon systems by West and Erbar[44]

Thermodynamic Property	K-value	Enthalpy	Liquid Density
Number of data points	3510	21 compositions	709
Temperature range, °F	−240 to 500	< -150 to >150	32 to 140
Pressure, psia	to 3707	to >1000	200 to 2000
Correlation	Absolute average deviation, %	Absolute average deviation, Btu/lb	Absolute average deviation, %
S-R-K	13.6	2.1	9.78
C-S	15.5	—	—
Starling-Han B-W-R	19.0	3.1	1.14

Source. E. W. West and J. H. Erbar, "An Evaluation of Four Methods of Predicting Thermodynamic Properties of Light Hydrocarbon Systems," paper presented at the 52nd Annual Meeting of NGPA, Dallas, Texas, March 26–28, 1973.

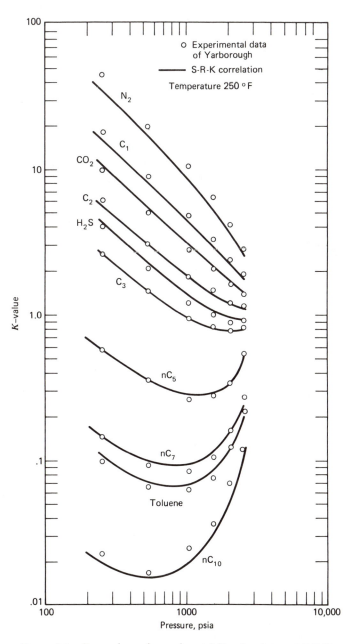

Figure 4.6.   Comparison of experimental K-value data and S-R-K correlation.

included are their evaluations of the Chao-Seader (C-S) correlation[13] described in Chapter 5 and the Starling and Han modification of the B-W-R equation.[10,11] The S-R-K equations appear to give the most reliable overall results for K-values and enthalpies over wide ranges of temperature and pressure. However, as indicated, the S-R-K correlation, like the R-K equation, still fails to predict liquid density with good accuracy. A more recent extension of the R-K equation by Peng and Robinson[45] is more successful in that respect.

Figure 4.6. shows the ability of the S-R-K correlation to predict K-values for the multicomponent system of 10 species studied experimentally by Yarborough.[46] The data cover more than a threefold range of volatility. Also, the S-R-K correlation appears to be particularly well suited for predicting K-values and enthalpies for natural gas systems at cryogenic temperatures, where the C-S correlation is not always adequate. Figures 4.7 and 4.8, which are based on the data of Cavett[47] and West and Erbar,[44] are comparisons of K-values computed

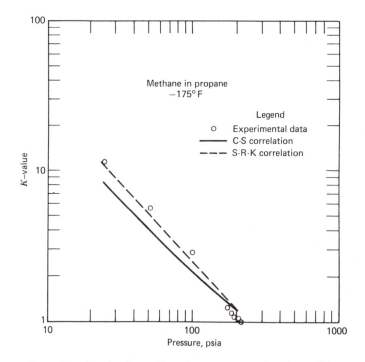

Figure 4.7. K-value for methane in propane at cryogenic conditions. (Data from R. H. Cavett, "Monsanto Physical Data System," paper presented at AIChE meeting, 1972, and E. W. West and J. H. Erbar, "An Evaluation of Four Methods of Predicting Thermodynamic Properties of Light Hydrocarbon Systems," paper presented at NGPA meeting, 1973.)

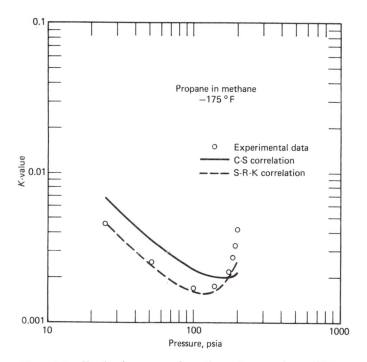

Figure 4.8. K-value for propane in methane at cryogenic conditions. (Data from R. H. Cavett, "Monsanto Physical Data System," paper presented at AIChE meeting, 1972, and E. W. West and J. H. Erbar, "An Evaluation of Four Methods of Predicting Thermodynamic Properties of Light Hydrocarbon Systems," paper presented at NGPA meeting, 1973.)

for the C-S and S-R-K correlations to experimental K-values of Wichterle and Kobayashi[48] for the methane–propane system at $-175°F$ ($-115°C$) over a pressure range of 25 to approximately 200 psia (0.172 to 1.379 MPa). While the S-R-K correlation follows the experimental data quite closely, the C-S correlation shows average deviations of approximately 16% and 32% for methane and propane, respectively.

Example 4.10. Wichterle and Kobayashi[48] measured equilibrium phase compositions for the methane–ethane–propane system at temperatures of -175 to $-75°F$ (158.15 to 213.71°K) and pressures to 875 psia (6.033 MPa). At $-175°F$ and 100 psia (0.689 MPa), one set of data is

Species	x_i	y_i	K_i
Methane	0.4190	0.9852	2.35
Ethane	0.3783	0.01449	0.0383
Propane	0.2027	0.000312	0.00154

Use the Soave–Redlich–Kwong correlation to estimate the compressibility factor, enthalpy (relative to zero-pressure vapor at 0°F) and K-values for the equilibrium phases. Necessary constants of pure species are in Appendix I. All values of k_{ij} are 0.0. Compare estimated K-values to experimental K-values.

Solution. By computer calculations, results are obtained as follows.

	Liquid Phase	Vapor Phase
Z	0.0275	0.9056
M, Lb/lbmole	27.03	16.25
v, Ft³/lbmole	0.8401	27.66
ρ, Lb/ft³	32.18	0.5876
H, Btu/lbmole	−4723.6	−1696.5
H/M, Btu/lb	−174.7	−104.4

	K-Values		
Species	Experimental	S-R-K	% Deviation
Methane	2.35	2.33	−0.85
Ethane	0.0383	0.0336	−12.27
Propane	0.00154	0.00160	3.90

As seen, agreement is quite good for methane and propane. Adjustment of the acentric factor for ethane would improve agreement for this species.
□

The Soave–Redlich–Kwong equation is rapidly gaining acceptance by the hydrocarbon processing industry. Further developments, such as that of Peng and Robinson,[45] are likely to improve predictions of liquid density and phase equilibria in the critical region. In general however, use of such equations appears to be limited to relatively small, nonpolar molecules. Calculations of phase equilibria with the S-R-K equations require initial estimates of the phase compositions.

References

1. Perry, R. H. and C. H. Chilton, Eds, *Chemical Engineers' Handbook*, 5th ed., McGraw–Hill Book Co., New York, 1973, 4-43 to 4-65.

2. Gully, A. J., *Refining Engineer*, **31**, C-34 to C-47 (May, 1959).

3. Stocking, M., J. H. Erbar, and R. N. Maddox, *Refining Engineer*, **32**, C-15 to C-18 (April, 1960).

4. Grayson, H. G., *Proc. API*, **42**, III, 62–71 (1962).

5. Hougen, O. A., K. M. Watson, and R. A. Ragatz, *Chemical Process Principles, Part II, Thermodynamics*, 2nd ed., John Wiley & Sons, Inc., New York, 1959, 892–893.

6. Prausnitz, J. M., *Molecular Thermodynamics of Fluid-Phase Equil-*

ibria, Prentice–Hall, Inc., Englewood Cliffs, N.J., 1969, 19-22.

7. Hougen, O. A., and K. M. Watson, *Chemical Process Principles, Part II, Thermodynamics*, John Wiley & Sons, Inc., New York, 1947, 663–665.

8. Mehra, V. S., G. M. Brown, and G. Thodos, *Chem. Eng. Sci.*, **17**, 33–43 (1962).

9. Benedict, M., G. B. Webb, and L. C. Rubin, *Chem. Eng. Progr.*, **47**, 449–454 (1951).

10. Starling, K. E., and M. S. Han, *Hydrocarbon Processing*, **51** (5), 129–132 (1972).

11. Starling, K. E., and M. S. Han, *Hydrocarbon Processing*, **51** (6), 107–115 (1972).

12. Soave, G., *Chem. Eng. Sci.*, **27**, 1197–1203 (1972).

13. Chao, K. C., and J. D. Seader, *AIChE J.*, **7**, 598–605 (1961).

14. Grayson, H. G., and C. W. Streed, Paper 20-PO7, Sixth World Petroleum Conference, Frankfurt, June, 1963.

15. Prausnitz, J. M., C. A. Eckert, R. V. Orye, and J. P. O'Connell, *Computer Calculations for Multicomponent Vapor-Liquid Equilibria*, Prentice–Hall, Inc., Englewood Cliffs, N.J., 1967.

16. Lee, B., J. H. Erbar, and W. C. Edmister, *AIChE J.*, **19**, 349–356 (1973).

17. Robinson, R. L., and K. C. Chao, *Ind. Eng. Chem., Process Design Devel.*, **10**, 221–229 (1971).

18. Prausnitz, J. M., and P. L. Chueh, *Computer Calculations for High-Pressure Vapor-Liquid Equilibria*, Prentice–Hall, Inc., Englewood Cliffs, New Jersey (1968).

19. Reid, R. C., and T. K. Sherwood, *The Properties of Gases and Liquids*, 2nd ed., McGraw–Hill Book Co., New York, 1966, 47-60.

20. O'Connell, J. P., and J. M. Prausnitz, *Ind. Eng. Chem., Process Des. Develop.*, **6**, 245–250 (1967).

21. Hayden, J. G., and J. P. O'Connell, *Ind. Eng. Chem., Process Des. Develop.*, **14**, 209-216 (1975).

22. Beattie, J. A., and O. C. Bridgeman, *J. Amer. Chem. Soc.*, **49**, 1665–1666 (1927).

23. Benedict, M., G. W. Webb, and L. C. Rubin, *J. Chem. Phys.*, **8**, 334–345 (1940),

24. Redlich, O., and J. N. S. Kwong, *Chem. Rev.*, 44, 233-244 (1949).

25. Shah, K. K., and G. Thodos, *Ind. Eng. Chem.*, **57** (3), 30–37 (1965).

26. Sage, B. H., and W. N. Lacey, *Ind. Eng. Chem.*, **30**, 673–681 (1938).

27. Edmister, W. C., *Hydrocarbon Processing*, **47** (9), 239–244 (1968).

28. Glanville, J. W., B. H. Sage, and W. N. Lacey, *Ind. Eng. Chem.*, **42**, 508-513 (1950).

29. Edmister, W. C., *Hydrocarbon Processing*, **47** (10), 145–149 (1968).

30. Yarborough, L., and W. C. Edmister, *AIChE J.*, **11**, 492–497 (1965).

31. Pitzer, K. S., D. Z. Lippman, R. F. Curl, Jr., C. M. Huggins, and D. E. Petersen, *J. Amer. Chem. Soc.*, **77**, 3433–3440 (1955).

32. Reid, R. C., J. M. Prausnitz, and T. K. Sherwood, *The Properties of Gases and Liquids*, 3rd ed., McGraw–Hill Book Company, New York, 1977, 184–185 and 193–197.

33. Redlich, O., and A. T. Kister, *J. Amer. Chem. Soc.*, **71**, 505-507 (1949).

34. Cavett, R. H., *Proc. API*, **42**, 351–366 (1962).

35. Frank. J. C., G. R. Geyer, and H. Kehde, *Chem. Eng. Prog.*, **65** (2), 79-86 (1969).

36. Chaiyavech, P., and M. Van Winkle, *J. Chem. Eng. Data*, **4**, 53–56 (1959).

37. Edmister, W. C., *Petroleum Refiner*, **28** (5), 149–160 (1949).

38. Prausnitz, J. M., *AIChE J.*, **6**, 78–82 (1961).

39. Benedict, M., C. A. Johnson, E. Solomon, and L. C. Rubin, *Trans. AIChE*, **41**, 371-392 (1945).

40. Goff, G. H., P. S. Farrington, and B. H. Sage, *Ind. Eng. Chem.*, **42**, 735–743 (1950).

41. Wilson, G. M., *Advan. Cryog. Eng.*, **11**, 392–400 (1966).

42. Redlich, O., *Ind. Eng. Chem., Fundam.*, **14**, 257 (1975).

43. GPA K and H Computer Program, Gas Processors Association, Tulsa, Oklahoma (August, 1974).

44. West, E. W., and J. H. Erbar, "An Evaluation of Four Methods of Predicting Thermodynamic Properties of Light Hydrocarbon Systems," paper presented at the 52nd Annual Meeting of NGPA, Dallas, Texas, March 26–28, 1973.

45. Peng, D. Y., and D. B. Robinson, *Ind. Eng. Chem., Fundam.*, **15**, 59–64 (1976).

46. Yarborough, L., *J. Chem. Eng. Data*, **17**, 129-133 (1972).

47. Cavett, R. H., "Monsanto Physical Data System." paper presented at the 65th Annual Meeting of AIChE, New York, November 26-30, 1972.

48. Wichterle, I., and R. Kobayashi, *J. Chem. Eng. Data*, **17**, 4–18 (1972).

Problems

4.1 For any component i of a multicomponent, multiphase system, derive (4-12), the equality of fugacity, from (4-8), the equality of chemical potential, and (4-11), the definition of fugacity.

4.2 Indicate which of the following K-value expressions, if any, is(are) rigorous. For those expressions, if any, that are not rigorous, cite the assumptions involved.

$$\text{(a) } K_i = \frac{\phi_{iL}}{\phi_{iv}} \quad \text{(b) } K_i = \frac{\nu_{iL}^o}{\nu_{iv}^o} \quad \text{(c) } K_i = \nu_{iL}^o \quad \text{(d) } K_i = \frac{\gamma_{iL}\nu_{iL}^o}{\phi_{iv}}$$

$$\text{(e) } K_i = \frac{P_i^s}{P} \quad \text{(f) } K_i = \left(\frac{\gamma_{iL}}{\gamma_{iv}}\right)\left(\frac{\nu_{iL}^o}{\nu_{iv}^o}\right) \quad \text{(g) } K_i = \frac{\gamma_{iL}P_i^s}{P}$$

4.3 Distribution coefficients for liquid–liquid equilibria can be calculated from

$$K_{D_i} \equiv \frac{x_i^{I}}{x_i^{II}} = \frac{\phi_{iL}^{II}}{\phi_{iL}^{I}}$$

(a) Derive this equation.
(b) Why is this equation seldom used for the prediction of K_{D_i}?

4.4 Repeat Example 4.2 for the given vapor mixture at 400°F and a pressure of 350 psia. Under these conditions, the vapor will be superheated.

4.5 Calculate the density in kilograms per cubic meter of isobutane at 93°C and 1723 kPa with (a) ideal gas law, (b) Redlich–Kwong equation of state.

4.6 For the Redlich-Kwong equation of state, derive an expression for

$$\left(\frac{\partial Z}{\partial T}\right)_P$$

4.7 Use the van der Waals equation of state to calculate the molal volume in cubic feet per pound-mole of isobutane at 190°F for:
(a) The vapor at 150 psia.
(b) The liquid at 3000 psia.
Compare your results to those in Fig. 4.1.

4.8 Use the R-K equation of state to predict the liquid molal volumes in cubic centimeters per gram-mole at 25°C and 1 atm of n-pentane, n-decane, and n-pentadecane. Compare your values to those in Appendix I and note any trend in the discrepancies.

4.9 Use the R-K equation of state to predict the density in kilograms per cubic meter of ethylbenzene at the critical point. Compare your value to that based on the value of Z_c given in Appendix I.

4.10 Use the R-K equation of state to predict the liquid molal volume of an equimolal mixture of ethane and n-decane at 100°F and 1000 psia. Compare your value to the experimental value of 2.13 ft³/lbmole [*J. Chem. Eng. Data*, **7**, 486 (1962)].

4.11 Estimate the vapor molal volume for 27.33 mole % nitrogen in ethane at 400°F and 2000 psia with the R-K equation of state. Compare your value to the experimental value of 4.33 ft³/lbmole [*Ind. Eng. Chem.*, **44**, 198 (1952)].

4.12 Repeat Example 4.2 for a mixture containing 0.4507 weight fraction propane at 400°F and 300 psia. The experimental value is 0.4582 ft³/lb.

4.13 Using the Antoine equation for vapor pressure, estimate the acentric factor for isobutane from Equation (4-68) and compare it to the value tabulated in Appendix I.

4.14 Derive Equations (4-66) and (4-67) from (4-38) and (4-51).

4.15 Estimate ϕ_{iv} for propane in the vapor mixture of Example 4.2 using the R-K equation of state.

4.16 Prove that the R-K equation predicts

$$Z_c = \frac{1}{3}$$

4.17 Derive an analytical expression for

$$\frac{\partial \phi_{iv}}{\partial T}$$

using the Redlich–Kwong equation.

4.18   Develop equations for computing the liquid-phase and vapor-phase activity coefficients from the Redlich–Kwong equation of state. Apply your equations to the propane–benzene system at 280°F and 400 psia where experimentally measured propane mole fractions are $x = 0.1322$ and $y = 0.6462$ [*Ind. Eng. Chem.*, **42**, 508 (1950)]. Do any difficulties arise in applying your equations?

4.19   Experimental measurements of Vaughan and Collins [*Ind. Eng. Chem.*, **34**, 885 (1942)] for the propane-isopentane system at 167°F and 147 psia show that a liquid mixture with a propane mole fraction of 0.2900 is in equilibrium with a vapor having a 0.6650 mole fraction of propane. Use the R-K equation to predict:
   (a) Vapor and liquid molal volumes in cubic meters per kilogram.
   (b) Vapor and liquid enthalpies in kilojoules per kilogram.
   (c) Vapor and liquid mixture fugacity coefficients for each component.
   (d) K-values for each component. Compare these to the experimental values.

4.20   For n-hexane at 280°F, use the R-K equation of state to predict:
   (a) The vapor pressure.
   (b) The latent heat of vaporization.
   The experimental values are 84.93 psia and 116.83 Btu/lb, respectively [*J. Chem. Eng. Data*, **9**, 223 (1964)].

4.21   For *trans*-2-butene at 130°F, use the R-K equation of state to predict:
   (a) The vapor pressure.
   (b) The saturated specific volumes of the liquid and vapor.
   (c) The latent heat of vaporization.
   The experimental values are 78.52 psia, 0.02852 ft³/lb, 1.249 ft³/lb, and 149.17 Btu/lb, respectively [*J. Chem. Eng. Data*, **9**, 536 (1964)].

4.22   For methane vapor at −100°F, predict:
   (a) The zero-pressure specific heat.
   (b) The specific heat at 800 psia using the R-K equation of state. The experimental value is 3.445 Btu/lb · °F [*Chem. Eng. Progr., Symp. Ser. No. 42*, **59** 52 (1963)].

4.23   Using the R-K equation of state, estimate the enthalpy of carbon dioxide relative to that of the ideal gas for the following conditions:
   (a) As a vapor at $T_r = 2.738$ and $P_r = 3.740$.
   (b) As a liquid at $T_r = 0.958$ and $P_r = 1.122$.
   The experimental values are −3.4 and −114.4 Btu/lb, respectively [*AIChE J.*, **11**, 334 (1965)].

4.24   For propylene at 100°F, use the R-K equation of state to estimate the fugacity and the pure component fugacity coefficient at:
   (a) 500 Psia.
   (b) 100 Psia.
   Values of the fugacity coefficients computed from the Starling modification of the B-W-R equation of state are 0.3894 and 0.9101, respectively (K. E. Starling, *Fluid Thermodynamic Properties for Light Petroleum Systems*, Gulf Publishing Co., Houston, Texas, 1973.)

4.25   Repeat Problem 4.24 (a) using Equation (4-71) in conjuction with the Antoine vapor pressure equation.

4.26 Using the Antoine vapor pressure equation, calculate relative volatilities from (4-76) for the isopentane/n-pentane system and compare the values on a plot with the following smoothed experimental values [*J. Chem. Eng. Data*, **8**, 504 (1963)].

Temperature, °F	α_{iC_5, nC_5}
125	1.26
150	1.23
175	1.21
200	1.18
225	1.16
250	1.14

4.27 Using (4-76) with the Antoine vapor pressure equation, calculate the relative volatility of the paraxylene–metaxylene system at a temperature of 138.72°C. A reported value is 1.0206 [*J. Chem. Eng. Japan*, **4**, 305 (1971)].

4.28 Assuming ideal solutions, as in Section 4.6, and using results in Example 4.7, predict the following for an equimolal liquid solution of styrene and ethylbenzene with its equilibrium vapor at 80°C.
(a) Total pressure, kilopascals.
(b) Vapor density, kilograms per cubic meter.
(c) Liquid density, kilograms per cubic meter.
(d) Vapor enthalpy, kilojoules per kilogram.
(e) Liquid enthalpy, kilojoules per kilogram.

4.29 Use Equation (4-35), the van der Waals equation of state, to derive equations similar to (4-91), (4-92), (4-98), and (4-101). Based on your results, calculate ideal K-values and the relative volatility for propylene/1–butene at 100°F and 125 psia. Compare your answer to that of Example 4.8.

4.30 Use the results of Problem 4.29 to compute the liquid-phase enthalpy for the conditions of Example 4.9 and compare your answer to the result of that example.

4.31 At 190°F and 600 psia, a methane/n-butane vapor mixture of 0.6037 mole fraction methane is in equilibrium with a liquid mixture containing 0.1304 mole fraction methane. Using physical property constants and correlation coefficients from Appendix I,
(a) Calculate the specific volumes in cubic meter per kilogram for the liquid and vapor mixtures using the R-K equation.
(b) Estimate the enthalpies of the liquid and vapor phases using the R-K equation.
(c) Calculate the values of the acentric factors using (4-68) and compare to the values listed in Appendix I.
(d) At 190°F, calculate the vapor pressure of methane and butane using (4-66) and (4-67) and compare to values computed from (4-69), the Antoine equation.
(e) Calculate the mixture fugacity coefficients ϕ_{iV} and ϕ_{iL} and the K-values for the R-K equation from (4-72), (4-73), and (4-27) and compare to the experimental values.
(f) Calculate the ideal K-value for n-butane from (4-75) and compare it to the experimental value. Why can't the ideal K-value of methane be computed?

(g) Calculate the K-value of n-butane from (4-92) and compare to experimental data.

4.32 Use the equations in Section 4.7 for ideal solutions at low to moderate pressures to predict the following at 0°F and 159 psia for the ethane–propane system with $x_{C_2} = 0.746$ and $y_{C_2} = 0.897$ [*J. Chem. Eng. Data*, **15**, 10 (1970)].
(a) Pure liquid fugacity coefficients.
(b) Pure vapor fugacity coefficients.
(c) K-values.
(d) Vapor density, pounds per cubic foot.
(e) Liquid density pounds per cubic foot.
(f) Vapor enthalpy British thermal units per pound.
(g) Liquid enthalpy, British thermal units per pound.

4.33 Use the equations in Section 4.7 to predict the K-values of the two butane isomers and the four butene isomers at 220°F and 276.5 psia. Compare these values with the following experimental results [*J. Chem. Eng. Data*, **7**, 331 (1962)].

Component	K-value
Isobutane	1.067
Isobutene	1.024
n-Butane	0.922
1-Butene	1.024
trans-2-Butene	0.952
cis-2-Butene	0.876

4.34 What are the advantages and disadvantages of the Peng–Robinson equation of state [*Ind. Eng. Chem., Fundam.*, **15**, 59 (1976); *AIChE J.*, **23**, 137 (1977); *Hydrocarbon Processing*, **57** (4), 95 (1978)] compared to the Soave–Redlich–Kwong equation of state?

4.35 What are the advantages and disadvantages of the Benedict–Webb–Rubin–Starling equation of state [K. E. Starling, *Fluid Thermodynamic Properties for Light Petroleum Systems*, Gulf Publishing Co., Houston, Texas, 1973; *Hydrocarbon Processing*, **51** (6), 107 (1972)] compared to the Soave–Redlich–Kwong equation of state?

4.36 Repeat problem 4.19 using the Soave–Redlich–Kwong equation of state.

4.37 Reamer, Sage, and Lacey [*Ind. Eng. Chem.*, **43**, 1436 (1951)] measured the following equilibrium phase compositions for the methane/n-butane/n-decane system at 280°F and 3000 psia.

Species	x_i	y_i
Methane	0.5444	0.9140
n-Butane	0.0916	0.0512
n-Decane	0.3640	0.0348

Use the Soave–Redlich–Kwong equation of state to predict for each phase the density, mixture fugacity coefficients, and enthalpy. Also predict the K-values and compare them to the experimental values derived from the above data.

5

Equilibrium Properties from Activity Coefficient Correlations

---

But, for strongly interacting molecules, regardless of size and shape, there are large deviations from random mixing (in liquid solutions); such molecules are far from 'color-blind' because their choice of neighbors is heavily influenced by differences in intermolecular forces. An intuitive idea toward describing this influence was introduced by (Grant M.) Wilson with his notion of local composition....

John M. Prausnitz, 1977

In Chapter 4, methods based on equations of state were presented for predicting thermodynamic properties of vapor and liquid mixtures. Alternatively, as developed in this chapter, predictions of liquid properties can be based on correlations for liquid-phase activity coefficients. Regular solution theory, which can be applied to mixtures of nonpolar compounds using only properties of the pure components, is the first type of correlation presented. This presentation is followed by a discussion of several correlations that can be applied to mixtures containing polar compounds, provided that experimental data are available to determine the binary interaction parameters contained in the correlations. If not, group-contribution methods, which have recently undergone extensive development, can be used to make estimates. All the correlations discussed can be applied to predict vapor–liquid phase equilibria; and some, as discussed in the final section of this chapter, can estimate liquid–liquid equilibria.

5.1 Regular Solutions and the Chao–Seader Correlation

For the more nonvolatile species of mixtures, dependency of K-values on composition is due primarily to nonideal solution behavior in the liquid phase. Prausnitz, Edmister, and Chao[1] showed that the relatively simple *regular solu-*

tion theory of Scatchard and Hildebrand[2] can be used to estimate deviations due to nonideal behavior for hydrocarbon–liquid mixtures. They expressed K-values in terms of (4-29), $K_i = \gamma_{iL} \nu_{iL}^0 / \phi_{iv}$. Chao and Seader[3] simplified and extended application of this equation to a general correlation for hydrocarbons and some light gases in the form of a compact set of equations especially suitable for use with a digital computer.

Simple correlations for the liquid-phase activity coefficient γ_{iL} based only on properties of pure species are not generally accurate. However, for hydrocarbon mixtures, regular solution theory is convenient and widely applied. The theory is based on the premise that nonideality is due to differences in van der Waals forces of attraction among the species present. Regular solutions have an endothermic heat of mixing and all activity coefficients are greater than one. These solutions are regular in the sense that molecules are assumed randomly dispersed. Unequal attractive forces between like and unlike molecule pairs tend to cause segregation of molecules. However, this segregation can be assumed to be counter-balanced by thermal energy with the result that local molecular concentrations are identical to overall solution concentrations. Therefore, the excess entropy is zero and entropy of regular solutions is identical to that of ideal solutions, in which the molecules are randomly dispersed. This is in contrast to an *athermal* solution, for which the excess enthalpy is zero.

For a real solution, the molal free energy g is the sum of the molal free energy for an ideal solution and an excess molal free energy g^E for nonideal effects. For a liquid solution

$$g = \sum_{i=1}^{C} x_i g_i + RT \sum_{i=1}^{C} x_i \ln x_i + g^E$$

$$= \sum_{i=1}^{C} x_i (g_i + RT \ln x_i + \bar{g}_i^E) \tag{5-1}$$

where excess molal free energy is the sum of partial excess molal free energies. The partial excess molal free energy is related by classical thermodynamics[4] to the liquid-phase activity coefficient by

$$\frac{\bar{g}_i^E}{RT} = \ln \gamma_i = \left[\frac{\partial (n_T g^E / RT)}{\partial n_i} \right]_{P,T,n_j} = \frac{g^E}{RT} - \sum x_k \left[\frac{\partial (g^E / RT)}{\partial x_k} \right]_{T,P,x_r} \tag{5-2}$$

where $j \neq i$, $r \neq k$, $r \neq i$, and $k \neq i$.

The relationship between excess molal free energy and excess molal enthalpy and entropy is

$$g^E = H^E - Ts^E = \sum_{i=1}^{C} x_i (\bar{H}_i^E - T\bar{s}_i^E) \tag{5-3}$$

For a multicomponent regular liquid solution, the excess molal free energy is

$$g^E = \sum_{i=1}^{C} (x_i v_{iL}) \left[\frac{1}{2} \sum_{i=1}^{C} \sum_{j=1}^{C} \Phi_i \Phi_j (\delta_i - \delta_j)^2 \right] \qquad (5\text{-}4)$$

where Φ is the volume fraction, assuming additive molal volumes, given by

$$\Phi_j = \frac{x_j v_{jL}}{\sum\limits_{i=1}^{C} x_i v_{iL}} = \frac{x_j v_{jL}}{v_L} \qquad (5\text{-}5)$$

and δ is the *solubility parameter*

$$\delta_j = \left[\frac{\lambda_j - RT}{v_{jL}} \right]^{1/2} \qquad (5\text{-}6)$$

Applying (5-2) to (5-4) gives an expression for the activity coefficient

$$\ln \gamma_{iL} = \frac{v_{iL} \left(\delta_i - \sum\limits_{j=1}^{C} \Phi_j \delta_j \right)^2}{RT} \qquad (5\text{-}7)$$

Because $\ln \gamma_{iL}$ varies almost inversely with absolute temperature, v_{iL} and δ_j are frequently taken as constants at some convenient reference temperature, such as 25°C. Thus, calculation of γ_{iL} by regular solution theory involves only the pure species constants v_L and δ. The latter parameter is often treated as an empirical constant determined by back calculation from experimental data. Chao and Seader[3] suggest that the solubility parameters of isomers be set equal. For species with a critical temperature below 25°C, v_L and δ at 25°C are hypothetical. However, they can be evaluated by back calculation from phase equilibria data. Recommended values of the solubility parameter are included in Appendix I.

When molecular size differences, as reflected by liquid molal volumes, are appreciable, the following Flory–Huggins size correction for athermal solutions can be added to the regular solution free energy contribution

$$g^E = RT \sum_{i=1}^{C} x_i \ln\left(\frac{\Phi_i}{x_i} \right) \qquad (5\text{-}8)$$

Substitution of (5-8) into (5-2) gives

$$\ln \gamma_{iL} = \ln\left(\frac{v_{iL}}{v_L} \right) + 1 - \frac{v_{iL}}{v_L} \qquad (5\text{-}9)$$

The complete expression for the activity coefficient of a species in a regular solution, including the Flory–Huggins correction, is

$$\gamma_{iL} = \exp \left\{ \frac{v_{iL} \left[\delta_i - \sum\limits_{j=1}^{C} \Phi_j \delta_j \right]^2}{RT} + \ln\left(\frac{v_{iL}}{v_L} \right) + 1 - \frac{v_{iL}}{v_L} \right\} \qquad (5\text{-}10)$$

The Flory–Huggins correction was not included in the treatment by Chao and Seader[3] but is contained in the correlation of Robinson and Chao.[5] The correction reduces the magnitude of the activity coefficient, and its use is recommended.

Example 5.1. Yerazunis, Plowright, and Smola[6] measured liquid-phase activity coefficients for the *n*-heptane–toluene system over the entire concentration range at 1 atm (101.3 kPa). Compute activity coefficients using regular solution theory both with and without the Flory–Huggins correction. Compare calculated values with experimental data.

Solution. Experimental liquid-phase compositions and temperatures for 7 of 19 points are as follows, where H denotes heptane and T denotes toluene.

$T, °C$	x_H	x_T
98.41	1.0000	0.0000
98.70	0.9154	0.0846
99.58	0.7479	0.2521
101.47	0.5096	0.4904
104.52	0.2681	0.7319
107.57	0.1087	0.8913
110.60	0.0000	0.0000

From (4-79) at 25°C, using liquid volume constants from Appendix I and the computation procedure of Example 4.8, $v_{HL} = 147.5$ cm^3/gmole and $v_{TL} = 106.8$ cm^3/gmole. As an example, consider mole fractions in the above table for 104.52°C. From (5-5), volume fractions are

$$\Phi_H = \frac{0.2681(147.5)}{0.2681(147.5) + 0.7319(106.8)} = 0.3359$$

$$\Phi_T = 1 - \Phi_H = 1 - 0.3359 = 0.6641$$

Substitution of these values, together with solubility parameters from Appendix I, into (5-7) gives

$$\gamma_H = \exp\left\{\frac{147.5[7.430 - 0.3359(7.430) - 0.6641(8.92)]^2}{1.987(377.67)}\right\} = 1.212$$

Values of γ_H and γ_T computed in this manner for all seven liquid-phase conditions are plotted in Fig. 5.1.

Applying the Flory–Huggins correction (5-10) to the same data point gives

$$\gamma_H = \exp\left[0.1923 + \ln\left(\frac{147.5}{117.73}\right) + 1 - \left(\frac{147.5}{117.73}\right)\right] = 1.179$$

Values of γ_H and γ_T computed in this manner are included in Fig. 5.1, which shows that theoretically calculated curves, especially those based on regular solution theory with the Flory–Huggins correction, are in reasonably good agreement with experimental values. Deviations from experiment are not greater than 12% for regular solution theory and not greater than 6% when the Flory–Huggins correction is included. Unfortunately, such good agreement is not always obtained with nonpolar hydrocarbon solutions as shown, for example, by Hermsen and Prausnitz,[7] who studied the cyclopentane–benzene system.

□

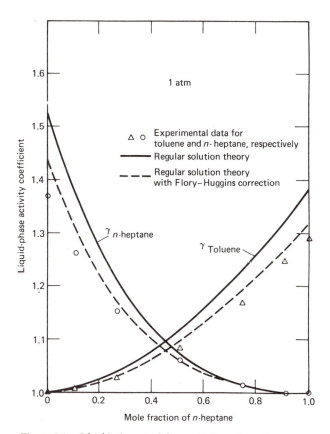

Figure 5.1.   Liquid-phase activity coefficients for n-heptane-toluene system at 1 atm.

In the Chao–Seader (C–S) correlation, the R–K equation of state (4-72) is used to compute ϕ_{iv}, which is close to unity at low pressures. As pressure increases, ϕ_{iv} remains close to one for very volatile components in the mixture. However, for components of low volatility, ϕ_{iv} will be much less than one as pressure approaches the convergence pressure of the mixture.

Chao and Seader developed an empirical expression for v_{iL}^o in terms of T_r, P_r, and ω using the generalized correlation of Pitzer et al.,[8] which is based on the equation of state given as (4-33). For hypothetical liquid conditions ($P < P_i^s$ or $T > T_{c_i}$), the correlation was extended by back calculating v_{iL}^o from vapor–liquid equilibrium data. The C–S equation for v_{iL}^o is

$$\log v_{iL}^o = \log v_{iL}^{(0)} + \omega_i \log v_{iL}^{(1)} \tag{5-11}$$

where

$$\log \nu_{iL}^{(0)} = A_0 + \frac{A_1}{T_{r_i}} + A_2 T_{r_i} + A_3 T_{r_i}^2 + A_4 T_{r_i}^3$$
$$+ (A_5 + A_6 T_{r_i} + A_7 T_{r_i}^2) P_{r_i} + (A_8 + A_9 T_{r_i}) P_{r_i}^2 - \log P_{r_i} \qquad (5\text{-}12)$$

and

$$\log \nu_{iL}^{(1)} = A_{10} + A_{11} T_{r_i} + \frac{A_{12}}{T_{r_i}} + A_{13} T_{r_i}^3 + A_{14}(P_{r_i} - 0.6) \qquad (5\text{-}13)$$

The constants for (5-13) are:

$$A_{10} = -4.23893 \qquad A_{12} = -1.22060$$
$$A_{11} = 8.65808 \qquad A_{13} = -3.15224 \qquad A_{14} = -0.025$$

Grayson and Streed[9] presented revised constants for A_0 through A_9 as follows.

	Simple Fluid, $\omega = 0$	Methane	Hydrogen
A_0	2.05135	1.36822	1.50709
A_1	−2.10899	−1.54831	2.74283
A_2	0	0	−0.02110
A_3	−0.19396	0.02889	0.00011
A_4	0.02282	−0.01076	0
A_5	0.08852	0.10486	0.008585
A_6	0	−0.02529	0
A_7	−0.00872	0	0
A_8	−0.00353	0	0
A_9	0.00203	0	0

Use of these revised constants, rather than the original constants of Chao and Seader, permits application of the C–S correlation to higher temperatures and pressures and gives improved predictions for hydrogen.

The empirical equations for ν_{iL}^o are applicable at reduced temperatures from 0.5 to 1.3. When the vapor is an ideal gas solution obeying the ideal gas law and the liquid solution is ideal, ν_{iL}^o is the ideal K-value.

Chao and Seader[3] tested their K-value correlation against 2696 experimental data points for paraffins, olefins, aromatics, naphthenes, and hydrogen and found an average deviation of 8.7%. For best results, they suggested that application of the C–S correlation be restricted to certain ranges of conditions. Lenoir and Koppany,[10] in a thorough study of the C–S correlation, added additional restrictions. The combined restrictions are as follows.

(a) $T < 500°F$ (260°C).

(b) $P < 1000$ psia (6.89 MPa).

(c) For hydrocarbons (except methane), $0.5 < T_{r_i} < 1.3$ and mixture critical pressure <0.8.

(d) For systems containing methane and/or hydrogen, molal average $T_r < 0.93$, and methane mole fraction < 0.3. Mole fraction of other dissolved gases < 0.2.

(e) When predicting K-values of paraffins or olefins, liquid-phase aromatic mole fraction should be <0.5. Conversely, when predicting K-values of aromatics, liquid-phase aromatic mole fraction should be >0.5.

In addition, as shown in Figs. 4.7 and 4.8, the C–S correlation may be unreliable at low temperatures and generally is not recommended at temperatures below about 0°F.

Example 5.2. Estimate the K-value of benzene in a solution with propane at 400°F (477.59°K) and 410.3 psia (2.829 MPa) by the Chao–Seader correlation. Experimental compositions of equilibrium phases and the corresponding K-values are given by Glanville et al.[11]

Solution. The C–S K-value expression is (4-29). The fugacity coefficient of benzene in the vapor mixture ϕ_{iV} was calculated in Example 4.6 from the R–K equation to be 0.7055 for the experimental vapor composition given in Example 4.2.

The fugacity coefficient of pure liquid benzene ν_{iL}^o depends on the values T_r, P_r, and ω for benzene. Using the critical constants from Appendix I, we have

$$T_r = \frac{859.67}{1012.7} = 0.8489, \qquad P_r = \frac{410.3}{714.226} = 0.5745$$

From (5-12), using the Grayson–Streed constants, we have

$$\log \nu_{iL}^{(0)} = 2.05135 - \frac{2.10899}{0.8489} - 0.19396(0.8489)^2 + 0.02282(0.8489)^3$$
$$+ [0.08852 - 0.00872(0.8489)^2]0.5745$$
$$+ [-0.00353 + 0.00203(0.8489)](0.5745)^2 - \log(0.5745)$$
$$= -0.271485$$

From (5-13)

$$\log \nu_{iL}^{(1)} = -4.23893 + 8.65808(0.8489) - 1.22060/0.8489 - 3.15224(0.8489)^3$$
$$- 0.025(0.5745 - 0.6) = -0.254672$$

From (5-11)

$$\log \nu_{iL}^o = -0.271485 + 0.2115696(-0.254672)$$
$$= -0.325366$$
$$\nu_{iL}^o = 10^{-0.325366} = 0.4728$$

For a vapor-phase mole fraction of benzene of 0.6051, as determined from the measured weight fraction in Example 4.2, the corresponding liquid-phase mole fraction is

0.6051/0.679 = 0.891, as computed from $y_i/K_i = x_i$, where K_i is 0.679 by interpolation from the data of Glanville et al.[11] for benzene at 400°F (204.4°C) and 410.3 psia (2.829 MPa). For this value of x_i for benzene, the liquid-phase activity coefficient γ_{iL} is 1.008 from (5-10) as computed in the manner of Example 5.1. In this case, the Flory–Huggins correction is negligible because propane and benzene have almost identical liquid volumes. From (4-29), $K = (1.008)(0.4278)/0.7055 = 0.676$, which is almost identical to the interpolated experimental value of 0.679.
☐

Other thermodynamic properties can be computed in a consistent manner with the C–S K-value correlation. For the vapor, the R–K equation of state is used to determine vapor mixture density from (4-38), as illustrated in Example 4.2 and vapor mixture enthalpy from (4-64).

Liquid mixture enthalpy is computed from the ν_{iL}^o and γ_{iL} equations used in the C–S correlation by classical procedures, as shown by Edmister, Persyn, and Erbar.[12] The starting equation is a combination of (4-55) and (4-57) with (4-62).

$$H_L = \sum_{i=1}^{C} x_i \left[H_{iV}^o - RT^2 \left(\frac{\partial \ln \nu_{iL}^o}{\partial T} \right)_P - RT^2 \left(\frac{\partial \ln \gamma_{iL}}{\partial T} \right)_{P,x_i} \right] \qquad (5\text{-}14)$$

Ideal gas enthalpy, H_{iV}^o, is obtained from (4-60). The derivative of pure component liquid fugacity coefficient with respect to temperature leads to the following relation for combined effects of pressure and latent heat of phase change from vapor to liquid.

$$RT^2 \left(\frac{\partial \ln \nu_{iL}^o}{\partial T} \right)_P = (H_{iV}^o - H_{iL}) = \frac{2.30258 RT^2}{T_{c_i}} \left[-\frac{A_1}{T_{r_i}^2} + A_2 + T_{r_i}(2A_3 + 3A_4 T_{r_i}) \right.$$
$$\left. + P_{r_i}(A_6 + 2A_7 T_{r_i}) + A_9 P_{r_i}^2 + \omega_i \left(A_{11} - \frac{A_{12}}{T_{r_i}^2} + 3A_{13} T_{r_i}^2 \right) \right] \qquad (5\text{-}15)$$

where the constants A_i are those of (5-12) and (5-13). The derivative of the liquid-phase activity coefficient leads to species excess enthalpy $\bar{H}_i^E$ (heat of mixing effect). For regular solutions, $\bar{H}_i^E > 0$ (endothermic).

$$RT^2 \left(\frac{\partial \ln \gamma_{iL}}{\partial T} \right)_{P,x_i} = (H_{iL} - \bar{H}_{iL}) = -\bar{H}_i^E = -\nu_{iL} \left(\delta_i - \sum_{j=1}^{C} \Phi_j \delta_j \right)^2 \qquad (5\text{-}16)$$

Example 5.3. Solve Example 4.4 using the liquid enthalpy equation of Edmister, Persyn, and Erbar,[12] which is based on the Chao–Seader correlation.

Solution. The liquid mixture contains 25.2 mole% propane in benzene at 400°F (477.59°K) and 750 psia (5.171 MPa). Denoting propane by P and benzene by B and applying (5-15) to propane with $\omega_P = 0.1538$, we have

$$T_{r_P} = \frac{859.67}{665.948} = 1.291 \quad \text{and} \quad P_{r_P} = \frac{750}{617.379} = 1.215$$

$$H^\circ_{PV} - H_{PL} = \frac{2.30258(1.987)(859.67)^2}{665.948}\left\{-\frac{(-2.10899)}{(1.291)^2} + 1.291[2(-0.19396) + 3(0.02282)(1.291)]\right.$$

$$+ 1.215[2(-0.00872)(1.291)] + 0.00203(1.215)^2$$

$$\left.+ 0.1538\left[8.65808 - \frac{(-1.22060)}{(1.291)^2} + 3(-3.15224)(1.291)^2\right]\right\}$$

$$= -637.35 \text{ Btu/lbmole } (-1.481 \text{ MJ/kgmole})$$

By a similar calculation,

$$H^\circ_{BV} - H_{BL} = 11290.8 \text{ Btu/lbmole } (26.2 \text{ MJ/kgmole})$$

Excess enthalpy for each species is obtained from (5-16). For propane, the liquid-phase volume fraction is computed from (5-7) as in Example 5.1.

$$\Phi_P = 0.252(84)/[0.252(84) + 0.748(89.4)] = 0.240$$

For benzene

$$\Phi_B = 1 - \Phi_P = 1 - 0.240 = 0.760$$

Therefore, for propane, using the solubility parameters from Appendix I

$$\bar{H}^E_P = 84[6.4 - 0.240(6.4) - 0.760(9.158)]^2(1.8)$$

$$= 664.30 \text{ Btu/lbmole } (1.544 \text{ MJ/kgmole})$$

Similarly, $\bar{H}^E_B = 70.50$ Btu/lbmole (163.9 kJ/kgmole).

The liquid mixture enthalpy relative to an ideal vapor at 400°F and 0 psia is obtained from the following equation, which is equivalent to (5-14).

$$H_L - H^\circ_V = \sum_{i=1}^{C} x_i[(H_{iL} - H^\circ_{iV}) + \bar{H}^E_i] = 0.252(637.35 + 664.3) \qquad (5\text{-}17)$$

$$+ 0.748(-11290.8 + 70.5) = -8065 \text{ Btu/lbmole of mixture } (-18.74 \text{ MJ/kgmole})$$

The mixture molecular weight from Example 4.4 is 69.54. Thus, the specific enthalpy difference is

$$\frac{H_L - H^\circ_V}{M} = \frac{-8065}{69.54} = -115.98 \text{ Btu/lb } (-269.5 \text{ kJ/kg})$$

This deviates by 8.5%, or approximately 11 Btu/lb (25.6 kJ/kg) from the measured value of -127.38 Btu/lb (-296 kJ/kg) of Yarborough and Edmister.[13] The computed excess enthalpy contribution is only 3.17 Btu/lb (7.37 kJ/kg).
□

An equation for liquid-phase molal volume that is consistent with the C–S correlation is derived by summing species molal volumes and correcting for excess volume (i.e., volume of mixing) from (4-58). Thus

$$v_L = \sum_{i=1}^{C} x_i(v_{iL} + \bar{v}^E_{iL}) = \sum_{i=1}^{C} x_i\left[v_{iL} + RT\left(\frac{\partial \ln \gamma_{iL}}{\partial P}\right)_{T,x_i}\right] \qquad (5\text{-}18)$$

For regular solutions, γ_{iL} can be considered independent of pressure. Thus, by

(5-18), $\bar{v}_{iL}^E = 0$. An equation for the pure species molal volume can then be obtained by combining (4-56) with (5-11), (5-12), and (5-13) with the result

$$v_{iL} = \frac{2.30258RT}{P_{c_i}}[A_5 + A_6 T_{r_i} + A_7 T_{r_i}^2 + 2P_{r_i}(A_8 + A_9 T_{r_i}) + \omega_i A_{14}] \qquad (5-19)$$

Example 5.4. Calculate the specific volume of a liquid-phase mixture containing 26.92 weight% propane (P) in benzene (B) at 400°F (477.59°K) and 1000 psia (6.895 MPa) using the C–S correlation. Compare the result with the measured value of Glanville et al.[11]

Solution. From Example 4.2, the mixture is 39.49 mole% propane and 60.51 mole% benzene with an average molecular weight of 64.58 lb/lbmole. Applying (5-19) to propane with $\omega_P = 0.1538$

$$T_{r_P} = \frac{859.67}{665.948} = 1.291 \qquad \text{and} \qquad P_{r_P} = \frac{1000}{617.379} = 1.620$$

$$v_{PL} = \frac{2.30258(10.731)(859.67)}{617.379}\{0.08852 - 0.00872(1.291)^2$$
$$+ 2(1.620)[-0.00353 + 0.00203(1.291)] + 0.1538(-0.025)\}$$
$$= 2.312 \text{ ft}^3/\text{lbmole} \ (0.1443 \text{ m}^3/\text{kgmole})$$

Similarly, $v_{BL} = 2.138$ ft³/lbmole (0.1335 m³/kgmole). From (5-18) with $\bar{v}_{iL}^E = 0$, the mixture molal volume is

$$\frac{v_L}{M} = \frac{2.207}{64.68} = 0.03412 \text{ ft}^3/\text{lb} \ (0.002130 \text{ m}^3/\text{kg})$$

This is 1.1% higher than the measured value of 0.03375 ft³/lb (0.002107 m³/kg).
□

The Chao–Seader correlation is widely used in the petroleum and natural gas industries. Waterman and Frazier[14] describe its use in the design of a wide variety of distillation separations involving light hydrocarbons. Correlations more sophisticated than the C–S correlation can give more accurate results in certain ranges of conditions. However, Lo[15] showed that computing requirements can become excessive and extrapolation more uncertain when more complex equations are utilized.

5.2 Nonideal Liquid Mixtures Containing Polar Species

When liquids contain dissimilar polar species, particularly those that can form or break hydrogen bonds, the ideal liquid solution assumption is almost always invalid. Ewell, Harrison, and Berg[16] provided a very useful classification of molecules based on the potential for association or solvation due to hydrogen bond formation. If a molecule contains a hydrogen atom attached to a donor atom (O, N, F, and in certain cases C), the active hydrogen atom can form a bond with another molecule containing a donor atom. The classification in Table

Table 5.1   Classification of molecules based on potential for forming hydrogen bonds

Class	Description	Examples
I	Molecules capable of forming three-dimensional networks of strong H-bonds	Water, glycols, glycerol, amino alcohols, hydroxylamines, hydroxyacids, polyphenols, and amides
II	Other molecules containing both active hydrogen atoms and donor atoms (O, N, and F)	Alcohols, acids, phenols, primary and secondary amines, oximes, nitro and nitrile compounds with α-hydrogen atoms, ammonia, hydrazine, hydrogen fluoride, and hydrogen cyanide
III	Molecules containing donor atoms but no active hydrogen	Ethers, ketones, aldehydes, esters, tertiary amines (including pyridine type), and nitro and nitrile compounds without α-hydrogen atoms
IV	Molecules containing active hydrogen atoms but no donor atoms that have two or three chlorine atoms on the same carbon atom as a hydrogen atom, or one chlorine on the same carbon atom and one or more chlorine atoms on adjacent carbon atoms	$CHCl_3$, CH_2Cl_2, CH_3CHCl_2, CH_2ClCH_2Cl, $CH_2ClCHClCH_2Cl$, and $CH_2ClCHCl_2$
V	All other molecules having neither active hydrogen atoms nor donor atoms	Hydrocarbons, carbon disulfide, sulfides, mercaptans, and halohydrocarbons not in Class IV

5.1 permits qualitative estimates of deviations from Raoult's law for binary pairs when used in conjunction with Table 5.2. Positive deviations correspond to values of $\gamma_{iL} > 1$. Nonideality results in a variety of variations of γ_{iL} with composition as shown in Fig. 5.2 for several binary systems, where the Roman numerals refer to classification groups in Tables 5.1 and 5.2. Starting with Fig. 5.2a and taking the other plots in order, we offer the following explanations for the nonidealities. Normal heptane (V) breaks ethanol (II) hydrogen bonds causing strong positive deviations. In Fig. 5.2b, similar but less positive deviations occur when acetone (III) is added to formamide (I). Hydrogen bonds are broken and formed with chloroform (IV) and methanol (II) in Fig. 5.2c, resulting in an unusual positive deviation curve for chloroform that passes through a maximum. In Fig. 5.2d, chloroform (IV) provides active hydrogen atoms that can form hydrogen bonds with oxygen atoms of acetone (III), thus causing negative deviations. For water (I) and n-butanol (II) in Fig. 5.2e,

Table 5.2 Molecule interactions causing deviations from Raoult's law

Type of Deviation	Classes	Effect on Hydrogen Bonding
Always negative	III + IV	H-bonds formed only
Quasi-ideal; always positive or ideal	III + III III + V IV + IV IV + V V + V	No H-bonds involved
Usually positive, but some negative	I + I I + II I + III II + II II + III	H-bonds broken and formed
Always positive	I + IV (frequently limited solubility) II + IV	H-bonds broken and formed, but dissociation of Class I or II is more important effect
Always positive	I + V II + V	H-bonds broken only

hydrogen bonds of both molecules are broken. Nonideality is sufficiently strong to cause phase separation over a wide region of overall composition. The trend toward strong nonideality in water–alcohol systems starting with methanol[17] is shown in Fig. 5.3. Not shown in Fig. 5.2 are curves for the methanol (II)–ethanol (II) system, which is almost an ideal solution.

Nonideal solution effects can be incorporated into K-value formulations in two different ways. Chapter 4 described the use of ϕ_i, the fugacity coefficient, in conjunction with an equation of state and adequate mixing rules. This is the method most frequently used for handling nonidealities in the vapor phase. However, ϕ_{iV} reflects the combined effects of a nonideal gas and a nonideal gas solution. At low pressures, both effects are negligible. At moderate pressures, a vapor solution may still be ideal even though the gas mixture does not follow the ideal gas law. Nonidealities in the liquid phase, however, can be severe even at low pressures. In Section 4.5, ϕ_{iL} was used to express liquid-phase nonidealities for nonpolar species. When polar species are present, mixing rules can be modified to include binary interaction parameters as in (4-113).

The other technique for handling solution nonidealities is to retain ϕ_{iV} but replace ϕ_{iL} by the product of γ_{iL} and ν_{iL}^o, where the former quantity accounts for

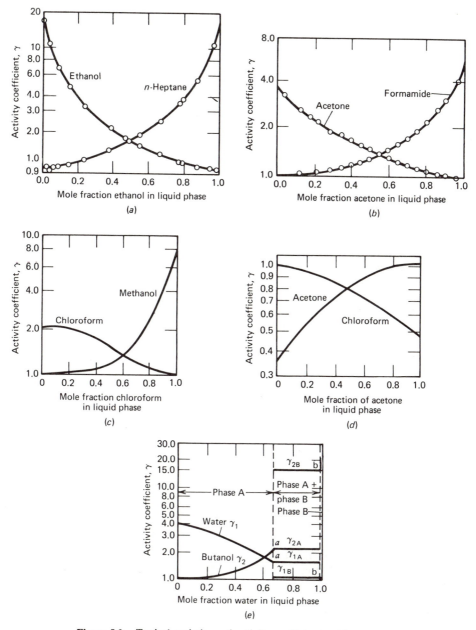

Figure 5.2. Typical variations of activity coefficients with composition in binary liquid systems. (*a*) Ethanol(II)-*n*-heptane(V). (*b*) Acetone(III)-Formamide(I). (*c*) Chloroform(IV)-methanol(II). (*d*) Acetone(III)-chloroform(IV). (*e*) Water(I)-*n*-butanol(II).

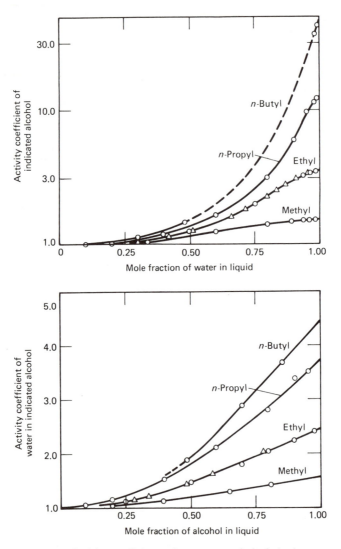

Figure 5.3. Activity coefficients of water-normal alcohol mixtures at 25°C. (*a*) Alcohols in water. (*b*) Water in alcohols.

deviations from nonideal solutions. Equation (4-27) then becomes

$$K_i = \frac{\gamma_{iL} \nu_{iL}^o}{\phi_{iV}} \qquad (5\text{-}20)$$

which was derived previously as (4-29). At low pressures, from Table 4.1,

$\nu_{iL}^{o} = P_{i}^{s}/P$ and $\phi_{iV} = 1.0$, so that (5-20) reduces to a modified Raoult's law K-value, which differs from (4-75) only in the γ_{iL} term.

$$K_{i} = \frac{\gamma_{iL}P_{i}^{s}}{P} \qquad (5\text{-}21)$$

Similarly, (4-77) becomes

$$P = \sum_{i=1}^{N} x_{i}\gamma_{iL}P_{i}^{s} \qquad (5\text{-}22)$$

At moderate pressures, assumption of an ideal vapor solution may still be valid. If so, (4-86) becomes

$$K_{i} = \gamma_{iL} \left(\frac{P_{i}^{s}}{P}\right)\left(\frac{\nu_{iV_{s}}^{o}}{\nu_{iV}^{o}}\right) \exp\left[\frac{\nu_{iL}(P - P_{i}^{s})}{RT}\right] \qquad (5\text{-}23)$$

For the general case, (5-20) is applied directly.

Many empirical and semitheoretical equations exist for estimating activity coefficients of binary mixtures containing polar and/or nonpolar species. These equations contain binary interaction parameters obtained from experimental data. Some of the more common equations are listed in Table 5.3 in binary-pair form. Of these, the recent universal quasi-chemical (UNIQUAC) equation of Abrams and Prausnitz[18] appears to be the most general; and all other equations in Table 5.3 are embedded in it. For a given activity coefficient correlation, (4-57) can be used to determine excess enthalpy. However, unless the dependency on pressure of the parameters and properties used in the equations for activity coefficient is known, excess liquid volumes cannot be determined directly from (4-58). Fortunately, the contribution of excess volume to total mixture volume is generally small for solutions of nonelectrolytes. For example, consider a 50 mole% solution of ethanol in n-heptane at 25°C. As shown in Fig. 5.2a, this is a highly nonideal, but miscible, liquid mixture. From the data of Van Ness, Soczek, and Kochar,[19] excess volume is only 0.465 cm³/gmmole compared to an estimated ideal solution molal volume of 106.3 cm³/gmmole.

5.3.   Van Laar Equation

Because of its flexibility, simplicity, and ability to fit many systems well, the van Laar equation[20] is widely used in practice. It can be derived from the general energy expansion of Wohl,[21] which considers effective volume fractions and molecular interactions. The so-called Carlson and Colburn[22] natural logarithm version of the van Laar equation is given in Table 5.3. However, a common logarithm form is more common. The Margules and Scatchard–Hamer equations in Table 5.3 can also be derived from the Wohl expansion by a set of different assumptions.

Table 5.3 Empirical and semitheoretical equations for correlating liquid phase activity coefficients of binary pairs

Name	Equation for Species 1	Equation for Species 2	Equation No.
Margules (one-constant, symmetric)	$\log \gamma_1 = A x_2^2$	$\log \gamma_2 = A x_1^2$	(5-24)
Margules (two-constant)	$\log \gamma_1 = x_2^2[\bar{A}_{12} + 2x_1(\bar{A}_{21} - \bar{A}_{12})]$	$\log \gamma_2 = x_1^2[\bar{A}_{21} + 2x_2(\bar{A}_{12} - \bar{A}_{21})]$	(5-25)
van Laar (two-constant)	$\ln \gamma_1 = \dfrac{A_{12}}{\left[1 + \dfrac{x_1 A_{12}}{x_2 A_{21}}\right]^2}$	$\ln \gamma_2 = \dfrac{A_{21}}{\left[1 + \dfrac{x_2 A_{21}}{x_1 A_{12}}\right]^2}$	(5-26)
Scatchard–Hamer (two-constant)	$\log \gamma_1 = \Phi_2^2\left[A + 2\Phi_1\left(B\dfrac{v_1}{v_2} - A\right)\right]$	$\log \gamma_2 = \Phi_1^2\left[B + 2\Phi_2\left(A\dfrac{v_2}{v_1} - B\right)\right]$	(5-27)
Wilson (two-constant)	$\ln \gamma_1 = -\ln(x_1 + \Lambda_{12}x_2)$ $\quad + x_2\left[\dfrac{\Lambda_{12}}{x_1 + \Lambda_{12}x_2} - \dfrac{\Lambda_{21}}{\Lambda_{21}x_1 + x_2}\right]$	$\ln \gamma_2 = -\ln(x_2 + \Lambda_{21}x_1)$ $\quad - x_1\left[\dfrac{\Lambda_{12}}{x_1 + \Lambda_{12}x_2} - \dfrac{\Lambda_{21}}{\Lambda_{21}x_1 + x_2}\right]$	(5-28)
NRTL (three-constant)	$\ln \gamma_1 = x_2^2\left\{\tau_{21}\left[\dfrac{\exp(-2\alpha_{12}\tau_{21})}{[x_1 + x_2\exp(-\alpha_{12}\tau_{21})]^2}\right] \right.$ $\left. + \tau_{12}\dfrac{\exp(-\alpha_{12}\tau_{12})}{[x_2 + x_1\exp(-\alpha_{12}\tau_{12})]^2}\right\}$	$\ln \gamma_2 = x_1^2\left\{\tau_{12}\left[\dfrac{\exp(-2\alpha_{12}\tau_{12})}{[x_2 + x_1\exp(-\alpha_{12}\tau_{12})]^2}\right] \right.$ $\left. + \tau_{21}\dfrac{\exp(-\alpha_{12}\tau_{21})}{[x_1 + x_2\exp(-\alpha_{12}\tau_{21})]^2}\right\}$	(5-29)
UNIQUAC (two-constant)	$\ln \gamma_1 = \ln\dfrac{\Psi_1}{x_1} + \dfrac{\bar{Z}}{2}q_1 \ln\dfrac{\theta_1}{\Psi_1}$ $\quad + \Psi_2\left(\ell_1 - \dfrac{r_1}{r_2}\ell_2\right) - q_1\ln(\theta_1 + \theta_2 T_{21})$ $\quad + \theta_2 q_1\left(\dfrac{T_{21}}{\theta_1 + \theta_2 T_{21}} - \dfrac{T_{12}}{\theta_2 + \theta_1 T_{21}}\right)$	$\ln \gamma_2 = \ln\dfrac{\Psi_2}{x_2} + \dfrac{\bar{Z}}{2}q_2 \ln\dfrac{\theta_2}{\Psi_2}$ $\quad + \Psi_1\left(\ell_2 - \dfrac{r_2}{r_1}\ell_1\right) - q_2\ln(\theta_2 + \theta_1 T_{12})$ $\quad + \theta_1 q_2\left(\dfrac{T_{12}}{\theta_2 + \theta_1 T_{12}} - \dfrac{T_{21}}{\theta_1 + \theta_2 T_{21}}\right)$	(5-30)

The van Laar interaction constants A_{ij} and A_{ji} are, in theory, only constant for a particular binary pair at a given temperature. In practice, they are frequently computed from isobaric data covering a range of temperature. The van Laar theory expresses the temperature dependence of A_{ij} to be

$$A_{ij} = \frac{A'_{ij}}{RT} \tag{5-31}$$

Regular solution theory and the van Laar equation are equivalent for a binary solution if

$$A_{ij} = \frac{v_{iL}}{RT}(\delta_i - \delta_j)^2 \tag{5-32}$$

The van Laar equation can fit activity coefficient–composition curves corresponding to both positive and negative deviations from Raoult's law, but cannot fit curves that exhibit minima or maxima such as those in Fig. 5.2c.

For a multicomponent mixture, it is common to neglect ternary and higher interactions and assume a pseudobinary system. The resulting van Laar expression for the activity coefficient depends only on composition and the binary constants. The following form given by Null[23] is preferred.

$$\ln \gamma_i = \frac{\sum\limits_{j=1}^{C} (x_j A_{ij})}{1 - x_i} \left[1 - \frac{x_i \sum\limits_{j=1}^{C} (x_j A_{ij})}{x_i \sum\limits_{j=1}^{C} (x_j A_{ij}) + (1 - x_i) \sum\limits_{j=1}^{C} (x_j A_{ji})} \right]^2 \tag{5-33}$$

This equation is restricted to conditions where all A_{ij} and A_{ji} pairs are of the same sign. If not and/or if some values of A_{ij} are large but complete miscibility still exists, a more complex form of (5-33) should be employed.[23] But most often (5-33) suffices. In using it, $A_{ii} = A_{jj} = 0$. For a multicomponent mixture of N species, $N(N - 1)/2$ binary pairs exist. For example, when $N = 5$, 10 binary pairs can be formed.

Extensive tabulations of van Laar binary constants are provided by Hála et al.[24] and Holmes and Van Winkle.[25] When $|A_{ij}| < 0.01$, γ_{iL} is within 1.00 ± 0.01 and it is reasonable to assume an ideal solution. When van Laar binary-pair constants are not available, the following procedure is recommended.

1. For isomers and close-boiling pairs of homologs that are assumed to form ideal solutions according to Table 5.2, $A_{ij} = A_{ji} = 0$.

2. For nonpolar hydrocarbon pairs known to follow regular solution theory, (5-32) can be used to estimate A_{ij} and A_{ji}.

3. For pairs containing polar or other species that do not follow regular solution theory, van Laar constants can be determined from activity coefficients computed from experimental data.

4. When data exist on closely related pairs, interpolation or extrapolation may be employed. For example, in Fig. 5.3a constants for the ethanol–water pair, if data were not available, could be interpolated from data on the other three alcohol–water pairs.

5. If no useful data exist, a procedure based on estimation of binary activity coefficients at infinite dilution suggested by Null[23] can be employed.

When data are isothermal, or isobaric over only a narrow range of temperature, determination of van Laar constants is conducted in a straightforward manner. The most accurate procedure is a nonlinear regression[26,27] to obtain the best fit to the data over the entire range of binary composition, subject to minimization of some objective function. A less accurate, but extremely rapid, hand-calculation procedure can be used when experimental data can be extrapolated to infinite dilution conditions. Modern experimental techniques are available for accurately and rapidly determining activity coefficients at infinite dilution. Applying (5-26) to the conditions $x_i = 0$ and then $x_j = 0$, we have

$$A_{ij} = \ln \gamma_i^\infty \qquad x_i = 0$$

and

$$A_{ji} = \ln \gamma_j^\infty \qquad x_j = 0 \qquad\qquad (5\text{-}34)$$

For practical applications, it is important that the van Laar equation correctly predict azeotrope formation. If activity coefficients are known or can be computed at the azeotropic composition, say from (5-21), ($\gamma_{iL} = P/P_i^s$ since $K_i = 1.0$), these coefficients can be used to determine the van Laar constants directly from the following equations obtained by solving (5-26) simultaneously for A_{12} and A_{21}

$$A_{12} = \ln \gamma_1 \left[1 + \frac{x_2 \ln \gamma_2}{x_1 \ln \gamma_1} \right]^2 \qquad\qquad (5\text{-}35)$$

$$A_{21} = \ln \gamma_2 \left[1 + \frac{x_1 \ln \gamma_1}{x_2 \ln \gamma_2} \right]^2 \qquad\qquad (5\text{-}36)$$

These equations are applicable in general to activity coefficient data obtained at any single composition.

The excess enthalpy due to liquid-phase nonideality can·be determined by applying (4-57) to (5-26), assuming the temperature dependence of (5-31). The result is the approximate relation

$$H^E = \sum_{i=1}^{C} x_i \bar{H}_i^E = RT \sum_{i=1}^{C} (x_i \ln \gamma_i) \qquad\qquad (5\text{-}37)$$

where the $\ln \gamma_{iL}$ term is estimated from (5-26).

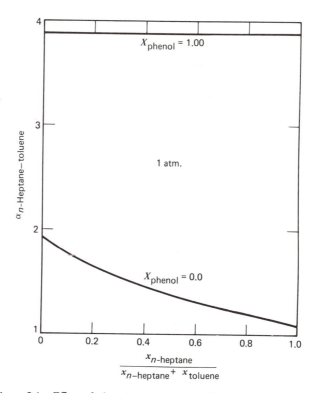

Figure 5.4. Effect of phenol on relative volatility between n-heptane and toluene.

Example 5.5. The relative volatility of n-heptane to toluene at atmospheric pressure, as computed from the experimental data of Yerazunis et al.,[6] is shown in Fig. 5.4 by the curve $x_{phenol} = 0$. When the mole fraction of toluene is low, the relative volatility is very low (approximately 1.10). To increase the relative volatility, a polar solvent, phenol, is added, as discussed by Dunn et al.,[28] and an extractive distillation separation (Table 1.1) can be carried out. With phenol, positive deviations from Raoult's law will occur according to Tables 5.1 and 5.2.

Use available experimental data at infinite dilution with the van Laar equations to estimate the relative volatility between n-heptane and toluene at atmospheric pressure for a liquid-phase mixture consisting of 5 mole% toluene, 15 mole% n-heptane, and 80 mole% phenol. Also compute excess enthalpy for this mixture.

Solution. Liquid-phase activity coefficients for n-heptane and toluene at infinite dilution in phenol were measured by Tassios[29] using gas–liquid chromatography. Experimental infinite-dilution activity coefficients for the n-heptane–toluene system are as shown in Fig. 5.1. Also, infinite-dilution activity coefficients for phenol in toluene, phenol in n-heptane, and toluene in phenol are available from Drickamer, Brown, and

White.[30] These data are summarized in the following table for a pressure of 1 atm.

Solute	Solvent	γ^∞_{solute}	T
n-Heptane	Phenol	12.5	70°C
Toluene	Phenol	2.92	70°C
n-Heptane	Toluene	1.372	110.6°C
Toluene	n-Heptane	1.293	98.41°C
Phenol	Toluene	2.6	231°F
Phenol	n-Heptane	15.8	210°F
Toluene	Phenol	2.15	360°F

Combining (5-31) with (5-34) gives

$$A'_{ij} = RT \ln \gamma^\infty_i$$

Let n-heptane be denoted by 1, toluene by 2, and phenol by 3. Then, the following values of A'_{ij} are obtained from the above values of γ^∞_i, where as an example

$$A'_{12} = 1.987(690.75) \ln 1.372 = 434 \text{ Btu/lbmole}$$

Similarly

$$A'_{21} = 341 \text{ Btu/lbmole} \qquad A'_{23} = 1315 \text{ Btu/lbmole}$$

$$A'_{13} = 3100 \text{ Btu/lbmole} \qquad A'_{32} = 1311 \text{ Btu/lbmole}$$

$$A'_{31} = 3668 \text{ Btu/lbmole} \qquad A'_{23} = 1247 \text{ Btu/lbmole}$$

The two values for A'_{23} are within 6% of each other. The value of 1315 Btu/lbmole from the measurements of Yerazunis et al.[6] will be used in the remaining calculations.

Before computing the relative volatility for the specified mixture, it is of interest to estimate the relative volatility for n-heptane and toluene at infinite dilution in phenol. Often, this represents the largest relative volatility obtainable for the given solvent at a given pressure. With essentially pure phenol, the temperature is the boiling point (819°R, 455°K) at the specified pressure of 1 atm (101.3 kPa). At this low pressure, (5-21) is applicable for the K-value. Combining this with (1-7), (5-31), and (5-34), we have

$$\alpha^\infty_{12} = \frac{P^s_1 \exp\left(\dfrac{A'_{13}}{RT}\right)}{P^s_2 \exp\left(\dfrac{A'_{23}}{RT}\right)}$$

where P^s_1 is obtained from the Antoine relation as in Example 4.5. Thus

$$\alpha^\infty_{12}(819°R, 1 \text{ atm}) = \frac{102.16 \exp\left[\dfrac{3100}{1.987(819)}\right]}{79.06 \exp\left[\dfrac{1311}{1.987(819)}\right]} = 3.88$$

This value, which is the upper line in Fig. 5.4, is considerably higher than the relative volatility in the absence of phenol.

In order to compute α_{12} for the 80 mole% phenol mixture, it is necessary to assume a temperature that will satisfy (5-22). By an iterative procedure, the correct temperature is found to be 215°F (101.7°C). The procedure is shown only for the final iteration at 215°F. For this temperature, van Laar constants are computed from (5-31).

$$A_{12} = \frac{434}{1.987(674.67)} = 0.324$$

Similarly

$$A_{21} = 0.254 \qquad A_{23} = 0.981$$
$$A_{13} = 2.31 \qquad A_{32} = 0.978$$
$$A_{31} = 2.74$$

From (5-33) applied to a ternary mixture

$$\ln \gamma_1 = (x_2 + x_3)(x_2 A_{12} + x_3 A_{13}) \left[\frac{x_2 A_{21} + x_3 A_{31}}{x_1(x_2 A_{12} + x_3 A_{13}) + (x_2 + x_3)(x_2 A_{21} + x_3 A_{31})} \right]^2$$

Using the above values of A_{ij} with $x_1 = 0.15$, $x_2 = 0.05$, and $x_3 = 0.80$, we have

$$\ln \gamma_1 = (0.05 + 0.80)[0.05(0.324) + 0.80(2.31)]$$
$$\times \left\{ \frac{0.05(0.254) + 0.80(2.74)}{0.15[0.05(0.324) + 0.80(2.31)] + (0.05 + 0.80)[0.05(0.324) + 0.80(2.31)]} \right\}^2$$

The result is $\gamma_1 = 5.27$.

Similarly, $\gamma_2 = 2.19$ and $\gamma_3 = 1.07$. To check the assumed temperature, total pressure is computed from (5-22) with vapor pressures from (4-69).

$$P = 0.15(16.15)(5.27) + 0.05(11.30)(2.19) + 0.80(0.847)(1.07) = 14.7 \text{ psia}$$

which is the specified pressure.

By combining (1-7) and (5-21), we have for α_{12}

$$\alpha_{12} = \frac{\gamma_1 P_1^s}{\gamma_2 P_2^s} = \frac{5.27(16.15)}{2.19(11.30)} = 3.44$$

From Fig. 5.4, this value is almost 200% higher than the value for a binary mixture without phenol, but with the same composition on a phenol-free basis.

By the above procedure, families of curves for different phenol mole fractions could be computed.

The excess enthalpy at 215°F (101.7°C) is obtained by applying (5-37) at 215°F.

$$H^E = 1.987(674.67)[0.15(\ln 5.27) + 0.05(\ln 2.19) + 0.80(\ln 1.07)]$$
$$= 459.3 \text{ Btu/lbmole } (1.07 \text{ MJ/kgmole})$$

☐

5.4 The Local Composition Concept and the Wilson Equation

Mixtures of self-associated polar molecules (Class II in Table 5.1) with nonpolar molecules such as hydrocarbons (Class V) can exhibit the strong nonideality of the positive deviation type shown in Fig. 5.2a. Figure 5.5 shows experimental data of Sinor and Weber[31] for ethanol (1)-n-hexane (2), a system of this type, at

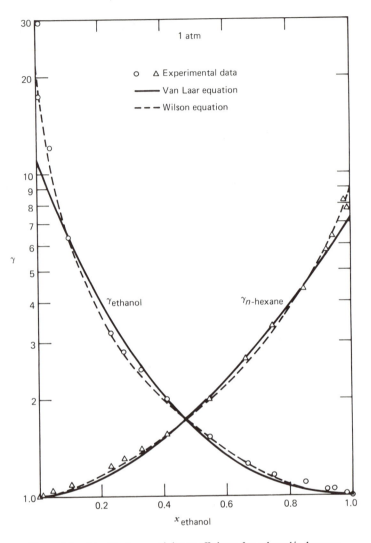

Figure 5.5. Liquid-phase activity coefficients for ethanol/n-hexane system. [Data from J. E. Sinor and J. H. Weber, *J. Chem. Eng. Data*, 5, 243–247 (1960).]

101.3 kPa. These data were correlated as shown in Fig. 5.5 with the van Laar equation by Orye and Prausnitz[32] to give $A_{12} = 2.409$ and $A_{21} = 1.970$. From $x_1 = 0.1$ to 0.9, the fit of the data to the van Laar equation is reasonably good; but, in the dilute regions, deviations are quite severe and the predicted activity coefficients for ethanol are low. An even more serious problem with these highly nonideal mixtures is that the van Laar equation may erroneously predict formation of two liquid phases (phase splitting).

Since its introduction in 1964, the Wilson equation,[33] shown in binary form in Table 5.3 as (5-28), has received wide attention because of its ability to fit strongly nonideal, but miscible, systems. As shown in Fig. 5.5, the Wilson equation, with the binary interaction constants of $\Lambda_{12} = 0.0952$ and $\Lambda_{21} = 0.2713$ determined by Orye and Prausnitz,[32] fits experimental data well even in dilute regions where variation of γ_1 becomes exponential. Corresponding infinite-dilution activity coefficients computed from the Wilson equation are $\gamma_1^\infty = 21.72$ and $\gamma_2^\infty = 9.104$.

In the Wilson equation, the effects of difference both in molecular size and intermolecular forces are incorporated by an extension of the Flory–Huggins relation (5-8). Overall solution volume fractions ($\bar{\Phi}_i = x_i v_{iL}/v_L$) are replaced by local volume fractions, $\bar{\Phi}_i$, which are related to local molecule segregations caused by differing energies of interaction between pairs of molecules. The concept of local compositions that differ from overall compositions is shown schematically for an overall equimolar binary solution in Fig. 5.6, which is taken

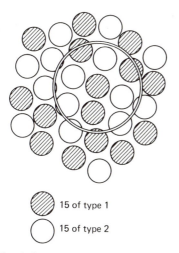

○ 15 of type 1

○ 15 of type 2

Overall mole fractions: $x_1 = x_2 = \frac{1}{2}$
Local mole fractions:

$$x_{21} = \frac{\text{Molecules of 2 about a central molecule 1}}{\text{Total molecules about a central molecule 1}}$$

$x_{21} + x_{11} = 1$, as shown

$x_{12} + x_{22} = 1$

$x_{11} \sim 3/8$

$x_{21} \sim 5/8$

Figure 5.6.   The concept of local compositions. [From P. M. Cukor and J. M. Prausnitz, *Intl. Chem. Eng. Symp. Ser. No.* 32, Instn. Chem. Engrs., London, **3**, 88 (1969).]

from Cukor and Prausnitz.[34] About a central molecule of type 1, the local mole fraction of molecules of type 2 is shown as 5/8.

For local volume fraction, Wilson proposed

$$\bar{\Phi}_i = \frac{v_{iL} x_i \exp(-\lambda_{ii}/RT)}{\displaystyle\sum_{j=1}^{C} v_{jL} x_j \exp(-\lambda_{ij}/RT)} \tag{5-38}$$

where energies of interaction $\lambda_{ij} = \lambda_{ji}$, but $\lambda_{ii} \neq \lambda_{jj}$. Following the treatment by Orye and Prausnitz,[32] substitution of the binary form of (5-38) into (5-8), and defining the binary interaction constants* as

$$\Lambda_{12} = \frac{v_{2L}}{v_{1L}} \exp\left[-\frac{(\lambda_{12} - \lambda_{11})}{RT}\right] \tag{5-39}$$

$$\Lambda_{21} = \frac{v_{1L}}{v_{2L}} \exp\left[-\frac{(\lambda_{12} - \lambda_{22})}{RT}\right] \tag{5-40}$$

leads to the following equation for a binary system

$$\frac{g^E}{RT} = -x_1 \ln(x_1 + \Lambda_{12} x_2) - x_2 \ln(x_2 + \Lambda_{21} x_1) \tag{5-41}$$

The Wilson equation is very effective for dilute composition where entropy effects dominate over enthalpy effects. The Orye–Prausnitz form of the Wilson equation for the activity coefficient, as given in Table 5.3, follows from combining (5-2) with (5-41). Values of $\Lambda_{ij} < 1$ correspond to positive deviations from Raoult's law, while values of $\Lambda_{ji} > 1$ correspond to negative deviations. Ideal solutions result from $\Lambda_{ij} = 1$. Studies indicate that λ_{ii} and λ_{ij} are temperature dependent. Values of v_{iL}/v_{jL} depend on temperature also, but the variation may be small compared to temperature effects on the exponential term.

The Wilson equation is readily extended to multicomponent mixtures. Like the van Laar equation (5-33), the following multicomponent Wilson equation involves only binary interaction constants.

$$\ln \gamma_k = 1 - \ln\left[\sum_{j=1}^{C} (x_j \Lambda_{kj})\right] - \sum_{i=1}^{C} \left[\frac{x_i \Lambda_{ik}}{\displaystyle\sum_{j=1}^{C} (x_j \Lambda_{ij})}\right] \tag{5-42}$$

where $\Lambda_{ii} = \Lambda_{jj} = \Lambda_{kk} = 1$. Unfortunately, Hála[35] showed that the binary interaction constants are not all independent. For example, in a ternary system, 1 of 6 binary constants depends on the other 5. In a quaternary system, only 9 of 12 binary constants are independent. However, Brinkman, Tao, and Weber[36] provide an example where the Hála constraint is not serious.

Binary and multicomponent forms of the Wilson equation were evaluated

---

* Wilson gives $\Lambda_{12} = 1 - A_{2/1}$ and $\Lambda_{21} = 1 - A_{1/2}$.

by Orye and Prausnitz,[32] Holmes and Van Winkle,[25] and Hudson and Van Winkle.[37] In the limit, as mixtures become only weakly nonideal, all the equations in Table 5.3 become essentially equivalent in form and, therefore, in accuracy. As mixtures become highly nonideal, but still miscible, the Wilson equation becomes markedly superior to the Margules, van Laar, and Scatchard–Hamer equations. The Wilson equation is consistently superior for multicomponent solutions. Values of the constants in the Wilson equation for a number of binary systems are tabulated in several sources.[25,32,37–40] Prausnitz et al.,[41] provide listings of FORTRAN computer programs for determining parameters of the Wilson equation from experimental data and for computing activity coefficients when parameters are known. Two limitations of the Wilson equation are its inability to predict immiscibility, as in Fig. 5.2e, and maxima and minima in the activity coefficient–mole fraction relationship, as shown in Fig. 5.2c.

When insufficient experimental data are available to determine binary Wilson parameters from a best fit of activity coefficients over the entire range of composition, infinite-dilution or single-point values can be used. At infinite dilution, the Wilson equation in Table 5.3 becomes

$$\ln \gamma_1^\infty = 1 - \ln \Lambda_{12} - \Lambda_{21} \qquad (5\text{-}43)$$

$$\ln \gamma_2^\infty = 1 - \ln \Lambda_{21} - \Lambda_{12} \qquad (5\text{-}44)$$

An iterative procedure[42] is required for obtaining Λ_{12} and Λ_{21}. If temperatures corresponding to γ_1^∞ and γ_2^∞ are not close or equal, (5-39) and (5-40) should be substituted into (5-43) and (5-44), with values of $(\lambda_{12} - \lambda_{11})$ and $(\lambda_{12} - \lambda_{22})$ determined from estimates of pure-component liquid molal volumes.

When the experimental data of Sinor and Weber[31] for n-hexane/ethanol shown in Fig. 5.5 are plotted as a y–x diagram in ethanol (Fig. 5.7) the equilibrium curve crosses the 45° line at an ethanol mole fraction $x = 0.332$. The measured temperature corresponding to this composition is 58°C. Ethanol has a normal boiling point of 78.33°C, which is higher than the normal boiling point of 68.75°C for n-hexane. Nevertheless, ethanol is more volatile than n-hexane up to an ethanol mole fraction of $x = 0.322$, the minimum-boiling azeotrope. This occurs because of the relatively close boiling points of the two species and the high activity coefficients for ethanol at low concentrations. At the azeotropic composition, $y_i = x_i$; therefore, $K_i = 1.0$. Applying (5-21) to both species, we have

$$\gamma_1 P_1^s = \gamma_2 P_2^s \qquad (5\text{-}45)$$

If species 2 is more volatile in the pure state ($P_2^s > P_1^s$), the criteria for formation of a minimum-boiling azeotrope are

$$\gamma_1 \geq 1 \qquad (5\text{-}46)$$

$$\gamma_2 \geq 1 \qquad (5\text{-}47)$$

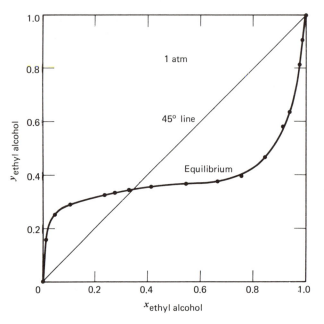

Figure 5.7. Equilibrium curve for n-hexane–ethanol system.

and

$$\frac{\gamma_1}{\gamma_2} > \frac{P_2^s}{P_1^s} \tag{5-48}$$

for x_1 less than the azeotropic composition. These criteria are most readily applied at $x_1 = 0$. For example, for the n-hexane (2)–ethanol (1) system at 1 atm (101.3 kPa), when the liquid-phase mole fraction of ethanol approaches zero, the temperature approaches 68.75°C (155.75°F), the boiling point of pure n-hexane. At this temperature, $P_1^s = 10$ psia (68.9 kPa) and $P_2^s = 14.7$ psia (101.3 kPa). Also from Fig. 5.5, $\gamma_1^\infty = 21.72$ when $\gamma_2 = 1.0$. Thus, $\gamma_1^\infty/\gamma_2 = 21.72$, but $P_2^s/P_1^s = 1.47$. Therefore, a minimum-boiling azeotrope will occur.

Maximum-boiling azeotropes are less common. They occur for relatively close-boiling mixtures when negative deviations from Raoult's law arise such that $\gamma_i < 1.0$. Criteria for their formation are derived in a manner similar to that of minimum-boiling azeotropes. At $x_1 = 1$, where species 2 is more volatile,

$$\gamma_1 = 1.0 \tag{5-49}$$

$$\gamma_2^\infty < 1.0 \tag{5-50}$$

and

$$\frac{\gamma_2^\infty}{\gamma_1} < \frac{P_1^s}{P_2^s} \tag{5-51}$$

For an azeotrope binary system, the two interaction constants Λ_{12} and Λ_{21} can be determined by solving (5-28) at the azeotropic composition as shown in the following example.

Example 5.6. From measurements by Sinor and Weber[31] of the azeotrope condition for the ethanol-n-hexane system at 1 atm (101.3 kPa, 14.696 psia), calculate Λ_{12} and Λ_{21}.

Solution. Let E denote ethanol and H n-hexane. The azeotrope occurs at $x_E = 0.332$, $x_H = 0.668$, and $T = 58°C$ (331.15°K). At 1 atm, (5-21) can be used to approximate K-values. Thus, at azeotropic conditions, $\gamma_i = P/P_i^s$. The vapor pressures at 58°C are $P_E^s = 6.26$ psia and $P_H^s = 10.28$ psia. Therefore

$$\gamma_E = \frac{14.696}{6.26} = 2.348$$

$$\gamma_H = \frac{14.696}{10.28} = 1.430$$

Substituting these values together with the above corresponding values of x_i into the binary form of the Wilson equation in (5-28) gives

$$\ln 2.348 = -\ln[0.332 + 0.668\Lambda_{EH}] + 0.668\left[\frac{\Lambda_{EH}}{0.332 + 0.668\Lambda_{EH}} - \frac{\Lambda_{HE}}{0.332\Lambda_{HE} + 0.668}\right]$$

$$\ln 1.430 = -\ln[0.668 + 0.332\Lambda_{HE}] - 0.332\left[\frac{\Lambda_{EH}}{0.332 + 0.668\Lambda_{EH}} - \frac{\Lambda_{HE}}{0.332\Lambda_{HE} + 0.668}\right]$$

Solving these two nonlinear equations simultaneously by an iterative procedure, we have $\Lambda_{EH} = 0.041$ and $\Lambda_{HE} = 0.281$. From these constants, the activity coefficient curves can be predicted if the temperature variations of Λ_{EH} and Λ_{HE} are ignored. The results are plotted in Fig. 5.8. The fit of experimental data is good except, perhaps, for ethanol near infinite-dilution conditions, where $\gamma_E^\infty = 49.82$ and $\gamma_H^\infty = 9.28$. The former value is considerably greater than the $\gamma_E^\infty = 21.72$ obtained by Orye and Prausnitz[32] from a fit of all experimental data points. However, if Figs. 5.5 and 5.8 are compared, it is seen that widely differing γ_E^∞ values have little effect on γ in the composition region $x_E = 0.15$ to 1.00, where the two sets of Wilson curves are almost identical. Hudson and Van Winkle[37] state that Wilson parameters based on data for a single liquid composition are sufficient for screening and preliminary design. Howeveruer, for accuracy over the entire composition range, commensurate with the ability of the Wilson equation, data for at least three well-spaced liquid compositions per binary are preferred.
□

A common procedure for screening possible mass separating agents for extractive distillation is to measure or estimate infinite-dilution activity coefficients for solutes in various polar solvents. However, the inverse determination for solvent at infinite dilution in the solute often is not feasible. In this

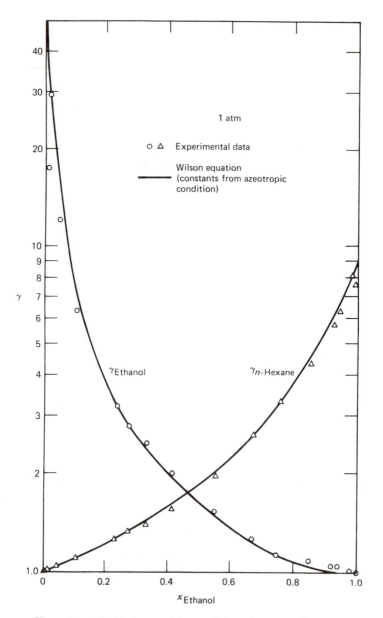

Figure 5.8. Liquid-phase activity coefficients for ethanol/n-hexane system.

case, the single-parameter modification of the Wilson equation by Tassios[43] is useful. From γ_i^∞ for a binary pair ij, the value of γ_j^∞ and the entire γ_i and γ_j curves can be predicted. Schreiber and Eckert[44] obtained good results using this technique provided both γ^∞ values were less than 10. The modification of the Wilson equation by Tassios consists of estimating λ_{ii} and λ_{jj} in (5-39) and (5-40) from the energy of vaporization by

$$\lambda_{ii} = -\eta(\lambda_i - RT) \tag{5-52}$$

Then, only the single parameter λ_{ij} remains to be determined per binary pair. Tassios used a value of $\eta = 1$, but Schreiber and Eckert suggested $\eta = 0.2$ on theoretical grounds.

The Wilson equation can be used also to determine excess enthalpy of a nonideal liquid solution. An approximate procedure is to apply (4-57) to (5-42), neglecting effects of temperature on $(\lambda_{ij} - \lambda_{ii})$ and v_{iL}/v_{jL} in (5-39) and (5-40). The result is

$$H^E = \sum_{i=1}^{C} x_i \bar{H}_i^E = \sum_{i=1}^{C} \left\{ x_i \left[\frac{\sum_{j,j\neq i}^{C} x_j \left(\frac{\partial \Lambda_{ij}}{\partial T}\right) RT^2}{x_i + \sum_{j,j\neq i}^{C} x_j \Lambda_{ij}} \right] \right\} \tag{5-53}$$

where

$$\left(\frac{\partial \Lambda_{ij}}{\partial T}\right) RT^2 = (\lambda_{ij} - \lambda_{ii})\Lambda_{ij} \tag{5-54}$$

More accurate estimates of H^E can be made at the expense of greatly added complexity if the temperature dependence of $(\lambda_{ij} - \lambda_{ii})$ and v_{iL}/v_{jL} are considered, as discussed by Duran and Kaliaguine[45] and Tai, Ramalho, and Kaliaguine.[46]

Example 5.7. Use (5-53) and (5-54) to estimate the excess enthalpy of a 40.43 mole% solution of ethanol (E) in n-hexane (H) at 25°C (298.15°K). Compare the estimate to the experimental value of 138.6 cal/gmmole reported by Jones and Lu.[47]

Solution. Smith and Robinson[48] determined the following Wilson parameters at 25°C.

$$\Lambda_{EH} = 0.0530 \qquad (\lambda_{EH} - \lambda_{EE}) = 2209.77 \text{ cal/gmmole}$$

$$\Lambda_{HE} = 0.2489 \qquad (\lambda_{EH} - \lambda_{HH}) = 354.79 \text{ cal/gmmole}$$

For a binary system, (5-53) and (5-54) combine to give

$$H^E = x_1 \left(\frac{x_2 \Lambda_{12}}{x_1 + x_2 \Lambda_{12}}\right)(\lambda_{12} - \lambda_{11}) + x_2 \left(\frac{x_1 \Lambda_{21}}{x_2 + x_1 \Lambda_{21}}\right)(\lambda_{12} - \lambda_{22}) \tag{5-55}$$

For $x_E = 0.4043$ and $x_H = 0.5957$

$$H^E = 0.4043 \left[\frac{0.5957(0.0530)}{0.4043 + 0.5957(0.0530)}\right](2209.77)$$

$$+ 0.5957 \left[\frac{0.4043(0.2489)}{0.5957 + 0.4043(0.2489)}\right](354.79) = 95.2 \text{ cal/gmole (398.3 kJ/kgmole)}$$

This value is 31.3% lower than the experimental value. This result indicates that the assumption of temperature independence of $(\lambda_{ij} - \lambda_{ii})$ and v_{iL}/v_{jL} is not valid. When temperature dependence is taken into account, predictions are improved. For best accuracy, however, Nagata and Yamada[49] showed that Wilson parameters should be determined by simultaneous fit of vapor–liquid equilibrium and heat of mixing data.

□

5.5 The NRTL Equation

The Wilson equation can be extended to immiscible liquid systems by multiplying the right-hand side of (5-41) by a third binary-pair constant evaluated from experimental data.[33] However, for multicomponent systems of three or more species, the third binary-pair constants must be the same for all constituent binary pairs. Furthermore, as shown by Hiranuma,[50] representation of ternary systems involving only one partially miscible binary pair can be extremely sensitive to the third binary-pair Wilson constant. For these reasons, application of the Wilson equation to liquid–liquid systems has not been widespread. Rather, the success of the Wilson equation for prediction of activity coefficients for miscible liquid systems greatly stimulated further development of the local composition concept in an effort to obtain more universal expressions for liquid-phase activity coefficients.

The nonrandom, two-liquid (NRTL) equation developed by Renon and Prausnitz,[51,52] as listed in Table 5.3, represents an accepted extension of Wilson's concept. The NRTL equation is applicable to multicomponent vapor–liquid, liquid–liquid, and vapor–liquid–liquid systems. For multicomponent vapor–liquid systems, only binary-pair constants from the corresponding binary-pair experimental data are required.

Starting with an equation similar to (5-38), but expressing local composition in terms of mole fractions rather than volume fractions, Renon and Prausnitz developed an equation for the local mole fraction of species i in a liquid cell occupied by a molecule of i at the center.

$$x_{ji} = \frac{x_j \exp(-\alpha_{ji}\tau_{ji})}{\sum\limits_{k=1}^{C} x_k \exp(-\alpha_{ki}\tau_{ki})} \tag{5-56}$$

For the binary pair ij, τ_{ji} and τ_{ij} are adjustable parameters, and $\alpha_{ji}(=\alpha_{ij})$ is a third parameter that can be fixed or adjusted. Excess free energy for the liquid system is expressed by an extension of Scott's cell theory, wherein only two-molecule interactions are considered.

$$\frac{g^E}{RT} = \sum_{i=1}^{C} x_i \left[\sum_{j=1}^{C} x_{ji}\tau_{ji} \right] \tag{5-57}$$

The expression for the activity coefficient is obtained by combining (5-2), (5-56) and (5-57) to give

$$\ln \gamma_i = \frac{\sum_{j=1}^{C}(\tau_{ji}G_{ji}x_j)}{\sum_{k=1}^{C}(G_{ki}x_k)} + \sum_{j=1}^{C}\left[\frac{(x_jG_{ij})}{\sum_{k=1}^{C}(G_{kj}x_k)}\left(\tau_{ij} - \frac{\sum_{k=1}^{C}(x_k\tau_{kj}G_{kj})}{\sum_{k=1}^{C}(G_{kj}x_k)}\right)\right] \tag{5-58}$$

where

$$G_{ji} = \exp(-\alpha_{ji}\tau_{ji}) \tag{5-59}$$

The τ coefficients are given by

$$\tau_{ij} = \frac{(g_{ij} - g_{jj})}{RT} \tag{5-60}$$

$$\tau_{ji} = \frac{(g_{ji} - g_{ii})}{RT} \tag{5-61}$$

where g_{ij}, g_{jj}, and so on are energies of interaction between molecule pairs. In the above equations, $G_{ji} \neq G_{ij}$, $\tau_{ij} \neq \tau_{ji}$, $G_{ii} = G_{jj} = 1$, and $\tau_{ii} = \tau_{jj} = 0$. Often $(g_{ij} - g_{jj})$ and other constants are linear in temperature. Hála[35] showed that not all values of $(g_{ij} - g_{jj})$ are independent for a multicomponent mixture.

The parameter α_{ji} characterizes the tendency of species j and species i to be distributed in a nonrandom fashion. When $\alpha_{ji} = 0$, local mole fractions are equal to overall solution mole fractions. Generally α_{ji} is independent of temperature and depends on molecule properties in a manner similar to the classifications in Tables 5.1 and 5.2. Values of α_{ji} usually lie between 0.2 and 0.47. When $\alpha_{ji} < 0.426$, phase immiscibility is predicted. Although α_{ji} can be treated as an adjustable parameter, to be determined from experimental binary-pair data, more commonly α_{ji} is set according to the following rules, which are occasionally ambiguous.

1. $\alpha_{ji} = 0.20$ for mixtures of saturated hydrocarbons and polar nonassociated species (e.g., n-heptane–acetone).

2. $\alpha_{ji} = 0.30$ for mixtures of nonpolar compounds (e.g., benzene–n-heptane), except fluorocarbons and paraffins; mixtures of nonpolar and polar nonassociated species (e.g., benzene–acetone); mixtures of polar species that exhibit negative deviations from Raoult's law (e.g., acetone–chloroform) and moderate positive deviations (e.g., ethanol–water); mixtures of water and polar nonassociated species (e.g., water–acetone).

3. $\alpha_{ji} = 0.40$ for mixtures of saturated hydrocarbons and homolog perfluorocarbons (e.g., n-hexane–perfluoro-n-hexane).

4. $\alpha_{ji} = 0.47$ for mixtures of an alcohol or other strongly self-associated species with nonpolar species (e.g., ethanol–benzene); mixtures of carbon tetrachloride with either acetonitrile or nitromethane; mixtures of water with either butyl–glycol or pyridine.

For a binary system, (5-58) reduces to (5-29) or the following expressions in G_{ij}.

$$\ln \gamma_1 = x_2^2 \left[\frac{\tau_{21} G_{21}^2}{(x_1 + x_2 G_{21})^2} + \frac{\tau_{12} G_{12}}{(x_2 + x_1 G_{12})^2} \right] \tag{5-62}$$

$$\ln \gamma_2 = x_1^2 \left[\frac{\tau_{12} G_{12}^2}{(x_2 + x_1 G_{12})^2} + \frac{\tau_{21} G_{21}}{(x_1 + x_2 G_{21})^2} \right] \tag{5-63}$$

For ideal solutions, $\tau_{ji} = 0$.

Binary and ternary forms of the NRTL equation were evaluated and compared to other equations for vapor–liquid equilibrium applications by Renon and Prausnitz,[51] Larson and Tassios,[53] Mertl,[54] Marina and Tassios,[55] and Tsuboka and Katayama.[56] In general, the accuracy of the NRTL equation is comparable to that of the Wilson equation. Although α_{ji} is an adjustable constant, there is little loss in accuracy over setting its value according to the rules described above. Methods for determining best values of NRTL binary parameters are considered in detail in the above references. Mertl[54] tabulated NRTL parameters obtained from 144 sets of data covering 102 different binary systems. Other listings of NRTL parameters are also available.[38,40,57]

As with the Wilson equation, the two NRTL parameters involving energy differences can be obtained from a single data point or from a pair of infinite-dilution activity coefficients using the above rules to set the value of α_{ji}. At infinite dilution, (5-62) and (5-63) reduce to

$$\ln \gamma_1^\infty = \tau_{21} + \tau_{12} \exp(-\alpha_{12} \tau_{12}) \tag{5-64}$$

$$\ln \gamma_2^\infty = \tau_{12} + \tau_{21} \exp(-\alpha_{12} \tau_{21}) \tag{5-65}$$

A one-parameter form of the NRTL equation was developed by Bruin and Prausnitz.[58]

The excess enthalpy of a nonideal liquid solution can best be estimated from the NRTL equation by applying (4-57) to (5-58) with the assumption that $(g_{ji} - g_{ii})$ and $(g_{ij} - g_{jj})$ vary linearly with temperature. For example, for a binary mixture

$$H^E = x_1 x_2 R \left[\frac{x_1 \tau_{21} G_{21}'}{(x_1 + x_2 G_{21})^2} + \frac{x_2 \tau_{12} G_{12}'}{(x_2 + x_1 G_{12})^2} \right] \tag{5-66}$$

where

$$G_{ij}' = \frac{dG_{ij}}{d(1/T)} \tag{5-67}$$

Nagata and Yamada[49] report that NRTL parameters must be determined from both vapor–liquid equilibrium and heat of mixing data for highly accurate predictions of H^E.

Example 5.8. For the ethanol (E)–n-hexane (H) system at 1 atm (101.3 kPa), a best fit of the Wilson equation using the experimental data of Sinor and Weber[31] leads to infinite-dilution activity coefficients of $\gamma_E^\infty = 21.72$ and $\gamma_H^\infty = 9.104$ as discussed in Example 5.6. Neglecting the effect of temperature, use these values to determine τ_{EH} and τ_{HE} in the NRTL equation. Then, estimate activity coefficients at the azeotropic composition $x_E = 0.332$. Compare the values obtained to those derived from experimental data in Example 5.6.

Solution. According to rules of Renon and Prausnitz,[51] α_{EH} is set at 0.47. Values of τ_{EH} and τ_{HE} are determined by solving (5-64) and (5-65) simultaneously.

$$\ln 21.72 = \tau_{HE} + \tau_{EH}\, \exp(-0.47\tau_{EH})$$

$$\ln 9.104 = \tau_{EH} + \tau_{HE}\, \exp(-0.47\tau_{HE})$$

By an iterative procedure, $\tau_{EH} = 2.348$ and $\tau_{HE} = 1.430$. Equation (5-59) gives

$$G_{EH} = \exp[-0.47(2.348)] = 0.3317$$

$$G_{HE} = \exp[-0.47(1.430)] = 0.5106$$

Equations (5-62) and (5-63) are solved for γ_E and γ_H at $x_E = 0.332$ and $x_H = 0.668$.

$$\gamma_E = \exp\left\{(0.668)^2\left[\frac{1.430(0.5106)^2}{[0.332 + 0.668(0.5106)]^2} + \frac{2.348(0.3317)}{[0.668 + 0.332(0.3317)]^2}\right]\right\}$$

$$= 2.563$$

Similarly, $\gamma_H = 1.252$.

The value of γ_E is 9.2% higher than the experimental value of 2.348 and the value of γ_H is 12.4% lower than the experimental value of 1.43. By contrast, from Fig. 5.5, the Wilson equation predicts values of $\gamma_E = 2.35$ and $\gamma_H = 1.36$, which are in closer agreement. Renon and Prausnitz[51] also show that solutions of alcohols and hydrocarbons represent the only case where the NRTL equation is not as good as or better than the Wilson equation.
□

5.6 The UNIQUAC Equation

In an attempt to place calculations of liquid-phase activity coefficients on a simpler, yet more theoretical basis, Abrams and Prausnitz[18,59,61] used statistical mechanics to derive a new expression for excess free energy. Their model, called UNIQUAC (*uni*versal *qua*si-*c*hemical), generalizes a previous analysis by Guggenheim and extends it to mixtures of molecules that differ appreciably in size and shape. As in the Wilson and NRTL equations, local concentrations are used. However, rather than local volume fractions or local mole fractions, UNIQUAC uses the local area fraction θ_{ij} as the primary concentration variable.

The local area fraction is determined by representing a molecule by a set of bonded segments. Each molecule is characterized by two structural parameters that are determined relative to a standard segment taken as an equivalent sphere of a mer unit of a linear, infinite-length polymethylene molecule. The two structural parameters are the relative number of segments per molecule r (volume parameter), and the relative surface area of the molecule q (surface parameter). Values of these parameters computed from bond angles and bond distances are given by Abrams and Prausnitz[18] and Gmehling and Onken[38] for a number of species. For other compounds, values can be estimated by the group-contribution method of Fredenslund et al.[60]

For a multicomponent liquid mixture, the UNIQUAC model gives the excess free energy as

$$\frac{g^E}{RT} = \sum_{i=1}^{C} x_i \ln\left(\frac{\Psi_i}{x_i}\right) + \frac{\bar{Z}}{2} \sum_{i=1}^{C} q_i x_i \ln\left(\frac{\theta_i}{\Psi_i}\right) - \sum_{i=1}^{C} q_i x_i \ln\left(\sum_{j=1}^{C} \theta_j T_{ji}\right) \qquad (5\text{-}68)$$

The first two terms on the right-hand side account for *combinatorial* effects due to differences in molecule size and shape; the last term provides a *residual* contribution due to differences in intermolecular forces, where

$$\Psi_i = \frac{x_i r_i}{\sum\limits_{i=1}^{C} x_i r_i} = \text{segment fraction} \qquad (5\text{-}69)$$

$$\theta_i = \frac{x_i q_i}{\sum\limits_{i=1}^{C} x_i q_i} = \text{area fraction} \qquad (5\text{-}70)$$

$\bar{Z}$ = lattice coordination number set equal to 10

$$T_{ji} = \exp\left(-\frac{u_{ji} - u_{ii}}{RT}\right) \qquad (5\text{-}71)$$

Equation (5-68) contains only two adjustable parameters for each binary pair, $(u_{ji} - u_{ii})$ and $(u_{ij} - u_{jj})$. Abrams and Prausnitz show that $u_{ji} = u_{ij}$ and $T_{ii} = T_{jj} = 1$. In general $(u_{ji} - u_{ii})$ and $(u_{ij} - u_{jj})$ are linear functions of temperature.

If (5-2) is combined with (5-68), an equation for the liquid-phase activity coefficient for a species in a multicomponent mixture is obtained as

$$\ln \gamma_i = \ln \gamma_i^C + \ln \gamma_i^R = \underbrace{\ln\left(\frac{\Psi_i}{x_i}\right) + \frac{\bar{Z}}{2} q_i \ln\left(\frac{\theta_i}{\Psi_i}\right) + \ell_i - \frac{\Psi_i}{x_i} \sum_{j=1}^{C} (x_j \ell_j)}_{C,\ \text{combinatorial}}$$

$$\underbrace{+ q_i \left[1 - \ln\left(\sum_{j=1}^{C} \theta_j T_{ji}\right) - \sum_{j=1}^{C} \left(\frac{\theta_j T_{ij}}{\sum\limits_{k=1}^{C} \theta_k T_{kj}} \right) \right]}_{R,\ \text{residual}} \qquad (5\text{-}72)$$

where

$$\ell_j = \left(\frac{\bar{Z}}{2}\right)(r_j - q_j) - (r_j - 1) \tag{5-73}$$

For a binary mixture of species 1 and 2, (5-72) reduces to (5-30) in Table 5.3 where ℓ_1 and ℓ_2 are given by (5-73), Ψ_1 and Ψ_2 by (5-69), θ_1 and θ_2 by (5-70), T_{12} and T_{21} by (5-71), and $\bar{Z} = 10$.

All well-known equations for excess free energy can be derived from (5-68) by making appropriate simplifying assumptions. For example, if q_i and r_i equal one, (5-30) for species 1 becomes

$$\ln \gamma_1 = -\ln(x_1 + x_2 T_{21}) + x_2\left(\frac{T_{21}}{x_1 + x_2 T_{21}} - \frac{T_{12}}{x_2 + x_1 T_{12}}\right)$$

which is identical to the Wilson equation if $T_{21} = \Lambda_{12}$ and $T_{12} = \Lambda_{21}$.

Abrams and Prausnitz found that for vapor–liquid systems. the UNIQUAC equation is as accurate as the Wilson equation. However, an important advantage of the UNIQUAC equation lies in its applicability to liquid–liquid systems, as discussed in Section 5.8. Abrams and Prausnitz also give a one-parameter form of the UNIQUAC equation. The methods previously described can be used to determine UNIQUAC parameters from infinite-dilution activity coefficients or from azeotropic or other single-point data.

Excess enthalpy can be estimated from the UNIQUAC equation by combining (4-57) with (5-72) using the assumption that $(u_{ji} - u_{ii})$ and $(u_{ij} - u_{jj})$ vary linearly with temperature.

Example 5.9. Solve Example 5.8 by the UNIQUAC equation using values of binary interaction constants given by Abrams and Prausnitz.[18]

Solution. At 1 atm (101.3 kPa), $(u_{HE} - u_{EE}) = 940.9$ cal/gmole and $(u_{EH} - u_{HH}) = -335.0$ cal/gmole. Corresponding values of the size and surface parameters from Abrams are $r_E = 2.17$, $r_H = 4.50$, $q_E = 2.70$, and $q_H = 3.86$. From Example 5.6, azeotropic conditions are 331.15°K, $x_E = 0.332$, and $x_H = 0.668$. From (5-69)

$$\Psi_E = \frac{0.332(2.17)}{0.332(2.17) + 0.668(4.50)} = 0.1933$$

$$\Psi_H = 1 - \Psi_E = 1 - 0.1933 = 0.8067$$

From (5-70)

$$\theta_E = \frac{0.332(2.70)}{0.332(2.70) + 0.668(3.86)} = 0.2580$$

$$\theta_H = 1 - \theta_E = 1 - 0.2580 = 0.7420$$

From (5-71)

$$T_{EH} = \exp\left[-\frac{(-335.0)}{1.987(331.15)}\right] = 1.664$$

$$T_{HE} = \exp\left[-\frac{(940.9)}{1.987(331.15)}\right] = 0.2393$$

From (5-73)

$$\ell_E = \frac{10}{2}(2.17 - 2.70) - (2.17 - 1) = -3.820$$

$$\ell_H = \frac{10}{2}(4.50 - 3.86) - (4.50 - 1) = -0.300$$

From (5-30) for a binary mixture

$$\ln \gamma_E = \ln\left(\frac{0.1933}{0.332}\right) + \left(\frac{10}{2}\right)(2.70) \ln\left(\frac{0.2580}{0.1933}\right)$$

$$+ 0.8067\left[-3.820 - \frac{2.17}{4.50}(-0.300)\right] - 2.70 \ln[0.2580 + 0.7420(0.2393)]$$

$$+ 0.7420(2.70)\left[\frac{0.2393}{0.2580 + 0.7420(0.2393)} - \frac{1.664}{0.7420 + 0.2580(1.664)}\right]$$

$$= 0.8904$$

Then $\gamma_E = \exp(0.8904) = 2.436$. Similarly, $\gamma_H = 1.358$. These values are, respectively, only 3.8% higher and 5.0% lower than experimental values.
□

5.7 Group Contribution Methods and the UNIFAC Model

Liquid-phase activity coefficients must be predicted for nonideal mixtures even when experimental phase equilibrium data are not available and when the assumption of regular solutions is not valid because polar compounds are present. For such predictions, Wilson and Deal[62] and then Derr and Deal,[63] in the 1960s, presented methods based on treating a solution as a mixture of functional groups instead of molecules. For example, in a solution of toluene and acetone, the contributions might be 5 aromatic CH groups, 1 aromatic C group, and 1 CH_3 group from toluene; and 2 CH_3 groups plus 1 CO carbonyl group from acetone. Alternatively, larger groups might be employed to give 5 aromatic CH groups and 1 CCH_3 group from toluene; and 1 CH_3 group and 1 CH_3CO group from acetone. As larger and larger functional groups are taken, the accuracy of molecular representation increases, but the advantage of the group-contribution method decreases because a larger number of groups is required. In practice, 50 functional groups can be used to represent literally thousands of multicomponent liquid mixtures.

To calculate the partial molal excess free energies $\bar{g}_i^E$ and from this the activity coefficients and the excess enthalpy, size parameters for each functional group and binary interaction parameters for each pair of functional groups are required. Size parameters can be calculated from theory. Interaction parameters are back-calculated from existing phase equilibrium data and then used with the size parameters to predict phase equilibrium properties of mixtures for which no data are available.

The UNIFAC (UNIQUAC *F*unctional-group *A*ctivity *C*oefficients) group-contribution method, first presented by Fredenslund, Jones, and Prausnitz[64] and further developed for use in practice by Fredenslund et al.[60] and Fredenslund, Gmehling and Rasmussen,[65] has several advantages over other group-contribution methods. (1) It is theoretically based on the UNIQUAC method;[18] (2) The parameters are essentially independent of temperature; (3) Size and binary interaction parameters are available for a wide range of types of functional groups;[60,65] (4) Predictions can be made over a temperature range of 275 to 425°K and for pressures up to a few atmospheres; (5) Extensive comparisons with experimental data are available.[65] All components in the mixture must be condensable.

The UNIFAC method for predicting liquid-phase activity coefficients is based on the UNIQUAC equation (5-72), wherein the molecular volume and area parameters in the combinatorial terms are replaced by

$$r_i = \sum_k \nu_k^{(i)} R_k \tag{5-74}$$

$$q_i = \sum_k \nu_k^{(i)} Q_k \tag{5-75}$$

where $\nu_k^{(i)}$ is the number of functional groups of type k in molecule i, and R_k and Q_k are the volume and area parameters, respectively, for the type-k functional group.

The residual term in (5-72), which is represented by $\ln \gamma_i^R$, is replaced by the expression

$$\ln \gamma_i^R = \sum_{\substack{k \\ \text{all functional} \\ \text{groups in the} \\ \text{mixture}}} \nu_k^{(i)} (\ln \Gamma_k - \ln \Gamma_k^{(i)}) \tag{5-76}$$

where Γ_k is the residual activity coefficient of the functional group k in the actual mixture, and $\Gamma_k^{(i)}$ is the same quantity but in a reference mixture that contains only molecules of type i. The latter quantity is required so that $\gamma_i^R \to 1.0$ as $x_i \to 1.0$. Both Γ_k and $\Gamma_k^{(i)}$ have the same form as the residual term in (5-72). Thus

$$\ln \Gamma_k = Q_k \left[1 - \ln \left(\sum_m \theta_m T_{mk} \right) - \sum_m \frac{\theta_m T_{km}}{\sum_n \theta_n T_{nm}} \right] \tag{5-77}$$

where θ_m is the area fraction of group m, given by an equation similar to (5-70)

$$\theta_m = \frac{X_m Q_m}{\sum_n X_n Q_n} \tag{5-78}$$

where X_m is the mole fraction of group m in the solution

$$X_m = \frac{\sum_j v_m^{(j)} x_j}{\sum_j \sum_n (v_n^{(j)} x_j)} \tag{5-79}$$

and T_{mk} is a group interaction parameter given by an equation similar to (5-71)

$$T_{mk} = \exp\left(-\frac{a_{mk}}{T}\right) \tag{5-80}$$

where $a_{mk} \neq a_{km}$. When $m = k$, then $a_{mk} = 0$ and $T_{mk} = 1.0$. For $\Gamma_k^{(i)}$, (5-77) also applies, where θ terms correspond to the pure component i.

Extensive tables of values for R_k, Q_k, a_{mk}, and a_{km} are available,[65,66] and undoubtedly will be updated as new experimental data are obtained for mixtures that contain functional groups not included in the current tables. Although values of R_k and Q_k are different for each functional group, values of a_{mk} are equal for all subgroups within a main group. For example, main group CH_2 consists of subgroups CH_3, CH_2, CH, and C. Accordingly, $a_{CH_3,CHO} = a_{CH_2,CHO} = a_{CH,CHO} = a_{C,CHO}$. Thus, the amount of experimental data required to obtain values of a_{mk} and a_{km} and the size of the corresponding bank of data for these parameters are not as large as might be expected.

Excess enthalpy can be estimated by the UNIFAC method in a manner analogous to the UNIQUAC equation except that values of a_{mk}, which are like $(u_{mk} - u_{km})/R$, are assumed to be independent of temperature.

Example 5.10. Solve Example 5.9 by the UNIFAC method using values of the volume, area, and binary interaction parameters from Fredenslund, Jones, and Prausnitz.[64]

Solution. Ethanol (CH_3CH_2OH) (E) is treated as a single functional group (17), while n-hexane (H) consists of two CH_3 groups (1) and four CH_2 groups (2) that are both contained in the same main group. The parameters are:

$$R_{17} = 2.1055 \qquad R_1 = 0.9011 \qquad R_2 = 0.6744$$

$$Q_{17} = 1.972 \qquad Q_1 = 0.848 \qquad Q_2 = 0.540$$

$$a_{17,1} = a_{17,2} = -87.93°K \qquad a_{1,17} = a_{2,17} = 737.5°K$$

Then $a_{1,2} = a_{2,1} = 0°K$ because groups 1 and 2 are in the same main group. Other conditions are

$$T = 331.15°K \qquad x_E = 0.332 \qquad x_H = 0.668$$

Combinatorial part. From (5-74)

$$r_E = (1)(2.1055) = 2.1055$$

$$r_H = (2)(0.9011) + (4)(0.6744) = 4.4998$$

From (5-75)

$$q_E = (1)(1.972) = 1.972$$

$$q_H = (2)(0.848) + (4)(0.540) = 3.856$$

From (5-69)

$$\Psi_E = \frac{(0.332)(2.1055)}{(0.332)(2.1055) + (0.668)(4.4998)} = 0.1887$$

$$\Psi_H = 1 - 0.1887 = 0.8113$$

From (5-70)

$$\theta_E = \frac{(0.332)(1.972)}{(0.332)(1.972) + (0.668)(3.856)} = 0.2027$$

$$\theta_H = 1 - 0.2027 = 0.7973$$

From (5-73), with $\bar{Z} = 10$

$$\ell_E = \left(\frac{10}{2}\right)(2.1055 - 1.972) - (2.1055 - 1) = -0.4380$$

$$\ell_H = \left(\frac{10}{2}\right)(4.4998 - 3.856) - (4.4998 - 1) = -0.280$$

From (5-72)

$$\ln \gamma_E^C = \ln\left(\frac{0.1887}{0.332}\right) + \left(\frac{10}{2}\right)(1.972) \ln\left(\frac{0.2027}{0.1887}\right) + (-0.4380)$$
$$- \frac{0.1887}{0.332}[0.332(-0.4380) + 0.668(-0.280)] = -0.1083$$

Similarly

$$\ln \gamma_H^C = -0.0175$$

Residual part. For pure ethanol, which is represented by a single functional group (17), $\ln \Gamma_{17}^{(E)} = 0$. For n-hexane with $a_{1,2} = a_{2,1}$, no difference in interactions exists and $\ln \Gamma_1^{(H)} = \ln \Gamma_2^{(H)} = 0$. For the actual mixture, from (5-79)

$$X_{17} = \frac{(1)(0.332)}{(1)(0.332) + (2)(0.668) + 4(0.668)} = 0.0765$$

Similarly,

$$X_1 = 0.3078 \quad \text{and} \quad X_2 = 0.6157$$

From (5-78)

$$\theta_{17} = \frac{(0.0765)(1.972)}{(0.0765)(1.972) + (0.3078)(0.848) + (0.6157)(0.540)} = 0.2027$$

Similarly

$$\theta_1 = 0.3506 \quad \text{and} \quad \theta_2 = 0.4467$$

From (5-80)

$$T_{17,1} = T_{17,2} = \exp\left[-\frac{(-87.93)}{331.15}\right] = 1.3041$$

Similarly,

$$T_{1,17} = T_{2,17} = \exp\left[-\frac{737.5}{331.15}\right] = 0.1078$$

$$T_{1,2} = T_{2,1} = 1.0$$

From (5-77)

$$\ln \Gamma_{17} = 1.972\{1 - \ln[(0.2027)(1) + (0.3506)(0.1078) + (0.4467)(0.1078)]$$

$$- \left[\frac{(0.2027)(1)}{(0.2027)(1) + (0.3506)(0.1078) + (0.4467)(0.1078)}\right.$$

$$+ \frac{(0.3506)(1.3041)}{(0.2027)(1.3041) + (0.3506)(1) + (0.4467)(1)}$$

$$\left.\left. + \frac{(0.4467)(1.3041)}{(0.2027)(1.3041) + (0.3506)(1) + (0.4467)(1)}\right]\right\} = 1.1061$$

Similarly

$$\ln \Gamma_1 = 0.0962 \quad \text{and} \quad \ln \Gamma_2 = 0.0612$$

From (5-76)

$$\ln \gamma_E^R = (1)(1.1061 - 0.0) = 1.1061$$

$$\ln \gamma_H^R = (2)(0.0962 - 0.0) + 4(0.0612 - 0.0) = 0.4372$$

$$\ln \gamma_E = \ln \gamma_E^C + \ln \gamma_E^R = -0.1083 + 1.1061 = 0.9978$$

$$\gamma_E = 2.71$$

Similarly,

$$\gamma_H = 1.52.$$

These values may be compared to results of previous examples for other local composition models.

	Expt.	Wilson	NRTL (constants from γ_i^∞)	UNIQUAC	UNIFAC
γ_E	2.348	2.35	2.563	2.436	2.71
γ_H	1.43	1.36	1.252	1.358	1.52

☐

In general, as stated by Prausnitz,[66] the choice of correlation (model) for the excess free energy is not as important as the procedure used to obtain the correlation parameters from limited experimental data.

5.8 Liquid–Liquid Equilibria

When species are notably dissimilar and activity coefficients are large, two and even more liquid phases may coexist at equilibrium. For example, consider the binary system of methanol (1) and cyclohexane (2) at 25°C. From measurements of Takeuchi, Nitta, and Katayama,[67] van Laar constants, as determined in Example 5.11 below, are $A_{12} = 2.61$ and $A_{21} = 2.34$, corresponding, respectively, to the infinite-dilution activity coefficients of 13.6 and 10.4 obtained using (5-34). These values of A_{12} and A_{21} can be used to construct an equilibrium plot of y_1 against x_1 assuming an isothermal condition. By combining (5-21), where $K_i = y_i/x_i$, and (5-22), one obtains the following relation for computing y_i from x_i.

$$y_1 = \frac{x_1 \gamma_1 P_1^s}{x_1 \gamma_1 P_1^s + x_2 \gamma_2 P_2^s} \tag{5-81}$$

Vapor pressures at 25°C are $P_1^s = 2.452$ psia (16.9 kPa) and $P_2^s = 1.886$ psia (13.0 kPa). Activity coefficients can be computed from the van Laar equation in Table 5.3. The resulting equilibrium plot is shown in Fig. 5.9, where it is observed that over much of the liquid-phase region three values of y_1 exist. This indicates phase instability. Experimentally, single liquid phases can exist only for cyclohexane-rich mixtures of $x_1 = 0.8248$ to 1.0 and for methanol-rich mixtures of $x_1 = 0.0$ to 0.1291. Because a coexisting vapor phase exhibits only a single composition, two coexisting liquid phases prevail at opposite ends of the dashed line in Fig. 5.9. The liquid phases represent solubility limits of methanol in cyclohexane and cyclohexane in methanol.

For two coexisting equilibrium liquid phases, from (4-31), the relation $\gamma_{iL}^{I} x_i^{I} = \gamma_{iL}^{II} x_i^{II}$ must hold. This permits determination of the two-phase region in Fig. 5.9 from the van Laar or other suitable activity coefficient equations for which the constants are known. Also shown in Fig. 5.9 is an equilibrium curve for the same binary system at 55°C based on data of Strubl et al.[68] At this higher temperature, methanol and cyclohexane are completely miscible. The data of Kiser, Johnson, and Shetlar[69] show that phase instability ceases to exist at 45.75°C, the critical solution temperature. Rigorous thermodynamic methods for determining phase instability and, thus, existence of two equilibrium liquid phases are generally based on free energy calculations, as discussed by Prausnitz.[70]

Most of the empirical and semitheoretical equations for liquid-phase activity coefficient listed in Table 5.3 apply to liquid–liquid systems. The Wilson equation is a notable exception. As examples, the van Laar equation will be discussed next, followed briefly by the NRTL, UNIQUAC, and UNIFAC equations.

Van Laar Equation

Starting with activity coefficients, which can be determined from (5-33), one can form distribution coefficients from (4-31). For partially miscible pairs, the preferred procedure for obtaining van Laar constants is a best fit of activity

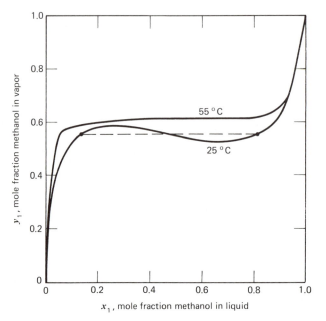

Figure 5.9. Equilibrium curves for methanol–cyclohexane systems. [Data from K. Strubl, V. Svoboda, R. Holub, and J. Pick, *Collect. Czech. Chem. Commun.*, **35**, 3004–3019 (1970)].

coefficient data for each species over the complete composition range. For example, in Fig. 5.2e, Phase A (water-rich phase) covers the range of water mole fraction from 0.0 to 0.66 and Phase B (*n*-butanol-rich phase) covers the range from 0.98 to 1.0. Frequently, only mutual solubility data for a binary pair are available. In this case, van Laar constants can be computed directly from the equations of Carlson and Colburn.[22] These are obtained by combining (5-26) with the liquid–liquid equilibrium condition of (4-31).

$$\frac{A_{ij}}{A_{ji}} = \frac{(x_i^I/x_j^I + x_i^{II}/x_j^{II})\left[\dfrac{\ln(x_i^{II}/x_i^I)}{\ln(x_j^I/x_j^{II})}\right] - 2}{(x_i^I/x_j^I + x_i^{II}/x_j^{II}) - \dfrac{2x_i^I x_i^{II} \ln(x_i^{II}/x_i^I)}{x_j^I x_j^{II} \ln(x_j^I/x_j^{II})}} \tag{5-82}$$

$$A_{ij} = \frac{\ln(x_i^{II}/x_i^I)}{(1 + A_{ij}x_i^I/A_{ji}x_j^I)^{-2} - (1 + A_{ij}x_i^{II}/A_{ji}x_j^{II})^{-2}} \tag{5-83}$$

Brian[71] showed that, when binary constants are derived from mutual solubility data, the van Laar equation behaves well over the entire single liquid-phase regions and is superior to the Margules and Scatchard–Hammer equations.

However, Joy and Kyle[72] suggest that caution be exercised in predicting ternary liquid–liquid equilibria from binary constants evaluated from mutual solubility data when only one of the binary pairs is partially miscible as in Fig. 3.11. The prediction is particularly difficult in the region of the plait point. Ternary systems involving two partially miscible binary pairs, such as those in Fig. 3.18b, can be predicted reasonably well.

Example 5.11. Experimental liquid–liquid equilibrium data for the methanol (1)–cylcohexane (2)–cyclopentane (3) system at 25°C were determined by Takeuchi, Nitta, and Katayama.[67] Use the van Laar equation to predict liquid–liquid distribution coefficients for the following liquid phases in equilibrium. Note that only methanol and cylcohexane form a binary of partial miscibility.

	I, Methanol-Rich Layer	II, Cyclohexane-Rich Layer
x_1	0.7615	0.1737
x_2	0.1499	0.5402
x_3	0.0886	0.2861

Solution. Assume cyclohexane and cyclopentane form an ideal solution.* Thus $A_{23} = A_{32} = 0.0$. Mutual solubility data for the methanol–cyclohexane system at 25°C are given by Takeuchi, Nitta, and Katayama[67] as

	I, Methanol-Rich Layer	II, Cyclohexane-Rich Layer
x_1	0.8248	0.1291
x_2	0.1752	0.8709

These binary mutual solubility data can be used in (5-82) and (5-83) to predict binary van Laar constants.

$$\frac{A_{12}}{A_{21}} = \frac{(0.8248/0.1752 + 0.1291/0.8709)\left[\frac{\ln(0.1291/0.8248)}{\ln(0.1752/0.8709)}\right] - 2}{(0.8248/0.1752 + 0.1291/0.8709) - \frac{2(0.8248)(0.1291)\ln(0.1291/0.8248)}{(0.1752)(0.8709)\ln(0.1752/0.8709)}} = 1.115$$

$$A_{12} = \frac{\ln(0.1291/0.8248)}{[1 + 1.115(0.8248/0.1752)]^{-2} - [1 + 1.115(0.1291/0.8709)]^{-2}} = 2.61$$

$$A_{21} = \frac{A_{12}}{(A_{12}/A_{21})} = \frac{2.61}{1.115} = 2.34$$

For the third binary pair, methanol (1) and cyclopentane (3), a critical solution temperature of 16.6°C with $x_3 = 0.58$ was reported by Kiser, Johnson, and Shetlar.[69] Thus

* Alternatively, modified regular solution theory (Equation 5-10) could be applied.

$x_1^I = x_1^{II} = 0.42$ and $x_3^I = x_3^{II} = 0.58$. Under these conditions, (5-82) and (5-83) become indeterminate. However, application of the incipient-phase instability conditions as discussed by Hildebrand and Scott[73] leads to the expressions

$$x_i^I \xrightarrow{\text{Lim}} x_i^{II} \quad \left(\frac{A_{ij}}{A_{ji}}\right) = \frac{1 - (x_i^I)^2}{2x_i^I - (x_i^I)^2} \tag{5-84}$$

$$x_i^I \xrightarrow{\text{Lim}} x_i^{II} \quad A_{ij} = \frac{13.50(1 - x_i^I)^2}{(2 - x_i^{II})^2[1 - (x_i^I)^2]} \tag{5-85}$$

From (5-85)* at 16.6°C

$$A_{13} = \frac{13.50(1 - 0.42)^2}{(2 - 0.42)^2[1 - (0.42)^2]} = 2.209$$

Applying (5-84)*

$$\frac{A_{13}}{A_{31}} = \frac{1 - (0.42)^2}{2(0.42) - (0.42)^2} = 1.241$$

$$A_{31} = \frac{A_{13}}{(A_{13}/A_{31})} = \frac{2.209}{1.241} = 1.780$$

Assuming temperature dependence of the van Laar constants as given by (5-31), at 25°C

$$A_{13} = 2.209\left(\frac{289.75}{298.15}\right) = 2.147$$

$$A_{31} = 1.780\left(\frac{289.75}{298.15}\right) = 1.730$$

In summary, binary constants at 25°C are

$$A_{12} = 2.61 \qquad A_{13} = 2.147 \qquad A_{23} = 0.0$$
$$A_{21} = 2.34 \qquad A_{31} = 1.730 \qquad A_{32} = 0.0$$

For the ternary system, activity coefficients for each species in each liquid phase are computed from (5-33) for the given phase compositions in the manner of Example 5.5. Results are

$$\gamma_1^I = 1.118 \qquad \gamma_2^I = 4.773 \qquad \gamma_3^I = 3.467$$
$$\gamma_1^{II} = 5.103 \qquad \gamma_2^{II} = 1.233 \qquad \gamma_3^{II} = 1.271$$

For methanol, using experimental data

$$(K_{D_1})_{\text{expt}} = \frac{x_1^I}{x_1^{II}} = \frac{0.7615}{0.1737} = 4.384$$

Using the activity coefficients from the van Laar equations

$$(K_{D_1})_{\text{van Laar}} = \frac{\gamma_1^{II}}{\gamma_1^I} = \frac{5.103}{1.118} = 4.565$$

---

*For a symmetrical system such as Fig. 5.2a, van Laar constants at incipient phase instability conditions ($x_i^I = 0.5$) are $A_{ij} = A_{ji} = 2.0$

Calculations for cyclohexane and cyclopentane are done in a similar manner. Results are

	K_D		
Species	Experimental	van Laar	% Deviation
Methanol	4.384	4.565	4.12
Cyclohexane	0.2775	0.2583	6.92
Cyclopentane	0.3097	0.3666	18.37

Corresponding relative selectivities of methanol for cyclopentane relative to cyclohexane are computed from (1-8)

$$(\beta_{3,2})_{\text{expt}} = \frac{0.3097}{0.2775} = 1.116$$

$$(\beta_{3,2})_{\text{predicted}} = \frac{0.3666}{0.2583} = 1.419$$

This represents a deviation of 27.2%, which is considerable.
□

NRTL Equation

The applicability of (5-58) to liquid–liquid system equilibria was studied by Renon and Prausnitz,[51,52] Joy and Kyle,[72] Mertl,[54] Guffey and Wehe,[74] Marina and Tassios,[55] and Tsuboka and Katayama.[56] In general, their studies show the NRTL equation to be superior to the van Laar or Margules equations. Multicomponent systems involving only one completely miscible binary pair can be predicted quite well, when the nonrandomness constants α_{ij} are set by the rules of Renon and Prausnitz and the binary interaction parameters $(g_{ij} - g_{jj})$ and $(g_{ji} - g_{ii})$ are computed from mutual solubility data for the partially miscible binary pairs. However, for multicomponent systems involving more than one completely miscible binary pair, predictions of liquid–liquid equilibrium, particularly in the plait-point region, are sensitive to values of α_{ij}, where i and j represent species of partially miscible binary pairs. Therefore, Renon et al.[57] recommended using ternary data to fix values of α_{ij}. Mertl,[54] however, suggested simultaneous evaluation of all sets of three NRTL binary parameters from binary data consisting of mutual solubilities and one vapor–liquid equilibrium point at the midcomposition range. In any case, for liquid–liquid equilibria, the NRTL equation must be treated as a three-parameter, rather than an effective two-parameter, equation.

Heidemann and Mandhane[75] and Katayama, Kato, and Yasuda[76] discovered complications that occasionally arise when the NRTL equation is used. The most serious problem is prediction of multiple miscibility gaps. For example, Heidemann and Mandhane cite the case of the n-butylacetate (1)–water (2) system. Experimental mutual solubilities are $x_1^I = 0.004564$ and $x_1^{II} = 0.93514$. NRTL parameters were computed from equilibrium data to be $\alpha_{12} = 0.391965$, $\tau_{12} = 3.00498$, and $\tau_{21} = 4.69071$. However, for these three parameters, two other sets of mutual solubilities satisfy the equilibrium conditions $\gamma_1^I x_1^I = \gamma_1^{II} x_1^{II}$ and $\gamma_2^I x_2^I = \gamma_2^{II} x_2^{II}$. These are

$$x_1^I = 0.004557 \qquad x_1^{II} = 0.59198$$

and

$$x_1^I = 0.59825 \qquad x_1^{II} = 0.93577$$

The existence of multiple miscibility gaps should be detected by following the procedure outlined by Heidemann and Mandhane. This includes seeking an alternate set of NRTL parameters that do not predict multiple sets of mutual solubilities.

UNIQUAC and UNIFAC Equations

The two adjustable UNIQUAC parameters $(u_{ij} - u_{jj})$ and $(u_{ji} - u_{ii})$ of (5-71) can be uniquely determined from mutual solubility data for each partially miscible binary pair. In this respect, the UNIQUAC equation has a distinct advantage over the three-parameter NRTL equation. Parameters for miscible binary pairs can be obtained in the usual fashion by fitting single or multiple vapor–liquid equilibrium data points, a pair of infinite-dilution activity coefficients, or an azeotropic condition. Prediction of ternary liquid–liquid equilibria from binary-pair data is reasonably good, even for systems containing only one partially miscible binary pair (1-2). For a ternary plait-point system, Abrams and Prausnitz[18] do not recommend that all six parameters be obtained by brute-force correlation when experimental ternary data are available. They suggest that for best estimates, binary-pair parameters for the two miscible binary pairs (1-3 and 2-3) be chosen so that binary vapor–liquid equilibrium data are reproduced within experimental uncertainty and, simultaneously, the experimental limiting liquid–liquid distribution coefficient $(K_D)_3 = (\gamma_3^\infty)^{II}/(\gamma_3^\infty)^I$ from ternary data is satisfied.

When accurate experimental data are available for a binary pair at several different temperatures under vapor–liquid and/or liquid–liquid conditions, plots of the best values for UNIQUAC binary interaction parameters appear to be smooth linear functions of temperature. Often, variations are small over moderate ranges of temperature as shown in Fig. 5.10 for ethanol–n-octane under vapor–liquid equilibrium conditions.

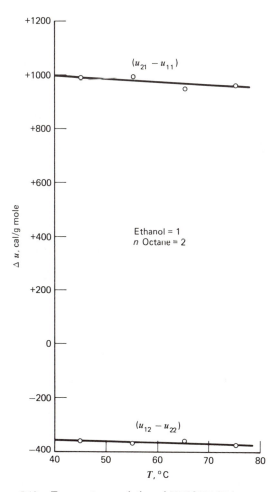

Figure 5.10. Temperature variation of UNIQUAC binary parameters for ethanol–n-octane system.

While not as quantitative for predicting liquid–liquid equilibrium as the UNIQUAC equation, the UNIFAC method can be used for order-of-magnitude estimates in the absence of experimental data. The UNIFAC method is almost always successful in predicting whether or not two liquid phases will form. Comparisons of experimental data with predictions by the UNIFAC method are given for a number of ternary systems by Fredenslund et al.[65] Two such comparisons are shown in Fig. 5.11.

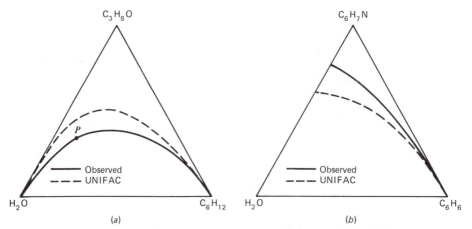

Figure 5.11. Comparison of UNIFAC predictions of liquid–liquid equilibrium with experimental data for two ternary systems. (*a*) Water–cyclohexane–2-propanol, type-I system. *P* = plait point. (*b*) Water–benzene–aniline, type-II system. (From A. Fredenslund, J. Gmehling, and P. Rasmussen, *Vapor–Liquid Equilibria Using UNIFAC, A Group Contribution Method*, Elsevier, Amsterdam, 1977.).

References

1. Prausnitz, J. M., W. C. Edmister, and K. C. Chao, *AIChE J.*, **6**, 214–219 (1960).

2. Hildebrand, J. H., J. M. Prausnitz, and R. L. Scott, *Regular and Related Solutions*, Van Nostrand Reinhold Company, New York, 1970.

3. Chao, K. C., and J. D. Seader, *AIChE J.*, **7**, 598–605 (1961).

4. Perry, R. H., and C. H. Chilton, Eds, *Chemical Engineers' Handbook*, 5th ed., McGraw-Hill Book Co., New York, 1973, 4–64.

5. Robinson, R. L., and K. C. Chao, *Ind. Eng. Chem., Process Des. Develop.*, **10**, 221–229 (1971).

6. Yerazunis, S., J. D. Plowright, and F. M. Smola, *AIChE J.*, **10**, 660–665 (1964).

7. Hermsen, R. W., and J. M. Praus-nitz, *Chem. Eng. Sci.*, **18**, 485–494 (1963).

8. Pitzer, K. S., D. Z. Lippman, R. F. Curl, Jr., C. M. Huggins, and D. E. Petersen, *J. Amer. Chem. Soc.*, **77**, 3433–3440 (1955).

9. Grayson, H. G., and C. W. Streed, Paper 20–P07, Sixth World Petroleum Conference, Frankfurt, June, 1963.

10. Lenoir, J. M., and C. R. Koppany, *Hydrocarbon Processing*, **46**(11), 249–252 (1967).

11. Glanville, J. W., B. H. Sage, and W. N. Lacey, *Ind. Eng. Chem.*, **42**, 508–513 (1950).

12. Edmister, W. C., C. L. Persyn, and J. H. Erbar, "Enthalpies of Hydro-carbon Mixtures in Coexisting Vapor–Liquid States," paper

presented at 42nd Annual Convention of NGPA, Houston, Texas, March 20–22, 1963.

13. Yarborough, L., and W. C. Edmister, *AIChE J.*, **11**, 492–497 (1965).

14. Waterman, W. W., and J. P. Frazier, *Hydrocarbon Processing*, **44**, 155–160 (1965).

15. Lo, C., *AIChE J.*, **18**, 866–867 (1972).

16. Ewell, R. H., J. M. Harrison, and L. Berg, *Ind. Eng. Chem.*, **36**, 871–875 (1944).

17. Butler, J. A. V., D. W. Thomson, and W. H. Maclennan, *J. Chem. Soc., Part I*, 674–686 (1933).

18. Abrams, D. S., and J. M. Prausnitz, *AIChE J.*, **21**, 116–128 (1975).

19. Van Ness, H. C., C. A. Soczek, and N. K. Kochar, *J. Chem. Eng. Data*, **12**, 346–351 (1967).

20. van Laar, J. J., *Z. Phys. Chem.*, **83**, 599–608 (1913).

21. Wohl, K., *Trans. AIChE*, **42**, 215–249 (1946).

22. Carlson, H. C., and A. P. Colburn, *Ind. Eng. Chem.*, **34**, 581–589 (1942).

23. Null, H. R., *Phase Equilibrium in Process Design*, Interscience Publishers, Inc., New York, 1970, 54–61.

24. Hála, E., I. Wichterle, J. Polák, and T. Boublik, *Vapour–Liquid Equilibrium Data at Normal Pressures*, Pergammon Press, Oxford, 1968.

25. Holmes, M. J., and M. Van Winkle, *Ind. Eng. Chem.*, **62**(1), 21–31 (1970).

26. Perry, R. H., and C. H. Chilton, Eds, *Chemical Engineers' Handbook*, 5th ed., McGraw–Hill Book Co., New York, 1973, 2-72 to 2-74.

27. Null, H. R., *Phase Equilibrium in Process Design*, Interscience Publishers, Inc., New York, 1970, 117–124.

28. Dunn, C. L., R. W. Millar, G. J. Pierotti, R. N. Shiras, and M. Souders, Jr., *Trans. AIChE*, **41**, 631–644 (1945).

29. Tassios, D. P., *Ind. Eng. Chem., Process Des. Develop.*, **11**, 43–46 (1972).

30. Drickamer, H. G., G. G. Brown, and R. R. White, *Trans. AIChE*, **41**, 555–605 (1945).

31. Sinor, J. E., and J. H. Weber, *J. Chem. Eng. Data*, **5**, 243–247 (1960).

32. Orye, R. V., and J. M. Prausnitz, *Ind. Eng. Chem.*, **57**(5), 18–26 (1965).

33. Wilson, G. M., *J. Amer. Chem. Soc.*, **86**, 127–130 (1964).

34. Cukor, P. M., and J. M. Prausnitz, *Instn. Chem. Eng. Symp. Ser. No. 32*, Instn. Chem. Engrs., London, **3**, 88 (1969).

35. Hála, E., *Collect. Czech. Chem. Commun.*, **37**, 2817–2819 (1972).

36. Brinkman, N. D., L. C. Tao, and J. H. Weber, *Ind. Eng. Chem., Fundam.*, **13**, 156–157 (1974).

37. Hudson, J. W., and M. Van Winkle, *Ind. Eng. Chem., Process Des. Develop.*, **9**, 466–472 (1970).

38. Gmehling, J., and U. Onken, "Vapor–Liquid Equilibrium Data Collection," *DECHEMA Chemistry Data Series*, **1–8**, Frankfurt (1977–1979).

39. Hirata, M., S. Ohe, and K. Nagahama, *Computer Aided Data Book of Vapor–Liquid Equilibria*, Elsevier, Amsterdam, 1975.

40. Nagata, I., *J. Chem. Eng. Japan*, **6**, 18–30 (1973).

41. Prausnitz, J. M., C. A. Eckert, R. V. Orye, and J. P. O'Connell, *Calculations for Multicomponent Vapor–Liquid Equilibria*, Prentice-Hall, Inc., Englewood Cliffs, N.J., 1967, Appendix A.

42. Null, H. R., *Phase Equilibrium in Process Design*, Interscience Publishers, Inc., New York, 1970, 147–151.

43. Tassios, D., *AIChE J.*, **17**, 1367–1371 (1971).

44. Schreiber, L. B., and C. A. Eckert, *Ind. Eng. Chem., Process Des. Develop.*, **10**, 572–576 (1971).

45. Duran, J. L., and S. Kaliaguine, *Can. J. Chem. Eng.*, **49**, 278–281 (1971).

46. Tai, T. B., R. S. Ramalho, and S. Kaliaguine, *Can. J. Chem. Eng.*, **50**, 771–776 (1972).

47. Jones, H. K. D., and B. C. Y. Lu, *J. Chem. Eng. Data*, **11**, 488–492 (1966).

48. Smith, V. C., and R. L. Robinson, Jr., *J. Chem. Eng. Data*, **15**, 391–395 (1970).

49. Nagata, I., and T. Yamada, *Ind. Eng. Chem., Process Des. Develop.*, **11**, 574–578 (1972).

50. Hiranuma, M., *J. Chem. Eng. Japan*, **8**, 162–163 (1957).

51. Renon, H., and J. M. Prausnitz, *AIChE J.*, **14**, 135–144 (1968).

52. Renon, H., and J. M. Prausnitz, *Ind. Eng. Chem., Process Des. Develop.*, **8**, 413–419 (1969).

53. Larson, C. D., and D. P. Tassios, *Ind. Eng. Chem., Process Des. Develop.*, **11**, 35–38 (1972).

54. Mertl, I., *Collect. Czech. Chem. Commun.*, **37**, 375–411 (1972).

55. Marina, J. M., and D. P. Tassios, *Ind. Eng. Chem., Process Des. Develop.*, **12**, 67–71 (1973).

56. Tsuboka, T., and T. Katayama, *J. Chem. Eng. Japan*, **8**, 181–187 (1975).

57. Renon, H., L. Asselineau, G. Cohen, and C. Raimbault, *Calculsur Ordinateur des Equilibres Liquide-Vapeur et Liquide-Liquide*, Technip, Paris, 1971.

58. Bruin, S., and J. M. Prausnitz, *Ind. Eng. Chem., Process Des. Develop.*, **10**, 562–572 (1971).

59. Prausnitz, J. M., et al., "Computer Calculations for Multicomponent Vapor–Liquid and Liquid–Liquid Equilibria", Prentice-Hall, Inc., Englewood Cliffs, N.J., 1980.

60. Fredenslund, A., J. Gmehling, M. L. Michelsen, P. Rasmussen, and J. M. Prausnitz, *Ind. Eng. Chem., Process Des. Develop.*, **16**, 450–462 (1977).

61. Abrams, D. S., Ph.D. thesis in chemical engineering, University of California, Berkeley, 1974.

62. Wilson, G. M., and C. H. Deal, *Ind. Eng. Chem., Fundam.*, **1**, 20–23 (1962).

63. Derr, E. L., and C. H. Deal, *Instn. Chem. Eng. Symp. Ser. No. 32*, Instn. Chem. Engrs., London, **3**, 40–51 (1969).

64. Fredenslund, A., R. L. Jones, and J. M. Prausnitz, *AIChE J.*, **21**, 1086–1099 (1975).

65. Fredenslund, A., J. Gmehling, and P. Rasmussen, *Vapor–Liquid Equilibria Using UNIFAC, A Group Contribution Method*, Elsevier, Amsterdam, 1977.

66. Prausnitz, J. M., "Phase Equilibria and Fluid Properties in the Chemical Industry, Estimation and Correlation," in T. S. Storvick and S. I. Sandler, Eds, *ACS Symp. Series 60*, 1977, 43.

67. Takeuchi, S., T. Nitta, and T. Katayama, *J. Chem. Eng. Japan*, **8**, 248–250 (1975).

68. Strubl, K., V. Svoboda, R. Holub, and J. Pick, *Collect. Czech. Chem. Commun.*, **35**, 3004–3019 (1970).

69. Kiser, R. W., G. D. Johnson, and M. D. Shetlar, *J. Chem. Eng. Data*, **6**, 338–341 (1961).

70. Prausnitz, J. M., *Molecular Thermodynamics of Fluid-Phase Equilibria*, Prentice–Hall, Inc., Englewood Cliffs, N.J., 1969.

71. Brian, P. L. T., *Ind. Eng. Chem., Fundam.*, **4**, 100–101 (1965).

72. Joy, D. S., and B. G. Kyle, *AIChE J.*, **15**, 298–300 (1969).

73. Hildebrand, J. H., and R. L. Scott, *The Solubility of Nonelectrolytes*, 3rd ed., Reinhold Publishing Corporation, New York, 1950.

74. Guffey, C. G., and A. H. Wehe, *AIChE J.*, **18**, 913–922 (1972).

75. Heidemann, R. A., and J. M. Mandhane, *Chem. Eng. Sci.*, **28**, 1213–1221 (1973).

76. Katayama, T., M. Kato, and M. Yasuda, *J. Chem. Eng. Japan*, **6**, 357–359 (1973).

Problems

5.1 For the conditions of Example 5.2, calculate γ_{iL}, v_{iL}^o, ϕ_{iV}, and K_i for propane. What is the deviation of K_i from the experimental value?

5.2 Derive Eq. (5-19).

5.3 Derive Eq. (5-7).

5.4 For the conditions of Problem 4.31:
(a) Calculate the liquid-phase activity coefficients using regular solution theory with the Flory–Huggins correction.
(b) Calculate the K-values using the C–S correlation and compare to the experimental values.
(c) Calculate the enthalpy of the liquid phase in kilojoules per kilogram-mole from (5-14).
(d) Calculate the specific volume of the liquid phase in cubic meters per kilogram-mole from (5-19).

5.5 Repeat Problem 4.19 using the Chao–Seader correlation.

5.6 Chao [Ph.D. thesis, Univ. Wisconsin (1956)] obtained the following vapor–liquid equilibrium data for the benzene–cyclohexane system at 1 atm.

T, °C	$x_{benzene}$	$y_{benzene}$	$\gamma_{benzene}$	$\gamma_{cyclohexane}$
79.7	0.088	0.113	1.300	1.003
79.1	0.156	0.190	1.256	1.008
78.5	0.231	0.268	1.219	1.019
78.0	0.308	0.343	1.189	1.032
77.7	0.400	0.422	1.136	1.056
77.6	0.470	0.482	1.108	1.075
77.6	0.545	0.544	1.079	1.102
77.6	0.625	0.612	1.058	1.138
77.8	0.701	0.678	1.039	1.178
78.0	0.757	0.727	1.025	1.221
78.3	0.822	0.791	1.018	1.263
78.9	0.891	0.863	1.005	1.328
79.5	0.953	0.938	1.003	1.369

Note that an azeotrope is formed at 77.6°C.

(a) Use the data to calculate and plot the relative volatility of benzene with respect to cyclohexane-versus-benzene composition in the liquid phase.

(b) Use the Chao–Seader correlation to predict α and plot the predictions on the plot of Part (a). Does the Chao–Seader correlation predict an azeotrope?

5.7 Hermsen and Prausnitz [*Chem. Eng. Sci.*, **18**, 485 (1963)] show that the mixture benzene–cyclopentane does not form a regular solution. Confirm this by computing liquid-phase activity coefficients for this system at 35°C, over the entire composition range, from regular solution theory, with and without the Flory–Huggins correction, and prepare a plot of your results similar to Fig. 5.1, compared to the following experimental values of Hermsen and Prausnitz.

$x_{benzene}$	$\gamma_{benzene}$	$\gamma_{cyclopentane}$
0.0	1.5541	1.0000
0.1	1.4218	1.0047
0.2	1.3150	1.0186
0.3	1.2294	1.0416
0.4	1.1611	1.0742
0.5	1.1075	1.1165
0.6	1.0664	1.1693
0.7	1.0362	1.2333
0.8	1.0157	1.3096
0.9	1.0038	1.3993
1.0	1.0000	1.5044

5.8 Use the C–S correlation to predict the liquid molal volume of an equimolal mixture of benzene and n-heptane at 25°C. Compare your result to that obtained from the following form of (5-18).

$$v_L = v_L^E + \sum_{i=1}^{C} x_i v_{iL}$$

where values of v_{iL} are taken from Appendix I and the value of v_L^E is taken as 0.57 cm³/gmole [*Ind. Eng. Chem. Fundam.*, **11**, 387 (1972)].

5.9 At −175°F and 25 psia, the experimental K-values for methane and propane in a binary mixture are 8.3 and 0.0066, respectively. Assuming that the vapor forms an ideal solution, estimate the liquid-phase activity coefficients for methane and propane. Do not assume that the gas phase obeys the ideal gas law. At − 175°F, the vapor pressures for methane and propane are 220 and 0.097 psia, respectively.

5.10 Use (5-23) with regular solution theory to predict K-values for the n-octane–ethylbenzene system at 270.3°F and 1 atm with experimental compositions of $x_{C_8} = 0.201$ and $y_{C_8} = 0.275$ [*Ind. Eng. Chem.*, **47**, 293 (1955)]. Compare your results with the experimental values computed from the experimental compositions.

5.11 Prove that the van Laar equation cannot predict activity coefficient curves with maxima of the type shown in Fig. 5.2c.

5.12 From the experimental infinite-dilution activity coefficients given in Problem 5.7 for the benzene–cyclopentane system, calculate the constants in the van Laar equation. With these constants, use the van Laar equation to compute the activity coefficients over the entire range of composition and compare them in a plot like Fig. 5.1 with the experimental points of Problem 5.7.

5.13 From the azeotropic composition for the benzene–cyclohexane system in Problem 5.6, calculate the constants in the van Laar equation. With these constants, use the van Laar equation to compute the activity coefficients over the entire range of composition and compare in a plot like Fig. 5.1 with the experimental data of Problem 5.6.

5.14 From Hála et al.,[24] the van Laar constants for the system n-hexane(1)-benzene(2) at approximately 70°C are determined to be $A_{12} = 0.6226$ and $A_{21} = 0.2970$.
(a) Predict the van Laar constants at 25°C.
(b) Predict the excess enthalpy at 25°C over the entire composition range and compare your values with the following experimental values of Jones and Lu [*J. Chem. Eng. Data*, **11**, 488 (1966)].

x_2	H^E, cal/gmole
0.1519	102.9
0.2136	129.8
0.3575	185.5
0.4678	209.7
0.5242	213.5
0.6427	207.5
0.6497	205.2
0.6957	196.1
0.7211	189.3
0.8075	155.3
0.9157	82.8

5.15 At 45°C, the van Laar constants for two of the pairs in the ternary system n-hexane(1)-isohexane(2)-methyl alcohol(3) are: $A_{13} = 2.35$, $A_{31} = 2.36$, $A_{23} = 2.14$, and $A_{32} = 2.22$. Assume that the two hexane isomers form an ideal solution. Use (5-33), the multicomponent form of the van Laar equation, to predict the liquid-phase activity coefficients of an equimolal mixture of the three components at 45°C. It is possible that application of the van Laar equation to this system may result in erroneous prediction of two liquid phases.

5.16 The Wilson constants for the ethanol(1)-benzene(2) system at 45°C are: $\Lambda_{12} = 0.124$ and $\Lambda_{21} = 0.523$. Use these constants with the Wilson equation to predict the liquid-phase activity coefficients for this system over the entire range of composition and compare them in a plot like Fig. 5.5 with following experimental results [*Aust. J. Chem.*, **7**, 264 (1954)].

x_1	$\ln \gamma_1$	$\ln \gamma_2$
0.0374	2.0937	0.0220
0.0972	1.6153	0.0519
0.3141	0.7090	0.2599
0.5199	0.3136	0.5392
0.7087	0.1079	0.8645
0.9193	0.0002	1.3177
0.9591	−0.0077	1.3999

5.17 At 1 atm (101.3 kPa), the acetone(1)-chloroform(2) system exhibits negative deviations from Raoult's law and a maximum-boiling azeotrope. Compute the constants
Λ_{12} and Λ_{21} in the Wilson equation for this system from the following data.
(a) Infinite-dilution activity coefficients $\gamma_1^\infty = 0.37$ and $\gamma_2^\infty = 0.46$.
(b) Azeotropic composition $x_1 = 0.345$, at 64.5°C.

From the results of Part (a), use the Wilson equation to predict activity
coefficients over the entire range of composition and plot the values after the
manner of Fig. 5.5. Repeat the calculations and plot using the results of Part (b).
Compare the two sets of predictions.

Use the results of Part (b) to predict and plot y–x and T–y–x equilibrium
curves for this system at 1 atm.

5.18 For the system methylethyl ketone (MEK)-n-hexane(H) at 60°C, the infinite-
dilution activity coefficient γ_{MEK}^∞ has been measured by chromatography to be 3.8.
(a) Predict γ_H^∞ and the γ_{MEK} and γ_H curves with composition by the single-
parameter modification of the Wilson equation.
(b) Compare your predictions with experimental data [*J. Chem. Eng. Data*, **12**, 319
(1967)].
(c) Use your predictions to calculate and plot y–x and T–y–x equilibrium curves
for this system at 1 atm.

5.19 For the system ethanol(1)-benzene(2), the Wilson constants at 45°C are: $\Lambda_{12} = 0.124$
and $\Lambda_{21} = 0.523$.
(a) Predict the Wilson constants at 25°C.
(b) Predict the excess enthalpy at 25°C over the entire composition range and
compare your values in a plot of H^E versus x_1 with the following experimental
values [*J. Chem. Eng. Data*, **11**, 480 (1966)].

x_1	H^E, cal/gmole
0.0250	75.2
0.0586	112.0
0.1615	190.0
0.2673	216.5
0.4668	197.0
0.5602	168.4
0.7074	114.8
0.9058	37.8

5.20 Infinite-dilution activity coefficients for the system methylacetate(1)-methanol(2) at
50°C are reported to be $\gamma_1^\infty = 2.79$ and $\gamma_2^\infty = 3.02$ [*Ind. Eng. Chem., Process Des.
Develop.*, **10**, 573 (1971)]. Using a value of $\alpha_{ji} = 0.30$, predict the values of τ_{12} and
τ_{21} in the NRTL equation. Compare your results with literature values of $\tau_{12} =
0.5450$ and $\tau_{21} = 0.5819$ [*Collect. Czech. Chem. Commun.*, **37**, 375 (1972)].

5.21 The NRTL constants for the chloroform(1)-methanol(2) system at 1 atm (53.5 to
63.0°C) are $\alpha_{12} = 0.30$, $\tau_{12} = 2.1416$, and $\tau_{21} = -0.1998$ [*Collect. Czech. Chem.
Commun.*, **37**, 375 (1972)]. Use these constants with the NRTL equation to predict
and plot α–x, y–x, and T–y–x equilibrium curves at 1 atm. Compare your results
with the following experimental data [*J. Chem. Eng. Data*, **7**, 367 (1962)].

x_1	y_1	$T, °C$
0.0400	0.1020	63.0
0.0950	0.2150	60.9
0.1960	0.3780	57.8
0.2870	0.4720	55.9
0.4250	0.5640	54.3
0.65	0.65	53.5 (azeotrope)
0.7970	0.7010	53.9
0.9040	0.7680	55.2
0.9700	0.8750	57.9

5.22 The UNIQUAC parameters for the acetone(1)-water(2) system at 1 atm are $(u_{21} - u_{11}) = -120.2$ cal/gmole, $(u_{12} - u_{22}) = 609.1$ cal/gmole, $r_1 = 2.57$, $r_2 = 0.92$, $q_1 = 2.34$, and $q_2 = 1.40$ [tabulations of J. M. Prausnitz as prepared by A. O. Lau (1975)]. Use these parameters with the UNIQUAC equation to predict and plot $\alpha-x$, $y-x$, and $T-y-x$ equilibrium curves at 1 atm. Compare your results with the following experimental data [*Ind. Eng. Chem.*, 37, 299 (1945)].

x_1	y_1	$T, °C$
0.015	0.325	89.6
0.036	0.564	79.4
0.074	0.734	68.3
0.175	0.800	63.7
0.259	0.831	61.1
0.377	0.840	60.5
0.505	0.849	59.9
0.671	0.868	59.0
0.804	0.902	58.1
0.899	0.938	57.4

5.23 An azeotropic condition for the system carbon tetrachloride(1)-n-propyl alcohol(2) occurs at $x_1 = y_1 = 0.75$, $T = 72.8°C$, and $P = 1$ atm. Predict the UNIQUAC equation constants $(u_{21} - u_{11})$ and $(u_{12} - u_{22})$, if $r_1 = 3.45$, $r_2 = 2.57$, $q_1 = 2.91$, and $q_2 = 3.20$. Compare your predictions to the following values obtained by fitting data over the entire range of composition: $(u_{21} - u_{11}) = -496.8$ cal/gmole and $(u_{12} - u_{22}) = 992.0$ cal/gmole [tabulations of J. M. Prausnitz as prepared by A. O. Lau (1975)].

5.24 At 20°C, estimate with the UNIFAC method the liquid-phase activity coefficients, equilibrium vapor composition, and total pressures for 25 mole% liquid solutions of the following hydrocarbons in ethanol.
(a) n-Pentane.
(b) n-Hexane.
(c) n-Heptane.
(d) n-Octane.
Necessary group-contribution parameters are given in Example 5.10.

5.25 For the binary system ethanol(1)-isooctane(2) at 50°C

$$\gamma_1^\infty = 21.17$$

$$\gamma_2^\infty = 9.84$$

Calculate the constants.
(a) A_{12} and A_{21} in the van Laar equations.
(b) Λ_{12} and Λ_{21} in the Wilson equations.
(c) Using these constants, calculate γ_1 and γ_2 over the entire composition range and plot the calculated points as log γ versus x_1.
(d) How well do the van Laar and Wilson predictions agree with the azeotropic point where

$$x_1 = 0.5941 \qquad \gamma_1 = 1.44$$

$$x_2 = 0.4059 \qquad \gamma_2 = 2.18$$

(e) Show that the van Laar equation erroneously predicts separation into two liquid phases over a portion of the composition range by calculating and plotting a y–x diagram like Fig. 5.9.

5.26 Mutual solubility data for the isooctane(1)-furfural(2) system at 25°C are [*Chem. Eng. Sci.*, **6**, 116 (1957)]:

	Liquid-Phase I	Liquid-Phase II
x_1	0.0431	0.9461
x_2	0.9569	0.0539

Predict and plot liquid-phase activity coefficients over the entire range of solubility using the NRTL equation with $\alpha_{12} = 0.30$.

5.27 Infinite-dilution, liquid-phase activity coefficients at 25°C are [*Chem. Eng. Sci.*, **6**, 116 (1957)]:

$$\gamma_3^\infty = 3.24 \qquad \gamma_1^\infty = 3.10 \text{ for the benzene(3)-isooctane(1) system}$$

and

$$\gamma_3^\infty = 2.51 \qquad \gamma_2^\infty = 3.90 \text{ for the benzene(3)-furfural(2) system}$$

Use this data, together with that of Problem 5.26, to predict by the van Laar equation the compositions and amounts of each of the two equilibrium liquid phases formed from 100, 100, and 50 kgmoles of isooctane, furfural, and benzene, respectively.

5.28 For binary mixtures that fit the van Laar equation such that $A_{12} = A_{21}$ in (5-26), determine the minimum value of A_{12} for the formation of two liquid phases. What is the mole fraction range of immiscibility if $A_{12} = A_{21} = 2.76$?

6

Specification of Design Variables

In the design of processes for physical separation of components by mechanisms involving mass and heat transfer, the first step usually consists of specification of process conditions or independent variables.

Mooson Kwauk, 1956

The solution to a multicomponent, multiphase, multistage separation problem is found in the simultaneous or iterative solution of, literally, hundreds of equations. This implies that a sufficient number of design variables is specified so that the number of unknown (output) variables exactly equals the number of (independent) equations. When this is done, a separation process is uniquely specified. If an incorrect number of design variables is chosen, multiple or inconsistent solutions or no solution at all will be found.

The computational difficulties attending the solution of large sets of frequently nonlinear equations are such that a judicious choice of design variables frequently ameliorates computational obstacles. In practice, however, the designer is not free to choose the design variables on the basis of computational convenience. More commonly he is confronted with a situation where the feed composition, the number of stages, and/or the product specifications are fixed and he must suitably arrange the equations so they can be solved.

An intuitively simple, but operationally complex, method of finding N_D, the number of *independent design variables, degrees of freedom*, or *variance* in the process, is to enumerate all pertinent variables N_V and to subtract from these the total number of independent equations N_E relating the variables.

$$N_D = N_V - N_E \qquad (6\text{-}1)$$

This approach to separation process design was developed by Kwauk,[1] and a modification of his methodology forms the basis for this chapter.

Typically, the variables in a separation process can be *intensive variables* such as composition, temperature, or pressure; *extensive variables* such as flow rate or heat transferred; or equipment parameters such as the number of equilibrium stages. Physical properties such as enthalpy or K-values are not counted. The variables are relatively easy to enumerate; but to achieve an unambiguous count of N_E it is necessary to carefully seek out all independent relationships due to material and energy conservations, phase equilibrium restrictions, process specifications, and equipment configurations.

Separation equipment consists of physically identifiable elements (equilibrium stages, condensers, reboilers, etc.) as well as stream dividers and stream mixers. It is helpful to examine each element separately, prior to synthesizing the complete system.

6.1   Stream Variables

For each single-phase stream containing C components, a complete specification of intensive variables consists of $C - 1$ mole fractions (or other concentration variables) plus temperature and pressure. This follows from the phase rule, which states that, for a single-phase system, the intensive variables are specified by $C - \mathscr{P} + 2 = C + 1$ variables. To this number can be added the total flow rate, an extensive variable. Finally, although the missing mole fractions are often treated implicitly, it is preferable for completeness to include these missing mole fractions in the list of stream variables and then to include in the list of equations the mole fraction constraint

$$\sum_{i=1}^{C} \text{mole fractions} = 1.0$$

Thus, associated with each stream are $C + 3$ variables. For example, for a liquid-phase stream, the variables might be

Liquid mole fractions $x_1, x_2, \ldots, x_C$.

Total molal flow rate L.

Temperature T.

Pressure P.

6.2   Adiabatic Equilibrium Stage

For a single adiabatic equilibrium stage with two entering streams and two exit streams, as shown in Fig. 6.1, the only variables are those associated with the streams. Thus

$$N_V = 4(C + 3) = 4C + 12$$

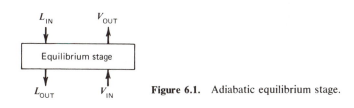

Figure 6.1. Adiabatic equilibrium stage.

The exit streams V_{OUT} and L_{OUT} are in equilibrium so there are equilibrium restrictions as well as component material balances, a total material balance, an enthalpy balance, and mole fraction constraints. Thus N_E, the number of equations relating these variables, is

Equations	Number of Equations
Pressure equality,	1
$P_{V_{OUT}} = P_{L_{OUT}}$	
Temperature equality,	1
$T_{V_{OUT}} = T_{L_{OUT}}$	
Phase equilibrium relationships,	C
$(y_i)_{V_{OUT}} = K_i (x_i)_{L_{OUT}}$	
Component material balances,	$C - 1$
$L_{IN}(x_i)_{L_{IN}} + V_{IN}(y_i)_{V_{IN}} = L_{OUT}(x_i)_{L_{OUT}} + V_{OUT}(y_i)_{V_{OUT}}$	
Total material balance,	1
$L_{IN} + V_{IN} = L_{OUT} + V_{OUT}$	
Adiabatic enthalpy balance,	1
$H_{L_{IN}}L_{IN} + H_{V_{IN}}V_{IN} = H_{L_{OUT}}L_{OUT} + H_{V_{OUT}}V_{OUT}$	
Mole fraction constraints,	4
e.g. $\sum_{i=1}^{C}(x_i)_{L_{IN}} = 1.0$	$N_E = 2C + 7$

Alternatively, C component material balances can be written. The total material balance is then a dependent equation obtained by summing the component material balances and applying the mole fraction constraints to eliminate the mole fractions. From (6-1)

$$N_D = (4C + 12) - (2C + 7) = 2C + 5$$

Several different sets of design variables can be specified. A typical set includes complete specification of the two entering streams as well as the stage pressure.

Variable Specification	Number of Variables
Component mole fractions, $(x_i)_{L_{IN}}$	$C - 1$
Total flow rate, L_{IN}	1
Component mole fractions, $(y_i)_{V_{IN}}$	$C - 1$
Total flow rate, V_{IN}	1
Temperature and pressure of L_{IN}	2
Temperature and pressure of V_{IN}	2
Stage pressure ($P_{V_{OUT}}$ or $P_{L_{OUT}}$)	1
	$\overline{N_D = 2C + 5}$

Specification of these $(2C + 5)$ variables permits calculation of the unknown variables L_{OUT}, V_{OUT}, $(x_C)_{L_{IN}}$, $(y_C)_{V_{IN}}$, all $(x_i)_{L_{OUT}}$, T_{OUT}, and all $(y_i)_{V_{OUT}}$, where C denotes the missing mole fraction.

6.3 Equilibrium Stage with Heat Addition, Feed Stream, and Sidestream

A more complex equilibrium stage is shown in Fig. 6.2. The feed stream has no variable values in common with L_{IN} and V_{OUT}, but the liquid side stream shown leaving the stage is identical in composition, T, and P to L_{OUT} though different in flow rate. Heat can be transferred to or from the stage at the rate Q (where a positive value denotes addition of heat to the stage). The number of total variables (including Q) is

$$N_V = 6(C + 3) + 1 = 6C + 19$$

The equations for this element are similar to those for an adiabatic equilibrium stage. But, in addition, component mole fractions of L_{OUT} and side stream S are identical. This situation is handled by $C - 1$ mole fraction equalities with the missing mole fractions accounted for by the usual mole fraction constraints.

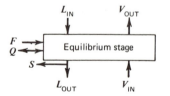

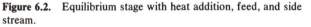

Figure 6.2. Equilibrium stage with heat addition, feed, and side stream.

Equations	Number of Equations
Pressure equalities,	2
$P_{V_{OUT}} = P_{L_{OUT}} = P_S$	
Temperature equalities,	2
$T_{V_{OUT}} = T_{L_{OUT}} = T_S$	
Phase equilibrium relationships,	C
$(y_i)_{V_{OUT}} = K_i(x_i)_{L_{OUT}}$	
Component material balances,	$C - 1$
$L_{IN}(x_i)_{L_{IN}} + V_{IN}(y_i)_{V_{IN}} + F(x_i)_F$	
$\quad = L_{OUT}(x_i)_{L_{OUT}} + S(x_i)_S + V_{OUT}(y_i)_{V_{OUT}}$	
Total material balance,	1
$F + L_{IN} + V_{IN} = L_{OUT} + S + V_{OUT}$	
Enthalpy balance,	1
$H_F F + H_{L_{IN}} L_{IN} + H_{V_{IN}} V_{IN} + Q = H_{L_{OUT}} L_{OUT}$	
$\quad + H_S S + H_{V_{OUT}} V_{OUT}$	
Mole fraction constraints,	6
e.g. $\displaystyle\sum_{i=1}^{C} (x_i)_S = 1.0$	
Mole fraction equalities,	$C - 1$
$(x_i)_{L_{OUT}} = (x_i)_S$	$\overline{N_E = 3C + 10}$

From (6-1)

$$N_D = (6C + 19) - (3C + 10) = 3C + 9$$

A typical set of design variables is as follows. Many other sets are possible.

Variable Specification	Number of Variables
Component mole fractions, $(x_i)_{L_{IN}}$	$C - 1$
Total flow rate, L_{IN}	1
Component mole fractions, $(y_i)_{V_{IN}}$	$C - 1$
Total flow rate, V_{IN}	1
Component mole fractions, $(x_i)_F$	$C - 1$
Total flow rate, F	1
T and P of L_{IN}, V_{IN}, F	6
Stage pressure ($P_{V_{OUT}}$, $P_{L_{OUT}}$, or P_S)	1
Stage temperature ($T_{V_{OUT}}$, $T_{L_{OUT}}$, or T_S)	1
Total side stream flow rate, S	1
	$\overline{3C + 9}$

These specifications differ from those given previously for an adiabatic stage in that the required heat transfer rate is an output variable. Alternatively, the heat transfer rate Q can be specified with stage temperature treated as an output variable. Also, an algebraic combination of variables can be specified in place of a single variable—for example, a value for S/L_{OUT} instead of a value for the total side stream flow rate S.

6.4 Condenser and Boiler

Figure 6.3 shows a boiler; if the flows of heat and mass are reversed, it is a condenser. Complete vaporization or condensation is assumed. The number of variables is

$$N_V = 2(C+3) + 1 = 2C + 7$$

The equations are

Equations	Number of Equations
Component material balances	$C - 1$
Total material balance	1
Enthalpy balance	1
Mole fraction constraints	2
	$N_E = C + 3$

The degrees of freedom are

$$N_D = (2C + 7) - (C + 3) = C + 4$$

By specifying, for example, $C - 1$ input flow component mole fractions, the total input flow rate, and T and P of both output and input streams, one can calculate Q.

If only partial condensation or vaporization occurs, equipment schematic diagrams are as shown in Fig. 6.4a and 6.4b.

The analysis for either element is identical. The total variables equal

$$N_V = 3(C + 3) + 1 = 3C + 10$$

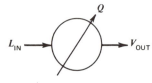

Figure 6.3.   Boiler.

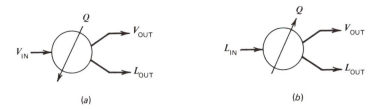

Figure 6.4. Partial condenser and vaporizer. (a) Partial condenser, (b) Partial vaporizer.

Relations among the variables include the following, where V_{OUT} and L_{OUT} are in equilibrium.

Equations	Number of Equations
Component material balances	$C - 1$
Total material balance	1
Enthalpy balance	1
Pressure equality, $P_{V_{OUT}} = P_{L_{OUT}}$	1
Temperature equality, $T_{V_{OUT}} = T_{L_{OUT}}$	1
Phase equilibrium relationships,	C
$\quad (y_i)_{V_{OUT}} = K_i(x_i)_{L_{OUT}}$	
Mole fraction constraints	3
	$N_E = 2C + 6$

Thus $N_D = C + 4$, which is identical to the result obtained for the total condenser or boiler.

6.5 Mixer, Divider, and Splitter

For mixers, dividers, and splitters involving three streams, as shown in Fig. 6.5, $N_V = 3(C + 3) + 1 = 3C + 10$. Equations for the mixer include $(C - 1)$ component material balances, a total material balance, an enthalpy balance, and three mole fraction constraints.

$$N_D = 3(C + 3) + 1 - (C + 4) = 2C + 6$$

Typical variable specifications are feed conditions for both inlet streams ($2C + 4$ variables), outlet stream pressure, and Q. All three streams are assumed to be of the same phase state (vapor or liquid).

In the stream divider, the relations include $[2(C - 1)]$ mole fraction equalities because all three streams have the same composition. There are also

Table 6.1 Degrees of freedom for separation operation elements and units

Schematic	Element or Unit Name	N_V, Total Variables	N_E, Independent Relationships	N_D, Degrees of Freedom
	Total boiler (Reboiler)	$(2C+7)$	$(C+3)$	$(C+4)$
	Total condenser	$(2C+7)$	$(C+3)$	$(C+4)$
	Partial (equilibrium) boiler (reboiler)	$(3C+10)$	$(2C+6)$	$(C+4)$
	Partial (equilibrium) condenser	$(3C+10)$	$(2C+6)$	$(C+4)$
	Adiabatic equilibrium stage	$(4C+12)$	$(2C+7)$	$(2C+5)$
	Equilibrium stage with heat transfer	$(4C+13)$	$(2C+7)$	$(2C+6)$
	Equilibrium feed stage with heat transfer and feed	$(5C+16)$	$(2C+8)$	$(3C+8)$

Unit	Diagram			
Equilibrium stage with heat transfer and side stream	(streams: Q, L_{in}, L_{out}, V_{out}, V_{in}, $^a S$)	$(5C+16)$	$(3C+9)$	$(2C+7)$
N connected equilibrium stages with heat transfer	(streams: Q_N, Q_{N-1}, Q_2, Q_1, Stage N ... Stage 1, L_{in}, L_{out}, V_{out}, V_{in})	$(7N+2NC+2C+7)$	$(5N+2NC+2)$	$(2N+2C+5)$
Stream mixer	(streams: Q, L_3, L_1, L_2)	$(3C+10)$	$(C+4)$	$(2C+6)$
Stream divider	(streams: L_2, L_3, Q, $^b L_1$)	$(3C+10)$	$(2C+5)$	$(C+5)$
Splitter, composition of $L^I_{OUT} \neq L^{II}_{OUT}$	(streams: L^I_{OUT}, Q, L^{II}_{OUT}, $^c L_{in}$)	$(3C+10)$	$(C+4)$	$(2C+6)$

[a] Side stream can be vapor or liquid.   [b] Alternatively, all streams can be vapor.   [c] Any stream can be vapor.

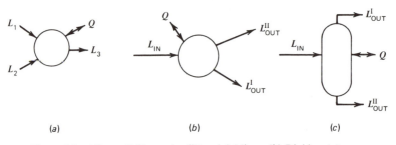

Figure 6.5.   Mixer, divider and splitter. (*a*) Mixer. (*b*) Divider. (*c*) Splitter.

pressure and temperature identities between L_{OUT}^{II} and L_{OUT}^{I}, plus one total material balance, one enthalpy balance, and three mole fraction constraints. Thus

$$N_D = (3C + 10) - (2C + 5) = C + 5$$

Variable specifications for the stream divider are typically $(C + 2)$ feed variables, outlet temperature (or heat transfer rate) and outlet pressure, and L_{OUT}^{II}, L_{OUT}^{I}/L_{OUT}^{II}, L_{OUT}^{I}/L_{IN}, and so on.

In the splitter, composition of L_{OUT}^{I} is not equal to L_{OUT}^{II} or L_{IN}. Also, stream compositions are not related by equilibrium constraints, nor need the outlet streams be at the same temperature or pressure. Thus, the only relationships are $(C - 1)$ component balances, one total material balance, an enthalpy balance, and three mole fraction constraints, so that

$$N_D = (3C + 10) - (C + 4) = 2C + 6$$

Examples of splitters are devices in which nonequilibrium separations are achieved by means of membranes, electrical fields, temperature changes, and others. The splitter can also be used to model any multistage chemical separator where stage details are not of interest.

Table 6.1 is a summary of the degrees of freedom in representative building-block elements for separation operations.

6.6   Combinations of Elements by an Enumeration Algorithm

An algorithm is easily developed for enumerating variables, equations, and degrees of freedom for combinations of elements to form units. The number of design variables for a separator (e.g., a distillation column) is obtained by summing the variables associated with the individual equilibrium stages, heat exchangers, and other elements *e* that comprise the separator. However, care must be taken to subtract from the total variables the $(C+3)$ variables for each of the N_R redundant interconnecting streams that arise when the output of one process element becomes the input to another. Also, if an unspecified number of repetitions of any element occurs within the unit, an additional variable is added, one for each group of repetitions, giving a total of N_A additional variables. In addition, N_R redundant mole fraction constraints are subtracted after summing the independent relationships of the individual elements. The number of degrees of freedom is obtained as before, from (6-1). Thus

$$(N_V)_{\text{unit}} = \sum_{\substack{\text{all} \\ \text{elements, } e}} (N_V)_e - N_R(C+3) + N_A \tag{6-2}$$

$$(N_E)_{\text{unit}} = \sum_{\substack{\text{all} \\ \text{elements, } e}} (N_E)_e - N_R \tag{6-3}$$

Combining (6-1), (6-2), and (6-3), we have

$$(N_D)_{\text{unit}} = \sum_{\substack{\text{all} \\ \text{elements, } e}} (N_D)_e - N_R(C+2) + N_A \tag{6-4}$$

or

$$(N_D)_{\text{unit}} = (N_V)_{\text{unit}} - (N_E)_{\text{unit}} \tag{6-5}$$

For the N-stage cascade unit of Fig. 6.6, with reference to the single

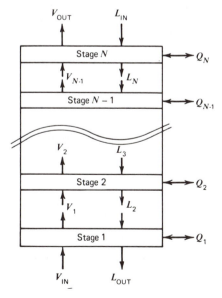

Figure 6.6. An N-stage cascade.

equilibrium stage with heat transfer in Table 6.1, the total variables from (6-2) are

$$(N_V)_{unit} = N(4C + 13) - [2(N - 1)](C + 3) + 1 = 7N + 2NC + 2C + 7$$

since $2(N - 1)$ interconnecting streams exist. The additional variable is the total number of stages (i.e., $N_A = 1$).

The number of independent relationships from (6-3) is

$$(N_E)_{unit} = N(2C + 7) - 2(N - 1) = 5N + 2NC + 2$$

since $2(N - 1)$ redundant mole fraction constraints exist.

The degrees of freedom from (6-5) are

$$(N_D)_{unit} = N_V - N_E = 2N + 2C + 5$$

One possible set of design variables is

Variable Specification	Number of Variables
Heat transfer rate for each stage (or adiabatic)	N
Stage pressures	N
Stream V_{IN} variables	$C + 2$
Stream L_{IN} variables	$C + 2$
Number of stages	$\dfrac{1}{2N + 2C + 5}$

Output variables for this specification include missing mole fractions for V_{IN} and L_{IN}, stage temperatures, and the variables associated with the V_{OUT} stream, L_{OUT} stream, and interstage streams. The results obtained in this example are included in Table 6.1. The N-stage cascade unit can represent simple absorbers, strippers, or liquid–liquid extractors.

6.7 Description Rule

An attractive alternative to counting variables and equations is to use the *Description Rule* of Hanson, Duffin, and Somerville.[2]

> To completely describe the separation operation, the number of independent variables which must be set, must equal the number that can be set by construction or controlled by external means.

To apply this rule it is necessary to identify the variables that "can be set by construction or controlled by external means." For the cascade shown in Fig.

6.6, this is easily done. They are

Variable	Number of Variables
Stage pressure	N
Stage temperature (or Q)	N
Feed stream V_{IN} variables	$C+2$
Feed stream L_{IN} variables	$C+2$
Number of stages	1
	$N_D = 2N + 2C + 5$

This is in agreement with the results obtained by enumeration.

6.8 Complex Units

In applying either the Description Rule of Section 6.7 or the enumeration algorithm of Section 6.6 to complex multistage separators that have auxiliary heat exchangers, boilers, stream mixers, stream dividers, and so on, a considerable amount of physical insight is required to develop a feasible list of design variables. This can best be illustrated by a few examples.

Example 6.1 Consider a multistage distillation column with one feed, one side stream, total condenser, partial reboiler, and provisions for heat transfer to or from any stage. This separator can be composed, as shown in Fig. 6.7, from the circled elements and units. Total variables are determined from (6-2) by summing the variables $(N_V)_e$ for each unit from Table 6.1 and then subtracting the redundant variables due to interconnecting flows. As before, redundant mole fraction constraints are subtracted from the summation of independent relationships for each element $(N_E)_e$. This problem was first treated by Gilliland and Reed[3] and more recently by Kwauk[1] and Smith.[4] Differences in N_D obtained by various authors are due, in part, to their method of numbering stages.

Solution. Here, the partial reboiler is the first equilibrium stage. From Table 6.1, element variables and relationships are obtained as follows.

Element or Unit	$(N_V)_e$	$(N_E)_e$
Total condenser	$(2C+7)$	$(C+3)$
Reflux divider	$(3C+10)$	$(2C+5)$
$(N-S)$ stages	$[7(N-S)+2(N-S)C+2C+7]$	$[5(N-S)+2(N-S)C+2]$
Side-stream stage	$(5C+16)$	$(3C+9)$
$(S-1)-(F)$ stages	$[7(S-1-F)+2(S-1-F)C+2C+7]$	$[5(S-1-F)+2(S-1-F)C+2]$
Feed stage	$(5C+16)$	$(2C+8)$
$(F-1)-1$ stages	$[7(F-2)+2(F-2)C+2C+7]$	$[5(F-2)+2(F-2)C+2]$
Partial reboiler	$(3C+10)$	$(2C+6)$
	$\Sigma(N_V)_e = 7N + 2NC + 18C + 59$	$\Sigma(N_E)_e = 5N + 2NC + 4C + 22$

Subtracting $(C+3)$ redundant variables for 13 interconnecting streams, according to (6-2) with $N_A = 0$ (no unspecified repetitions) gives

$$(N_V)_{\text{unit}} = \sum (N_V)_e - 13(C + 3) = 7N + 2NC + 5C + 20$$

Subtracting the corresponding 13 redundant mole fraction constraints, according to (6-3), we have

$$(N_E)_{\text{unit}} = \sum (N_E)_e - 13 = 5N + 2NC + 4C + 9$$

Therefore, from (6-5)

$$N_D = (7N + 2NC + 5C + 20) - (5N + 2NC + 4C + 9) = 2N + C + 11$$

A set of feasible design variable specifications is

Variable Specification	*Number of Variables*
1. Pressure at each stage (including partial reboiler)	N
2. Pressure at reflux divider outlet	1
3. Pressure at total condenser outlet	1
4. Heat transfer rate for each stage (excluding partial reboiler)	$(N - 1)$
5. Heat transfer rate for divider	1
6. Feed mole fractions and total feed rate	C
7. Feed temperature	1
8. Feed pressure	1
9. Condensate temperature (e.g., saturated liquid)	1
10. Total number of stages N	1
11. Feed stage location	1
12. Side-stream stage location	1
13. Side-stream total flow rate S	1
14. Total distillate flow rate, D or D/F	1
15. Reflux flow rate L_R or reflux ratio L_R/D	1
	$N_D = (2N + C + 11)$

In most separation operations, variables related to feed conditions, stage heat transfer rates, and stage pressures are known or set. Remaining specifications have proxies, provided that the variables are mathematically independent of each other and of those already known. Thus, in the above list the first 9 entries are almost always known or specified. Variables 10 to 15, however, have surrogates. Some of these are:

    16.   Condenser heat duty Q_C
    17.   Reboiler heat duty Q_R
    18.   Recovery or mole fraction of one component in bottoms
    19.   Recovery or mole fraction of one component in distillate
    20.   Maximum vapor rate in column

The combination 1 to 9, 10, 11, 12, 14, 16, and 20 is convenient if the problem is one of calculating the performance of an existing column on a new feed. Here the maximum vapor rate is known as is the condenser heat duty, and the product compositions are calculated.

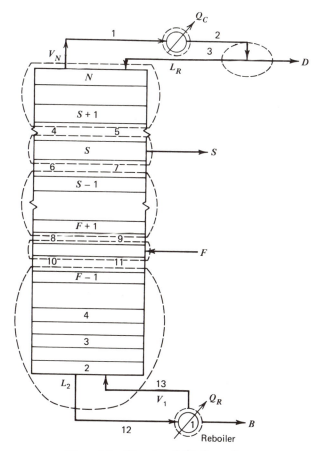

Figure 6.7. Complex distillation unit.

Heat duties Q_C and Q_R are not good design variables because they are difficult to specify. Condenser duty Q_C, for example, must be specified so that the condensate temperature lies between that corresponding to a saturated liquid and the freezing point of the condensate. Otherwise, a physically unrealizable (or no) solution to the problem is obtained. Similarly, it is much easier to calculate Q_R knowing the total flow rate and enthalpy of the bottom streams than vice versa. In general Q_R and the maximum vapor rate are so closely related that it is not advisable to specify both of them. The same is true of Q_C and Q_R.

Other proxies are possible—stage temperatures, for example, or a flow for each stage, or any independent variable that characterizes the process. The problem of independence of variables requires careful consideration. Distillate product rate, Q_C, and L_R/D, for example, are not independent. It should be noted also that, for the design case, we specify recoveries of no more than two species (items 18 and 19). These species are referred to as key components. Attempts to specify recoveries of three or four species will usually result in unsuccessful solutions of the equations.

□

Example 6.2. Consider a liquid–liquid extraction separator with central feed and extract reflux, as shown in Fig. 6.8. The five elements or units circled are a set of stages above the feed stage, the feed stage, a set of stages below the feed stage, the splitter, in which solvent is recovered, and a divider that sends reflux to the bottom stage. Suggest a feasible set of design variables.

Solution. The degrees of freedom for this complex separator unit can be determined as in Example 6.1. An alternative method is to apply (6-4) by summing the degrees of freedom for each element $(N_D)_e$ and subtracting $(C + 2)$ redundant independent variables for each interconnecting stream. Using this alternative procedure with values of $(N_D)_e$ from Table 6.1, the elements and design variables are:

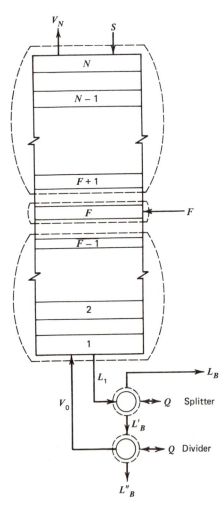

Figure 6.8. Liquid–liquid extraction unit.

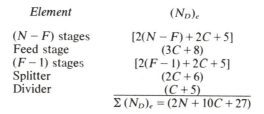

Element	$(N_D)_e$
$(N - F)$ stages	$[2(N - F) + 2C + 5]$
Feed stage	$(3C + 8)$
$(F - 1)$ stages	$[2(F - 1) + 2C + 5]$
Splitter	$(2C + 6)$
Divider	$(C + 5)$
	$\Sigma (N_D)_e = (2N + 10C + 27)$

There are seven interconnecting streams. Thus, the number of design variables from (6-4) is:

$$(N_D)_{\text{unit}} = (2N + 10C + 27) - 7(C + 2)$$
$$= 2N + 3C + 13$$

A feasible set of variable specifications is:

Variable Specification	Number of Variables
Pressure at each stage	N
Temperature or heat transfer rate for each stage	N
Solvent feed flow rate, composition, temperature, and pressure	$(C + 2)$
Feed stream flow rate, composition, temperature, and pressure	$(C + 2)$
Total number of stages N	1
Feed stage location	1
Splitter:	
Component recovery	C
T, P of L_B and L'_B	4
Divider:	
P, Q	2
Reflux ratio, V_0/L''_B	1
	$N_D = 2N + 3C + 13$

$\square$

Example 6.3. Is the following problem from Henley and Staffin[5] completely specified?

A mixture of maleic anhydride and benzoic acid containing 10 mole percent acid is a product of the manufacture of phthalic anhydride. The mixture is to be distilled continuously at a pressure of 13.2 kPa [100 torr] to give a product of 99.5 mole percent maleic anhydride and a bottoms of 0.5 mole percent anhydride. Using the data below [omitted here], calculate the number of plates (stages) using an L/D of 1.6 times the minimum.

Solution. The degrees of freedom for this distillation operation are determined as follows by using Table 6.1, assuming the partial reboiler is the first stage.

Table 6.2 Typical variable specifications for design cases

Unit Operation	N_D	Variable Specification[a]	
		Case I, Component Recoveries Specified	Case II, Number of Equilibrium Stages Specified
(a) Absorption (two inlet streams)	$2N + 2C + 5$	1. Recovery of one key component.	1. Number of stages.
(b) Distillation (one inlet stream, total condenser, partial reboiler)	$2N + C + 9$	1. Condensate at saturation temperature. 2. Recovery of light-key component. 3. Recovery of heavy-key component. 4. Reflux ratio (> minimum). 5. Optimum feed stage.[b]	1. Condensate at saturation temperature. 2. Number of stages above feed stage. 3. Number of stages below feed stage. 4. Reflux ratio. 5. Distillate flow rate.

(c) Distillation
(one inlet stream,
partial condenser,
partial reboiler,
vapor distillate only)

$2N + C + 6$

1. Recovery of light-key component.
2. Recovery of heavy-key component.
3. Reflux ratio ($>$ minimum).
4. Optimum feed stage.[b]

1. Number of stages above feed stage.
2. Number of stages below feed stage.
3. Reflux ratio.
4. Distillate flow rate.

(d) Liquid–liquid extraction with two solvents (three inlet streams)

$2N + 3C + 8$

1. Recovery of key component one.
2. Recovery of key component two.

1. Number of stages above feed.
2. Number of stages below feed.

257

Table 6.2, continued

Unit Operation	N_D	Variable Specification[a]	
		Case I, Component Recoveries Specified	**Case II, Number of Equilibrium Stages Specified**
(e) Reboiled absorption (two inlet streams)	$2N + 2C + 6$	1. Recovery of light-key component. 2. Recovery of heavy-key component. 3. Optimum feed stage.[b]	1. Number of stages above feed. 2. Number of stages below feed. 3. Bottoms flow rate.
(f) Reboiled stripping (one inlet stream)	$2N + C + 3$	1. Recovery of one key component. 2. Reboiler heat duty.[d]	1. Number of stages. 2. Bottoms flow rate.

(g) Distillation (one inlet stream, partial condenser, partial reboiler, both liquid and vapor distillates)

$2N + C + 9$

1. Ratio of vapor distillate to liquid distillate.
2. Recovery of light-key component.
3. Recovery of heavy-key component.
4. Reflux ratio (> minimum).
5. Optimum feed stage.[b]

1. Ratio of vapor distillate to liquid distillate.
2. Number of stages above feed stage.
3. Number of stages below feed stage.
4. Reflux ratio.
5. Liquid distillate flow rate.

(h) Extractive distillation (two inlet streams, total condenser, partial reboiler, single-phase condensate)

$2N + 2C + 12$

1. Condensate at saturation temperature.
2. Recovery of light-key component.
3. Recovery of heavy-key component.
4. Reflux ratio (> minimum)
5. Optimum feed stage.[b]
6. Optimum MSA stage.[b]

1. Condensate at saturation temperature.
2. Number of stages above MSA stage.
3. Number of stages between MSA and feed stages.
4. Number of stages below feed stage.
5. Reflux ratio.
6. Distillate flow rate.

Table 6.2, continued

Unit Operation	N_D	Variable Specification[a]	
		Case I, Component Recoveries Specified	Case II, Number of Equilibrium Stages Specified
(i) Liquid–liquid extraction (two inlet streams)	$2N + 2C + 5$	1. Recovery of one key component.	1. Number of stages.
(j) Stripping (two inlet streams)	$2N + 2C + 5$	1. Recovery of one key component.	1. Number of stages.

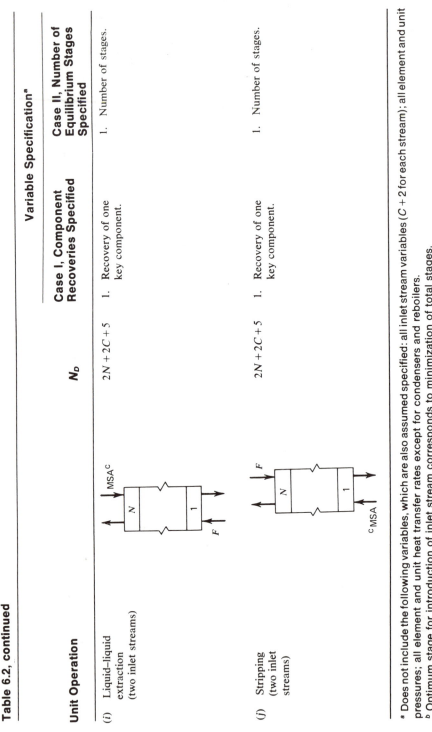

[a] Does not include the following variables, which are also assumed specified: all inlet stream variables ($C + 2$ for each stream); all element and unit pressures; all element and unit heat transfer rates except for condensers and reboilers.
[b] Optimum stage for introduction of inlet stream corresponds to minimization of total stages.
[c] For Case I variable specifications, MSA flow rates must be greater than minimum values for specified recoveries.
[d] For Case I variable specifications, reboiler heat duty must be greater than minimum value for specified recovery.

Element or Unit	$(N_D)_e$
Reflux divider	$(C+5)$
Total condenser	$(C+4)$
Stages above feed stage	$[2(N-F)+2C+5]$
Stages below feed stage	$[2(F-2)+2C+5]$
Feed stage	$(3C+8)$
Partial reboiler	$(C+4)$
	$\Sigma\,(N_D)_e = 2N+10C+27$

There are nine interconnecting streams. Thus, the degrees of freedom from (6-4) are

$$N_D = 2N+10C+27-9(C+2) = 2N+C+9$$

The only variables specified in the problem statement are

Variable Specification	Number of Variables
Stage pressures (including reboiler)	N
Condenser pressure	1
Reflux divider pressure	1
L_R/D	1
Feed composition	$C-1$
Mole fraction of maleic anhydride in distillate	1
Mole fraction of maleic anhydride in bottoms	1
	$\overline{C+N+4}$

The problem is underspecified by $(N+5)$ variables. It can be solved if we assume:

Additional Variable Specification	Number of Variables
Feed T and P	2
Total condenser giving saturated reflux	1
Heat transfer rate (loss) in divider	1
Adiabatic stages (excluding boiler, which is assumed to be a partial reboiler	$N-1$
Feed stage location (assumed to be optimum)	1
Feed rate	1
	$\overline{N+5}$

☐

6.9 Variable Specifications for Typical Design Cases

The design of multistage separation operations involves solving the variable relationships for output variables after selecting values of design variables to satisfy the degrees of freedom. Two cases are commonly encountered. In Case I, recovery specifications are made for one or two key components and the number of required equilibrium stages is determined. In Case II, the number of equilibrium stages is specified and component separations are computed. For multicomponent feeds, the second case is more widely employed because less computational complexity is involved. Table 6.2 is a summary of possible variable specifications for each of these two cases for a number of separator types previously discussed in Chapter 1 and shown in Table 1.1. For all separators in Table 6.2, it is assumed that all inlet streams are completely specified (i.e., $C - 1$ mole fractions, total flow rate, temperature, and pressure) and all element and unit pressures and heat transfer rates (except for condensers and reboilers) are specified. Thus, only variables to satisfy the remaining degrees of freedom are listed.

References

1. Kwauk, M., *AIChE J.*, **2**, 240–248 (1956).

2. Hanson, D. N., J. H. Duffin, and G. F. Somerville, *Computation of Multistage Separation Processes*, Reinhold Publishing Corporation, New York, 1962, Chapter 1.

3. Gilliland, E. R., and C. E. Reed, *Ind. Eng. Chem.*, **34**, 551–557 (1942).

4. Smith, B., *Design of Equilibrium Stage Processes*, McGraw–Hill Book Co., New York, 1963, Chapter 3.

5. Henley, E. J., and H. K. Staffin, *Stagewise Process Design*, John Wiley & Sons, Inc., New York, 1963, 198.

Problems

6.1 Consider the equilibrium stage shown in Fig. 1.12. Conduct a degrees-of-freedom analysis by performing the following steps.
(a) List and count the variables.
(b) Write and count the equations relating the variables.
(c) Calculate the degrees of freedom.
(d) List a reasonable set of design variables.

6.2 Can the following problems be solved uniquely?
(a) The feed streams to an adiabatic equilibrium stage consist of liquid and vapor streams of known composition, flow rate, temperature, and pressure. Given the stage (outlet) temperature and pressure, calculate the composition and amounts of equilibrium vapor and liquid leaving the stage.
(b) The same as Part (a), except that the stage is not adiabatic.
(c) The same as Part (a), except that, in addition to the vapor and liquid streams leaving the stage, a vapor side stream, in equilibrium with the vapor leaving the stage, is withdrawn.
(d) A multicomponent vapor of known temperature, pressure, and composition is to be partially condensed in a condenser. The pressure in the condenser and the inlet cooling water temperature are fixed. Calculate the cooling water required.
(e) A mixture of ^{235}U and ^{238}U is partially diffused through a porous membrane

barrier to effect isotope enrichment. The process is adiabatic. Given the separation factor and the composition and conditions of the feed, calculate the pumping requirement.

6.3 Consider an adiabatic equilibrium flash. The variables are all as indicated in the sketch below.

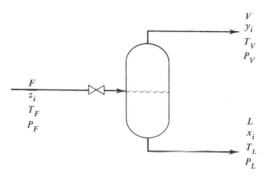

(a) Determine N_V = number of variables.
(b) Write all the independent equations that relate the variables.
(c) Determine N_E = number of equations.
(d) Determine the number of degrees of freedom.
(e) What variables would you prefer to specify in order to solve a typical adiabatic flash problem?

6.4 Determine the number of degrees of freedom for a nonadiabatic equilibrium flash for one liquid feed, one vapor stream product, and two immiscible liquid stream products.

6.5 Determine N_D for the following unit operations in Table 6.2: (b), (c), and (g).

6.6 Determine N_D for unit operations (e) and (f) in Table 6.2.

6.7 Determine N_D for unit operation (h) in Table 6.2. How would N_D change if a liquid side stream were added to a stage that was located between the feed F and stage 2?

6.8 The following are not listed as design variables for the distillation unit operations in Table 6.2.
(a) Condenser heat duty.
(b) Stage temperature.
(c) Intermediate stage vapor rate.
(d) Reboiler heat load.
Under what conditions might these become design variables? If so, which variables listed in Table 6.2 would you eliminate?

6.9 Show for distillation that, if a total condenser is replaced by a partial condenser, the degrees of freedom are reduced by three, provided that the distillate is removed solely as a vapor.

6.10 Determine the number of independent variables and suggest a reasonable set for (a) a new column, and (b) an existing column, for the crude oil distillation column with side stripper shown below. Assume that water does not condense.

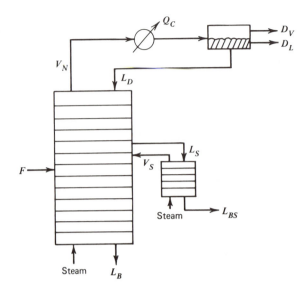

6.11 Show that the degrees of freedom for a liquid–liquid extraction column with two feeds and raffinate reflux is $3C + 2N + 13$.

6.12 Determine the degrees of freedom and a reasonable set of specifications for an azeotropic distillation problem wherein the formation of one minimum-boiling azeotrope occurs.

6.13 A distillation column consisting of four equilibrium trays, a reboiler, a partial condenser, and a reflux divider is being used to effect a separation of a five-component stream.

   The feed is to the second tray, the trays and divider are adiabatic, and the pressure is fixed throughout the column. The feed is specified.

   The control engineer has specified three control loops that he believes to be independent. One is to control the reflux/distillate ratio, the second to control the distillate/feed ratio, and the third to maintain top tray temperature. Comment on this proposed control scheme.

6.14 (a) Determine for the distillation column below the number of independent design variables.
   (b) It is suggested that a feed consisting of 30% A, 20% B, and 50% C at 37.8°C and 689 kPa be processed in an existing 15-plate, 3-m-diameter column that is designed to operate at vapor velocities of 0.3 m/sec and an L/V of 1.2. The pressure drop per plate is 373 Pa at these conditions, and the condenser is cooled by plant water, which is at 15.6°C.

   The product specifications in terms of the concentration of A in the distillate and C in the bottoms have been set by the process department, and the plant manager has asked you to specify a feed rate for the column.

   Write a memorandum to the plant manager pointing out why you can't do this and suggest some alternatives.

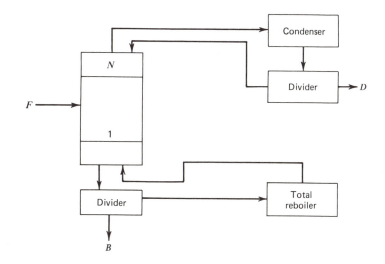

6.15 Unit operation (*b*) in Table 6.2 is to be heated by injecting live steam directly into the bottom plate of the column instead of by using a reboiler, for a separation involving ethanol and water. Assuming a fixed feed, an adiabatic operation, atmospheric pressure throughout, and a top alcohol concentration specification:
   (a) What is the total number of design variables for the general configuration?
   (b) How many design variables will complete the design? Which variables do you recommend?

6.16 Calculate the degrees of freedom of the mixed-feed triple-effect evaporator shown below. Assume the steam and all drain streams are at saturated conditions and the feed is an aqueous solution of dissolved organic solids (two-component streams). Also, assume that all overhead streams are pure water vapor with no entrained solids (one-component streams).

   If this evaporator is used to concentrate a feed containing 2% solids to a product with 25% solids using 689 kPa saturated steam, calculate the number of unspecified design variables and suggest likely candidates. Assume perfect insulation in each effect.

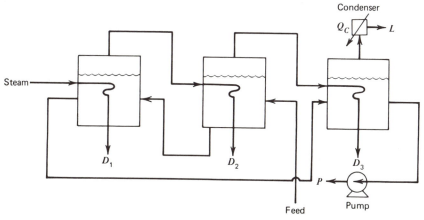

6.17 A reboiled stripper is to be designed for the task shown below. Determine:
(a) The number of variables.
(b) The number of equations relating the variables.
(c) The number of degrees of freedom.
and indicate:
(d) Which additional variables, if any, need to be specified.

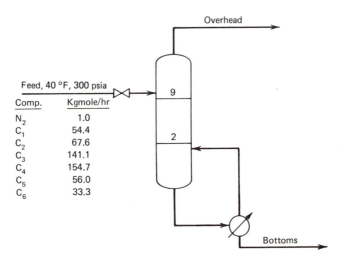

Feed, 40 °F, 300 psia	
Comp.	Kgmole/hr
N_2	1.0
C_1	54.4
C_2	67.6
C_3	141.1
C_4	154.7
C_5	56.0
C_6	33.3

6.18 The thermally coupled distillation system shown below is to be used to separate a mixture of three components into three products. Determine for the system:
(a) The number of variables.
(b) The number of equations relating the variables.
(c) The number of degrees of freedom.
and propose:
(d) A reasonable set of design variables.

6.19 When the feed to a distillation column contains a small amount of impurities that are much more volatile than the desired distillate, it is possible to separate the volatile impurities from the distillate by removing the distillate as a liquid side stream from a stage located several stages below the top stage. As shown below, this additional top section of stages is referred to as a *pasteurizing section*.
(a) Determine the number of degrees of freedom for the unit.
(b) Determine a reasonable set of design variables.

6.20 A system for separating a mixture into three products is shown below. For it, determine:
(a) The number of variables.
(b) The number of equations relating the variables.
(c) The number of degrees of freedom.
and propose:
(d) A reasonable set of design variables.

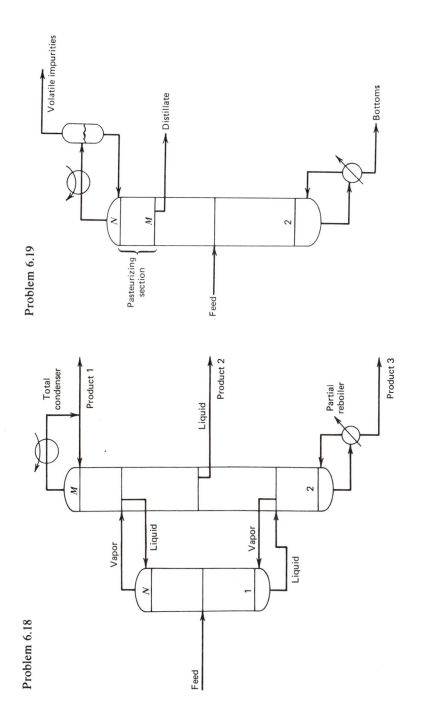

Problem 6.18

Problem 6.19

Problem 6.20

Problem 6.21

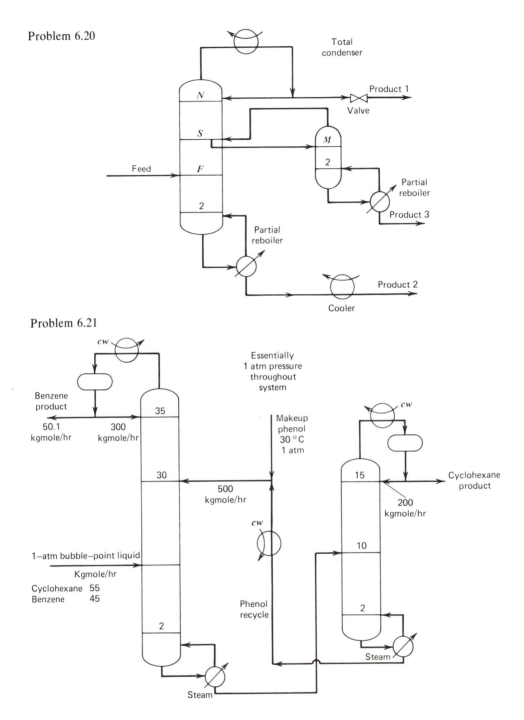

6.21 A system for separating a binary mixture by extractive distillation, followed by ordinary distillation for recovery and recycle of the solvent, is shown on previous page. Are the design variables shown sufficient to completely specify the problem? If not, what additional design variable(s) would you select?

6.22 A single distillation column for separating a three-component mixture into three products is shown below. Are the design variables shown sufficient to specify the problem completely? If not, what additional design variable(s) would you select?

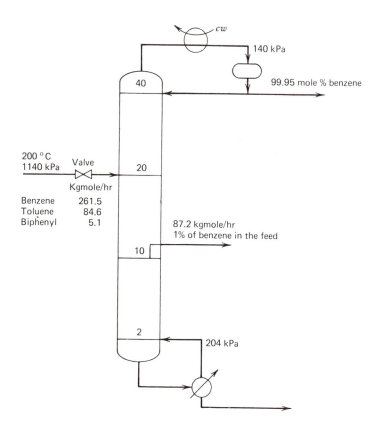

cw

140 kPa

40

99.95 mole % benzene

200 °C
1140 kPa   Valve

20

Kgmole/hr

Benzene	261.5
Toluene	84.6
Biphenyl	5.1

87.2 kgmole/hr
1% of benzene in the feed

10

2

204 kPa

7
Equilibrium Flash
Vaporization and
Partial Condensation

The equilibrium flash separator is the simplest equilibrium-stage process with which the designer must deal. Despite the fact that only one stage is involved, the calculation of the compositions and the relative amounts of the vapor and liquid phases at any given pressure and temperature usually involves a tedious trial-and-error solution.

Buford D. Smith, 1963

A *flash* is a single-stage distillation in which a feed is partially vaporized to give a vapor that is richer in the more volatile components. In Fig. 7.1*a*, a liquid feed is heated under pressure and flashed adiabatically across a valve to a lower pressure, the vapor being separated from the liquid residue in a flash drum. If the valve is omitted, a low-pressure liquid can be partially vaporized in the heater and then separated into two phases. Alternatively, a vapor feed can be cooled and partially condensed, with phase separation in a flash drum as in Fig. 7.1*b* to give a liquid that is richer in the less volatile components. In both cases, if the equipment is properly designed, the vapor and liquid leaving the drum are in equilibrium.[1]

Unless the relative volatility is very large, the degree of separation achievable between two components in a single stage is poor; so flashing and partial condensation are usually auxiliary operations used to prepare feed streams for further processing. However, the computational methods used in single-stage calculation are of fundamental importance. Later in this chapter we show that stages in an ordinary distillation column are simply adiabatic flash chambers, and columns can be designed by an extension of the methods developed for a single-stage flash or partial condensation. Flash calculations are also widely used

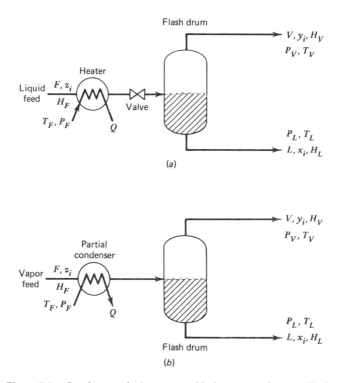

Figure 7.1. Continuous single-stage equilibrium separation. (*a*) Flash vaporization. (Adiabatic flash with valve; isothermal flash without valve when T_V is specified. (*b*) Partial condensation. (Analogous to isothermal flash when T_V is specified.

to determine the phase condition of a stream of known composition, temperature, and pressure.

For the single-stage equilibrium operations shown in Fig. 7.1, with one feed stream and two product streams, the $(2C + 6)$ equations listed in Table 7.1 relate the $(3C + 10)$ variables (F, V, L, $\mathbf{z}$, $\mathbf{y}$, $\mathbf{x}$, T_F, T_V, T_L, P_F, P_V, P_L, Q), leaving $(C + 4)$ degrees of freedom. Assuming that $(C + 2)$ feed variables—F, T_F, P_F, and $(C - 1)$ values of z_i—are known, two additional variables can be specified. In this chapter, computational procedures are developed for several different sets of specifications, the most important being:

1. P_L (or P_V) and V/F (or L/F): percent vaporization (percent condensation).

2. P_L (or P_V) and T_L (or T_V): isothermal flash.

3. P_L (or P_V) and $Q = 0$: adiabatic flash.

Table 7.1 Equations for single-stage flash and partial condensation operations

Equation[a]	Number of Equations	Equation Number
$P_V = P_L$ (Mechanical equilibrium)	1	(7-1)
$T_V = T_L$ (Thermal equilibrium)	1	(7-2)
$y_i = K_i x_i$ (Phase equilibrium)	C	(7-3)
$Fz_i = Vy_i + Lx_i$ (Component material balance)	$C-1$	(7-4)
$F = V + L$ (Total material balance)	1	(7-5)
$H_F F + Q = H_V V + H_L L$ (Enthalpy balance)	1	(7-6)
$\sum z_i = 1$ (Summation)	1	(7-7)
$\sum y_i = 1$ (Summation)	1	(7-8)
$\sum x_i = 1$ (Summation)	1	(7-9)
	$N_E = \overline{2C + 6}$	

$$K_i = K_i\{T_V, P_V, \mathbf{y}, \mathbf{x}\} \qquad H_F = H_F\{T_F, P_F, \mathbf{z}\}$$

$$H_V = H_V\{T_V, P_V, \mathbf{y}\} \qquad H_L = H_L\{T_L, P_L, \mathbf{x}\}$$

[a]These equations are restricted to two equilibrium phases. For a treatment of three-phase flash calculations, see E. J. Henley and E. M. Rosen, *Material and Energy Balance Computations*, J. Wiley & Sons, Inc., 1968, Chapter 8.

7.1 Graphical Methods for Binary Mixtures

For binary mixtures, percent vaporization or condensation is conveniently determined from graphical construction by methods similar to those of Section 3.5. Figure 7.2 shows the equilibrium curve and the $y = x$ (45° line) curve for n-hexane in a mixture with n-octane at 101 kPa. Operating lines representing various percentages of feed vaporized are obtained by combining (7-4) and (7-5) to eliminate L to give for n-hexane (the more volatile component)

$$y = -\frac{(1-\psi)}{\psi}x + \frac{z}{\psi} \tag{7-10}$$

where $\psi = V/F$, the fraction vaporized. This line has a slope of $-(1-\psi)/\psi$ in y-x coordinates. When (7-10) is solved simultaneously with $x = y$, we find that $x = y = z$.

A graphical method for obtaining the composition y and x of the exit equilibrium streams as a function of ψ consists of finding the intersection of the operating line with the equilibrium line. Assume, for instance, that a feed of 60 mole% n-hexane in n-octane enters a flash drum at 1 atm. If $\psi = 0.5$, then $-(1-\psi)/\psi = -1$. A line of slope -1 passing through ($z = 0.6$) intersects the

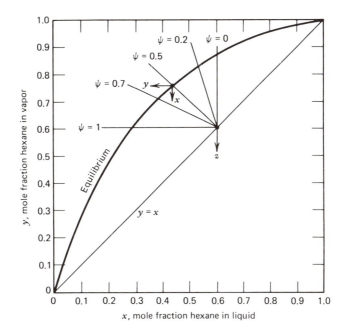

Figure 7.2. Flash vaporization for n-hexane and n-octane at 1 atm (101.3 kPa).

equilibrium curve of Fig. 7.2 at approximately $y = 0.77$, $x = 0.43$, the compositions of the vapor and liquid leaving the drum. Values of y and x corresponding to other values of ψ are also shown. For $\psi = 0$, the feed is at its bubble point; for $\psi = 1$, the feed is at its dew point. If the flash temperature rather than ψ is specified, a T-x-y diagram like Fig. 3.4 becomes a more convenient tool than a y-x diagram.

7.2 Multicomponent Isothermal Flash and Partial Condensation Calculations

If the equilibrium temperature T_V (or T_L) and the equilibrium pressure P_V (or P_L) of a multicomponent mixture are specified, values of the remaining $2C + 6$ variables are determined from the equations in Table 7.1 by an *isothermal flash calculation*. However, the computational procedure is not straightforward because (7-4) is a nonlinear equation in the unknowns V, L, y_i, and x_i. Many solution strategies have been developed, but the generally preferred procedure, as given in Table 7.2, is that of Rachford and Rice[2] when K-values are independent of composition.

First, equations containing only a single unknown are solved. Thus, (7-1), (7-2), and (7-7) are solved respectively for P_L, T_L, and the unspecified value of z_i. Next, because the unknown Q appears only in (7-6), Q is computed only after all other equations have been solved. This leaves (7-3), (7-4), (7-5), (7-8), and (7-9) to be solved for V, L, y, and x. These equations can be partitioned so as to solve for the unknowns in a sequential manner by substituting (7-5) into (7-4) to eliminate L; combining the result with (7-3) to obtain two equations, (7-12) and (7-13), one in x_i but not y_i and the other in y_i but not x_i; and summing these two equations and combining them with (7-8) and (7-9) in the form $\Sigma y_i - \Sigma x_i = 0$ to eliminate y_i and x_i and give a nonlinear equation (7-11) in V (or ψ) only. Upon solving this equation in an iterative manner for V, one obtains the remaining unknowns directly from (7-12) to (7-15). When T_F and/or P_F are not specified, (7-15) is not solved for Q. In this case the equilibrium phase condition of a mixture at a known temperature ($T_V = T_L$) and pressure ($P_V = P_L$) is determined.

Equation (7-11) can be solved by trial-and-error by guessing values of ψ between 0 and 1 until $f\{\psi\} = 0$. The form of the function, as computed for Example 7.1, is shown in Fig. 7.3

The most widely employed computer methods for solving (7-11) are false position and Newton's method.[3] In the latter, a predicted value of the ψ root for

Table 7.2   Rachford–Rice procedure for isothermal flash calculations when K-values are independent of composition

Specified variables: F, T_F, P_F, z_1, z_2, ..., z_{C-1}, T_V, P_V

Steps	Equation Number
(1) $T_L = T_V$	
(2) $P_L = P_V$	
(3) $z_C = 1 - \sum_{i=1}^{C-1} z_i$	
(4) Solve	
$$f\{\psi\} = \sum_{i=1}^{C} \frac{z_i(1 - K_i)}{1 + \psi(K_i - 1)} = 0$$	(7-11)
for $\psi = V/F$, where $K_i = K_i(T_V, P_V)$.	
(5) $V = F\psi$	
(6) $x_i = \dfrac{z_i}{1 + \psi(K_i - 1)}$	(7-12)
(7) $y_i = \dfrac{z_i K_i}{1 + \psi(K_i - 1)} = x_i K_i$	(7-13)
(8) $L = F - V$	(7-14)
(9) $Q = H_V V + H_L L - H_F F$	(7-15)

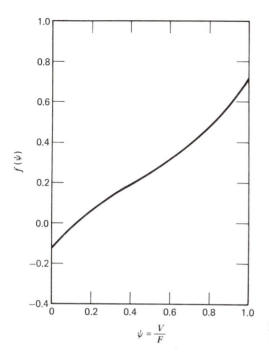

$$\psi = \frac{V}{F}$$

Figure 7.3.   Rachford–Rice function for Example 7.1.

iteration $k + 1$ is computed from the recursive relation

$$\psi^{(k+1)} = \psi^{(k)} - \frac{f\{\psi^{(k)}\}}{f'\{\psi^{(k)}\}} \tag{7-16}$$

where the derivative in (7-16) is

$$f'\{\psi^{(k)}\} = \sum_{i=1}^{c} \frac{z_i(1 - K_i)^2}{[1 + \psi^{(k)}(K_i - 1)]^2} \tag{7-17}$$

The iteration can be initiated by assuming $\psi^{(1)} = 0.5$. Sufficient accuracy will be achieved by terminating the iterations when $|\psi^{(k+1)} - \psi^{(k)}|/\psi^{(k)} < 0.0001$. Values of $\psi^{(k+1)}$ should be constrained to lie between 0 and 1. Thus, if $\psi^{(k)} = 0.10$ and $\psi^{(k+1)}$ is computed from (7-16) to be -0.05, $\psi^{(k+1)}$ should be reset to, say, one half of the interval from $\psi^{(k)}$ to 0 or 1, whichever is closer to $\psi^{(k+1)}$. In this case, $\psi^{(k+1)}$ would be reset to 0.05. One should check the existence of a valid root ($0 \le \psi \le 1$) before employing the procedure of Table 7.2 by testing to see if the equilibrium condition corresponds to subcooled liquid or superheated vapor. This check is discussed in the next section.

If K-values are functions of phase compositions, the procedure for solving (7-11) is more involved. Two widely used algorithms are shown in Fig. 7.4, where **x**, **y**, and **K** are vectors of the x, y, and K-values. In procedure Fig. 7.4a, (7-11) is

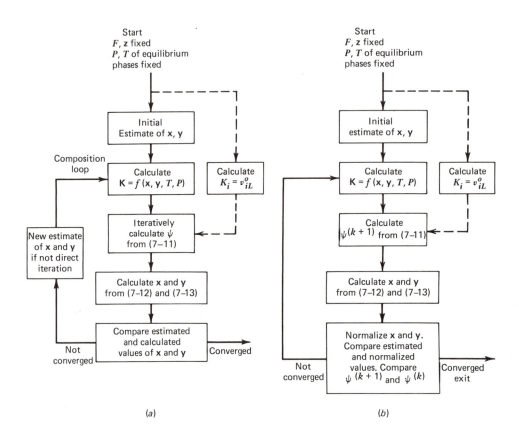

Figure 7.4. Algorithm for isothermal flash calculation when K-values are composition dependent. (a) Separate nested iterations on ψ and $(\mathbf{x}, \mathbf{y})$. (b) Simultaneous iteration on ψ and $(\mathbf{x}, \mathbf{y})$.

iterated to convergence for *each* set of $\mathbf{x}$ and $\mathbf{y}$. For each converged ψ, a new set of $\mathbf{x}$ and $\mathbf{y}$ is generated and used to compute a new set of $\mathbf{K}$. Iterations on $\mathbf{x}$ and $\mathbf{y}$ are continued in the composition loop until no appreciable change occurs on successive iterations. This procedure is time consuming, but generally stable.

In the alternative procedure, Fig. 7.4b, $\mathbf{x}$, $\mathbf{y}$, and $\mathbf{K}$ are iterated simultaneously with each iteration on ψ using (7-11). Values of $\mathbf{x}$ and $\mathbf{y}$ from (7-12) and (7-13) are normalized ($x_i = x_i / \Sigma\, x_i$ and $y_i = y_i / \Sigma\, y_i$) before computing new K-values. This procedure is very rapid but may not always converge. In general, both procedures require an initial estimate of $\mathbf{x}$ and $\mathbf{y}$. An alternative is indicated by the dashed line, where K-values are computed from (4-29), assuming $K_i = v_{IL}^o$.

In both procedures, direct iteration on $\mathbf{x}$ and $\mathbf{y}$ is generally satisfactory.

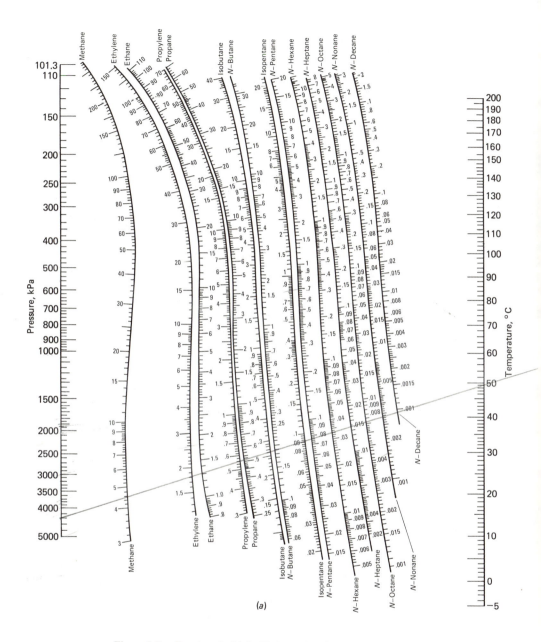

(a)

Figure 7.5. K-values in Light-Hydrocarbon Systems. (a) High-temperature range. (b) Low-temperature range. [D. B. Dadyburjor, *Chem. Eng. Progr.*, **74** (4), 85–86 (1978). The SI version of charts of C. L. DePriester, *Chem. Eng. Progr.*, *Symp.* Ser., **49** (7), 1 (1953).]

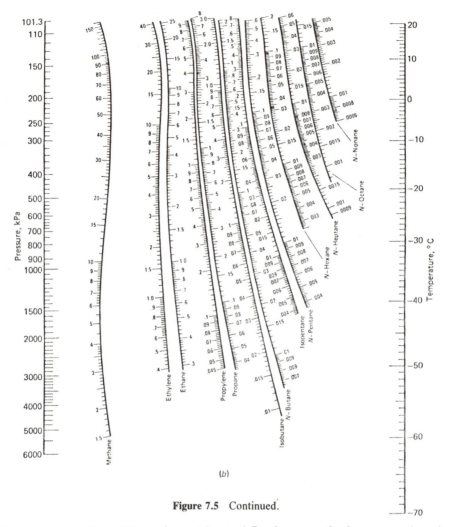

(b)

Figure 7.5 Continued.

However, sometimes Wegstein or Newton–Raphson methods are employed as convergence acceleration promoters.

Example 7.1 is an application of Table 7.2 to the problem of obtaining outlet compositions and flow rates for an isothermal flash using K-values that are concentration independent and are taken from Fig. 7.5. Example 7.2 involves an isothermal flash using the Chao–Seader correlation for K-values in conjunction with the algorithm of Fig. 7.4b.

Example 7.1. A 100-kgmole/hr feed consisting of 10, 20, 30, and 40 mole% of propane (3), n-butane (4), n-pentane (5), and n-hexane (6), respectively, enters a distillation column at 100 psia (689.5 kPa) and 200°F (366.5°K). Assuming equilibrium, what fraction of the feed enters as liquid and what are the liquid and vapor compositions?

Solution. At flash conditions, from Fig. 7.5*a*, $K_3 = 4.2$, $K_4 = 1.75$, $K_5 = 0.74$, $K_6 = 0.34$. Thus (7-11) for the root ψ becomes

$$0 = \frac{0.1(1-4.2)}{1+\psi(4.2-1)} + \frac{0.2(1-1.75)}{1+\psi(1.75-1)} + \frac{0.3(1-0.74)}{1+\psi(0.74-1)} + \frac{0.4(1-0.34)}{1+\psi(0.34-1)}$$

Solution of this equation by Newton's method using an initial guess for ψ of 0.50 gives the following iteration history.

k	$\psi^{(k)}$	$f\{\psi^{(k)}\}$	$f'\{\psi^{(k)}\}$	$\psi^{(k+1)}$	$\left\|\dfrac{\psi^{(k+1)}-\psi^{(k)}}{\psi^{(k)}}\right\|$
1	0.5000	0.2515	0.6259	0.0982	0.8037
2	0.0982	−0.0209	0.9111	0.1211	0.2335
3	0.1211	−0.0007	0.8539	0.1219	0.0065
4	0.1219	0.0000	0.8521	0.1219	0.0000

For this example, convergence is very rapid. The equilibrium vapor flow rate is $0.1219(100) = 12.19$ kgmole/hr, and the equilibrium liquid flow rate is $(100 - 12.19) = 87.81$ kgmole/hr. Alternatively, ψ could be determined by trial and error to obtain the curve in Fig. 7.3 from which the proper ψ root corresponds to $f\{\psi\} = 0$. The liquid and vapor compositions computed from (7-12) and (7-13) are:

	x	*y*
Propane	0.0719	0.3021
n-Butane	0.1833	0.3207
n-Pentane	0.3098	0.2293
n-Hexane	0.4350	0.1479
	1.0000	1.0000

□

The degree of separation achieved by a flash when relative volatilities are very large is illustrated by the following industrial application.

Example 7.2. In the high-pressure, high-temperature thermal hydrodealkylation of toluene to benzene $(C_7H_8 + H_2 \rightarrow C_6H_6 + CH_4)$,[4] excess hydrogen is used to minimize cracking of aromatics to light gases. In practice, conversion of toluene is only 70%. To separate and recycle hydrogen, hot reactor effluent vapor of 5597 kgmole/hr is partially condensed with cooling water to 100°F (310.8°K), with product phases separated in a flash drum as in Fig. 7.1*b*. If the flow rate and composition of the reactor effluent is as follows, and the flash drum pressure is 485 psia (3344 kPa), calculate equilibrium compositions and flow rates of vapor and liquid leaving the flash drum.

Component	*Mole Fraction*
Hydrogen (H)	0.31767
Methane (M)	0.58942
Benzene (B)	0.07147
Toluene (T)	0.02144
	1.00000

Table 7.3   Iteration results for Example 7.2

Iteration, k	Assumed Fraction Vaporized, $\psi^{(k)}$	x				Calculated Fraction Vaporized	$\left\|\dfrac{\psi^{(k-1)} - \psi^{(k)}}{\psi^{(k)}}\right\|$
		H	M	B	T		
^{a}1	0.500	—	—	—	—	0.98933	0.9787
2	0.98933	0.0797	0.5023	0.3211	0.0969	0.97137	0.0181
3	0.97137	0.0019	0.0246	0.6490	0.3245	0.94709	0.0250
4	0.94709	0.0016	0.0216	0.7118	0.2650	0.92360	0.0248
5	0.92360	0.0025	0.0346	0.7189	0.2440	0.91231	0.0122
6	0.91231	0.0035	0.0482	0.7147	0.2336	0.91065	0.0018
7	0.91065	0.0040	0.0555	0.7108	0.2297	0.91054	0.0001
8	0.91054	0.0041	0.0569	0.7099	0.2291	0.91053	0.0000
9	0.91053	0.0041	0.0571	0.7097	0.2291	0.91053	0.0000
10	0.91053	0.0041	0.0571	0.7097	0.2291	0.91053	0.0000

Iteration, k	Assumed Fraction Vaporized, $\psi^{(k)}$	y				Calculated Fraction Vaporized	$\left\|\dfrac{\psi^{(k+1)} - \psi^{(k)}}{\psi^{(k)}}\right\|$
		H	M	B	T		
^{a}1	0.500	—	—	—	—	0.98933	0.9787
2	0.98933	0.3850	0.6141	0.0008	0.0001	0.97137	0.0181
3	0.97137	0.3317	0.6144	0.0459	0.0080	0.94709	0.0250
4	0.94709	0.3412	0.6316	0.0239	0.0033	0.92360	0.0248
5	0.92360	0.3455	0.6383	0.0144	0.0018	0.91231	0.0122
6	0.91231	0.3476	0.6409	0.0102	0.0013	0.91065	0.0018
7	0.91065	0.3484	0.6416	0.0089	0.0011	0.91054	0.0001
8	0.91054	0.3485	0.6417	0.0088	0.0010	0.91053	0.0000
9	0.91053	0.3485	0.6417	0.0088	0.0010	0.91053	0.0000
10	0.91053	0.3485	0.6417	0.0088	0.0010	0.91053	0.0000

Iteration, k	Assumed Fraction Vaporized, $\psi^{(k)}$	K-Values				Calculated Fraction Vaporized	$\left\|\dfrac{\psi^{(k+1)} - \psi^{(k)}}{\psi^{(k)}}\right\|$
		H	M	B	T		
^{a}1	0.500	17.06	4.32	0.00866	0.00302	0.98933	0.9787
2	0.98933	41.62	6.07	0.01719	0.00600	0.97137	0.0181
3	0.97137	85.01	11.59	0.01331	0.00501	0.94709	0.0250
4	0.94709	87.00	11.71	0.01268	0.00473	0.92360	0.0248
5	0.92360	86.20	11.54	0.01244	0.00461	0.91231	0.0122
6	0.91231	84.99	11.36	0.01235	0.00456	0.91065	0.0018
7	0.91065	84.30	11.26	0.01233	0.00455	0.91054	0.0001
8	0.91054	84.16	11.25	0.01233	0.00454	0.91053	0.0000
9	0.91053	84.15	11.24	0.01233	0.00454	0.91053	0.0000
10	0.91053	84.15	11.24	0.01233	0.00454	0.91053	0.0000

^{a} K-values for Iteration 1 were computed from $K_i = \nu^0_{iL}$.

Solution.   The Chao–Seader K-value correlation discussed in Section 5.1 is suitable for this mixture. Calculations of K-values are made in the manner of Example 5.1. Because the K-values are composition dependent, one of the algorithms of Fig. 7.4 applies. Choosing Fig. 7.4a and using the dashed-line option with the Chao–Seader correlation, we solve (7-11) by Newton's procedure to generate values of $\psi^{(k+1)}$. Results are listed in Table 7.3. For the first three iterations, K-values change markedly as phase mole fractions vary significantly. Thereafter, changes are much less. After seven iterations, the fraction vaporized is converged; but three more iterations are required before successive sets of mole fractions are identical to four significant figures. Final liquid and vapor flows are 500.76 and 5096.24 kgmole/hr, respectively, with compositions from Iteration 10. Final converged values of thermodynamic properties and the K-values are

Component	v_L°	γ_L	ϕ_V	K
Hydrogen	17.06	5.166	1.047	84.15
Methane	4.32	2.488	0.956	11.24
Benzene	0.00866	1.005	0.706	0.01233
Toluene	0.00302	1.000	0.665	0.00454

Because the relative volatility between methane and benzene is $(11.24/0.01233) = 911.6$, a reasonably sharp separation occurs. Only 0.87% of methane in the feed appears in the equilibrium liquid, and 11.15% of benzene in the feed appears in the equilibrium vapor. Only 0.12% of hydrogen is in the liquid.
□

7.3 Bubble- and Dew-Point Calculations

A first estimate of whether a multicomponent feed gives a two-phase equilibrium mixture when flashed at a given temperature and pressure can be made by inspecting the K-values. If all K-values are greater than one, the exit phase is superheated vapor above the *dew point* (the temperature and pressure at which the first drop of condensate forms). If all K-values are less than one, the single exit phase is a subcooled liquid below the *bubble point* (at which the first bubble of vapor forms).

A precise indicator is the parameter ψ, which must lie between zero and one. If $f\{\psi\} = f\{0\}$ in (7-11) is greater than zero, the mixture is below its bubble point; and if $f\{0\} = 0$, the mixture is at its bubble point. Since $\Sigma z_i = 1$, (7-11) becomes

$$f\{\psi\} = f\{0\} = 1 - \sum_{i=1}^{C} z_i K_i \qquad (7\text{-}18)$$

The bubble-point criterion, therefore, is

$$1 = \sum_{i=1}^{C} z_i K_i \qquad (7\text{-}19)$$

with $x_i = z_i$ and $y_i = K_i x_i$.
Equation (7-19) is useful for calculating bubble-point temperature at a specified pressure or bubble-point pressure at a specified temperature.

If $f\{\psi\} = f\{1\}$, (7-11) becomes

$$f\{\psi\} = f\{1\} = \sum_{i=1}^{C} \frac{z_i}{K_i} - 1 \qquad (7\text{-}20)$$

If $f\{1\} < 0$, the mixture is above its dew point (superheated vapor). If $f\{1\} = 1$, the mixture is at its dew point. Accordingly, the dew point criterion is

$$\sum_{i=1}^{C} \frac{z_i}{K_i} = 1 \tag{7-21}$$

with $y_i = z_i$ and $x_i = y_i/K_i$.

The bubble- and dew-point criteria, (7-19) and (7-21), are generally highly nonlinear in temperature but only moderately nonlinear in pressure, except in the region of the convergence pressure where K-values of very light or very heavy species can change radically with pressure, as in Fig. 4.6. Therefore, iterative procedures are required to solve for bubble- and dew-point conditions. One exception is where Raoult's law K-values are applicable as described in

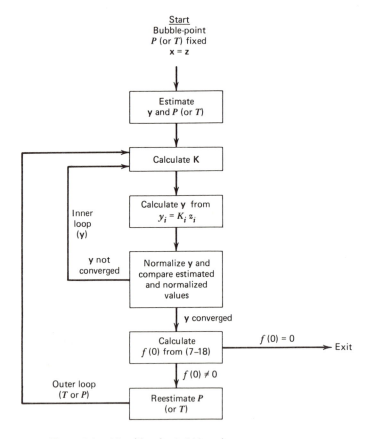

Figure 7.6. Algorithm for bubble-point temperature or pressure when K-values are composition dependent.

Section 4.6. Substitution of (4-75) into (7-18) leads to an equation for the direct calculation of bubble-point pressure

$$P_{\text{bubble}} = \sum_{i=1}^{C} z_i P_i^s \tag{7-22}$$

where P_i^s is the temperature-dependent vapor pressure of species i. Similarly, from (4-75) and (7-21), the dew-point pressure is

$$P_{\text{dew}} = \left[\sum_{i=1}^{C} \frac{z_i}{P_i^s} \right]^{-1} \tag{7-23}$$

Another useful exception occurs for mixtures at the bubble point when K-values can be expressed by the modified Raoult's law (5-21). Substituting this

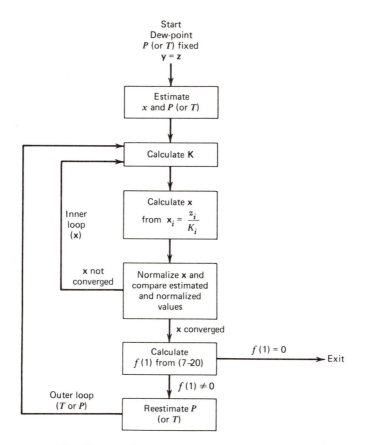

Figure 7.7. Algorithm for dew-point temperature or pressure when K-values are composition dependent.

equation into (7-18), we have

$$P_{bubble} = \sum_{i=1}^{C} \gamma_i z_i P_i^s \qquad (7\text{-}24)$$

where liquid-phase activity coefficients can be computed for known temperatures and compositions by methods in Chapter 5.

Where K-values are nonlinear in pressure and temperature and are composition dependent, algorithms such as those in Figs. 7.6 and 7.7 can be employed. For solving (7-19) and (7-21), the Newton–Raphson method is convenient if K-values can be expressed analytically in terms of temperature or pressure. Otherwise, the method of false position can be used. Unfortunately, neither method is guaranteed to converge to the correct solution. A more reliable but tedious numerical method, especially for bubble-point temperature calculations involving strongly nonideal liquid solutions, is Muller's method.[5]

Bubble- and dew-point calculations are useful to determine saturation conditions for liquid and vapor streams, respectively. It is important to note that when vapor–liquid equilibrium is established, the vapor is at *its* dew point and the liquid is at *its* bubble point.

Example 7.3. Cyclopentane is to be separated from cyclohexane by liquid–liquid extraction with methanol at 25°C. Calculate the bubble-point pressure using the equilibrium liquid-phase compositions given in Example 5.11.

Solution. Because the pressure is likely to be low, the modified Raoult's law is suitable for computing K-values. Therefore, the bubble point can be obtained directly from (7-24). The activity coefficients calculated from the van Laar equation in Example 5.11 can be used. Vapor pressures of the pure species are computed from the Antoine relation (4-69) using constants from Appendix I. Equation (7-24) applies to either the methanol-rich layer or the cyclohexane-rich layer, since from (4-31) $\gamma_i^I x_i^I = \gamma_i^{II} x_i^{II}$. Results will differ depending on the accuracy of the activity coefficients. Choosing the methanol-rich layer, we find

	$x^I = z$	γ^I	Vapor Pressure, psia
Methanol	0.7615	1.118	2.45
Cyclohexane	0.1499	4.773	1.89
Cyclopentane	0.0886	3.467	6.14

$$P_{bubble} = 1.118(0.7615)(2.45) + 4.773(0.1499)(1.89)$$
$$+ 3.467(0.0886)(6.14) = 5.32 \text{ psia}(36.7 \text{ kPa})$$

A similar calculation based on the cyclohexane-rich layer gives 5.66 psia (39.0 kPa), which is 6.4% higher. A pressure higher than this value will prevent formation of vapor at this location in the extraction process.

□

Example 7.4. Propylene is to be separated from 1-butene by distillation into a distillate vapor containing 90 mole% propylene. Calculate the column operating pressure assuming the exit temperature from the partial condenser is 100°F (37.8°C), the minimum attainable with cooling water. Determine the composition of the liquid reflux.

Solution. The operating pressure corresponds to a dew-point condition for the vapor distillate composition. The composition of the reflux corresponds to the liquid in equilibrium with the vapor distillate at its dew point. As shown in Example 4.8, propylene (P) and 1-butene (B) form an ideal solution. Ideal K-values are plotted for 100°F in Fig. 4.4. The method of false position[3] can be used to perform the iterative calculations by rewriting (7-21) in the form

$$f\{P\} = \sum_{i=1}^{C} \frac{z_i}{K_i} - 1 \qquad (7\text{-}25)$$

The recursion relationship for the method of false position is based on the assumption that $f\{P\}$ is linear in P such that

$$P^{(k+2)} = P^{(k+1)} - f\{P^{(k+1)}\}\left[\frac{P^{(k+1)} - P^{(k)}}{f\{P^{(k+1)}\} - f\{P^{(k)}\}}\right] \qquad (7\text{-}26)$$

Two values of P are required to initialize this formula. Choose 100 psia and 190 psia. At 100 psia

$$f\{P\} = \frac{0.90}{1.97} + \frac{0.10}{0.675} - 1.0 = -0.3950$$

Subsequent iterations give

k	$P^{(k)}$, psia	K_P	K_B	$f\{P^{(k)}\}$
1	100	1.97	0.675	−0.3950
2	190	1.15	0.418	0.0218
3	185.3	1.18	0.425	−0.0020
4	185.7	1.178	0.424	−0.0001

Iterations are terminated when $|P^{(k+2)} - P^{(k+1)}|/P^{(k+1)} < 0.0001$.

An operating pressure of 185.7 psia (1279.9 kPa) at the partial condenser outlet is indicated. The composition of the liquid reflux is obtained from $x_i = z_i/K_i$ with the result:

	Equilibrium Mole Fraction	
Component	**Vapor Distillate**	**Liquid Reflux**
Propylene	0.90	0.764
1-Butene	0.10	0.236
	1.00	1.000

☐

7.4 Adiabatic Flash

When the pressure of a liquid stream of known composition, flow rate, and temperature (or enthalpy) is reduced adiabatically across a valve as in Fig. 7.1a, an adiabatic flash calculation can be made to determine the resulting temperature, compositions, and flow rates of the equilibrium liquid and vapor streams for a specified flash drum pressure. In this case, the procedure of Fig. 7.4a is applied in an iterative manner, as in Fig. 7.8, by choosing the flash temperature T_V as the iteration or tear variable whose value is guessed. Then ψ, **x**, **y**, and L are determined as for an isothermal flash. The guessed value of T_V (equal to T_L) is checked by an enthalpy balance obtained by combining (7-15) for $Q = 0$ with (7-14) to give

$$f\{T_V\} = \frac{\psi H_V + (1-\psi)H_L - H_F}{1000} = 0 \tag{7-27}$$

where the division by 1000 is to make the terms in (7-27) of the order of one. The method of false position can be used to iterate on temperature T_V until $|T_V^{(k+1)} - T_V^{(k)}|/T_V^{(k)} < 0.0001$. The procedure is facilitated when mixtures form ideal solutions, or when initial K-values can be estimated from $K_i = \nu_{iL}^o$.

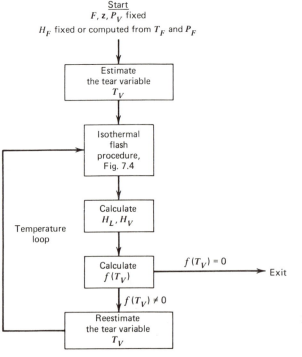

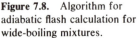

Figure 7.8. Algorithm for adiabatic flash calculation for wide-boiling mixtures.

The algorithm of Fig. 7.8 is very successful when (7-11) is not sensitive to T_V. This is the case for wide-boiling mixtures such as those in Example 7.2. For close-boiling mixtures (e.g., isomers), the algorithm may fail because (7-11) may become extremely sensitive to the value of T_V. In this case, it is preferable to select ψ as the tear variable and solve (7-11) iteratively for T_V.

$$f\{T_V\} = \sum_{i=1}^{C} \frac{z_i(1 - K_i)}{1 + \psi(K_i - 1)} = 0 \qquad (7\text{-}28)$$

then solve (7-12) and (7-13) for x and y, respectively, and then (7-27) directly for ψ noting that

$$f\{\psi\} = \frac{\psi H_V + (1 - \psi)H_L - H_F}{1000} = 0 \qquad (7\text{-}29)$$

from which

$$\psi = \frac{H_F - H_L}{H_V - H_L} \qquad (7\text{-}30)$$

If ψ from (7-30) is not equal to the value of ψ guessed to solve (7-28), the new value of ψ is used to repeat the loop starting with (7-28).

In rare cases, (7-11) and (7-27) may both be very sensitive to ψ and T_V, and neither of the above two tearing procedures may converge. Then, it is necessary to combine (7-12) and (7-13) with (7-27) to give

$$f_2\{\psi, T_V\} = \frac{\psi H_V\{T_V, \psi\} + (1 - \psi)H_L\{T_V, \psi\} - H_F}{1000} = 0 \qquad (7\text{-}31)$$

and solve this nonlinear equation simultaneously with (7-11), another nonlinear equation, in the form

$$f_1\{\psi, T_V\} = \sum_{i=1}^{C} \frac{z_i[1 - K_i\{T_V\}]}{1 + \psi[K_i\{T_V\} - 1]} = 0 \qquad (7\text{-}32)$$

Example 7.5. The equilibrium liquid from the flash drum at 100°F and 485 psia in Example 7.2 is fed to a stabilizer to remove the remaining hydrogen and methane. Pressure at the feed plate of the stabilizer is 165 psia (1138 kPa). Calculate the percent vaporization of the feed if the pressure is decreased adiabatically from 485 to 165 psia by valve and pipe line pressure drop.

Solution. This problem is solved by making an adiabatic flash calculation using the Chao–Seader K-value correlation and the corresponding enthalpy equations of Edmister, Persyn, and Erbar in Section 5.1. Computed enthalpy H_F of the liquid feed at 100°F and 485 psia is -5183 Btu/lbmole (-12.05 MJ/kgmole), calculated in the manner of Example 5.3. The adiabatic flash algorithm of Fig. 7.8 applies because the mixture is wide boiling. For the isothermal flash step, the algorithm of Fig. 7.4a is used with the Rachford–Rice and Newton procedures. Two initial assumptions for the equilibrium temperature are required to begin the method of false position in the temperature loop of Fig. 7.8. These are taken as $T_1 = 554.7°R$ and $T_2 = (1.01)T_1$. Results for the temperature loops are:

m, Temp. Loop Iteration	$T^{(m-1)}$, °F	$T^{(m)}$, °F	Vapor Fraction, ψ	H_V, Btu lbmole	H_L, Btu lbmole	$f\{T_V\}$, (7-27)	$T^{(m+1)}$, °F	$\left\| \dfrac{T^{(m+1)} - T^{(m)}}{T^{(m)} + 459.7} \right\|$
1	—	95.0	0.03416	4365	−5589	−0.0660	—	—
2	95.0	100.547	0.03462	4419	−5484	0.0418	98.397	0.00384
3	100.547	98.397	0.03444	4398	−5525	0.0002	98.407	0.00020

Convergence is obtained after only three iterations in the temperature loop because (7-11) is quite insensitive to the assumed value of T_V. From six to seven iterations were required for each of the three passes through the isothermal flash procedure. Final equilibrium streams at 98.4°F (a 1.6°F drop in temperature) and 165 psia are

Component	Equilibrium Vapor, lbmole/hr	Equilibrium Liquid, lbmole/hr	K-value
Hydrogen	1.87	0.20	255.8
Methane	15.00	13.58	30.98
Benzene	0.34	355.07	0.0268
Toluene	0.04	114.66	0.00947
	17.25	483.51	

Only 3.44% of the feed is vaporized.

$\square$

7.5   Other Flash Specifications

A schematic representation of an equilibrium flash or partial condensation is shown in Fig. 7.9. Pressure may or may not be reduced across the valve. In practice, heat transfer, when employed, is normally done in a separate upstream heat exchanger, as in Fig. 7.1a, rather than in the flash drum. In the schematic, Q may be 0, positive, or negative.

In general, a large number of sets of specifications is possible for flash calculations. Some useful ones are listed in Table 7.4. In all cases, the feed flow rate F, composition z_i, pressure P_F, and temperature T_F (or enthalpy H_F) are assumed known. Cases 6, 7, and 8 should be used with extreme caution because valid solutions are generally possible only for very restricted regions of mole fractions in the equilibrium vapor and/or liquid. A preferred approach for determining the approximate degree of separation possible between two key components in a single-stage flash is to utilize case 5, where V is set equal to the amounts of light-key and lighter species in the feed. An algorithm for this case

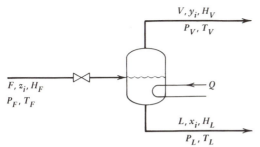

Figure 7.9.   General equilibrium flash.

Table 7.4   Flash calculation specifications

Case	Specified Variables	Type of Flash	Output Variables
1	P_V, T_V	Isothermal	Q, V, y_i, L, x_i
2	$P_V, Q = 0$	Adiabatic	T_V, V, y_i, L, x_i
3	$P_V, Q \neq 0$	Nonadiabatic	T_V, V, y_i, L, x_i
4	P_V, L	Percent liquid	Q, T_V, V, y_i, x_i
5	$P_V(\text{or } T_V), V$	Percent vapor	$Q, T_V(\text{or } P_V), y_i, L, x_i$
6	$P_V, x_j(\text{or } x_j L)$	Liquid purity	$Q, T_V, V, y_i, L, x_{i_{i \neq j}}$
7	$P_V(\text{or } T_V), y_j(\text{or } y_j V)$	Vapor purity	$Q, T_V(\text{or } P_V), V, y_{i_{i \neq j}}, L, x_i$
8	$y_j(\text{or } y_j V), x_k(\text{or } x_k L)$	Separation	$Q, P_V, T_V, V, y_{i_{i \neq j}}, L, x_{i_{i \neq k}}$

can be constructed by embedding one of the isothermal flash procedures of Fig. 7.4 into an iterative loop where values of T_V are assumed until the specified V (or ψ) is generated.

7.6   Multistage Flash Distillation

A single-stage flash distillation generally produces a vapor that is only somewhat richer in the lower-boiling constituents than the feed. Further enrichment can be achieved by a series of flash distillations where the vapor from each stage is condensed, then reflashed. In principle, any desired product purity could be obtained by such a multistage flash technique provided a suitable number of stages is employed. However, in practice, recovery of product would be small, heating and cooling requirements high, and relatively large quantities of various liquid products would be produced.

As an example, consider Fig. 7.10a where n-hexane (H) is separated from n-octane by a series of three flashes at 1 atm (pressure drop and pump needs are ignored). The feed to the first flash stage is an equimolal bubble-point liquid at a flow rate of 100 lbmole/hr. A bubble-point temperature calculation yields 192.3°F. Using Case 5 of Table 7.4, where the vapor rate leaving stage 1 is set

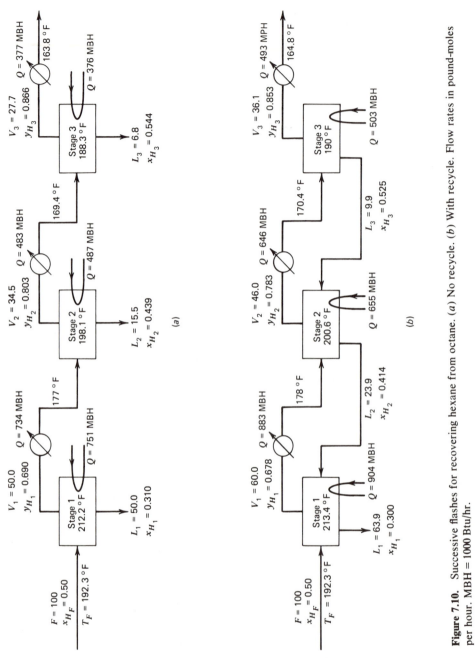

Figure 7.10. Successive flashes for recovering hexane from octane. (*a*) No recycle. (*b*) With recycle. Flow rates in pound-moles per hour. MBH = 1000 Btu/hr.

equal to the amount of n-hexane in the feed to stage 1, the calculated equilibrium exit phases are as shown. The vapor V_1 is enriched to a hexane mole fraction of 0.690. The heating requirement is 751,000 Btu/hr. Equilibrium vapor from stage 1 is condensed to bubble-point liquid with a cooling duty of 734,000 Btu/hr. Repeated flash calculations for stages 2 and 3 give the results shown. For each stage, the leaving molal vapor rate is set equal to the moles of hexane in the feed to the stage. The purity of n-hexane is increased from 50 mole% in the feed to 86.6 mole% in the final condensed vapor product, but the recovery of hexane is only 27.7(0.866)/50 = 48%. Total heating requirement is 1,614,000 Btu/hr and liquid products total 72.3 lbmole/hr.

In comparing feed and liquid products from two contiguous stages, we note that liquid from the later stage and the feed to the earlier stage are both leaner in hexane, the more volatile species, than the feed to the later stage. Thus, if intermediate streams are recycled, intermediate recovery of hexane should be improved. This processing scheme is depicted in Fig. 7.10b, where again the molar fraction vaporized in each stage equals the mole fraction of hexane in combined feeds to the stage. The mole fraction of hexane in the final condensed vapor product is 0.853, just slightly less than that achieved by successive flashes without recycle. However, the use of recycle increases recovery of hexane from 48 to 61.6%. As shown in Fig. 7.10b, increased recovery of hexane is accompanied by approximately 28% increased heating and cooling requirements. If the same degree of heating and cooling is used for the no-recycle scheme in Fig. 7.10a as is used in 7.10b, the final hexane mole fraction y_{H_3} is reduced from 0.866 to 0.815, but hexane recovery is increased to 36.1(0.815)/50 = 58.8%.

Both successive flash arrangements in Fig. 7.10 involve a considerable number of heat exchangers and pumps. Except for stage 1, the heaters in Fig. 7.10a can be eliminated if the two intermediate total condensers are converted to partial condensers with duties of 247 MBH[a] (734–487) and 107 MBH (483–376). Total heating duty is now only 751,000 Btu/hr, and total cooling duty is 731,000 Btu/hr. Similarly, if heaters for stages 2 and 3 in Fig. 7.10b are removed by converting the two total condensers to partial condensers, total heating duty is 904,000 Btu/hr (20% greater than the no-recycle case), and cooling duty is 864,000 Btu/hr (18% greater than the no-recycle case).

A considerable simplification of the successive flash technique with recycle is shown in Fig. 7.11a. The total heating duty is provided by a feed boiler ahead of stage 1. The total cooling duty is utilized at the opposite end to condense totally the vapor leaving stage 3. Condensate in excess of distillate is returned as reflux to the top stage, from which it passes successively from stage to stage countercurrently to vapor flow. Vertically arranged adiabatic stages eliminate the need for interstage pumps, and all stages can be contained within a single piece

[a] MBH = 1000 Btu/hr.

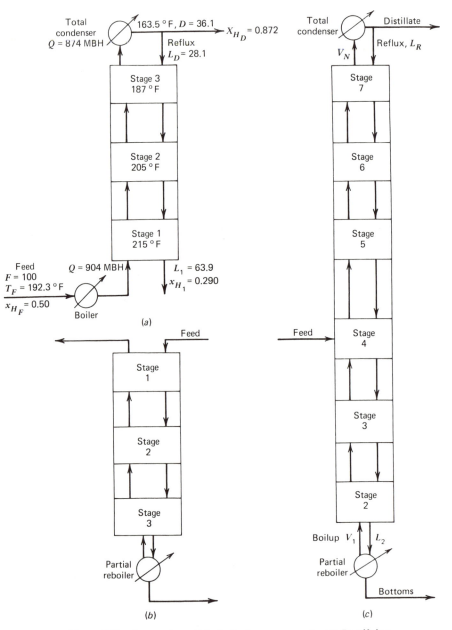

Figure 7.11. Successive adiabatic flash arrangements. (*a*) Rectifying section. (*b*) Stripping section. (*c*) Multistage distillation.

of equipment. Such a set of stages is called a *rectifying section*. As discussed in Chapter 17, such an arrangement may be inefficient thermodynamically because heat is added at the highest temperature level and removed at the lowest temperature level.

The number of degrees of freedom for the arrangement in Fig. 7.11a is determined by the method of Chapter 5 to be $(C + 2N + 10)$. If all independent feed conditions, number of stages (3), all stage and element pressures (1 atm.), bubble-point liquid leaving the condenser, and adiabatic stages and divider are specified, two degrees of freedom remain. These are specified to be a heating duty for the boiler and a distillate rate equal to that of Fig. 7.10b. Calculations result in a mole fraction of 0.872 for hexane in the distillate. This is somewhat greater than that shown in Fig. 7.10b.

The same principles by which we have concluded that the adiabatic multi-stage countercurrent flow arrangement is advantageous for concentrating a light component in an overhead product can be applied to the concentration of a heavy component in a bottoms product, as in Fig. 7.11b. Such a set of stages is called a *stripping section*.

7.7 Combinations of Rectifying and Stripping Stages (MSEQ Algorithm)

Figure 7.11c, a combination of Fig. 7.11a and 7.11b with a liquid feed, is a complete column for rectifying and stripping a feed to effect a sharper separa-tion between light key and heavy key components than is possible with either a stripping or an enriching section alone. Adiabatic flash stages are placed above and below the feed. Recycled liquid reflux L_R is produced in the condenser and vapor boilup in the reboiler. The reflux ratios are L_R/V_N and L_2/V_1 at the top and the bottom of the apparatus, respectively. All flows are countercurrent. This type of arrangement is widely used in industry for multistage distillation. Modifications are used for most other multistage separation operations depicted in Table 1.1.

The rectifying stages above the point of feed introduction purify the light product by contacting it with successively richer liquid reflux. Stripping stages below the feed increase light product recovery because vapor relatively low in volatile constituents strips them out of the liquid. For heavy product, the functions are reversed; the stripping section increases purity; the enriching section, recovery.

Algorithms based on those developed for single-stage flash distillations can be applied to multistage distillation with reflux and boilup. Although these are not widely used industrially, because they are less efficient than the methods to be given in Chapter 15, the stage-to-stage flash algorithm (MSEQ), which will now be developed, is highly instructive, computationally stable, and very versatile. Consider Fig. 7.12, which is a four-stage column with a partial reboiler as a first stage and a total condenser. Stage 3 receives a feed of 100 lbmole/hr of a saturated liquid at column pressure of 100 psi (68.94 kPa) and 582°F (323.3°K) containing mole percentages of 30, 30, and 40 propane, n-butane, and n-pentane, respectively. Other specifications are adiabatic stages and dividers, isobaric conditions, a B of 50 lbmole/hr, saturated reflux, and a reflux ratio $L_R/D = 2$ to account for the remainder of the $(C + 2N + 9)$ degrees of freedom listed in Table 6.2.

To calculate the output variables, repeated flash calculations are made. However, an iterative procedure is required because initially all inlet streams to a particular stage or element are not known. A method discussed by McNeil and Motard[6] is to proceed repeatedly down and up the column until convergence of the output variables on successive iterations is achieved to an acceptable tolerance. This procedure is analogous to the start-up of a distillation column.

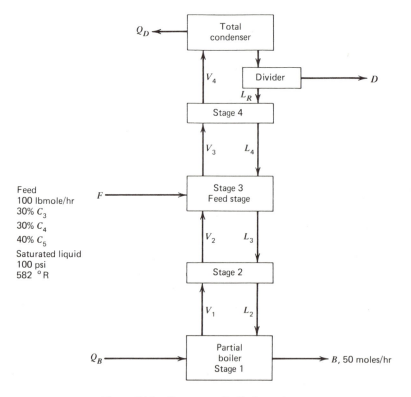

Figure 7.12.   Four-stage distillation column.

Let us begin at stage 3 where the known feed is introduced. Unknown variables are initially set to zero but are continually replaced with the latest computed values as the iterations proceed.

Iteration 1 (Down from the Feed and up to the Reflux Divider)

(a) *Stage 3.* With streams L_4 and V_2 unknown, an adiabatic flash simply gives L_3, x_3, and T_3 equal to corresponding values for the feed.

(b) *Stage 2.* With stream V_1 unknown, an adiabatic flash simply gives L_2, x_2, and T_2 equal to L_3, x_3, and T_3, respectively.

(c) *Stage 1, Partial Reboiler.* With stream L_2 known and the flow rate of stream B specified, a total material balance gives $V_1 = (L_2 - B) = (100 - 50) = 50$ lbmole/hr. To obtain this flow rate for V_1, a percent vapor flash (Case 5, Table 7.4) is performed to give Q_B, T_1, y_1, and x_B. The algorithm for this type of flash is shown in Fig. 7.13.

(d) *Stage 2.* With initial estimates of streams L_3 and V_1 now available, an

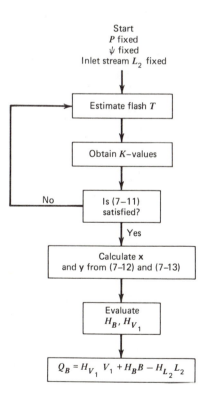

Figure 7.13 Percent vapor flash algorithm for partial reboiler.

adiabatic flash is again performed at stage 2 using an algorithm similar to that of Fig. 7.8, where "F fixed" refers here to L_3 plus V_1. Initial output variables for stream V_2 and a revised set of output variables for stream L_2 are determined.

(e) *Stage 3.* With stream L_4 still unknown, but stream F known and an initial estimate now available for stream V_2, an adiabatic flash is again performed at stage 3.

(f) *Stage 4.* With only an initial estimate of stream V_3 available from the previous stage 3 calculation, an adiabatic flash simply results in stream L_4 remaining at zero and an initial estimate of stream V_4 as identical to stream V_3.

(g) *Total Condenser.* Stream V_4 is condensed to a saturated liquid with temperature T_D computed by a bubble point and condenser duty Q_D computed by an enthalpy balance.

(h) *Reflux Divider.* Condensed stream V_4 is divided according to the specified reflux ratio of $(L_R/D) = 2$. Thus, $D = V_4/[1 + (L_R/D)]$ and $L_R = (V_4 - D)$. During the first iteration, boilup and reflux are generated, but in amounts less than the eventual converged steady-state quantities.

Iteration 2. (Down from Stage 4 and up to the Reflux Divider)

The second iteration starts with initial estimates for almost all streams. An exception is stream L_4, for which no estimate is needed. Initial estimates of streams V_3 and L_R are used in an adiabatic flash calculation for stage 4 to determine an initial estimate for stream L_4. Subsequently, flash calculations are performed in order for stages 3, 2, and 1, and then back up the column for stages 2, 3, and 4 followed by the total condenser and reflux divider. At the conclusion of the second iteration, generally all internal vapor and liquid flow rates are increased over values generated during the first iteration.

Subsequent Iterations

Iterations are continued until successive values for Q_B, Q_D, all V, all L, all T, all $\mathbf{y}$, and all $\mathbf{x}$ do not differ by more than a specified tolerance. In particular, the value of the flow rate for D must satisfy the overall total material balance $F = (D + B)$.

The above flash cascade technique provides a rigorous, assumption-free, completely stable, iterative calculation of unknown stream flow rates, temperatures, and compositions, as well as reboiler and condenser duties. The rate of convergence is relatively slow, however, particularly at high reflux ratios, because the initial vapor and liquid flow rates for all streams except F and B are initially zero, and only $(F - B - D)$ moles are added to the column at each iteration. Another drawback of this multistage flash (MSEQ) method is that an adiabatic flash is required to obtain each stage temperature. Most of the methods to be discussed in Chapter 15 are faster because they start with guesses for all flow rates and stage temperatures, and these are then updated at each iteration.

References

1. Hughes, R. R., H. D. Evans, and C. V. Sternling, *Chem. Eng. Progr.*, **49**, 78–87 (1953).

2. Rachford, H. H., Jr., and J. D. Rice, *J. Pet. Tech.*, **4** (10), Section 1, p. 19, and Section 2, p. 3 (October, 1952).

3. Perry, R. H., and C. W. Chilton, Eds, *Chemical Engineers' Handbook*, 5th ed, McGraw–Hill Book Co., New York, 1973, 2-53 to 2-55.

4. Seader, J. D., and D. E. Dallin, *Chemical Engineering Computing*, AIChE Workshop Series, **1**, 87–98 (1972).

5. Muller, D. E., *Math. Tables Aids Comput.*, **10**, 205–208 (1956).

6. McNeil, L. J., and R. L. Motard, "Multistage Equilibrium Systems," *Proceedings of GVC/AIChE Meeting at Munich*, Vol. II, C-5, 3 (1974).

Problems

7.1 A liquid containing 60 mole% toluene and 40 mole% benzene is continuously fed to and distilled in a single-stage unit at atmospheric pressure. What percent of benzene in the feed leaves in the vapor if 90% of the toluene entering in the feed leaves in the liquid? Assume a relative volatility of 2.3 and obtain the solution graphically.

7.2 Using the y-x and T-y-x diagrams in Figs. 3.3 and 3.4, determine the temperature, amounts, and compositions of the equilibrium vapor and liquid phases at 101 kPa for the following conditions with a 100-kgmole mixture of $nC_6(H)$ and $nC_8(C)$.
(a) $z_H = 0.5$, $\psi = 0.2$.
(b) $z_H = 0.4$, $y_H = 0.6$.
(c) $z_H = 0.6$, $x_C = 0.7$.
(d) $z_H = 0.5$, $\psi = 0$.
(e) $z_H = 0.5$, $\psi = 1.0$.
(f) $z_H = 0.5$, $T = 200°F$.

7.3 A liquid mixture consisting of 100 kgmoles of 60 mole% benzene, 25 mole% toluene, and 15 mole% o-xylene is flashed at 1 atm and 100°C.
(a) Compute the amounts of liquid and vapor products and their composition.
(b) Repeat the calculation at 100°C and 2 atm.
(c) Repeat the calculation at 105°C and 0.1 atm.
(d) Repeat the calculation at 150°C and 1 atm.
Assume ideal solutions and use the Antoine equation.

7.4 The system shown below is used to cool the reactor effluent and separate the light gases from the heavier hydrocarbons. Calculate the composition and flow rate of the vapor leaving the flash drum. Does the flow rate of liquid quench influence the result?

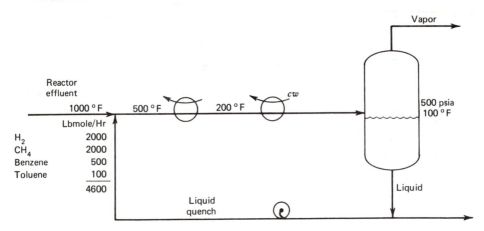

K-values for the components at 500 psia and 100°F are:

	K
H_2	80
CH_4	10
Benzene	0.010
Toluene	0.004

7.5 The mixture shown below is partially condensed and separated into two phases. Calculate the amounts and compositions of the equilibrium phases, V and L.

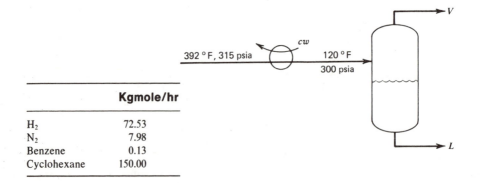

	Kgmole/hr
H_2	72.53
N_2	7.98
Benzene	0.13
Cyclohexane	150.00

7.6 Determine the phase condition of a stream having the following composition at 7.2°C and 2620 kPa.

Component	Kgmole/hr
N_2	1.0
C_1	124.0
C_2	87.6
C_3	161.6
nC_4	176.2
nC_5	58.5
nC_6	33.7

7.7 An equimolal solution of benzene and toluene is totally evaporated at a constant temperature of 90°C. What are the pressures at the beginning and the end of the vaporization process? Assume an ideal solution and use the Antoine equation with constants from Appendix I for vapor pressures.

7.8 For a binary mixture, show that at the bubble point with one liquid phase present

$$x_1 = \frac{1 - K_2}{K_1 - K_2}$$

7.9 Consider the basic flash equation

$$\sum_{i=1}^{C} \frac{z_i(1 - K_i)}{1 + \left(\dfrac{V}{F}\right)(K_i - 1)} = 0$$

Under what conditions can this equation be satisfied?

7.10 The following mixture is introduced into a distillation column as saturated liquid at 1.72 MPa. Calculate the bubble-point temperature using the K-values of Fig. 7.5.

Compound	Kgmole/hr
Ethane	1.5
Propane	10.0
n-Butane	18.5
n-Pentane	17.5
n-Hexane	3.5

7.11 A liquid containing 30 mole% toluene, 40 mole% ethylbenzene, and 30 mole% water is subjected to a continuous flash distillation at a total pressure of 0.5 atm. Assuming that mixtures of ethylbenzene and toluene obey Raoult's law and that the hydrocarbons are completely immiscible in water and vice versa, calculate the temperature and composition of the vapor phase at the bubble-point temperature.

7.12 A seven-component mixture is flashed at a specified temperature and pressure.
(a) Using the K-values and feed composition given below, make a plot of the Rachford–Rice flash function.

$$f\{\psi\} = \sum_{i=1}^{C} \frac{z_i(1 - K_i)}{1 + \psi(K_i - 1)}$$

at intervals of ψ of 0.1, and from the plot estimate the correct root of ψ.

(b) An alternate form of the flash function is:

$$f\{\psi\} = \sum_{i=1}^{C} \frac{z_i K_i}{1 + \psi(K_i - 1)} - 1$$

Make a plot of this equation also at intervals of ψ of 0.1 and explain why the first function is preferred.

Component	z_i	K_i
1	0.0079	16.2
2	0.1321	5.2
3	0.0849	2.6
4	0.2690	1.98
5	0.0589	0.91
6	0.1321	0.72
7	0.3151	0.28

(c) Assume the K-values are for 150°F and 50 psia and that they obey Raoult's and Dalton's laws. Calculate the bubble-point and dew-point pressures.

7.13 The following equations are given by Sebastiani and Lacquaniti [Chem. Eng. Sci., 22, 1155 (1967)] for the vapor–liquid equilibria in the water (W)–acetic acid (A) system.

$$\log \gamma_w = x_A^2 [A + B(4x_w - 1) - C(x_w - x_A)(4x_A - 1)^2]$$

$$\log \gamma_A = x_w^2 [A + B(4x_w - 3) + C(x_w - x_A)(4x_w - 3)^2]$$

$$A = 0.1182 + \frac{64.24}{T(°K)}$$

$$B = 0.1735 - \frac{43.27}{T(°K)}$$

$$C = 0.1081$$

Find the dew point and bubble point of a mixture of composition $x_w = 0.5$, $x_A = 0.5$ at 1 atm. Flash the mixture at a temperature halfway between the dew point and the bubble point.

7.14 Find the bubble-point and dew-point temperatures of a mixture of 0.4 mole fraction toluene (1) and 0.6 mole fraction iso-butanol (2) at 101.3 kPa. The K-values can be calculated from (5-21) using the Antoine equation for vapor pressure and γ_1 and γ_2 from the van Laar equation (5-26), with $A_{12} = 0.169$ and $A_{21} = 0.243$. If the same mixture is flashed at a temperature midway between the bubble point and dew point and 101.3 kPa, what fraction is vaporized, and what are the compositions of the two phases?

7.15 (a) For a liquid solution having a molar composition of ethyl acetate (A) of 80% and ethyl alcohol (E) of 20%, calculate the bubble-point temperature at 101.3 kPa and the composition of the corresponding vapor using (5-21) with the Antoine equation and the van Laar equation (5-26) with $A_{AE} = 0.144$, $A_{EA} = 0.170$.

(b) Find the dew point of the mixture.

(c) Does the mixture form an azeotrope? If so, predict the temperature and composition.

7.16 The overhead system for a distillation column is as shown below. The composition of the *total* distillates is indicated, with 10 mole% of it being taken as vapor. Determine the pressure in the reflux drum, if the temperature is 100°F. Use the K-values given below at any other pressure by assuming that K is inversely proportional to pressure.

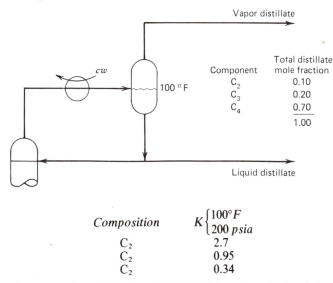

Vapor distillate

Component	Total distillate mole fraction
C_2	0.10
C_3	0.20
C_4	0.70
	1.00

100 °F

Liquid distillate

Composition	$K \begin{cases} 100°F \\ 200\,psia \end{cases}$
C_2	2.7
C_2	0.95
C_2	0.34

7.17 The following stream is at 200 psia and 200°F. Determine whether it is a subcooled liquid or a superheated vapor, or whether it is partially vaporized, without making a flash calculation.

Component	Lbmole/hr	K-value
C_3	125	2.056
nC_4	200	0.925
nC_5	175	0.520
	500	

7.18 Prove that the vapor leaving an equilibrium flash is at its dew point and that the liquid leaving an equilibrium flash is at its bubble point.

7.19 In the sketch below, 150 kgmole/hr of a saturated liquid L_1 at 758 kPa of molar composition—propane 10%, n-butane 40%, and n-pentane 50%—enters the reboiler from stage 1. What are the composition and amounts of V_B and B? What is Q_R, the reboiler duty?

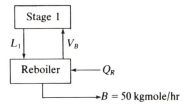

Stage 1

L_1 V_B

Reboiler ← Q_R

→ B = 50 kgmole/hr

7.20 Streams entering stage F of a distillation column are as shown. What is the temperature of stage F and the composition and amount of streams V_F and L_F if the pressure is 758 kPa for all streams?

Bubble-point feed, 160 kgmole/hr

Mole percent
C_3 20
nC_4 40
nC_5 40

Stream	Total Flow Rate, kg mole/hr	Composition Mole%		
		C_3	nC_4	nC_5
L_{F-1}	100	15	45	40
V_{F+1}	196	30	50	20

7.21 Flash adiabatically a stream composed of the six hydrocarbons given below. Feed enthalpy H_F is 13,210 Btu and the pressure is 300 psia.

$$P = 300 \text{ psia}$$

Feed composition and K-values: $\left(\dfrac{K_i}{T}\right)^{1/3} = a_{1,i} + a_{2,i}T + a_{3,i}T^2 + a_{4,i}T^3 (T = °R)$

Component	z_i	$a_1 \times 10^2$	$a_2 \times 10^5$	$a_3 \times 10^8$	$a_4 \times 10^{12}$
C_2H_4	0.02	-5.177995	62.124576	-37.562082	8.0145501
C_2H_6	0.03	-9.8400210	67.545943	-37.459290	-9.0732459
C_3H_6	0.05	-25.098770	102.39287	-75.221710	153.84709
C_3H_8	0.10	-14.512474	53.638924	-5.3051604	-173.58329
nC_4	0.60	-14.181715	36.866353	16.521412	-248.23843
iC_4	0.20	-18.967651	61.239667	-17.891649	-90.855512

Enthalpy (T in °R)

$$P = 300 \text{ psia}$$

$$(H_{L_i})^{1/2} = c_{1,i} + c_{2,i}T + c_{3,i}T^2 \qquad (H_{V_i})^{1/2} = e_{1,i} + e_{2,i}T + e_{3,i}T^2$$

Component	c_1	$c_2 \times 10$	$c_3 \times 10^5$	e_1	$e_2 \times 10^4$	$e_3 \times 10^6$
C_2H_4	-7.2915	1.5411962	-1.6088376	56.79638	615.93154	2.4088730
C_2H_6	-8.4857	1.6286636	-1.9498601	61.334520	588.7543	11.948654
C_3H_6	-12.4279	1.8834652	-2.4839140	71.828480	658.5513	11.299585
C_3H_8	-14.50006	1.9802223	-2.9048837	81.795910	389.81919	36.47090
nC_4	-20.29811	2.3005743	-3.8663417	152.66798	-1153.4842	146.64125
iC_4	-16.553405	2.1618650	-3.1476209	147.65414	-1185.2942	152.87778

Note. Basis for enthalpy: Saturated liquid at $-200°F$.

7.22 As shown below, a hydrocarbon mixture is heated and expanded before entering a distillation column. Calculate the mole percentage vapor phase and vapor and liquid phase mole fractions at each of the three locations indicated by a pressure specification.

Component	Mole Fraction
C_2	0.03
C_3	0.20
nC_4	0.37
nC_5	0.35
nC_6	0.05
	1.00

7.23 One hundred kilogram-moles of a feed composed of 25 mole% n-butane, 40 mole% n-pentane, and 35 mole% n-hexane are flashed at steady-state conditions. If 80% of the hexane is to be recovered in the liquid at 240°F, what pressure is required, and what are the liquid and vapor compositions?

7.24 For a mixture consisting of 45 mole% n-hexane, 25 mole% n-heptane, and 30 mole% n-octane at 1 atm
(a) Find the bubble- and dew-point temperatures.
(b) The mixture is subjected to a flash distillation at 1 atm so that 50% of the feed is vaporized. Find the flash temperature, and the composition and relative amounts of the liquid and vapor products.
(c) Repeat Parts (a) and (b) at 5 atm and 0.5 atm.
(d) If 90% of the hexane is taken off as vapor, how much of the octane is taken off as vapor?

7.25 An equimolal mixture of ethane, propane, n-butane, and n-pentane is subjected to a flash vaporization at 150°F and 205 psia. What are the expected liquid and vapor products? Is it possible to recover 70% of the ethane in the vapor by a single-stage flash at other conditions without losing more than 5% of nC_4 to the vapor?

7.26 (a) Use Newton's method to find the bubble-point temperature of the mixture given below at 50 psia.

Pertinent data at 50 psia are:

Component	z_i	$a_1 \times 10^1$	$a_2 \times 10^3$	$a_3 \times 10^4$	$a_4 \times 10^5$
Methane	0.005	5.097584	0.2407971	−0.5376841	0.235444
Ethane	0.595	−7.578061	3.602315	−3.955079	1.456571
n-Butane	0.400	−6.460362	2.319527	−2.058817	0.6341839

where $(K_i/T)^{1/3} = a_i + a_2 T + a_3 T^2 + a_4 T^3 (T, °R)$

(b) Find the temperature that results in 25% vaporization at this pressure. Determine the corresponding liquid and vapor compositions.

7.27 Derive algorithms for carrying out the adiabatic flash calculations given below, assuming that expressions for K-values are available.
(a) As functions of T and P only.
(b) As functions of T, P, and liquid (but not vapor) composition.
     Expressions for component enthalpies as a function of T are also available at specified pressure and excess enthalpy is negligible.

Given	Find
H_F, P	ψ, T
H_F, T	ψ, P
H_F, ψ	T, P
ψ, T	H_F, P
ψ, P	H_F, T
T, P	ψ, H_F

7.28 (a) Consider the three-phase flash.

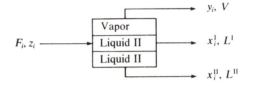

Derive equations of the form $f\{z_i, \psi, K_i\}$ for this three-phase system, where $\psi = V/L^I$, $\xi = L^I/(L^I + L^{II})$. Suggest an algorithm to solve these equations.
(b) Isobutanol (1) and water (2) form an immiscible mixture at the bubble point. The van Laar constants are $A_{12} = 0.74$ and $A_{21} = 0.30$ and the vapor pressures are given by the Antoine equation. For the overall composition, $z_1 = 0.2$ and $z_2 = 0.8$, find the composition of the two liquid phases and the vapor at 1 atm at the bubble point.
(c) For the same overall composition, what are the liquids in equilibrium at 80°C?

7.29 Propose a detailed algorithm for the Case 5 flash in Table 7.4 where the percent vaporized and the flash pressure are to be specified.

7.30 Solve Problem 7.1 by assuming an ideal solution and using the Antoine equation with constants from Appendix I for the vapor pressure. Also determine the temperature.

7.31 In Fig. 7.10, is anything gained by totally condensing the vapor leaving each stage? Alter the processes in Fig. 7.10a and 7.10b so as to eliminate the addition of heat to stages 2 and 3 and still achieve the same separations.

7.32 Develop an algorithm similar to MSEQ in Section 7.7 to design liquid–liquid extraction columns such as the one of Fig. 6.8, Example 6.2.
     The equilibrium data are available in the form of Fig. 3.11 for a ternary system. The specification of variables is as in Example 6.2.

7.33 Saturated vapor is fed to the second stage of the column shown below. The feed rate, bottoms rate, and reflux ratio are as given.
     Assuming that vapor and liquid rates leaving an adiabatic stage are equal,

respectively, to the same rates entering the stage (constant molal overflow assumption), insert in the table below the values calculated using the MSEQ algorithm. Start at the boiler and work up and down the column until you are almost converged. How could you obtain the final compositions and temperatures after the flows have converged?

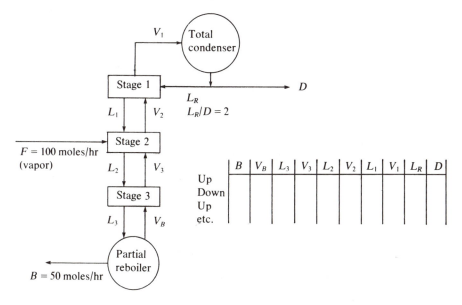

7.34 Develop an algorithm similar to MSEQ to design a liquid–liquid extraction column such as the one of Fig. 6.8, Example 6.2.

Use your algorithm to solve for V_N and L_B'' given the following data: $V_0/L_B'' = 0.3$, L_B is pure solvent, $F = 100$ kg with 95% water and 5% ethylene glycol, S is pure furfural at 100 kg/hr. Equilibrium data are given in Table 3.2.

8
Graphical Multistage Calculations by the McCabe-Thiele Method

However, it is felt that the following graphical method is simpler, exhibits its results in a plainer manner than any method, analytical or graphical, so far proposed, and is accurate enough for all practical use.

Warren L. McCabe and
Edwin W. Thiele, 1925

Graphical methods are exceedingly useful for visualizing the relationships among a set of variables, and as such are commonly used in chemical engineering. They are useful in stagewise contactor design because design procedures involve the solution of equilibrium relationships simultaneously with material and enthalpy balances. The equilibrium relationships, being complex functions of the system properties, are frequently presented in the graphical forms discussed in Chapter 3. Material and enthalpy balance equations can be plotted together on these same graphs. By proper choice of coordinates and appropriate geometric constructions, it is then possible to achieve graphical solutions to design problems. It is also true that any problem amenable to solution by graphical techniques can also be handled analytically. In fact, the increasing availability of high-speed computing facilities and software has, perhaps, antiquated graphical methods for rigorous design purposes. However, they remain useful for preliminary designs and particularly for easy visualization of the interrelationships among variables.

We begin the development of graphical design procedures by deriving generalized material balance equations for countercurrent cascades. These are first applied to simple, single-section distillation, extraction, absorption, and ion-exchange columns and then to increasingly complex binary distillation systems. Incorporation of enthalpy balances into graphical design procedures is deferred until Chapter 10.

8.1   Countercurrent Multistage Contacting

In a countercurrent multistage section, the phases to be contacted enter a series of ideal or equilibrium stages from opposite ends. A contactor of this type is diagramatically represented by Fig. 8.1, which could be a series of stages in an absorption, a distillation, or an extraction column. Here L and V are the molal (or mass) flow rates of the heavier and lighter phases, and x_i and y_i the corresponding mole (or mass) fractions of component i, respectively. This chapter focuses on binary or pseudobinary systems so the subscript i is seldom required. Unless specifically stated, y and x will refer to mole (or mass) fractions of the lighter component in a binary mixture, or the species that is transferred between phases in three-component systems.

The development begins with a material balance about stage $n + 1$, the top stage of the cascade in Fig. 8.1, as indicated by the top envelope to $n + 1$. Streams L_{n+2} and V_n enter the stage, and L_{n+1} and V_{n+1} leave.

$$L_{n+2}x_{n+2} + V_ny_n = V_{n+1}y_{n+1} + L_{n+1}x_{n+1} \tag{8-1}$$

or

$$y_n = \frac{L_{n+1}}{V_n} x_{n+1} + \frac{V_{n+1}y_{n+1} - L_{n+2}x_{n+2}}{V_n} \tag{8-2}$$

Letting Δ represent $V_{n+1}y_{n+1} - L_{n+2}x_{n+2}$, the net flow of the light component out the top of the section, we obtain

$$y_n = \frac{L_{n+1}}{V_n} x_{n+1} + \frac{\Delta}{V_n} \tag{8-3}$$

A similar material balance is performed around the top of the column and stage n, the second stage from the top, as indicated by the envelope to n in Fig. 8.1. Streams L_{n+2} and V_{n-1} enter the boundary, and streams L_n and V_{n+1} leave.

$$L_{n+2}x_{n+2} + V_{n-1}y_{n-1} = V_{n+1}y_{n+1} + L_nx_n$$

Solving for y_{n-1}, we have

$$y_{n-1} = \frac{L_n}{V_{n-1}} x_n + \left[\frac{\Delta}{V_{n-1}}\right] \tag{8-4}$$

Equations (8-3) and (8-4) can be used to locate the points (y_n, x_{n+1}) and (y_{n-1}, x_n), and others on an x-y diagram. The line passing through these points is called

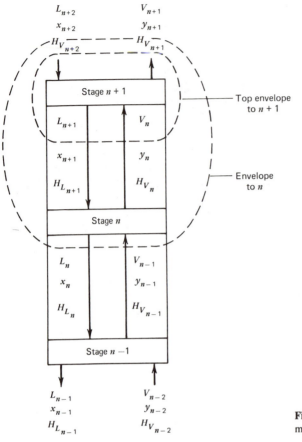

Figure 8.1. Countercurrent multistage contactor.

the *operating line*. All countercurrently passing streams in the column, (L_{n+2}, V_{n+1}), (L_{n+1}, V_n), (L_n, V_{n-1}), and so on lie on this operating line, which may be curved or straight. If the ratio of phase flows is constant throughout the section of stages, then $L/V = L_{n+1}/V_n = L_n/V_{n-1} = \ldots = L_{n-1}/V_{n-2}$; and the slopes of the lines defined by (8-3) and (8-4) are identical. Furthermore, if V and L are constant, all passing streams in the column lie on the same *straight* operating line of slope L/V, which may be drawn if we know either:

1. The concentrations for only one set of passing streams—for example, (y_{n-1}, x_n) or (y_n, x_{n+1}), and L/V, the ratio of phase flows in the contactor. A common statement of this problem is: Given L/V and the inlet and outlet composition at one end, calculate the inlet and outlet compositions at the other end and the number of stages.

2. The concentrations of any two pairs of passing streams. The two most convenient streams to analyze are those entering and leaving the cascade (L_{n+2}, V_{n+1} and L_{n-1}, V_{n-2}). These points lie at the ends of the operating line. A common statement of this problem is: Given inlet and exit compositions, calculate L/V and the number of stages.

A relationship between the ratios of the flow rates and the compositions of passing streams may be developed if it is assumed that the flow rates of liquid and vapor are constant. Then, subtraction of (8-4) from (8-3) yields for any stage in the section

$$\frac{y_n - y_{n-1}}{x_{n+1} - x_n} = \frac{L}{V} \tag{8-5}$$

The number of theoretical stages required to effect the transfer of a specified amount of light component from phase L to phase V can be determined by using the material balance operating line in conjunction with a phase equilibrium curve on an x-y diagram. An example of a graphical construction for a simple countercurrent multistage section is shown in Fig. 8.2. The compositions of entering and discharging streams are the specified points A and B, which are located as (y_0, x_1) and (y_n, x_{n+1}) in Fig. 8.2a, which also shows the phase equilibrium curve for the system. If L/V is constant throughout the

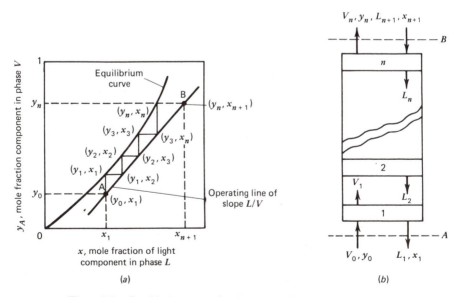

Figure 8.2. Graphical construction for a two-plase countercurrent multistage separator. (a) McCabe–Thiele construction. (b) Column.

section of stages, then the straight line connecting point A with point B is the operating line (the locus of all passing streams). To determine the number of stages required to achieve the change of composition from A to B, step off the stages as shown by the staircase construction. Starting at A (the composition of passing streams under stage 1) move vertically on x_1 to the equilibrium curve to find y_1 (the composition of the vapor leaving stage 1). Next, move horizontally on y_1 to the operating line point (y_1, x_2)—the composition of the passing streams between stages 1 and 2. Continue vertically and then horizontally between points on the operating line and the equilibrium line until point B is reached or overshot. Four equilibrium stages, as represented by points on the equilibrium curve, are required for the separation. Note that this procedure is as accurate as is the draftsman and that there is rarely an integer number of stages.

8.2 Application to Rectification of Binary Systems

The graphical construction shown in Fig. 8.2 was applied to distillation calculations for binary systems by McCabe and Thiele in 1925.[1] Since then, the x-y plot with operating line(s) and equilibrium curve has come to be known as the McCabe–Thiele diagram. The method will be developed here in the context of the rectifying column of Fig. 8.3 where the vapor feed to the bottom of the column provides the energy to maintain a vapor phase V rising through the column. The countercurrent liquid phase L is provided by totally condensing the overhead vapor and returning a portion of this distillate to the top of the column as saturated liquid reflux. The column is assumed to operate at constant pressure, and the stages and divider are adiabatic. An *internal reflux ratio* is defined as L_n/V_n, and an *external reflux ratio* as L_R/D.

The unit of Fig. 8.3 has $C + 2N + 8$ degrees of freedom. Suppose the following design variables are specified.

Adiabatic stages and divider	$N + 1$
Constant pressure (1 atm) in stages, condenser, divider	$N + 2$
Feed T, P, composition and rate (saturated vapor, 40 kgmole/hr, 20 mole% hexane in octane	$C + 2$
Product (hexane) composition, $x_D = 0.9$	1
Product rate, 5 kgmole/hr	1
Saturated reflux	$\dfrac{1}{C + 2N + 8}$

The design problem is fully specified, so it is possible to calculate the number of

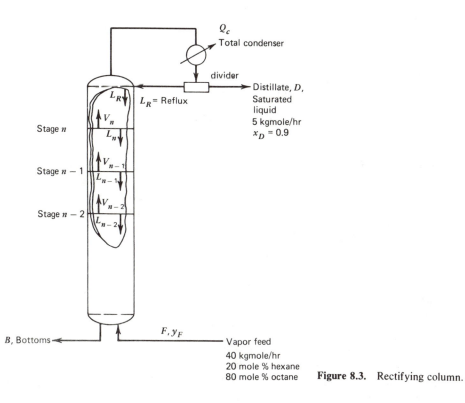

Figure 8.3. Rectifying column.

stages required for the separation as well as all other column parameters (stage compositions, temperatures, reflux ratios, etc.).

Before demonstrating the graphical McCabe–Thiele solution, we examine what is involved in an analytical stage-by-stage solution. The first step is to write material and heat balances on the basis of 1 hr (40 kgmole of feed).

A. An overall material balance, input = output.

$$F = B + D \qquad (8\text{-}6)$$

Since $F = 40$ and $D = 5$, $B = 35$.

B. A material balance for hexane. Let x_D = mole fraction of hexane in D, and x_B = the mole fraction of hexane in B.

$$y_F F = D x_D + B x_B$$

Therefore,

$$40(0.2) = 5(0.9) + 35(x_B) \qquad (8\text{-}7)$$

and

$$x_B = 0.1$$

C.   An overall enthalpy balance. Let

$$Q_C = \text{heat removed in condenser, J}$$

$$H = \text{enthalpy, J/kgmole}$$

$$FH_F = DH_D + BH_B + Q_C \tag{8-8}$$

All quantities in this equation except Q_C are known from the specification or can be computed from information obtained from (8-6) and (8-7). (Dew point and bubble point calculations are necessary to determine stream temperatures before determining enthalpies.) Thus, from (8-8), Q_C can be computed.

D.   Enthalpy balance around condenser and divider.

$$V_n H_{V_n} = DH_D + L_R H_R + Q_C \tag{8-9}$$

with

$$V_n = D + L_R \tag{8-10}$$

Since $y_n = x_D$ and the reflux is saturated, its temperature T_D can be calculated by a bubble point on D, and $H_R = H_D$ are then fixed. Then T_n is obtained by a dew point on V_n, which establishes H_{V_n}. Then (8-9) and (8-10) can be solved for L_R and V_n.

E.   Stage-to-stage calculations. For the top stage there is
(a) A total material balance

$$V_{n-1} + L_R = V_n + L_n \tag{8-11}$$

(b) A material balance for hexane

$$y_{n-1} V_{n-1} + L_R x_R = V_n y_n + L_n x_n \tag{8-12}$$

where $y_n = x_R$
(c) A phase equilibrium relationship. Figure 3.3 can be considered an equation of the form

$$x_n = f\{y_n\} \tag{8-13}$$

(d) An enthalpy balance

$$H_{V_{n-1}} V_{n-1} + H_R L_R = H_{V_n} V_n + H_{L_n} L_n \tag{8-14}$$

The known variables in (8-9) through (8-12) are x_R (since $x_R = x_D$), L_R, y_n (since $y_n = x_D$), and V_n. Also properties H_R (since $H_R = H_D$) and H_{V_n} are known if it is assumed that for a saturated stream the enthalpy is known if the composition is known; thus, the four unknowns are y_{n-1}, V_{n-1}, L_n, and x_n. Properties $H_{V_{n-1}}$ and H_{L_n} are related to y_{n-1} and x_n and the respective dew-point and bubble-point temperatures. Hence, solution of the four equations gives the unknown conditions about stage n. By developing additional energy and material balances analogous to (8-11), (8-12), and (8-14) about stage $n - 1$ and applying the phase equilibrium relation (8-13) to this stage, one can ascertain unknown conditions about stage $n - 1$. One continues in this manner to stage $n - m$, on which $x_{n-m} \leq 0.1 = x_B$, the bottoms product composition computed from (8-7).

Both analytical and graphical solutions are greatly simplified if $L_R = L_n = L = $ constant, and $V_n = V_{n-1} = V = $ constant. In this case it is possible to dispense with one equation for each stage, namely the enthalpy balance. This is called the *constant molal overflow assumption* (which was also embodied in (8-5)) and is valid if, over the temperature and pressure operating range of the separator:

1. The molar heats of vaporization of both species of the binary system are equal.

2. Heats of mixing, stage heat losses, and sensible heat changes of both liquid and vapor are negligible.

Then, every mole of condensing vapor vaporizes exactly 1 mole of liquid. Since it is the molar latent heats that are presumed to be equal, the flows must be specified in terms of moles, and the concentrations in terms of mole fractions for use with the constant molal overflow assumption.

Often the assumption of constant L/V in a section of stages causes no significant error. However, it is important to understand how deviations from this condition arise.

1. For homologous series of compounds, the molal heat of vaporization generally increases with increasing molecular weight. When conditions are close to isothermal, this causes a decrease in the molal vapor rate as we move down the stages.

2. The temperature decreases as we move up the stages. This results in an increase in molal heat of vaporization, but a decrease in sensible heat of both vapor and liquid for a given species. The net result depends on the particular mixture.

In general, the importance of energy effects is determined largely by the magnitude of the difference in passing vapor and liquid flows. In rectifying sections where $L < V$, a relatively small value of external reflux may be reduced to zero before reaching the bottom of the section. In stripping sections where

$V < L$, a relatively small amount of boilup from the reboiler may be reduced to zero before reaching the top of the section. Heat effects can also be severe in separators with large temperature differences between top and bottom sections.

Example 8.1. Using graphical (McCabe–Thiele) methods, calculate the number of stages required to make the separation described in Section 8.2 and Fig. 8.3. Assume constant molal overflow ($V = 40$ kgmole/hr).

Solution. The McCabe–Thiele diagram is shown in Fig. 8.4. A logical start is at point A, where $y_n = x_D = x_R = 0.9$. Since x_R and y_n are passing streams, point A must lie on the operating line, which from (8-2) for constant molal overflow is:

$$y = \frac{L}{V} x + \frac{V y_n - L x_R}{V}$$

Here, x and y are the compositions of any two passing streams, and L/V = the slope of the operating line; $35/40 = 0.875$ in this example. With reference to Fig. 8.1 ($V y_n - L x_R$) is Δ, the difference in net flow of the two phases at the top of the column. In the nomenclature of this example, the difference is

$$(V_n y_n - L_R x_R) = D x_D$$

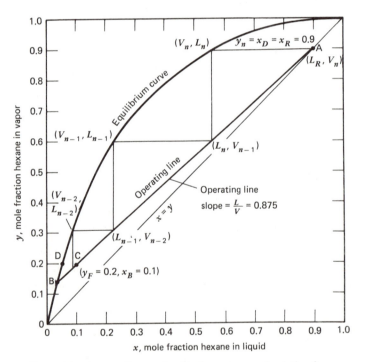

Figure 8.4. An x-y diagram for hexane–octane and solution for Example 8.1, $P = 101$ kPa.

Thus, the operating line equation becomes

$$y = \frac{L}{V} x + \frac{D x_D}{V}$$

The line $\overline{ACB}$ in Fig. 8.4 is the operating line, which passes through the point A ($y_n = x_D = 0.9$) and point C ($x_B = 0.1$, $y_F = 0.2$), both of which represent passing streams. Its slope is $L/V = (0.9 - 0.2)/(0.9 - 0.1) = 0.875$ as stated above. The corresponding external reflux ratio $= L_R/D = (L/V)/(1 - L/V) = 7$.

Proceeding down the column of stages by going from the passing streams (L_R, V_n) to the equilibrium streams leaving plate n (V_n, L_n) to the passing streams below n (L_n, V_{n-1}, etc.) the number of plates required to reach $x_B = 0.1$ is seen to be less than three. □

The contactor discussed in Example 8.1 is a rectifying or enriching column since its main function is to purify the lower-boiling constituent, hexane. The device is incapable of producing an octane bottoms product much richer in octane than the feed. For example, if the distillate rate were increased from 5 to approximately 7.44 so the operating line intersected the equilibrium curve at point D ($y = 0.2$, $x = 0.04$) and if an infinite number of equilibrium stages were employed, a maximum octane purity in the bottoms product of 96 mole% would be achieved. Almost any purity of hexane in the distillate can be attained, however. In order to produce a higher concentration of octane in the bottoms, stages below the feed point are required—the so-called stripping section. Methods for designing columns having both stripping and enriching sections will be developed in Section 8.6.

8.3 Application to Extraction

Liquid–liquid extraction introduces an added complexity, since the material balance must take cognizance of at least two components in the feed and at least one additional component as a mass separating agent. In distillation, energy transfer makes the separation of the mixture possible. In extraction (Fig. 8.5) a solvent S is used instead of heat to extract a solute from a mixture of the solute and another solvent W. The two solvents are immiscible in each other, but the solute is miscible in both. We denote solvent-rich extract phase flow by E and raffinate phase flow by R.

The nomenclature adopted in Fig. 8.5 is:

E_a, E_{n-2}, etc. = Total mass flow rates of extract phase.

E_S = Mass flow rate of solvent in extract phase (assumed constant).

y_a, y_{n-2}, etc. = Mass fraction solute in phase E.

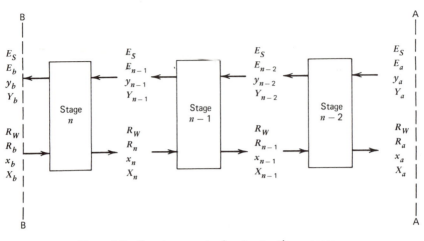

Figure 8.5. Countercurrent solvent extraction process.

Y_a, Y_{n-2}, etc. = Mass flow solute/mass flow solvent in phase E.
R_n, R_{n-1}, etc. = Total mass flow of raffinate phase.
R_W = Mass flow rate of inert W in raffinate phase (assumed constant).
x_n, x_{n-1}, etc. = Mass fraction solute in phase R.
X_n, X_{n-1}, etc. = Mass flow of solute/mass flow of inerts in phase R.

The overall total and component mass balances per unit time are, in terms of total mass flow rates and mass fractions

$$E_a + R_b = E_b + R_a$$

$$R_b x_b + E_a y_a = E_b y_b + R_a x_a$$

or in terms of mass ratios with R_W and E_S constant

$$R_W(X_b - X_a) = E_S(Y_b - Y_a) \tag{8-15}$$

Equation (8-15) is analogous to (8-5) and gives the slope of an operating line on an X-Y diagram such as Fig. 8.6 in terms of the difference in composition of two passing streams in the column.

$$\frac{R_W}{E_S} = \frac{Y_b - Y_a}{X_b - X_a} = \frac{\Delta Y}{\Delta X} \tag{8-16}$$

Similarly, by taking a mass balance about stage $n-2$, we develop an expression for the operating line.

$$E_S Y_a + R_W X_{n-1} = E_S Y_{n-2} + R_W X_a$$

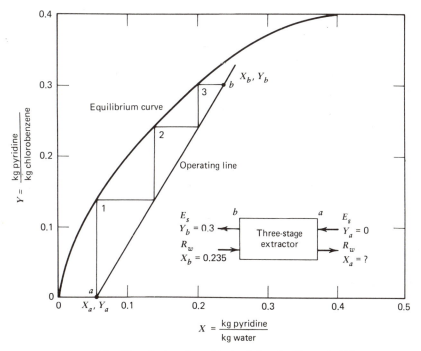

Figure 8.6.   Flowsheet and graphical solution for Example 8.2.

Therefore

$$Y_{n-2} = \frac{R_W}{E_S}(X_{n-1}) + \frac{E_S Y_a - R_W X_a}{E_S} \tag{8-17}$$

Equation (8-17) is the relationship for passing streams between stages $n-1$ and $n-2$ and represents a straight operating line on a Y-X phase equilibrium diagram for all pairs of passing streams between stages, when R_W and E_S are constant from stage to stage. This is analogous to the assumption of constant molal overflow in distillation. Note, however, that the simplifying assumption of constant R_W/E_S permits us to use either mole or mass ratios. Characteristic of a system that permits the assumption of a constant R_W/E_S with a relatively small error is complete immiscibility of the solvent S and inert W.

   If the equilibrium curve and the material balance equations are expressed in other than the mass ratio concentrations X and Y and mass of solvent and mass of inert F_S and R_W, the operating line will not be straight, even under conditions of complete immiscibility, since the ratio of total phases is not constant because of transfer of solute between phases.

Example 8-2. Pyridine may be removed from water by extraction with chlorobenzene with the equilibrium data at process conditions being as shown in Fig. 8.6.

An extract solution E_b containing 0.3 kg pyridine/kg chlorobenzene (Y_b) is to be produced by contacting 100 kg/hr of aqueous pyridine solution R_b with pure chlorobenzene, solution E_b, of mass ratio concentration $Y_a = 0$, in a countercurrent extraction column containing the equivalent of three theoretical stages. Pyridine concentration in the feed is 0.235 kg pyridine/kg water (X_b).

Assuming that chlorobenzene and water are mutually insoluble, what is the required ratio of pure solvent to feed and what is the concentration of pyridine in the final raffinate (X_a)?

Solution. First, a degrees-of-freedom analysis is conducted to determine whether the problem is specified adequately. From Table 6.2, for N connected equilibrium stages, there are $2N + 2C + 5$ degrees of freedom. Since there are three stages and three components, the number of design variables $= 6 + 6 + 5 = 17$.

Specified are:

Stage pressures	3
Stage heat transfer rates (equivalent to specifying stage temperatures)	3
Compositions, T, and P of E_S at a and R_w at b (the two feed streams)	8
Composition of stream E_S at b	2
Feed rate, 100 kg/hr	$\underline{1}$
	17

T and P specifications are implied in the equilibrium Y-X graph. Thus, exactly the number of independent design variables has been specified to solve the problem.

The point $Y_b = 0.3$, $X_b = 0.235$ is plotted. Point a with $Y_a = 0$ is then located by a trial-and-error procedure since there is only one operating line that will result in exactly three stages between point a and point b. The result is $X_a = 0.05$.

From (8-16) or the slope of the operating line, R_w/E_S is 1.62. Thus, on a basis of 100 kg of feed, since $S_b = 0.235$ kg pyridine/kg water, there are 81.0 kg of H_2O in the feed; and 50.0 kg of solvent must be used. This result should be checked by a material balance

$$E_S Y_a + R_w X_b = E_S Y_b + R_w X_a$$

$$50(0) + 81(0.235) = 50(0.3) + 81(0.05) \approx 19.04$$

☐

8.4 Application to Absorption

Gas absorption and liquid–liquid extraction are analogous in that in each there are two carrier streams and at least one solute to be partitioned between them. The following example illustrates the application of the McCabe–Thiele graphical method to a gas absorption problem. Mole units are used.

Example 8.3. Ninety-five percent of the acetone vapor in an 85 vol% air stream is to be absorbed by countercurrent contact with a stream of pure water in a valve–tray column with an expected overall tray efficiency of 30%. The column will operate essentially at 20°C and 101 kPa pressure. Equilibrium data for acetone–water at these conditions are

Mole percent acetone in water	3.30	7.20	11.7	17.1
Acetone partial pressure in air, torr	30.00	62.80	85.4	103.0

Calculate:

(a)   Minimum value of L/G, the ratio of moles of water per mole of air.

(b)   The number of equilibrium stages required, using a value of L/G of 1.25 times the minimum.

(c)   The concentration of acetone in the exit water. From Table 6.2, for N connected equilibrium stages, there are $2N + 2C + 5$ degrees of freedom. Specified in this problem are

Stage pressures (101 kPa)	N
Stage temperatures (20°C)	N
Feed stream composition	$C - 1$
Water stream composition	$C - 1$
Feed stream T, P	2
Water stream T, P	2
Acetone recovery	1
L/G	$\dfrac{1}{2N + 2C + 4}$

The remaining specification will be the feed flow rate.

Solution.   Assumptions: (1) No water is vaporized. (2) No air dissolves in the water. (3) No acetone is in the entering water. Basis: 100 kgmole/hr of entering gas.

Conditions at column bottom:

$$\text{Acetone in gas} = 15 \text{ kgmole/hr}$$

$$\text{Air in gas} = 85 \text{ kgmole/hr}$$

$$Y_B = 15/85 = 0.176 \text{ mole acetone/mole air}$$

Conditions at column top:

$$\text{Acetone in gas} = (0.05)(15) = 0.75 \text{ kgmole/hr}$$

$$\text{Air in gas} = 85.00 \text{ kgmole/hr}$$

$$Y_T = 0.75/85 = 0.00882 \text{ mole acetone/mole air}$$

Converting the equilibrium data to mole ratios we have

Equilibrium Curve Data ($P = 101$ kPa)

x	$1 - x$	X $x/(1 - x)$	y (p/P)	$1 - y$	Y $y/(1 - y)$
0		0	0		0
0.033	0.967	0.0341	0.0395	0.9605	0.0411
0.072	0.928	0.0776	0.0826	0.9174	0.0901
0.117	0.883	0.1325	0.1124	0.8876	0.1266
0.171	0.829	0.2063	0.1355	0.8645	0.1567

These data are plotted in Fig. 8.7 as Y versus X, the coordinates that linearize the operating line, since the slope $\Delta Y/\Delta X =$ moles water/mole air is constant throughout the column.

(a) **Minimum solvent rate.** The operating line, which must pass through (X_T, Y_T), cannot go below the equilibrium curve (or acetone would be desorbed rather than absorbed). Thus, the slope of the dashed operating line in Fig. 8.7, which is tangent to the equilibrium curve, represents the minimum ratio of water to air that can be used when $Y_B = 0.176$ is required. The corresponding X'_B is the highest X_B possible. At this minimum solvent rate, $(L/G)_{min} = 1.06$ moles H_2O/mole air, an infinite number of stages would be required.

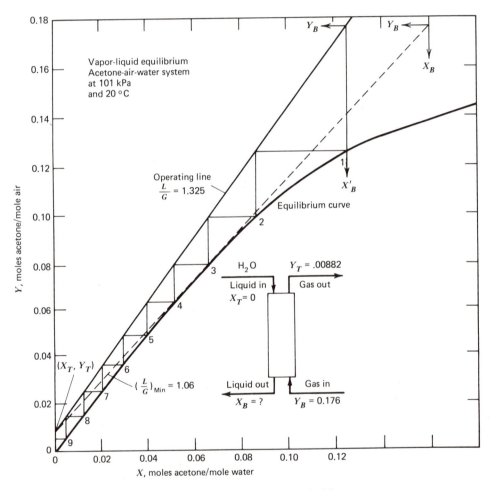

Figure 8.7. Solution to Example 8.3.

(b) **Actual plates required**

$$L/G = 1.25(L/G)_{min} = 1.325 \text{ moles } H_2O/\text{mole air}$$

Intersection of the operating line with $Y_B = 0.176$ occurs at $X_B = 0.126$.

Theoretical stages = 8.7, where, for convenience, stage numbering in Fig. 8.7 is down from the top.
Actual plates = 8.7/0.30 = 29.

☐

8.5 Application to Ion Exchange

By proper choice of coordinate system, McCabe–Thiele diagrams can be applied to exchange of ion G in an ion-exchange resin, with ion A in solution.

$$A + G \cdot resin = G + A \cdot resin$$

Example 8.4. If y and x are used to denote the equivalent ionic fraction of A on the resin and in solution, respectively, and the separation factor is $\alpha = (y_A/x_A)/(y_G/x_G) = 3.00$, then the equilibrium data can be presented graphically, as in Fig. 8.8, by the relation $y = \alpha x/[1 - x(1 - \alpha)]$, where the unsubscripted y and x refer to ion A.

Assume that the ionic fraction of component A in the solution entering at the top of the continuous countercurrent ion exchanger of Fig. 8.8 is 1 and that the y leaving is 0.7. At the bottom of the ion exchanger, $y = 0$ and $x = 0.035$.
(a) How many equivalent stages are in the column?

(b) If the exchange process is equimolal and subject to the following process conditions, how many cubic feet per minute of solution L can the column process?

ρ = bulk density of resin = 0.65 g/cm^3 of bed

Q = total capacity = 5 milligram equivalents (meq)/g (dry)

R = resin flow rate = 10 ft^3 bed/min (0.283 m^3/min)

C = total ionic level in solution = 1 meq/cm^3

Solution. Basis of 1 minute:
(a) The straight operating line in Fig. 8.8 passes through the two specified terminal points. It is seen that less than three stages are required.

(b) The slope of the operating line is 0.721 and the material balance is

$$\text{Decrease of A in solution} = \text{Increase of A on resin}$$

$$C(\Delta x)L = Q\rho R(\Delta y)$$

or

$$\frac{\Delta y}{\Delta x} = 0.72 = \frac{CL}{Q\rho R} = \frac{(1)L}{(5)(0.65)(10)}$$

Therefore, $L = 23.43$ ft^3/min (0.663 m^3/min).

☐

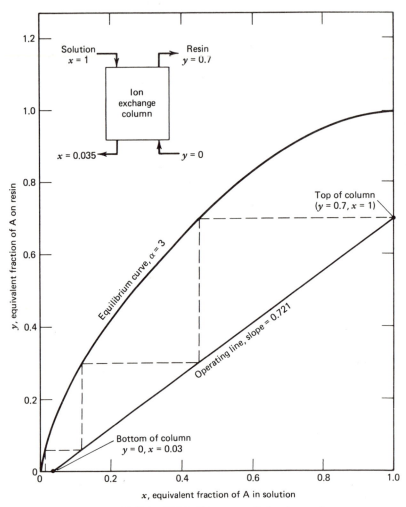

Figure 8.8. McCabe–Thiele diagram applied to ion exchange.

8.6 Application to Binary Distillation

Once the principle of the operating line and the method of stepping off stages between it and the equilibrium curve are clearly understood, the procedures can be applied to more complex separators. In the case of the distillation operation in Fig. 8.9, this might include:

1. Feeding the column at some intermediate feed plate while simultaneously

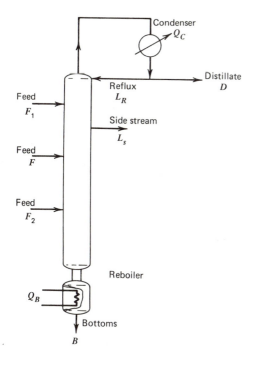

Figure 8.9.   Various distillation operations.

returning reflux to the top of the column. The thermal condition of the feed may be a liquid below its boiling point, a saturated liquid, part liquid–part vapor, saturated vapor, or superheated vapor.

2. A multiple feeding arrangement with feeds entering intermediate plates (feeds F_1 and F_2).

3. Operation approaching the *minimum reflux ratio*. As will be seen, this results in a maximum amount of product per unit heat input, with the number of plates approaching infinity.

4. Operation at *total reflux* ($L/V = 1$). This gives the minimum number of plates required to achieve a separation when no feed enters and no product is withdrawn.

5. Withdrawal of an intermediate side stream as a product.

The fractionation column in Fig. 8.10 contains both an enriching section and a stripping section. In Section 8.2, the operating line for the enriching section was developed, and the same approach can be used to derive the operating line for the stripping section, again assuming constant molal overflow. Let $\bar{L}$ and $\bar{V}$ denote liquid and vapor flows in the stripping section, noting they may differ

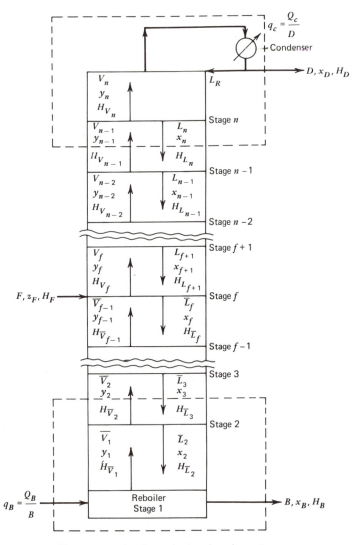

Figure 8.10.   Sections of a fractionation column.

from enriching section values because of flow changes across the feed stage. Taking a balance about the lower dotted section of the fractionation column in Fig. 8.10, we have

Total balance: $\qquad\qquad \bar{V}_2 + B = \bar{L}_3$ $\qquad\qquad$ (8-18)

Component balance: $y_2\bar{V}_2 + Bx_B = \bar{L}_3 x_3$ $\qquad$ (8-19)

$$y_2 = \frac{\bar{L}_3}{\bar{V}_2} x_3 - \frac{Bx_B}{\bar{V}_2} \tag{8-20}$$

or

$$y_2 = \frac{\bar{L}}{\bar{V}} x - \frac{Bx_B}{\bar{V}} \tag{8-21}$$

The enthalpy balance is

$$\bar{H}_{\bar{V}_2}\bar{V}_2 + H_B B = H_{\bar{L}_3}\bar{L}_3 + Bq_B \tag{8-22}$$

where

$$Bq_B = Q_B \tag{8-23}$$

The intersection of the stripping section operating line (8-21) with the $(x = y)$ line is given by

$$x\bar{V} + Bx_B = \bar{L}x$$

Since

$$\bar{L} - \bar{V} = B$$

the intersection is at $x = x_B$.

The locus of the intersections of the operating lines for the enriching section and the stripping section of Fig. 8.11 is obtained by simultaneous solution of the generalized enriching section operating line (8-24) and the stripping section operating line (8-25), which is a generalization of (8-21).

$$y = \frac{L}{V} x + \frac{D}{V} x_D \tag{8-24}$$

$$y = \frac{\bar{L}}{\bar{V}} x - \frac{B}{\bar{V}} x_B \tag{8-25}$$

Subtracting (8-25) from (8-24), we have

$$y(V - \bar{V}) = x(L - \bar{L}) + (Dx_D + Bx_B) \tag{8-26}$$

Substituting (8-27), an overall component balance, for the last bracketed term in (8-26),

$$Fz_F = Dx_D + Bx_B \tag{8-27}$$

and for $(V - \bar{V})$ a total material balance around the feed stage,

$$(V - \bar{V}) = F - (\bar{L} - L)$$

and combining (8-26) and (8-27), we obtain

$$y = \left(\frac{q}{q-1}\right) x - \left(\frac{z_F}{q-1}\right) \tag{8-28}$$

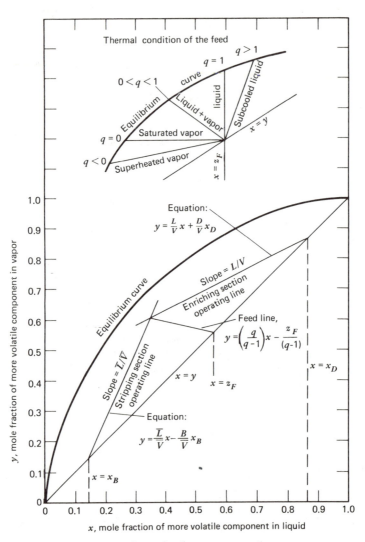

Figure 8.11. Operating lines on an x-y diagram.

where

$$q = \frac{\bar{L} - L}{F} \tag{8-29}$$

Equation (8-28) is the equation of a straight line, the so-called q-line. Its slope, $q/(q-1)$, marks the intersection of the two operating lines. It is easily shown that it also intersects the $x = y$ line at $x = z_F$.

Thermal Condition of the Feed

The magnitude of q is related to the thermal condition of the feed. This can be demonstrated by writing enthalpy and material balance equations around the feed stage in Fig. 8.10, assuming constant molal overflow conditions

$$H_F F + H_{\bar{V}_{f-1}} \bar{V} + H_{L_{f+1}} L = H_{\bar{L}_f} + H_{V_f} V \qquad (8\text{-}30)$$

$$F + \bar{V} + L = \bar{L} + V \qquad (8\text{-}31)$$

Assuming $H_{L_{f+1}} = H_{\bar{L}_f}$ and $H_{\bar{V}_{f-1}} = H_{V_f}$ we can combine (8-30) and (8-31) and substitute the result into the definition of q given by (8-29) to obtain

$$q = \frac{H_{V_f} - H_F}{H_{V_f} - H_{\bar{L}_f}} \qquad (8\text{-}32)$$

Equation (8-32) states that the value of q can be established by dividing the enthalpy required to bring the feed to saturated vapor by the latent heat of vaporization of the feed. The effect of thermal condition of the feed on the slope of the q-line described by (8-32) is summarized and shown schematically in the insert of Fig. 8.11. Included are examples of q-lines for cases where the feed is above the dew point and below the bubble point.

Determination of the Number of Theoretical Stages

After the operating lines in the enriching section and the stripping section have been located, the theoretical stages are determined by stepping off stages in accordance with the procedure described in Section 8.1.

If the distillation column is equipped with a *partial reboiler*, this reboiler is equivalent to a single theoretical contact since, in effect, the reboiler accepts a liquid feed stream and discharges liquid and vapor streams in equilibrium: B and $\bar{V}_1$ in Fig. 8.10. However, not all reboilers operate this way. Some columns are equipped with *total reboilers* in which either all of the entering liquid from the bottom stage is vaporized and the bottoms are discharged as a vapor, or the liquid from the bottom stage is divided into a bottoms product and reboiler feed that is totally or partially vaporized.

If the overhead vapor is only partially condensed in a *partial condenser*, rather than being totally condensed, the condenser functions as another stage to the column. In Fig. 8.12, L_R and D leaving the partial condenser are in equilibrium.

Taking a material balance about stage n and the top of the column, we have

$$V y_{n-1} = L x_n + D y_D$$

and, making the usual simplifying assumptions, the operating line becomes

$$y = \frac{L}{V} x + \frac{D}{V} y_D$$

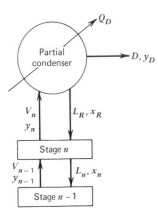

Figure 8.12. A partial condenser.

This operating line intersects the $(x = y)$ line at $x = y_D$. Figure 8.16, which is discussed later, shows the corresponding McCabe–Thiele construction.

Location of the Feed Stage

The feed stage location is at the changeover point from stepping-off stages between the enriching section operating line and the equilibrium curve to stepping-off stages between the stripping section operating line and the equilibrium curve (stage 3 in Fig. 8.13a). If for the same x_B, x_D, L/V, and $\bar{L}/\bar{V}$ a feed location below the optimum stage is chosen (stage 5 in Fig. 8.13b), more equilibrium contacts are required to effect the separation.

If a feed stage location above the optimum is chosen as shown in Fig. 8.13c, again more equilibrium contacts will be required to effect the separation than if the optimum feed point were used.

Limiting Operating Conditions

(a) *Minimum Number of Stages.* In the enriching section, the steepest slope any operating line through point x_D can have is $L/V = 1$. Under these conditions there is no product and the number of stages is a minimum for any given separation. For the stripping section, the lowest slope an operating line through x_B can have is $\bar{L}/\bar{V} = 1$, in which case no bottoms stream is withdrawn.

At $L = V$, and with no products, the operating line equations (8-24) and (8-25) become simply $y = x$. This situation, termed *total reflux*, affords the greatest possible area between the equilibrium curve and the operating lines, thus fixing the minimum number of equilibrium stages required to produce x_B and x_D. It is possible to operate a distillation column at total reflux, as shown in Fig. 8.14.

(b) *Minimum Reflux* Ratio (L_R/D or L/V). Algebraically, $L_R/D = (L/V)/(1 - L/V)$. Thus, L/D is a minimum when L/V is a minimum. Since no

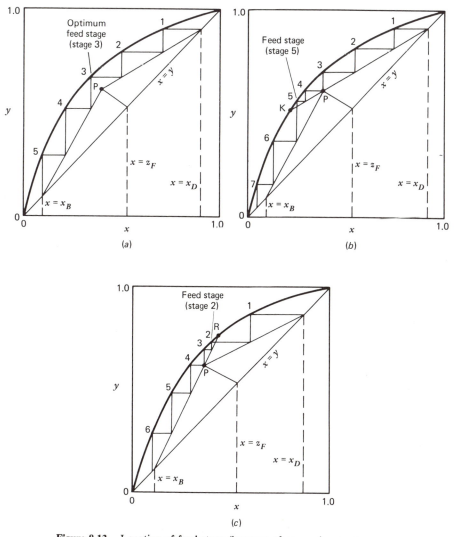

Figure 8.13. Location of feed stage (because of convenience, stages are stepped off and counted in top-down direction. (*a*) Optimum feed-stage location. (*b*) Feed-stage location below optimum stage. (*c*) Feed-stage location above optimum stage.

point on the operating line can lie above the equilibrium curve, the minimum slope of the operating line is determined by an intersection of an operating line with the equilibrium curve. The two minimum reflux situations that normally occur in practice are shown in Fig. 8.15.

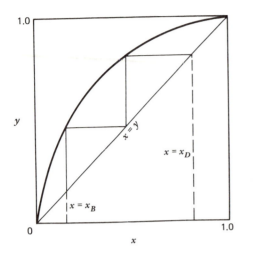

Figure 8.14.   Minimum stages, total reflux.

1. The intersection of the two operating lines is seen to fall on the equilibrium curve in Fig. 8.15a. This case is typical of ideal and near-ideal binary mixtures and was encountered in Example 8.4.

2. The slope of the enriching section operating line is tangent to the equilibrium line at the so-called pinch point R in Fig. 8.15b. This case can occur with nonideal binary mixtures.

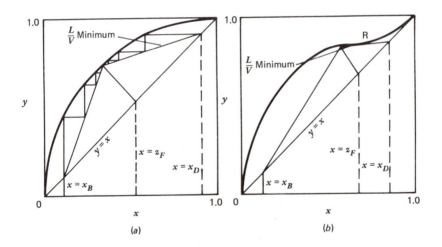

(a)   (b)

Figure 8.15.   Minimum reflux ratio conditions. (a) Intersection of operating lines at equilibrium curve. (b) Operating line tangent to equilibrium curve.

In either case, an infinite number of stages is required to accomplish a separation at minimum reflux. This is the exact converse of the minimum stage–total reflux case.

Large Number of Stages

When conditions are such that the number of theoretical stages involved is very large, but finite, the McCabe–Thiele method is difficult to apply because graphical construction can be inaccurate, unless special coordinate systems are used.[2] However, for such conditions, relatively close-boiling mixtures are generally involved and the assumption of constant relative volatility is reasonable. If, in addition, constant molar overflow can be assumed, then a direct analytical solution for binary mixtures devised by Smoker[3] can be applied to determine the stage requirements for a specified separation. Although not developed here, the general analytical approach is illustrated in Part (b) of the next example.

Example 8.5. One hundred kilogram-moles per hour of a feed containing 30 mole% n-hexane and 70% n-octane is to be distilled in a column consisting of a partial reboiler, a theoretical plate, and a partial condenser, all operating at 1 atm (101.3 kPa). The feed, a bubble-point liquid, is fed to the reboiler, from which a liquid bottoms product is continuously withdrawn. Bubble-point reflux is returned to the plate. The vapor distillate contains 80 mole% hexane, and the reflux ratio (L_R/D) is 2. Assume the partial reboiler, plate, and partial condenser each function as an equilibrium stage.

(a) Using the McCabe–Thiele method, calculate the bottoms composition and moles of distillate produced per hour.

(b) If the relative volatility α is assumed constant at a value of 5 over the composition range for this example (the relative volatility actually varies from 4.3 at the reboiler to 6.0 at the condenser) calculate the bottoms composition analytically.

Solution. Following the procedure of Chapter 6, we have $N_D = C + 2N + 5$ degrees of freedom. With two stages and two species, $N_D = 11$. Specified in this problem are

Feed stream variables	4
Plate and reboiler pressures	2
Condenser pressure	1
Q for plate	1
Number of stages	1
Reflux ratio, L_R/D	1
Distillate composition	1
	11

The problem is fully specified and can be solved.

(a) **Graphical solution.** A diagram of the separator is given in Fig. 8.16 as is the graphical solution, which is constructed in the following manner.

1. The point $y_D = 0.8$ is located on the $x = y$ line.

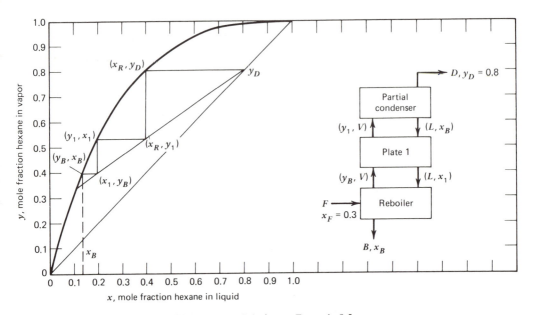

Figure 8.16. Solution to Example 8.5.

2. Conditions in the condenser are fixed, because x_R is in equilibrium with y_D, hence the point x_R, y_D is located.

3. The operating line with slope $L/V = 2/3$ is now drawn through the point $y_D = 0.8$. Note that $(L/V) = (1/D)/[1 + (L/D)]$.

4. Three theoretical stages are stepped off and the bottoms composition $x_B = 0.135$ is read.

The amount of distillate is determined from overall material balances. For hexane

$$z_F F = y_D D + x_B B$$

Therefore

$$(0.3)(100) = (0.8)D + (0.135)B$$

and for the total flow

$$B = 100 - D$$

Solving these equations simultaneously for D, we have

$$D = 24.8 \text{ kgmole/hr}$$

(b) **Analytical solution.** From (3-10) for constant α, equilibrium compositions are given by

$$x = \frac{y}{y + \alpha(1 - y)} \qquad (8\text{-}33)$$

The steps in the solution are as follows.

1. The liquid leaving the partial condenser at x_R is calculated from (7-33).

$$x_R = \frac{0.8}{0.8 + 5(1 - 0.8)} = 0.44$$

2. Then y_1 is determined by a material balance about the partial condenser.

$$Vy_1 = Dy_D + Lx_R$$

$$y_1 = (1/3)(0.8) + (2/3)(0.44) = 0.56$$

3. From (8.33), for plate 1

$$x_1 = \frac{0.56}{0.56 + 5(1 - 0.56)} = 0.203$$

4. By material balance around the top two stages

$$Vy_B = Dx_D + Lx_1$$

$$y_B = 1/3(0.8) + 2/3(0.203) = 0.402$$

5. $$x_B = \frac{0.402}{0.402 + 5(1 - 0.402)} = 0.119$$

By approximating the equilibrium curve with $\alpha = 5$, an answer of $0.119 = x_B$ rather than $x_B = 0.135$ has been obtained.

☐

Example 8.6. (a) Solve Example 8.5 graphically, assuming the feed is introduced on plate 1, rather than into the reboiler. (b) Determine the minimum number of stages required to carry out the calculated separation.

Solution. (a) The flowsheet and solution given in Fig. 8.17 are obtained as follows.

1. The point x_B, y_D is located on the equilibrium line.

2. The operating line for the enriching section is drawn through the point $y = x = 0.8$, with a slope of $L/V = 2/3$.

3. Intersection of the q-line, $x_F = 0.3$ (saturated liquid) with the enriching section operating line is located at point P. The stripping section operating line must also go through this point.

4. The slope of the stripping section operating line is found by trial and error, since there are three equilibrium contacts in the column with the middle stage involved in the switch from one operating line to the other. The result is $x_B = 0.07$, and the amount of distillate is obtained from the combined total and hexane overall material balances (as in Example 8.5a).

$$(0.3)(100) = (0.8D) + 0.07(100 - D)$$

Therefore

$$D = 31.5 \text{ mole/hr}$$

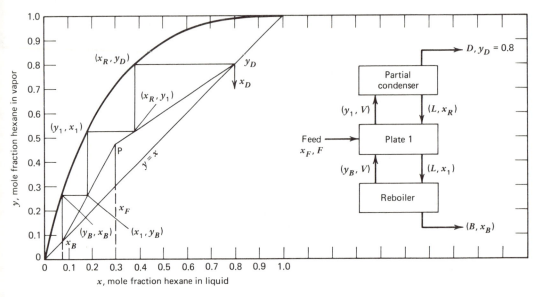

Figure 8.17. Solution to Example 8.7.

Comparing this result to that obtained in Example 8.5, we find that the bottoms purity and distillate yield are improved by introduction of the feed on plate 1, rather than into the reboiler.

(b) The construction corresponding to total reflux ($L/V = 1$, no products, no feed, minimum plates) is shown in Fig. 8.18. Slightly more than two stages are required for an x_B of 0.07, compared to the three stages previously required.
□

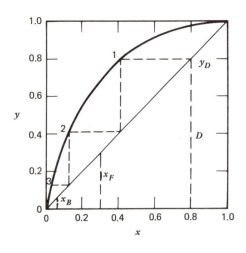

Figure 8.18. Solution to total reflux case in Example 8.6.

8.7   Stage Efficiencies

In previous discussions of stage contacts, it was assumed that the phases leaving the stages are in thermodynamic equilibrium. In industrial countercurrent multistage equipment, it is not always practical to provide the combination of residence time and intimacy of contact required to accomplish equilibrium. Hence, concentration changes for a given stage are less than predicted by equilibrium considerations. One way of characterizing this is by *overall stage efficiency* for a particular separator.

$$\text{Overall stage efficiency} = \frac{\begin{array}{c}\text{Theoretical contacts required for}\\ \text{given separation}\end{array}}{\begin{array}{c}\text{Actual number of contacts required}\\ \text{for the same separation}\end{array}} \times 100\%$$

This definition has the advantage of being very simple to use; however, it does not take into account the variation in efficiency from one stage to another due to changes in physical properties of the system caused by changes in composition and temperature. In addition, efficiencies for various species and for various tray designs may differ.

A stage efficiency frequently used to describe individual tray performance for individual components is the *Murphree plate efficiency*.[4] This efficiency can be defined on the basis of either phase and, for a given component, is equal to the change in actual composition in the phase, divided by the change predicted by equilibrium considerations. This definition as applied to the vapor phase is expressed mathematically as:

$$E_V = \frac{y_n - y_{n-1}}{y_n^* - y_{n-1}} \times 100\% \tag{8-34}$$

where E_V is the Murphree plate efficiency based on the vapor phase, and y^* is the composition in the hypothetical vapor phase that would be in equilibrium with the liquid composition leaving the actual stage. Note that, in large-diameter trays, appreciable differences in liquid composition exist, so that the vapor generated by the first liquid to reach the plate will be richer in the light component than the vapor generated where the liquid exits. This may result in an *average* vapor richer in the light component than would be in equilibrium with the exit liquid, and a Murphree efficiency greater than 100%. A liquid phase efficiency E_L analogous to E_V can be defined using liquid phase concentrations.

In stepping-off stages, the Murphree plate efficiency dictates the percentage of the distance taken from the operating line to the equilibrium line; only E_V or E_L of the total vertical or horizontal path is traveled. This is shown in Fig. 8.19*a* for the case of Murphree efficiencies based on the vapor phase, and in Fig. 8.19*b* for the liquid phase. In effect the dashed curves for actual exit phase com-

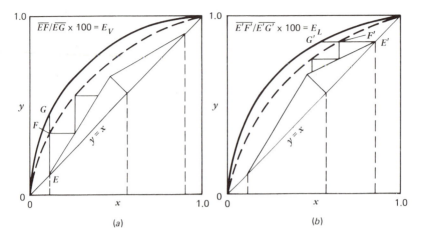

Figure 8.19.   Murphree plate efficiencies. (a) Based on vapor phase. (b) Based on liquid phase.

positions replace the thermodynamic equilibrium curves for a particular set of operating lines.

8.8   Application to Complex Distillation Columns

A multiple feed arrangement is shown in Fig. 8.9. In the absence of side stream L_s, this arrangement has no effect on the material balance equations associated with the enriching section of the column above the upper feed point F_1. The section of column between the upper feed point and lower feed point F_2 (in the absence of feed F) is represented by an operating line of slope L'/V', this line intersecting the enriching section operating line. A similar argument holds for the stripping section of the column. Hence it is possible to apply the McCabe–Thiele graphical principles as shown in Fig. 8.20a. The operating condition for Fig. 8.9, with $F_1 = F_2 = 0$ but $L_s \neq 0$, is represented graphically in Fig. 8.20b.

For certain types of distillation, an inert hot gas is introduced directly into the base of the column. Open steam, for example, can be used if one of the components in the mixture is water, or if the water can form a second phase thereby reducing the boiling point, as in the steam distillation of fats where heat is supplied by live superheated steam, no reboiler being used. In this application, Q_B of Fig. 8.9 is a stream of composition $y = 0$, which with $x = x_B$ becomes a point on the operating line, since the passing streams at this point actually exist at the end of the column. The use of open steam rather than a reboiler for the operating condition $F_1 = F_2 = L_s = 0$ is represented graphically in Fig. 8.20c.

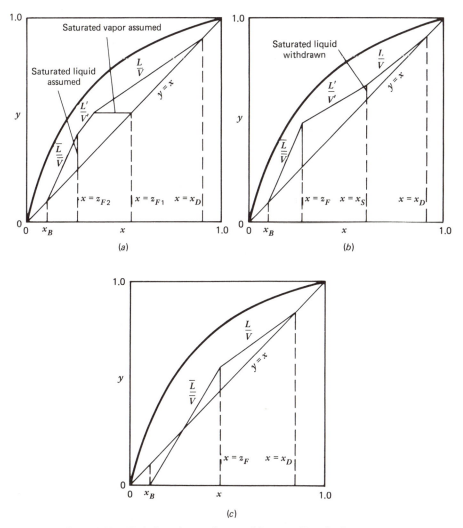

Figure 8.20. Variation of operating conditions. (*a*) Two feeds (saturated liquid and saturated vapor). (*b*) One feed, one side stream (saturated liquid). (*c*) Open steam system.

Example 8.7 A complex distillation column, equipped with a partial reboiler and total condenser and operating at steady state with a saturated liquid feed, has a liquid side stream draw-off in the enriching section. Making the usual simplifying assumptions:

(a) Derive an equation for the two operating lines in the enriching section.

(b) Find the point of intersection of these operating lines.

(c) Find the intersection of the operating line between F and L_s with the diagonal.

(d) Show the construction on an x-y diagram.

    Solution. (a) Taking a material balance over section 1 in Fig. 8.21, we have

$$V_{n-1}y_{n-1} = L_n x_n + D x_D$$

About section 2,

$$V_{s-2}y_{s-2} = L'_{s-1}x_{s-1} + L_s x_s + D x_D$$

Invoking the usual simplifying assumptions, the two operating lines are

$$y = \frac{L'}{V}x + \frac{L_s x_s + D x_D}{V} \qquad \text{and} \qquad y = \frac{L}{V}x + \frac{D}{V}x_D$$

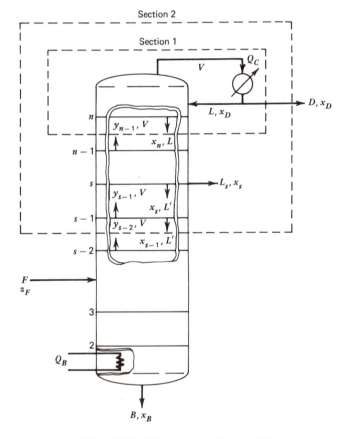

Figure 8.21. Flowsheet for Example 8.7.

(b) Equating the two operating lines, we find that the intersection occurs at

$$(L - L')x = L_s x_s$$

and, since $L \quad L' = L_s$, the point of intersection becomes $x = x_s$.

(c) The intersection of the lines

$$y = \frac{L'}{V}x + \frac{L_s x_s + Dx_D}{V}$$

and

$$y = x$$

occurs at

$$x = \frac{L_s x_s + Dx_D}{L_s + D}$$

(d) The x-y diagram is shown in Fig. 8.22.

□

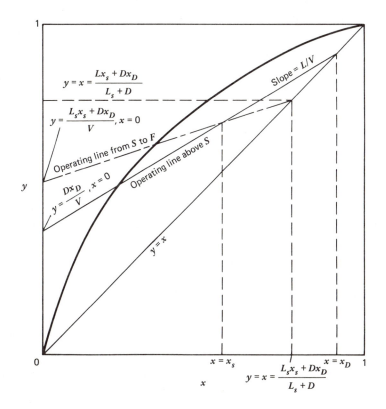

Figure 8.22. Graphical solution to Example 8.7.

8.9   Cost Considerations

For the stagewise contact design problems considered in this chapter, solutions were found for given operating conditions, reflux ratios, number of stages or plates, feed locations, and other specifications. There are an infinite number of possible solutions for a given separation requirement because an infinite combination of, for example, stages and reflux ratios can be used. The final selection is based primarily on cost considerations, which include capital costs for the equipment and installation and operating costs of utilities, labor, raw materials, and maintenance.

Increased reflux ratio has the effect of decreasing the required number of theoretical contacts, but it increases the internal flow, equipment diameter, and energy requirements. It is frequently necessary to solve the separation problem a considerable number of times, imposing various operating conditions and carefully ascertaining the effect on the total cost of the particular solution proposed. Optimum design conditions are discussed in Chapter 13.

References

1. McCabe, W. L., and E. W. Thiele, *Ind. Eng. Chem.*, **17**, 605–611 (1925).

2. Horvath, P. J., and R. F. Schubert, *Chem. Eng.*, **65** (3), 129–132 (February 10, 1958).

3. Smoker, E. H., *Trans. AIChE*, **34**, 165–172 (1938).

4. Murphree, E. V., *Ind. Eng. Chem.*, **17**, 747–750, 960–964 (1925).

Problems

Unless otherwise stated, the usual simplifying assumptions of saturated liquid feed and reflux, optimum feed plate, no heat losses, steady state, and constant molar liquid and vapor flows apply to each of the following problems. Additional problems can be formulated readily from many of the problems in Chapter 10.

8.1   A plant has a batch of 100 kgmole of a liquid mixture containing 20 mole% benzene and 80 mole% chlorobenzene. It is desired to rectify this mixture at 1 atm to obtain bottoms containing only 0.1 mole% benzene. The relative volatility may be assumed constant at 4.13. There are available a suitable still to vaporize the feed and a column containing the equivalent of four perfect plates. The run is to be made at total reflux. While the steady state is being approached, a finite amount of distillate is held in a reflux trap. When the steady state is reached, the bottoms contain 0.1 mole% benzene. With this apparatus, what yield of bottoms can be obtained? The holdup of the column is negligible compared to that in the still and in the reflux trap.

8.2 (a) For the cascade (a) shown below, calculate the composition of streams V_4 and L_1. Assume constant molar overflow, atmospheric pressure, saturated liquid and vapor feeds, and the vapor–liquid equilibrium data given below. Compositions are in mole percents.

(b) Given the feed compositions in cascade (a), how many stages would be required to produce a V_4 containing 85% alcohol?

(c) For the configuration in cascade (b), with $D = 50$ moles what are the compositions of D and L_1?

(d) For the configuration of cascade (b), how many stages are required to produce a D of 50% alcohol?

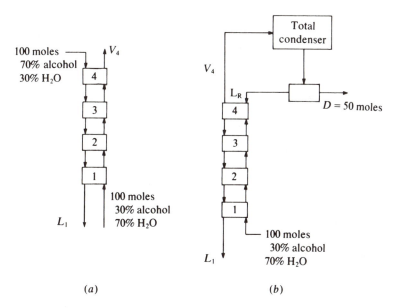

(a)   (b)

Equilibrium data, mole fraction alcohol

x	0.1	0.3	0.5	0.7	0.9
y	0.2	0.5	0.68	0.82	0.94

8.3 A distillate containing 45 wt% isopropyl alcohol, 50 wt% diisopropyl ether, and 5 wt% water is obtained from the heads column of an isopropyl alcohol finishing unit. The company desires to recover the ether from this stream by liquid–liquid extraction with water entering the top and the feed entering the bottom so as to produce an ether containing no more than 2.5 wt% alcohol and to obtain the extracted alcohol at a concentration of at. least 20 wt%. The unit will operate at 25°C and 1 atm. Using a McCabe–Thiele diagram, find how many theoretical stages are required assuming constant phase ratios.

Is it possible to obtain an extracted alcohol composition of 25 wt%?

Phase equilibrium data at 25°C, 1 atm

Ether Phase			Water Phase		
Wt% Alcohol	Wt% Ether	Wt% Water	Wt% Alcohol	Wt% Ether	Wt% Water
2.4	96.7	0.9	8.1	1.8	90.1
3.2	95.7	1.1	8.6	1.8	89.6
5.0	93.6	1.4	10.2	1.5	88.3
9.3	88.6	2.1	11.7	1.6	86.7
24.9	69.4	5.7	17.5	1.9	80.6
38.0	50.2	11.8	21.7	2.3	76.0
45.2	33.6	21.2	26.8	3.4	69.8

Additional points on phase boundary

Wt% Alcohol	Wt% Ether	Wt% Water
45.37	29.70	24.93
44.55	22.45	33.00
39.57	13.42	47.01
36.23	9.66	54.11
24.74	2.74	72.52
21.33	2.06	76.61
0	0.6	99.4
0	99.5[a]	0.5[a]

[a] Estimated.

8.4 Solve graphically the following problems from Chapter 1: (a) 1.12; (b) 1.13; (c) 1.14; (d) 1.15; (e) 1.16; and (f) 1.19.

8.5 An aromatic compound is to be stripped from oil by direct contact with counterflowing superheated steam in a plate column. The operation is isothermal at 130°C and the total pressure is 108.3 kPa.

> Oil feed rate: 3175 kg/hr.
> Concentration of aromatic in feed: 5 wt%.
> Concentration of aromatic in exit oil: 0.5 wt%.
> Moles steam/mole oil = 3.34.
> Vapor pressure of aromatic at 130°C = 150 kPa.
> Molecular weight of oil = 220.
> Molecular weight of aromatic = 78.

Calculate the number of theoretical stages required, assuming the oil and steam are not miscible and the ratio of H_2O/oil is constant in the column.

8.6 Liquid air is fed to the top of a perforated tray reboiled stripper operated at substantially atmospheric pressure. Sixty percent of the oxygen in the feed is to be drawn off in the bottoms vapor product from the still. This product is to contain 0.2 mole% nitrogen. Based on the assumptions and data given below, calculate:

(a) the mole% of nitrogen in the vapor leaving the top plate
(b) the moles of vapor generated in the still per 100 moles of feed
(c) the number of theoretical plates required.

Notes: To simplify the problem, assume constant molal overflow equal to the moles of feed. Liquid air contains 20.9 mole% of oxygen and 79.1 mole% of nitrogen. The equilibrium data [*Chem. Met. Eng.*, **35**, 622 (1928)] at atmospheric pressure are:

Temperature, °K	Mole percent N_2 in liquid	Mole percent N_2 in vapor
77.35	100.00	100.00
77.98	90.00	97.17
78.73	79.00	93.62
79.44	70.00	90.31
80.33	60.00	85.91
81.35	50.00	80.46
82.54	40.00	73.50
83.94	30.00	64.05
85.62	20.00	50.81
87.67	10.00	31.00
90.17	0.00	0.00

8.7 The exit gas from an alcohol fermenter consists of an air–CO_2 mixture containing 10 mole% CO_2 that is to be absorbed in a 5.0–N solution of triethanolamine, which contains 0.04 moles of carbon dioxide per mole of amine. If the column operates isothermally at 25°C, if the exit liquid contains 0.8 times the maximum amount of carbon dioxide, and if the absorption is carried out in a six-theoretical-plate column, calculate:

(a) Moles of amine solution/mole of feed gas.
(b) Exit gas composition.

Equilibrium Data

Y	0.003	0.008	0.015	0.023	0.032	0.043	0.055	0.068	0.083	0.099	0.12
X	0.01	0.02	0.03	0.04	0.05	0.06	0.07	0.08	0.09	0.10	0.11

Y = mole CO_2/mole air; X = mole CO_2/mole amine solution

8.8 A straw oil used to absorb benzene from coke oven gas is to be steam stripped in a bubble plate column at atmospheric pressure to recover the dissolved benzene. Equilibrium conditions are approximated by Henry's law such that, when the oil phase contains 10 mole% C_6H_6, the C_6H_6 partial pressure above it is 5.07 kPa. The oil may be considered nonvolatile, and the operation adiabatic. The oil enters containing 8 mole% benzene, 75% of which is to be recovered. The steam leaving contains 3 mole% C_6H_6.

(a) How many theoretical stages are required in the column?

(b) How many moles of steam are required per 100 moles of oil–benzene mixture?
(c) If 85% of the benzene is to be recovered with the same oil and steam rates, how many theoretical stages are required?

8.9 A solvent recovery plant consists of a plate column absorber and a plate column stripper. Ninety percent of the benzene (B) in the gas stream is recovered in the absorption column. Concentration of benzene in the inlet is $Y_1 = 0.06$ mole B/mole B-free gas. The oil entering the top of the absorber contains $X_2 = 0.01$ mole B/mole pure oil. In the leaving stream, $X_1 = 0.19$ mole B/mole pure oil. Operation temperature is 77°F (250°C).

Open, superheated steam is used to strip the benzene out of the benzene-rich oil, at 110°C. Concentration of benzene in the oil = 0.19 and 0.01 (mole ratios) at inlet and outlet, respectively. Oil (pure)-to-gas (benzene-free) flow rate ratio = 2.0. Vapors are condensed, separated, and removed.

$$MW \ oil = 200 \quad MW \ benzene = 78 \quad MW \ gas = 32$$

Equilibrium data at column pressures

X in oil	0	0.04	0.08	0.12	0.16	0.20	0.24	0.28
Y in gas, 25°C	0	0.011	0.0215	0.032	0.0405	0.0515	0.060	0.068
Y in steam, 110°C	0	0.1	0.21	0.33	0.47	0.62	0.795	1.05

Calculate:
(a) Molar flow rate ratio in absorber.
(b) Number of theoretical plates in absorber.
(c) Minimum steam flow rate required to remove the benzene from 1 mole oil under given terminal conditions, assuming an infinite plate column.

8.10 A dissolved gas A is to be stripped from a stream of water by contacting the water with air in three ideal stages.

The incoming air G_o contains no A; and the liquid feed L_o contains 0.09 kg A/kg water. The weight ratio of water to air is 10/1 in the feed streams for each case.

Using the equilibrium diagram below, calculate the unknowns listed for each of the configurations shown below and compare the recovery of A for the three cases.

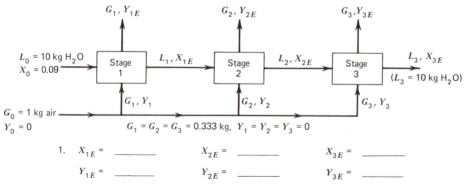

1. $X_{1E} = $ _____ $X_{2E} = $ _____ $X_{3E} = $ _____

     $Y_{1E} = $ _____ $Y_{2E} = $ _____ $Y_{3E} = $ _____

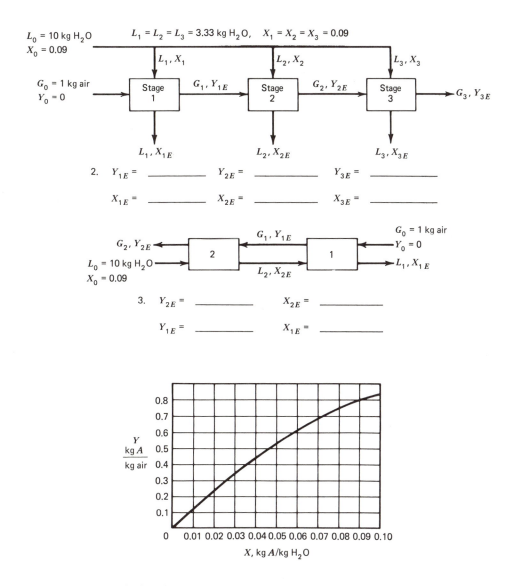

2. $Y_{1E} =$ _____ $Y_{2E} =$ _____ $Y_{3E} =$ _____

   $X_{1E} =$ _____ $X_{2E} =$ _____ $X_{3E} =$ _____

3. $Y_{2E} =$ _____ $X_{2E} =$ _____

   $Y_{1E} =$ _____ $X_{1E} =$ _____

8.11 A saturated liquid mixture of 69.4 mole% benzene in toluene is to be continuously distilled at atmospheric pressure to produce a distillate containing 90 mole% benzene, with a yield of 25 moles distillate per 100 moles of feed. The feed is sent to a steam-heated still (reboiler), where residue is to be withdrawn continuously. The vapors from the still pass directly to a partial condenser. From a liquid separator following the condenser, reflux is returned to the still. Vapors from the separator, which are in equilibrium with the liquid reflux, are sent to a total condenser and are continuously withdrawn as distillate. At equilibrium the mole ratio of benzene to toluene in the vapor may be taken as equal to 2.5 times the mole ratio of benzene to toluene in the liquid. Calculate analytically and graphically the total moles of vapor generated in the still per 100 moles of feed.

8.12 A mixture of A (more volatile) and B is being separated in a plate distillation column. In two separate tests run with a saturated liquid feed of 40 mole% A, the following compositions, in mole percent A, were obtained for samples of liquid and vapor streams from three consecutive stages between the feed and total condenser at the top.

Stage	Test 1		Test 2	
	V	**L**	**V**	**L**
$M+2$	80.2	72.0	75.0	68.0
$M+1$	76.5	60.0	68.0	60.5
M	66.5	58.5	60.5	53.0

Determine the reflux ratio and overhead composition in each case, assuming the column has more than three stages.

8.13 The McCabe–Thiele diagram below refers to the usual distillation column. What is the significance of x_1 (algebraic value and physical significance)?

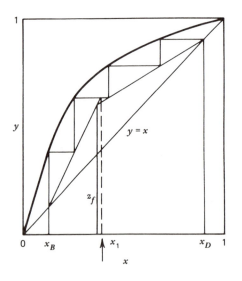

8.14 A saturated liquid mixture containing 70 mole% benzene and 30 mole% toluene is to be distilled at atmospheric pressure in order to produce a distillate of 80 mole% benzene. Five procedures, described below, are under consideration. For each of the procedures, calculate and tabulate the following.

(a) Moles of distillate per 100 moles of feed.
(b) Moles of total vapor generated per mole of distillate.
(c) Mole percent benzene in the residue.
(d) For each part make a y-x diagram. On this, indicate the compositions of the overhead product, the reflux, and the composition of the residue.
(e) If the objective is to maximize total benzene recovery, which, if any, of these procedures is preferred?

Note: Assume the relative volatility equals 2.5.

Description of Procedures

1. Continuous distillation followed by partial condensation. The feed is sent to the direct-heated still pot, from which the residue is continuously withdrawn. The vapors enter the top of a helically coiled partial condenser that discharges into a trap. The liquid is returned (refluxed) to the still, while the residual vapor is condensed as a product containing 80 mole% benzene. The molar ratio of reflux to product is 0.5.

2. Continuous distillation in a column containing one perfect plate. The feed is sent to the direct-heated still, from which residue is withdrawn continuously. The vapors from the plate enter the top of a helically coiled partial condenser that discharges into a trap. The liquid from the trap is returned to the plate, while the uncondensed vapor is condensed to form a distillate containing 80 mole% benzene. The molar ratio of reflux to product is 0.5.

3. Continuous distillation in a column containing the equivalent of two perfect plates. The feed is sent to the direct-heated still, from which residue is withdrawn continuously. The vapors from the top plate enter the top of a helically coiled partial condenser that discharges into a trap. The liquid from the trap is returned to the top plate (refluxed) while the uncondensed vapor is condensed to form a distillate containing 80 mole% benzene. The molar ratio of reflux to product is 0.5.

4. The operation is the same as that described in Part 3 with the exception that the liquid from the trap is returned to the bottom plate.

5. Continuous distillation in a column containing the equivalent of one perfect plate. The feed at its boiling point is introduced on the plate. The residue is withdrawn continuously from the direct-heated still pot. The vapors from the plate enter the top of a helically coiled partial condenser that discharges into a trap. The liquid from the trap is returned to the plate while the uncondensed vapor is condensed to form a distillate containing 80 mole% benzene. The molar ratio of reflux to product is 0.5.

8.15 A saturated liquid mixture of benzene and toluene containing 50 mole% benzene is to be distilled in an apparatus consisting of a still pot, one perfect plate, and a total condenser. The still pot is equivalent to one stage, and the pressure is 101 kPa.

   The still is supposed to produce a distillate containing 75 mole% benzene. For each of the procedures given, calculate, if possible,

1. Moles of distillate per 100 moles of feed.

Assume a constant relative volatility of 2.5.

The procedures are:
(a) No reflux with feed to the still pot.
(b) Feed to the still pot, reflux ratio $L/D = 3$.
(c) Feed to the plate with a reflux ratio of 3.
(d) Feed to the plate with a reflux ratio of 3. However, in this case, a partial condenser is employed.
(e) A preheater partially vaporizes the feed to 25 mole% vapor, which is fed to the plate. The reflux ratio is 3.
(f) Solve part (b) using minimum reflux.
(g) Solve part (b) using total reflux.

8.16 A fractionation column operating at 101 kPa is to separate 30 kg/hr of a solution of benzene and toluene containing 0.6 mass fraction toluene into an overhead product containing 0.97 mass fraction benzene and a bottoms product containing 0.98 mass fraction toluene. A reflux ratio of 3.5 is to be used. The feed is liquid at its boiling point, feed is on the optimum tray, and the reflux is at saturation temperature.
(a) Determine the quantity of top and bottom products.
(b) Determine the number of stages required.

Equilibrium data in mole fraction benzene, 101 kPa

y	0.21	0.37	0.51	0.64	0.72	0.79	0.86	0.91	0.96	0.98
x	0.1	0.2	0.3	0.4	0.5	0.6	0.7	0.8	0.9	0.95

8.17 A mixture of 54.5 mole% benzene in chlorobenzene at its bubble point is fed continuously to the bottom plate of a column containing two theoretical plates. The column is equipped with a partial reboiler and a total condenser. Sufficient heat is supplied to the reboiler to make V/F equal to 0.855, and the reflux ratio L/V in the top of the column is kept constant at 0.50. Under these conditions, what quality of product and bottoms (x_D, x_W) can be expected?

Equilibrium data at column pressure, mole fraction benzene

x	0.100	0.200	0.300	0.400	0.500	0.600	0.700	0.800
y	0.314	0.508	0.640	0.734	0.806	0.862	0.905	0.943

8.18 A certain continuous distillation operation operating with a reflux ratio (L/D) of 3.5 yielded a distillate containing 97 wt% B (benzene) and bottoms containing 98 wt% T (toluene).

Due to an accident, the 10 plates in the bottom section of the column were ruined; but the 14 upper plates were left intact.

It is suggested that the column still be used, with the feed (F) as saturated vapor at the dew point, with $F = 13,600$ kg/hr containing 40 wt% B and 60 wt% T.

Assuming that the plate efficiency remains unchanged at 50%,
(a) Can this column still yield a distillate containing 97 wt% B?
(b) How much distillate can we get?

(c) What will the composition (in mole percent) of the residue be?
For vapor–liquid equilibrium data, see Problem 8.16.

8.19 A distillation column having eight theoretical stages (six + partial reboiler + partial condenser) is being used to separate 100 mole/hr of a saturated liquid feed containing 50 mole% A into a product stream containing 90 mole% A. The liquid-to-vapor molar ratio at the top plate is 0.75. The saturated liquid feed is introduced on plate 5.
(a) What is the composition of the bottoms?
(b) What is the L/V ratio in the stripping section?
(c) What are the moles of bottoms/hour?
Unbeknown to the operators, the bolts holding plates 5, 6, and 7 rust through, and the plates fall into the still pot. If no adjustments are made, what is the new composition of the bottoms?

It is suggested that, instead of returning reflux to the top plate, an equivalent amount of liquid product from another column be used as reflux. If this product contains 80 mole% A, what now is the composition of
(a) The distillate?
(b) The bottoms?

Equilibrium data, mole fraction of A

y	0.19	0.37	0.5	0.62	0.71	0.78	0.84	0.9	0.96
x	0.1	0.2	0.3	0.4	0.5	0.6	0.7	0.8	0.9

8.20 A distillation unit consists of a partial reboiler, a column with seven perfect plates, and a total condenser. The feed consists of a 50 mole% mixture of benzene in toluene. It is desired to produce a distillate containing 96 mole% benzene.
(a) With saturated liquid feed fed to the fifth plate from the top, calculate:

1. Minimum reflux ratio $(L_R/D)_{min}$.

2. The bottoms composition, using a reflux ratio (L_R/D) of twice the minimum.

3. Moles of product per 100 moles of feed.

(b) Repeat part (a) for a saturated vapor feed fed to the fifth plate from the top.
(c) With saturated vapor feed fed to the reboiler and a reflux ratio (L/V) of 0.9, calculate

1. Bottoms composition.

2. Moles of product per 100 moles of feed.

8.21 A valve–tray fractionating column containing eight theoretical plates, a partial reboiler equivalent to one theoretical plate, and a total condenser has been in operation separating a benzene–toluene mixture containing 36 mole% benzene at 101 kPa. Under normal operating conditions, the reboiler generates 100 kgmole of vapor per hour. A request has been made for very pure toluene, and it is proposed to operate this column as a stripper, introducing the feed on the top plate as a saturated liquid, employing the same boilup at the still, and returning no reflux to the column.
(a) What is the minimum feed rate under the proposed conditions, and what is the corresponding composition of the liquid in the reboiler at the minimum feed?

(b) At a feed rate 25% above the minimum, what is the rate of production of toluene, and what are the compositions of product and distillate?

For equilibrium data, see Problem 8.16.

8.22 A solution of methanol and water at 101 kPa containing 50 mole% methanol is continuously rectified in a seven-plate, perforated tray column, equipped with a total condenser and a partial reboiler heated by steam.

During normal operation, 100 kgmole/hr of feed is introduced on the third plate from the bottom. The overhead product contains 90 mole% methanol, and the bottoms product contains 5 mole% methanol. One mole of liquid reflux is returned to the column for each mole of overhead product.

Recently it has been impossible to maintain the product purity in spite of an increase in the reflux ratio. The following test data were obtained.

Stream	Kgmole/hr	Mole% alcohol
Feed	100	51
Waste	62	12
Product	53	80
Reflux	94	—

What is the most probable cause of this poor performance? What further tests would you make to establish definitely the reason for the trouble? Could some 90% product be obtained by increasing the reflux ratio still farther while keeping the vapor rate constant?

Vapor-liquid equilibrium data at 1 atm [*Chem. Eng. Progr.*, 48, 192 (1952)] in mole fraction methanol

x	0.0321	0.0523	0.0750	0.1540	0.2250	0.3490	0.8130	0.9180
y	0.1900	0.2940	0.3520	0.5160	0.5930	0.7030	0.9180	0.9630

8.23 A fractionating column equipped with a partial reboiler heated with steam, as shown below, and with a total condenser, is operated continuously to separate a mixture of 50 mole% A and 50 mole% B into an overhead product containing 90 mole% A and a bottoms product containing 20 mole% A. The column has three theoretical plates, and the reboiler is equivalent to one theoretical plate. When the system is operated at an $L/V = 0.75$ with the feed as a saturated liquid to the bottom plate of the column, the desired products can be obtained. The system is instrumented as shown. The steam to the reboiler is controlled by a flow controller so that it remains constant. The reflux to the column is also on a flow controller so that the *quantity of reflux* is constant. The feed to the column is normally 100 mole/hr, but it was inadvertently cut back to 25 mole/hr. What would be the composition of the reflux, and what would be the composition of the vapor leaving the reboiler under these new conditions? Assume the vapor leaving the reboiler is not superheated. Relative volatility for the system is 3.0.

8.24 A saturated vapor mixture of maleic anhydride and benzoic acid containing 10 mole% acid is a by-product of the manufacture of phthalic anhydride. This

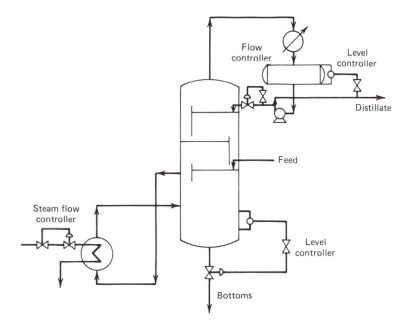

mixture is to be distilled continuously at 13.3 kPa to give a product of 99.5 mole% maleic anhydride and a bottoms of 0.5 mole% anhydride. Using the data below, calculate the number of plates using an L/D of 1.6 times the minimum.

Vapor pressure (torr)

Temperatures (°C)	10	50	100	200	400
Maleic anhydride	80.0	122.5	144.0	167.8	181
Benzoic acid	131.6	167.8	185.0	205.8	227

8.25 A bubble-point binary mixture containing 5 mole% A in B is to be distilled to give a distillate containing 35 mole% A and bottoms product containing 0.2 mole% A. If the relative volatility is constant at a value of 6, calculate the following assuming the column will be equipped with a partial reboiler and a partial condenser.
(a) The minimum number of equilibrium stages.
(b) The minimum boilup ratio V/B leaving the reboiler.
(c) The actual number of equilibrium stages for an actual boilup ratio equal to 1.2 times the minimum value.

8.26 Methanol (M) is to be separated from water (W) by distillation as shown below. The feed is subcooled such that $q = 1.12$. Determine the feed stage location and the number of theoretical stages required. Vapor–liquid equilibrium data are given in Problem 8.22.

8.27 A mixture of acetone and isopropanol containing 50 mole% acetone is to be

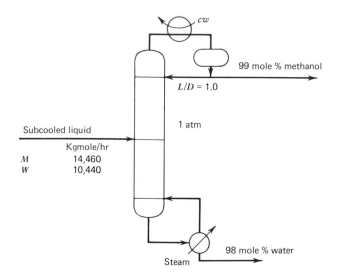

distilled continuously to produce an overhead product containing 80 mole% acetone and a bottoms containing 25 mole% acetone. If a saturated liquid feed is employed, if the column is operated with a reflux ratio of 0.5, and if the Murphree vapor efficiency is 50%, how many plates will be required? Assume a total condenser, partial reboiler, saturated reflux, and optimum feed stage. The vapor–liquid equilibrium data for this system are:

Equilibrium data, mole% acetone

Liquid	0	2.6	5.4	11.7	20.7	29.7	34.1	44.0	52.0
Vapor	0	8.9	17.4	31.5	45.6	55.7	60.1	68.7	74.3
Liquid	63.9	74.6	80.3	86.5	90.2	92.5	95.7	100.0	
Vapor	81.5	87.0	89.4	92.3	94.2	95.5	97.4	100.0	

8.28 A mixture of 40 mole% carbon disulfide (CS_2) in carbon tetrachloride (CCl_4) is continuously distilled. The feed is 50% vaporized ($q = 0.5$). The top product from a total condenser is 95 mole% CS_2, and the bottoms product from a partial reboiler is a liquid of 5 mole% CS_2.

   The column operates with a reflux ratio L/D of 4 to 1. The Murphree vapor efficiency is 80%.
(a) Calculate graphically the minimum reflux, the minimum boilup ratio from the reboiler V/B, and the minimum number of stages (including reboiler).
(b) How many trays are required for the actual column at 80% efficiency by the McCabe–Thiele method.
The coordinates of the x-y diagram at column pressure for this mixture in terms of CS_2 mole fraction are:

x	0.05	0.1	0.2	0.3	0.4	0.5	0.6	0.7	0.8	0.9
y	0.135	0.245	0.42	0.545	0.64	0.725	0.79	0.85	0.905	0.955

8.29 A distillation unit consists of a partial reboiler, a bubble cap column, and a total condenser. The overall plate efficiency is 65%. The feed is a liquid mixture, at its bubble point, consisting of 50 mole% benzene in toluene. This liquid is fed to the optimum plate. The column is to produce a distillate containing 95 mole% benzene and a bottoms of 95 mole% toluene. Calculate for an operating pressure of 1 atm:
(a) Minimum reflux ratio $(L/D)_{min}$.
(b) Minimum number of actual plates to carry out the desired separation.
(c) Using a reflux ratio (L/D) of 50% more than the minimum, the number of actual plates needed.
(d) The kilograms per hour of product and residue, if the feed is 907.3 kg/hr.
(e) The saturated steam at 273.7 kPa required per hour to heat the reboiler using enthalpy data below and any assumptions necessary.
(f) A rigorous enthalpy balance on the reboiler, using the enthalpy data tabulated below and assuming ideal solutions.

H_L and H_V in Btu/lbmole at reboiler temperature

	H_L	H_V
benzene	4,900	18,130
toluene	8,080	21,830

For vapor/liquid equilibrium data, see Problem 8.16.

8.30 A continuous distillation unit, consisting of a perforated-tray column together with a partial reboiler and a total condenser, is to be designed to operate at atmospheric pressure to separate ethanol and water. The feed, which is introduced into the column as liquid at its boiling point, contains 20 mole% alcohol. The distillate is to contain 85% alcohol, and the alcohol recovery is to be 97%.
(a) What is the molar concentration of the bottoms?
(b) What is the minimum value of:

1. The reflux ratio L/V?
2. Of L/D?
3. Of the boilup ratio V/B from the reboiler?

(c) What is the minimum number of theoretical stages and the corresponding number of actual plates if the overall plate efficiency is 55%?
(d) If the reflux ratio L/V used is 0.80, how many actual plates will be required?
   Vapor–liquid equilibrium for ethanol–water at 1 atm in terms of mole fractions of ethanol are [*Ind. Eng. Chem.*, **24**, 882 (1932)]:

x	y	$T, °C$
0.0190	0.1700	95.50
0.0721	0.3891	89.00
0.0966	0.4375	86.70
0.1238	0.4704	85.30
0.1661	0.5089	84.10
0.2337	0.5445	82.70
0.2608	0.5580	82.30
0.3273	0.5826	81.50
0.3965	0.6122	80.70
0.5079	0.6564	79.80
0.5198	0.6599	79.70
0.5732	0.6841	79.30
0.6763	0.7385	78.74
0.7472	0.7815	78.41
0.8943	0.8943	78.15

8.31 A solvent A is to be recovered by distillation from its water solution. It is necessary to produce an overhead product containing 95 mole% A and to recover 95% of the A in the feed. The feed is available at the plant site in two streams, one containing 40 mole% A and the other 60 mole% A. Each stream will provide 50 mole/hr of component A, and each will be fed into the column as saturated liquid. Since the less volatile component is water, it has been proposed to supply the necessary heat in the form of open steam. For the preliminary design, it has been suggested that the operating reflux ratio L/D be 1.33 times the minimum value. A total condenser will be employed. For this system, it is estimated that the overall plate efficiency will be 70%. How many plates will be required, and what will be the bottoms composition? The relative volatility may be assumed to be constant at 3.0. Determine analytically the points necessary to locate the operating lines. Each feed should enter the column at its optimum location.

8.32 A saturated liquid feed stream containing 40 mole% n-hexane (H) and 60 mole% n-octane is fed to a plate column. A reflux ratio L/D equal to 0.5 is maintained at the top of the column. An overhead product of 0.95 mole fraction H is required, and the column bottoms is to be 0.05 mole fraction H. A cooling coil submerged in the liquid of the second plate from the top removes sufficient heat to condense 50 mole% of the vapor rising from the third plate down from the top.
(a) Derive the equations needed to locate the operating lines.
(b) Locate the operating lines and determine the required number of theoretical plates if the optimum feed plate location is used.

8.33 One hundred kilogram/moles per hour of a saturated liquid mixture of 12 mole% ethyl alcohol in water is distilled continuously by direct steam at 1 atm. Steam is introduced directly to the bottom plate. The distillate required is 85 mole% alcohol, representing 90% recovery of the alcohol in the feed. The reflux is saturated liquid with $L/D = 3$. Feed is on the optimum stage. Vapor–liquid equilibrium data are given in Problem 8.30. Calculate:
(a) Steam requirement (kgmole/hr).
(b) Number of theoretical stages.
(c) The feed stage (optimum).
(d) Minimum reflux ratio $(L/D)_{min}$.

8.34 A water–isopropanol mixture at its bubble point containing 10 mole% isopropanol is to be continuously rectified at atmospheric pressure to produce a distillate containing 67.5 mole% isopropanol. Ninety-eight percent of the isopropanol in the feed must be recovered. If a reflux ratio L/D of 1.5 of minimum is to be used, how many theoretical stages will be required.
(a) If a partial reboiler is used?
(b) If no reboiler is used and saturated steam at 101 kPa is introduced below the bottom plate?
How many stages are required at total reflux?

Vapor–liquid equilibrium data, mole fraction of isopropanol at 101 kPa

°C	93.00	84.02	82.12	81.25	80.62	80.16	80.28	81.51
y	0.2195	0.4620	0.5242	0.5516	0.5926	0.6821	0.7421	0.9160
x	0.0118	0.0841	0.1978	0.3496	0.4525	0.6794	0.7693	0.9442

Notes: Composition of the azcotrope: $x = y = 0.6854$. Boiling point of azeotrope: 80.22°C.

8.35 An aqueous solution containing 10 mole% isopropanol is fed at its bubble point to the top of a continuous stripping column, operated at atmospheric pressure, to produce a vapor containing 40 mole% isopropanol. Two procedures are under consideration, both involving the same heat expenditure; that is, V/F (moles of vapor generated/mole of feed) = 0.246 in each case.
Scheme (1) uses a partial reboiler at the bottom of a plate-type stripping column, generating vapor by the use of steam condensing inside a closed coil. In Scheme (2) the reboiler is omitted and live steam is injected directly below the bottom plate. Determine the number of stages required in each case.

Problem 8.36

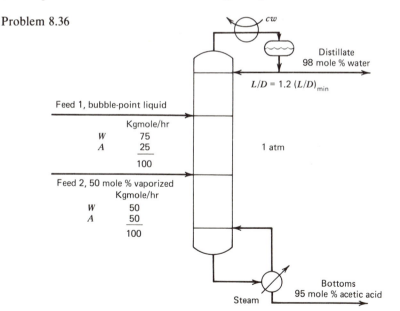

Equilibrium data for the system isopropanol–water are given in Problem 8.34. The usual simplifying assumptions may be made.

8.36 Determine the optimum stage location for each feed and the number of theoretical stages required for the distillation separation shown above using the following equilibrium data in mole fractions.

Water (W)/acetic acid (A) at 1 atm

x_W	0.0055	0.0530	0.1250	0.2060	0.2970	0.5100	0.6490	0.8030	0.9594
y_W	0.320	0.1330	0.2400	0.3380	0.4370	0.6300	0.7510	0.8660	0.9725

8.37 Determine the number of theoretical stages required and the optimum stage locations for the feed and liquid side stream for the distillation process shown below assuming that methanol (M) and ethanol (E) form an ideal solution. Use the Antoine equation for vapor pressures.

8.38 A mixture of n-heptane and toluene (T) is separated by extractive distillation with

Problem 8.37

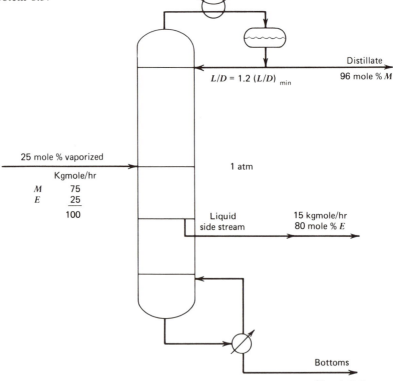

$L/D = 1.2 \, (L/D)_{min}$

Distillate
96 mole % M

25 mole % vaporized

	Kgmole/hr
M	75
E	25
	100

1 atm

Liquid
side stream

15 kgmole/hr
80 mole % E

Bottoms
95 mole % E

phenol (P). Distillation is then used to recover the phenol for recycle as shown in sketch (a) below, where the small amount of n-heptane in the feed is ignored. For the conditions shown in sketch (a), determine the number of theoretical stages required.

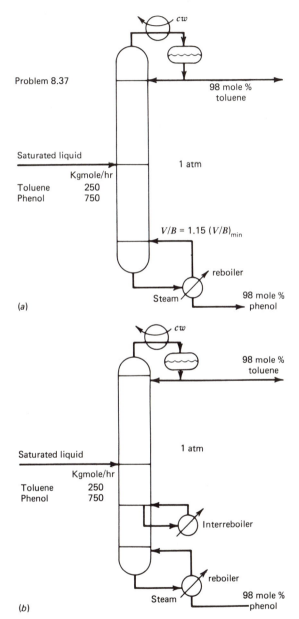

Problem 8.37

98 mole % toluene

Saturated liquid

	Kgmole/hr
Toluene	250
Phenol	750

1 atm

$V/B = 1.15 \, (V/B)_{min}$

reboiler

Steam

98 mole % phenol

(a)

cw

98 mole % toluene

1 atm

Saturated liquid

	Kgmole/hr
Toluene	250
Phenol	750

Interreboiler

reboiler

Steam

98 mole % phenol

(b)

Note that heat will have to be supplied to the reboiler at a high temperature because of the high boiling point of phenol. Therefore, consider the alternative scheme in sketch (b), where an interreboiler to be located midway between the bottom plate and the feed stage is used to provide 50% of the boilup used in sketch (a). The remainder of the boilup is provided by the reboiler. Determine the number of theoretical stages required for the case with the interreboiler and the temperature of the interreboiler stage. Vapor–liquid equilibrium data at 1 atm are [*Trans. AIChE*, **41**, 555 (1945)]:

x_T	y_T	$T, °C$
0.0435	0.3410	172.70
0.0872	0.5120	159.40
0.1186	0.6210	153.80
0.1248	0.6250	149.40
0.2190	0.7850	142.20
0.2750	0.8070	133.80
0.4080	0.8725	128.30
0.4800	0.8901	126.70
0.5898	0.9159	122.20
0.6348	0.9280	120.20
0.6512	0.9260	120.00
0.7400	0.9463	119.70
0.7730	0.9536	119.40
0.8012	0.9545	115.60
0.8840	0.9750	112.70
0.9108	0.9796	112.20
0.9394	0.9861	113.30
0.9770	0.9948	111.10
0.9910	0.9980	111.10
0.9939	0.9986	110.50
0.9973	0.9993	110.50

8.39 A distillation column for the separation of *n*-butane from *n*-pentane was recently put into operation in a petroleum refinery. Apparently, an error was made in the design because the column fails to make the desired separation as shown in the following table [*Chem. Eng. Prog.*, **61** (8), 79 (1965)].

	Design specification	Actual operation
Mole% nC_5 in distillate	0.26	13.49
Mole% nC_4 in bottoms	0.16	4.28

In order to correct the situation, it is proposed to add an intercondenser in the rectifying section to generate more reflux and an interreboiler in the stripping section to produce additional boilup. Show by use of a McCabe–Thiele diagram how such a proposed change can improve the operation.

8.40 In the production of chlorobenzenes by the chlorination of benzene, the two close-boiling isomers paradichlorobenzene (P) and orthodichlorobenzene (O) are separated by distillation. The feed to the column consists of 62 mole% of the para isomer and 38 mole% of the ortho isomer. Assume the pressures at the bottom and top of the column are 20 psia (137.9 kPa) and 15 psia (103.43 kPa), respectively. The distillate is a liquid containing 98 mole% para isomer. The bottoms product is to contain 96 mole% ortho isomer. At column pressure, the feed is slightly vaporized with $q = 0.9$. Calculate the number of theoretical stages required for a reflux ratio equal to 1.15 times the minimum reflux ratio. Base your calculations on a constant relative volatility obtained as the arithmetic average between the column top and column bottom using the Antoine vapor pressure equation and the assumption of Raoult's and Dalton's laws. The McCabe–Thiele construction should be carried out with special log–log coordinates as described by P. J. Horvath and R. F. Schubert [*Chem. Eng.*, **65** (3), 129 (Feb. 10, 1958)].

8.41 Relatively pure oxygen and nitrogen can be obtained by the distillation of air using the Linde double column that, as shown below, consists of a column operating at elevated pressure surmounted by an atmospheric pressure column. The boiler of the upper column is at the same time the reflux condenser for both columns. Gaseous air plus enough liquid to take care of heat leak into the column (more liquid, of course, if liquid-oxygen product is withdrawn) enters the exchanger at the

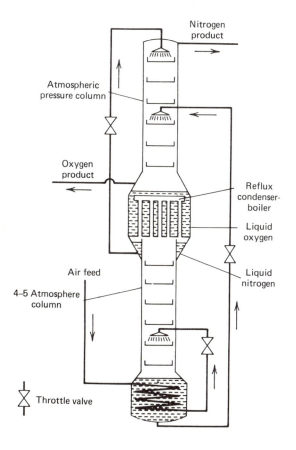

base of the lower column and condenses, giving up heat to the boiling liquid and thus supplying the vapor flow for this column. The liquid air enters an intermediate point in this column, as shown. The vapors rising in this column are partially condensed to form the reflux, and the uncondensed vapor passes to an outer row of tubes and is totally condensed, the liquid nitrogen collecting in an annulus, as shown. By operating this column at 4 to 5 atm, the liquid oxygen boiling at 1 atm is cold enough to condense pure nitrogen. The liquid that collects in the bottom of the lower column contains about 45% O_2 and forms the feed for the upper column. Such a double column can produce a very pure oxygen with high oxygen recovery and relatively pure nitrogen.

On a *single* McCabe–Thiele diagram—using equilibrium lines, operating lines, *q*-lines, 45° line, stepped-off stages, and other illustrative aids—show *qualitatively* how the stage requirements of the double column could be computed.

9
Batch Distillation

All the vapour rising from the liquid must be condensed in the specially provided Liebig condenser and be collected as distillate. Subject to this condition, and in view of the rapid stirring effected by the rising vapour, it would seem safe to assume that the distillate really represents the vapour which is in equilibrium with the liquid at the time in question. The compositions of the liquid and vapour are of course continually changing as the batch distillation proceeds.

Lord Rayleigh, 1902

In batch operations, an initial quantity of material is charged to the equipment and, during operation, one or more phases are continuously withdrawn. A familiar example is ordinary laboratory distillation, where liquid is charged to a still and heated to boiling. The vapor formed is continuously removed and condensed.

In batch separations there is no steady state, and the composition of the initial charge changes with time. This results in an increase in still temperature and a decrease in the relative amount of lower boiling components in the charge as distillation proceeds.

Batch operation is used to advantage if:

1. The required operating capacity of a proposed facility is too small to permit continuous operation at a practical rate. Pumps, boilers, piping, instrumentation, and other auxiliary equipment generally have a minimum industrial operating capacity.

2. The operating requirements of a facility fluctuate widely in characteristics of feed material as well as processing rate. Batch equipment usually has

considerably more operating flexibility than continuous equipment. This is the reason for the predominance of batch equipment for multipurpose solvent recovery or pilot-plant applications.

9.1 Differential Distillation

The simplest case of batch distillation corresponds to use of the apparatus shown in Fig. 9.1. There is no reflux; at any instant, vapor leaving the still pot with composition y_D is assumed to be in equilibrium with the liquid in the still, and $y_D = x_D$. Thus, there is only a single stage. The following nomenclature is used assuming that all compositions refer to a particular species in the multicomponent mixture.

$$D = \text{distillate rate, mole/hr}$$

$$y = y_D = x_D = \text{distillate composition, mole fraction}$$

$$W = \text{amount of liquid in still}$$

$$x = x_W = \text{composition of liquid in still}$$

Also, subscript o refers to initial charge condition. For the more volatile component

$$\text{Rate of output} = D y_D$$

$$\left.\begin{array}{l}\text{Rate of depletion}\\ \text{in the still}\end{array}\right\} = -\frac{d}{dt}(W x_W) = -W\frac{dx_W}{dt} - x_W\frac{dW}{dt}$$

Thus, by material balance at any instant

$$W\frac{dx_W}{dt} + x_W\frac{dW}{dt} = -D y_D \qquad (9\text{-}1)$$

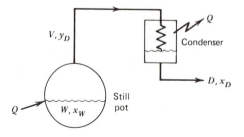

Figure 9.1. Differential distillation.

Thus

$$Wdx_W + x_W dW = y_D(-Ddt) = y_D dW$$

since by total balance $-Ddt = dW$. Integrating from the initial charge condition

$$\int_{x_{W_o}}^{x_W} \frac{dx_W}{y_D - x_W} = \int_{W_o}^{W} \frac{dW}{W} \tag{9-2}$$

This is the well-known Rayleigh equation,[1] as first applied to the separation of wide-boiling mixtures such as $HCl-H_2O$, $H_2SO_4-H_2O$, and NH_3-H_2O. Without reflux, y_D and x_W are in equilibrium and (9-2) can be written as:

$$\int_{x_o}^{x} \frac{dx}{y - x} = \int_{W_o}^{W} \frac{dW}{W} \tag{9-3}$$

Equation (9-3) is easily integrated for the case where pressure is constant, temperature change in the still pot is relatively small (close-boiling mixture), and K-values are composition independent. Then $y = Kx$, where K is approximately constant, and (9-3) becomes

$$\ln \left(\frac{W}{W_o} \right) = \frac{1}{K-1} \ln \left(\frac{x}{x_o} \right) \tag{9-4}$$

For a binary mixture, if the relative volatility α can be assumed constant, substitution of (3-9) into (9-3) followed by integration and simplification gives

$$\ln \left(\frac{W_o}{W} \right) = \frac{1}{\alpha - 1} \left[\ln \left(\frac{x_o}{x} \right) + \alpha \ln \left(\frac{1-x}{1-x_o} \right) \right] \tag{9-5}$$

If the equilibrium relationship $y = f(x)$ is in graphical or tabular form, for which no analytical relationship is available, integration of (9-3) can be performed graphically.

Example 9.1. A batch still is loaded with 100 kgmole of a liquid containing a binary mixture of 50 mole% benzene in toluene. As a function of time, make plots of (a) still temperature, (b) instantaneous vapor composition, (c) still pot composition, and (d) average total distillate composition. Assume a constant boilup rate of 10 kgmole/hr and a constant relative volatility of 2.41 at a pressure of 101.3 kPa (1 atm).

Solution. Initially, $W_o = 100$, $x_o = 0.5$. Solving (9-5) for W at values of x from 0.5 in increments of 0.05 and determining corresponding values of time from $t = (W_o - W)/10$, we generate the following table.

t, hr	2.12	3.75	5.04	6.08	6.94	7.66	8.28	8.83	9.35
W, kgmole	78.85	62.51	49.59	39.16	30.59	23.38	17.19	11.69	6.52
x	0.45	0.40	0.35	0.30	0.25	0.20	0.15	0.10	0.05

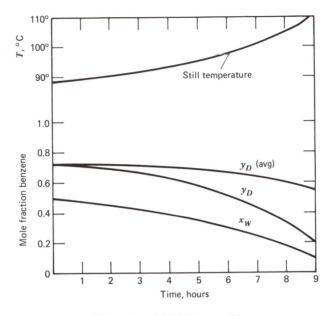

Figure 9.2. Distillation conditions.

The instantaneous vapor composition y is obtained from (3-9), $y = 2.41x/(1 + 1.41x)$, the equilibrium relationship for constant α. The average value of y_D is related to x and W by combining overall component and total material balances to give

$$(y_D)_{avg} = \frac{W_o x_o - W x}{W_o - W} \qquad (9\text{-}6)$$

To obtain the temperature in the still, it is necessary to have experimental T-x-y data for benzene–toluene at 101.3 kPa as given in Fig. 3.4. The temperature and compositions as a function of time are shown in Fig. 9.2.

If the equilibrium relationship between the components is in graphical form, (9-3) may be integrated graphically.

□

9.2 Rectification with Constant Reflux

A batch column with plates above a still pot functions as a rectifier, which can provide a sharper separation than differential distillation. If the reflux ratio is fixed, distillate and still bottoms compositions will vary with time. Equation (9-2) applies with $y_D = x_D$. Its use is facilitated with the McCabe–Thiele diagram as described by Smoker and Rose.[2]

Initially, the composition of the liquid in the reboiler of the column in Fig.

9.3 is the charge composition, x_o. If there are two theoretical stages and there is no appreciable liquid holdup except in the still pot, the initial distillate composition x_{D_o} at Time 0 can be found by constructing an operating line of slope L/V, such that exactly two stages are stepped off from x_o to the $y = x$ line as in Fig. 9.3. At an arbitrary time, say Time 1, at still pot composition x_W, the distillate composition is x_D. A time-dependent series of points is thus established by trial and error, L/V and the stages being held constant.

Equation (9-2) cannot be integrated directly if the column has more than one stage, because the relationship between y_D and x_W depends on the liquid-to-vapor ratio and number of stages, as well as the phase equilibrium relationship. Thus, as shown in the following example, (9-2) is integrated graphically with pairs of values for x_D and x_W obtained from the McCabe–Thiele diagram.

The time t required for batch rectification at constant reflux ratio and negligible holdup in the trap can be computed by a total material balance based on a constant boilup rate V, as shown by Block.[3]

$$t = \frac{W_o - W}{V\left(1 - \dfrac{L}{V}\right)} \tag{9-7}$$

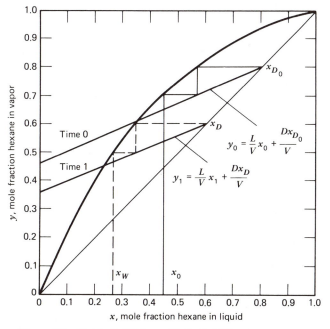

Figure 9.3.   Batch distillation with fixed L/V and two stages.

Example 9.2. A three-stage batch column (first stage is the still pot) is charged with 100 kgmole of a 20 mole% n-hexane in n-octane mix. At an L/V ratio of 0.5, how much material must be distilled if an average product composition of 70 mole% C_6 is required? The phase equilibrium curve at column pressure is as given in Fig. 9.4. If the boilup rate is 10 kgmole/hr, calculate the distillation time.

Solution. A series of operating lines and, hence, values of x_W are obtained by the trial-and-error procedure described above and indicated on Fig. 9.4 for x_o and $x_w = 0.09$. It is then possible to construct Table 9.1. The graphical integration is shown in Fig. 9.5. Assuming a final value of $x_W = 0.1$, for instance

$$\ln \frac{100}{W} = \int_{0.1}^{0.2} \frac{dx_W}{y_D - x_W} = 0.162$$

Hence $W = 85$ and $D = 15$. From (8-6)

$$(x_D)_{\text{avg}} = \frac{100(0.20) - 85(0.1)}{(100 - 85)} = 0.77$$

The $(x_D)_{\text{avg}}$ is higher than 0.70; hence, another final x_W must be chosen. By trial, the correct answer is found to be $x_W = 0.06$, $D = 22$, and $W = 78$ corresponding to a value of 0.25 for the integral.

From (9-7), the distillation time is

$$t = \frac{100 - 78}{10(1 - 0.5)} = 4.4 \text{ hr}$$

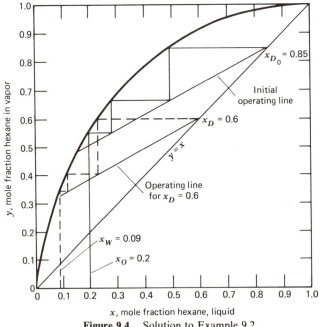

Figure 9.4. Solution to Example 9.2.

Table 9.1   Graphical integration, example 9.2

y_D	x_W	$\dfrac{1}{y_D - x_W}$
0.85	0.2	1.54
0.60	0.09	1.96
0.5	0.07	2.33
0.35	0.05	3.33
0.3	0.035	3.77

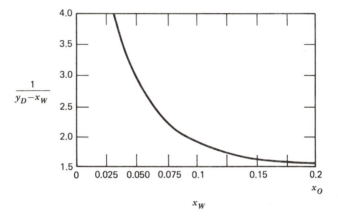

Figure 9.5.   Graphical integration for Example 9.2.

If differential distillation is used, Fig. 9.4 shows that a 70 mole% hexane distillate is impossible.

☐

If a constant distillate composition is required, this can be achieved by increasing the reflux ratio as the system is depleted in the more volatile material. Calculations are again made with the McCabe–Thiele diagram as described by Bogart[4] and illustrated by the following example. Other methods of operating batch columns are described by Ellerbe.[5]

Example 9.3.   A three-stage batch still is loaded initially with 100 kgmole of a liquid containing a mixture of 50 mole% n-hexane in n-octane. The boilup rate is 20 kgmole/hr. A liquid distillate of 0.9 mole fraction hexane is to be maintained by continuously adjusting the reflux ratio. What should the reflux ratio be one hour after start-up? Theoretically, when must the still be shut down? Assume negligible holdup on the plates.

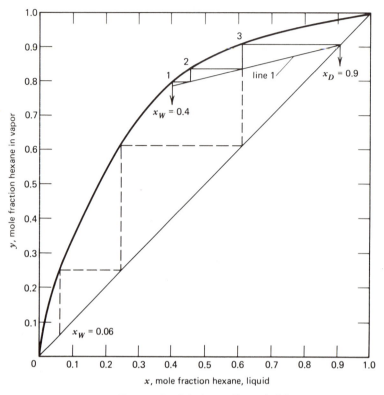

Figure 9.6. Solution to Example 9.3.

Solution. After one hour, the still residue composition is

$$x_W = [50 - 20(0.9)]/80 = 0.4$$

Since $y_D = 0.9$, the operating line is located so that there are three stages between x_W and x_D. The slope of line 1, Fig. 9.6, is $L/V = 0.22$.

At the highest reflux rate possible, $L/V = 1$, $x_W = 0.06$ according to the dash-line construction shown in Fig. 9.6. The corresponding time is

$$0.06 = \frac{50 - 20t(0.9)}{100 - 20t}$$

Solving, $t = 2.58$ hrs.

This chapter has presented only a brief introduction to batch distillation. A more complete treatment including continuous, unsteady-state distillation is given by Holland.[6]

References

1. Rayleigh, J. W. S., *Phil. Mag. and J. Sci.*, Series 6, **4** (23), 521–537 (1902).

2. Smoker, E. H., and A. Rose, *Trans. AIChE*, **36**, 285–293 (1940).

3. Block, B., *Chem. Eng.*, **68** (3), 87–98 (1961).

4. Bogart, M. J. P., *Trans. AIChE*, **33**, 139–152 (1937).

5. Ellerbe, R. W., *Chem. Eng.*, **80**, 110–116 (1973).

6. Holland, C. D., *Unsteady State Processes with Applications in Multicomponent Distillation*, Prentice–Hall, Inc., Englewood Cliffs, N.J., 1966.

Problems

9.1 (a) A bottle of pure n-heptane is accidentally poured into a drum of pure toluene in a commercial laboratory. One of the laboratory assistants, with almost no background in chemistry, suggests that, since heptane boils at a lower temperature than does toluene, the following purification procedure can be used.

> Pour the mixture (2 mole% n-heptane) into a simple still pot. Boil the mixture at 1 atm and condense the vapors until all heptane is boiled away. Obtain the pure toluene from the residue in the still pot.

You, being a chemical engineer, immediately realize that the above purification method will not work. Please indicate this by a curve showing the composition of the material remaining in the pot after various quantities of the liquid have been distilled. What is the composition of the residue after 50 wt% of the original material has been distilled? What is the composition of the cumulative distillate?

(b) When one half of the heptane has been distilled, what is the composition of the cumulative distillate and of the residue? What weight percent of the original material has been distilled?

Vapor–liquid equilibrium data at 1 atm [*Ind. Eng. Chem.*, **42**, 2912 (1949)] are:

Mole fraction n-heptane

Liquid	Vapor	Liquid	Vapor	Liquid	Vapor
0.025	0.048	0.286	0.396	0.568	0.637
0.062	0.107	0.354	0.454	0.580	0.647
0.129	0.205	0.412	0.504	0.692	0.742
0.185	0.275	0.448	0.541	0.843	0.864
0.235	0.333	0.455	0.540	0.950	0.948
0.250	0.349	0.497	0.577	0.975	0.976

9.2 A mixture of 40 mole% isopropanol in water is to be distilled at 1 atm by a simple batch distillation until 70% of the charge (on a molal basis) has been vaporized (equilibrium data are given in Problem 8.36). What will be the compositions of the liquid residue remaining in the still pot and of the collected distillate?

9.3 A 30 mole% feed of benzene in toluene is to be distilled in a batch operation. A product having an average composition of 45 mole% benzene is to be produced. Calculate the amount of residue, assuming $\alpha = 2.5$, and $W_o = 100$.

9.4 A charge of 250 lb of 70 mole% benzene and 30 mole% toluene is subjected to batch differential distillation at atmospheric pressure. Determine composition of distillate and residue after one third of the original mass is distilled off. Assume the mixture forms an ideal solution and apply Raoult's and Dalton's laws with the Antoine equation.

9.5 A mixture containing 60 mole% benzene and 40 mole% toluene is subjected to batch differential distillation at 1 atm, under three different conditions.
(a) Until the distillate contains 70 mole% benzene.
(b) Until 40 mole% of the feed is evaporated.
(c) Until 60 mole% of the original benzene leaves in the vapor phase.
Using $\alpha = 2.43$, determine for each of the three cases:
1. The number of moles in the distillate for 100 moles feed.
2. The composition of distillate and residue.

9.6 A mixture consisting of 15 mole% phenol in water is to be batch distilled at 260 torr. What fraction of the original batch remains in the still when the total distillate contains 98 mole% water? What is the residue concentration?
Vapor–liquid equilibrium data at 260 torr [*Ind. Eng. Chem.*, **17**, 199 (1925)] are:

x, % (H_2O)	1.54	4.95	6.87	7.73	19.63	28.44	39.73	82.99	89.95	93.38	95.74
y, % (H_2O)	41.10	79.72	82.79	84.45	89.91	91.05	91.15	91.86	92.77	94.19	95.64

9.7 A still is charged with 25 gmole of a mixture of benzene and toluene containing 0.35 mole fraction benzene. Feed of the same composition is supplied at a rate of 7 gmole/hr, and the heat rate is adjusted so that the liquid level in the still remains constant. No liquid leaves the still pot, and $\alpha = 2.5$. How long will it be before the distillate composition falls to 0.45 mole fraction benzene?

9.8 Repeat Problem 9.2 for the case of a batch distillation carried out in a two-stage column with a reflux ratio of $L/V = 0.9$.

9.9 Repeat Problem 9.3 assuming the operation is carried out in a three-stage still with an $L/V = 0.6$.

9.10 An acetone–ethanol mixture of 0.5 mole fraction acetone is to be separated by batch distillation at 101 kPa.

Vapor–liquid equilibrium data at 101 kPa are:

Mole fraction acetone

y	0.16	0.25	0.42	0.51	0.60	0.67	0.72	0.79	0.87	0.93
x	0.05	0.10	0.20	0.30	0.40	0.50	0.60	0.70	0.80	0.90

(a) Assuming an L/D of 1.5 times the minimum, how many stages should this column have if we want the composition of the distillate to be 0.9 mole fraction acetone at a time when the residue contains 0.1 mole fraction acetone?

(b) Assume the column has eight stages and the reflux rate is varied continuously so that the top product is maintained constant at 0.9 mole fraction acetone. Make a plot of the reflux ratio versus the still pot composition and the amount of liquid left in the still.

(c) Assume now that the same distillation is carried out at constant reflux ratio (and varying product composition). We wish to have a residue containing 0.1 and an (average) product containing 0.9 mole fraction acetone, respectively. Calculate the total vapor generated. Which method of operation is more energy intensive? Can you suggest operating policies other than constant reflux ratio and constant distillate compositions that might lead to equipment and/or operating cost savings?

9.11 One kilogram-mole of an equimolar mixture of benzene and toluene is fed to a batch still containing three equivalent stages (including the boiler). The liquid reflux is at its bubble point, and $L/D = 4$. What is the average composition and amount of product at a time when the instantaneous product composition is 55 mole% benzene? Neglect holdup, and assume $\alpha = 2.5$.

9.12 A distillation system consisting of a reboiler and a total condenser (no column) is to be used to separate A and B from a trace of nonvolatile material. The reboiler initially contains 20 lbmole of feed of 30 mole% A. Feed is to be supplied to the reboiler at the rate of 10 lbmole/hr, and the heat input is so adjusted that the total moles of liquid in the reboiler remains constant at 20. No residue is withdrawn from the still. Calculate the time required for the composition of the overhead product to fall to 40 mole % A. The relative volatility may be assumed to be constant at 2.50.

9.13 The fermentation of corn produces a mixture of 3.3 mole% ethyl alcohol in water. If 20 mole% of this mixture is distilled at 1 atm by a simple batch distillation, calculate and plot the instantaneous vapor composition as a function of mole percent of batch distilled. If reflux with three theoretical stages were used, what is the maximum purity of ethyl alcohol that could be produced by batch distillation?
Equilibrium data are given in Problem 8.30.

10

Graphical Multistage Calculations by the Ponchon-Savarit Method

In general the Ponchon-Savarit diagram is somewhat more difficult to use than the constant O/V (McCabe–Thiele) diagram, but it is the exact solution for theoretical plates assuming that the enthalpy data employed are correct.

Clark S. Robinson and Edwin R. Gilliland, 1950

The McCabe–Thiele constructions described in Chapter 8 embody rather restrictive tenets. The assumptions of constant molal overflow in distillation and of interphase transfer of solute only in extraction seriously curtail the general utility of the method. Continued use of McCabe–Thiele procedures can be ascribed to the fact that (a) they often represent a fairly good engineering approximation and (b) sufficient thermodynamic data to justify a more accurate approach is often lacking. In the case of distillation, enthalpy-concentration data needed for making stage-to-stage enthalpy balances are often unavailable, while, in the case of absorption or extraction, complete phase equilibrium data may not be at hand.

In this chapter, the Ponchon–Savarit graphical method[1,2] for making stage-to-stage calculations is applied to distillation and extraction. Like the McCabe–Thiele procedure, the Ponchon–Savarit method is restricted to binary distillation and ternary extraction systems. The method does, however, obviate the need for the constant phase-ratio flow assumptions.

10.1 Mass and Energy Balances on Enthalpy-Concentration Diagrams

According to the first law of thermodynamics, the energy conservation per unit mass for a steady-state flow process is

$$\delta q - \delta W_s = dH + d(U^2/2g_c) + dZ$$

If the system is adiabatic, $\delta q = 0$; if no shaft work is done, $\delta W_s = 0$; if the kinetic energy effects are negligible, $d(U^2/2g_c) = 0$; and, if the elevation above the datum plane is constant, $dZ = 0$. Hence only the enthalpy term remains

$$dH = 0 \tag{10-1}$$

Equation (10-1) applies to a mixing process where streams A and B are combined to form C—H_A, H_B, and H_C being the respective enthalpies per unit mass of the stream designated by the subscript. If A, B, and C denote mass flow rates, then

$$H_A A + H_B B = H_C(A + B) \tag{10-2}$$

and, letting x denote mass fraction of one of the components in A, B, or C, we have

$$x_A A + x_B B = x_c C = x_C(A + B) \tag{10-3}$$

Simultaneous solution of (10-2) and (10-3) yields

$$\frac{H_B - H_C}{x_B - x_C} = \frac{H_C - H_A}{x_C - x_A} \quad \text{or} \quad \frac{x_B - x_C}{H_B - H_C} = \frac{x_C - x_A}{H_C - H_A} \tag{10-4}$$

Equation (10-4) is the three-point form of a straight line and is shown in a one-phase field on an $H - x$ diagram in Fig. 10.1.

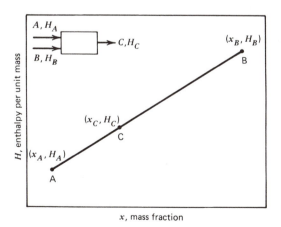

Figure 10.1. Adiabatic mixing process on an enthalpy-concentration diagram.

By (10-4) the point (x_C, H_C) must lie on a straight line connecting the points (x_A, H_A) and (x_B, H_B), since the slopes between point A and C, and C and B are equal. As usual, the ratio of line segments represents the weight ratios of the corresponding streams.

In addition to representing the addition process $A + B = C$, the material balance line $\overline{ABC}$ is also the locus of the equivalent subtraction $B = C - A$. This point is raised to introduce the concept of B as a difference point. If a mixture (x_A, H_A) is subtracted from C, a stream of composition x_B and enthalpy H_B results. It is also useful to recall the geometric relationship between two right triangles having parallel sides; namely, that the ratios of all parallel sides are equal. For instance, in Fig. 10.1,

$$\frac{\overline{BC}}{\overline{CA}} = \frac{x_B - x_C}{x_C - x_A} = \frac{H_B - H_C}{H_C - H_B} = \frac{A}{B} \tag{10-5}$$

10.2   Nonadiabatic Mass and Enthalpy Balances

Under nonadiabatic conditions, q is not zero and, adopting the standard convention of heat transferred out of the system as negative, (10-1) becomes

$$q = \Delta H$$

For multicomponent systems, q may be defined on the basis of a unit mass of any of the streams. To wit, if Q is total heat transfer, the thermal absorption per unit mass of A and B becomes

$$q_A = \frac{Q}{A}; \quad q_B = \frac{Q}{B}; \quad q_C = \frac{Q}{A + B} = \frac{Q}{C}$$

If q_A is factored into (10-2)

$$(H_A + q_A)A + H_B B = H_C C \tag{10-6}$$

Analogously, (10-4) becomes

$$\frac{x_B - x_C}{H_B - H_C} = \frac{x_C - x_A}{H_C - (H_A + q_A)} \tag{10-7}$$

In Fig. 10.2, the graphical representation of (10-7), the point H_A is replaced by $(H_A + q_A)$. Alternatively, if (10-6) were written in terms of q_B, then H_B in (10-7) would be replaced by $(H_B + q_B)$; if it were written in terms of q_C, H_C would become $(H_C - q_C)$.

Figure 10.3 shows the three equivalent ways of representing the nonadiabatic mixing process. By assigning the entire energy effect to stream A, B, or C, we create, respectively, the virtual streams A', B', or C'. These points must lie on a

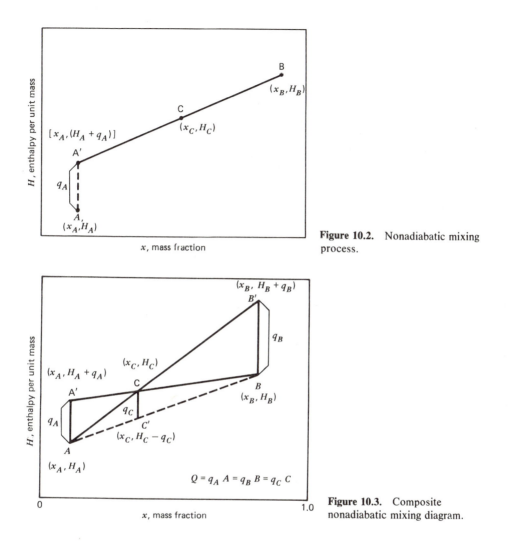

Figure 10.2. Nonadiabatic mixing process.

Figure 10.3. Composite nonadiabatic mixing diagram.

straight line drawn through the other two actual points; i.e., $\overline{A'CB}$, $\overline{AC'B}$, and $\overline{ACB'}$ must fall on different straight lines.

10.3 Application to Binary Distillation

Stage $n-1$ of the column shown in Fig. 10.4 is a mixing device where streams L_n, V_{n-2} enter and the equilibrated streams V_{n-1}, L_{n-1} leave. In Fig. 10.5, the mixing–equilibrating action of the stage is shown in two steps. The vapor V_{n-2}

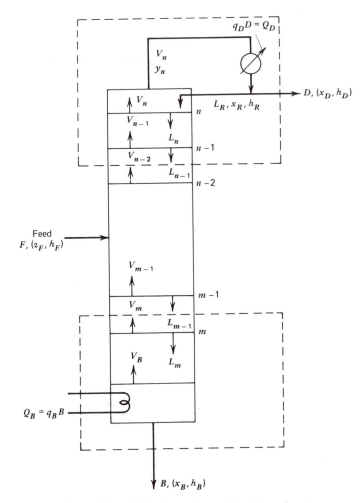

Figure 10.4. Distillation column with total condenser.

and liquid L_n are mixed to give overall composition z, which then separates into two equilibrium vapor and liquid phases V_{n-1}, L_{n-1} that are connected by a tie line through z.

Figure 10.5 demonstrates the basic concept behind the Ponchon–Savarit method. In the McCabe–Thiele construction, material balance equations are plotted on an x-y phase equilibrium diagram, the stages being calculated by alternate use of the material balance and equilibrium relationships. The Ponchon diagram embodies both enthalpy and material balance relationships as well as

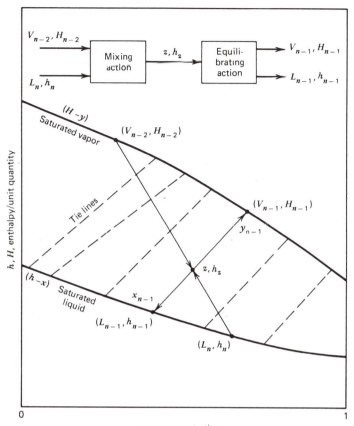

Figure 10.5.   Two-phase mixing on an enthalpy-concentration diagram.

phase equilibrium conditions. Since it is unnecessary to assume constant molal overflow, the calculations can be done on a per mole or per pound basis. Any set of consistent units may be used.

The design equations are developed by making material and enthalpy balances about the portion of the enriching section of the column in Fig. 10.4 enclosed within the dotted line. For the more volatile component, the material balance is

$$y_{n-2}V_{n-2} = x_{n-1}L_{n-1} + Dx_D \tag{10-8}$$

the total material balance is

$$V_{n-2} = L_{n-1} + D \tag{10-9}$$

and the enthalpy balance is

$$q_D D + H_{n-2} V_{n-2} = h_{n-1} L_{n-1} + h_D D \qquad (10\text{-}10)$$

where, to simplify the notation, H = vapor enthalpy and h = liquid enthalpy.

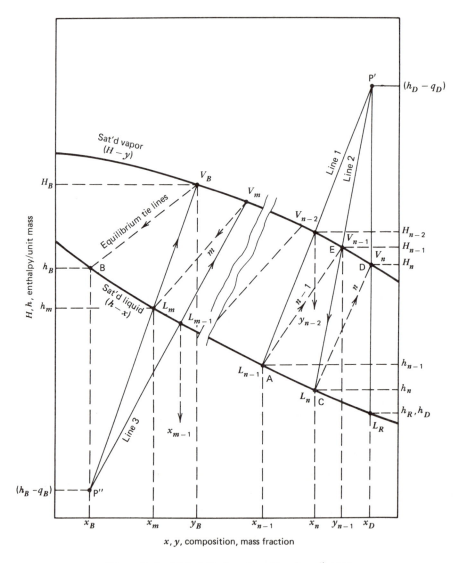

Figure 10.6.   Material balances on Ponchon diagram.

Solving (10-8) and (10-9) simultaneously for L_{n-1}/D,

$$\frac{L_{n-1}}{D} = \frac{x_D - y_{n-2}}{y_{n-2} - x_{n-1}} \tag{10-11}$$

while simultaneous solution of (10-9) and (10-10) gives

$$\frac{L_{n-1}}{D} = \frac{(h_D - q_D) - H_{n-2}}{H_{n-2} - h_{n-1}} \tag{10-12}$$

Combining (10-11) and (10-12) and rearranging, we have

$$\frac{(h_D - q_D) - H_{n-2}}{x_D - y_{n-2}} = \frac{H_{n-2} - h_{n-1}}{y_{n-2} - x_{n-1}} \tag{10-13}$$

Equation (10-13) is an operating line for the two passing streams V_{n-2} and L_{n-1}. The equation, which is a three-point equation of a straight line on a Ponchon diagram, states that the points $(h_D - q_D, x_D)$, (H_{n-2}, y_{n-2}), and (h_{n-1}, x_{n-1}) lie on the same straight line, since the slopes between these pairs of points are equal, and the two lines have one point (H_{n-2}, y_{n-2}) in common. If all phases are saturated, then the points denoting streams V_{n-1} and L_{n-1} lie on the saturated vapor and liquid lines, respectively. In Fig. 10.6 the line satisfying the conditions of (10-13) is labeled Line 1.

The point P', whose coordinates are $(h_D - q_D, x_D)$ is called the *difference* or *net flow point*, since from (10-10), $(h_D - q_D) = (H_{n-2}V_{n-2} - h_{n-1}L_{n-1})/D$, which involves the net enthalpy of passing streams.

Ideal Stages in the Enriching Section

If the lower bound of the material balance loop in the enriching section of the column of Fig. 10.4 had cut the column between plates n and $(n$-1) instead of between $(n - 1)$ and $(n - 2)$, an operating-line equation equivalent to (10-13) would result

$$\frac{(h_D - q_D) - H_{n-1}}{x_D - y_{n-1}} = \frac{H_{n-1} - h_n}{y_{n-1} - x_n} \tag{10-14}$$

Equation (10-14) is shown as Line 2 on Fig. 10.6. This line also contains the difference point P'.

If the material balance loop encloses only the overhead condenser and reflux divider, the operating line is

$$\frac{(h_D - q_D) - H_n}{x_D - y_n} = \frac{H_n - h_R}{y_n - x_D} \tag{10-15}$$

But with a total condenser, $y_n = x_D$. Therefore, the line passing through (H_n, y_n), (h_R, x_D), $(h_D - q_D, x_D)$ is a vertical line passing through the difference point P'.

To illustrate the method of stepping off stages, consider V_n the vapor leaving stage n. This stream is found on the vertical operating line. Since L_n is in equilibrium with V_n, proceed from V_n to L_n along equilibrium tie line n (path $\overline{DC}$). In a manner similar to the McCabe–Thiele method, obtain stream V_{n-1}, which passes countercurrently to L_n, by moving along the operating line $\overline{CEP'}$.

Notice that, once the difference point P' is located, all passing streams between any two stages lie on a line that cuts the phase envelopes and terminates at point P'. If equilibrium liquid and vapor are saturated, their compositions are determined by the points at which the material balance line cuts the liquid and vapor phase envelopes.

The internal reflux ratios L/V for each stage of the column can be expressed in terms of the line segments and coordinates of Fig. 10.6. The ratio of liquid to vapor between stages $n-1$ and $n-2$, for instance, is

$$\frac{L_{n-1}}{V_{n-2}} = \frac{\overline{V_{n-2}P'}}{\overline{L_{n-1}P'}} = \frac{x_D - y_{n-2}}{x_D - x_{n-1}} = \frac{(h_D - q_D) - H_{n-2}}{(h_D - q_D) - h_{n-1}} \qquad (10\text{-}16)$$

while the ratio of reflux to distillate on the top stage is, as in (10-12)

$$\frac{L_R}{D} = \frac{\overline{P'V_n}}{\overline{V_nL_R}} = \frac{(h_D - q_D) - H_n}{H_n - h_R} \qquad (10\text{-}17)$$

These relationships apply even if one of the streams is not saturated. Reflux L_R, for instance, might be subcooled. Note that from (10-17) the difference point P' may be located if the reflux ratio at the top of the column and the composition and thermal condition of the product are known.

The Stripping Section

The principles developed for the enriching section also apply to the stripping section of the column in Fig. 10.4.

Simultaneous material, enthalpy, and component balances about the portion of the column enclosed in the dotted envelope of Fig. 10.6 yield

$$\frac{(h_B - q_B) - H_m}{x_B - y_m} = \frac{H_m - h_{m-1}}{y_m - x_{m-1}} \qquad (10\text{-}18)$$

where q_B is the heat added to the reboiler per unit mass (or mole) of bottoms. Equation (10-18), like (10-13) is an operating-line equation for the two passing streams V_m and L_{m-1}. The material balance line defined by (10-18) is shown as Line 3 in Fig. 10.6. Point P'' is the difference point for the stripping section, the ratio of liquid to vapor between stages m and $m-1$ being

$$\frac{L_{m-1}}{V_m} = \frac{\overline{P''V_m}}{\overline{P''L_{m-1}}} \qquad (10\text{-}19)$$

Stages in the stripping section are stepped off in the same manner as that for the enriching section. From L_{m-1} go to V_m, hence to L_m along tie line m, to passing stream V_B, and hence to B, a partial reboiler acting as a stage.

As before, note that all operating lines in the stripping section pass through P″ and that the compositions of any two saturated passing streams are marked by the points where the straight lines through P″ cut the phase envelopes.

The Overall Column

Considering the entire column in Fig. 10.4, a total overall material balance is

$$F = D + B \qquad (10\text{-}20)$$

a component balance is

$$z_F F = x_D D + x_B B \qquad (10\text{-}21)$$

and an enthalpy balance is

$$Fh_F + q_B B + q_D D = h_D D + h_B B \qquad (10\text{-}22)$$

Algebraic manipulations involving (10-20), (10-21), and (10-22) produce a composite balance line passing through (h_F, z_F) and the two difference points, $(h_B - q_B, x_B)$ and $(h_D - q_D, x_D)$.

$$\frac{D}{B} = \frac{z_F - x_B}{x_D - z_F} = \frac{h_F - (h_B - q_B)}{(h_D - q_D) - h_F} \qquad (10\text{-}23)$$

The line $\overline{\text{P}'\text{P}''}$ in Fig. 10.7 is a plot of (10-23). Presented also in Fig. 10.7 are other pertinent constructions whose relationships to column parameters are summarized in Table 10.1.

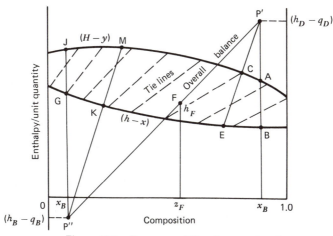

Figure 10.7. Summary of Ponchon construction.

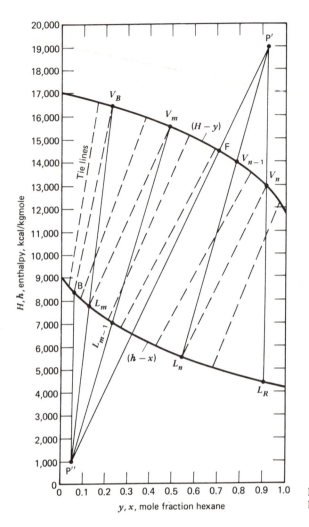

Figure 10.8. Solution to Example 10.1.

Example 10.1. One hundred kilogram-moles per hour of saturated n-hexane-n-octane vapor containing 69 mole% hexane is separated by distillation at atmospheric pressure into product containing 90 mole% hexane and bottoms containing 5 mole% hexane. The total condenser returns 42.5 mole% of the condensate to the column as saturated liquid. Using the graphical Ponchon method and the enthalpy-concentration data of Fig. 10.8 calculate:

(a) The rate of production of bottoms and overhead.

(b) The kilocalories per hour supplied to the boiler and removed at the condenser.

(c) The compositions of streams V_{n-1}, L_n, V_m, and L_{m-1} of Fig. 10.4.

Table 10.1 Summary of Ponchon diagram construction in Fig. 10.7 (All phases assumed saturated)

Section of Column	Line Segments (Fig. 10.7)	Significance
Enriching	$\overline{P'B}$	Heat removed in condenser per pound of distillate
Enriching	$\overline{CP'}/\overline{EP'}$	L/V general
Enriching	$\overline{AP'}/\overline{BP'}$	L/V on top plate (internal reflux ratio)
Enriching	$\overline{AP'}/\overline{AB}$	L/D on top plate (external reflux ratio)
Stripping	$\overline{P''G}$	Heat added in reboiler per pound of bottoms
Stripping	$\overline{MP''}/\overline{KP''}$	L/V general
Stripping	$\overline{MP''}/\overline{MK}$	L/B general
Overall	$\overline{P'P''}/\overline{FP''}$	F/D
Overall	$\overline{FP''}/\overline{FP'}$	D/B

Solution (*basis one hour*). The point F is located on the saturated vapor line at $z_F = 0.69$. Next P′ is located. From Table 10.1, $L_R/V_n = \overline{P'V_n}/\overline{P'L_R} = 0.425$. The overall balance line P′FP″ is drawn, and the point P″ located at the intersection of the balance line with $x_B = 0.05$.

(a) $D/F = \overline{FP''}/\overline{P'P''} = 0.75$
   $D = 75$ kgmole/hr
   $B = 25$ kgmole/hr

Check via an overall hexane balance:

$$\text{Input} = 100(0.69) = 69$$

$$\text{Output} = (0.9)(75) + (0.05)(25) = 68.75$$

which is close to 69.

(b) $q_D = \overline{P'L_R} = (4{,}500 - 19{,}000) = -14{,}500$ kcal/kgmole of product $(-60{,}668$ J/gmole)

$$Q_D = -q_D D = \frac{14{,}500 \text{ kcal}}{\text{kgmole}} \frac{(75 \text{ kgmole})}{\text{hr}} = -1{,}087{,}500 \text{ kcal/hr (4.55 GJ/hr)}$$

$$Q_B = \overline{P''B} = (8{,}400 - 1000)(25) = 185{,}000 \text{ kcal/hr (774 MJ/hr)}$$

(c) The point L_n is on a tie line with V_n. Then V_{n-1} is located by drawing an operating line from L_n to P′ and noting its intersection with the $H - y$ phase envelope.

Starting at B in the stripping section, we move to V_B along a tie line, to L_m along an operating line, hence to V_m along a tie line, and finally to L_{m-1} along an operating line. At

$$V_{n-1}; y_{n-1} = 0.77$$

$$L_n; x_n = 0.54$$

$$V_m; y_m = 0.47$$

$$L_{m-1}; x_{m-1} = 0.23$$

Feed Stage Location

Example 10.1 demonstrated the method of stepping off stages in both the stripping and enriching sections of the column. Since point P′ is the appropriate difference point for stages above the feed stage and P″ for stages below the feed stage, it follows that the transition from one difference point to the other occurs at the feed stage. Hence the problem of transition from enriching to stripping stages is synonymous with the feed stage location problem.

Section 8.5 showed that in the McCabe–Thiele method the optimum transition point lies at the intersection of the enriching and stripping section operating lines, point P in Fig. 8.13a. The analogous location on the Ponchon diagram is the line $\overline{P'P''}$ in Fig. 10.9. Hence the optimum feed stage location is where the equilibrium tie line, stage 5, crosses $\overline{P'P''}$; hence stage 5 becomes the feed stage.

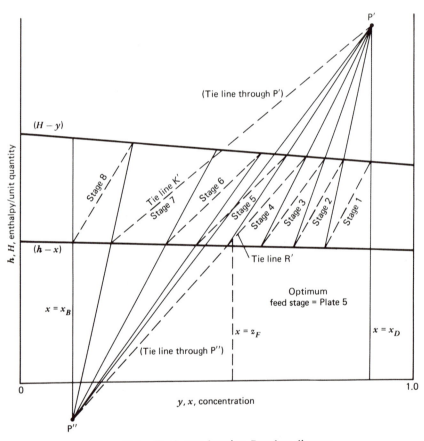

Figure 10.9.   Feed stage location, Ponchon diagram.

Points K and R in Fig. 8.13*b* and 8.13*c* represent, respectively, lower and upper limit feed stage locations where operating lines intersect the equilibrium curve. The exactly equivalent situation occurs in the Ponchon diagram at tie lines K′ and R′. These tie lines coincide with operating lines; hence they represent pinch points just as points K and R did. Using the difference point P′, it is impossible to move beyond tie line K′, just as it is impossible to move past point K on the enriching section operating line in Fig. 8.13*b*. In order to move farther down the column, we must shift from the enriching to the stripping section operating line.

In Fig. 10.9 the phase envelope lines are straight. If the saturated vapor (H-y) and saturated liquid (h-x) lines are both straight and parallel, then the L/V ratios of Fig. 10.9 would be constant throughout the column; hence the Ponchon diagram gives answers equivalent to those of the McCabe–Thiele diagram at constant molal overflow.

Limiting Reflux Ratios

Total reflux corresponds to a situation where there is no feed, distillate, or bottoms and where the minimum number of stages is required to achieve a desired separation. It will be recalled that in the McCabe–Thiele method this situation corresponded to having the operating line coincide with the $y = x$ line.

Figure 10.10 shows the total reflux condition on the Ponchon diagram.

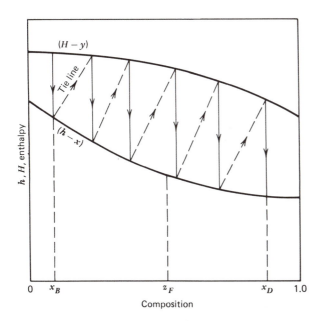

Figure 10.10. Total reflux by Ponchon method.

Difference points P' and P" lie at + and − infinity, respectively, since $y = x$, $B = 0$, $D = 0$, and $(h_D - q_D) = +\infty$, and $(h_B - q_B) = -\infty$.

Minimum reflux conditions correspond to a situation of minimum L/V, maximum product, and infinite stages. In the McCabe–Thiele constructions minimum reflux was determined either by feed conditions or an equilibrium line pinch condition as in Fig. 8.15.

Figure 10.11 demonstrates the analogous Ponchon constructions. In Fig. 10.11a where the minimum reflux ratio is determined by feed conditions, the difference points P' and P" are located by extending the equilibrium tie line through z_F to its intersections with the two difference points.

In Fig. 10.11b, extension of the feed point tie line places the difference points at P'₁ and P"₁. The pinch point, however, manifests itself at tie line T. This tie line, which intersects the line $x = x_D$ at P', gives higher L/V ratios than the difference point P'₁. This suggests that the minimum reflux ratio in the enriching section is set by the highest intersection made by the steepest tie line in the section with the line $x = x_D$. Use of difference point P' in stepping off stages gives an infinite number of stages with the pinch occurring at x_T. A differentially higher L/V ratio results in an operable design.

Point P" is found by the intersection of $\overline{P'z_F}$ with $x = x_B$. If there is a tie line in the stripping section that gives a lower P", this must be used; otherwise, a pinch region will develop in the stripping section.

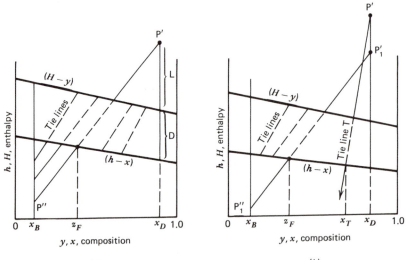

(a)   (b)

Figure 10.11 Minimum reflux by Ponchon method. (a) Minimum reflux determined by the line through feed point, liquid feed. (b) Minimum reflux ratio determined by tie line between feed composition and overhead product.

Partial Condenser

The construction of Fig. 10.12 corresponds to an enriching section with a partial condenser, as shown in the insert. Here V_n and L_R are on the same operating line, and L_R is in equilibrium with D.

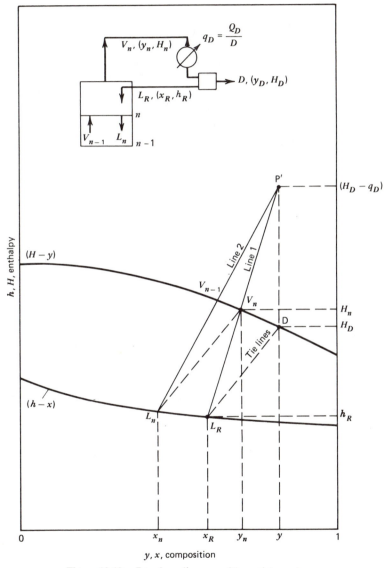

Figure 10.12.   Ponchon diagram with partial condenser.

Example 10.2.   A 59 mole% saturated vapor feed stream containing a 50 mole% mixture of
n-hexane-n-octane is to be separated into a distillate containing 95 mole% hexane and a
bottoms containing 5 mole% hexane. The operation is to be carried out in a column with a
reboiler and partial condenser. Reflux is saturated liquid, and feed enters at the optimum
stage.

(a)   What is the minimum number of stages the column must have to carry out the
separation?

(b)   At an L_R/D of $1.5\,(L/D)_{min}$ for the top stage, how many stages are required? Solve
this part of the problem using both Ponchon and McCabe–Thiele diagrams.

(c)   Make a plot of number of theoretical stages versus the L/D ratio on the top stage
and heat required in kilocalories per kilogram-mole of bottoms.

Solution.   (a) The minimum number of stages is established at total reflux. In this case
the operating lines are vertical. Three theoretical stages are required as seen in Fig. 10.13.

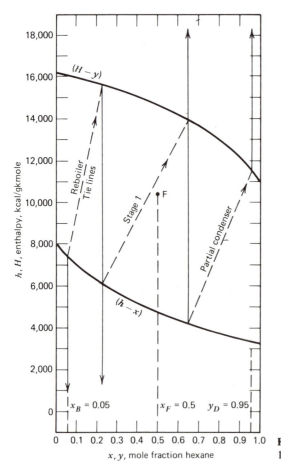

Figure 10.13.   Solution to Example
10.2—Part (a).

(b) There being no tie line that intersects the line $x = x_D$ above P_1', the equilibrium tie line through F serves to establish the difference points. From Fig. 10.14, (L/D) at the top stage is $\overline{P_1' V_{n1}}/\overline{V_{n1} L_R} = 0.7$.

The point P' is located by drawing the line $\overline{L_R V_n P'}$ such that $\overline{P' V_n}/\overline{V_n L_R} = 0.7(1.5) = 1.05$. Point P'' is located at the intersection of the extension of line $\overline{P'F}$ with $x = x_B$.

The stages are stepped off in Fig. 10.15a. Three stages are required in addition to the

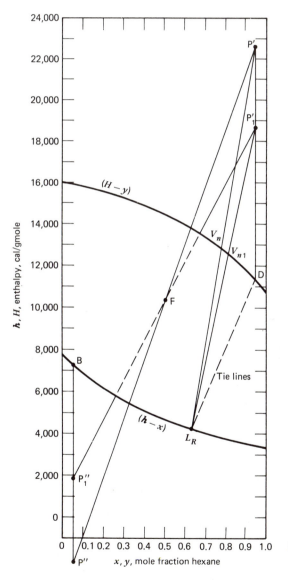

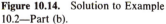

Figure 10.14. Solution to Example 10.2—Part (b).

partial reboiler and partial condenser stages. The feed is introduced on stage 2. Equivalent results are obtained by the McCabe–Thiele diagram in Fig. 10.15b, where, the (curved) operating lines were obtained using Fig. 10.15a and the intersections of Ponchon operating lines originating at P' or P" and the phase envelopes.

If the number of stages is determined on the McCabe–Thiele diagram with the simplifying assumption of constant molal overflow, the straight (dashed) operating lines in

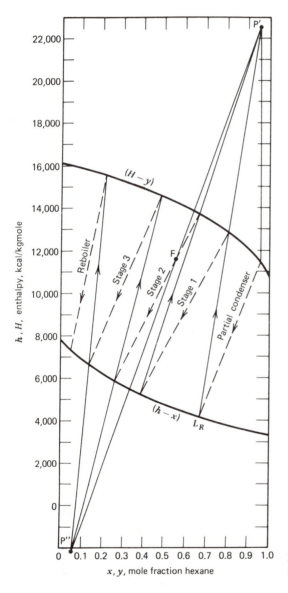

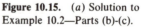

Figure 10.15. (a) Solution to Example 10.2—Parts (b)-(c).

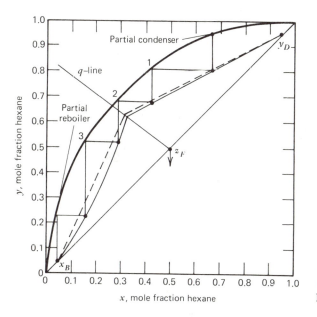

Figure 10.15. (b) Solution to Example 10.2—Part (b).

Fig. 10.15*b* apply. Although not shown, the corresponding number of stages is 5.5 including the partial reboiler and partial condenser. This is 10% higher than the exact result of 5.0 stages.

(c) A plot of q_B and L/D versus the number of stages (including reboiler and condenser) is given in Fig. 10.16. Note that L/D and q_B go to infinity as $n \rightarrow 3$ and that n goes to infinity as $L/D \rightarrow 0.7$ and $q_B \rightarrow 5,500$.

□

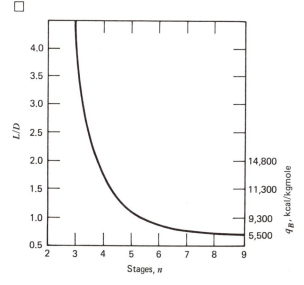

Figure 10.16. Solution to Example 10.2—Part (d).

Side Streams and Multiple Feeds

Although no new principles are involved, it is of interest to examine the type of constructions required in situations involving multiple feeds and/or side streams.

Consider a column of the type shown in Fig. 8.9, but having feed streams F_1 and F_2 and no side stream. Assuming that the distillate and bottoms compositions as well as the reflux ratio on the top plate are specified, it is possible to locate the point P' on Fig. 10.17. Next consider the two feed streams. If F_1 and F_2 are mixed, the resultant stream $\mathscr{F} = F_1 + F_2$ can be located by material and enthalpy balance methods, or by line segment ratios, since $\overline{F_1\mathscr{F}}/\overline{F_2\mathscr{F}} = F_2/F_1$.

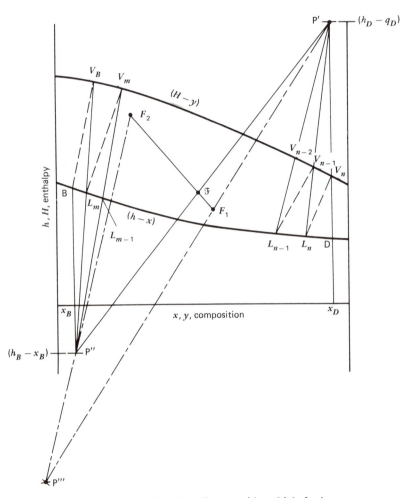

Figure 10.17. Ponchon diagram with multiple feeds.

The difference point P″, which lies on the intersection of $x = x_B$ and the extension of the line $\mathscr{F}P′$, is then located. It is now possible to step off stages by using the upper difference point for the section of the column above F_1 and the lower one for the section below F_2.

To handle the section of the column between F_1 and F_2 it is necessary to find a new difference point. The construction is shown in Fig. 10.17. For the section of the column between the top and a stage between F_1 and F_2, the net flow point

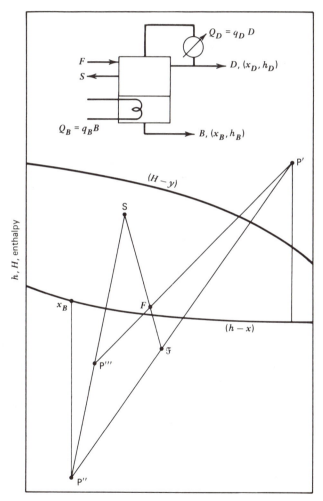

Figure 10.18.   Ponchon diagram with side stream.

must lie on a line through F_1 and P'. Similarly, the line through F_2 and P''
represents the net flow between a stage in the section between F_1 and F_2 and the
bottom of the column. The intersection of the two balance lines occurs at P''',
which is the difference point to be used in stepping off plates between F_1 and F_2.
This may, of course, be verified by material and enthalpy balances.

Another illustration may further clarify the procedures. Figure 10.18 shows
a column having one side stream S along with the usual feed, bottoms, and
distillate. Again, it is assumed that point P' has been located.

Since in this case $\mathscr{F} = F - S$, the point $\mathscr{F}$ in Fig. 10.18 is located so that
$\overline{\mathscr{F}S}/\overline{\mathscr{F}F} = \overline{F/S}$. Point P'' is located by extending the line $\overline{P'\mathscr{F}}$ to its intersection
with $x = x_B$.

The difference points P' and P'' are used in stepping off stages between the
feed and top and between side stream and bottoms, respectively. Between F and
S, we use the difference point P''', which must be located at the intersection of
$\overline{SP''}$ and the extension of $\overline{FP'}$.

10.4   Stage Efficiencies and Actual Number of Plates

The overall and Murphree efficiencies were defined in Chapter 8.

$$\text{Overall efficiency} = \frac{\text{Theoretical contacts}}{\text{Actual contacts required}} \times 100\%$$

$$E_V = \text{Murphree (vapor–phase) efficiency} = \frac{y_n - y_{n-1}}{y_n^* - y_{n-1}} \times 100\% \qquad (10\text{-}24)$$

Figure 10.19 is a graphical representation of (10-24) on a Ponchon diagram.
If the stage efficiency is 100%, move from y_{n-1} to x_n; hence to y_n^*. If E_V is less
than 100%, locate y_n by (10-24). In Fig. 10.19 it is assumed that y_n is a saturated
vapor.

The same procedure used to establish the pseudoequilibrium line on the
McCabe–Thiele plot can be used in conjunction with the Ponchon method. If the
Murphree efficiency is constant, a pseudoequilibrium curve could be drawn
through y_n' and other points identically. This line $\overline{AB}$, in conjunction with the
difference points, could be used to step off stages.

10.5   Application to Extraction

It was seen how readily the McCabe–Thiele diagram can be applied to extraction, provided the coordinates are adjusted to allow for the inert carriers. The
same is true of the Ponchon diagram.

Figure 10.20 is a countercurrent extraction apparatus with both extract and

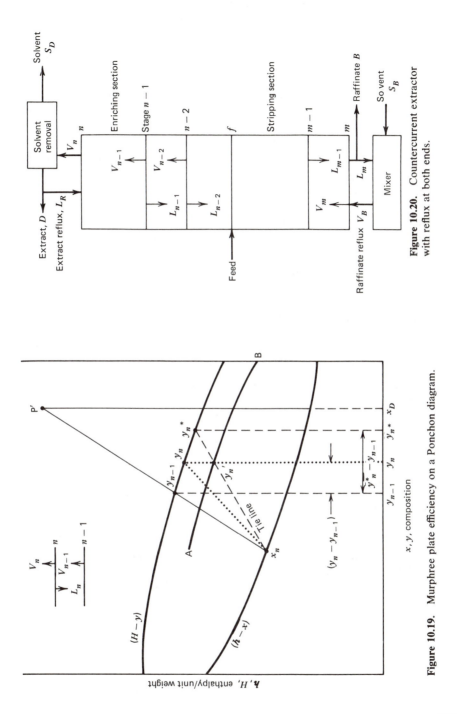

Figure 10.20. Countercurrent extractor with reflux at both ends.

Figure 10.19. Murphree plate efficiency on a Ponchon diagram.

raffinate reflux. The extract reflux is obtained by removing solvent from V_n and returning a portion L_R of relatively solvent-free extract to the apparatus. Thus, the extract phase undergoes added enrichment in the enriching section and it is possible to obtain extract concentrations much richer in solute than the extract in equilibrium with the feed. At the opposite end of the extractor, a portion of the raffinate L_m is diverted and mixed with incoming solvent S_B to produce V_B.

To construct and use a Ponchon diagram for extraction problems it is convenient to consider the solvent in extraction as analogous to enthalpy in distillation. Thus, the solvent-free extract phase in extraction becomes analogous to the enthalpy-rich vapor phase in distillation, as discussed by Smith.[3] Similarities between the two processes are noted in Table 10.2. The most important difference is the choice of coordinates. In Fig. 10.21, Janecke coordinates, as described in Chapter 3, are used. The abscissa is on a solvent-free basis $A/(A + C)$ and the ordinate is $S/(A + C)$. Thus, the coordinates of pure solvent are at $X = 0$, $Y = \text{infinity}$, and the flows L and V are solvent free.

Other points of importance are shown in Fig. 10.21. If the solvent removed in the solvent removal step converts the extract V_n from the top stage into saturated streams, then D and L_R are located on the saturated extract line. If all solvent is removed, they are on the abscissa ($Y = 0$).

The two net flow points P' and P'' are found as in distillation. Point P' is an

Table 10.2   Equivalent parameters in distillation and extraction

Distillation	Extraction
D = distillate	D = extract product (solvent-free basis)
Q = heat	S = mass solvent
Q_D = heat withdrawn in condenser	S_D = solvent withdrawn at top of column
$q_D = Q_D/D$	S_D/D
Q_B = heat added in reboiler	S_B = solvent added in mixer
$q_B = Q_B/B$	S_B/B
B = bottoms	B = raffinate (solvent-free basis)
L = saturated liquid	L = saturated raffinate (solvent-free)
V = saturated vapor	V = saturated extract (solvent-free)
A = more volatile component	A = solute to be recovered
C = less volatile component	C = component from which A is extracted
F = feed	F = feed
x = mole fraction A in liquid	X = mole or wt ratio of A (solvent-free), $A/(A + C)$
y = mole fraction A in vapor	$Y = S/(A + C)$
$P' = h_D + Q_D/D$	$P' = Y_D + S_D/D$
$P'' = h_B - Q_B/B$	$P'' = Y_B - S_B/B$

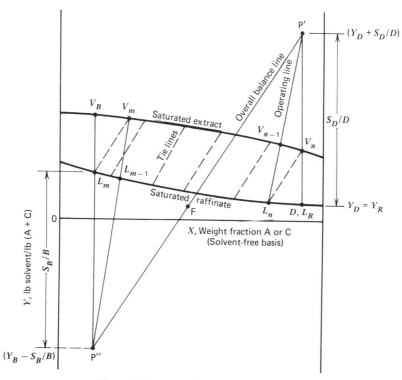

Figure 10.21. Janecke diagram for extraction.

imaginary mixture of solvent S_D and extract D. Hence it is located by adding S_D/D to the extract coordinate Y_D. Likewise, the net flow point P″ is obtained by taking the Y coordinate of B (or L_m) and subtracting the solvent S_B/B.

Since the feed is split as in distillation, the overall balance line will be through the point P′, P″, and the feed point F.

The ratios of line segments have their usual significance. A (solvent-free) material balance above the solvent removal step (assuming S_D is pure solvent) gives

$$V_n = D + L_R \tag{10-25}$$

Equation (10-25) coupled with the solvent balance

$$V_n Y_n - S_D = D Y_D + L_R Y_R \tag{10-26}$$

yields

$$\frac{L_R}{D} = \frac{(Y_D + S_D/D) - Y_n}{Y_n - Y_R} = \frac{\overline{P'V_n}}{\overline{V_n L_R}} \tag{10-27}$$

Other relationships that are easily derivable include

$$\frac{V_B}{B} = \frac{\overline{L_m P''}}{\overline{V_B L_m}} \qquad \frac{L_n}{V_{n-1}} = \frac{\overline{P' V_{n-1}}}{\overline{P' L_n}}$$

$$\frac{L_R}{V_n} = \frac{\overline{P' V_n}}{\overline{P' L_R}} \qquad \frac{D}{B} = \frac{\overline{FP''}}{\overline{P'F}} \qquad \frac{L_{m-1}}{V_m} = \frac{\overline{P'' V_m}}{\overline{P'' L_{m-1}}}$$

Stages are stepped off exactly as in distillation. Intersections of straight lines through P' with phase equilibrium lines mark the composition of two passing streams in the enriching section. Likewise, a line through P'' cuts the envelope at points that denote the composition of passing streams in the stripping section. The significance of total and minimum reflux is also analogous. At total reflux, S_D/D and S_B/B are located at $+$ and $-$ infinity, so the operating lines are vertical. At minimum reflux, S_D/D and S_B/B are reduced (numerically) to the point where one of the operating lines becomes parallel to a tie line, causing a pinch condition.

The significance of line segments and the general methodology apply to extraction columns with or without extract and/or raffinate reflux. Raffinate reflux, in fact, is seldom employed and frequently results in greater, not lower stage requirements, as discussed in Chapter 11.

Example 10.3.   As shown in Fig. 10.22, a countercurrent extraction cascade equipped with a solvent separator to provide extract reflux is used to separate methylcyclopentane A and n-hexane C into a final extract and raffinate containing 95 wt% and 5 wt% A, respectively. The feed rate is 1000 kg/hr with 55 wt% A, and the mass ratio of aniline, the solvent S, to feed is 4.0. The feed contains no aniline and the fresh solvent is pure. Recycle solvent is also assumed pure. Determine the reflux ratio and number of stages. Equilibrium data at column temperature and pressure are shown in Fig. 10.23. Feed is to enter at the optimum stage.

Solution.   Based on the cascade schematic shown in Fig. 10.22, a degrees-of-freedom analysis based on information in Table 6-1 gives:

$$N_D = 3C + 2N + 15$$

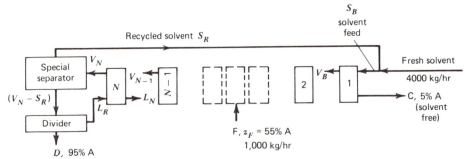

Figure 10.22.   Flowsheet, Example 10.3.

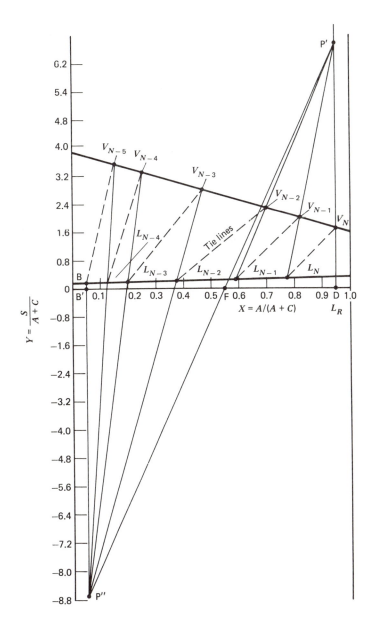

Figure 10.23. Solution to Example 10.3.

The specifications satisfy the degrees of freedom as follows.

Stage pressures	N
Adiabatic stages	N
T and P of stream S_R	2
T and P of stream $(V_N - S_R)$	2
Heat leak and P of divider and solvent mixer	4
S (feed rate not given)	$C + 1$
F	$C + 2$
Feed stage location	1
Solvent/feed ratio S_B/F	1
Recovery of each component in the special separator	C
Concentration of A in product and bottoms	2

$$3C + 2N + 15$$

The points F, L_R, D, B' (solvent-free), and B (solvent-saturated) are located in Fig. 10.23. The points P' and P″, which lie on the verticals through D and B, are found by solving for D (or B) by material balances. An overall material balance for A is

$$0.95D + 0.05B = (1000)(0.55)$$

A total overall balance is

$$D + B = 1000$$

Thus $D = 556$, $B = 444$, and, since $\underline{S_B} = 4{,}000$ kg, $S_B/B = 9.0$. Now the point P″ can be located and then P', by constructing $\overline{\text{P'FP''}}$.

Next, step off stages starting with V_N. Slightly less than six are required with the feed introduced as indicated in Fig. 10.23.

The reflux ratio L_R/D on the top stage is $\overline{\text{P'}V_N}/\overline{V_N D} = 2.95$.

□

References

1. Ponchon, M., *Tech. Moderne*, **13**, 20, 55 (1921).

2. Savarit, R., *Arts et Mètiers*, pp. 65, 142, 178, 241, 266, 307 (1922).

3. Smith, B. D., *Design of Equilibrium Stage Processes*, McGraw–Hill Book Co., New York, 1962, 193.

Problems

10.1 An equimolal mixture of *n*-hexane in *n*-octane having an enthalpy of 4000 cal/gmole is (1) pumped from 1 to 5 atm, (2) passed through a heat exchanger, and (3) flashed to atmospheric pressure. Sixty mole percent of the feed is converted to vapor in the process. Using Fig. 10.8, determine the composition of liquid and vapor leaving the flash drum and the total heat added in the heat exchanger.

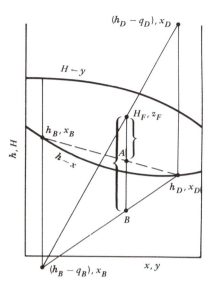

10.2 In the H-y-x diagram above, what is the significance of the lines $H_F - A$ and $H_F - B$?

10.3 An equimolal mixture of carbon tetrachloride and toluene is to be fractionated so as to produce an overhead containing 4 mole% toluene and a bottoms containing 4 mole% carbon tetrachloride. Calculate by the Ponchon method the theoretical minimum reflux ratio, the theoretical minimum number of stages, and the number of theoretical stages when $L/D = 2.5$. The thermal condition of the feed is saturated liquid, which is sent to the optimum stage.

The assumption may be made that the enthalpies of the liquid and the vapor are linear functions of composition.

	Normal Boiling Point, °C	Average Liquid Specific Heat	Latent Heat of Vaporization
CCl₄	76.4	0.225 cal/gm °C	46.42 cal/gm at 76.4°C
Toluene	110.4	0.500 cal/gm °C	86.8 cal/gm at 110.4°C

Equilibrium data (mole fractions CCl₄)

y	0.37	0.62	0.79	0.92
x	0.2	0.4	0.6	0.8

10.4 A mixture of 45 mole% isobutane in n-pentane, at conditions such that 40 mole% is vapor, is to be rectified into a distillate containing only 2 mole% n-pentane. The pressure on the system will be 308 kPa (3.04 atm absolute). The reflux is saturated liquid.

Using the data below, construct an enthalpy-concentration diagram based on an enthalpy datum of liquid at 68°F and determine the minimum number of stages required to make the separation. Also, calculate the condenser duty.

Equilibrium constants for isobutane and n-pentane.

$P = 308$ kPa (3.04 atm abs)

T, °F	K_{nC_5}	K_{iC_4}
100	0.36	1.7
140	0.70	2.6
160	0.90	3.1
150	0.80	2.9
120	0.50	2.1
80	0.25	1.3
70	0.10	1.1

Boiling point at 308 kPa (3.04 atm abs): isobutane = 20°C (68°F), n-pentane = 73.9°C (165°F).

Heat of mixing = negligible.

Heat capacity of liquid isobutane = $0.526 + 0.725 \times 10^{-3} T$ Btu/lb · °F ($T = $°R)

Heat capacity of liquid n-pentane = $0.500 + 0.643 \times 10^{-3} T$ Btu/lb · °F ($T = $°R)

Latent heat of vaporization at boiling point (308 kPa): isobutane = 141 Btu/lb (3.28×10^5 J/kg); n-pentane = 131 Btu/lb (3.4×10^5 J/kg).

Average heat capacity of isobutane vapor at 308 kPa (3.04 atm) = 27.6 Btu/lbmole · °F (1.15×10^5 J/kgmole · °K).

Average heat capacity of the n-pentane vapor at 308 kPa (3.04 atm) = 31 Btu/lbmole · °F (1.297×10^{-5} J/kgmole · °K)

10.5 A saturated liquid feed containing 40 mole% n-hexane and 60 mole% n-octane is fed to a distillation column at a rate of 100 gmole/hr.

A reflux ratio $L/D = 1.5 \, (L/D)_{min}$ is maintained at the top of the column. The overhead product is 95 mole% hexane, and the bottoms product is 10 mole% hexane. If each theoretical plate section loses 80,000 cal/hr (3.35×10^5 J/hr), step off the theoretical plates on the Ponchon diagram, taking into account the column heat losses.

See Fig. 10.8 for H-x-y data.

10.6 A mixture of 80 mole% isopropanol in isopropyl ether is to be fractionated to produce an overhead product containing 77 mole% of the ether and a bottoms product containing 5 mole% of the ether. If the tower is to be designed to operate at 1 atm and a reflux ratio L/D of 1.3 $(L/D)_{min}$, how many theoretical plates will be required?

Determine the number of plates by means of an enthalpy concentration diagram. Assume that the enthalpies of the saturated liquid and the saturated vapor are linear functions of composition. The feed is introduced at its bubble point.

	Normal Boiling Point, °F	H_L, Btu/lbmole	H_v, Btu/lbmole
Isopropyl ether	155	6,580	18,580
Isopropyl alcohol	180	6,100	23,350

Equilibrium data at 1 atm with mole fractions of ether

y	0.315	0.465	0.560	0.615	0.660	0.205	0.750	0.79
x	0.1	0.2	0.3	0.4	0.5	0.6	0.7	0.8

10.7 A refluxed stripper is to operate as shown below. The system is benzene–toluene at 1 atm. The stripping vapor is introduced directly below the bottom plate, and the liquid from this plate is taken as bottoms product. Using the Ponchon method, determine:

(a) The reflux ratio (L/D) at the top of the tower and the condenser heat duty.
(b) Rates of production of distillate and bottoms product.
(c) Total number of theoretical stages required.
(d) Optimum locations for introducing the feed stream and withdrawing the side stream.

Equilibrium data are given in Problem 10.16.

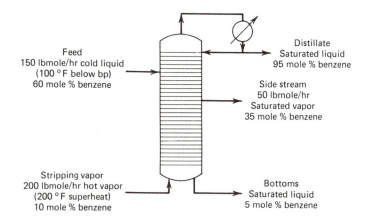

Feed
150 lbmole/hr cold liquid
(100 °F below bp)
60 mole % benzene

Distillate
Saturated liquid
95 mole % benzene

Side stream
50 lbmole/hr
Saturated vapor
35 mole % benzene

Stripping vapor
200 lbmole/hr hot vapor
(200 °F superheat)
10 mole % benzene

Bottoms
Saturated liquid
5 mole % benzene

10.8 An equimolal mixture of carbon tetrachloride and toluene is to be fractionated so as to produce an overhead containing 6 mole% toluene and a bottoms containing 4 mole% carbon tetrachloride and a side stream from the third theoretical plate from the top containing 20 mole% toluene. The thermal conditions of the feed and side stream are saturated liquid.

The rate of withdrawal of the side stream is 25% of the column feed rate. External reflux ratio is $L/D = 2.5$. Using the Ponchon method, determine the number of theoretical plates required. However, if the specifications are excessive, make revisions before obtaining a solution.

The assumption may be made that the enthalpies of the liquid and vapor are linear functions of composition. Equilibrium data are given in Problem 10.3.

Methanol–water vapor–liquid equilibrium and enthalpy data for 1 atm (MeOH = Methyl alcohol)

Mole%	Enthalpy above 0°C Btu/lbmole solution				Vapor-Liquid Equilibrium Data J. G. Dunlop, M.S. thesis, Brooklyn Polytechn. Inst., 1948)		
MeOH	**Saturated Vapor**		**Saturated Liquid**		**Mole% MeOH in**		
y or x	**t°C**	**H_V**	**t°C**	**H_L**	**Liquid, x**	**Vapor, y**	**Boiling Point, °C**
0	100	20,720	100	3240	0	0	100
5	98.9	20,520	92.8	3070	2.0	13.4	96.4
10	97.7	20,340	87.7	2950	4.0	23.0	93.5
15	96.2	20,160	84.4	2850	6.0	30.4	91.2
20	94.8	20,000	81.7	2760	8.0	36.5	89.3
30	91.6	19,640	78.0	2620	10.0	41.8	87.7
40	88.2	19,310	75.3	2540	15.0	51.7	84.4
50	84.9	18,970	73.1	2470	20.0	57.9	81.7
60	80.9	18,650	71.2	2410	30.0	66.5	78.0
70	76.6	18,310	69.3	2370	40.0	72.9	75.3
80	72.2	17,980	67.6	2330	50.0	77.9	73.1
90	68.1	17,680	66.0	2290	60.0	82.5	71.2
100	64.5	17,390	64.5	2250	70.0	87.0	69.3
					80.0	91.5	67.6
					90.0	95.8	66.0
					95.0	97.9	65.0
					100.0	100.0	64.5

10.9   Using the above equilibrium data and the enthalpy data at the top of page 405, solve Problem 8.26 by the Ponchon method.

10.10   One hundred pound-moles per hour of a mixture of 60 mole% methanol in water at 30°C and 1 atm is to be separated by distillation at the same pressure into a distillate containing 98 mole% methanol and a bottoms product containing 96 mole% water. The overhead condenser will produce a subcooled reflux at 40°C. Determine by the Ponchon method:

(a) The minimum cold external reflux in moles per mole distillate.

(b) The number of theoretical stages required for total reflux.

(c) The number of theoretical stages required for a cold external reflux of 1.3 times the minimum.

(d) The internal reflux ratio leaving the top stage, the stage above the feed stage, the stage below the feed stage, and the bottom stage leading to the reboiler.

(e) The duties in British thermal units per hour of the condenser and reboiler.

(f) The temperatures of the top stage and the feed stage.

(g) All the items in Parts (c) to (f) if an interreboiler is inserted on the second stage from the bottom with a duty equal to half that determined for the reboiler in Part (e).

Enthalpy of the liquid, btu/lb mole of solution

Mole % MeOH	Temperature, °C										
	0	10	20	30	40	50	60	70	80	90	100
0	0	324	648	972	1296	1620	1944	2268	2592	2916	3240
5	−180	167	533	887	1235	1592	1933	2291	2646	2997	
10	−297	50	432	810	1181	1564	1922	2300	2673		
15	−373	−18	364	751	1145	1548	1915	2304	2686		
20	−410	−58	328	718	1129	1541	1908	2304	2693		
25	−428	−76	310	706	1123	1527	1901	2304			
30	−427	−79	308	704	1120	1537	1901	2304			
40	−410	−65	320	713	1123	1543	1910	2304			
50	−380	−36	340	731	1138	1557	1930	2318			
60	−335	7	380	765	1174	1577	1953	2345			
70	−279	63	434	812	1220	1600	1985				
80	−209	130	495	869	1260	1638	2016				
90	−121	211	562	940	1310	1678	2048				
100	0	333	675	1022	1375	1733	2092				

(h) All the items in Parts (c) to (f) if a boilup ratio of 1.3 times the minimum value is used and an intercondenser is inserted on the third stage from the top with a duty equal to half that for the condenser in Part (e).

Equilibrium and enthalpy data are given in Problem 10.9.

10.11 An equimolal bubble-point mixture of propylene and 1-butene is to be distilled at 200 psia (1.379 MPa) into 95 mole% pure products with a column equipped with a partial condenser and a partial reboiler.

(a) Construct y-x and H-y-x diagrams using the method of Section 4.7.

(b) Determine by both the McCabe–Thiele and Ponchon methods the number of theoretical stages required at an external reflux ratio equal to 1.3 times the minimum value.

10.12 A mixture of ethane and propane is to be separated by distillation at 475 psia. Explain in detail how a series of isothermal flash calculations using the Soave–Redlich–Kwong equation of state can be used to establish y-x and H-y-x diagrams so that the Ponchon–Savarit method can be applied to determine the stage and reflux requirements.

10.13 One hundred kilogram-moles per hour of a 30 mole% bubble-point mixture of acetone (1) in water (2) is to be distilled at 1 atm to obtain 90 mole% acetone and 95 mole% water using a column with a partial reboiler and a total condenser. The van Laar constants at this pressure are (E. Hale et al., *Vapour-Liquid Equilibrium Data at Normal Pressures*, Pergammon Press, Oxford, 1968) $A_{12} = 2.095$ and $A_{21} = 1.419$.

(a) Construct y-x and H-y-x diagrams at 1 atm.

(b) Use the Ponchon–Savarit method to determine the equilibrium stages required for an external reflux ratio of 1.5 times the minimum value.

10.14 A feed at 21.1°C, 101 kPa (70°F, 1 atm) containing 50.0 mass% ethanol in water is to be stripped in a reboiled stripper to produce a bottoms product containing 1.0 mass% ethanol. Overhead vapors are withdrawn as a top product.
(a) What is the minimum heat required in the reboiler per pound of bottoms product to effect this separation?
(b) What is the composition of the distillate vapor for Part (a)?
(c) If V/B at the reboiler is 1.5 times the minimum and the Murphree plate efficiency (based on vapor compositions) is 70%, how many plates are required for the separation?
Equilibrium data for this system at 1 atm are as follows.

Saturation Temp., °F	Ethanol Concentration		Enthalpy of Mixture Btu/lb	
	Mass fraction in liquid	Mass fraction in vapor	Liquid	Vapor
212	0	0	180.1	1150
208.5	0.020	0.192		
204.8	0.040	0.325		
203.4	0.050	0.377	169.3	1115
197.2	0.100	0.527	159.8	1072
189.2	0.200	0.656	144.3	1012.5
184.5	0.300	0.713	135.0	943
179.6	0.500	0.771	122.9	804
177.8	0.600	0.794	117.5	734
176.2	0.700	0.822	111.1	664
174.3	0.800	0.858	103.8	596
174.0	0.820	0.868		
173.4	0.860	0.888		
173.0	0.900	0.912	96.6	526
173.0	1.000	0.978	89.0	457.5

Note: Reference states for enthalpy = pure liquids, 32°F.

10.15 An equimolal mixture of acetic acid and water is to be separated into a distillate containing 90 mole% water and a bottoms containing 20 mole% water with a plate column having a partial reboiler and a partial condenser. Determine the minimum reflux, and, using a reflux L/D 1.5 times the minimum, calculate the theoretical plates. Assume linear H-x-y, feed on the optimum plate, and operation at 1 atm.
If the Murphree efficiency is 85%, how many stages are required?

Equilibrium data (mole fraction water)

y	0.17	0.3	0.42	0.53	0.63	0.72	0.79	0.86	0.93
x	0.1	0.2	0.3	0.4	0.5	0.6	0.7	0.8	0.9

Notes: Acetic Acid: Liquid $C_p = 31.4$ Btu/lbmole · °F(1.31×10^{-5} J/kgmole · °K).
At normal bp, heat of vap. = 10,430 Btu/lbmole (2.42×10^7 J/kgmole).
Water: Liquid $C_p = 18.0$ Btu/lbmole · °F (7.53×10^4 J/kgmole · °K).
At normal bp, heat of vap. = 17,500 Btu/lbmole (4.07×10^7 J/kgmole).

10.16 An equimolal mixture of benzene and toluene is to be distilled in a plate column at atmospheric pressure. The feed, saturated vapor, is to be fed to the optimum plate. The distillate is to contain 98 mole% benzene, while the bottoms is to contain 2 mole% benzene. Using the Ponchon method and data below [*Ind. Eng. Chem.*, **39**, 752 (1947)], calculate:
(a) Minimum reflux ratio (L/D).
(b) The number of theoretical plates needed and the duties of the reboiler and condenser, using a reflux ratio (L/V) of 0.80.
(c) To which actual plate the feed should be sent, assuming an overall plate efficiency of 65%.

Enthalpy data (1 Atm, 101 kPa)

Composition, mole fraction benzene		Enthalpy, Btu/lbmole	
x	y	Saturated Liquid	Saturated Vapor
0	0.00	8,075	21,885
0.1	0.21	7,620	21,465
0.2	0.38	7,180	21,095
0.3	0.51	6,785	20,725
0.4	0.62	6,460	20,355
0.5	0.72	6,165	19,980
0.6	0.79	5,890	19,610
0.7	0.85	5,630	19,240
0.8	0.91	5,380	18,865
0.9	0.96	5,135	18,500
1.0	1.00	4,900	18,130

10.17 A feed stream containing 35 wt% acetone in water is to be extracted at 25°C in a countercurrent column with extract and raffinate reflux to give a raffinate containing 12% acetone and an extract containing 55% acetone. Pure 1,1,2,-trichloroethane, which is to be the solvent, is removed in the solvent separator, leaving solvent-free product. Raffinate reflux is saturated. Determine
(a) The minimum number of stages.
(b) Minimum reflux ratios.
(c) The number of stages for an extract solvent rate twice that at minimum reflux.
Repeat using a feed containing 50 wt% acetone. Was reflux useful in this case? Feed is to the optimum stage.

System acetone-water-1, 1, 2-trichloroethane, 25°C, composition on phase boundary [*Ind. Eng. Chem.*, 38, 817 (1946)]

	Acetone, weight fraction	Water, weight fraction	Trichloroethane, weight fraction
	0.60	0.13	0.27
	0.50	0.04	0.46
	0.40	0.03	0.57
Extract	0.30	0.02	0.68
	0.20	0.015	0.785
	0.10	0.01	0.89
	0.55	0.35	0.10
	0.50	0.43	0.07
	0.40	0.57	0.03
Raffinate	0.30	0.68	0.02
	0.20	0.79	0.01
	0.10	0.895	0.005

Tie-line data

Raffinate, weight fraction acetone	Extract, weight fraction acetone
0.44	0.56
0.29	0.40
0.12	0.18

Note: This problem is more easily solved using the techniques of Chapter 11.

10.18 A feed mixture containing 50 wt% *n*-heptane and 50 wt% methyl cyclohexane (MCH) is to be separated by liquid–liquid extraction into one product containing 92.5 wt% methylcyclohexane and another containing 7.5 wt% methylcyclohexane. Aniline will be used as the solvent.
(a) What is the minimum number of theoretical stages necessary to effect this separation?
(b) What is the minimum extract reflux ratio?
(c) If the reflux ratio is 7.0, how many theoretical contacts will be required?

Liquid–liquid equilibrium data for the system *n*-heptane–methyl cyclohexane–aniline at 25°C and at 1 atm (101 kPa)

Hydrocarbon Layer		Solvent Layer	
Weight percent MCH, solvent-free basis	Pounds aniline/ pound solvent-free mixture	Weight percent MCH, solvent-free basis	Pounds aniline/ pound solvent-free mixture
0.0	0.0799	0.0	15.12
9.9	0.0836	11.8	13.72
20.2	0.087	33.8	11.5
23.9	0.0894	37.0	11.34
36.9	0.094	50.6	9.98
44.5	0.0952	60.0	9.0
50.5	0.0989	67.3	8.09
66.0	0.1062	76.7	6.83
74.6	0.1111	84.3	6.45
79.7	0.1135	88.8	6.0
82.1	0.116	90.4	5.9
93.9	0.1272	96.2	5.17
100.0	0.135	100.0	4.92

11

Extraction Calculations by Triangular Diagrams

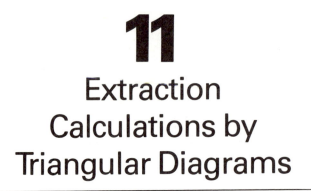

Representation of (liquid–liquid) equilibrium in an isothermal ternary system by a simple mathematical expression is almost impossible and the best representation of such a case is a graphical one employing triangular co-ordinates.

T. G. Hunter and
A. W. Nash, 1934

The triangular phase equilibrium diagrams introduced in Section 3.10 are commonly used for multistage extraction calculations involving ternary systems. The equilateral-triangle diagram was developed by Hunter and Nash,[1] and the extension to right-triangle diagrams was proposed by Kinney.[2] In this chapter we show how they are used in the design of countercurrent contactors. In general they are not as convenient as Ponchon diagrams because they tend to become too cluttered if there are more than a few stages. However, for highly immiscible Class I equilibrium diagrams, Ponchon-type constructions are not possible.

11.1 Right-Triangle Diagrams

In Fig. 11.1, the horizontal, vertical, and diagonal axes represent weight fractions of glycol (G), furfural (F), and water (H), respectively. Because there are only two independent composition variables, any point, such as A, can be located if two compositions are specified ($x_G = 0.43$, $x_F = 0.28$). The point A falls in the two-phase region; hence, at equilibrium the mixture separates into streams A' and A'' whose compositions are fixed by the intersection of the tie line with the phase envelope. The extract A'' is richer in glycol (the solute) and in furfural (the solvent).

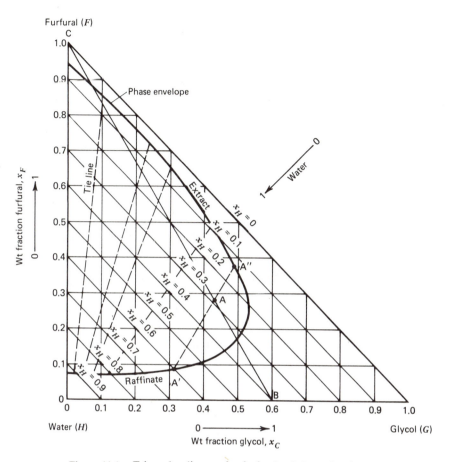

Figure 11.1.   Triangular diagram for furfural–ethylene glycol–water.

Material Balances

In stage n of Fig. 11.2, the streams V and L are termed the *overflow* and *underflow*. A component balance for any one of the three components is

$$L_{n-1}x_{n-1} + V_{n+1}y_{n+1} = V_n y_n + L_n x_n \qquad (11\text{-}1)$$

where y and x are the component mass fractions for the overflow and underflow, respectively. In the following discussion, y and x will refer to the solute only.

The mixing process in Fig. 11.2 can be shown on Fig. 11.1. Suppose a feed stream L_{n-1} consisting of 60 wt% glycol in water, ($x_{n-1} = 0.6$, point B) is mixed with pure solvent (V_{n+1}, point C) in the ratio of 2.61 kg to 1 kg solvent; then $\overline{AC}/\overline{AB} =$ feed/solvent $= 2.61$. The resulting mixture splits into equilibrium

Figure 11.2. An ideal stage.

extract phase A'' and raffinate phase A', where A'' corresponds to 48% G and A' to 32% G.

Overall Column Balances

Figure 11.3a represents a portion of a cascade for which the saturated terminal glycol (solute) compositions $x_{n+1} = 0.35$, $x_W = 0.05$, and $y_W = 0$ (pure solvent) are specified. The ratio of solvent to feed is given as 0.56; that is, $(V_W/L_{n+1} = 100/180)$.

To make an overall balance, we define the mixing point M by

$$M = V_W + L_{n+1} = V_n + L_W \tag{11-2}$$

Since we know the compositions of V_W and L_{n+1} and their mass ratio, M can be located by the lever rule described in Section 3.10. The point on Fig. 11.4 corresponding to solvent/feed = 0.56 is M_1, and the point of intersection of a line through M_1 and L_W with the saturated extract defines V_n at $y_n = 0.33$ because M, V_n, and L_W must lie on the same straight line. The compositions of all streams are now established.

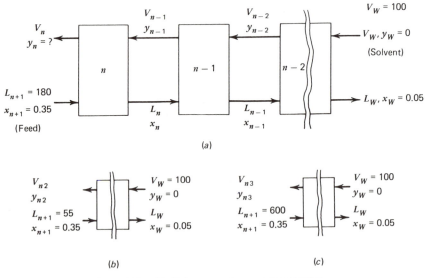

Figure 11.3. Multistage countercurrent contactor.

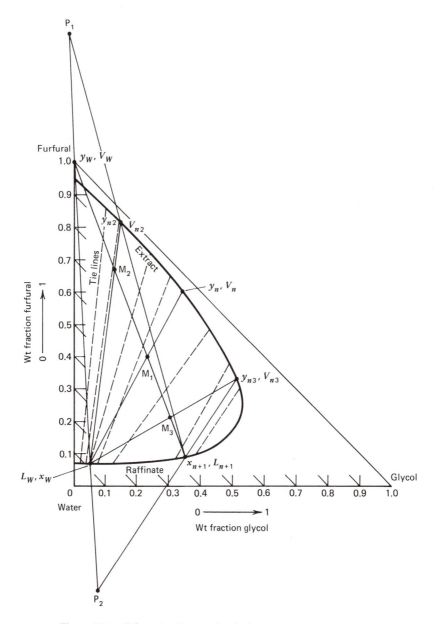

Figure 11.4. Triangular diagram for furfural–ethylene glycol–water.

Stripping and Enriching Section Analogies

As in the Ponchon diagram analysis, we now invoke the concept of a net flow difference point P, which is the terminus of the operating line that marks the location of any two passing streams. In Fig. 11.3a

$$P = V_n - L_{n+1} = V_{n-1} - L_n = V_{n-2} - L_{n-1} = \ldots \qquad (11\text{-}3)$$

In accordance with this nomenclature, the contactor of Fig. 11.3 functions as an enriching section if $V > L$, or as a stripping section if $L > V$.

Case 1. *Enriching Section.*   Difference Point P_1: In Fig. 11.3b, the solvent-to-feed ratio is given as $V_w/L_{n+1} = 100/55 = 1.82$ and $x_w = 0.05$ and $x_{n+1} = 0.35$. This allows point M_2 to be located by the lever rule. From L_w, a line through M_2 locates V_{n2} with the terminal composition $y_{n2} = 0.14$. We can solve for L_w and V_{n2} by the lever rule or by material balances.

Total balance: $L_w + V_n = 155$

Furfural balance: $(0)(100) + (0.35)(55) = 0.5 L_w + 0.14 V_n$

Solving,

$$V_n = 128 \qquad L_w = 27$$

In accordance with (11-3), the net flow point P_1 in Fig. 11.4 is located at the intersection of the material balance lines, which are extensions of the lines drawn from L_w to V_w and from L_{n+1} to V_{n2}.

Case 2. *Stripping Section.*   Difference Point P_2: If, as shown in Fig. 11.3c, the solvent-to-feed ratio is 100/600, the mixing point in Fig. 11.4 is M_3, and y_{n3} can be located. By a material balance, $V_{n3} = 380$ and $L_w = 320$. The net flow point, by (11-3), is fixed at P_2 by intersection of the two material balance lines. Note that, by definition, this is a stripping section because $L > V$.

Stepping Off Stages

Stepping off stages in a countercurrent cascade entails alternate use of equilibrium and material balance lines. In Case 1, Fig. 11.3b, the difference point is at P_1 in Fig. 11.4; and the terminal solute compositions are $x_{n+1} = 0.35$, $y_{n2} = 0.14$, $x_w = 0.05$, and $y_w = 0$. If we start at the top of the column, we are on operating line $\overline{P_1 x_{n+1}}$. We then move down to stage n along the equilibrium tie line through y_{n2}. Less than one stage is required to reach x_w, since the raffinate in equilibrium with an extract phase of $y_{n2} = 0.14$ is $x_w = 0.04$, where these are solute weight fractions.

Minimum Stages

The minimum number of stages in distillation is at total reflux when $L = V$, the composition of the passing streams are equal, and the heat addition per pound of distillate is infinite. For the simple configuration in Fig. 11.3a, because there is no reflux, the condition of minimum stages corresponds to an infinite V_n/L_{n+1} (solvent/feed) ratio. Thus the mixing point M must lie on the pure solvent apex, providing the apex is in the two-phase region. Otherwise, we move as far as possible toward the apex.

Infinite Stages

Case 2 in Figs. 11.3c and 11.4 (difference point P_2) represents a pinch point on plate n, since an equilibrium tie line and operating line coincide along $\overline{P_2 x_{n+1} y_{n3}}$. If less solvent is used, point M_3 is lowered, P_2 is raised, and the contactor is inoperable.

An analogous minimum-solvent, maximum-stage situation exists for the $V > L$ enriching section situation. Here, the net flow point P_1 would be lowered until an operating line–equilibrium line pinch is encountered; that is no further change in composition takes place as we add stages.

11.2   Equilateral-Triangle Diagrams

Countercurrent extraction calculations are readily made on equilateral-triangle phase equilibrium diagrams, no new principles being involved. In Fig. 11.5, we see the solution to Case 1 of Fig. 11.3b.

The point y_{n2} is found by dividing the line between $y_w = 0$, $x_{n+1} = 0.35$ into the section $\overline{Mx_{n+1}}/\overline{My_w} = V_w/L_{n+1} = 1.82$ and locating y_{n2} on the intersection of the tie line $\overline{Mx_w}$ with the saturated extract phase envelope. The difference point P is at the intersection of the material balance lines through $\overline{x_{n+1}y_{n2}}$ and $\overline{x_w y_w}$.

It is seen that the separation can be carried out with one stage.

11.3   Extract and Raffinate Reflux

The extraction cascade in Fig. 11.6 has both extract and raffinate reflux. Raffinate reflux is not processed through the solvent recovery unit since additional solvent would have to be added in any case. It is necessary, however, to remove solvent from extract reflux.

The use of raffinate reflux has been judged to be of little, if any, benefit by Wehner[3] and Skelland.[4] Since only the original raffinate plus extra solvent is refluxed, the amount of material, not the compositions, is affected.

Analysis of a complex operation such as this involves relatively straightforward extensions of the methodology already developed. Results, however, depend critically on the equilibrium phase diagram and it is very difficult to draw any general conclusions with respect to the effect (or even feasibility) of reflux. Most frequently, column parameters are dictated by the necessity of operating in

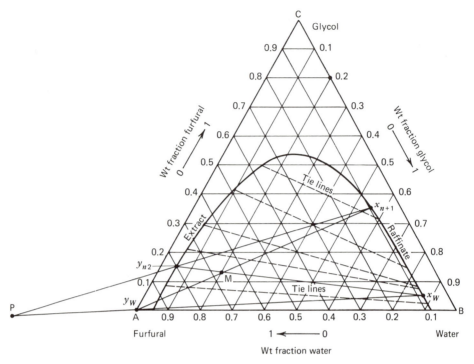

Figure 11.5. Equilateral equilibrium diagram, furfural–ethylene glycol–water. $T = 25°C$. $P = 101$ kPa.

a region where the two phases are rapidly separable and emulsion and foaming difficulties are minimal.

Some of these factors are considered below in Example 11.1, which demonstrates the benefits (or lack thereof) of using reflux in the extraction of acetone from ethyl acetate with water. We will first investigate the effect of solvent ratios on stage requirements in a simple countercurrent cascade and then the effect of extract and raffinate reflux on the same system. The data (at 30°C and 101 kPa) are from Venkataratnam and Rao;[5] and portions of this analysis are from Sawistowski and Smith.[6] A right-triangle, rather than an equilateral-triangle, diagram is frequently preferred for this type of analysis because the ordinate scale can be expanded or contracted to prevent crowding of the construction. Another alternative is to transfer the operating line and equilibrium data information to an x-y McCabe–Thiele diagram, for stepping off stages. In most parts of this problem, so few stages are required that an equilateral-triangle diagram can be used.

Example 11.1. (a) Calculate the weight ratio of water (S) to feed required to reduce the

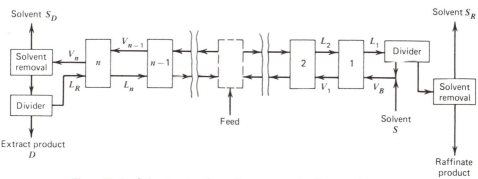

Figure 11.6.   Solvent extraction with extract and raffinate reflux.

acetone (A) concentration in a feed mixture of 30 wt% acetone (A) and 70 wt% ethyl acetate (E) to 5% by weight acetone (water-free) as a function of the number of stages in the countercurrent cascade show in Fig. 11.7. (b) Examine advantages, if any, of extract reflux.

Solution.   (a) Prior to investigating the maximum and minimum solvent/feed ratios, which occur, respectively, at the minimum and maximum number of stages, we develop the solution for $S/F = 1.5$.

In Fig. 11.7, the feed is at F and the water-free raffinate product at B. The saturated raffinate is located at B' on the saturated raffinate envelope on a line connecting B with the 100% water vertex at S.

The composition of the saturated extract D' is obtained from a material balance about the extractor.

$$S + F = D' + B' = M \tag{11-4}$$

Since the ratio of $S/F = 1.5$, point M can be located such that $\overline{FM}/\overline{MS} = 1.5$. A straight line must also pass through D', B', and M. Therefore, D' can be located by extending B'M to the extract envelope.

The net flow difference point P is $P = S - B' = D' - F$. It is located at the inter-section of extensions of lines $\overline{FD'}$ and $\overline{B'S}$.

Stepping off stages poses no problem. Starting at D', we follow a tie line to L_1. Then V_2 is located by noting the intersection of the operating line L_1P with the phase envelope. Additional stages are stepped off in the same manner by alternating between the tie lines and operating lines. For the sake of clarity, only the first stage is shown; three are required.

Minimum stages.   The mixing point M must, by (11-4), lie on a line joining S and F, such that $\overline{FM}/\overline{MS} = S/F$. Since the minimum number of stages corresponds to maximum solvent flow, we move toward S as far as possible. In Fig. 11.8 the point $M_{max} = D'_{max}$ on the extract envelope represents the point of maximum possible solvent addition. If more were added, two phases could not exist. The difference point P_{max} is also at $M_{max} = D'_{max}$ since this is the intersection of $\overline{D'B'}$ and $\overline{FS}$. By chance, the line $\overline{D'B'}$ coincides with a tie line so only one stage is required. Note that this represents a hypothetical situation since removal of solvent from D' gives a product having the feed composition $x_D = 0.3$, $\overline{B'M_{max}}/\overline{D'M_{max}} = \infty$, and B' equals essentially zero.

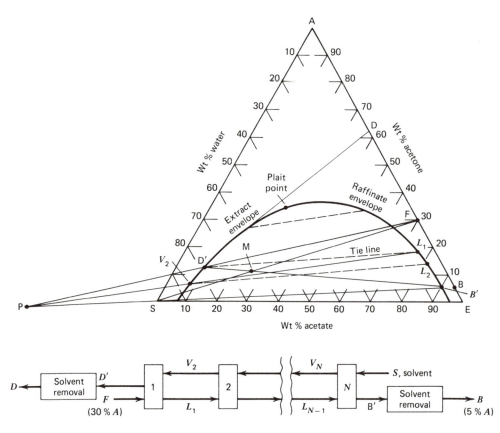

Figure 11.7.　Simple cascade, $S/F = 1.5$.

Infinite stages (minimum solvent).　When an operating line coincides with a tie line, composition of successive extract and raffinate streams remains constant; and we have encountered a *pinch point* or *pinch zone*. In Fig. 11.8, point D'_{min} has been chosen so that the operating line through D'_{min} and F coincides with a tie line through D'_{min} and L_1. This gives a pinch zone at the feed end (top) of the extractor. This is not always the minimum solvent point; the pinch could occur at other locations in the extractor. If so, it is readily apparent from the diagram. The mixing point M_{min} is located, as before, by the intersection of $\overline{B'_{min}D'_{min}}$ and $\overline{SF}$ so $S_{min}/F = 0.76$.

Results of computations for other S/F ratios are summarized in Table 11.1.

Table 11.1　Results for Example 11.1(a)

Countercurrent cascade (without reflux)	S/F (solvent/feed ratio)	0.76	1.5	3	8.46
	N (stages)	∞	3	1.9	1
	x_D (wt% acetone)	65	62	48	30

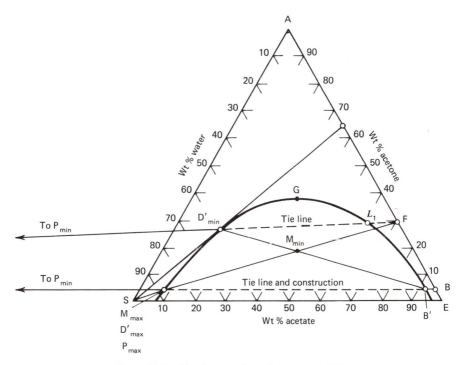

Figure 11.8. Simple cascade—min and max S/F.

(b) The maximum possible concentration of acetone in the solvent-free extract is 65 wt%, corresponding to minimum solvent of $S/F = 0.76$. Since an S/F of 1.5 results in a 62 wt% acetone product, use of extract reflux for the purpose of producing a purer (solvent-free) product is not very attractive, given the particular phase equilibrium diagram and feedstock in this example. To demonstrate the technique, the calculation for extract reflux is carried out nevertheless.

For the case of the simple countercurrent cascade, the extract pinch point is at 58 wt% H_2O, 27 wt% A, and 15 wt% E as shown in Fig. 11.8 at D'_{min}. If there are stages above the feed, as in the refluxed cascade of Fig. 11.9, it is possible to reduce the water content to (theoretically) about 32 wt% (point G) in Fig. 11.8. However, the solvent-free product would not be as rich in acetate (53 wt%).

Assume that a saturated extract containing 50 wt% water is required elsewhere in the process. This is shown as point D' in Fig. 11.9. The configuration is as in Fig. 11.6 but without raffinate reflux and with the product removed prior to solvent separation. The ratio of S/F is taken to be 1.5, and the raffinate again is 5 wt% acetone (water-free) at point B.

Since there are both stripping and enriching sections, there are two net flow points, P' and P'', above and below the feed stage, respectively.

First the mixing point $M = F + S$, with $S/F = 1.5$, is located. Also, since $M = S_D + D' + B'$ and $P' = V_n - L_R = D' + S_D$ we can locate P' by $P' = M - B' = V_n - L_R = V_{n-1} - L_n$.

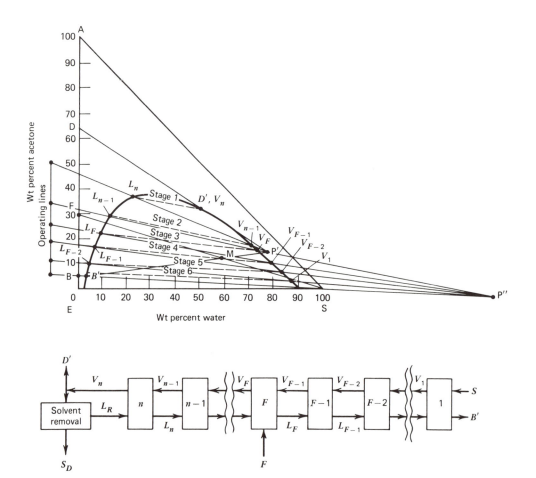

Figure 11.9. Extract reflux for Example 11.1.

Thus, P′ must lie on the line $\overline{\text{B}'\text{M}}$ extended. Furthermore, V_n has the same composition as D', and L_R is simply D' with the solvent removed (point D). Hence P′ in Fig. 11.9, is located at the intersection of $\overline{\text{D}'\text{D}}$ and $\overline{\text{B}'\text{M}}$.

It is now possible to step off stages for the enriching section, stopping when the feed line $\overline{\text{FP}'}$ is crossed. At this point, the stripping section net flow point P″ must be located. Since $P'' = B' - S = F - P'$, this is easily done.

Six stages are required with two above the feed. The extract reflux ratio $(V_n - D')/D'$ is 2.39.

Consider the case of total reflux (minimum stages) with the configuration and feed streams in Fig. 11.9. With reference to Fig. 11.10, $L_R = 62$ wt%, $F = 30$ wt%, $D' = 33$ wt%, and $B' = 4.9$ wt%, acetone. As in the case of the simple cascade of Fig. 11.8, as the solvent-to-feed ratio is increased, the mixing point $M = F + S$ moves towards the pure solvent axis. At maximum possible solvent addition M, P′ and P″ all lie at the intersection

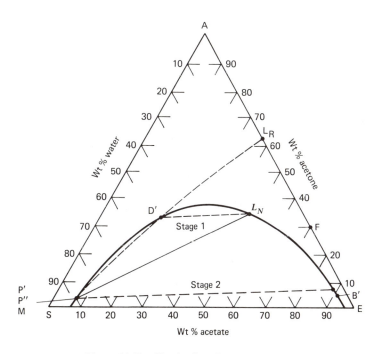

Figure 11.10. Total reflux, for Example 11.1.

of the line through F and S with the extract phase envelope, since $P' = S_D$ (D' being zero) and $P'' = F - P' = -P'$, since $F = 0$. No product is removed at either end of the apparatus, there is no feed, and solvent moves straight through the extractor with $S_D = S$. A little more than two stages is required.

Consider the case of minimum reflux ratio (infinite stages). As the amount of solvent is reduced, point M (equal to $S + F$) in Fig. 11.9 moves towards F, and P' (equal to $D' + S_D$) moves towards D'. Point P'' (equal to $B' - S$) moves away from the equilibrium curve. The maximum distance that points M, P', and P'' can be moved is determined by the slope of the tie lines. The minimum solvent ratio, which corresponds to the minimum reflux ratio, is reached when a tie line and an operating line coincide. A pinch point can occur either in the enriching or in the stripping section of the column, so it is necessary to seek the highest value of the minimum reflux ratio by trial and error. In this example, it occurs at the feed stage. The minimum reflux ratio is 0.58 and the corresponding minimum solvent ratio is 0.74.

□

References

1. Hunter, T. G., and A. W. Nash, *J. Soc. Chem. Ind.*, **53**, 95T–102T (1934).

2. Kinney, G. F., *Ind. Eng. Chem.*, **34**, 1102–1104 (1942).

3. Wehner, J. F., *AIChE J.*, **5**, 406 (1959).

4. Skelland, A. H. P., *Ind. Eng. Chem.*, **53**, 799–800 (1961).

5. Venkataratnam, A., and R. J. Rao, *Chem. Eng. Sci.*, **7**, 102–110 (1957).

6. Sawistowski, H., and W. Smith, *Mass Transfer Process Calculations*, Interscience Publishers, Inc., New York, 1963.

Problems

11.1 Benzene and trimethylamine (TMA) are to be separated in a three-stage liquid–liquid extraction column using water as the solvent. If the solvent-free extract and raffinate products are to contain, respectively, 70 and 3 wt% TMA, find the original feed composition and the water-to-feed ratio with a right-triangle diagram. There is no reflux and the solvent is pure water.

Trimethylamine–water–benzene compositions on phase boundary

	Extract, wt%			Raffinate, wt%	
TMA	**H$_2$O**	**Benzene**	**TMA**	**H$_2$O**	**Benzene**
5.0	94.6	0.04	5.0	0.0	95.0
10.0	89.4	0.06	10.0	0.0	90.0
15.0	84.0	1.0	15.0	1.0	84.0
20.0	78.0	2.0	20.0	2.0	78.0
25.0	72.0	3.0	25.0	4.0	71.0
30.0	66.4	3.6	30.0	7.0	63.0
35.0	58.0	7.0	35.0	15.0	50.0
40.0	47.0	13.0	40.0	34.0	26.0

Tie-line data

Extract, wt% TMA	Raffinate, wt% TMA
39.5	31.0
21.5	14.5
13.0	9.0
8.3	6.8
4.0	3.5

11.2 One thousand kilograms per hour of a 45 wt% acetone-in-water solution is to be extracted at 25°C in a continuous countercurrent system with pure 1,1,2-trichloroethane to obtain a raffinate containing 10 wt% acetone. Using the equilibrium data in Problem 10.17, determine with a right-triangle diagram:
(a) The minimum flow rate of solvent.

(b) The number of stages required for a solvent rate equal to 1.5 times the minimum.

(c) The flow rate and composition of each stream leaving each stage.

11.3 The system docosane–diphenylhexane–furfural is representative of more complex systems encountered in the solvent refining of lubricating oil. Five hundred kilograms per hour of a 40 wt% mixture of diphenylhexane in docosane are to be continuously extracted in a countercurrent system with 500 kg/hr of a solvent containing 98 wt% furfural and 2 wt% diphenylhexane to produce a raffinate that contains only 5 wt% diphenylhexane. Calculate with a right-triangle diagram the number of theoretical stages required and the kilograms per hour of diphenylhexane in the extract at 45°C and at 80°C. Compare the results.

Equilibrium Data [Ind. Eng. Chem., 35, 711 (1943)]—
Binodal curves in docosane–diphenylhexane–furfural system

Wt% at 45°C			Wt% at 80°C		
Docosane	Diphenylhexane	Furfural	Docosane	Diphenylhexane	Furfural
96.0	0.0	4.0	90.3	0.0	9.7
84.0	11.0	5.0	50.5	29.5	20.0
67.0	26.0	7.0	34.2	35.8	30.0
52.5	37.5	10.0	23.8	36.2	40.0
32.6	47.4	20.0	16.2	33.8	50.0
21.3	48.7	30.0	10.7	29.3	60.0
13.2	46.8	40.0	6.9	23.1	70.0
7.7	42.3	50.0	4.6	15.4	80.0
4.4	35.6	60.0	3.0	7.0	90.0
2.6	27.4	70.0	2.2	0.0	97.8
1.5	18.5	80.0			
1.0	9.0	90.0			
0.7	0.0	99.3			

Tie lines in docosane–diphenylhexane–furfural system

Docosane Phase Composition, wt%			Furfural Phase Composition, wt%		
Docosane	Diphenylhexane	Furfural	Docosane	Diphenylhexane	Furfural
Temperature, 45°C					
85.2	10.0	4.8	1.1	9.8	89.1
69.0	24.5	6.5	2.2	24.2	73.6
43.9	42.6	13.3	6.8	40.9	52.3
Temperature, 80°C					
86.7	3.0	10.3	2.6	3.3	94.1
73.1	13.9	13.0	4.6	15.8	79.6
50.5	29.5	20.0	9.2	27.4	63.4

11.4 Solve each of the following liquid–liquid extraction problems.
(a) Problem 10.17 using a right-triangle diagram.
(b) Problem 10.18 using an equilateral-triangle diagram.
(c) Problem 10.19 using a right-triangle diagram.

11.5 For each of the ternary systems given below indicate whether
(a) Simple countercurrent extraction

or (b) Countercurrent extraction with extract reflux
or (c) Countercurrent extraction with raffinate reflux
or (d) Countercurrent extraction with both extract and raffinate reflux

would be expected to yield the most economical process.

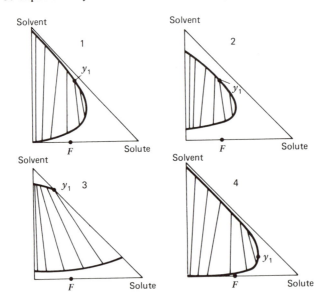

11.6 Two solutions, feed F at the rate of 75 kg/hr containing 50 wt% acetone and 50 wt% water, and feed F' at the rate of 75 kg/hr containing 25 wt% acetone and 75 wt% water, are to be extracted in a countercurrent system with 37.5 kg/hr of 1,1,2-trichloroethane at 25°C to give a raffinate containing 10 wt% acetone. Calculate the number of stages required and the stage to which each feed should be introduced. Equilibrium data are given in Problem 10.17.

11.7 The three-stage extractor shown below is used to extract the amine from a fluid consisting of 40 wt% benzene and 60 wt% trimethylamine. The solvent (water) flow to stage 3 is 750 kg/hr and the feed flow rate is 1000 kg/hr. Determine the required solvent flow rates S_1 and S_2.

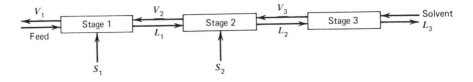

11.8 The extraction process shown below is conducted in a multiple-feed countercurrent unit without extract or raffinate reflux. Feed F' is composed of solvent and solute and is an extract phase feed. Feed F'' is composed of unextracted raffinate and solute and is a raffinate phase feed.

Derive the equations required to establish the three reference points needed to step off the theoretical stages in the extraction column. Show the graphical determination of these points on a right-triangle graph.

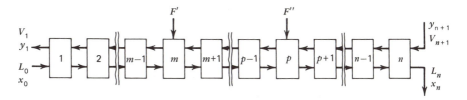

11.9 A mixture containing 50 wt% methylcyclohexane (MCH) in n-heptane is fed to a countercurrent stage-type extractor at 25°C. Aniline is used as solvent. Reflux is used on both ends of the column.

An extract containing 95 wt% MCH and a raffinate containing 5 wt% MCH (both on solvent-free basis) are required.

The minimum reflux ratio of extract is 3.49. Using a right-triangle diagram, calculate:
(a) The reflux ratio of raffinate.
(b) How much aniline must be removed at the separator "on top" of the column.
(c) How much solvent must be added to the solvent mixer at the bottom.
Equilibrium data are given in Problem 10.18.

11.10 Two liquids A and B, which have nearly identical boiling points, are to be separated by liquid–liquid extraction with a solvent C. The data below represent the equilibrium between the two liquid phases at 95°C.

Equilibrium data

Extract Layer			Raffinate Layer		
A, %	B, %	C, %	A, %	B, %	C, %
0	7	93.0	0	92.0	8.0
1.0	6.1	92.9	9.0	81.7	9.3
1.8	5.5	92.7	14.9	75.0	10.1
3.7	4.4	91.9	25.3	63.0	11.7
6.2	3.3	90.5	35.0	51.5	13.5
9.2	2.4	88.4	42.0	41.0	17.0
13.0	1.8	85.2	48.1	29.3	22.6
18.3	1.8	79.9	52.0	20.0	28.0
24.5	3.0	72.5	47.1	12.9	40.0
31.2	5.6	63.2	Plait point		

Adapted from McCabe and Smith, *Unit Operations of Chemical Engineering*, 3rd ed., McGraw–Hill Book Co., New York, 1976, 639.

Determine the minimum amount of reflux that must be returned from the extract product and from the raffinate product to produce an extract containing 83% A and 17% B (on a solvent-free basis) and a raffinate product containing 10% A and 90% B (solvent-free). The feed contains 35% A and 65% B on a solvent-free basis and is a saturated raffinate. The raffinate is the heavy liquid. Determine the number of ideal stages on both sides of the feed required to produce the same end products from the same feed when the reflux ratio of the extract, expressed as pounds of extract reflux per pound of extract product (including solvent), is twice the minimum. Calculate the masses of the various streams per 1000 pounds of feed. Solve the problem, using equilateral-triangle coordinates, right-triangle coordinates, and solvent-free coordinates.

Approximate Methods for Multicomponent, Multistage Separations

> In the design of any [distillation] column it is important to know at least two things. One is the minimum number of plates required for the separation if no product, or practically no product, is withdrawn from the column. This is the condition of total reflux. The other point is the minimum reflux that can be used to accomplish the desired separation. While this case requires the minimum expenditure of heat, it necessitates a column of infinite height. Obviously, all other cases of practical operation lie in between these two conditions.
>
> Merrill R. Fenske, 1932

Although rigorous computer methods are available for solving multicomponent separation problems, approximate methods continue to be used in practice for various purposes, including preliminary design, parametric studies to establish optimum design conditions, and process synthesis studies to determine optimal separation sequences.

This chapter presents three useful approximate methods: the Fenske–Underwood–Gilliland method and variations thereof for determining reflux and stage requirements of multicomponent distillation; the Kremser group method and variations thereof for separations involving various simple countercurrent cascades such as absorption, stripping, and liquid–liquid extraction; and the Edmister group method for separations involving countercurrent cascades with intermediate feeds such as distillation. These methods can be applied readily by hand calculations if physical properties are composition independent. However, since they are iterative in nature, computer solutions are recommended.

12.1 Multicomponent Distillation by Empirical Method

An algorithm for the empirical method that is commonly referred to as the Fenske–Underwood–Gilliland method, after the authors of the three important steps in the procedure, is shown in Fig. 12.1 for a distillation column of the type shown in Table 1.1. The column can be equipped with a partial or total condenser. From Table 6.2, the degrees of freedom with a total condenser are $2N + C + 9$. In this case, the following variables are generally specified with the partial reboiler counted as a stage.

	Number of specifications
Feed flow rate	1
Feed mole fractions	$C - 1$
*Feed temperature	1
*Feed pressure	1
Adiabatic stages (excluding reboiler)	$N - 1$
Stage pressures (including reboiler)	N
Split of light-key component	1
Split of heavy-key component	1
Feed stage location	1
Reflux ratio (as multiple of minimum reflux)	1
Reflux temperature	1
Adiabatic reflux divider	1
Pressure of total condenser	1
Pressure at reflux divider	$\dfrac{1}{2N + C + 9}$

Similar specifications can be written for columns with a partial condenser.

Selection of Two Key Components

For multicomponent feeds, specification of two key components and their distribution between distillate and bottoms is accomplished in a variety of ways. Preliminary estimation of the distribution of nonkey components can be sufficiently difficult as to require the iterative procedure indicated in Fig. 12.1. However, generally only two and seldom more than three iterations are necessary.

Consider the multicomponent hydrocarbon feed in Fig. 12.2. This mixture is typical of the feed to the recovery section of an alkylation plant.[1] Components are listed in order of decreasing volatility. A sequence of distillation columns

---

* Feed temperature and pressure may correspond to known stream conditions leaving the previous piece of equipment.

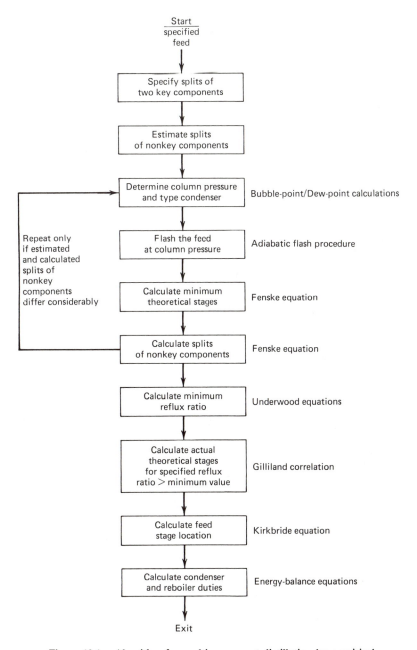

Figure 12.1.   Algorithm for multicomponent distillation by empirical method.

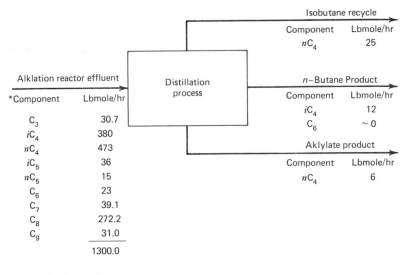

Figure 12.2. Separation specifications for alkylation reactor effluent.

including a deisobutanizer and a debutanizer is to be used to separate this mixture into the three products indicated. In Case 1 of Table 12.1, the deisobutanizer is selected as the first column in the sequence. Since the allowable quantities of n-butane in the isobutane recycle, and isobutane in the n-butane product, are specified, isobutane is the light key and n-butane is the heavy key. These two keys are adjacent in the order of volatility. Because a fairly sharp separation between these two keys is indicated and the nonkey components are not close in volatility to the butanes, as a preliminary estimate, we can assume the separation of the nonkey components to be perfect.

Alternatively, in Case 2, if the debutanizer is placed first in the sequence, specifications in Fig. 12.2 require that n-butane be selected as the light key. However, selection of the heavy key is uncertain because no recovery or purity is specified for any component less volatile than n-butane. Possible heavy-key components for the debutanizer are iC_5, nC_5, or C_6. The simplest procedure is to select iC_5 so that the two keys are again adjacent.

For example, suppose we specify that 13 lbmole/hr of iC_5 in the feed is allowed to appear in the distillate. Because the split of iC_5 is not sharp and nC_5 is close in volatility to iC_5, it is probable that the quantity of nC_5 in the distillate will not be negligible. A preliminary estimate of the distributions of the nonkey components

Table 12.1 Specifications of key component splits and preliminary estimation of nonkey component splits for alkylation reactor effluent

Component	Feed, lbmole/hr	Case 1, Deisobutanizer Column First, lbmole/hr		Case 2, Debutanizer Column First (iC_5 is HK), lbmole/hr		Case 3, Debutanizer Column First (C_6 is HK), lbmole/hr	
		Distillate	Bottoms	Distillate	Bottoms	Distillate	Bottoms
C_3	30.7	(30.7)	(0)	(30.7)	(0)	(30.7)	(0)
iC_4	380	[b]368	[a]12	(380.0)	(0)	(380.0)	(0)
nC_4	473	[a]25	[b]448	[b]467	[a]6	[b]467	[a]6
iC_5	36	(0)	(36)	[a]13	[b]23	(13)	(23)
nC_5	15	(0)	(15)	(1)	(14)	(1)	(14)
C_6	23	(0)	(23)	(0)	(23)	[a]0.01	[b]22.99
C_7	39.1	(0)	(39.1)	(0)	(39.1)	(0)	(39.1)
C_8	272.2	(0)	(272.2)	(0)	(272.2)	(0)	(272.2)
C_9	31.0	(0)	(31.0)	(0)	(31.0)	(0)	(31.0)
	1300.0	423.7	876.3	891.7	408.3	891.71	408.29

[a] Specification.
[b] By material balance.
(Preliminary estimate.)

431

for Case 2 is given in Table 12.1. Although iC_4 may also distribute, a preliminary estimate of zero is made for the bottoms quantity.

Finally, in Case 3, we select C_6 as the heavy key for the debutanizer at a specified rate of 0.01 lbmole/hr in the distillate as shown in Table 12.1. Now iC_5 and nC_5 will distribute between the distillate and bottoms in amounts to be determined; as a preliminary estimate, we assume the same distribution as in Case 2.

In practice, the deisobutanizer is usually placed first in the sequence. In Table 12.1, the bottoms for Case 1 then becomes the feed to the debutanizer, for which, if nC_4 and iC_5 are selected as the key components, component separation specifications for the debutanizer are as indicated in Fig. 12.3 with preliminary estimates of the separation of nonkey components shown in parentheses. This separation has been treated by Bachelor.[2] Because nC_4 and C_8 comprise 82.2 mole% of the feed and differ widely in volatility, the temperature difference between distillate and bottoms is likely to be large. Furthermore, the light-key split is rather sharp; but the heavy-key split is not. As will be shown later, this case provides a relatively severe test of the empirical design procedure discussed in this section.

Column Operating Pressure and Type Condenser

For preliminary design, column operating pressure and type condenser can be established by the procedure shown in Fig. 12.4, which is formulated to achieve, if possible, reflux drum pressures P_D between 0 and 415 psia (2.86 MPa) at a minimum temperature of 120°F (49°C) (corresponding to the use of water as the coolant in the overhead condenser). The pressure and temperature limits are representative only and depend on economic factors. Both column and condenser pressure drops of 5 psia are assumed. However, when column tray requirements are known, more refined computations should allow at least 0.1 psi/tray for atmospheric or superatmospheric column operation and 0.05 psi/tray pressure drop for vacuum column operation together with a 5 to 2 psia condenser pressure drop. Column bottom temperature must not result in bottoms decomposition or correspond to a near-critical condition. A total condenser is used for reflux drum pressures to 215 psia. A partial condenser is used from 215 psia to 365 psia. A refrigerant is used for overhead condenser coolant if pressure tends to exceed 365 psia.

With column operating pressures established, the column feed can be adiabatically flashed at an estimated feed tray pressure of $P_D + 7.5$ psia to determine feed-phase condition.

Example 12.1. Determine column operating pressures and type of condenser for the debutanizer of Fig. 12.3.

Solution. Using the estimated distillate composition in Fig. 12.3, we compute the

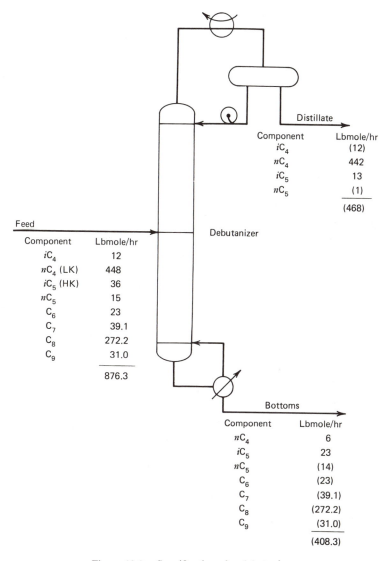

Figure 12.3. Specifications for debutanizer.

distillate bubble-point pressure at 120°F (48.9°C) iteratively from (7-18) in a manner similar to Example 7.4. This procedure gives 79 psia as the reflux drum pressure. Thus, a total condenser is indicated. Allowing a 5-psi condenser pressure drop, column top pressure is $(79 + 5) = 84$ psia; and, allowing a 5-psi pressure drop through the column, the bottoms pressure is $(84 + 5) = 89$ psia.

Bachelor[2] sets column pressure at 80 psia throughout. He obtains a distillate tem-

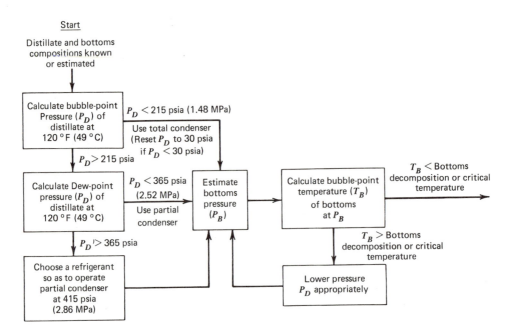

Figure 12.4. Algorithm for establishing distillation column pressure and type condenser.

perature of 123°F. A bubble-point calculation for the bottoms composition at 80 psia gives 340°F. This temperature is sufficiently low to prevent decomposition.

Feed to the debutanizer is presumably bottoms from a deisobutanizer operating at a pressure of perhaps 100 psia or more. Results of an adiabatic flash of this feed to 80 psia are given by Bachelor[2] as follows.

	Pound-moles per hour	
Component	**Vapor Feed**	**Liquid Feed**
iC_4	3.3	8.7
nC_4	101.5	346.5
iC_5	4.6	31.4
nC_5	1.6	13.4
nC_6	1.3	21.7
nC_7	1.2	37.9
nC_8	3.2	269.0
nC_9	0.2	30.8
	116.9	759.4

The temperature of the flashed feed is 180°F (82.2°C). From above, the feed mole fraction vaporized is $(116.9/876.3) = 0.1334$.

☐

Minimum Equilibrium Stages

For a specified separation between two key components of a multicomponent mixture, an exact expression is easily developed for the required minimum number of equilibrium stages, which corresponds to total reflux. This condition can be achieved in practice by charging the column with feedstock and operating it with no further input of feed and no withdrawal of distillate or bottoms, as illustrated in Fig. 12.5. All vapor leaving stage N is condensed and returned to stage N as reflux. All liquid leaving stage 1 is vaporized and returned to stage 1 as boilup. For steady-state operation within the column, heat input to the reboiler and heat output from the condenser are made equal (assuming no heat losses). Then, by a material balance, vapor and liquid streams passing between any pair of stages have equal flow rates and compositions; for example, $V_{N-1} = L_N$ and $y_{i,N-1} = x_{i,N}$. However, vapor and liquid flow rates will change from stage to stage unless the assumption of constant molal overflow is valid.

Derivation of an exact equation for the minimum number of equilibrium stages involves only the definition of the K-value and the mole fraction equality between stages. For component i at stage 1 in Fig. 12.5

$$y_{i,1} = K_{i,1}x_{i,1} \tag{12-1}$$

But for passing streams

$$y_{i,1} = x_{i,2} \tag{12-2}$$

Combining these two equations,

$$x_{i,2} = K_{i,1}x_{i,1} \tag{12-3}$$

Similarly, for stage 2

$$y_{i,2} = K_{i,2}x_{i,2} \tag{12-4}$$

Combining (12-3) and (12-4), we have

$$y_{i,2} = K_{i,2}K_{i,1}x_{i,1} \tag{12-5}$$

Equation (12-5) is readily extended in this fashion to give

$$y_{i,N} = K_{i,N}K_{i,N-1} \cdots K_{i,2}K_{i,1}x_{i,1} \tag{12-6}$$

Similarly, for component j

$$y_{j,N} = K_{j,N}K_{j,N-1} \cdots K_{j,2}K_{j,1}x_{j,1} \tag{12-7}$$

Combining (12-6) and (12-7), we find that

$$\frac{y_{i,N}}{y_{j,N}} = \alpha_N \alpha_{N-1} \cdots \alpha_2 \alpha_1 \left(\frac{x_{i,1}}{x_{j,1}}\right) \tag{12-8}$$

or

$$\left(\frac{x_{i,N+1}}{x_{i,1}}\right)\left(\frac{x_{j,1}}{x_{j,N+1}}\right) = \prod_{k=1}^{N_{min}} \alpha_k \tag{12-9}$$

Total condenser

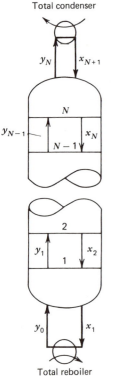

Total reboiler   **Figure 12.5.** Distillation column operation at total reflux.

where $\alpha_k = K_{i,k}/K_{j,k}$, the relative volatility between components i and j. Equation (12-9) relates the relative enrichments of any two components i and j over a cascade of N theoretical stages to the stage relative volatilities between the two components. Although (12-9) is exact, it is rarely used in practice because the conditions of each stage must be known to compute the set of relative volatilities. However, if the relative volatility is constant, (12-9) simplifies to

$$\left(\frac{x_{i,N+1}}{x_{i,1}}\right)\left(\frac{x_{j,1}}{x_{j,N+1}}\right) = \alpha^N \tag{12-10}$$

or

$$N_{\min} = \frac{\log\left[\left(\frac{x_{i,N+1}}{x_{i,1}}\right)\left(\frac{x_{j,1}}{x_{j,N+1}}\right)\right]}{\log \alpha_{i,j}} \tag{12-11}$$

Equation (12-11) is extremely useful. It is referred to as the *Fenske equation.*[3] When $i =$ the light key and $j =$ the heavy key, the minimum number of equilibrium stages is influenced by the nonkey components only by their effect (if any) on the value of the relative volatility between the key components.

Equation (12-11) permits a rapid estimation of minimum equilibrium stages. A more convenient form of (12-11) is obtained by replacing the product of the mole fraction ratios by the equivalent product of mole distribution ratios in terms of component distillate and bottoms flow rates d and b, respectively,* and by replacing the relative volatility by a geometric mean of the top-stage and bottom-stage values. Thus

$$N_{min} = \frac{\log\left[\left(\frac{d_i}{d_j}\right)\left(\frac{b_j}{b_i}\right)\right]}{\log \alpha_m} \tag{12-12}$$

where the mean relative volatility is approximated by

$$\alpha_m = [(\alpha_{i,j})_N (\alpha_{i,j})_1]^{1/2} \tag{12-13}$$

Thus, the minimum number of equilibrium stages depends on the degree of separation of the two key components and their relative volatility, but is independent of feed-phase condition. Equation (12-12) in combination with (12-13) is exact for two minimum stages. For one stage, it is equivalent to the equilibrium flash equation. In practice, distillation columns are designed for separations corresponding to as many as 150 minimum equilibrium stages.

When relative volatility varies appreciably over the cascade and when more than just a few stages are involved, the Fenske equation, although inaccurate, generally predicts a conservatively high value for N. Under varying volatility conditions, the Winn equation[4] is more accurate if the assumption that

$$K_i = \zeta_{i,j} K_j^{\varphi_{i,j}} \tag{12-14}$$

is valid, where ζ and φ are empirical constants determined for the pressure and temperature range of interest. Dividing each side of (12-6) by the φ power of (12-7) and combining with (12-14) gives the Winn equation

$$N_{min} = \frac{\log\left[\left(\frac{x_{i,N+1}}{x_{i,1}}\right)\left(\frac{x_{j,1}}{x_{j,N+1}}\right)^{\varphi_{i,j}}\right]}{\log \zeta_{i,j}} \tag{12-15}$$

If (12-14) does not apply (e.g., with very nonideal mixtures), the Winn equation can also be in error.

Example 12.2. For the debutanizer shown in Fig. 12.3 and considered in Example 12.1, estimate the minimum equilibrium stages by (a) the Fenske equation and (b) the Winn equation. Assume uniform operating pressure of 80 psia (552 kPa) throughout and utilize the ideal K-values given by Bachelor[2] as plotted after the method of Winn[4] in Fig. 12.6.

---

* This substitution is valid even though no distillate or bottoms products are withdrawn at total reflux.

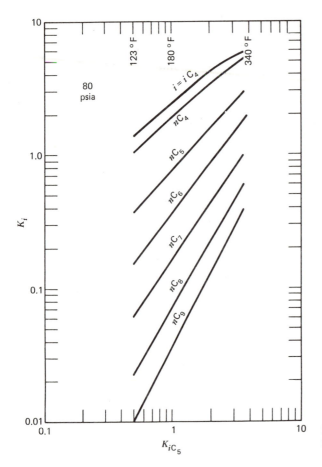

Figure 12.6. Ideal K-values for hydrocarbons at 80 psia.

Solution. The two key components are n-butane and isopentane. Distillate and bottoms conditions based on the estimated product distributions for nonkey components in Fig. 12.3 are

Component	$x_{N+1} = x_D$	$x_1 = x_B$
iC_4	0.0256	~0
nC_4 (LK)	0.9445	0.0147
iC_5 (HK)	0.0278	0.0563
nC_5	0.0021	0.0343
nC_6	~0	0.0563
nC_7	~0	0.0958
nC_8	~0	0.6667
nC_9	~0	0.0759
	1.0000	1.0000

(a) From Fig. 12.6, at 123°F, the assumed top-stage temperature,

$$(\alpha_{nC_4,iC_5})_N = 1.03/0.495 = 2.08$$

At 340°F, the assumed bottom-stage temperature,

$$(\alpha_{nC_4,iC_5})_1 = 5.20/3.60 = 1.44$$

From (12-13)

$$\alpha_m = [(2.08)(1.44)]^{1/2} = 1.73$$

Noting that $(d_i/d_j) = (x_{D_i}/x_{D_j})$ and $(b_i/b_j) = (x_{B_i}/x_{B_j})$, (12-12) becomes

$$N_{min} = \frac{\log[(0.9445/0.0147)(0.0563/0.0278)]}{\log 1.73} = 8.88 \text{ stages}$$

(b) From Fig. 12.6, a straight line fits the slightly curved line for nC_4 according to the equation

$$K_{nC_4} = 1.862 K_{iC_5}^{0.843}$$

so that $\zeta = 1.862$ and $\phi = 0.843$ in (12-14). From (12-15), we have

$$N_{min} = \frac{\log[(0.9445/0.0147)(0.0563/0.0278)]^{0.843}}{\log 1.862} = 7.65 \text{ stages}$$

The Winn equation gives approximately one less stage than the Fenske equation.

☐

Distribution of Nonkey Components at Total Reflux

The Fenske and Winn equations are not restricted to the two key components. Once N_{min} is known, they can be used to calculate mole fractions x_{N+1} and x_1 for all nonkey components. These values provide a first approximation to the actual product distribution when more than the minimum stages are employed. A knowledge of the distribution of nonkey components is also necessary when applying the method of Winn because (12-15) cannot be converted to a mole ratio form like (12-12).

Let $i = $ a nonkey component and $j = $ the heavy-key or reference component denoted by r. Then (12-12) becomes

$$\left(\frac{d_i}{b_i}\right) = \left(\frac{d_r}{b_r}\right)(\alpha_{i,r})_m^{N_{min}} \tag{12-16}$$

Substituting $f_i = d_i + b_i$ in (12-16) gives

$$b_i = \frac{f_i}{1 + \left(\frac{d_r}{b_r}\right)(\alpha_{i,r})_m^{N_{min}}} \tag{12-17}$$

or

$$d_i = \frac{f_i\left(\frac{d_r}{b_r}\right)(\alpha_{i,r})_m^{N_{min}}}{1 + \left(\frac{d_r}{b_r}\right)(\alpha_{i,r})_m^{N_{min}}} \tag{12-18}$$

Equations (12-17) and (12-18) give the distribution of nonkey components at total reflux as predicted by the Fenske equation.

The Winn equation is manipulated similarly to give

$$b_i = \frac{f_i}{1 + \left[\dfrac{\zeta_{i,r}^{N_{min}}}{\left(\dfrac{b_r}{d_r}\right)^{\varphi_{i,r}}\left(\dfrac{B}{D}\right)^{1-\varphi_{i,r}}}\right]} \tag{12-19}$$

and

$$d_i = \frac{f_i}{1 + \left[\dfrac{\left(\dfrac{b_r}{d_r}\right)^{\varphi_{i,r}}\left(\dfrac{B}{D}\right)^{1-\varphi_{i,r}}}{\zeta_{i,r}^{N_{min}}}\right]} \tag{12-20}$$

where

$$B = \sum_i b_i \tag{12-21}$$

$$D = \sum_i d_i \tag{12-22}$$

To apply (12-19) and (12-20), values of B and D are assumed and then checked with (12-21) and (12-22).

For accurate calculations, (12-17), (12-18), (12-19), and/or (12-20) should be used to compute the smaller of the two quantities b_i and d_i. The other quantity is best obtained by overall material balance.

Example 12.3. Estimate the product distributions for nonkey components by the Fenske equation for the problem of Example 12.2.

Solution. All nonkey relative volatilities are calculated relative to isopentane using the K-values of Fig. 12.6.

Component	α_{i,iC_5}		
	123°F	340°F	Geometric mean
iC_4	2.81	1.60	2.12
nC_5	0.737	0.819	0.777
nC_6	0.303	0.500	0.389
nC_7	0.123	0.278	0.185
nC_8	0.0454	0.167	0.0870
nC_9	0.0198	0.108	0.0463

Based on $N_{min} = 8.88$ stages from Example 12.2 and the above geometric-mean relative volatilities, values of $(\alpha_{i,r})_m^{N_{min}}$ are computed relative to isopentane as tabulated below.

From (12-17), using the feed rate specifications in Fig. 12.3 for f_i,

$$b_{iC_4} = \frac{12}{1 + \left(\frac{13}{23}\right)790} = 0.0268 \text{ lbmole/hr}$$

$$d_{iC_4} = f_{iC_4} - b_{iC_4} = 12 - 0.0268 = 11.9732 \text{ lbmole/hr}$$

Results of similar calculations for the other nonkey components are included in the following table.

Component	$(\alpha_{i,iC_5})_m^{N_{min}}$	d_i	b_i
iC_4	790	11.9732	0.0268
nC_4	130	442.0	6.0
iC_5	1.00	13.0	23.0
nC_5	0.106	0.851	14.149
nC_6	0.000228	0.00297	22.99703
nC_7	3.11×10^{-7}	6.87×10^{-6}	39.1
nC_8	3.83×10^{-10}	5.98×10^{-8}	272.2
nC_9	1.41×10^{-12}	2.48×10^{-11}	31.0
		467.8272	408.4728

□

Minimum Reflux

Minimum reflux is based on the specifications for the degree of separation between two key components. The minimum reflux is finite and feed product withdrawals are permitted. However, a column can not operate under this condition because of the accompanying requirement of infinite stages. Nevertheless, minimum reflux is a useful limiting condition.

For binary distillation at minimum reflux, as shown in Fig. 8.15a, most of the stages are crowded into a constant-composition zone that bridges the feed stage. In this zone, all vapor and liquid streams have compositions essentially identical to those of the flashed feed. This zone constitutes a single *pinch point* or *point of infinitude* as shown in Fig. 12.7a. If nonideal phase conditions are such as to create a point of tangency between the equilibrium curve and the operating line in the rectifying section, as shown in Fig. 8.15b, the pinch point will occur in the rectifying section as in Fig. 12.7b. Alternatively, the single pinch point can occur in the stripping section.

Shiras, Hanson, and Gibson[5] classified multicomponent systems as having one (Class 1) or two (Class 2) pinch points. For Class 1 separations, all components in the feed distribute to both the distillate and bottoms products. Then the single pinch point bridges the feed stage as shown in Fig. 12.7c. Class 1 separations can occur when narrow-boiling-range mixtures are distilled or when the degree of separation between the key components is not sharp.

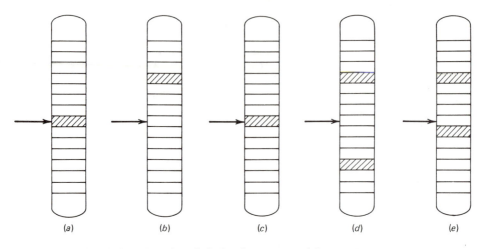

Figure 12.7.   Location of pinch-point zones at minimum reflux. (a)
Binary system. (b) Binary system; nonideal conditions giving point of
tangency. (c) Multicomponent system; all components distributed
(Class 1). (d) Multicomponent system; not all LLK and HHK
distributing (Class 2). (e) Multicomponent system; all LLK, if any,
distributing but not all HHK distributing (Class 2).

For Class 2 separations, one or more of the components appear in only one
of the products. If neither the distillate nor the bottoms product contains all feed
components, two pinch points occur away from the feed stage as shown in Fig.
12.7d. Stages between the feed stage and the rectifying section pinch point
remove heavy components that do not appear in the distillate. Light components
that do not appear in the bottoms are removed by the stages between the feed
stage and the stripping section pinch point. However, if all feed components
appear in the bottoms, the stripping section pinch moves to the feed stage as
shown in Fig. 12.7e.

Consider the general case of a rectifying section pinch point at or away
from the feed stage as shown in Fig. 12.8. A component material balance over all
stages gives

$$y_{i,\infty} V_\infty = x_{i,\infty} L_\infty + x_{i,D} D \tag{12-23}$$

A total balance over all stages is

$$V_\infty = L_\infty + D \tag{12-24}$$

Since phase compositions do not change in the pinch zone, the phase equilibrium
relation is

$$y_{i,\infty} = K_{i,\infty} x_{i,\infty} \tag{12-25}$$

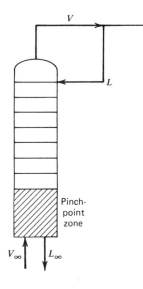

Figure 12.8. Rectifying section pinch-point zone.

Combining (12-23) and (12-25) for components i and j to eliminate $y_{i,\infty}$ and V_∞, solving for the internal reflux ratio at the pinch point, and substituting $(\alpha_{i,j})_\infty = K_{i,\infty}/K_{j,\infty}$, we have

$$\frac{L_\infty}{D} = \frac{\left[\dfrac{x_{i,D}}{x_{i,\infty}} - (\alpha_{i,j})_\infty \dfrac{x_{j,D}}{x_{j,\infty}}\right]}{(\alpha_{i,j})_\infty - 1} \tag{12-26}$$

For Class 1 separations, flashed feed and pinch zone compositions are identical.* Therefore, $x_{i,\infty} = x_{i,F}$ and (12-26) for the light key (LK) and the heavy key (HK) becomes

$$\frac{(L_\infty)_{min}}{F} = \frac{\dfrac{L_F}{F}\left[\dfrac{Dx_{LK,D}}{L_F x_{LK,F}} - (\alpha_{LK,HK})_F \dfrac{Dx_{HK,D}}{L_F x_{HK,F}}\right]}{(\alpha_{LK,HK})_F - 1} \tag{12-27}$$

This equation is attributed to Underwood[6] and can be applied to subcooled liquid or superheated vapor feeds by using fictitious values of L_F and $x_{i,F}$ computed by making a flash calculation outside of the two-phase region. As with the Fenske equation, (12-27) applies to components other than the key components. Therefore, for a specified split of two key components, the distribution of nonkey components is obtained by combining (12-27) with the analogous equation for

---

* Assuming the feed is neither subcooled nor superheated.

component i in place of the LK to give

$$\frac{Dx_{i,D}}{L_F x_{i,F}} = \left[\frac{(\alpha_{i,HK})_F - 1}{(\alpha_{LK,HK})_F - 1}\right]\left(\frac{Dx_{LK,D}}{L_F x_{LK,F}}\right) + \left[\frac{(\alpha_{LK,HK})_F - (\alpha_{i,HK})F}{(\alpha_{LK,HK})_F - 1}\right]\left(\frac{Dx_{HK,D}}{L_F x_{HK,F}}\right) \quad (12\text{-}28)$$

For a Class 1 separation

$$0 < \left(\frac{Dx_{i,D}}{Fx_{i,F}}\right) < 1$$

for all nonkey components. If so, the external reflux ratio is obtained from the internal reflux by an enthalpy balance around the rectifying section in the form

$$\frac{(L_{min})_{external}}{D} = (R_{min})_{external} = \frac{(L_\infty)_{min}(H_{V_\infty} - H_{L_\infty}) + D(H_{V_\infty} - H_V)}{D(H_V - H_L)} \quad (12\text{-}29)$$

where subscripts V and L refer to vapor leaving the top stage and external liquid reflux, respectively. For conditions of constant molal overflow

$$(R_{min})_{external} = (L_\infty)_{min}/D$$

Even when (12-27) is invalid, it is useful because, as shown by Gilliland,[7] the minimum reflux ratio computed by assuming a Class 1 separation is equal to or greater than the true minimum. This is because the presence of distributing nonkey components in the pinch-point zones increases the difficulty of the separation, thus increasing the reflux requirement.

Example 12.4.   Calculate the minimum internal reflux for the problem of Example 12.2 assuming a Class 1 separation. Check the validity of this assumption.

Solution.   From Fig. 12.6, the relative volatility between $nC_4(LK)$ and $iC_5(HK)$ at the feed temperature of 180°F is 1.93. Feed liquid and distillate quantities are given in Fig. 12.3 and Example 12.1. From (12-27)

$$(L_\infty)_{min} = \frac{759.4\left[\dfrac{442}{346.5} - 1.93\left(\dfrac{13}{31.4}\right)\right]}{1.93 - 1} = 389 \text{ lbmole/hr}$$

Distribution of nonkey components in the feed is determined by (12-28). The most likely nonkey component to distribute is nC_5 because its volatility is close to that of $iC_5(HK)$, which does not undergo a sharp separation. For nC_5, using data for K-values from Fig. 12.6, we have

$$\frac{Dx_{nC_5,D}}{L_F x_{nC_5,F}} = \left[\frac{0.765 - 1}{1.93 - 1}\right]\left(\frac{442}{346.5}\right) + \left[\frac{1.93 - 0.765}{1.93 - 1}\right]\left(\frac{13}{31.4}\right)$$
$$= 0.1963$$

Therefore, $Dx_{nC_5,D} = 0.1963(13.4) = 2.63$ lbmole/hr of nC_5 in the distillate. This is less than the quantity of nC_5 in the total feed. Therefore, nC_5 distributes between distillate and bottoms. However, similar calculations for the other nonkey components give negative distillate flow rates for the other heavy components and, in the case of iC_4, a distillate flow rate greater than the feed rate. Thus, the computed reflux rate is not valid. However, as expected, it is greater than the true internal value of 298 lbmole/hr reported by Bachelor.[2]

□

For Class 2 separations, (12-23) to (12-26) still apply. However, (12-26) cannot be used directly to compute the internal minimum reflux ratio because values of $x_{i,\infty}$ are not simply related to feed composition for Class 2 separations. Underwood[8] devised an ingenious algebraic procedure to overcome this difficulty. For the rectifying section, he defined a quantity Φ by

$$\sum \frac{(\alpha_{i,r})_\infty x_{i,D}}{(\alpha_{i,r})_\infty - \Phi} = 1 + (R_\infty)_{min} \qquad (12\text{-}30)$$

Similarly, for the stripping section, Underwood defined Φ' by

$$\sum \frac{(\alpha'_{i,r})_\infty x_{i,B}}{(\alpha'_{i,r})_\infty - \Phi'} = 1 - (R'_\infty)_{min} \qquad (12\text{-}31)$$

where $R'_\infty = L'_\infty/B$ and the prime refers to conditions in the stripping section pinch-point zone. In his derivation, Underwood assumed that relative volatilities are constant in the region between the two pinch-point zones and that $(R_\infty)_{min}$ and $(R'_\infty)_{min}$ are related by the assumption of constant molal overflow in the region between the feed entry and the rectifying section pinch point and in the region between the feed entry and the stripping section pinch point. Hence

$$(L'_\infty)_{min} - (L_\infty)_{min} = qF \qquad (12\text{-}32)$$

With these two critical assumptions, Underwood showed that at least one common root θ (where $\theta = \Phi = \Phi'$) exists between (12-30) and (12-31).

Equation (12-30) is analogous to the following equation derived from (12-25), and the relation $\alpha_{i,r} = K_i/K_r$

$$\sum \frac{(\alpha_{i,r})_\infty x_{i,D}}{(\alpha_{i,r})_\infty - L_\infty/[V_\infty(K_r)_\infty]} = 1 + (R_\infty)_{min} \qquad (12\text{-}33)$$

where $L_\infty/[V_\infty(K_r)_\infty]$ is called the *absorption factor* for a reference component in the rectifying section pinch-point zone. Although Φ is analogous to the absorption factor, a different root of Φ is used to solve for $(R_\infty)_{min}$ as discussed by Shiras, Hanson, and Gibson.[5]

The common root θ may be determined by multiplying (12-30) and (12-31) by D and B, respectively, adding the two equations, substituting (12-31) to eliminate $(R'_\infty)_{min}$ and $(R_\infty)_{min}$, and utilizing the overall component balance $z_{i,F}F = x_{i,D}D + x_{i,B}B$ to obtain

$$\sum \frac{(\alpha_{i,r})_\infty z_{i,F}}{(\alpha_{i,r})_\infty - \theta} = 1 - q \qquad (12\text{-}34)$$

where q is the thermal condition of the feed from (8-32) and r is conveniently taken as the heavy key, HK. When only the two key components distribute, (12-34) is solved iteratively for a root of θ that satisfies $\alpha_{LK,HK} > \theta > 1$. The following modification of (12-30) is then solved for the internal reflux ratio $(R_\infty)_{min}$.

$$\sum \frac{(\alpha_{i,r})_\infty x_{i,D}}{(\alpha_{i,r})_\infty - \theta} = 1 + (R_\infty)_{min}. \qquad (12\text{-}35)$$

If any nonkey components are suspected of distributing, estimated values of $x_{i,D}$ cannot be used directly in (12-35). This is particularly true when nonkey components are intermediate in volatility between the two key components. In this case, (12-34) is solved for m roots of θ where m is one less than the number of distributing components. Furthermore, each root of θ lies between an adjacent pair of relative volatilities of distributing components. For instance, in Example 12.4, it was found the nC_5 distributes at minimum reflux, but nC_6 and heavier do not and iC_4 does not. Therefore, two roots of θ are necessary where

$$\alpha_{nC_4,iC_5} > \theta_1 > 1.0 > \theta_2 > \alpha_{nC_5,iC_5}.$$

With these two roots, (12-35) is written twice and solved simultaneously to yield $(R_\infty)_{min}$ and the unknown value of $x_{nC_5,D}$. The solution must, of course, satisfy the condition $\sum x_{i,D} = 1.0$.

With the internal reflux ratio $(R_\infty)_{min}$ known, the external reflux ratio is computed by enthalpy balance with (12-29). This requires a knowledge of the rectifying section pinch-point compositions. Underwood[8] shows that

$$x_{i,\infty} = \frac{\theta x_{i,D}}{(R_\infty)_{min}[(\alpha_{i,r})_\infty - \theta]} \qquad (12\text{-}36)$$

with $y_{i,\infty}$ given by (12-23). The value of θ to be used in (12-36) is the root of (12-35) satisfying the inequality

$$(\alpha_{HNK,r})_\infty > \theta > 0$$

where HNK refers to the heaviest nonkey in the distillate at minimum reflux. This root is equal to $L_\infty/[V_\infty(K_r)]$ in (12-33). With wide-boiling feeds, the external reflux can be significantly higher than the internal reflux. Bachelor[2] cites a case where the external reflux rate is 55% greater than the internal reflux.

For the stripping section pinch-point composition, Underwood obtains

$$x'_{i,\infty} = \frac{\theta x_{i,B}}{[(R'_\infty)_{min} + 1][(\alpha_{i,r})_\infty - \theta]} \qquad (12\text{-}37)$$

where, in this case, θ is the root of (12-35) satisfying the inequality

$$(\alpha_{HNK,r})_\infty > \theta > 0$$

where HNK refers to the heaviest nonkey in the bottoms product at minimum reflux.

Because of their relative simplicity, the Underwood minimum reflux equations for Class 2 separations are widely used, but too often without examining the possibility of nonkey distribution. In addition, the assumption is frequently made that $(R_\infty)_{min}$ equals the external reflux ratio. When the assumptions of constant relative volatility and constant molal overflow in the regions between the two pinch-point zones are not valid, values of the minimum reflux ratio computed from the Underwood equations for Class 2 separations can be appreciably in error because of the sensitivity of (12-34) to the value of q as will be shown in Example 12.5. When the Underwood assumptions appear to be valid and a negative minimum reflux ratio is computed, this may be interpreted to mean that a rectifying section is not required to obtain the specified separation. The Underwood equations show that the minimum reflux depends mainly on the feed condition and relative volatility and, to a lesser extent, on the degree of separation between the two key components. A finite minimum reflux ratio exists even for a perfect separation.

An extension of the Underwood method for distillation columns with multiple feeds is given by Barnes, Hanson, and King[9]. Exact computer methods for determining minimum reflux are available.[2,10] For making rigorous distillation calculations at actual reflux conditions by the computer methods of Chapter 15, knowledge of the minimum reflux is not essential; but the minimum number of equilibrium stages is very useful.

Example 12.5. Repeat Example 12.4 assuming a Class 2 separation and utilizing the corresponding Underwood equations. Check the validity of the Underwood assumptions. Also calculate the external reflux ratio.

Solution. From the results of Example 12.4, assume the only distributing nonkey component is n-pentane. Assuming the feed temperature of 180°F is reasonable for computing relative volatilities in the pinch zone, the following quantities are obtained from Figs. 12.3 and 12.6.

Species i	$z_{i,F}$	$(\alpha_{i,HK})_\infty$
iC_4	0.0137	2.43
$nC_4(LK)$	0.5113	1.93
$iC_5(HK)$	0.0411	1.00
nC_5	0.0171	0.765
nC_6	0.0262	0.362
nC_7	0.0446	0.164
nC_8	0.3106	0.0720
nC_9	0.0354	0.0362
	1.0000	

The q for the feed is assumed to be the mole fraction of liquid in the flashed feed. From

Example 12.1, $q = 1 - 0.1334 = 0.8666$. Applying (12-34), we have

$$\frac{2.43(0.0137)}{2.43 - \theta} + \frac{1.93(0.5113)}{1.93 - \theta} + \frac{1.00(0.0411)}{1.00 - \theta} + \frac{0.765(0.0171)}{0.765 - \theta}$$

$$+ \frac{0.362(0.0262)}{0.362 - \theta} + \frac{0.164(0.0446)}{0.164 - \theta} + \frac{0.072(0.3106)}{0.072 - \theta} + \frac{0.0362(0.0354)}{0.0362 - \theta}$$

$$= 1 - 0.8666$$

Solving this equation by a bounded Newton method for two roots of θ that satisfy

$$\alpha_{nC_4, iC_5} > \theta_1 > \alpha_{iC_5, iC_5} > \theta_2 > \alpha_{nC_5, iC_5}$$

or

$$1.93 > \theta_1 > 1.00 > \theta_2 > 0.765$$

$\theta_1 = 1.04504$ and $\theta_2 = 0.78014$. Because distillate rates for nC_4 and iC_5 are specified (442 and 13 lbmole/hr, respectively), the following form of (12-35) is preferred.

$$\sum_i \frac{(\alpha_{i,r})_\infty (x_{i,D}D)}{(\alpha_{i,r})_\infty - \theta} = D + (L_\infty)_{min} \tag{12-38}$$

with the restriction that

$$\sum_i (x_{i,D}D) = D \tag{12-39}$$

Assuming $x_{i,D}D$ equals 0.0 for components heavier than nC_5 and 12.0 lbmole/hr for iC_4, we find that these two relations give the following three linear equations.

$$D + (L_\infty)_{min} = \frac{2.43(12)}{2.43 - 1.04504} + \frac{1.93(442)}{1.93 - 1.04504} + \frac{1.00(13)}{1.00 - 1.04504}$$

$$+ \frac{0.765(x_{nC_5,D}D)}{0.765 - 1.04504}$$

$$D + (L_\infty)_{min} = \frac{2.43(12)}{2.43 - 0.78014} + \frac{1.93(442)}{1.93 - 0.78014} + \frac{1.00(13)}{1.00 - 0.78014}$$

$$+ \frac{0.765(x_{nC_5,D}D)}{0.765 - 0.78014}$$

$$D = 12 + 442 + 13 + (x_{nC_5,D}D)$$

Solving these three equations gives

$$x_{nC_5,D}D = 2.56 \text{ lbmole/hr}$$

$$D = 469.56 \text{ lbmole/hr}$$

$$(L_\infty)_{min} = 219.8 \text{ lbmole/hr}$$

The distillate rate for nC_5 is very close to the value of 2.63 computed in Example 12.4, if we assume a Class 1 separation. The internal minimum reflux ratio at the rectifying pinch point is considerably less than the value of 389 computed in Example 12.4 and is also much less than the true internal value of 298 reported by Bachelor.[2] The main reason for the discrepancy between the value of 219.8 and the true value of 298 is the invalidity of the assumption of constant molal overflow. Bachelor[2] computed the pinch-point region flow rates and temperatures shown in Fig. 12.9. The average temperature of the region between the two pinch regions is 152°F (66.7°C), which is appreciably lower than the

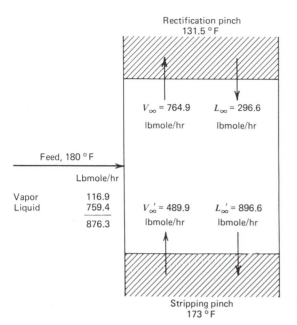

Rectification pinch
131.5 °F

$V_\infty = 764.9$

lbmole/hr

$L_\infty = 296.6$

lbmole/hr

Feed, 180 °F

Lbmole/hr

Vapor   116.9
Liquid   759.4
    ———
    876.3

$V_\infty' = 489.9$

lbmole/hr

$L_\infty' = 896.6$

lbmole/hr

Stripping pinch
173 °F

Figure 12.9. Pinch-point region conditions for Example 12.5 from computations by Bachelor. [J. B. Bachelor, *Petroleum Refiner*, **36** (6), 161–170 (1957).]

flashed feed temperature. The relatively hot feed causes additional vaporization across the feed zone. The effective value of q in the region between the pinch points is obtained from (8-29)

$$q_{\text{eff}} = \frac{L_\infty' - L_\infty}{F} = \frac{896.6 - 296.6}{876.3} = 0.685$$

This is considerably lower than the value of 0.8666 for q based on the flashed feed condition. On the other hand, the value of $\alpha_{LK,HK}$ at 152°F (66.7°C) is not much different from the value at 180°F (82.2°C). If this example is repeated using q equal to 0.685, the resulting value of $(L_\infty)_{\min}$ is 287.3 lbmole/hr, which is only 3.6% lower than the true value of 298. Unfortunately, in practice, this corrected procedure cannot be applied because the true value of q cannot be readily determined.

To compute the external reflux ratio from (12-29), rectifying pinch-point compositions must be calculated from (12-36) and (12-23). The root of θ to be used in (12-36) is obtained from the version of (12-35) used above. Thus

$$\frac{2.43(12)}{2.43 - \theta} + \frac{1.93(442)}{1.93 - \theta} + \frac{1.00(13)}{1.00 - \theta} + \frac{0.765(2.56)}{0.765 - \theta} = 469.56 + 219.8$$

where $0.765 > \theta > 0$. Solving, $\theta = 0.5803$. Liquid pinch-point compositions are obtained from the following form of (12-36)

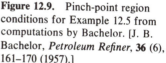

$$x_{i,\infty} = \frac{\theta(x_{i,D}D)}{(L_\infty)_{\min}[(\alpha_{i,r})_\infty - \theta]}$$

with $(L_\infty)_{\min} = 219.8$ lbmole/hr.

For iC_4,

$$x_{iC_4,\infty} = \frac{0.5803(12)}{219.8(2.43 - 0.5803)} = 0.0171$$

From a combination of (12-23) and (12-24)

$$y_{i,\infty} = \frac{x_{i,\infty}L_\infty + x_{i,D}D}{L_\infty + D}$$

For iC_4

$$y_{iC_4,\infty} = \frac{0.0171(219.8) + 12}{219.8 + 469.56} = 0.0229$$

Similarly, the mole fractions of the other components appearing in the distillate are

Component	$x_{i,\infty}$	$y_{i,\infty}$
iC_4	0.0171	0.0229
nC_4	0.8645	0.9168
iC_5	0.0818	0.0449
nC_5	0.0366	0.0154
	1.0000	1.0000

The temperature of the rectifying section pinch point is obtained from either a bubble-point temperature calculation on $x_{i,\infty}$ or a dew-point temperature calculation on $y_{i,\infty}$. The result is 126°F. Similarly, the liquid-distillate temperature (bubble point) and the temperature of the vapor leaving the top stage (dew point) are both computed to be approximately 123°F. Because rectifying section pinch-point temperature and distillate temperatures are very close, it would be expected that $(R_\infty)_{min}$ and $(R_{min})_{external}$ would be almost identical. Bachelor[2] obtained a value of 292 lbmole/hr for the external reflux rate compared to 298 lbmole/hr for the internal reflux rate.

☐

Actual Reflux Ratio and Theoretical Stages

To achieve a specified separation between two key components, the reflux ratio and the number of theoretical stages must be greater than their minimum values. The actual reflux ratio is generally established by economic considerations at some multiple of minimum reflux. The corresponding number of theoretical stages is then determined by suitable analytical or graphical methods or, as discussed in this section, by an empirical equation. However, there is no reason why the number of theoretical stages cannot be specified as a multiple of minimum stages and the corresponding actual reflux computed by the same empirical relationship. As shown in Fig. 12.10 from studies by Fair and Bolles,[11] the optimum value of R/R_{min} is approximately 1.05. However, near-optimal conditions extend over a relatively broad range of mainly larger values of R/R_{min}. In practice, superfractionators requiring a large number of stages are frequently designed for a value of R/R_{min} of approximately 1.10, while separations requiring a small number of stages are designed for a value of R/R_{min} of approximately

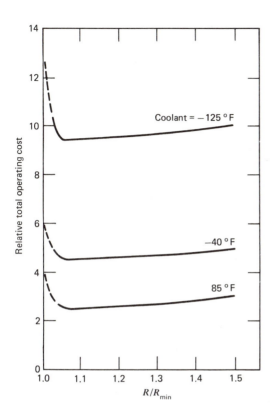

Figure 12.10. Effect of reflux ratio on cost. [J. R. Fair and W. L. Bolles, *Chem. Eng.*, **75** (9), 156–178 (April 22, 1968).

1.50. For intermediate cases, a commonly used rule of thumb is R/R_{min} equal to 1.30.

The number of equilibrium stages required for the separation of a binary mixture assuming constant relative volatility and constant molal overflow depends on $z_{i,F}$, $x_{i,D}$, $x_{i,B}$, q, R, and α. From (12-11) for a binary mixture, N_{min} depends on $x_{i,D}$, $x_{i,B}$, and α, while R_{min} depends on $z_{i,F}$, $x_{i,D}$, q, and α. Accordingly, a number of investigators have assumed empirical correlations of the form

$$N = N\{N_{min}\{x_{i,D}, x_{i,B}, \alpha\}, R_{min}\{z_{i,F}, x_{i,D}, q, \alpha\}, R\}$$

Furthermore, they have assumed that such a correlation might exist for nearly-ideal multicomponent systems where the additional feed composition variables and nonkey relative volatilities also influence the value of R_{min}.

The most successful and simplest empirical correlation of this type is the one developed by Gilliland[12] and slightly modified in a later version by Robinson and Gilliland.[13] The correlation is shown in Fig. 12.11, where the three sets of data points, which are based on accurate calculations, are the original points

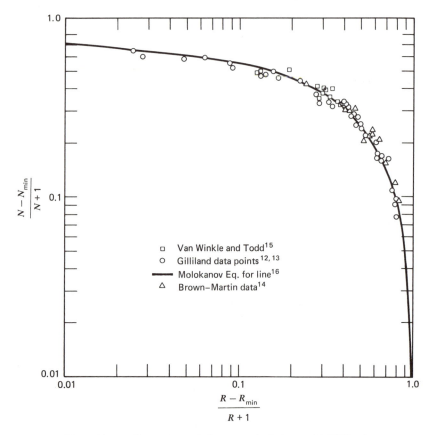

Figure 12.11. Comparison of rigorous calculations with Gilliland correlation.

from Gilliland[12] and the multicomponent data points of Brown and Martin[14] and Van Winkle and Todd.[15] The 61 data points cover the following ranges of conditions.

1. Number of components: 2 to 11.

2. q: 0.28 to 1.42.

3. Pressure: vacuum to 600 psig.

4. α: 1.11 to 4.05.

5. R_{min}: 0.53 to 9.09.

6. N_{min}: 3.4 to 60.3.

The line drawn through the data represents the equation developed by Molokanov et al.[16]

$$Y = \frac{N - N_{min}}{N + 1} = 1 - \exp\left[\left(\frac{1 + 54.4X}{11 + 117.2X}\right)\left(\frac{X - 1}{X^{0.5}}\right)\right] \qquad (12\text{-}40)$$

where

$$X = \frac{R - R_{min}}{R + 1}$$

This equation satisfies the end points ($Y = 0$, $X = 1$) and ($Y = 1$, $X = 0$). At a value of R/R_{min} near the optimum of 1.3, Fig. 12.11 predicts an optimum ratio for N/N_{min} of approximately 2. The value of N includes one stage for a partial reboiler and one stage for a partial condenser, if any.

The Gilliland correlation is very useful for preliminary exploration of design variables. Although never intended for final design, the Gilliland correlation was used to design many existing distillation columns for multicomponent separations without benefit of accurate stage-by-stage calculations. In Fig. 12.12, a replot of the correlation in linear coordinates shows that a small initial increase in R above R_{min} causes a large decrease in N, but further changes in R have a

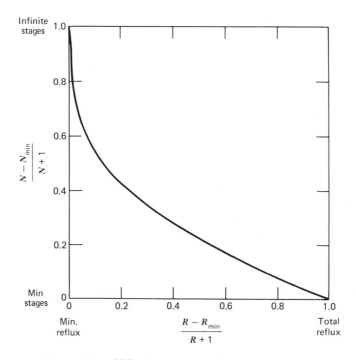

Figure 12.12. Gilliland correlation with linear coordinates.

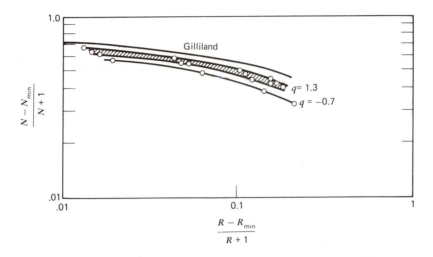

Figure 12.13. Effect of feed condition on Gilliland correlation. [G. Guerreri, *Hydrocarbon Processing*, **48** (8), 137–142 (August, 1969).]

much smaller effect on N. The knee in the curve of Fig. 12.12 corresponds closely to the optimum value of R/R_{min} in Fig. 12.10.

Robinson and Gilliland[13] state that a more accurate correlation should utilize a parameter involving the feed condition q. This effect is shown in Fig. 12.13 using data points for the sharp separation of benzene–toluene mixtures from Guerreri.[17] The data, which cover feed conditions ranging from subcooled liquid to superheated vapor (q equals 1.3 to -0.7), show a trend toward decreasing theoretical stage requirements with increasing feed vaporization. The Gilliland correlation would appear to be conservative for feeds having low values of q. Donnell and Cooper[18] state that this effect of q is only important when the α between the key components is high or when the feed is low in volatile components.

A serious problem with the Gilliland correlation can occur when stripping is much more important than rectification. For example, Oliver[19] considers a fictitious binary case with specifications of $z_F = 0.05$, $x_D = 0.40$, $x_B = 0.001$, $q = 1$, $\alpha = 5$, $R/R_{min} = 1.20$, and constant molal overflow. By exact calculations, $N = 15.7$. From the Fenske equation, $N_{min} = 4.04$. From the Underwood equation, $R_{min} = 1.21$. From (12-40) for the Gilliland correlation, $N = 10.3$. This is 34% lower than the exact value. This limitation, which is caused by ignoring boilup, is discussed further by Strangio and Treybal,[20] who present a more accurate method for such cases.

Example 12.6. Use the Gilliland correlation to estimate the theoretical stage requirements for the debutanizer of Examples 12.1, 12.2, and 12.5 for an external reflux of 379.6

lbmole/hr (30% greater than the exact value of the minimum reflux rate from Bachelor) by the following schemes.

(a) Fenske equation for N_{min}; Underwood equation for R_{min}.

(b) Winn equation for N_{min}; Underwood equation for R_{min}.

(c) Winn equation for N_{min}; Exact value for R_{min}.

Solution. From the examples cited, values of R_{min} and $[(R - R_{min})/(R + 1)]$ for the various cases are obtained using a distillate rate from Example 12.5 of 469.56 lbmole/hr. Thus, $R = 379.6/469.56 = 0.808$.

$$\frac{R - R_{min}}{R + 1} = X$$

Case	R_{min}	
(a)	0.479	0.182
(b)	0.479	0.182
(c)	0.622	0.103

From (12-40) for Case (a)

$$\frac{N - N_{min}}{N + 1} = 1 - \exp\left[\left(\frac{1 + 54.4(0.182)}{11 + 117.2(0.182)}\right)\left(\frac{0.182 - 1}{0.182^{0.5}}\right)\right] = 0.476$$

$$N = \frac{8.88 + 0.476}{1 - 0.476} = 17.85$$

Similarly, for the other cases, the following results are obtained, where $N - 1$ corresponds to the equilibrium stages in the tower allowing one theoretical stage for the reboiler, but no stage for the total condenser.

Case	N_{min}	N	$N - 1$
(a) (Fenske–Underwood)	8.88	17.85	16.85
(b) (Winn–Underwood)	7.65	15.51	14.51
(c) (Winn–Exact R_{min})	7.65	18.27	17.27

Although values for N from Cases (a) and (c) are close, it should be kept in mind that, had the exact value of R_{min} not been known and a value of R equal to 1.3 times R_{min} from the Underwood method been used, the value of R would have been 292 lbmole/hr. But this, by coincidence, is only the true minimum reflux. Therefore, the desired separation would not be achieved.

□

Feed Stage Location

Implicit in the application of the Gilliland correlation is the specification that the theoretical stages be distributed optimally between the rectifying and stripping sections. As suggested by Brown and Martin,[14] the optimum feed stage can be located by assuming that the ratio of stages above the feed to stages below the feed is the same as the ratio determined by simply applying the Fenske equation to the separate sections at total reflux conditions to give

$$\frac{N_R}{N_S} \simeq \frac{(N_R)_{min}}{(N_S)_{min}} = \frac{\log\left[(x_{LK,D}/z_{LK,F})(z_{HK,F}/x_{HK,D})\right]}{\log\left[(z_{LK,F}/x_{LK,B})(x_{HK,B}/z_{HK,F})\right]} \frac{\log\left[(\alpha_B\alpha_F)^{1/2}\right]}{\log\left[(\alpha_D\alpha_F)^{1/2}\right]} \qquad (12\text{-}41)$$

Unfortunately, (12-41) is not reliable except for fairly symmetrical feeds and separations.

A reasonably good approximation of optimum feed stage location can be made by employing the empirical equation of Kirkbride[21]

$$\frac{N_R}{N_S} = \left[\left(\frac{z_{HK,F}}{z_{LK,F}} \right) \left(\frac{x_{LK,B}}{x_{HK,D}} \right)^2 \left(\frac{B}{D} \right) \right]^{0.206} \tag{12-42}$$

An extreme test of both these equations is provided by the fictitious binary mixture problem due to Oliver[19] cited in the previous section. Exact calculations by Oliver and calculations by (12-41) and (12-42) give the following results.

Method	N_R/N_S
Exact	0.08276
Kirkbride (12-42)	0.1971
Fenske ratio (12-41)	0.6408

Although the result from the Kirkbride equation is not very satisfactory, the Fenske ratio method is much worse.

Example 12.7.   Use the Kirkbride equation to determine the feed stage location for the debutanizer of Example 12.1, assuming an equilibrium stage requirement of 18.27.

Solution.   Assume that the product distribution computed in Example 12.2 based on the Winn equation for total reflux conditions is a good approximation to the distillate and bottom compositions at actual reflux conditions. Therefore

$$x_{nC_4,B} = \frac{6.0}{408.5} = 0.0147$$

$$x_{iC_5,D} = \frac{13}{467.8} = 0.0278$$

$$D = 467.8 \text{ lbmole/hr}$$

$$B = 408.5 \text{ lbmole/hr}$$

From Fig. 12.3

$$z_{nC_4,F} = 448/876.3 = 0.5112$$

$$z_{iC_5,F} = 36/876.3 = 0.0411$$

From (12-42)

$$\frac{N_R}{N_S} = \left[\left(\frac{0.0411}{0.5112} \right) \left(\frac{0.0147}{0.0278} \right)^2 \left(\frac{408.5}{467.8} \right) \right]^{0.206} = 0.445$$

Therefore,

$$N_R = \frac{0.445}{1.445} (18.27) = 5.63 \text{ stages}$$

$$N_S = 18.27 - 5.63 = 12.64 \text{ stages}$$

Rounding the estimated stage requirements leads to 1 stage as a partial reboiler, 12 stages below the feed, and 6 stages above the feed.

□

Distribution of Nonkey Components at Actual Reflux

For multicomponent mixtures, all components distribute to some extent between distillate and bottoms at total reflux conditions. However, at minimum reflux conditions none or only a few of the nonkey components distribute. Distribution ratios for these two limiting conditions are shown in Fig. 12.14 for the debutanizer example. For total reflux conditions, results from the Fenske equation in Example 12.3 plot as a straight line for the log–log coordinates. For minimum reflux, results from the Underwood equation in Example 12.5 are shown as a dashed line.

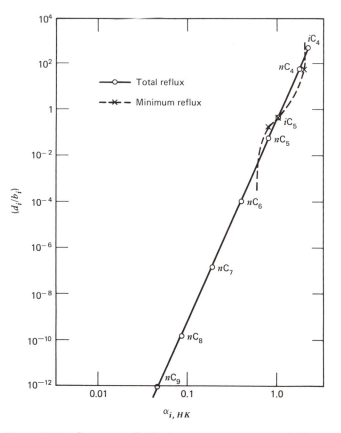

Figure 12.14. Component distribution ratios at extremes of distillation operating conditions.

It might be expected that a product distribution curve for actual reflux conditions would lie between the two limiting curves. However, as shown by Stupin and Lockhart,[22] product distributions in distillation are complex. A typical result is shown in Fig. 12.15. For a reflux ratio near minimum, the product distribution (Curve 3) lies between the two limits (Curves 1 and 4). However, for a high reflux ratio, the product distribution for a nonkey component (Curve 2) may actually lie outside of the limits and an inferior separation results.

For the behavior of the product distribution in Fig. 12.15, Stupin and Lockhart provide an explanation that is consistent with the Gilliland correlation of Fig. 12.11. As the reflux ratio is decreased from total reflux, while maintaining the specified splits of the two key components, equilibrium stage requirements increase only slowly at first, but then rapidly as minimum reflux is approached. Initially, large decreases in reflux cannot be adequately compensated for by increasing stages. This causes inferior nonkey component distributions. However, as minimum reflux is approached, comparatively small decreases in reflux are more than compensated for by large increases in equilibrium stages; and the separation of nonkey components becomes superior to that at total reflux. It appears reasonable to assume that, at a near-optimal reflux ratio of 1.3,

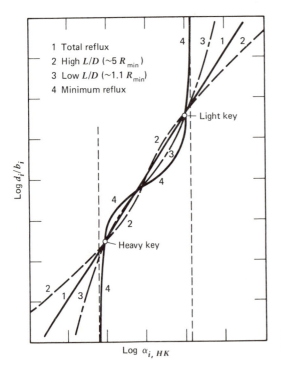

Figure 12.15. Component distribution ratios at various reflux ratios.

nonkey component distribution is close to that estimated by the Fenske or Winn equations for total reflux conditions.

12.2 Multistage Countercurrent Cascades—Group Methods

Many multicomponent separators are cascades of stages where the two contacting phases flow countercurrently. Approximate calculation procedures have been developed to relate compositions of streams entering and exiting cascades to the number of equilibrium stages required. These approximate procedures are called *group methods* because they provide only an overall treatment of the stages in the cascade without considering detailed changes in temperature and composition in the individual stages. In this section, single cascades used for absorption, stripping, liquid–liquid extraction, and leaching are considered. An introductory treatment of the countercurrent cascade for liquid–liquid extraction was given in Section 1.5.

Kremser[23] originated the group method. He derived overall species material balances for a multistage countercurrent absorber. Subsequent articles by Souders and Brown,[24] Horton and Franklin,[25] and Edmister[26] improved the method. The treatment presented here is similar to that of Edmister[27] for general application to vapor–liquid separation operations. Another treatment by Smith and Brinkley[28] emphasizes liquid–liquid separations.

Consider first the countercurrent cascade of N adiabatic equilibrium stages used, as shown in Fig. 12.16a, to absorb species present in the entering vapor. Assume these species are absent in the entering liquid. Stages are numbered from top to bottom. A material balance around the top of the absorber, including stages 1 through $N - 1$, for any absorbed species gives

$$v_N = v_1 + l_{N-1} \tag{12-43}$$

where

$$v = yV \tag{12-44}$$

$$l = xL \tag{12-45}$$

and $l_0 = 0$. From equilibrium considerations for stage N

$$y_N = K_N x_N \tag{12-46}$$

Combining (12-44), (12-45), and (12-46), v_N becomes,

$$v_N = \frac{l_N}{(L_N/K_N V_N)} \tag{12-47}$$

An *absorption factor* A for a given stage and component is defined by

$$A = \frac{L}{KV} \tag{12-48}$$

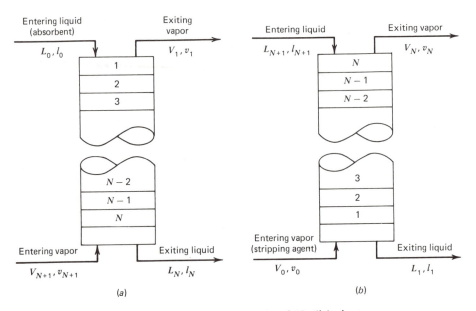

Figure 12.16. Countercurrent cascades of N adiabatic stages. (a) Absorber. (b) Stripper.

Combining (12-47) and (12-48), we have

$$v_N = l_N/A_N \qquad (12\text{-}49)$$

Substituting (12-49) into (12-43),

$$l_N = (l_{N-1} + v_1)A_N \qquad (12\text{-}50)$$

The internal flow rate l_{N-1} can be eliminated by successive substitution using material balances around successively smaller sections of the top of the cascade. For stages 1 through $N-2$,

$$l_{N-1} = (l_{N-2} + v_1)A_{N-1} \qquad (12\text{-}51)$$

Substituting (12-51) into (12-50), we have

$$l_N = l_{N-2}A_{N-1}A_N + v_1(A_N + A_{N-1}A_N) \qquad (12\text{-}52)$$

Continuing this process to the top stage, where $l_1 = v_1A_1$, ultimately converts (12-52) into

$$l_N = v_1(A_1A_2A_3\ldots A_N + A_2A_3\ldots A_N + A_3\ldots A_N + \cdots + A_N) \qquad (12\text{-}53)$$

A more useful form is obtained by combining (12-53) with the overall component balance

$$l_N = v_{N+1} - v_1 \tag{12-54}$$

to give an equation for the exiting vapor in terms of the entering vapor and a recovery fraction

$$v_1 = v_{N+1}\phi_A \tag{12-55}$$

where, by definition, the recovery fraction is

$$\phi_A = \frac{1}{A_1 A_2 A_3 \dots A_N + A_2 A_3 \dots A_N + A_3 \dots A_N + \dots + A_N + 1} = \begin{array}{l} \text{fraction of} \\ \text{species in} \\ \text{entering vapor} \\ \text{that is not} \\ \text{absorbed.} \end{array} \tag{12-56}$$

In the group method, an *average effective absorption factor* A_e replaces the separate absorption factors for each stage. Equation (12-56) then becomes

$$\phi_A = \frac{1}{A_e^N + A_e^{N-1} + A_e^{N-2} + \dots + A_e + 1} \tag{12-57}$$

When multiplied and divided by $(A_e - 1)$, (12-57) reduces to

$$\phi_A = \frac{A_e - 1}{A_e^{N+1} - 1} \tag{12-58}$$

Figure 12.17 from Edmister[27] is a plot of (12-58) with a probability scale for ϕ_A, a logarithmic scale for A_e, and N as a parameter. This plot in linear coordinates was first developed by Kremser.[23]

Consider next the countercurrent stripper shown in Fig. 12.16b. Assume that the components stripped from the liquid are absent in the entering vapor, and ignore condensation or absorption of the stripping agent. In this case, stages are numbered from bottom to top to facilitate the derivation. The pertinent stripping equations follow in a manner analogous to the absorber equation. The results are

$$l_1 = l_{N+1}\phi_S \tag{12-59}$$

where

$$\phi_S = \frac{S_e - 1}{S_e^{N+1} - 1} \tag{12-60}$$

$$S = \frac{KV}{L} = \frac{1}{A} \tag{12-61}$$

Figure 12.17 also applies to (12-60).

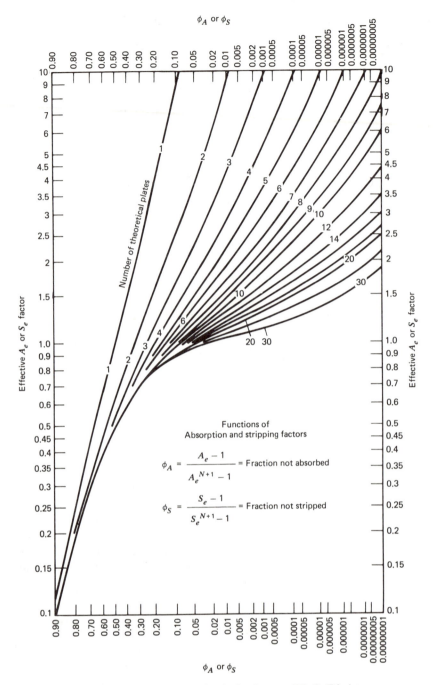

Figure 12.17. Absorption and stripping factors. [W. C. Edmister, *AIChE J.*, **3**, 165–171 (1957)].

Absorbers

As shown in Fig. 12.18, absorbers are frequently coupled with strippers or distillation columns to permit regeneration and recycle of absorbent. Since stripping action is not perfect, absorbent entering the absorber contains species present in the vapor entering the absorber. Vapor passing up through the absorber can strip these as well as the absorbed species introduced in the makeup absorbent. A general absorber equation is obtained by combining (12-55) for absorption of species from the entering vapor with a modified form of (12-59) for stripping of the same species from the entering liquid. For stages numbered from top to bottom, as in Fig. 12.16a, (12-59) becomes

$$l_N = l_0 \phi_S \tag{12-62}$$

or, since

$$l_0 = v_1 + l_N$$

$$v_1 = l_0(1 - \phi_S) \tag{12-63}$$

The total balance in the absorber for a component appearing in both entering vapor and entering liquid is obtained by adding (12-55) and (12-63)* to give

$$v_1 = v_{N+1} \phi_A + l_0(1 - \phi_S) \tag{12-64}$$

which is generally applied to each component in the vapor entering the absorber. Equation (12-62) is used for species appearing only in the entering liquid.

To obtain values of ϕ_A and ϕ_S for use in (12-62) and (12-64), expressions are required for A_e and S_e. These are conveniently obtained from equations derived by Edmister.[26]

$$A_e = [A_N(A_1 + 1) + 0.25]^{1/2} - 0.5 \tag{12-65}$$

$$S_e = [S_1(S_N + 1) + 0.25]^{1/2} - 0.5 \tag{12-66}$$

where stage numbers refer to Fig. 12.16a. These equations are exact for an adiabatic absorber with two stages and are reasonably good approximations for adiabatic absorbers containing more than two stages.

Values of A and S at the top and bottom stages are based on assumed temperatures and on total molal vapor and liquid rates leaving the stages. Total flow rates can be estimated by the following equations of Horton and Franklin.[25]

$$V_2 = V_1 \left(\frac{V_{N+1}}{V_1}\right)^{1/N} \tag{12-67}$$

$$L_1 = L_0 + V_2 - V_1 \tag{12-68}$$

$$V_N = V_{N+1} \left(\frac{V_1}{V_{N+1}}\right)^{1/N} \tag{12-69}$$

---

* Equation (12-63) neglects the presence of the species in the entering vapor since it is already considered in (12-55).

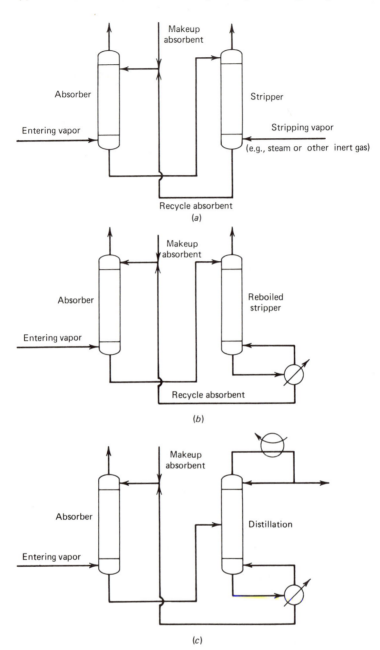

Figure 12.18. Various coupling schemes for absorbent recovery. (*a*)
Use of steam or inert gas stripper. (*b*) Use of reboiled stripper. (*c*)
Use of distillation.

These equations are not exact, but assume that the molal vapor contraction per stage is the same percentage of the molal vapor flow to the stage in question. Assuming that the temperature change of the liquid is proportional to the volume of gas absorbed, we have

$$\frac{T_N - T_1}{T_N - T_0} = \frac{V_{N+1} - V_2}{V_{N+1} - V_1} \tag{12-70}$$

This equation is solved simultaneously with an overall enthalpy balance for T_1 and T_N. Generally, if $T_0 \approx T_{N+1}$, $(T_1 - T_0)$ ranges from 0 to 20°F, depending on the fraction of entering gas absorbed.

The above system of equations is highly useful for studying the effects of variables during preliminary design studies. In general, for a given feed gas, the fractional absorption of only one key component can be specified. The key species will often have an effective absorption factor greater than one. The degree of absorption of the other components in the feed can be controlled to a limited extent by selection of absorber pressure, feed gas temperature, absorbent composition, absorbent temperature, and either absorbent flow rate or number of equilibrium stages. Alternatively, as in Example 12.8 below, the degree of absorption for all species can be computed for a specified absorbent flow rate and a fixed number of equilibrium stages. For maximum absorption, temperatures should be as low as possible and pressure as high as possible; but expensive refrigeration and gas compression generally place limits on the operating conditions. As can be seen from Fig. 12.17, maximum selectivity (ratios of ϕ_A values) in absorption are achieved more readily by increasing the number of stages, rather than by increasing the absorbent flow rate. The minimum absorbent flow rate, corresponding to an infinite number of stages, can be estimated from the following equation obtained from (12-58) with $N = \infty$

$$(L_0)_{\min} = K_K V_{N+1}(1 - \phi_{A_K}) \tag{12-71}$$

where subscript K refers to the key component. Equation (12-71) assumes that the key component does not appear in the entering liquid absorbent, $A_K < 1$, and that the fraction of the feed gas absorbed is small.

In processes such as the recovery of gasoline components from natural gas, where only a small fraction of the feed gas is absorbed, calculation of absorption and stripping factors is greatly simplified by assuming $L_1 \approx L_N \approx L_0$, $V_1 \approx V_N \approx V_{N+1}$, and $T_1 \approx T_N \approx (T_0 + T_{N+1})/2$. These assumptions constitute the Kremser approximation. In cases where an appreciable fraction of the entering gas is absorbed, these assumptions are still useful in obtaining an initial estimate of species material balances prior to the application of the calculation procedure of Edmister.

Example 12.8. As shown in Fig. 12.19, the heavier components in a slightly superheated

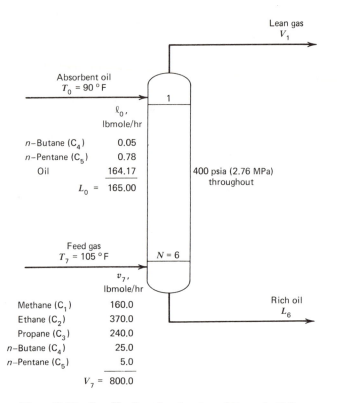

Figure 12.19. Specifications for absorber of Example 12.8.

hydrocarbon gas are to be removed by absorption at 400 psia (2.76 MPa) with a relatively high-molecular-weight oil. Estimate exit vapor and exit liquid flow rates and compositions by the group method if the entering absorbent flow rate is 165 lbmole/hr and 6 theoretical stages are used. Assume the absorbent is recycled from a stripper and has the temperature and composition shown.

Thermodynamic properties of the six species at 400 psia from 0°F to 300°F have been correlated by the following equations, whose constants are tabulated in Table 12.2.

$$K_i = \alpha_i + \beta_i T + \gamma_i T^2 + \delta_i T^3$$
$$H_{V_i} = A_i + B_i T + C_i T^2$$
$$H_{L_i} = a_i + b_i T + c_i T^2$$

where $T = °F$ and $H = Btu/lbmole$.

Solution. For the first iteration, the Kremser approximation is used at a temperature midway between the temperatures of the two entering streams and with $V = V_7$ and $L = L_0$. For each species, from (12-48),

$$A_i = \frac{1}{S_i} = \frac{165}{(K_i)_{97.5°F}(800)} = \frac{0.2063}{(K_i)_{97.5°F}}$$

Table 12.2 Constants for thermodynamic properties in Example 12.8—condition: 400 psia and 0°F to 300°F

Species	α_i	β_o	γ_i	δ_i
C_1	4.35	2.542×10^{-2}	-2.0×10^{-5}	8.333×10^{-9}
C_2	0.65	8.183×10^{-3}	2.25×10^{-5}	-2.333×10^{-8}
C_3	0.15	2.383×10^{-3}	2.35×10^{-5}	-2.333×10^{-8}
nC_4	0.0375	5.725×10^{-4}	1.075×10^{-5}	-2.5×10^{-10}
nC_5	0.0105	2.692×10^{-4}	2.55×10^{-6}	1.108×10^{-8}
Abs. oil	1.42×10^{-5}	3.64×10^{-7}	3.44×10^{-9}	1.50×10^{-11}

Species	A_i	B_i	C_i	a_i	b_i	c_i
C_1	1604	9.357	1.782×10^{-3}	0	14.17	-1.782×10^{-3}
C_2	4661	15.54	3.341×10^{-3}	0	16.54	3.341×10^{-3}
C_3	5070	26.45	0	0	22.78	4.899×10^{-3}
nC_4	5231	33.90	5.812×10^{-3}	0	31.97	5.812×10^{-3}
nC_5	5411	42.09	8.017×10^{-3}	0	39.68	8.017×10^{-3}
Abs. oil	8000	74.67	3.556×10^{-2}	0	69.33	3.556×10^{-2}

Corresponding values of ϕ_A and ϕ_S for $N = 6$ are obtained from Fig. 12.17 or (12-58) and (12-60), $(v_i)_1$ from (12-64), and $(l_i)_6$ from an overall component material balance

$$(l_i)_6 = (l_i)_0 + (v_i)_7 - (v_i)_1 \tag{12-72}$$

The results are

Component	$K_{97.5°F}$	A	S	ϕ_A	ϕ_S	v_1	l_6
						Pound-moles per hour	
C_1	6.646	0.0310	—	0.9690	—	155.0	5.0
C_2	1.640	0.1258	—	0.8743	—	323.5	46.5
C_3	0.5841	0.3531	—	0.6474	—	155.4	84.6
C_4	0.1953	1.056	0.9468	0.1205	0.1673	3.05	22.0
C_5	0.07126	2.894	0.3455	1.114×10^{-3}	0.6549	0.28	5.50
Abs. oil	0.000096	—	4.669×10^{-4}	—	0.99953	0.075	164.095
					TOTALS	637.305	327.695

The second iteration is begun using results from the first iteration to estimate terminal stage conditions. From (12-67) to (12-69)

$$V_2 = 637.3 \left(\frac{800}{637.3} \right)^{1/6} = 661.9 \text{ lbmole/hr}$$

$$L_1 = 165 + 661.9 - 637.295 = 189.6 \text{ lbmole/hr}$$

$$V_6 = 800 \left(\frac{637.3}{800} \right)^{1/6} = 770.3 \text{ lbmole/hr}$$

Terminal stage temperatures are estimated from (12-70) and from an overall enthalpy balance based on exit vapor and exit liquid rates from the first iteration. Thus

$$\frac{T_6 - T_1}{T_6 - 90} = \frac{800 - 661.9}{800 - 637.3} = 0.849$$

or

$$T_1 - 0.151 T_6 = 76.41$$

The overall energy balance is

$$V_7 H_{V_7} + L_0 H_{L_0} = V_1 H_{V_1} + L_6 H_{L_6}$$

These two equations are solved iteratively by assuming T_1, computing T_6, and checking the enthalpy balance. Assume $T_1 = T_0 + 10 = 90 + 10 = 100°F$ (37.8°C). Then

$$T_6 = \frac{100 - 76.41}{0.151} = 156°F \ (68.9°C)$$

The corresponding enthalpy balance check gives

$$V_7 H_{V_7} + L_0 H_{L_0} - V_1 H_{V_1} - L_6 H_{L_6} = \Delta Q = -170,800 \ \text{Btu/hr}$$

If T_1 is assumed to be 105°F, then $T_6 = 189°F$ and $\Delta Q = -798,600$ Btu/hr. By extrapolation, to make $\Delta Q = 0$; $T_1 = 98.6°F$ (37.0°C) and $T_6 = 147°F$ (63.9°C).

A successive substitution procedure is now employed wherein the terminal stage conditions computed from values of $(v_i)_1$ and $(l_i)_6$ in the previous iteration are used to estimate new values of $(v_i)_1$ and $(l_i)_6$. Absorption and stripping factors for the terminal stages are computed as follows using

$$\frac{L_1}{V_1} = \frac{189.6}{637.3} = 0.298$$

$$\frac{L_6}{V_6} = \frac{327.7}{770.3} = 0.425$$

The results are as follows.

| Component | 98.6°F | | 147°F | |
	A_1	S_1	A_6	S_6
C_1	0.0446	—	0.0554	—
C_2	0.1800	—	0.1879	—
C_3	0.5033	—	0.4556	—
C_4	1.501	0.6664	1.205	0.8298
C_5	4.105	0.2436	3.032	0.3298
Abs. oil	—	3.292×10^{-3}	—	4.457×10^{-4}

Effective absorption and stripping factors are computed from (12-65) and (12-66). Values of ϕ_A and ϕ_S are then read from Fig. 12.17 or computed from (12-58) and (12-60). Values for $(v_i)_1$ and $(l_i)_6$ are obtained from (12-64) and (12-72), respectively.

Component	A_e	S_e	ϕ_A	ϕ_S	Pound-moles per hour v_1	l_6
C_1	0.0549	—	0.9451	—	151.2	8.8
C_2	0.1868	—	0.8132	—	300.9	69.1
C_3	0.4669	—	0.5357	—	128.6	111.4
C_4	1.307	0.7122	0.0557	0.3173	1.43	23.62
C_5	3.466	0.2576	4.105×10^{-4}	0.7425	0.20	5.58
Abs. oil	—	3.292×10^{-3}	—	0.99967	0.054	164.116
				TOTALS	582.384	382.616

The value of $L_6 = 382.6$ lbmole/hr computed in the second iteration is considerably higher than the value of 327.7 lbmole/hr from the initial iteration. More importantly, the quantities absorbed in the first and second iterations are 162.7 and 217.6 lbmole/hr, respectively, the second value being 34% higher than the first. To obtain converged results, additional iterations are conducted in the same manner as the second iteration. The results for L_6 are as follows, where the difference between successive values of L_6 is monotonically reduced.

k, iteration number	L_6, lbmole/hr	$\dfrac{(L_6)_k - (L_6)_{k-1}}{(L_6)_{k-1}} \times 100$
1	327.7	—
2	382.6	16.8
3	402.8	5.3
4	410.3	1.9
5	413.1	0.68
6	414.2	0.27
7	414.6	0.097

Component	$(v_i)_1$, lbmole/hr Kremser Approximation	Edmister Group Method	Exact Solution	$(l_i)_6$, lbmole/hr Kremser Approximation	Edmister Group Method	Exact Solution
C_1	155.0	149.1	147.64	5.0	10.9	12.36
C_2	323.5	288.2	276.03	46.5	81.8	94.97
C_3	155.4	112.0	105.42	84.6	128.0	134.58
C_4	3.05	0.84	1.18	22.0	24.21	23.87
C_5	0.28	0.17	0.21	5.50	5.61	5.57
Abs. oil	0.075	0.045	0.05	164.095	164.125	164.12
TOTALS	637.305	550.355	530.53	327.695	414.645	434.47

After seven iterations, successive values of L_6 are within 0.1%, which is a satisfactory criterion for convergence. The final results for $(v_i)_1$ and $(l_i)_6$ above are compared to the results from the first iteration (Kremser approximation) and to the results of an exact solution to be described in Chapter 15.

The Kremser approximation gives product distributions that deviate considerably from the exact solution, but the Edmister group method is in reasonably good agreement with the exact results. However, terminal stage temperatures of $T_1 = 100.4°F$ (38.0°C) and $T_6 = 163.3°F$ (72.9°C) predicted by the Edmister group method are in poor agreement with values of $T_1 = 150.8°F$ (66.0°C) and $T_6 = 143.7$ (62.1°C) from an exact solution.

☐

Example 12.8 illustrates the ineffectiveness of absorption for making a split between two species that are adjacent in volatility. The exact solution predicts that 530 lbmole/hr of vapor exits from the top of the absorber. This rate corresponds to the sum of the flow rates of the two most volatile compounds, methane and ethane. Thus we might examine the ability of the absorber to separate ethane from propane. The fractions of each vapor feed component that appear in the exit vapor and in the exit liquid are:

Component	v_1/v_7	l_6/v_7
C_1	0.923	0.077
C_2	0.746	0.254
C_3	0.439	0.561
C_4	0.046^a	0.954^a
C_5	0.008^a	0.992^a

[a] Corrected to eliminate effect of stripping of this component due to presence of a small amount of it in the absorbent.

These values indicate a very poor separation between ethane and propane. A relatively good separation is achieved between methane and butane, but both ethane and propane distribute between exit vapor and exit liquid to a considerable extent. The absorber is mainly effective at absorbing butane and pentane, but only at the expense of considerable absorption of ethane and propane.

Strippers

The vapor entering a stripper is often steam or another inert gas. When the stripping agent contains none of the species in the feed liquid, is not present in the entering liquid, and is not absorbed or condensed in the stripper, the only direction of mass transfer is from the liquid to the gas phase. Then, only values of S_e are needed to apply the group method via (12-59) and (12-60). The equations for strippers are analogous to (12-65) to (12-71) for absorbers. Thus, with the stages of a stripper numbered as in Fig. 12.16b

$$S_e = [S_N(S_1 + 1) + 0.25]^{1/2} - 0.5 \qquad (12\text{-}73)$$

To estimate S_1 and S_N, total flow rates can be approximated by

$$L_2 = L_1\left(\frac{L_{N+1}}{L_1}\right)^{1/N} \qquad (12\text{-}74)$$

$$V_1 = V_0 + L_2 - L_1 \qquad (12\text{-}75)$$

$$L_N = L_{N+1}\left(\frac{L_1}{L_{N+1}}\right)^{1/N} \qquad (12\text{-}76)$$

Taking the temperature change of the liquid to be proportional to the liquid contraction, we have

$$\frac{T_{N+1} - T_N}{T_{N+1} - T_1} = \frac{L_{N+1} - L_N}{L_{N+1} - L_1} \qquad (12\text{-}77)$$

This equation is solved simultaneously with an overall enthalpy balance for T_1 and

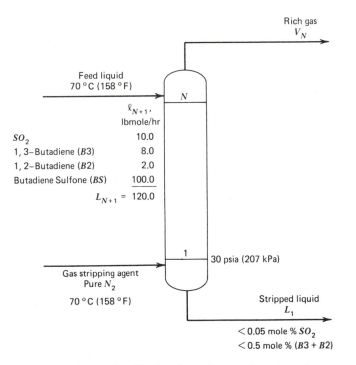

Figure 12.20. Specifications for stripper of Example 12.9.

T_N, the terminal stage temperatures. Often $(T_{N+1} - T_N)$ ranges from 0 to 20°F, depending on the fraction of entering liquid stripped.

For optimal stripping, temperatures should be high and pressures low. However, temperatures should not be so high as to cause decomposition, and vacuum should be used only if necessary. The minimum stripping agent flow rate, for a specified value of ϕ_S for a key component K corresponding to an infinite number of stages, can be estimated from an equation obtained from (12-60) with $N = \infty$

$$(V_0)_{min} = \frac{L_{N+1}}{K_K}(1 - \phi_{S_K}) \tag{12-78}$$

This equation assumes that $A_K < 1$ and the fraction of liquid feed stripped is small.

Example 12.9.   Sulfur dioxide and butadienes ($B3$ and $B2$) are to be stripped with nitrogen from the liquid stream given in Fig. 12.20 so that butadiene sulfone (BS) product will contain less than 0.05 mole% SO_2 and less than 0.5 mole% butadiene. Estimate the flow rate of nitrogen, N_2, and the number of equilibrium stages required.

Solution.   Neglecting stripping of BS, the stripped liquid must have the following component flow rates, and corresponding values for ϕ_S:

Species	l_1, lbmole/hr	$\phi_S = \dfrac{l_1}{l_{N+1}}$
SO_2	< 0.0503	<0.00503
$B3 + B2$	< 0.503	<0.0503
BS	100.0	—

Thermodynamic properties can be computed based on ideal solutions at low pressures as described in Section 4.6. For butadiene sulfone, which is the only species not included in Appendix I, the vapor pressure is

$$P^s_{BS} = \exp\left[17.30 - \frac{11142}{T + 459.67}\right]$$

where P^s_{BS} is in pounds force per square inch absolute and T is in degrees Fahrenheit. The liquid enthalpy of BS is

$$(H_L)_{BS} = 50T$$

where $(H_L)_{BS}$ is in British thermal units per pound-mole and T is in degrees Fahrenheit.

The entering flow rate of the stripping agent V_0 is not specified. The minimum rate at infinite stages can be computed from (12-78), provided that a key component is selected. Suppose we choose $B2$, which is the heaviest component to be stripped to a specified extent. At 70°C, the vapor pressure of $B2$ is computed from (4-83) to be 90.4 psia. From (4-75) at 30 psia total pressure

$$K_{B2} = \frac{90.4}{30} = 3.01$$

From (12-78), using $(\phi_S)_{B2} = 0.0503$, we have

$$(V_0)_{min} = \frac{120}{3.01}(1 - 0.0503) = 37.9 \text{ lbmole/hr}$$

For this value of $(V_0)_{min}$, (12-78) can now be used to determine ϕ_S for $B3$ and SO_2. The K-values for these two species are 4.53 and 6.95, respectively. From (12-78), at infinite stages with $V_0 = 37.9$ lbmole/hr,

$$(\phi_S)_{B3} = 1 - \frac{4.53(37.9)}{120} = -0.43$$

$$(\phi_S)_{SO_2} = 1 - \frac{6.95(37.9)}{120} = -1.19$$

These negative values indicate complete stripping of $B3$ and SO_2. Therefore, the total butadienes in the stripped liquid would only be $(0.0503)(2.0) = 0.1006$ compared to the specified value of 0.503. We can obtain a better estimate of $(V_0)_{min}$ by assuming all of the butadiene content of the stripped liquid is due to $B2$. Then $(\phi_S)_{B2} = 0.503/2 = 0.2515$, and $(V_0)_{min}$ from (12-78) is 29.9 lbmole/hr. Values of $(\phi_S)_{B3}$ and $(\phi_S)_{SO_2}$ are still negative.

The actual entering flow rate for the stripping vapor must be greater than the minimum value. To estimate the effect of V_0 on the theoretical stage requirements and ϕ_S values for the nonkey components, the Kremser approximation is used with K-values at 70°C and 30 psia, $L = L_{N+1} = 120$ lbmole/hr, and $V = V_0$ equal to a series of multiples of 29.9 lbmole/hr. The calculations are greatly facilitated if values of N are selected and values of V are determined from (12-61), where S is obtained from Fig. 12.17. Because $B3$ will be found to some extent in the stripped liquid, $(\phi_S)_{B2}$ will be held below 0.2515. By making iterative calculations, one can choose $(\phi_S)_{B2}$ so that $(\phi_S)_{B2+B3}$ satisfies the specification of, say, 0.05. For 10 theoretical stages, assuming essentially complete stripping of $B3$ such that $(\phi_S)_{B2} \approx 0.25$, $S_{B2} = 0.76$ from Fig. 12.17. From (12-61),

$$V = V_0 = \frac{(120)(0.76)}{3.01} = 30.3 \text{ lbmole/hr}$$

For $B3$, from (12-61),

$$S_{B3} = \frac{(4.53)(30.3)}{120} = 1.143$$

From Fig. 12.17, $(\phi_S)_{B3} = 0.04$. Thus

$$(\phi_S)_{B2+B3} = \frac{0.25(2) + 0.04(8)}{10} = 0.082$$

This is considerably above the specification of 0.05. Therefore, repeat the calculations

N	V_0, lbmole/hr	$V_0/(V_0)_{min}$	**Fraction Not Stripped**				
			ϕ_{SO_2}	ϕ_{B3}	ϕ_{B2}	ϕ_{B2+B3}	ϕ_{BS}
∞	29.9	1.00	0.0	0.0	0.2515	0.0503	0.9960
10	33.9	1.134	0.0005	0.017	0.18	0.050	0.9955
5	45.4	1.518	0.0050	0.029	0.117	0.040	0.9940
3	94.3	3.154	0.0050	0.016	0.045	0.022	0.9874

with, say, $(\phi_S)_{B2} = 0.09$ and continue to repeat until the specified value of $(\phi_S)_{B2+B3}$ is obtained. In this manner, calculations for various numbers of theoretical stages are carried out with converged results as above.

These results show that the specification on SO_2 cannot be met for $N < 6$ stages. Therefore, for 5 and 3 stages, SO_2 must be the key component for determining the value of V_0.

From the above table, initial estimates of L_1 can be determined from (12-59) for given values of V_0 and N. For example, for $N = 5$

	Pound-moles per hour			
Component	l_{N+1}	v_0	l_1	v_5
SO_2	10.0	0.0	0.05	9.95
$B3$	8.0	0.0	0.23	7.77
$B2$	2.0	0.0	0.23	1.77
BS	100.0	0.0	99.40	0.60
N_2	0.0	45.4	0.00	45.40
	120.0	45.4	99.91	64.59

The Edmister group method can now be employed to refine the estimates of l_1. From (12-74) through (12-76)

$$L_2 = 99.91\left(\frac{120}{99.91}\right)^{1/5} = 103.64 \text{ lbmole/hr}$$

$$V_1 = 45.4 + 103.64 - 99.91 = 49.13 \text{ lbmole/hr}$$

$$L_5 = 120\left(\frac{99.91}{120}\right)^{1/5} = 115.68 \text{ lbmole/hr}$$

From (12-77), with T in degrees Fahrenheit,

$$\frac{158 - T_5}{158 - T_1} = \frac{120 - 115.68}{120 - 99.91} = 0.215$$

or

$$T_5 - 0.215 T_1 = 124$$

Solving this equation simultaneously with the overall enthalpy balance, as in Example 12.8, we have $T_1 = 124.6°F$ (51.4°C) and $T_5 = 150.8°F$ (66.0°C).

Stripping factors can now be computed for the two terminal stages, where

$$\frac{V_1}{L_1} = \frac{49.13}{99.91} = 0.4917$$

$$\frac{V_5}{L_5} = \frac{65.49}{115.68} = 0.5661.$$

Also, it will be assumed that stripper top pressure is 29 psia (199.9 kPa).

Component	Bottom Stage, 124.6°F, 30 psia		Top Stage, 150.8°F, 29 psia	
	K_1	S_1	K_5	S_5
SO_2	4.27	2.10	6.50	3.68
$B3$	2.89	1.42	4.27	2.42
$B2$	1.84	0.905	2.82	1.60
BS	0.0057	0.0028	0.0133	0.0075

Effective stripping factors are computed from (12-73). Values of ϕ_S are read from Fig. 12.17 or computed from (12-60), followed by calculation of $(l_i)_1$ from (12-59) and $(v_i)_5$ from an overall species material balance like (12-72).

Component	S_e	ϕ_S	Pound-Moles per Hour	
			l_1	v_5
SO_2	2.91	0.0032	0.032	9.968
$B3$	1.97	0.017	0.136	7.864
$B2$	1.32	0.075	0.150	1.850
BS	0.0075	0.9925	99.24	0.75
N_2	—	—	0.00	45.40
			99.568	65.832

The calculated values of L_1 and V_5 are very close to the assumed values from the Kremser approximation. Therefore, the calculations need not be repeated. However, the stripped liquid contains 0.032 mole% SO_2 and 0.29 mole% butadienes, which are considerably less than limiting specifications. The entering stripping gas flow rate could therefore be reduced somewhat for five theoretical stages.
□

Liquid–Liquid Extraction

A schematic representation of a countercurrent extraction cascade is shown in Fig. 12.21, with stages numbered from the top down and solvent V_{N+1} entering at the bottom.* The group method of calculation can be applied with the equations written by analogy to absorbers. In place of the K-value, the distribution coefficient is used

$$K_{D_i} = \frac{y_i}{x_i} = \frac{v_i/V}{l_i/L} \tag{12-79}$$

* In a vertical extractor, solvent would have to enter at the top if of greater density than the feed.

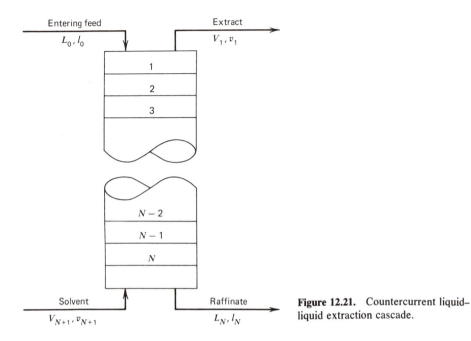

Entering feed

L_0, l_0

Extract

V_1, v_1

1

2

3

$N-2$

$N-1$

N

Solvent

V_{N+1}, v_{N+1}

Raffinate

L_N, l_N

Figure 12.21. Countercurrent liquid–liquid extraction cascade.

Here, y_i is the mole fraction of i in the solvent or extract phase and x_i is the mole fraction in the feed or raffinate phase. Also, in place of the absorption factor, an *extraction factor E* is used where

$$E_i = \frac{K_{D_i} V}{L} \qquad (12\text{-}80)$$

The reciprocal of E is

$$U_i = \frac{1}{E_i} = \frac{L}{K_{D_i} V} \qquad (12\text{-}81)$$

The working equations for each component are

$$v_1 = v_{N+1}\phi_U + l_0(1 - \phi_E) \qquad (12\text{-}82)$$

$$l_N = l_0 + v_{N+1} - v_1 \qquad (12\text{-}83)$$

where

$$\phi_U = \frac{U_e - 1}{U_e^{N+1} - 1} \qquad (12\text{-}84)$$

$$\phi_E = \frac{E_e - 1}{E_e^{N+1} - 1} \qquad (12\text{-}85)$$

$$V_2 = V_1 \left(\frac{V_{N+1}}{V_1} \right)^{1/N} \tag{12-86}$$

$$L_1 = L_0 + V_2 - V_1 \tag{12-87}$$

$$V_N = V_{N+1} \left(\frac{V_1}{V_{N+1}} \right)^{1/N} \tag{12-88}$$

$$E_e = [E_1(E_N + 1) + 0.25]^{1/2} - 0.5 \tag{12-89}$$

$$U_e = [U_N(U_1 + 1) + 0.25]^{1/2} - 0.5 \tag{12-90}$$

If desired, (12-79) through (12-90) can be applied with mass units rather than mole units. No enthalpy balance equations are required because ordinarily temperature changes in an adiabatic extractor are not great unless the feed and solvent enter at appreciably different temperatures or the heat of mixing is large. Unfortunately, the group method is not always reliable for liquid–liquid extrac-

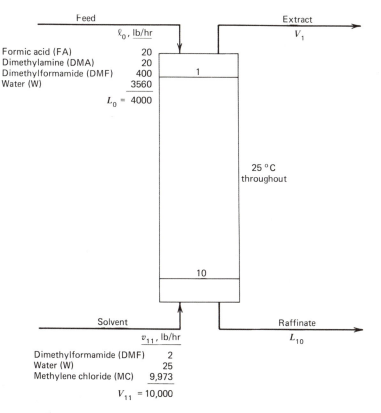

Figure 12.22. Specifications for extractor of Example 12.10.

tion cascades because the distribution coefficient, as discussed in Chapter 5, is a ratio of activity coefficients, which can vary drastically with composition.

Example 12.10. Countercurrent liquid–liquid extraction with methylene chloride is to be used at 25°C to recover dimethylformamide from an aqueous stream as shown in Fig. 12.22. Estimate flow rates and compositions of extract and raffinate streams by the group method using mass units. Distribution coefficients for all components except DMF are essentially constant over the expected composition range and on a mass fraction basis are:

Component	K_{D_i}
MC	40.2
FA	0.005
DMA	2.2
W	0.003

The distribution coefficient for DMF depends on concentration in the water-rich phase as shown in Fig. 12.23.

Solution. Although the Kremser approximation could be applied for the first trial calculation, the following values will be assumed from guesses based on the magnitudes of the K_D-values.

	Pounds per Hour			
Component	**Feed,** l_0	**Solvent,** v_{11}	**Raffinate,** l_{10}	**Extract,** v_1
FA	20	0	20	0
DMA	20	0	0	20
DMF	400	2	2	400
W	3560	25	3560	25
MC	0	9,973	88	9,885
	4000	10,000	3670	10,330

From (12-86) through (12-88), we have

$$V_2 = 10,330 \left(\frac{10,000}{10,330} \right)^{1/10} = 10,297 \text{ lb/hr}$$

$$L_1 = 4000 + 10,297 - 10,330 = 3967 \text{ lb/hr} \tag{12-91}$$

$$V_{10} = 10,000 \left(\frac{10,330}{10,000} \right)^{1/10} = 10,033 \text{ lb/hr}$$

From (12-80), (12-81), (12-89), and (12-90), assuming a mass fraction of 0.09 for DMF in L_1 in order to obtain $(K_D)_{DMF}$ for stage 1, we have

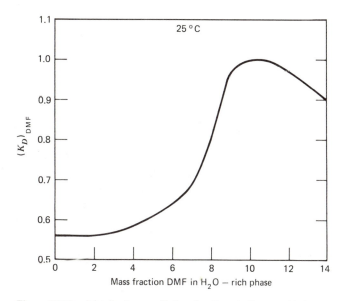

Figure 12.23. Distribution coefficient for dimethylformamide between water and methylene chloride.

Component	E_1	E_{10}	U_1	U_{10}	E_e	U_e
FA	0.013	0.014	—	—	0.013	—
DMA	5.73	6.01	—	—	5.86	—
DMF	2.50	1.53	0.400	0.653	2.06	0.579
W	0.0078	0.0082	128	122	0.0078	125
MC	—	—	0.0096	0.0091	—	0.0091

From (12-85), (12-84), (12-82), and (12-83), we have

Component	ϕ_E	ϕ_U	Pounds per Hour Raffinate, l_{10}	Extract, v_1
FA	0.9870	—	19.7	0.3
DMA	0.0	—	0.0	20.0
DMF	0.000374	0.422	1.3	400.7
W	0.9922	0.0	3557.2	37.8
MC	—	0.9909	90.8	9,882.2
			3669.0	10,331.0

The calculated total flow rates L_{10} and V_1 are almost exactly equal to the assumed rates. Therefore, an additional iteration is not necessary. The degree of extraction of DMF is very high. It would be worthwhile to calculate additional cases with less solvent and/or fewer equilibrium stages.

☐

12.3 Complex Countercurrent Cascades—Edmister Group Method

Edmister[27] applied the group method to complex separators where cascades are coupled to condensers, reboilers, and/or other cascades. Some of the possible combinations, as shown in Fig. 12.24, are fractionators (distillation columns), reboiled strippers, reboiled absorbers, and refluxed inert gas strippers. In Fig. 12.24, five separation zones are delineated: (1) partial condensation, (2) absorption cascade, (3) feed stage flash, (4) stripping cascade, and (5) partial reboiling.

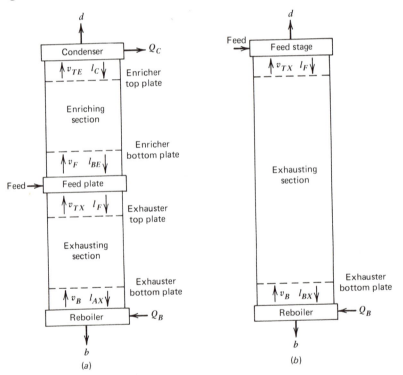

Figure 12.24. Complex countercurrent cascades. (*a*) Fractionator. (*b*) Reboiled stripper. (*c*) Reboiled absorber. (*d*) Refluxed inert gas stripper.

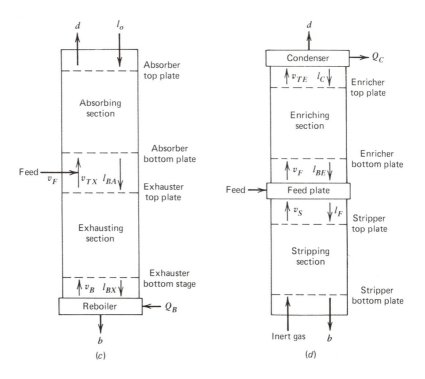

Figure 12.24. Continued.

The combination of an absorption cascade topped by a condenser is referred to as an *enricher*. A partial reboiler topped by a stripping cascade is referred to as an *exhauster*. As shown in Fig. 12.25 stages for an enricher are numbered from the top down and the overhead product is distillate, while for an exhauster stages are numbered from the bottom up. Feed to an enricher is vapor, while feed to an exhauster is liquid. The recovery equations for an enricher are obtained from (12-64) by making the following substitutions, which are obtained from material balance and equilibrium considerations.

$$v_1 = l_0 + d \tag{12-92}$$

$$v_{N+1} = l_N + d \tag{12-93}$$

and

$$l_0 = dA_0 \tag{12-94}$$

where

$$A_0 = \frac{L_0}{DK_0} \quad \text{(if partial condenser)} \tag{12-95}$$

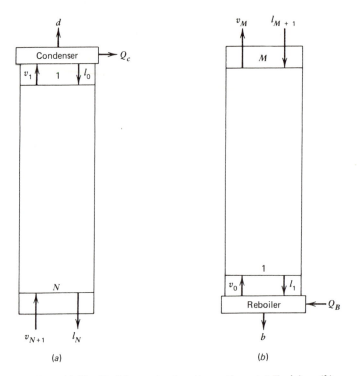

Figure 12.25.   Enricher and exhauster sections. (*a*) Enricher. (*b*) Exhauster.

$$A_0 = \frac{L_0}{D} \quad \text{(if total condenser)} \tag{12-96}$$

The resulting enricher recovery equations for each species are

$$\frac{l_N}{d} = \frac{A_0 \phi_{SE} + 1}{\phi_{AE}} - 1 \tag{12-97}$$

or

$$\frac{v_{N+1}}{d} = \frac{A_0 \phi_{SE} + 1}{\phi_{AE}} \tag{12-98}$$

where the additional subscript E on ϕ denotes an enricher.

The recovery equations for an exhauster are obtained in a similar manner*
as:

$$\frac{v_M}{b} = \frac{S_0 \phi_{AX} + 1}{\phi_{SX}} - 1 \tag{12-99}$$

* In this case the full stripper equation $l_1 = l_{M+1}\phi_s + v_0(1 - \phi_A)$ is used rather than (12-59).

or

$$\frac{l_{M+1}}{b} = \frac{S_0\phi_{AX} + 1}{\phi_{SX}} \tag{12-100}$$

where $S_0 = \dfrac{K_0 V_0}{B}$ for a partial reboiler and subscript X denotes an exhauster.

For either an enricher or exhauster, ϕ_A and ϕ_E are given, as before, by (12-58) and (12-60), respectively, where

$$A_e = [A_B(A_T + 1) + 0.25]^{1/2} - 0.5 \tag{12-101}$$

$$S_e = [S_T(S_B + 1) + 0.25]^{1/2} - 0.5 \tag{12-102}$$

Subscripts B and T designate the bottom and top stages in the section, respectively. For example, for the exhauster of Fig. 12.25b, T refers to stage M and B refers to stage 1.

When cascade sections are coupled, a feed stage is employed and an adiabatic flash is carried out on the combination of feed, vapor rising from the cascade below, and liquid falling from the cascade above. The absorption factor for the feed stage is related to the flashed streams leaving by

$$A_F = \frac{L_F}{K_F V_F} = \frac{L_F}{(v_F L_F / V_F l_F) V_F} = \frac{l_F}{v_F} \tag{12-103}$$

Equations for absorber, stripper, enricher, and exhauster cascades plus the feed stage are summarized in Table 12.3. The equations are readily combined to obtain product distribution equations for the types of separators in Fig. 12.24. For a fractionator, we combine (12-98), (12-100), and (12-103) to eliminate l_F and v_F, noting that $v_{N+1} = v_F$ and $l_{M+1} = l_F$. The result is

$$\frac{b}{d} = A_F \left(\frac{A_0\phi_{SE} + 1}{\phi_{AE}}\right) \bigg/ \left(\frac{S_0\phi_{AX} + 1}{\phi_{SX}}\right) \tag{12-104}$$

Table 12.3   Cascade equations for group method

Cascade Type	Figure Number	Stage Group Equation	Equation Number
Absorber	12.16a	$v_1 = v_{N+1}\phi_A + l_0(1 - \phi_S)$	(12-105)
Stripper	12.16b (with stages $1 - M$)	$l_1 = l_{M+1}\phi_S + v_0(1 - \phi_A)$	(12-106)
Enricher	12.25a	$\dfrac{v_{N+1}}{d} = \dfrac{A_0\phi_{SE} + 1}{\phi_{AE}}$	(12-107)
Exhauster	12.25b	$\dfrac{l_{M+1}}{b} = \dfrac{S_0\phi_{AX} + 1}{\phi_{SX}}$	(12-108)
Feed Stage	12.24a	$\dfrac{l_F}{v_F} = A_F$	(12-109)

Application of (12-104) requires iterative calculation procedures to establish values of necessary absorption factors, stripping factors, and recovery fractions.

Distillation

Convenient specifications for applying the Edmister group method to distillation are those of Table 6.2, Case II, that is, number of equilibrium stages (N) above the feed stage, number of equilibrium stages (M) below the feed stage, external reflux rate (L_0) or reflux ratio (L_0/D), and distillate flow rate (D). The first iteration is initiated by assuming the split of the feed into distillate and bottoms and determining the corresponding product temperatures by appropriate dew-point and/or bubble-point calculations. Vapor and liquid flow rates at the top of the enricher are set by the specified L_0 and D. This enables the condenser duty and the reboiler duty to be computed. From the reboiler duty, vapor and liquid flow rates at the bottom of the exhauster section are computed. Vapor and liquid flow rates in the feed zone are taken as averages of values computed from the constant molal overflow assumption as applied from the top of the column down and from the bottom of the column up. After a feed-zone temperature is assumed, (12-104) is applied to calculate the split of each component between distillate and bottoms products. The resulting total distillate rate D is compared to the specified value. Following the technique of Smith and Brinkley,[28] subsequent iterations are carried out by adjusting the feed-zone temperature until the computed distillate rate is essentially equal to the specified value. Details are illustrated in the following example.

Example 12.11. The hydrocarbon gas of Example 12.8 is to be distilled at 400 psia (2.76 MPa) to separate ethane from propane, as shown in Fig. 12.26. Estimate the distillate and bottoms compositions by the group method. Enthalpies and K-values are obtained from the equations and constants in Example 12.8.

Solution. The first iteration is made by assuming a feed-stage temperature equal to the feed temperature at column pressure. An initial estimate of distillate and bottoms composition for the first iteration is conveniently made by assuming the separation to be as perfect as possible.

Component	Feed	Assumed Distillate, *d*	Assumed Bottoms, *b*
		Pound-Moles per Hour	
C_1	160	160	0
C_2	370	370	0
C_3	240	0	240
C_4	25	0	25
C_5	5	0	5
	800	530	270

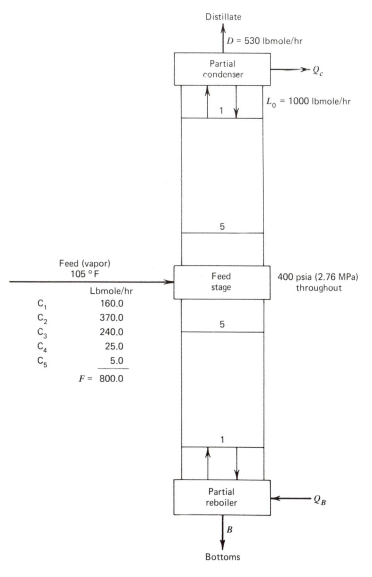

Figure 12.26. Specifications for fractionator of Example 12.11.

For these assumed products, the distillate temperature is obtained from a dew-point calculation as 11.5°F (−11.4°C) and the bottoms temperature is obtained from a bubble-point calculation as 164.9°F (73.8°C). The dew-point calculation also gives values of $(x_i)_0$, from which values of $(y_i)_1$ for the top stage of the enricher are obtained from a component material balance around the partial condenser stage

$$(y_i)_1 = \frac{d_i + (x_i)_0 L_0}{D + L_0} \tag{12-110}$$

A dew-point calculation on $(y_i)_1$ gives 26.1°F (−3.3°C) as the enricher top-stage temperature. The condenser duty is determined as 4,779,000 Btu/hr from an enthalpy balance around the partial condenser stage

$$Q_C = V_1 H_{V_1} - D H_D - L_0 H_{L_0} \tag{12-111}$$

A reboiler duty of 3,033,000 Btu/hr is determined from an overall enthalpy balance

$$Q_B = D H_D + B H_B + Q_C - F H_F \tag{12-112}$$

where for this example an equilibrium flash calculation for the feed composition at 105°F (40.6°C) indicates that the feed is slightly superheated.

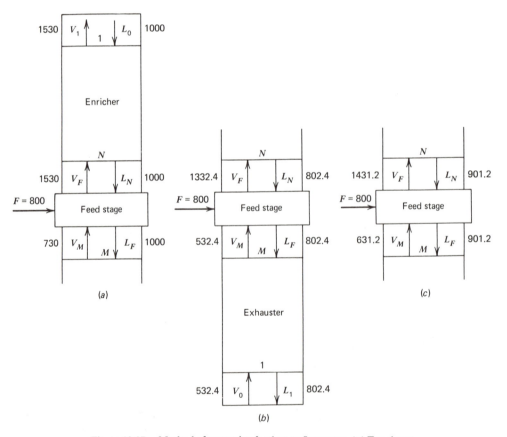

Figure 12.27. Method of averaging feed-zone flow rates. (*a*) Top down. (*b*) Bottom up. (*c*) Averaged rates.

The vapor rate leaving the partial reboiler and entering the bottom stage of the exhauster is obtained from an enthalpy balance around the partial reboiler stage

$$V_0 = \frac{Q_B - B H_B + L_1 H_{L_1}}{H_{v_0}} \qquad (12\text{-}113)$$

In this equation, enthalpy H_{v_0} is computed for the vapor composition $(y_i)_0$ obtained from the bottoms bubble-point calculation for T_B. The equation is then applied iteratively with B, H_B, and H_{v_0} known, by assuming a value for V_0 and computing L and $(x_i)_1$ from

$$L_1 = V_0 + B \qquad (12\text{-}114)$$

$$(x_i)_1 = \frac{(y_i)_0 V_0 + b_i}{L_1} \qquad (12\text{-}115)$$

A bubble-point calculation on $(x_i)_1$ gives T_1, from which H_{L_1} can be determined. The vapor rate V_0 computed from (12-113) is compared to the assumed value. The method of direct substitution is applied until successive values of V_0 are essentially identical. Convergence is rapid to give $V_0 = 532.4$ lbmole/hr, $L_1 = 802.4$ lbmole/hr, and T_1 (the exhaust bottom-stage-temperature) = 169.8°F (76.6°C).

If the constant molal overflow assumption is applied separately to the enricher and exhauster sections, feed-stage flows do not agree and are, therefore, averaged as shown in Fig. 12.27, where $L_N = (1000 + 802.4)/2 = 901.2$ and $V_M = (730 + 532.4)/2 = 631.2$. If the above computation steps involving the partial reboiler, (12-111) through (12-115), were not employed and the constant molal overflow assumption were extended all the way from the top of the enricher section to the bottom of the exhauster section, the vapor rate from the partial reboiler would be estimated as 730 lbmole/hr. This value is considerably larger than the reasonably accurate value of 532.4 lbmole/hr obtained from the partial reboiler stage equations.

Effective absorption and stripping factors for each component are computed for the enricher and exhauster sections from (12-101) and (12-102) for terminal-stage conditions based on the above calculations.

	Enricher		Exhauster	
	Top Stage	Bottom Stage	Top Stage	Bottom Stage
V, lbmole/hr	1530	1431.2	631.2	532.4
L, lbmole/hr	1000	901.2	901.2	802.4
T, °F	26.1	105	105	169.8

Feed-stage absorption factors are based on $V = 1431.2$ lbmole/hr, $L = 901.2$ lbmole/hr, and $T = 105$°F (40.6°C).

Distillate and bottoms compositions are computed from (12-104) together with

$$d_i = \frac{f_i}{1 + (b_i/d_i)} \qquad (12\text{-}116)$$

$$b_i = f_i - d_i. \qquad (12\text{-}117)$$

The results are

Component	Pound-Moles per Hour	
	d	*b*
C_1	160.0	0.0
C_2	363.8	6.2
C_3	3.4	236.6
C_4	0.0	25.0
C_5	0.0	5.0
TOTAL	527.2	272.8

The first iteration is repeated with this new estimate of product compositions to give

Component	Pound-Moles per Hour	
	d	*b*
C_1	160.0	0.0
C_2	363.5	6.5
C_3	4.3	235.7
C_4	0.0	25.0
C_5	0.0	5.0
TOTAL	527.8	272.2

Thus, for an assumed feed-zone temperature of 105°F (40.6°C), the computed distillate rate of 527.8 lbmole/hr is less than the specified value of 530.

A second iteration is performed in a manner similar to the first iteration, but using the last estimate of product compositions and estimating the feed-zone temperature from a linear temperature profile between the distillate and bottoms. This temperature is 88.2°F (31.2°C) and results in a distillate rate of 519.7 lbmole/hr.

The method of false position is now employed for subsequent iterations involving adjustment of the feed-zone temperature until the specified distillate rate is approached. The iteration results are

Iteration	Feed-Zone Temperature, °F	Distillate Rate, lbmole/hr
1	105	527.8
2	88.2	519.7
3	109.5	531.1
4	107.4	530.2

The final results for d_i and b_i after the fourth iteration are adjusted to force d_i to the specified value of D. The adjustment is made by the method of Lyster et al.,[29] which involves finding the value of Θ in the relation

$$D = \sum_i \frac{f_i}{1 + \Theta(b_i/d_i)} \qquad (12\text{-}118)$$

Then

$$d_i = \frac{f_i}{1 + \Theta(b_i/d_i)} \qquad (12\text{-}119)$$

and b_i is given by (12-117). The results of the Edmister group method are compared to the results of an exact solution to be described in Chapter 15.

Component	d, lbmole/hr		b, lbmole/hr	
	Edmister Group Method	Exact Method	Edmister Group Method	Exact Solution
C_1	160.00	160.00	0.00	0.00
C_2	364.36	365.39	5.64	4.61
C_3	5.64	4.61	234.36	235.39
C_4	0.00	0.00	25.00	25.00
C_5	0.00	0.00	5.00	5.00
TOTAL	530.00	530.00	270.00	270.00

The Edmister method predicts ethane and propane recoveries of 98.5% and 97.7%, respectively, compared to values of 98.8% and 98.1% by the exact method.

It is also of interest to compare the other important results given in the following tabulation.

	Edmister Group Method	Exact Solution
Temperatures, °F		
Distillate	15.5	14.4
Bottoms	161.4	161.6
Feed zone	107.4	99.3
Exchanger duties, Btu/hr		
Condenser	4,996,660	4,947,700
Reboiler	3,254,400	3,195,400
Vapor flow rates, lbmole/hr		
V_0 (exhauster)	564.3	555.5
V_M	647.2	517.9
V_{N+1}	1447.2	1358.2

Except in the vicinity of the feed stage, the results from the Edmister group method compare quite well with the exact solution.

☐

The separation achieved by distillation in this example is considerably different from the separation achieved by absorption in Example 12.8. Although the overhead total exit vapor flow rates are approximately the same (530 lbmole/hr) in this example and in Example 12.8, a reasonably sharp split between ethane and propane occurs for distillation, while the absorber allows appreciable quantities of both ethane and propane to appear in the overhead exit vapor and the bottoms exit liquid. If the absorbent rate in Example 12.8 is doubled, the recovery of propane in the bottoms exit liquid approaches 100%, but more than 50% of the ethane also appears in the bottoms exit liquid.

References

1. Kobe, K. A., and J. J. McKetta, Jr., Eds, *Advances in Petroleum Chemistry and Refining*, Vol. 2, Interscience Publishers, Inc., New York, 1959, 315–355.

2. Bachelor, J. B., *Petroleum Refiner*, 36 (6), 161–170 (1957).

3. Fenske, M. R., *Ind. Eng. Chem.*, 24, 482–485 (1932).

4. Winn, F. W., *Petroleum Refiner*, 37 (5), 216–218 (1958).

5. Shiras, R. N., D. N. Hanson, and C. H. Gibson, *Ind. Eng. Chem.*, 42, 871–876 (1950).

6. Underwood, A. J. V., *Trans. Inst. Chem. Eng. (London)*, 10, 112–158 (1932).

7. Gilliland, E. R., *Ind. Eng. Chem.*, 32, 1101–1106 (1940).

8. Underwood, A. J. V., *J. Inst. Petrol.*, 32, 614–626 (1946).

9. Barnes, F. J., D. N. Hanson, and C. J. King, *Ind. Eng. Chem., Process Des. Develop.*, 11, 136–140 (1972).

10. Tavana, M., and D. N. Hanson, *Ind. Eng. Chem., Process Des. Develop.*, 18, 154–156 (1979).

11. Fair, J. R., and W. L. Bolles, *Chem. Eng.*, 75 (9), 156–178 (April 22, 1968).

12. Gilliland, E. R., *Ind. Eng. Chem.*, 32, 1220–1223 (1940).

13. Robinson, C. S., and E. R. Gilliland, *Elements of Fractional Distillation*, 4th ed, McGraw–Hill Book Co., New York, 1950, 347–350.

14. Brown, G. G., and H. Z. Martin, *Trans. AIChE*, 35, 679–708 (1939).

15. Van Winkle, M., and W. G. Todd, *Chem. Eng.*, 78 (21), 136–148 (September 20, 1971).

16. Molokanov, Y. K., T. P. Korablina, N. I. Mazurina, and G. A. Nikiforov, *Int. Chem. Eng.*, 12 (2), 209–212 (1972).

17. Guerreri, G., *Hydrocarbon Processing*, 48 (8), 137–142 (August, 1969).

18. Donnell, J. W., and C. M. Cooper, *Chem. Eng.*, 57, 121–124 (June, 1950).

19. Oliver, E. D., *Diffusional Separation Processes: Theory, Design, and Evaluation*, John Wiley & Sons, Inc., New York, 1966, 104–105.

20. Strangio, V. A., and R. E. Treybal,

Ind. Eng. Chem., Process Des. Develop., **13**, 279–285 (1974).

21. Kirkbride, C. G., *Petroleum Refiner*, **23** (9), 87–102 (1944).

22. Stupin, W. J., and F. J. Lockhart, "The Distribution of Non-Key Components in Multicomponent Distillation," paper presented at the 61st Annual Meeting of AIChE, Los Angeles, California, December 1–5, 1968.

23. Kremser, A., *Nat. Petroleum News*, **22** (21), 43–49 (May 21, 1930).

24. Souders, M., and G. G. Brown, *Ind. Eng. Chem.*, **24**, 519–522 (1932).

25. Horton, G., and W. B. Franklin, *Ind. Eng. Chem.*, **32**, 1384–1388 (1940).

26. Edmister, W. C., *Ind. Eng. Chem.*, **35**, 837–839 (1943).

27. Edmister, W. C., *AIChE J.*, **3**, 165–171 (1957).

28. Smith, B. D., and W. K. Brinkley, *AIChE J.*, **6**, 446–450 (1960).

29. Lyster, W. N., S. L. Sullivan, Jr., D. S. Billingsley, and C. D. Holland, *Petroleum Refiner*, **38** (6), 221–230 (1959).

Problems

12.1 A mixture of propionic and *n*-butyric acids, which can be assumed to form ideal solutions, is to be separated by distillation into a distillate containing 95 mole% propionic acid and a bottoms product containing 98 mole% *n*-butyric acid. Determine the type of condenser to be used and estimate the distillation column operating pressure.

12.2 A sequence of two distillation columns is to be used to produce the products indicated below. Establish the type of condenser and an operating pressure for each column for:
(a) The direct sequence (C_2/C_3 separation first).
(b) The indirect sequence (C_3/nC_4 separation first).
Use *K*-values from Fig. 7.5.

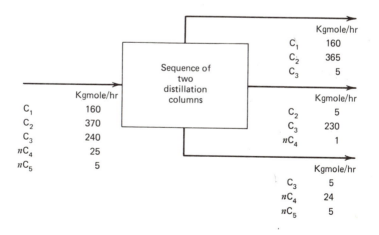

12.3 For each of the two distillation separations ($D-1$ and $D-2$) indicated below, establish the type condenser and an operating pressure.

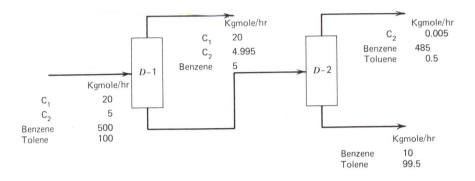

12.4 A deethanizer is to be designed for the separation indicated below. Estimate the number of equilibrium stages required assuming it is equal to 2.5 times the minimum number of equilibrium stages at total reflux.

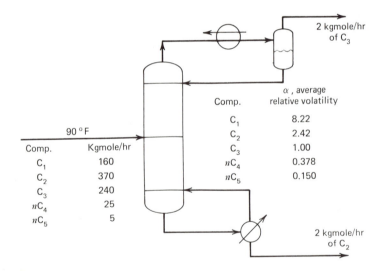

12.5 For the complex distillation operation shown below, use the Fenske equation to determine the minimum number of stages required between:
(a) The distillate and feed.
(b) The feed and side stream.
(c) The side stream and bottoms.
The K-values can be obtained from Raoult's law.

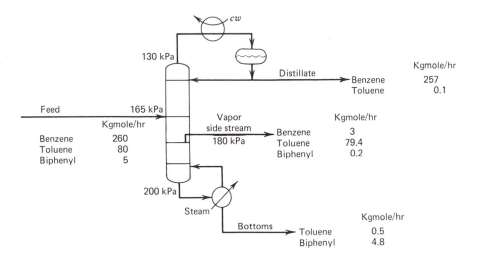

12.6 A 25 mole% mixture of acetone A in water W is to be separated by distillation at an average pressure of 130 kPa into a distillate containing 95 mole% acetone and a bottoms containing 2 mole% acetone. The infinite dilution activity coefficients are

$$\gamma_A^\infty = 8.12 \qquad \gamma_W^\infty = 4.13$$

Calculate by the Fenske equation the number of equilibrium stages required. Compare the result to that calculated from the McCabe–Thiele method.

12.7 For the distillation operation indicated below, calculate the minimum number of equilibrium stages and the distribution of the nonkey components by the Fenske equation using Fig. 7.5 for K-values.

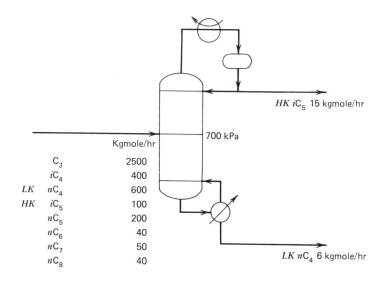

12.8 For the distillation operation shown below, establish the type condenser and an operating pressure, calculate the minimum number of equilibrium stages, and estimate the distribution of the nonkey components. Obtain K-values from Fig. 7.5

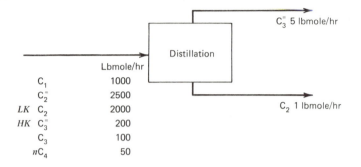

12.9 For 15 minimum equilibrium stages at 250 psia, calculate and plot the percent recovery of C_3 in the distillate as a function of distillate flow rate for the distillation of 1000 lbmole/hr of a feed containing 3% C_2, 20% C_3, 37% nC_4, 35% nC_5, and 5% nC_6 by moles. Obtain K-values from Fig. 7.5.

12.10 Use the Underwood equation to estimate the minimum external reflux ratio for the separation by distillation of 30 mole% propane in propylene to obtain 99 mole% propylene and 98 mole% propane, if the feed condition at a column operating pressure of 300 psia is:
(a) Bubble-point liquid.
(b) Fifty mole percent vaporized.
(c) Dew-point vapor.
Use K-values from Fig. 7.5.

12.11 For the conditions of Problem 12.7, compute the minimum external reflux rate and the distribution of the nonkey components at minimum reflux by the Underwood equation if the feed is a bubble-point liquid at column pressure.

12.12 Calculate and plot the minimum external reflux ratio and the minimum number of equilibrium stages against percent product purity for the separation by distillation of an equimolal bubble-point liquid feed of isobutane/n-butane at 100 psia. The distillate is to have the same iC_4 purity as the bottoms is to have nC_4 purity. Consider percent purities from 90 to 99.99%. Discuss the significance of the results.

12.13 Use the Fenske–Underwood–Gilliland shortcut method to determine the reflux ratio required to conduct the distillation operation indicated below if $N/N_{min} = 2.0$, the average relative volatility = 1.11, and the feed is at the bubble-point temperature at column feed-stage pressure. Assume external reflux equals internal reflux at the upper pinch zone. Assume a total condenser and a partial reboiler.

12.14 A feed consisting of 62 mole% paradichlorobenzene in orthodichlorobenzene is to be separated by distillation at near atmospheric pressure into a distillate containing 98 mole% para isomer and bottoms containing 96 mole% ortho isomer.
 If a total condenser and partial reboiler are used, $q = 0.9$, average relative

volatility = 1.154, and reflux/minimum reflux = 1.15, use the Fenske–Underwood–Gilliland procedure to estimate the number of theoretical stages required.

Problem 12.13

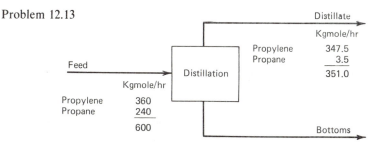

12.15 Explain why the Gilliland correlation can give erroneous results for cases where the ratio of rectifying to stripping stages is small.

12.16 The hydrocarbon feed to a distillation column is a bubble-point liquid at 300 psia with the mole fraction composition: $C_2 = 0.08$, $C_3 = 0.15$, $nC_4 = 0.20$, $nC_5 = 0.27$, $nC_6 = 0.20$, and $nC_7 = 0.10$.
(a) For a sharp separation between nC_4 and nC_5, determine the column pressure and type condenser if condenser outlet temperature is 120°F.
(b) At total reflux, determine the separation for eight theoretical stages overall, specifying 0.01 mole fraction nC_4 in the bottoms product.
(c) Determine the minimum reflux for the separation in Part (b).
(d) Determine the number of theoretical stages at $L/D = 1.5$ times minimum using the Gilliland correlation.

12.17 The following feed mixture is to be separated by ordinary distillation at 120 psia so as to obtain 92.5 mole% of the nC_4 in the liquid distillate and 82.0 mole% of the iC_5 in the bottoms.

Component	Lbmole/hr
C_3	5
iC_4	15
nC_4	25
iC_5	20
nC_5	35
	100

(a) Estimate the minimum number of equilibrium stages required by applying the Fenske equation. Obtain K-values from Fig. 7.5.
(b) Use the Fenske equation to determine the distribution of nonkey components between distillate and bottoms.
(c) Assuming that the feed is at its bubble point, use the Underwood method to estimate the minimum reflux ratio.
(d) Determine the number of theoretical stages required by the Gilliland correlation assuming $L/D = 1.2(L/D)_{min}$, a partial reboiler, and a total condenser.
(e) Estimate the feed-stage location.

12.18 Consider the separation by distillation of a chlorination effluent to recover C_2H_5Cl. The feed is bubble-point liquid at the column pressure of 240 psia with the following composition and K-values for the column conditions.

Component	Mole Fraction	K
C_2H_4	0.05	5.1
HCl	0.05	3.8
C_2H_6	0.10	3.4
C_2H_5Cl	0.80	0.15

Specifications are:

$$(x_D/x_B) \text{ for } C_2H_5Cl = 0.01$$
$$(x_D/x_B) \text{ for } C_2H_6 = 75$$

Calculate the product distribution, the minimum theoretical stages, the minimum reflux, and the theoretical stages at 1.5 times minimum L/D and locate the feed stage. The column is to have a partial condenser and a partial reboiler.

12.19 One hundred kilogram-moles per hour of a three-component bubble-point mixture to be separated by distillation has the following composition.

Component	Mole Fraction	Relative Volatility
A	0.4	5
B	0.2	3
C	0.4	1

(a) For a distillate rate of 60 kgmole/hr, five theoretical stages, and total reflux, calculate by the Fenske equation the distillate and bottoms compositions.
(b) Using the separation in Part (a) for components B and C, determine the minimum reflux and minimum boilup ratio by the Underwood equation.
(c) For an operating reflux ratio of 1.2 times the minimum, determine the number of theoretical stages and the feed-stage location.

12.20 For the conditions of Problem 12.6, determine the ratio of rectifying to stripping equilibrium stages by:
(a) Fenske equation.
(b) Kirkbride equation.
(c) McCabe–Thiele diagram.

12.21 Solve Example 12.8 by the Kremser method, but for an absorbent flow rate of 330 lbmole/hr and three theoretical stages. Compare your results to the Kremser results of Example 12.8 and discuss the effect of trading stages for absorbent flow.

12.22 Derive (12-65) in detail starting with (12-56) and (12-57).

12.23 Estimate the minimum absorbent flow rate required for the separation calculated in Example 12.8 assuming that the key component is propane, whose flow rate in the exit vapor is to be 105.4 lbmole/hr.

12.24 Solve Example 12.8 with the addition of a heat exchanger at each stage so as to maintain isothermal operation of the absorber at:
(a) 100°F.
(b) 125°F.
(c) 150°F.

12.25 One million pound-moles per day of a gas of the following composition is to be absorbed by n-octane at $-30°F$ and 550 psia in an absorber having 10 theoretical stages so as to absorb 50% of the ethane. Calculate by the group method the required flow rate of absorbent and the distribution of all the components between the lean gas and rich oil.

Component	Mole% in feed gas	K-value @ $-30°F$ and 550 psia
C_1	94.9	2.85
C_2	4.2	0.36
C_3	0.7	0.066
nC_4	0.1	0.017
nC_5	0.1	0.004

12.26 Determine by the group method the separation that can be achieved for the absorption operation indicated below for the following combinations of conditions.
(a) Six equilibrium stages and 75 psia operating pressure.
(b) Three equilibrium stages and 150 psia operating pressure.
(c) Six equilibrium stages and 150 psia operating pressure.

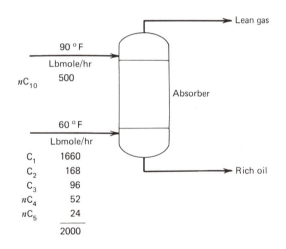

12.27 One thousand kilogram-moles per hour of rich gas at 70°F with 25% C_1, 15% C_2, 25% C_3, 20% nC_4, and 15% nC_5 by moles is to be absorbed by 500 kgmole/hr of nC_{10} at 90°F in an absorber operating at 4 atm. Calculate by the group method the percent absorption of each component for:
(a) Four theoretical stages.
(b) Ten theoretical stages.
(c) Thirty theoretical stages.
Use Fig. 7.5 for K-values.

12.28 For the flashing and stripping operation indicated below, determine by the group method the kilogram-moles per hour of steam if the stripper is operated at 2 atm and has five theoretical stages.

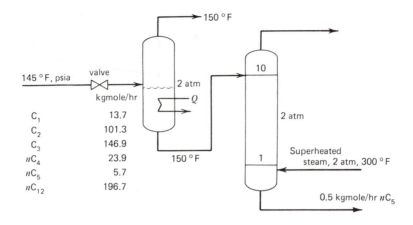

12.29 A stripper operating at 50 psia with three equilibrium stages is used to strip 1000 kgmole/hr of liquid at 250°F having the following molar composition: 0.03% C_1, 0.22% C_2, 1.82% C_3, 4.47% nC_4, 8.59% nC_5, 84.87% nC_{10}. The stripping agent is 100 kgmole/hr of superheated steam at 300°F and 50 psia. Use the group method to estimate the compositions and flow rates of the stripped liquid and rich gas.

12.30 One hundred kilogram-moles per hour of an equimolar mixture of benzene B, toluene T, n-hexane C_6, and n-heptane C_7 is to be extracted at 150°C by 300 kgmole/hr of diethylene glycol (DEG) in a countercurrent liquid–liquid extractor having five equilibrium stages. Estimate the flow rates and compositions of the extract and raffinate streams by the group method. In mole fraction units, the distribution coefficients for the hydrocarbon can be assumed essentially constant at the following values.

Component	$K_{D_i} = y$(solvent phase)$/x$(raffinate phase)
B	0.33
T	0.29
C_6	0.050
C_7	0.043

For diethylene glycol, assume $K_D = 30$ [E. D. Oliver, *Diffusional Separation Processes*, John Wiley & Sons, Inc., New York, 1966, 432].

12.31 A reboiled stripper in a natural gas plant is to be used to remove mainly propane and lighter components from the feed shown below. Determine by the group method the compositions of the vapor and liquid products.

12.32 Repeat Example 12.11 for external reflux flow rates L_0 of:
(a) 1500 lbmole/hr.
(b) 2000 lbmole/hr.
(c) 2500 lbmole/hr.
Plot d_{C_3}/b_{C_3} as a function of L_0 from 1000 to 2500 lbmole/hr. In making the calculations, assume that stage temperatures do not change from the results of Example 12.11. Discuss the validity of this assumption.

Problem 12.31

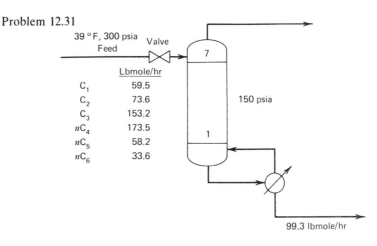

12.33 Repeat Example 12.11 for the following numbers of equilibrium stages (see Fig. 12.25).
(a) $M = 10$, $N = 10$.
(b) $M = 15$, $N = 15$.

Plot d_{C_3}/b_{C_3} as a function of $M + N$ from 10 to 30 stages. In making the calculations, assume that stage temperatures and total flow rates do not change from the results of Example 12.11. Discuss the validity of these assumptions.

12.34 Use the Edmister group method to determine the compositions of the distillate and bottoms for the distillation operation below.

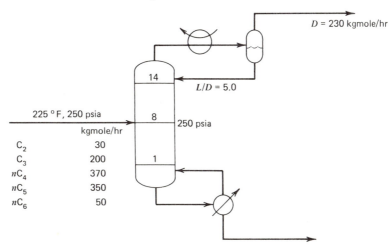

At column conditions, the feed is approximately 23 mole% vapor.

12.35 A bubble-point liquid feed is to be distilled as shown. Use the Edmister group method to estimate the compositions of the distillate and bottoms. Assume initial overhead and bottoms temperatures are 150 and 250°F, respectively.

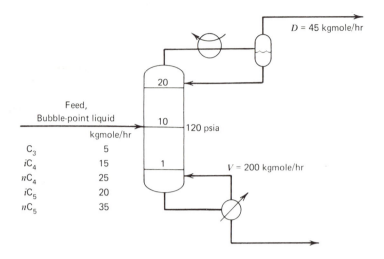

12.36 A mixture of ethylbenzene and xylenes is to be distilled as shown below. Assuming the applicability of Raoult's and Dalton's laws:

(a) Use the Fenske–Underwood–Gilliland method to estimate the number of stages required for a reflux-to-minimum reflux ratio of 1.10. Estimate the feed stage location by the Kirkbride equation.

(b) From the results of Part (a) for reflux, stages, and distillate rate, use the Edmister group method to predict the compositions of the distillate and bottoms. Compare the results with the specifications.

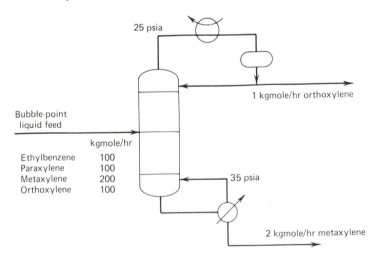

13

Stage Capacity and Efficiency

The capacity of a fractionating column may be limited by the maximum quantity of liquid that can be passed downward or by the maximum quantity of vapor that can be passed upward, per unit time, without upsetting the normal functioning of the column.

Mott Souders, Jr., and
George Granger Brown, 1934

The plate efficiency of fractionating columns and absorbers is affected by both the mechanical design of the column and the physical properties of the solution.

Harry E. O'Connell, 1946

Equipment for multistage separations frequently consists of horizontal phase-contacting trays arranged in a vertical column. The degree of separation depends upon the number of trays, their spacing, and their efficiency. The cross sectional area of the column determines the capacity of the trays to pass the streams being contacted.

In this chapter, methods are presented for determining capacity (or column diameter) and efficiency of some commonly used devices for vapor–liquid and liquid–liquid contacting. Emphasis is placed on approximate methods that are suitable for preliminary process design. Tray selection, sizing, and cost estimation of separation equipment are usually finalized after discussions with equipment vendors.

13.1 Vapor–Liquid Contacting Trays

In Section 2.5, bubble cap trays, sieve trays, and valve trays are cited as the most commonly used devices for contacting continuous flows of vapor and liquid phases; however, almost all new fabrication is with sieve or valve trays. For all three devices, as shown in Fig. 13.1, vapor, while flowing vertically upward, contacts liquid in crossflow on each tray. When trays are properly designed and operated, vapor flows only through perforated or open regions of the trays, while liquid flows downward from tray to tray only by means of

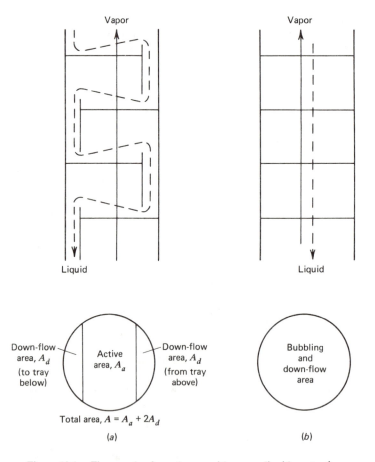

Figure 13.1. Flow modes for columns with vapor–liquid contacting trays. (*a*) Cross-flow mode. (*b*) Counter-flow mode.

downcomers. This cross-flow mode is preferred over a true counter-flow mode where both phases pass through the same open regions of a tray, because the former mode is capable of a much wider operating range and better reliability.

Capacity and Column Diameter

For a given ratio of liquid-to-vapor flow rates, as shown in Fig. 2.4, a maximum vapor velocity exists beyond which incipient column flooding occurs because of backup of liquid in the downcomer. This condition, if sustained, leads to carryout of liquid with the overhead vapor leaving the column. *Down-flow flooding* takes place when liquid backup is caused by downcomers of inadequate cross-sectional area A_d, but rarely occurs if downcomer cross-sectional area is at least 10% of total column cross-sectional area and if tray spacing is at least 24 in. The usual design limit is *entrainment flooding*, which is caused by excessive carry up of liquid e by vapor entrainment to the tray above. At incipient flooding, $(e + L) \gg L$ and downcomer cross-sectional area is inadequate for the excessive liquid load $(e + L)$.

Entrainment of liquid may be due to carry up of suspended droplets by rising vapor or to throw up of liquid particles by vapor jets formed at tray perforations, valves, or bubble cap slots. Souders and Brown[1] successfully correlated entrainment flooding data for 10 commercial tray columns by assuming that carry up of suspended droplets controls the quantity of entrainment. At low vapor velocity, a droplet settles out; at high vapor velocity, it is entrained. At the flooding or incipient entrainment velocity U_f, the droplet is suspended such that the vector sum of the gravitational, buoyant, and drag forces acting on the droplet, as shown in Fig. 13.2, are zero. Thus,

$$\sum F = 0 = F_g - F_b - F_d \tag{13-1}$$

Therefore, in terms of droplet diameter d_p

$$\rho_L\left(\frac{\pi d_p^3}{6}\right)g - \rho_V\left(\frac{\pi d_p^3}{6}\right)g - C_D\left(\frac{\pi d_p^2}{4}\right)\frac{U_f^2}{2}\rho_V = 0 \tag{13-2}$$

where C_D is the drag coefficient. Solving for flooding velocity, we have

$$U_f = C\left(\frac{\rho_L - \rho_V}{\rho_V}\right)^{1/2} \tag{13-3}$$

where C = capacity parameter of Souders and Brown. According to the above theory

$$C = \left(\frac{4 d_p g}{3 C_D}\right)^{1/2} \tag{13-4}$$

Parameter C can be calculated from (13-4) if the droplet diameter d_p is

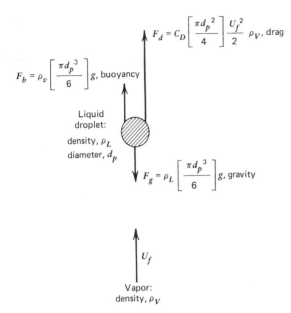

$$F_d = C_D \left[\frac{\pi d_p^2}{4} \right] \frac{U_f^2}{2} \rho_V, \text{drag}$$

$$F_b = \rho_v \left[\frac{\pi d_p^3}{6} \right] g, \text{buoyancy}$$

Liquid droplet:

density, ρ_L

diameter, d_p

$$F_g = \rho_L \left[\frac{\pi d_p^3}{6} \right] g, \text{gravity}$$

U_f

Vapor:

density, ρ_V

Figure 13.2. Forces acting on suspended liquid droplet.

known. In practice, however, C is treated as an empirical quantity that is determined from experimental data obtained from operating equipment. Souders and Brown considered all the important variables that could influence the value of C and obtained a correlation for commercial-size columns with bubble cap trays. Data used to develop the correlation covered column pressures from 10 mmHg to 465 psia, plate spacing from 12 to 30 in., and liquid surface tension from 9 to 60 dynes/cm. In accordance with (13-4), they found that the value of C increased with increasing surface tension, which would increase d_p. Also, C increased with increasing tray spacing, since this allowed more time for agglomeration to a larger d_p.

Using additional commercial column operating data, Fair[2] produced the more general correlation of Fig. 13.3, which is applicable to columns with bubble cap and sieve trays. Whereas Souders and Brown based the vapor velocity on the entire column cross-sectional area, Fair utilized a net vapor flow area equal to the total inside column cross-sectional area minus the area blocked off by the downcomer(s) bringing liquid down to the tray underneath, that is, $(A - A_d)$ in Fig. 13.1a. The value of C_F in Fig. 13.3 is seen to depend on the tray spacing and on the ratio $F_{LV} = (LM_L/VM_V)(\rho_V/\rho_L)^{0.5}$ (where flow rates are in molal units), which is a kinetic energy ratio that was first used by Sherwood, Shipley, and Holloway[3] to correlate packed-column flooding data. The value of C for use in (13-3) is obtained from Fig. 13.3 by correcting C_F for surface tension, foaming tendency, and the ratio of vapor hole area A_h to tray active area A_a, according to

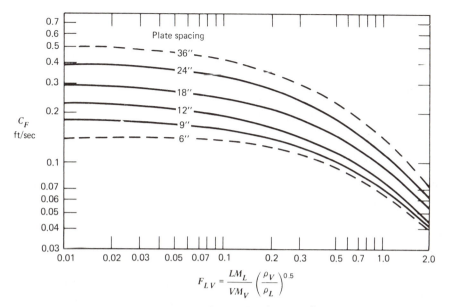

$$F_{LV} = \frac{LM_L}{VM_V}\left(\frac{\rho_V}{\rho_L}\right)^{0.5}$$

Figure 13.3.   Entrainment flooding capacity.

the empirical relationship

$$C = F_{ST}F_F F_{HA} C_F \tag{13-5}$$

where

F_{ST} = surface tension factor = $(\sigma/20)^{0.2}$

F_F = foaming factor

F_{HA} = 1.0 for $A_h/A_a \geq 0.10$ and $5(A_h/A_a) + 0.5$ for $0.06 \leq A_h/A_a \leq 0.1$

σ = liquid surface tension, dynes/cm

For nonfoaming systems, $F_F = 1.0$; for many absorbers, F_F may be 0.75 or even less. The quantity A_h is taken to be the area open to the vapor as it penetrates into the liquid on a tray. It is the total cap slot area for bubble cap trays and the perforated area for sieve trays.

Figure 13.3 appears to be applicable to valve trays also. This is shown in Fig. 13.4 where entrainment flooding data of Fractionation Research, Inc. (FRI)[4,5], for a 4-ft-diameter column with 24-in. tray spacing are compared to the correlation in Fig. 13.3. As seen, the correlation is conservative for these tests. For valve trays, the slot area A_h is taken as the full valve opening through which vapor enters the frothy liquid on the tray at a 90° angle with the axis of the column.

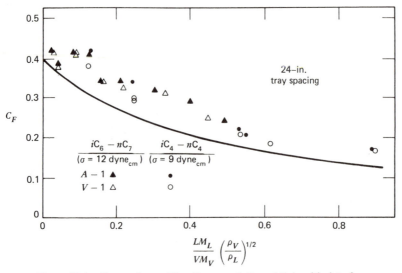

Figure 13.4. Comparison of flooding correlation of Fair with data for valve trays.

When using a well-designed tray, based on geometrical factors such as those listed by Van Winkle,[6] one finds that application of Fig. 13.3 will generally give somewhat conservative values of C_F for columns with bubble cap, sieve, and valve trays. Typically, column diameter D_T is based on 85% of the flooding velocity U_f, calculated from (13-3), using C from (13-5), based on C_F from Fig. 13.3. By the continuity equation, the molal vapor flow rate is related to the flooding velocity by

$$V = (0.85 U_f)(A - A_d)\frac{\rho_V}{M_V} \qquad (13\text{-}6)$$

where A = total column cross-sectional area = $\pi D_T^2/4$. Thus

$$D_T = \left[\frac{4 V M_V}{0.85 U_f \pi (1 - A_d/A)\rho_V}\right]^{0.5} \qquad (13\text{-}7)$$

Oliver[7] suggests that A_d/A be estimated from F_{LV} in Fig. 13.3 by

$$\frac{A_d}{A} = \begin{cases} 0.1; \ F_{LV} \leq 0.1 \\[2mm] 0.1 + \dfrac{(F_{LV} - 0.1)}{9}; \ 0.1 \leq F_{LV} \leq 1.0 \\[2mm] 0.2; \ F_{LV} \geq 1.0 \end{cases}$$

Because of the need for internal access to columns with trays, a packed column is generally used if the calculated diameter from (13-7) is less than 2.5 ft.

Tray Spacing and Turndown Ratio

The tray spacing must be specified to compute column diameter using Fig. 13.3. As spacing is increased, column height is increased but column diameter is reduced. A spacing of 24 in., which is considered minimum for ease of maintenance, is considered optimum for a wide range of conditions; however, a smaller spacing may be desirable for small-diameter columns with a large number of stages and a larger spacing is frequently used for large-diameter columns with a small number of stages.

As shown in Fig. 2.4, a minimum vapor rate exists below which liquid may weep or dump through tray perforations or risers instead of flowing completely across the active area and into the downcomer to the tray below. Below this minimum, the degree of contacting of liquid with vapor is reduced, causing tray efficiency to decline. The ratio of the vapor rate at flooding to the minimum vapor rate is the *turndown ratio*, which is approximately 10 for bubble cap and valve trays but only about 3 for sieve trays.

When vapor and liquid flow rates change appreciably from tray to tray, column diameter, tray spacing, or hole area can be varied to reduce column cost and insure stable operation at high efficiency. Variation of tray spacing is particularly applicable to columns with sieve trays because of their low turndown ratio.

Example 13.1. Determine the column diameter for the reboiled absorber of Example 15.8 using a solution based on the Soave–Redlich–Kwong equation of state for thermodynamic properties. Computed temperatures and molal flow rates, densities, and average molecular weights for vapor and liquid leaving each theoretical stage are as follows.

Stage	T, °F	Pound-moles per hour		Pound per cubic foot		Molecular weight	
		V	L	ρ_V	ρ_L	M_V	M_L
1	128.1	530.0	888.6	1.924	41.1	26.6	109.1
2	136.3	918.6	962.7	2.067	40.4	28.5	104.0
3	142.3	992.7	1007.8	2.133	40.0	29.5	101.4
4	147.6	1037.8	1044.1	2.180	39.6	30.3	99.7
5	152.3	1074.1	1076.8	2.221	39.2	30.9	98.2
6	155.7	1106.8	1115.5	2.263	39.0	31.6	96.5
7	154.0	1145.5	1219.6	2.339	38.2	32.2	92.1
8	162.8	449.6	1309.1	2.525	37.2	34.3	88.8
9	170.5	539.1	1379.1	2.645	36.7	35.9	86.8
10	179.3	609.1	1450.9	2.754	36.2	37.4	85.2
11	191.4	680.9	1505.2	2.860	36.0	39.2	84.5
12	221.5	735.2	1366.0	2.832	36.2	41.2	91.7

Solution. To determine the limiting column diameter, (13-7) is applied to each stage. From Fig. 13.3, values of C_F corresponding to a 24-in. tray spacing and computed values of the kinetic energy ratio F_{LV} are obtained using the given values of V, L, ρ_V, ρ_L, M_V, and M_L as follows. Values of A_d/A, based on recommendations given by Oliver[7] are included.

Stage	F_{LV}	C_F, ft/sec	A_d/A
1	1.49	0.082	0.2
2	0.866	0.131	0.185
3	0.806	0.138	0.178
4	0.776	0.142	0.175
5	0.758	0.144	0.173
6	0.741	0.147	0.171
7	0.754	0.145	0.173
8	1.96	0.063	0.2
9	1.66	0.074	0.2
10	1.50	0.081	0.2
11	1.34	0.090	0.2
12	1.16	0.103	0.2

Many of the F_{LV} values are large, indicating rather high liquid loads.

Values of C_F are used in (13-5) to compute the Souders and Brown capacity parameter C. Because the liquid flows contain a large percentage of absorber oil, a foaming factor, F_F of 0.75 is assumed. Also assume $A_h/A_a > 0.1$ so that F_{HA} is 1.0. The value of F_{ST} is determined from surface tension σ, which can be estimated for mixtures by various methods. For paraffin hydrocarbon mixtures, a suitable estimate in dynes per centimeter, based on densities, can be obtained from the equation

$$\sigma = \left(\frac{\rho_L - \rho_V}{18.5}\right)^4$$

From the computed value of C, (13-3) is used to obtain the flooding velocity U_f, from which the column diameter corresponding to 85% of flooding is determined from (13-7). For example, computations for stage 1 give

$$\sigma = \left(\frac{41.1 - 1.924}{18.5}\right)^4 = 20.1 \text{ dynes/cm}$$

$$F_{ST} = \left(\frac{20.1}{20}\right)^{0.2} = 1.001$$

$$C = (1.001)(0.75)(1.0)(0.082) = 0.062 \text{ ft/sec}$$

$$U_f = 0.062\left(\frac{41.1 - 1.924}{1.924}\right)^{0.5} = 0.28 \text{ ft/sec}$$

$$D_T = \left[\frac{(4)(530/3600)(26.6)}{0.85(0.28)(3.14)(1 - 0.2)(1.924)}\right]^{0.5} = 3.7 \text{ ft}$$

For other stages, values of these quantities are computed in a similar manner with the following results.

Stage	dyne/cm	F_{ST}	C, ft/sec	U_f, ft/sec	D_T, ft	Percent Flooding for 5 ft D_T
1	20.1	1.001	0.062	0.28	3.7	46.6
2	18.4	0.983	0.097	0.42	3.9	51.7
3	17.5	0.974	0.101	0.43	4.0	54.4
4	16.7	0.965	0.103	0.43	4.1	57.1
5	16.0	0.956	0.103	0.42	4.2	60.0
6	15.6	0.952	0.105	0.42	4.3	62.9
7	14.1	0.932	0.101	0.40	4.5	68.9
8	12.3	0.907	0.043	0.16	4.5	68.9
9	11.5	0.895	0.050	0.18	4.6	71.9
10	10.7	0.882	0.054	0.19	4.8	78.3
11	10.3	0.876	0.059	0.20	4.9	81.6
12	10.6	0.881	0.068	0.23	4.9	81.6

The maximum diameter computed is 4.9 ft. If a column of constant diameter equal to 5 ft is selected, the percentage of flooding at each stage, as tabulated above, is obtained by replacing the quantity 0.85 in (13-7) by the fraction flooding FF and solving for FF using a D_T of 5 ft. A maximum turndown ratio of $100/46.6 = 2.15$ occurs at stage 1. This should be acceptable even with sieve trays. Alternatively, reduction of tray spacing for the top six trays in order to increase the percent of flooding is possible. Because liquid flow rates are relatively high, this reduction might be undesirable. Also, because the diameter at 85% of flooding gradually increases from 3.7 to 4.9 feet down the column, a two-diameter column with swedging from 4 to 5 ft does not appear attractive.
□

Efficiency

For a given separation, the ratio of the required number of equilibrium stages N to the number of actual trays N_a defines an overall tray efficiency.

$$E_o = \frac{N}{N_a} \tag{13-8}$$

This efficiency is a complex function of tray design, fluid properties, and flow patterns. Theoretical approaches to the estimation of E_o have been developed but are based on point calculations of mass transfer[8] that are not discussed here. However, for well-designed bubble cap, sieve, and valve trays, the available empirical correlations described here permit reasonable predictions of E_o.

Using average liquid-phase viscosity as the sole correlating variable, Drickamer and Bradford[9] developed a remarkable correlation that showed good agreement with 60 operating data points for commercial fractionators, absorbers, and strippers. The data covered ranges of average temperature from 60 to 507°F and pressures from 14.7 to 485 psia; E_o varied from 8.7 to 88%. As can be shown by theoretical considerations, liquid viscosity significantly influences the mass transfer resistance in the liquid phase.

Mass transfer theory indicates that, when the volatility covers a wide range, the relative importance of liquid-phase and gas-phase mass transfer resistances can shift. Thus, as might be expected, O'Connell[10] found that the Drickamer–Bradford correlation inadequately correlated data for fractionators operating on key components with large relative volatilities and for absorbers and strippers involving a wide range of volatility of key components. Separate correlations in terms of a viscosity–volatility product were developed for fractionators and for absorbers and strippers by O'Connell[10] using data for columns with bubble cap trays. However, as shown in Fig. 13-5, Lockhart and Leggett[11] were able to obtain a single correlation by using the product of liquid viscosity and an appropriate volatility as the correlating variable. For fractionators, the relative volatility of the key components was used; for hydrocarbon absorbers, the volatility was taken as 10 times the K-value of the key component, which must be reasonably distributed between top and bottom products. The data used by

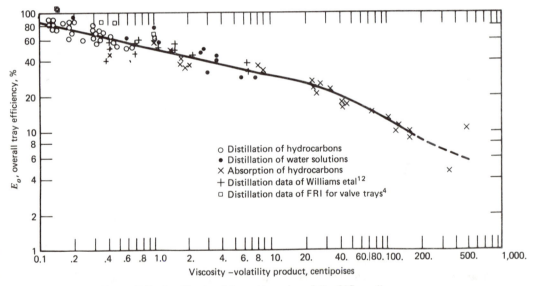

Figure 13.5. Lockhart and Leggett version of the O'Connell correlation for overall tray efficiency of fractionators, absorbers, and strippers. (Adapted from F. J. Lockhart and C. W. Leggett, *Advances in Petroleum Chemistry and Refining*, Vol. 1, eds. K. A. Kobe and John J. McKetta, Jr., Interscience Publishers, Inc., New York, © 1958, 323–326.)

O'Connell covered a range of relative volatility from 1.16 to 20.5 and are shown in Fig. 13.5. A comprehensive study of the effect on E_o of the ratio of liquid-to-vapor molal flow rates L/V for eight different binary systems in a 10-in.-diameter column with bubble cap trays was reported by Williams, Stigger, and Nichols.[12] The systems included water, hydrocarbons, and other organic compounds. For fractionation with L/V nearly equal to 1.0, their distillation data, which are included in Fig. 13.5, are in reasonable agreement with the O'Connell correlation. For the distillation of hydrocarbons in a column having a diameter of 0.45 m, Zuiderweg, Verburg, and Gilissen[13] found differences in E_o among bubble cap, sieve, and valve trays to be insignificant when operating at 85% of flooding. Accordingly, Fig. 13.5 is assumed to be applicable to all three tray types, but may be somewhat conservative for well-designed trays. For example, data of FRI for valve trays operating with the cyclohexane/n-heptane and isobutane/n-butane systems are also included in Fig. 13.5 and show efficiencies 10 to 20% higher than the correlation.

In using Fig. 13.5 to predict E_o, we compute the viscosity and relative volatility for fractionators at the arithmetic average of values at column top and bottom temperatures and pressures for the composition of the feed. For absorbers and strippers, both viscosity and the K-value are evaluated at rich-oil conditions.

Most of the data used to develop the correlation of Fig. 13.5 are for columns having a liquid flow path across the active tray area of from 2 to 3 ft. Gautreaux and O'Connell,[14] using theory and experimental data, showed that higher efficiencies are achieved for longer flow paths. For short liquid flow paths, the liquid flowing across the tray is usually completely mixed. For longer flow paths, the equivalent of two or more completely mixed, successive liquid zones may be present. The result is a greater average driving force for mass transfer and, thus, a higher efficiency; perhaps even greater than 100%. Provided that the viscosity–volatility product lies between 0.1 and 1.0, Lockhart and Leggett[11] recommend addition of the increments in Table 13.1 to the value of E_o from Fig. 13.5 when liquid flow path is greater than 3 ft. However, at large liquid rates, long liquid path lengths are undesirable because they lead to excessive liquid gradients. When the effective height of a liquid on a tray is appreciably higher at the inflow side than at the overflow weir, vapor may prefer to enter the tray in the latter region, leading to nonuniform bubbling action. Multipass trays, as shown in Fig. 13.6, are used to prevent excessive liquid gradients. Estimation of the desired number of flow paths can be made with Fig. 13.7, which was derived from recommendations by Koch Engineering Company.[15]

Based on estimates of the number of actual trays and tray spacing, we can compute the height of column between the top tray and the bottom tray. By adding an additional 4 ft above the top tray for removal of entrained liquid and 10 ft below the bottom tray for bottoms surge capacity, we can estimate total column height. If the height is greater than 212 ft (equivalent to 100 trays on 24-in.

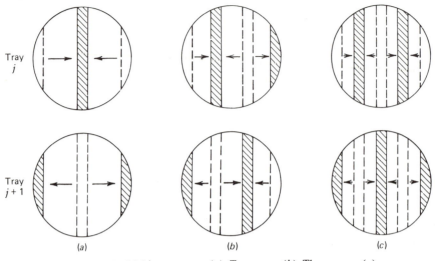

Figure 13.6. Multipass trays. (*a*) Two-pass. (*b*) Three-pass. (*c*) Four-pass.

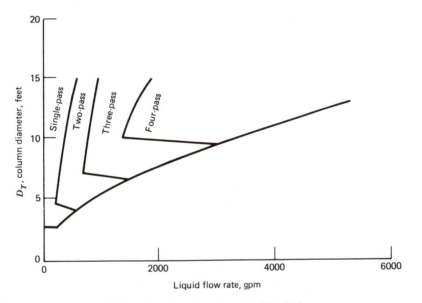

Figure 13.7. Estimation of number of required liquid flow passes. (Derived from Koch Flexitray Design Manual, Bulletin 960, Koch Engineering Co., Inc., Wichita, Kansas, 1960.)

spacing), two or more columns arranged in series may be preferable to a single column.*

Table 13.1 Correction to overall tray efficiency for length of liquid flow path ($0.1 \leq \mu\alpha \leq 1.0$)

Length of Liquid Flow Path, ft	Factor to be Added to E_o from Fig. 13.5, %
3	0
4	10
5	15
6	20
8	23
10	25
15	27

Example 13.2. Estimate the overall tray efficiency, number of actual trays, and column height for the reboiled absorber considered in Examples 15.8 and 13.1.

Solution. The Lockhart-Leggett version of the O'Connell correlation given in Fig. 13.5 can be used to estimate E_o. Because absorber oil is used, it is preferable to compute relative volatility and viscosity at average liquid conditions. This is most conveniently done for this example by averaging between the top and bottom stages. From the solution based on the Soave–Redlich–Kwong correlation, with ethane and propane as light key and heavy key, respectively

$$\alpha_{avg} = \frac{\alpha_{top} + \alpha_{bottom}}{2} = \frac{2.60 + 1.98}{2} = 2.29$$

Liquid mixture viscosity is estimated from the pure paraffin hydrocarbon of equivalent molecular weight. The result is

$$\mu_{avg} = \frac{\mu_{top} + \mu_{bottom}}{2} = \frac{0.36 + 0.18}{2} = 0.27 \text{ cp}$$

Thus

$$\mu_{avg}\alpha_{avg} = (2.29)(0.27) = 0.62$$

From Fig. 13.5, $E_o = 55\%$. From Fig. 13.7, for $D = 5$ ft and a maximum liquid flow rate from Example 13.1 of

$$\frac{(1505.2)(84.5)(7.48)}{(60)(36.0)} = 440 \text{ gpm}$$

it is seen that two liquid flow passes are required. Consequently, the liquid flow path is less than 3 ft and, according to Table 13.1, a correction to E_o is not required.

---

*The tallest distillation column in the world is located at the Shell Chemical Company complex in Deer Park, Texas. Fractionating ethylene, the tower is 338 ft tall and 18 ft in diameter [*Chem. Eng.*, **84** (26), 84 (1977)].

The number of actual trays, exclusive of the reboiler, is obtained from (13-8). From Example 13.1, with $N = 12$

$$N_a = 12/0.55 = 22 \text{ actual trays}$$

Column height (not including top and bottom heads) is

$$4 + 21(2) + 10 = 56 \text{ ft}$$

☐

13.2 Flash and Reflux Drums

A vertical vessel (drum), as shown in Fig. 7.1, can be used to separate vapor from liquid following equilibrium flash vaporization or partial condensation.* A reasonable estimate of the minimum drum diameter D_T, to prevent liquid carryover by entrainment can be made by using (13-7) in conjunction with the curve for 24-in. tray spacing in Fig. 13.3 and a value of $F_{HA} = 1.0$ in (13-5). To absorb process upsets and fluctuations and otherwise facilitate control, vessel volume V_V is determined on the basis of liquid residence time t, which should be at least 5 min with the vessel half full of liquid.[16] Thus

$$V_V = 2LM_L t/\rho_L \qquad (13\text{-}9)$$

Assuming a vertical, cylindrical vessel and neglecting the volume associated with the heads, we find that the height H of the vessel is

$$H = \frac{4V_V}{\pi D_T^2} \qquad (13\text{-}10)$$

However, if $H > 4D_T$, it is generally preferable to increase D_T and decrease H to give $H = 4D$. Then

$$D_T = \frac{H}{4} = \left(\frac{V_V}{\pi}\right)^{1/3} \qquad (13\text{-}11)$$

A height above the liquid level of at least 4 ft is necessary for feed entry and disengagement of liquid droplets from the vapor.

When vapor is totally condensed, a cylindrical, horizontal reflux drum is commonly employed to receive the condensate. Equations (13-9) and (13-11) permit estimates of the drum diameter and length L_V by assuming an optimum L_V/D of four[16] and the same liquid residence time suggested for a vertical drum.

Example 13.3. Equilibrium vapor and liquid streams leaving a flash drum are as follows.

---

* A horizontal drum is preferred by some if liquid loads are appreciable.

Component	Vapor	Liquid
Pound-moles per hour		
HCl	49.2	0.8
Benzene	118.5	81.4
Monochlorobenzene	71.5	178.5
TOTAL	239.2	260.7
Pounds per hour	19,110	26,480
T, °F	270	270
P, psia	35	35
Density, lb/ft³	0.371	57.08

Determine the dimensions of the flash drum.

Solution

$$F_{LV} = \frac{26,480}{19,110}\left(\frac{0.371}{57.08}\right)^{0.5} = 0.112$$

From Fig. 13.3, C_F at a 24-in. tray spacing is 0.34. Assume $C = C_F$. From (13-3)

$$U_f = 0.34\left(\frac{57.08 - 0.37}{0.371}\right)^{0.5} = 4.2 \text{ ft/sec} = 15,120 \text{ ft/hr}$$

From (13-7) with $A_d/A = 0$

$$D_T = \left[\frac{(4)(19,110)}{(0.85)(15,120)(3.14)(1)(0.371)}\right]^{0.5} = 2.26 \text{ ft}$$

From (13-9) with $t = 5 \text{ min} = 0.0833 \text{ hr}$

$$V_V = \frac{(2)(26,480)(0.0833)}{(57.08)} = 77.3 \text{ ft}^3$$

From (13-10)

$$H = \frac{(4)(77.3)}{(3.14)(2.26)^2} = 19.3 \text{ ft}$$

However, $H/D_T = 19.3/2.26 = 8.54 > 4$. Therefore, redimension V_V for $H/D_T = 4$. From (13-11)

$$D_T = \left(\frac{77.3}{3.14}\right)^{1/3} = 2.91 \text{ ft}$$

$$H = 4D_T = (4)(2.91) = 11.64 \text{ ft}$$

Height above the liquid level is $11.64/2 = 5.82$ ft, which is adequate. Alternatively, with a height of twice the minimum disengagement height, $H = 8$ ft and $D_T = 3.5$ ft.
□

13.3   Liquid–Liquid Contactors

A wide variety of equipment is available for countercurrent contacting of continuous flows of two essentially immiscible liquid phases. Such equipment includes devices that create interfacial contacting area solely by liquid head or jets (e.g., plate and spray columns) and other devices that incorporate mechanical agitation (e.g., pulsed, rotating disk, and reciprocating plate columns). Because mass transfer rates for liquid–liquid contacting are greatly increased when mechanical agitation is provided, the latter devices are the ones most commonly used. A popular device for liquid–liquid extraction is the rotating-disc contactor (RDC), which, according to Reman and Olney,[17] offers ease of design, construction, and maintenance; provides flexibility of operation; and has been thoroughly tested on a commercial scale. The total volumetric throughput per volume of a theoretical stage for the RDC is equaled only by the reciprocating-plate column (RPC),[18] which also offers many desirable features. The RDC has been constructed in diameters up to at least 9 ft and is claimed to be suitable for diameters up to 20 ft, while the RPC has been fabricated in diameters up to 3 ft with at least 6-ft-diameter units being possible. In this section, we consider capacity and efficiency only for the RDC and RPC.

Capacity and Column Diameter

Because of the larger number of important variables, estimation of column diameter for liquid–liquid contacting devices can be far more complex and is more uncertain than for the vapor–liquid contactors. These variables include individual phase flow rates, density difference between the two phases, interfacial tension, direction of mass transfer, viscosity and density of the continuous phase, rotating or reciprocating speed (and amplitude and plate hole size for the RPC), and compartment geometry. Column diameter is best determined by scale-up from tests run in standard laboratory or pilot plant test units, which have a diameter of 1 in. or larger. The sum of the measured superficial velocities of the two liquid phases in the test unit can be assumed to hold for the larger commercial unit. This sum is often expressed in total gallons per hour per square foot of empty column cross section and has been measured to be as large as 1837 in the RPC and 1030 in the RDC.[18] In the absence of laboratory data, preliminary estimates of column diameter can be made by a simplification of the theory of Logsdail, Thornton, and Pratt,[19] which has been compared recently to other procedures by Landau and Houlihan[20] in the case of the RDC. Because the relative motion between a dispersed droplet phase and a continuous phase is involved, this theory is based on a concept that is similar to that developed in Section 13.1 for liquid droplets dispersed in a vapor phase.

Consider the case of liquid droplets of the phase of smaller density rising through the denser, downward-flowing continuous liquid phase, as shown in Fig.

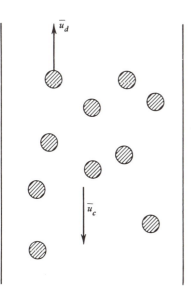

Figure 13.8.   Countercurrent flows of liquid phases in a column.

13.8. If the average superficial velocities of the discontinuous (droplet) phase and the continuous phase are U_d in the upward direction and U_c in the downward direction, respectively, the corresponding average actual velocities relative to the column wall are

$$\bar{u}_d = \frac{U_d}{\phi_d} \tag{13-12}$$

and

$$\bar{u}_c = \frac{U_c}{1 - \phi_d} \tag{13-13}$$

where ϕ_d is the average fractional volumetric holdup of dispersed (droplet) phase in the column. The average droplet rise velocity relative to the continuous phase is the sum of (13-12) and (13-13) or

$$\bar{u}_r = \frac{U_d}{\phi_d} + \frac{U_c}{1 - \phi_d} \tag{13-14}$$

This relative velocity can be expressed in terms of a modified form of (13-3) where the continuous phase density in the buoyancy term is replaced by the density of the two-phase mixture ρ_m. Thus, after noting for the case here that F_d and F_g act downward while F_b acts upward, we obtain

$$\bar{u}_r = C \left(\frac{\rho_m - \rho_d}{\rho_c} \right)^{1/2} f\{1 - \phi_d\} \tag{13-15}$$

where C is in the same parameter as in (13-4) and $f\{1 - \phi_d\}$ is a factor that allows for the hindered rising effect of neighboring droplets. The density ρ_m is a volumetric mean given by

$$\rho_m = \phi_d \rho_d + (1 - \phi_d)\rho_c \tag{13-16}$$

Therefore

$$\rho_m - \rho_d = (1 - \phi_d)(\rho_c - \rho_d) \tag{13-17}$$

Substitution of (13-17) into (13-15) yields

$$\bar{u}_r = C\left(\frac{\rho_c - \rho_d}{\rho_c}\right)^{1/2}(1 - \phi_d)^{1/2}f\{1 - \phi_d\} \tag{13-18}$$

From experimental data, Gayler, Roberts, and Pratt[21] found that, for a given liquid–liquid system, the right-hand side of (13-18) could be expressed empirically as

$$\bar{u}_r = u_o(1 - \phi_d) \tag{13-19}$$

where u_o is a characteristic rise velocity for a single droplet, which depends upon all the variables discussed above, except those on the right-hand side of (13-14). Thus, for a given liquid–liquid system, column design, and operating conditions, the combination of (13-14) and (13-19) gives

$$\frac{U_d}{\phi_d} + \frac{U_c}{1 - \phi_d} = u_o(1 - \phi_d) \tag{13-20}$$

where u_o is a constant. Equation (13-20) is cubic in ϕ_d, with a typical solution as shown in Fig. 13.9 for $U_c/u_o = 0.1$. Thornton[22] argues that, with U_c fixed, an increase in U_d results in an increased value of the holdup ϕ_d until the flooding point is reached, at which $(\partial U_d/\partial \phi_d)_{U_c} = 0$. Thus, in Fig. 13.9, only that portion of the curve for $\phi_d = 0$ to $(\phi_d)_f$, the holdup at the flooding point, is realized in practice. Alternatively, with U_d fixed, $(\partial U_c/\partial \phi_d)_{U_d} = 0$ at the flooding point. If these two derivatives are applied to (13-20), we obtain, respectively

$$U_c = u_o[1 - 2(\phi_d)_f][1 - (\phi_d)_f]^2 \tag{13-21}$$

$$U_d = 2u_o[1 - (\phi_d)_f](\phi_d)_f^2 \tag{13-22}$$

Combining (13-21) and (13-22) to eliminate u_o gives the following expression for $(\phi_d)_f$.

$$(\phi_d)_f = \frac{\left(1 + 8\dfrac{U_c}{U_d}\right)^{0.5} - 3}{4\left(\dfrac{U_c}{U_d} - 1\right)} \tag{13-23}$$

This equation predicts values of $(\phi_d)_f$ ranging from zero at $U_d/U_c = 0$ to 0.5 at

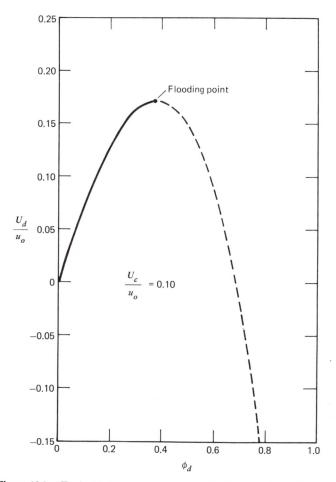

Figure 13.9.   Typical holdup curve in liquid–liquid extraction column.

$U_c/U_d = 0$. At $U_d/U_c = 1$, $(\phi_d)_f = 1/3$. The simultaneous solution of (13-20) and (13-23) results in Fig. 13.10 for the variation of total capacity as a function of phase flow ratio. The largest total capacities are achieved, as might be expected, at the smallest ratios of dispersed phase flow rate to continuous phase flow rate.

For fixed values of column geometry and rotor speed, experimental data of Logsdail et al.[19] for a laboratory scale RDC indicate that the dimensionless group $(u_o \mu_c \rho_c/\sigma \Delta \rho)$ is approximately constant, where σ is the interfacial tension. Data of Reman and Olney[17] and Strand, Olney, and Ackerman[23] for well-designed and efficiently operated commercial RDC columns ranging from 8 in. to 42 in. in diameter indicate that this dimensionless group has a value of roughly 0.01 for

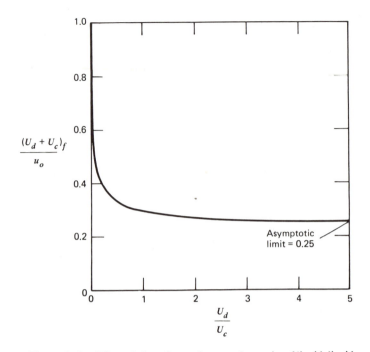

Figure 13.10. Effect of phase flow ratio on total capacity of liquid–liquid extraction columns.

systems involving water as either the continuous or dispersed phase. This value is suitable for preliminary calculations of RDC and RPC column diameters, when the sum of the actual superficial phase velocities is taken as 50% of the estimated sum at flooding conditions.

Example 13.4. Estimate the diameter of an RDC to extract acetone from a dilute toluene–acetone solution into water at 20°C. The flow rates for the dispersed organic and continuous aqueous phases are 27,000 and 25,000 lb/hr, respectively.

Solution. The necessary physical properties are: $\mu_c = 1.0 \, \text{cp}$ (0.000021 lbf/sec/ft²); $\rho_c = 1.0 \, \text{gm/cm}^3$; $\Delta\rho = 0.14 \, \text{gm/cm}^3$; $\sigma = 32 \, \text{dynes/cm}$ (0.00219 lbf/ft).

$$\frac{U_d}{U_c} = \left(\frac{27,000}{25,000}\right)\left(\frac{\rho_c}{\rho_d}\right) = \left(\frac{27,000}{25,000}\right)\left(\frac{1.0}{0.86}\right) = 1.26$$

From Fig. 13.10, $(U_d + U_c)_f / u_o = 0.29$.
Assume

$$\frac{u_o \mu_c \rho_c}{\sigma \Delta\rho} = 0.01$$

Therefore

$$u_o = \frac{(0.01)(0.00219)(0.14)}{(0.000021)(1.0)} = 0.146 \text{ ft/sec}$$

$$(U_d + U_c)_f = 0.29(0.146) = 0.0423 \text{ ft/sec}$$

$$(U_d + U_c)_{50\% \text{ of flooding}} = \left(\frac{0.0423}{2}\right)(3600) = 76.1 \text{ ft/hr}$$

$$\text{Total ft}^3/\text{hr} = \frac{27,000}{(0.86)(62.4)} + \frac{25,000}{(1.0)(62.4)} = 904 \text{ ft}^3/\text{hr}$$

$$\text{Column cross-sectional area} = A = \frac{904}{76.1} = 11.88 \text{ ft}^2$$

$$\text{Column diameter} = D_T = \left(\frac{4A}{\pi}\right)^{0.5} = \left[\frac{(4)(11.88)}{3.14}\right]^{0.5} = 3.9 \text{ ft}$$

☐

Efficiency

Despite their compartmentalization, mechanically assisted liquid–liquid extraction columns, such as the RDC and RPC, operate more nearly like differential contacting devices than like staged contactors. Therefore, it is more common to consider stage efficiency for such columns in terms of HETS (height equivalent to a theoretical stage) or as some function of mass transfer parameters, such as HTU (height of a transfer unit). Although not on as sound a theoretical basis as the HTU, the HETS is preferred here because it can be applied directly to determine column height from the number of equilibrium stages.

Unfortunately, because of the great complexity of liquid–liquid systems and the large number of variables that influence contacting efficiency, general correlations for HETS have not been developed. However, for well-designed and efficiently operated columns, the available experimental data indicate that the dominant physical properties influencing HETS are the interfacial tension, the phase viscosities, and the density difference between the phases. In addition, it has been observed by Reman[24] for RDC units and by Karr and Lo[25] for RPC columns that HETS decreases with increasing column diameter because of axial mixing effects.

It is preferred to obtain values of HETS by conducting small-scale laboratory experiments with systems of interest. These values are scaled to commercial-size columns by assuming that HETS varies inversely with column diameter D, raised to an exponent, which may vary from 0.2 to 0.4 depending on the system.

In the absence of experimental data, the crude correlation of Fig. 13.11 can be used for preliminary design if phase viscosities are no greater than 1 cp. The data points correspond to essentially minimum reported HETS values for RDC and RPC units with the exponent on column diameter set arbitrarily at one third.

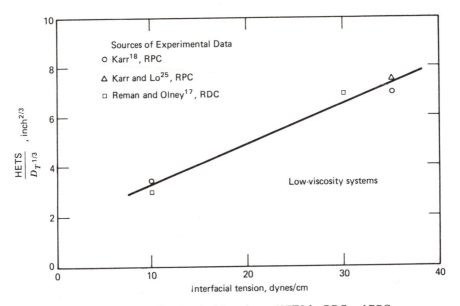

Figure 13.11. Effect of interfacial tension on HETS for RDC and RPC.

The points represent values of HETS that vary from as low as 6 in. for a 3-in. diameter laboratory-size column operating with a low interfacial tension–low viscosity system such as methylisobutylketone–acetic acid–water, to as high as 25 in. for a 36-in. diameter commercial column operating with a high interfacial tension–low viscosity system such as xylene–acetic acid–water. For systems having one phase of high viscosity, values of HETS can be 24 in. or more even for a small laboratory-size column.

Example 13.5. Estimate HETS for the conditions of Example 13.4.

Solution. Because toluene has a viscosity of approximately 0.6 cp, this is a low-viscosity system. From Example 13.4, the interfacial tension is 32 dynes/cm. From Fig. 13.11

$$HETS/D^{1/3} = 6.9$$

For $D = 3.9\,ft$

$$HETS = 6.9[(3.9)(12)]^{1/3} = 24.8\ in.$$

□

References

1. Souders, M., and G. G. Brown, *Ind. Eng. Chem.*, **26**, 98–103 (1934).

2. Fair, J. R., *Petro/Chem Eng.*, **33**, 211–218 (September, 1961).

3. Sherwood, T. K., G. H. Shipley, and F. A. L. Holloway, *Ind. Eng. Chem.*, **30**, 765–769 (1938).

4. Glitsch Ballast Tray, Bulletin No. 159, Fritz W. Glitsch and Sons, Inc., Dallas, Texas (from FRI report of September 3, 1958).

5. Glitsch V-1 Ballast Tray, Bulletin No. 160, Fritz W. Glitsch and Sons, Inc., Dallas, Texas (from FRI report of September 25, 1959).

6. Van Winkle, M., *Distillation*, McGraw–Hill Book Co., New York, 1967, 572–573.

7. Oliver, E. D., *Diffusional Separation Processes: Theory, Design, and Evaluation*, John Wiley & Sons, Inc., New York, 1966, 320–321.

8. Bubble-Tray Design Manual, Prediction of Fractionation Efficiency, AIChE, New York, 1958.

9. Drickamer, H. G., and J. R. Bradford, *Trans. AIChE*, **39**, 319–360 (1943).

10. O'Connell, H. E., *Trans. AIChE*, **42**, 741–755 (1946).

11. Lockhart, F. J., and C. W. Leggett in *Advances in Petroleum Chemistry and Refining*, Vol. *1*, ed. by K. A. Kobe and John J. McKetta, Jr., Interscience Publishers, Inc., New York, 1958, 323–326.

12. Williams, G. C., E. K. Stigger, and J. H. Nichols, *Chem. Eng. Progr.*, **46** (1), 7–16 (1950).

13. Zuiderweg, F. J., H. Verburg, and F. A. H. Gilissen, *Proc. International Symposium on Distillation*, Inst.

Chem. Eng., London, 202–207 (1960).

14. Gautreaux, M. F., and H. E. O'Connell, *Chem. Eng. Progr.*, **51** (5), 232–237 (1955).

15. Koch Flexitray Design Manual, Bulletin 960, Koch Engineering Co., Inc., Wichita, Kansas, 1960.

16. Younger, A. H., *Chem. Eng.*, **62** (5), 201–202 (1955).

17. Reman, G. H., and R. B. Olney, *Chem. Eng. Progr.*, **52** (3), 141–146 (1955).

18. Karr, A. E., *AIChE J.*, **5**, 446–452 (1959).

19. Logsdail, D. H., J. D. Thornton, and H. R. C. Pratt, *Trans. Inst. Chem. Eng.*, **35**, 301–315 (1957).

20. Landau, J., and R. Houlihan, *Can. J. Chem. Eng.*, **52**, 338–344 (1974).

21. Gayler, R., N. W. Roberts, and H. R. C. Pratt, *Trans. Inst. Chem. Eng.*, **31**, 57–68 (1953).

22. Thornton, J. D., *Chem. Eng. Sci.*, **5**, 201–208 (1956).

23. Strand, C. P., R. B. Olney, and G. H. Ackerman, *AIChE J.*, **8**, 252–261 (1962).

24. Reman, G. H., *Chem. Eng. Progr.*, **62** (9), 56–61 (1966).

25. Karr, A. E., and T. C. Lo, "Performance of a 36″ Diameter Reciprocating-Plate Extraction Column," paper presented at the 82nd National Meeting of AIChE, Atlantic City (Aug. 29–Sept. 1, 1976).

Problems

13.1 Conditions for the top tray of a distillation column are as shown below. Determine the column diameter corresponding to 85% of flooding if a valve tray is used. Make whatever assumptions are necessary.

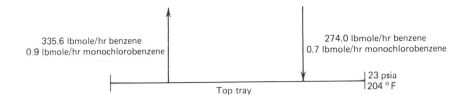

335.6 lbmole/hr benzene
0.9 lbmole/hr monochlorobenzene

274.0 lbmole/hr benzene
0.7 lbmole/hr monochlorobenzene

23 psia
204 °F

Top tray

13.2 For the bottom tray of a reboiled stripper, conditions are as shown below. Estimate the column diameter corresponding to 85% of flooding for a valve tray.

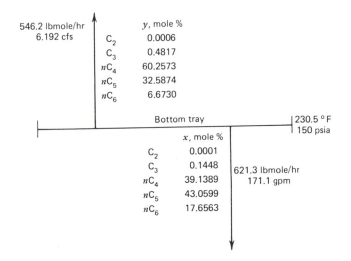

546.2 lbmole/hr
6.192 cfs

y, mole %

C_2	0.0006
C_3	0.4817
nC_4	60.2573
nC_5	32.5874
nC_6	6.6730

Bottom tray

230.5 °F
150 psia

x, mole %

C_2	0.0001
C_3	0.1448
nC_4	39.1389
nC_5	43.0599
nC_6	17.6563

621.3 lbmole/hr
171.1 gpm

13.3 Determine the column diameter, tray efficiency, number of actual trays, and column height for the distillation column of Example 12.11 or 15.2 if perforated trays are used.

13.4 Determine the column diameter, tray efficiency, number of actual trays, and column height for the absorber of Example 12.8 or 15.4 if valve trays are used.

13.5 A separation of propylene from propane is achieved by distillation as shown below, where two columns in series are used because a single column would be too tall. The tray numbers refer to equilibrium stages. Determine the column diameters, tray efficiency, number of actual trays, and column heights if perforated trays are used.

13.6 Determine the height and diameter of a vertical flash drum for the conditions shown below.

13.7 Determine the length and diameter of a horizontal reflux drum for the conditions shown below.

13.8 Results of design calculation for a methanol–water distillation operation are given below.

Problem 13.5

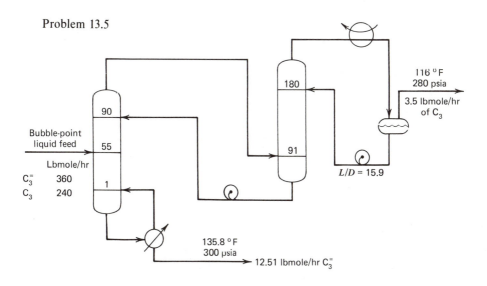

Bubble-point
liquid feed

Lbmole/hr

$C_3^=$ 360
C_3 240

90

55

1

180

91

$L/D = 15.9$

116 °F
280 psia

3.5 lbmole/hr
of C_3

135.8 °F
300 psia
→ 12.51 lbmole/hr $C_3^=$

Problem 13.6

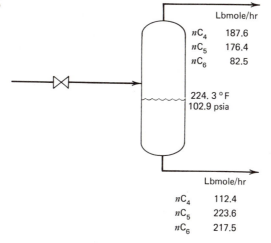

Lbmole/hr

nC_4 187.6
nC_5 176.4
nC_6 82.5

224. 3 °F
102.9 psia

Lbmole/hr

nC_4 112.4
nC_5 223.6
nC_6 217.5

Problem 13.7

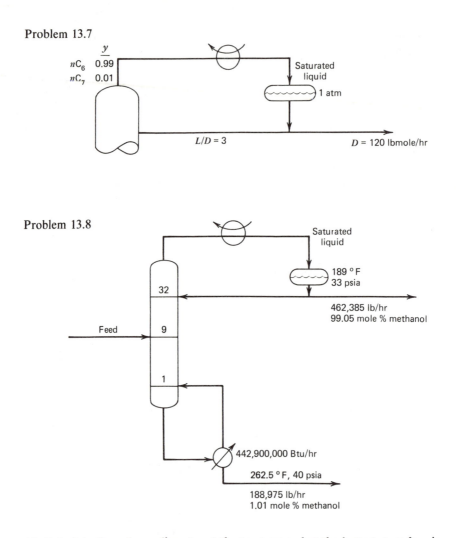

Problem 13.8

(a) Calculate the column diameter at the top tray and at the bottom tray for sieve trays. Should the column be swedged?
(b) Calculate the length and diameter of the horizontal reflux drum.

13.9 Estimate the diameter and height of an RDC to carry out the liquid–liquid extraction operations of:
(a) Example 12.10.
(b) Example 15.5.

14

Synthesis of Separation Sequences

A new general heuristic which should be of value to the design engineer, whether or not he is using the computer, is that the next separator to be incorporated into the separator sequence at any point is the one that is cheapest.

Leaving the most difficult separations, both in terms of separation factor and in terms of split fraction, until all nonkeys have been removed is a natural consequence of cheapest first.

Roger W. Thompson and
C. Judson King, 1972

Previous chapters have dealt mainly with the design of simple separators that produce two products. However, as discussed in Chapter 1, industrial separation problems generally involve the separation of multicomponent mixtures into more than two products. Although one separator of complex design often can be devised to produce all the desired products, a sequence of simple separators is more commonly used because it is frequently more economical than one complex separator.

A sequence may be simple as in Fig. 14.1 or complex as in Fig. 14.2. It is simple if each separator performs a relatively sharp split between two key components and if neither products nor energy is recycled between separators. In this chapter, methods for the synthesis of simple sequences containing simple separators are presented.

A cost that is a combination of capital and operating expenses can be computed for each separator in a sequence. The sequence cost is the sum of the separator costs. In general, one seeks the optimal or lowest-cost sequence; and, perhaps, several near-optimal sequences. However, other factors such as operability, reliability, and safety must be deliberated before a final sequence is selected.

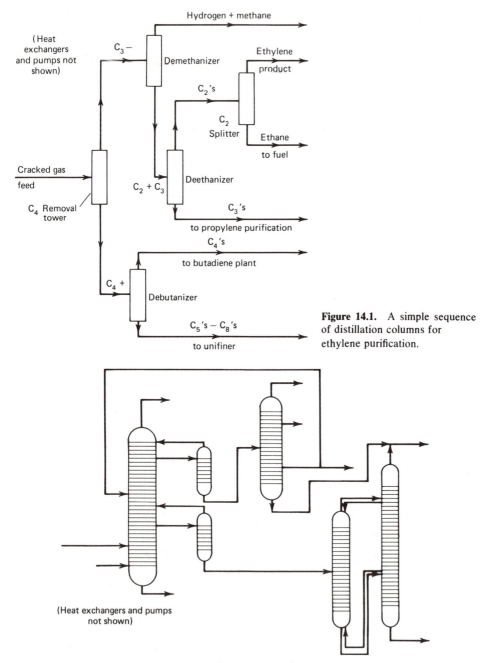

Figure 14.1. A simple sequence of distillation columns for ethylene purification.

Figure 14.2. A complex sequence of separators.

14.1   The Combinatorial Problem and Forbidden Splits

The creation of even a simple sequence involves decisions concerning separation methods and arrangement of separators. Suppose the sequence is restricted to single-feed, two-product separators that all employ the same method of separation. Energy-separating agents are used; no mass-separating agents are employed. Such sequences are commonly observed in industrial plants, where ordinary distillation is the only separation method. Let the process feed contain R components arranged in the order of decreasing relative volatility. Assume that the order of relative volatility remains constant as the process feed is separated into R products, which are nearly pure.

If the process feed is comprised of A, B, and C, two sequences exist of two separators each, as shown in Fig. 14.3. If the separation point is between A and B in the first separator, sequence Fig. 14.3a is obtained. Sequence Fig. 14.3b results if the split is between B and C in the first separator. A split between A

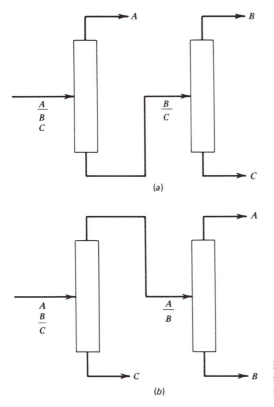

(a)

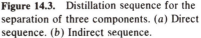

(b)

Figure 14.3.   Distillation sequence for the separation of three components. (a) Direct sequence. (b) Indirect sequence.

and C with B present is not permitted in simple sequences containing simple separators.

A recursion formula for the number of sequences S corresponding to the separation of a mixture of R components into R products can be developed in the following manner.[1] For the first separator in the sequence, $(R-1)$ separation points are possible. Let j be the number of components appearing in the overhead product; then $(R-j)$ equals the number of components appearing in the bottoms product. If S_i is the number of possible sequences for i components, then, for a given split in the first separator, the number of sequences is the product $S_j S_{R-j}$. But in the first separator, $(R-1)$ different splits are possible. Therefore, the total number of sequences for R components is the sum

$$S_R = \sum_{j=1}^{R-1} S_j S_{R-j} = \frac{[2(R-1)]!}{R!(R-1)!} \qquad (14\text{-}1)$$

For R equal to two, only one sequence comprised of a single separator is possible. Thus, from (14-1), $S_2 = S_1 S_1 = 1$ and $S_1 = 1$. Similarly, for $R = 3$, $S_3 = S_1 S_2 + S_2 S_1 = 2$, as shown in Fig. 14.3. Values for R up to 11, as given in Table 14.1, are derived in a progressive manner. The five sequences for a four-component feed are shown in Fig. 14.4.

The separation of a multicomponent feed produces subgroups or streams of adjacent, ordered components that are either separator feeds or final products. For example, a four-component process feed comprised of four ordered components can produce the 10 different subgroups listed in Table 14.2 from the five different sequences in Fig. 14.4. In general, the total number of different subgroups G, including the process feed, is simply the arithmetic progression

Table 14.1   Number of separators, sequences, subgroups, and unique splits for simple sequences using one simple method of separation

R, Number of Components	Number of Separators in a Sequence	S, Number of Sequences	G, Number of Subgroups	U, Number of Unique Splits
2	1	1	3	1
3	2	2	6	4
4	3	5	10	10
5	4	14	15	20
6	5	42	21	35
7	6	132	28	56
8	7	429	36	84
9	8	1,430	45	120
10	9	4,862	55	165
11	10	16,796	66	220

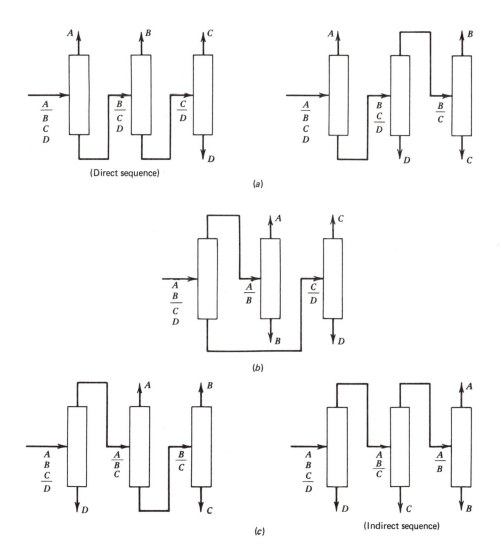

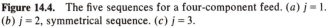

Figure 14.4. The five sequences for a four-component feed. (a) $j = 1$.
(b) $j = 2$, symmetrical sequence. (c) $j = 3$.

$$G = \sum_{j=1}^{R} j = \frac{R(R+1)}{2} \tag{14-2}$$

For process feeds containing more than three components, some splits will
be common to two or more sequences. For a four-component process feed, 3
separators are needed for each of five sequences. Thus, the total number of
separators for all sequences is 15. However, as shown in Fig. 14.4 and Table

Table 14.2 Subgroups for a four-component process feed

Process Feed First Separator	Feeds to Subsequent Separators		Products
$\begin{pmatrix} A \\ B \\ C \\ D \end{pmatrix}$	$\begin{pmatrix} A \\ B \\ C \end{pmatrix}$	$\begin{pmatrix} A \\ B \end{pmatrix}$	(A)
	$\begin{pmatrix} B \\ C \\ D \end{pmatrix}$	$\begin{pmatrix} B \\ C \end{pmatrix}$	(B)
		$\begin{pmatrix} C \\ D \end{pmatrix}$	(C)
			(D)

14.3, only 10 of these correspond to unique splits U given by the relation

$$U = \sum_{j=1}^{R-1} j(R-j) = \frac{R(R-1)(R+1)}{6} \tag{14-3}$$

Values of G and U for R up to 11 are included in Table 14.1, from which it is apparent that, as R increases, S, G, and U also increase. However, in the limit as $R \to \infty$, $G\{R+1\}/G\{R\} \to 1$, $U\{R+1\}/U\{R\} \to 1$, but $S\{R+1\}/S\{R\} \to 4$. While it

Table 14.3 Unique splits for a four-component process feed

Splits for First Separator	Splits for Subsequent Separators	
$\begin{pmatrix} \frac{A}{B} \\ C \\ D \end{pmatrix}$	$\begin{pmatrix} \frac{A}{B} \\ C \end{pmatrix}$	$\begin{pmatrix} \frac{A}{B} \end{pmatrix}$
$\begin{pmatrix} A \\ \frac{B}{C} \\ D \end{pmatrix}$	$\begin{pmatrix} A \\ \frac{B}{C} \end{pmatrix}$	$\begin{pmatrix} \frac{B}{C} \end{pmatrix}$
$\begin{pmatrix} A \\ B \\ \frac{C}{D} \end{pmatrix}$	$\begin{pmatrix} \frac{B}{C} \\ D \end{pmatrix}$	$\begin{pmatrix} \frac{C}{D} \end{pmatrix}$
	$\begin{pmatrix} B \\ \frac{C}{D} \end{pmatrix}$	

may be feasible to design all separators and examine all sequences for small values of R, it is prohibitive to do so for large values of R.

When a process feed comprised of R components is separated into P products, where $R > P$, (14-1), (14-2), and (14-3) still apply with R replaced by P if multicomponent products are not produced by blending and consist of adjacent, ordered components. For example, in Fig. 1.18, where one multicomponent product (iC_5, nC_5) is produced, $R = 5$, $P = 4$, and $S = 5$.

Example 14.1. Ordinary distillation is to be used to separate the ordered mixture C_2, $C_3^=$, C_3, $1-C_4^=$, nC_4 into the three products C_2; ($C_3^=$, $1-C_4^=$); (C_3, nC_4). Determine the number of possible sequences.

Solution. Neither multicomponent product contains adjacent components in the ordered list. Therefore, the mixture must be completely separated with subsequent blending to produce the ($C_3^=$, $1-C_4^=$) and C_3, nC_4) products. Thus, from Table 14.1 with R taken as 5, $S = 14$.
□

The combinatorial problems of (14-1), (14-2), and (14-3) summarized in Table 14.1 are magnified greatly when more than one separation method is considered. For computing the number of possible sequences, Thompson and King[1] give the following equation, which is, however, restricted to sequences like that shown in Fig. 1.20, where a mass separating agent (MSA) is recovered for recycle in the separator following the separator into which it is introduced. In effect, these two separators are considered to be one separation problem and the mass separating agent is not counted as a component. Thus, extending (14-1),

$$S = T^{R-1} S_R \qquad (14\text{-}4)$$

where T is the number of different separation methods to be considered. For example, if $R = 4$ and ordinary distillation, extractive distillation with phenol, extractive distillation with aniline, and liquid–liquid extraction with methanol are the separation methods to be considered, then from (14-1) or Table 14.1, and (14-4), $S = 4^{4-1}(5) = 64(5) = 320$. The number of possible sequences is 64 times that when only ordinary distillation is considered. Implicit in this calculation is the assumption that the production of multicomponent products by blending is prohibited. This restriction is violated in the sequence shown in Fig. 1.20, where mixed butenes are produced by blending 1-butene with 2-butenes. When blending is permitted and the MSA need not be recovered in the separator directly following the separator into which it is introduced, the number of possible sequences is further increased. However, if any separators are forbidden for obvious technical or economic reasons, the number of sequences is decreased.

In order to reduce the magnitude of the combinatorial problem, it is desirable to make a preliminary screening of separation methods based on an examination of various factors. While not rigorous, the graphical method of

Souders[2] is simple and convenient. It begins with an examination of the technical feasibility of ordinary distillation, which, in principle, is applicable over the entire region of coexisting vapor and liquid phases. This region extends from the crystallization temperature to the convergence pressure, provided that species are thermally stable at the conditions employed. The column operating pressure is determined by the method discussed in Section 12.1 and illustrated in Fig. 12.4. If refrigeration is required for the overhead condenser, alternatives to ordinary distillation, such as absorption and reboiled absorption, might be considered. At the other extreme, if vacuum operation of ordinary distillation is indicated, liquid–liquid extraction with various solvents might be considered. Between these extremes, ordinary distillation is generally not feasible economically when the relative volatility between key components is less than approximately 1.05. Even when this separation index is exceeded, extractive distillation and liquid–liquid extraction may be attractive alternatives provided that relative separation indices (α and β, respectively) for these methods lie above the respective curves plotted in Fig. 14.5 or if an order of volatility (or other separation index) for these alternative methods is achieved that permits production of multicomponent products without blending. To develop these curves, Souders assumed a solvent concentration of 67 mole% and a liquid rate four times that used in ordinary distillation. In general, extractive distillation need not be considered when the relative volatility for ordinary distillation is greater than about two.

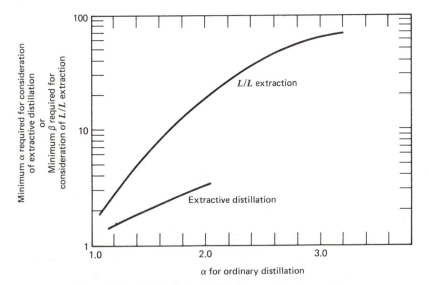

Figure 14.5.   Relative selectivities for equal-cost separators.

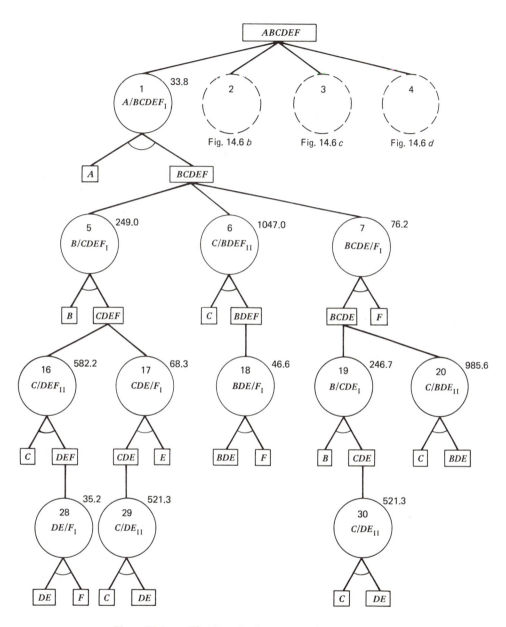

Figure 14.6. *a.* First branch of sequences for Example 14.2.

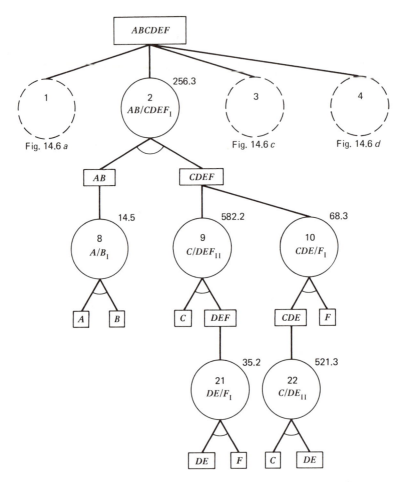

Figure 14.6. *b.* Second branch of sequences for Example 14.2.

Example 14.2. Consider the separation problem studied by Hendry and Hughes[3] as shown in Fig. 1.16. Determine the feasibility of two-product, ordinary distillation (method I), select an alternative separation technique (method II), if necessary, and forbid impractical splits. Determine the feasible separation subproblems and the number of possible separation sequences that incorporate only these subproblems.

Solution. The normal boiling points of the six species are listed in Table 1.5. Because both *trans-* and *cis-*butene-2 are contained in the butenes product and are adjacent when species are ordered by relative volatility, they need not be separated. Using the method of Section 12.1, we find that all ordinary distillation columns could be operated above atmospheric pressure and with cooling-water condensers. Approximate relative volatilities assuming ideal solutions at 150°F (65.6°C) are as follows for all

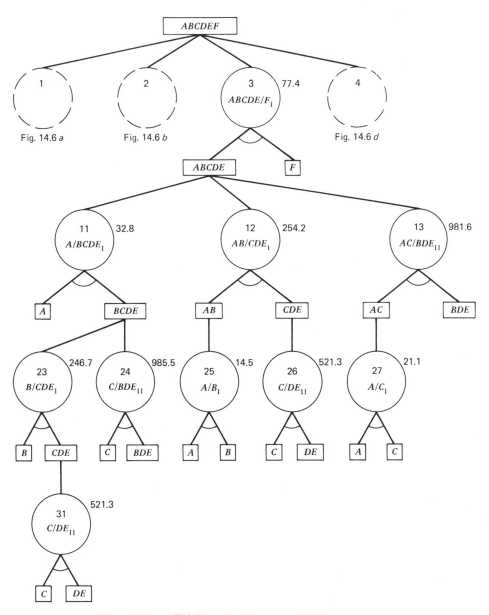

Figure 14.6. *c.* Third branch of sequences for Example 14.2.

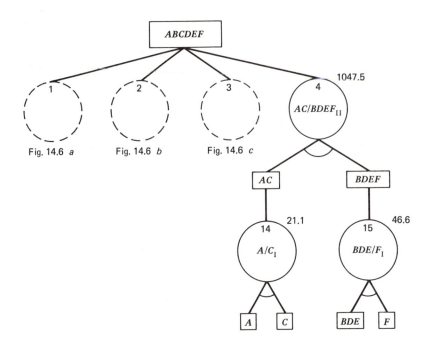

Figure 14.6. *d.* Fourth branch of sequences for Example 14.2. (Data for Fig. 14.6 from J. E. Hendry, Ph.D. thesis in chemical engineering, University of Wisconsin, Madison, 1972.)

adjacent binary pairs except *trans*-butene-2 and *cis*-butene-2, which need not be separated because they are contained in the butenes final product.

Adjacent Binary Pair	*Approximate Relative Volatility at 150°F (65.6°C)*
Propane/1-butene (*A/B*)	2.45
1-Butene/*n*-butane (*B/C*)	1.18
n-Butane/*trans*-butene-2 (*C/D*)	1.03
cis-Butene-2/*n*-pentane (*E/F*)	2.50

Because of their high relative volatilities, splits *A/B* and *E/F*, even in the presence of the other hydrocarbons, should be by ordinary distillation only. Split *C/D* is considered infeasible by ordinary distillation; split *B/C* is feasible, but an alternative method might be more attractive.

According to Buell and Boatright,[4] the use of approximately 96 wt% aqueous furfural as a solvent for extractive distillation increases the volatility of paraffins relative to olefins, causing a reversal in volatility between 1-butene and *n*-butane and giving the separation order (*A C B D E F*). Thus, the three olefins, which are specified as one product, are grouped together. They give an approximate relative volatility of 1.17 for split (*C/B*). In the presence of *A*, this would add the additional split (*A/C*), which by ordinary distillation has a very desirable relative volatility of 2.89. Also, split

$(\ldots C/D \ldots)_{\text{II}}$, with a relative volatility of 1.70, is more attractive than split $(\ldots C/D \ldots)_{\text{I}}$ according to Fig. 14.5.

In summary, the splits to be considered, with all others forbidden, are $A/B \ldots)_{\text{I}}$, $(\ldots B/C \ldots)_{\text{I}}$, $(\ldots E/F)_{\text{I}}$, $(A/C)_{\text{I}}$, $(\ldots C/B \ldots)_{\text{II}}$, and $(\ldots C/D \ldots)_{\text{II}}$. All feasible subgroups, separations, and sequences can be generated by developing an and/or-directed graph,[5,6] as shown for this example in Fig. 14.6, where rectangular nodes designate subgroups and numbered, circular nodes represent separations. A separation includes the separator for recovering the MSA when separation method II is used. A separation sequence is developed by starting at the process feed node and making path decisions until all products are produced. From each rectangular (*or*) node, one path is taken; if present, both paths from a circular (*and*) node must be taken. Figure 14.6 is divided into four main branches, one for each of the four possible separations that come first in the various sequences. Figure 14.6 contains the 31 separations and 12 sequences listed by separation number in Table 14.4.

☐

The decimal numbers in Fig. 14.6 are annual costs for the separations in thousands of dollars per year as derived from data given by Hendry.[7] The sequence costs are listed in Table 14.4. The lowest-cost sequence is illustrated in Fig. 1.20; the highest-cost sequence is 31% greater in cost than the lowest. If only the $(A/C \ldots)_{\text{II}}$ and $(\ldots E/F)_{\text{II}}$ splits are prohibited as by Hendry and Hughes,[3] the consequence is 64 unique separation subproblems and 227 sequences. However, every one of the additional 215 sequences is more than 350% greater in cost than the lowest-cost sequence.

Table 14.4   Sequences for Example 14.2

Sequence	Cost, $/yr.
1–5–16–28	900,200
1–5–17–29	872,400
1–6–18	1,127,400
1–7–19–30	878,000
1–7–20	1,095,600
2 ⟨ 8 / 9–21	888,200
2 ⟨ 8 / 10–22	860,400
3–11–23–31	878,200
3–11–24	1,095,700
3–12 ⟨ 25 / 26	867,400
3–13–27	1,080,100
4–14–15	1,115,200

14.2 Heuristic and Evolutionary Synthesis Techniques

Heuristic methods, which seek the solution to a problem by means of plausible but fallible rules, are used widely to overcome the need to examine all possible sequences in order to find the optimal and near-optimal arrangements. By the use of heuristics, good sequences can be determined quickly, even when large numbers of components are to be separated, without designing or costing equipment. A number of heuristics have been proposed for simple sequences comprised of ordinary distillation columns.[8,9,10] The most useful of these heuristics can be selected, as described by Seader and Westerberg,[11] in the following manner to develop sequences, starting with the process feed.

1. When the adjacent ordered components in the process feed vary widely in relative volatility, sequence the splits in the order of decreasing relative volatility.

2. Sequence the splits to remove components in the order of decreasing molar percentage in the process feed when that percentage varies widely but relative volatility does not vary widely.

3. When neither relative volatility nor molar percentage in the feed varies widely, remove the components one by one as overhead products. This is the direct sequence, shown in Fig. 14.3a and 14.4a.

These three heuristics are consistent with several observations concerned with the effect of the presence of nonkey components on the cost of splitting two key components. The observations are derived from the approximate design methods described in Chapter 12 by assuming constant molal overflow and constant relative volatility for each pair of adjacent components. The cost of ordinary distillation depends mainly on the required number of equilibrium stages and the required boilup rate, which influences column diameter and reboiler duty. Equation (12-12) shows that the minimum stage requirement for a given key-component split is independent of the presence of nonkey components. However, the equations of Underwood for minimum reflux can be used to show that the minimum boilup rate, while not influenced strongly by the extent of separation of the key components, may be increased markedly when amounts of nonkey components are present in the separator feed. Then, from the graphical relationship of Gilliland shown in Fig. 12.11, for a given ratio of actual-to-minimum-theoretical stages, the actual boilup ratio is greater in the presence of nonkey components than in their absence.

Example 14.3. Consider the separation problem shown in Fig. 1.18, except that separate isopentane and *n*-pentane products are also to be obtained with 98% recoveries. Use heuristics to determine a good sequence of ordinary distillation units.

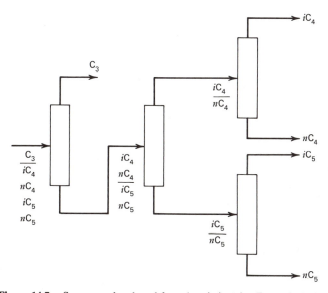

Figure 14.7. Sequence developed from heuristic 1 for Example 14.3.

Solution. Approximate relative volatilities for all adjacent pairs except iC_5/nC_5 are given in Table 1.6. The latter pair, with a normal boiling-point difference of 8.3°C, has an approximate relative volatility of 1.35 from Fig. 1.17. For this example, we have wide variations in both relative volatility and molar percentages in the process feed. The choice is heuristic 1, which dominates over heuristic 2 and leads to the sequence shown in Fig. 14.7, where the first split is between the pair with the highest relative volatility. This sequence also corresponds to the optimal arrangement.

□

When separation methods other than ordinary distillation (particularly those employing mass separating agents) are considered, two additional heuristics[3,12] are useful.

4. When an MSA is used, remove it in the separator following the one into which it is introduced.

5. When multicomponent products are specified, favor sequences that produce these products directly or with a minimum of blending unless relative volatilities are appreciably lower than for a sequence that requires additional separators and blending.

Example 14.4. Use heuristics to develop a good sequence for the problem considered in Example 14.2.

Solution. From Example 14.2, splits $(A/B\ldots)_1$ and $(\ldots E/F)_1$ have the largest values of relative volatility. Because these values are almost identical, either split can be placed first in the sequence. We will choose the former, noting that the small amounts of C_3 and nC_5 in the feed render this decision of minor importance. Following removal of propane A and n-pentane F by ordinary distillation, we can apply heuristic 5 by conducting the extractive distillation separation $(C/BDE)_{II}$, followed by a separator to recover the MSA. The resulting sequence produces the multicomponent product BDE directly; however, the relative volatility of $(C/BDE)_{II}$ is only 1.17. Alternatively, the split $(C/DE)_{II}$, with a much higher relative volatility of 1.70, preceded by the split $(B/CDE)_I$ to remove B and followed by removal of the MSA can be employed, with the product BDE formed by blending of B with DE. Because of the very strong effect of relative volatility on cost, the alternative sequence, shown in Fig. 14.8, may be preferable despite the need for one additional ordinary distillation separation with a low relative volatility. While this sequence is not optimal, it is near-optimal at a cost, according to Fig. 14.6a and Table 14.4, of \$878,000/yr compared to \$860,400/yr for the optimal sequence in Fig. 1.20. If the first two separations in Fig. 14.8 are interchanged, the cost of the new sequence is increased to \$878,200/yr.

☐

Once a good separation sequence has been developed from heuristics, improvements can be attempted by evolutionary synthesis as discussed by Seader and Westerberg.[11] This method involves moving from a starting sequence to better and better sequences by a succession of small modifications. Each new sequence must be costed to determine if it is better. The following evolutionary rules of Stephanopoulos and Westerberg[13] provide a systematic approach.

1. Interchange the relative positions of two adjacent separators.

2. For a given separation using separation method I, substitute separation method II.

Often evolutionary synthesis can lead to the optimal sequence without

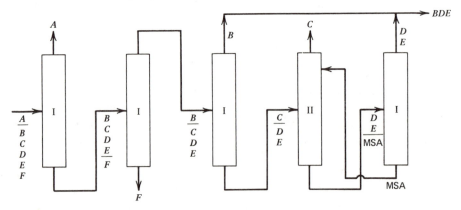

Figure 14.8. Sequence developed by heuristics for Example 14.4.

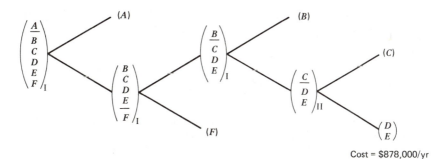

Cost = $878,000/yr

(a)

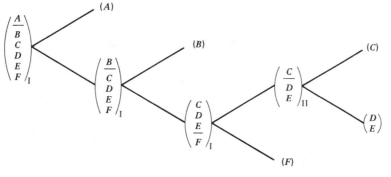

Cost = $872,400/yr

(b)

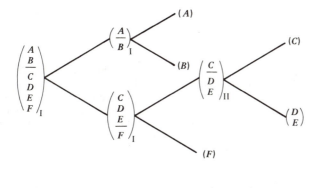

Cost = $860,400/yr

(c)

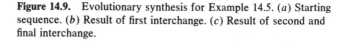

Figure 14.9. Evolutionary synthesis for Example 14.5. (*a*) Starting sequence. (*b*) Result of first interchange. (*c*) Result of second and final interchange.

requiring examination of all sequences. The efficiency depends on the strategy employed to direct the modifications. The heuristics already cited are useful in this regard.

Example 14.5. Use evolutionary synthesis to seek an improvement to the sequence that was developed from heuristics in Example 14.4.

Solution. The starting sequence in Fig. 14.8 can be represented more conveniently by the binary tree in Fig. 14.9a. Evolutionary rule 2 cannot be applied if the forbidden splits of Example 14.4 are still prohibited. If evolutionary rule 1 is applied, three interchanges are possible.

(a) $(A/B...)_I$ with $(...E/F)_I$.

(b) $(...E/F)_I$ with $(B/C...)_I$.

(c) $(C/D...)_{II}$ with $(B/C...)_I$.

Interchange (a) is not likely to cause any appreciable change in sequence cost when heuristics 1 and 2 are considered. The relative volatilities of the key components for the two splits are almost identical and neither of the two compounds removed (A and F) are present in appreciable amounts. Therefore, do not make this interchange.

Interchange (b) may have merit. From Fig. 1.16, the process feed contains a larger amount of 1-butene B than n-pentane F. According to heuristic 2, components present in larger quantities should be removed early in the sequence. However, heuristic 2 was ignored and heuristic 1 was applied in developing the sequence in Example 14.4. By making this interchange, we can remove B earlier in the sequence.

Interchange (c) will result in the sequence that produces BDE directly. But, as discussed in Example 14.4, this sequence is likely to be more costly because of the small relative volatility for the extractive distillation separation $(C/BDE)_{II}$. Therefore, we make interchange (b) only, with the result shown in Fig. 14.9b at a lower cost of $872,400/yr.

If we apply evolutionary rule 1 to the sequence in Figure 14.9b, the possible new interchanges are:

(d) $(A/B...)_I$ with $(...B/C...)_I$.

(e) $(...E/F)_I$ with $(C/D...)_{II}$.

Interchange (d) opposes heuristic 1 and supports heuristic 2. The latter heuristic may dominate here because of the very small amount of propane A, and larger amount of 1-butene B in the process feed. Interchange (e) would appear to offer less promise than interchange (d) because of the difficulty of the $(C/D...)_{II}$ split. The result of making interchange (d) is shown in Fig. 14.9c at a cost of $860,400/yr. No further interchanges are possible; the final sequence happens to be the optimum.
□

14.3 Algorithmic Synthesis Techniques

Neither heuristic nor evolutionary synthesis procedures are guaranteed to generate the optimal sequence. They are useful in preliminary design because, generally, good sequences are developed. For final design, it may be desirable to generate optimal and near-optimal sequences from which a final selection can be made. The true optimum for a simple sequence can be determined only if optimization is conducted with respect to separator design variables as well as arrangement of separators in the manner of Hendry and Hughes[3] or Westerberg and Stephanopoulos.[14]

Design variables for ordinary distillation include pressure, reflux rate and degree of subcooling, extent of feed preheating or precooling, feed-stage location, tray selection, choice of utilities, type of condenser operation, extent of approach to flooding, and approach temperatures in exchangers. For other separators, additional design variables include for the MSA: inlet flow rate, extent of preheating or precooling, and inlet stage location. Unless separation cost is dominated by utility cost, as shown in studies by Heaven,[15] Hendry,[7] King,[16] and Tedder,[17] the minimization of cost is not particularly sensitive to these design variables over reasonable ranges of their values. Then, the ordered branch search procedure of Rodrigo and Seader[6] can be employed to find the optimal and near-optimal sequences, frequently without examining all sequences or designing all separators. The procedure is particularly efficient for large combinatorial problems if impractical splits are forbidden, and if the assumption can be made that the presence of relatively small amounts of nonkey components has only a slight effect on the cost. In that case, only the key components are allowed to distribute between the overhead and bottoms with the result that this assumption, suggested by Hendry and Hughes,[3] results in duplicate and even multiplicate separations. For example, Fig. 14.6 includes the following duplicate separations—8, 25; 9, 16; 10, 17; 14, 27; 15, 18; 19, 23; 20, 24; 21, 28—and one multiplicate separation—22, 26, 29, 30, 31. Thus, of 31 separations, only 19 are unique and need be designed. Separator costs in Fig. 14.6 are based on the assumption of duplicate and multiplicate separators.

The ordered branch search technique consists of two steps, best explained by reference to an *and/or* graph such as Fig. 14.6. The search begins from the process feed by branching to and costing all alternative first-generation separations. The heuristic of Thompson and King,[12] which selects the separation having the lowest cost ("cheapest first"), is used to determine which branch to take. The corresponding separation is made to produce the two subgroups from which the branching and selection procedure is continued until subgroups corresponding to all products appear. However, in the case of multicomponent products, it is permissible to form them by blending. The total cost of the initial sequence developed in this manner is referred to as the *initial upper bound*, and is obtained by accumulating costs of the separations as they are selected. This "cheapest-first" sequence frequently is the optimal or one of the near-optimal separation schemes.

The second step involves backtracking and then branching to seek lower cost sequences. When one is found, it becomes the new upper bound. In this phase, branching does not continue necessarily until a sequence is completed. Branching is discontinued and backtracking resumed until the cost of a partially completed sequence exceeds the upper bound. Thus, all sequences need not be developed and not all unique separators need be designed and costed.

The backtracking procedure begins at the last separator of a completed or partially completed sequence by a backward move to the previous subgroup. If alternative separators are generated from that subgroup, they become parts of alternative sequences and are considered for branching. When all these alternative sequences have been completed or partially completed by comparing their costs to the latest upper bound, backtracking is extended backward another step. The backtracking and branching procedure is repeated until no further sequences remain to be developed.

Near-optimal sequences having costs within a specified percentage of the optimal sequence can be developed simultaneously in phase two by delaying the switch from branching to backtracking until the cost of a partially completed sequence exceeds the product of the upper bound and a specified factor (say, 1.10).

Example 14.6. Use the ordered branch search method to determine the optimal sequence for the problem of Example 14.2.

Solution. Using the information in Fig. 14.6, we develop the initial sequence in step one as shown in Fig. 14.10, where the numbers refer to separator numbers in Fig. 14.6 and an asterisk beside a number designates the lowest-cost separator for a given branching step. Thus, the "cheapest-first" sequence is 1–7–19–30 with a cost of $878,000/yr. This sequence, which is shown in Fig. 14.8, is identical to that developed from heuristics in Example 14.4.

Step two begins with backtracking in two steps from separator 30 to subgroup *BCDE*, which is the bottoms product from separator 7. Branching from this subgroup to separator 20 completes a second sequence, 1–7–20, with a total cost of $1,095,600/yr, which is in excess of the initial upper bound. Backtracking to subgroup *BCDEF* and

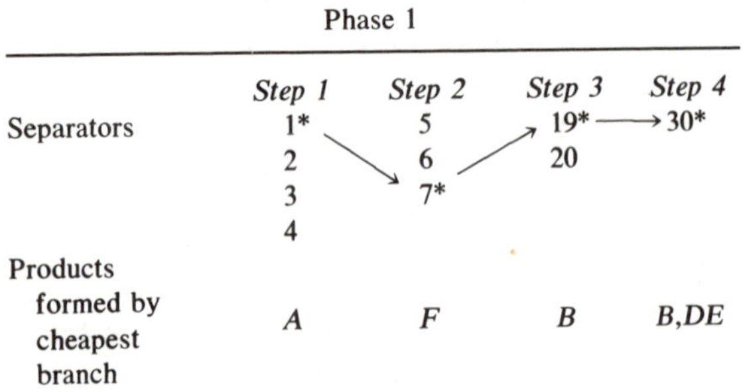

	Phase 1			
	Step 1	*Step 2*	*Step 3*	*Step 4*
Separators	1*	5	19* ⟶ 30*	
	2	6	20	
	3	7*		
	4			
Products formed by cheapest branch	A	F	B	B,DE

Figure 14.10. Development of the "cheapest-first" sequence for example 14.6. (Asterisk denotes the lowest-cost separator for a given branching step.)

Table 14.5 Sequences examined in Example 14.6

Sequence (or Partial Sequence) by Separator Numbers from Fig. 14.6	Total Cost (or Partial Cost), $/yr	Comments
1–7–19–30	878,000	Initial upper bound, ("cheapest-first" sequence)
1–7–20	1,095,600	
1–5–17–29	872,400	New upper bound
1–5–16–28	900,200	
(1–6)	(1,080,800)	
3–11–23–31	878,200	
3–11–24	1,095,700	
3–12 ⟨25/26⟩	867,400	New upper bound
(3–13)	(1,059,000)	
2 ⟨8/10–22⟩	860,400	New upper bound (optimal sequence)
2 ⟨8/9–21⟩	888,200	
(4)	(1,047,500)	

subsequent branching to separators 5, 17, and 20 develops the sequence 1–5–17–29, which is a new upper bound at $872,400/yr. Subsequent backtracking and branching develops the sequences and partial sequences in the order shown in Table 14.5. In order to obtain the optimal sequence, 9 of 12 sequences are completely developed and 27 of 31 separators (17 of 19 unique separators) are designed and costed. The optimal sequence is only 2% lower in cost than the "cheapest-first" sequence.

□

References

1. Thompson, R. W., and C. J. King, "Synthesis of Separation Schemes," Technical Report No. LBL-614, Lawrence Berkeley Laboratory (July, 1972).

2. Souders, M., *Chem. Eng. Progr.*, **60** (2), 75–82 (1964).

3. Hendry, J. E., and R. R. Hughes, *Chem. Eng. Progr.*, **68** (6), 71–76 (1972).

4. Buell, C. K., and R. G. Boatright, *Ind. Eng. Chem.*, **39**, 695–705 (1947).

5. Nilsson, N. J., *Problem-Solving Methods in Artificial Intelligence*, McGraw–Hill Book Co., New York, 1971.

6. Rodrigo, B. F. R., and J. D. Seader, *AIChE J.*, **21**, 885–894 (1975).

7. Hendry, J. E., Ph.D. thesis in chem-

ical engineering, University of Wisconsin, Madison, 1972.

8. Lockhart, F. J., *Petroleum Refiner*, **26** (8), 104–108 (1947).

9. Rod, V., and J. Marek, *Collect. Czech. Chem. Commun.*, **24**, 3240–3248 (1959).

10. Nishimura, H., and Y. Hiraizumi, *Int. Chem. Eng.*, **11**, 188–193 (1971).

11. Seader, J. D., and A. W. Westerberg, *AIChE J.*, **23**, 951–954 (1977).

12. Thompson, R. W., and C. J. King, *AIChE J.*, **18**, 941–948 (1972).

13. Stephanopoulos, G., and A. W.

Westerberg, *Chem. Eng. Sci.*, **31**, 195–204 (1976).

14. Westerberg, A. W., and G. Stephanopoulos, *Chem. Eng. Sci.*, **30**, 963–972 (1975).

15. Heaven, D. L., M.S. thesis in chemical engineering, University of California, Berkeley, 1969.

16. King, C. J., *Separation Processes*, McGraw–Hill Book Co., New York, 1971.

17. Tedder, D. W., Ph.D. thesis in chemical engineering, University of Wisconsin, Madison, 1975.

Problems

14.1 Stabilized effluent from a hydrogenation unit, as given below, is to be separated by ordinary distillation into five relatively pure products. Four distillation columns will be required. According to (14-1), these four columns can be arranged into 14 possible sequences. Draw sketches, as in Fig. 14.4, for each of these sequences.

Component	Feed Flow Rate, lbmole/hr	Approximate Relative Volatility Relative to $C5$
Propane ($C3$)	10.0	8.1
Butene-1 ($B1$)	100.0	3.7
n-Butane (NB)	341.0	3.1
Butene-2 isomers ($B2$)	187.0	2.7
n-Pentane ($C5$)	40.0	1.0

14.2 The feed to a separation process consists of the following species.

Species Number	Species
1	Ethane
2	Propane
3	Butene-1
4	n-Butane

It is desired to separate this mixture into essentially pure species. The use of two types of separators is to be explored.
1. Ordinary distillation.
2. Extractive distillation with furfural (Species 5).

The separation orderings are:

	Separator Type	
	1	**2**
Species number	1	1
	2	2
	3	4
	4	3
	5	5

(a) Determine the number of possible separation sequences.
(b) What splits would you forbid so as to greatly reduce the number of possible sequences?

14.3 Thermal cracking of naphtha yields the following gas, which is to be separated by a distillation train into the products indicated. If reasonably sharp separations are to be achieved, determine by heuristics two good sequences.

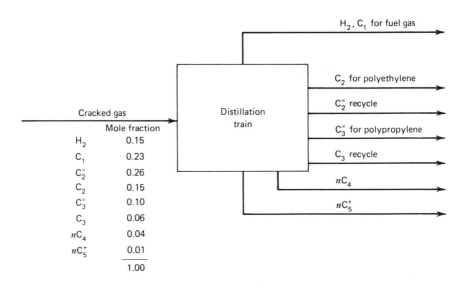

Cracked gas	Mole fraction
H_2	0.15
C_1	0.23
$C_2^=$	0.26
C_2	0.15
$C_3^=$	0.10
C_3	0.06
nC_4	0.04
nC_5^+	0.01
	1.00

H_2, C_1 for fuel gas
C_2 for polyethylene
$C_2^=$ recycle
$C_3^=$ for polypropylene
C_3 recycle
nC_4
nC_5^+

14.4 Investigators at the University of California at Berkeley have studied all 14 possible sequences for separating the following mixture at a flow rate of 200 lbmole/hr into its five components at about 98% purity each.[15]

Species	Symbol	Feed, mole fraction	Approximate relative volatility relative to n-pentane
Propane	A	0.05	8.1
Isobutane	B	0.15	4.3
n-Butane	C	0.25	3.1
Isopentane	D	0.20	1.25
n-Pentane	E	0.35	1.0
		1.00	

For each sequence, they determined the annual operating cost, including depreciation of the capital investment. Cost data for the *best three* sequences and the *worst* sequence are as follows.

Best sequence

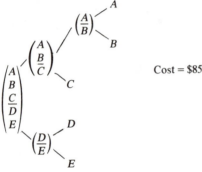

Cost = $858,780/yr

Second-best sequence

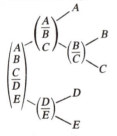

Cost = $863,580/yr

Third-best sequence

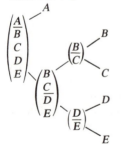

Cost = $871,460/yr

Worst sequence

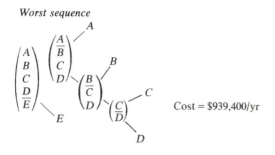

Cost = $939,400/yr

Explain in detail, as best you can, why the best sequences are best and the worst sequence is worst using the heuristics. Which heuristics appear to be most important?

14.5 Apply heuristics to determine the two most favorable sequences for Problem 14.1.

14.6 The effluent from a reactor contains a mixture of various chlorinated derivatives of the hydrocarbon RH_3, together with the hydrocarbon itself and HCl. Based on the following information and the heuristics, devise the best two feasible sequences. Explain your reasoning. Note that HCl may be corrosive.

Species	Lbmole/hr	α, relative to RCl_3	Purity desired
HCl	52	4.7	80%
RH_3	58	15.0	85%
RCl_3	16	1.0	98%
RH_2Cl	30	1.9	95%
$RHCl_2$	14	1.2	98%

14.7 The following stream at 100°F and 480 psia is to be separated into the four indicated products. Determine the best distillation sequence using the heuristics.

		Percent Recovery			
Species	Feed, lbmole/hr	Product 1	Product 2	Product 3	Product 4
H_2	1.5	~100			
CH_4	19.3	99			
Benzene	262.8		98		
Toluene	84.7			98	
Biphenyl	5.1				90

14.8 The following stream at 100°F and 20 psia is to be separated into the four indicated products. Determine the best distillation sequence by the heuristics.

Species	Feed, lbmole/hr	Percent Recovery			
		Product 1	Product 2	Product 3	Product 4
Benzene	100	98			
Toluene	100		98		
Ethylbenzene	200			98	
p-Xylene	200			98	
m-Xylene	200			98	
o-Xylene	200				98

14.9 The following cost data, which include operating cost and depreciation of capital investment, pertain to Problem 14.1. Determine by the ordered branch search technique:
(a) The best sequence.
(b) The second-best sequence.
(c) The worst sequence.

Split	Cost, $/yr
C3/B1	15,000
B1/NB	190,000
NB/B2	420,000
B2/C5	32,000
C3, B1/NB	197,000
C3/B1, NB	59,000
B1, NB/B2	500,000
B1/NB, B2	247,000
NB, B2/C5	64,000
NB/B2, C5	460,000
C3, B1, NB/B2	510,000
C3, B1/NB, B2	254,000
C3/B1, NB, B2	85,000
B1, NB, B2/C5	94,000
B1, NB/B2, C5	530,000
B1/NB, B2, C5	254,000
C3, B1, NB, B2/C5	95,000
C3, B1, NB/B2, C5	540,000
C3, B1/NB, B2, C5	261,000
C3/B1, NB, B2, C5	90,000

14.10 A hypothetical mixture of four species, A, B, C, and D, is to be separated into the four separate components. Two different separator types are being considered, neither of which requires a mass separating agent. The separation orders for the two types are:

Separator Type I	Separator Type II
A	B
B	A
C	C
D	D

Annual cost data for all the possible splits are given below. Use the ordered branch search technique to determine:
(a) The best sequence.
(b) The second-best sequence.
(c) The worst sequence.
For each answer, draw a diagram of the separation scheme, being careful to label each separator as to whether it is Type I or II.

Subgroup	Split	Type Separator	Annual Cost × $10,000
(A, B)	A/B	I	8
		II	15
(B, C)	B/C	I	23
		II	19
(C, D)	C/D	I	10
		II	18
(A, C)	A/C	I	20
		II	6
(A, B, C)	A/B, C	I	10
	B/A, C	II	25
	A, B/C	I	25
		II	20
(B, C, D)	B/C, D	I	27
		II	22
	B, C/D	I	12
		II	20
(A, C, D)	A/C, D	I	23
		II	10
	A, C/D	I	11
		II	20
(A, B, C, D)	A/B, C, D	I	14
	B/A, C, D	II	20
	A, B/C, D	I	27
		II	25
	A, B, C/D	I	13
		II	21

14.11 The following stream at 100°F and 250 psia is to be separated into the four

indicated products. Also given is the cost of each of the unique separators. Use the ordered branch search technique to determine:
(a) The best sequence.
(b) The second-best sequence.

Species	Symbol	Feed Rate, lbmole/hr	Percent Recovery			
			Product 1	Product 2	Product 3	Product 4
Propane	A	100	98			
i-Butane	B	300		98		
n-Butane	C	500			98	
i-Pentane	D	400				98

Unique Separator	Cost, $/yr
A/B	26,100
B/C	94,900
C/D	59,300
A/BC	39,500
AB/C	119,800
B/CD	112,600
BC/D	76,800
A/BCD	47,100
AB/CD	140,500
ABC/D	94,500

14.12 The following stream at 100°F and 300 psia is to be separated into four essentially pure products. Also given is the cost of each unique separator.
(a) Draw an *and/or* directed graph of the entire search space.
(b) Use the ordered branch search method to determine the best sequence.

Species	Symbol	Feed Rate, lb/mole/hr
i-Butane	A	300
n-Butane	B	500
i-Pentane	C	400
n-Pentane	D	700

Unique Separator	Cost, $/yr
A/B	94,900
B/C	59,300
C/D	169,200
A/BC	112,600
AB/C	76,800
B/CD	78,200
BC/D	185,300
A/BCD	133,400
AB/CD	94,400
ABC/D	241,800

14.13 Consider the problem of separation, by ordinary distillation, of propane A, isobutane B, n-butane C, isopentane D, and n-pentane E.
(a) What is the total number of sequences and unique splits?
Using heuristics only, develop flowsheets for:
(b) Equimolal feed with product streams A, (B, C), and (D, E) required.
(c) Feed consisting of $A = 10$, $B = 10$, $C = 60$, $D = 10$, and $E = 20$ (relative moles) with products A, B, C, D, and E.
(d) Given below [A. Gomez and J. D. Seader, *AIChE J.*, **22**, 970 (1976)] is a plot of the effect of nonkey components on the cost of the separation. Determine the optimal, or a near-optimal, separation sequence for products (A, B, C, D, E) and an equimolal feed.

Component	Boiling-pt pressure at 100°F	α at Boiling pt. pressure
A	210	
B	65	2.2
C	35	1.44
D	5	2.73
E	0	1.25

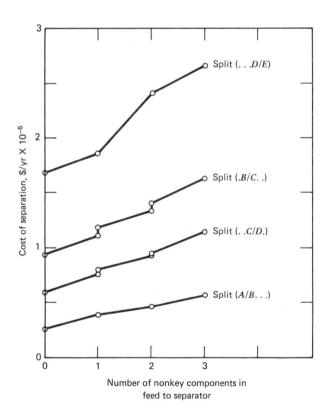

Rigorous Methods for Multicomponent, Multistage Separations

The availability of large electronic computers has made possible the rigorous solution of the equilibrium-stage model for multicomponent, multistage distillation column to an exactness limited only by the accuracy of the phase-equilibrium and enthalpy data utilized.

Buford D. Smith, 1973

Previous chapters have considered graphical, empirical, and approximate group methods for the solution of multistage separation problems. Except for simple cases, such as binary distillation, these methods are suitable only for preliminary design studies. Final design of multistage equipment for conducting multicomponent separations requires rigorous determination of temperatures, pressures, stream flow rates, stream compositions, and heat transfer rates at each stage.* This determination is made by solving material balance, energy (enthalpy) balance, and equilibrium relations for each stage. Unfortunately, these relations are nonlinear algebraic equations that interact strongly. Consequently, solution procedures are relatively difficult and tedious. However, once the procedures are programmed for a high-speed digital computer, solutions are achieved fairly rapidly and almost routinely. Such programs are readily available and widely used.

---

* However, rigorous calculational procedures may not be justified when multicomponent physical properties or stage efficiencies are not reasonably well known.

15.1 Theoretical Model for an Equilibrium Stage

Consider a general, continuous, steady-state vapor–liquid or liquid–liquid separator consisting of a number of stages arranged in a countercurrent cascade. Assume that phase equilibrium is achieved at each stage and that no chemical reactions occur. A general schematic representation of an equilibrium stage j is shown in Fig. 15.1 for a vapor–liquid separator, where the stages are numbered down from the top. The same representation applies to a liquid–liquid separator if the higher-density liquid phases are represented by liquid streams and the lower-density liquid phases are represented by vapor streams.

Entering stage j can be one single- or two-phase feed of molal flow rate F_j, with overall composition in mole fractions $z_{i,j}$ of component i, temperature T_{F_j}, pressure P_{F_j}, and corresponding overall molal enthalpy H_{F_j}. Feed pressure is assumed equal to or greater than stage pressure P_j. Any excess feed pressure $(P_F - P_j)$ is reduced to zero adiabatically across valve F.

Also entering stage j can be interstage liquid from stage $j - 1$ above, if any, of molal flow rate L_{j-1}, with composition in mole fractions $x_{i,j-1}$, enthalpy $H_{L_{j-1}}$, temperature T_{j-1}, and pressure P_{j-1}, which is equal to or less than the pressure of stage j. Pressure of liquid from stage $j - 1$ is increased adiabatically by hydrostatic head change across head L.

Similarly, from stage $j + 1$ below, interstage vapor of molal flow rate V_{j+1}, with composition in mole fractions $y_{i,j+1}$, enthalpy $H_{V_{j+1}}$, temperature T_{j+1}, and pressure P_{j+1} can enter stage j. Any excess pressure $(P_{j+1} - P_j)$ is reduced to zero adiabatically across valve V.

Leaving stage j is vapor of intensive properties $y_{i,j}$, H_{V_j}, T_j, and P_j. This stream can be divided into a vapor side stream of molal flow rate W_j and an interstage stream of molal flow rate V_j to be sent to stage $j - 1$ or, if $j = 1$, to leave the separator as a product. Also leaving stage j is liquid of intensive properties $x_{i,j}$, H_{L_j}, T_j, and P_j, which is in equilibrium with vapor $(V_j + W_j)$. This liquid can be divided also into a liquid side stream of molal flow rate U_j and an interstage or product stream of molal flow rate L_j to be sent to stage $j + 1$ or, if $j = N$, to leave the multistage separator as a product.

Heat can be transferred at a rate Q_j from ($+$) or to ($-$) stage j to simulate stage intercoolers, interheaters, condensers, or reboilers as shown in Fig. 1.7. The model in Fig. 15.1 does not allow for pumparounds of the type shown in Fig. 15.2. Such pumparounds are often used in columns having side streams in order to conserve energy and balance column vapor loads.

Associated with each general theoretical stage are the following indexed equations expressed in terms of the variable set in Fig. 15.1. However, variables other than those shown in Fig. 15.1 can be used. For example, component flow rates can replace mole fractions, and side-stream flow rates can be expressed as fractions of interstage flow rates. The equations are similar to those of Section

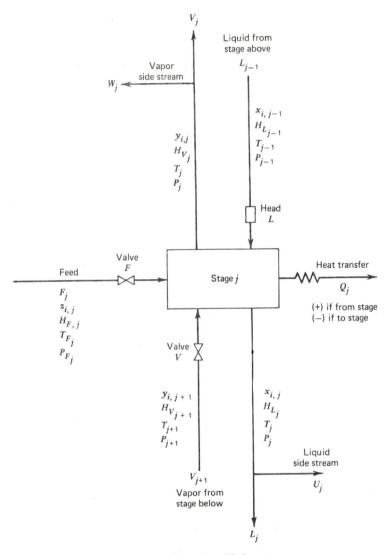

Figure 15.1.   General equilibrium stage.

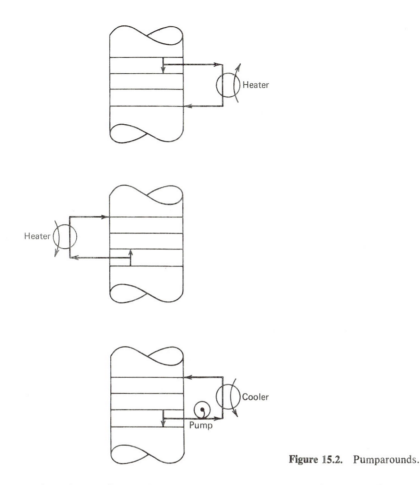

Figure 15.2. Pumparounds.

6.3* and are often referred to as the MESH equations after Wang and Henke.[1]

1. **M** equations—Material balance for each component (C equations for each stage).

$$M_{i,j} = L_{j-1}x_{i,j-1} + V_{j+1}y_{i,j+1} + F_j z_{i,j} - (L_j + U_j)x_{i,j} - (V_j + W_j)y_{i,j} = 0 \quad (15\text{-}1)$$

2. **E** equations—phase Equilibrium relation for each component (C equations for each stage).

$$E_{i,j} = y_{i,j} - K_{i,j}x_{i,j} = 0 \qquad (15\text{-}2)$$

where $K_{i,j}$ is the phase equilibrium ratio.

---

* Unlike the treatment in Section 6.3, all C component material balances are included here, and the total material balance is omitted. Also, the separate but equal temperature and pressure of the equilibrium phases are replaced by the stage temperature and pressure.

3. S equations—mole fraction Summations (one for each stage).

$$(S_y)_j = \sum_{i=1}^{C} y_{i,j} - 1.0 = 0 \qquad (15\text{-}3)$$

$$(S_x)_j = \sum_{i=1}^{C} x_{i,j} - 1.0 = 0 \qquad (15\text{-}4)$$

4. H equation—energy balance (one for each stage).

$$H_j = L_{j-1}H_{L_{j-1}} + V_{j+1}H_{V_{j+1}} + F_jH_{F_j} - (L_j + U_j)H_{L_j} - (V_j + W_j)H_{V_j} - Q_j = 0 \quad (15\text{-}5)$$

where kinetic and potential energy changes are ignored.

A total material balance equation can be used in place of (15-3) or (15-4). It is derived by combining these two equations and $\sum_i z_{i,j} = 1.0$ with (15-1) summed over the C components and over stages 1 through j to give

$$L_j = V_{j+1} + \sum_{m=1}^{j} (F_m - U_m - W_m) - V_1 \qquad (15\text{-}6)$$

In general, $K_{i,j} = K_{i,j}\{T_j, P_j, x_j, y_j\}$, $H_{V_j} = H_{V_j}\{T_j, P_j, y_j\}$, and $H_{L_j} = H_{L_j}\{T_j, P_j, x_j\}$. If these relations are not counted as equations and the three properties are not counted as variables, each equilibrium stage is defined only by the $2C + 3$ MESH equations. A countercurrent cascade of N such stages, as shown in Fig. 15.3, is represented by $N(2C + 3)$ such equations in $[N(3C + 10) + 1]$ variables. If N and all F_j, $z_{i,j}$, T_{F_j}, P_{F_j}, P_j, U_j, W_j, and Q_j are specified, the model is represented by $N(2C + 3)$ simultaneous algebraic equations in $N(2C + 3)$ unknown (output) variables comprised of all $x_{i,j}$, $y_{i,j}$, L_j, V_j, and T_j, where the M, E, and H equations are nonlinear. If other variables are specified, as they often are, corresponding substitutions are made to the list of output variables. Regardless of the specifications, the result is a set of nonlinear equations that must be solved by iterative techniques.

15.2 General Strategy of Mathematical Solution

A wide variety of iterative solution procedures for solving nonlinear algebraic equations has appeared in the literature. In general, these procedures make use of equation partitioning in conjunction with equation tearing and/or linearization by Newton–Raphson techniques, which are described in detail by Myers and Seider.[2] The equation-tearing method was applied in Section 7.4 for computing an adiabatic flash.

Early attempts to solve (15-1) to (15-5) or equivalent forms of these equations resulted in the classical *stage-by-stage, equation-by-equation* calculational procedures of Lewis–Matheson[3] in 1932 and Thiele–Geddes[4] in 1933 based on equation tearing for solving simple fractionators with one feed and two

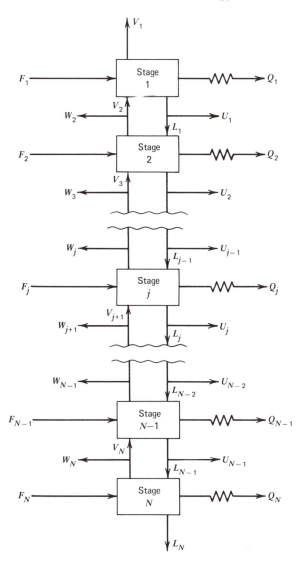

Figure 15.3.   General counter-current cascade of N stages.

products. Composition-independent K-values and component enthalpies were generally employed. The Thiele–Geddes method was formulated to handle the Case II variable specification in Table 6.2 wherein the number of equilibrium stages above and below the feed, the reflux ratio, and the distillate flow rate are specified, and stage temperatures and interstage vapor (or liquid) flow rates are the iteration (tear) variables. Although widely used for hand calculations in the years immediately following its appearance in the literature, the Thiele–Geddes method was found often to be numerically unstable when attempts were made to program it for a digital computer. However, Holland[5] and co-workers developed an improved Thiele–Geddes procedure called the *theta method*, which in various versions has been applied with considerable success.

The Lewis–Matheson method is also an equation-tearing procedure. It was formulated according to the Case I variable specification in Table 6.2 to determine stage requirements for specifications of the separation of two key components, a reflux ratio and a feed-stage location criterion. Both outer and inner iterations are required. The outer loop tear variables are the mole fractions or flow rates of nonkey components in the products. The inner loop tear variables are the interstage vapor (or liquid) flow rates. The Lewis–Matheson method was widely used for hand calculations, but it also proved often to be numerically unstable when implemented on a digital computer.

Rather than using an equation-by-equation solution procedure, Amundson and Pontinen,[6] in a significant development, showed that (15-1), (15-2), and (15-6) of the MESH equations for a Case II specification could be combined and solved component-by-component from simultaneous linear equation sets for all N stages by an equation-tearing procedure using the same tear variables as the Thiele–Geddes method. Although too tedious for hand calculations, such equation sets are readily solved with a digital computer.

In a classic study, Friday and Smith[7] systematically analyzed a number of tearing techniques for solving the MESH equations. They carefully considered the choice of output variable for each equation. They showed that no one technique could solve all types of problems. For separators where the feed(s) contains only components of similar volatility (narrow-boiling case), a modified Amundson–Pontinen approach termed the bubble-point (BP) method was recommended. For a feed(s) containing components of widely different volatility (wide-boiling case) or solubility, the BP method was shown to be subject to failure and a so-called sum-rates (SR) method was suggested. For intermediate cases, the equation-tearing technique may fail to converge; in that case, Friday and Smith indicated that either a Newton–Raphson method or a combined tearing and Newton–Raphson technique was necessary. Current practice is based mainly on the BP, SR, and Newton–Raphson methods, all of which are treated in this chapter. The latter method permits considerable flexibility in the choice of specified variables and generally is capable of solving all problems.

15.3 Equation-Tearing Procedures

In general the modern equation-tearing procedures are readily programmed, are rapid, and require a minimum of computer storage. Although they can be applied to a much wider variety of problems than the classical Thiele–Geddes tearing procedure, they are usually limited to the same choice of specified variables. Thus, neither product purities, species recoveries, interstage flow rates, nor stage temperatures can be specified.

Tridiagonal Matrix Algorithm

The key to the success of the BP and SR tearing procedures is the tridiagonal matrix that results from a modified form of the M equations (15-1) when they are torn from the other equations by selecting T_j and V_j as the tear variables, which leaves the modified M equations linear in the unknown liquid mole fractions. This set of equations for each component is solved by a highly efficient and reliable algorithm due to Thomas[8] as applied by Wang and Henke.[1] The modified M equations are obtained by substituting (15-2) into (15-1) to eliminate y and by substituting (15-6) into (15-1) to eliminate L. Thus, equations for calculating y and L are partitioned from the other equations. The result for each component and each stage is as follows where the i subscripts have been deleted from the B, C, and D terms.

$$A_j x_{i,j-1} + B_j x_{i,j} + C_j x_{i,j+1} = D_j \qquad (15\text{-}7)$$

where

$$A_j = V_j + \sum_{m=1}^{j-1} (F_m - W_m - U_m) - V_1 \qquad 2 \le j \le N \qquad (15\text{-}8)$$

$$B_j = -\left[V_{j+1} + \sum_{m=1}^{j} (F_m - W_m - U_m) - V_1 + U_j + (V_j + W_j)K_{i,j} \right] \qquad 1 \le j \le N \tag{15-9}$$

$$C_j = V_{j+1} K_{i,j+1} \qquad 1 \le j \le N - 1 \qquad (15\text{-}10)$$

$$D_j = -F_j z_{i,j} \qquad 1 \le j \le N \qquad (15\text{-}11)$$

with $x_{i,o} = 0$, $V_{N+1} = 0$, $W_1 = 0$, and $U_N = 0$, as indicated in Fig. 15-3. If the modified M equations are grouped by component, they can be partitioned by writing them as a series of C separate tridiagonal matrix equations where the output variable for each matrix equation is x_i over the entire countercurrent cascade of N stages.

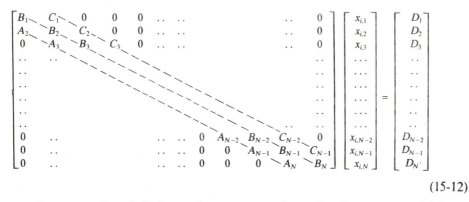

$$(15\text{-}12)$$

Constants B_j and C_j for each component depend only on tear variables T

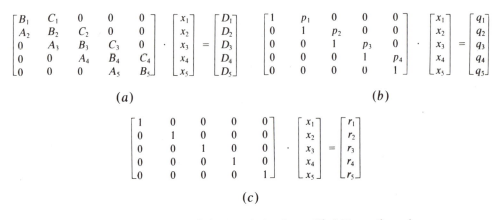

(a)   (b)

(c)

Figure 15.4. The coefficient matrix for the modified M-equations of a given component at various steps in the Thomas algorithm for five equilibrium stages (Note that the i subscript is deleted from x) (a) Initial matrix. (b) Matrix after forward elimination. (c) Matrix after backward substitution.

and V provided that K-values are composition independent. If not, compositions from the previous iteration may be used to estimate the K-values.

The Thomas algorithm for solving the linearized equation set (15-12) is a Gaussian elimination procedure that involves forward elimination starting from stage 1 and working toward stage N to finally isolate $x_{i,N}$. Other values of $x_{i,j}$ are then obtained starting with $x_{i,N-1}$ by backward substitution. For five stages, the matrix equations at the beginning, middle, and end of the procedure are as shown in Fig. 15.4.

The equations used in the Thomas algorithm are as follows:

For stage 1, (15-7) is $B_1 x_{i,1} + C_1 x_{i,2} = D_1$, which can be solved for $x_{i,1}$ in terms of unknown $x_{i,2}$ to give

$$x_{i,1} = \frac{D_1 - C_1 x_{i,2}}{B_1}$$

Let

$$p_1 = \frac{C_1}{B_1} \quad \text{and} \quad q_1 = \frac{D_1}{B_1}$$

Then

$$x_{i,1} = q_1 - p_1 x_{i,2} \tag{15-13}$$

Thus, the coefficients in the matrix become $B_1 \leftarrow 1$, $C_1 \leftarrow p_1$, and $D_1 \leftarrow q_1$, where $\leftarrow$ means "is replaced by." Only values for p_1 and q_1 need be stored.

For stage 2, (15-7) can be combined with (15-13) and solved for $x_{i,2}$ to give

$$x_{i,2} = \frac{D_2 - A_2 q_1}{B_2 - A_2 p_1} - \left(\frac{C_2}{B_2 - A_2 p_1}\right) x_{i,3}$$

Let

$$q_2 = \frac{D_2 - A_2 q_1}{B_2 - A_2 p_1} \quad \text{and} \quad p_2 = \frac{C_2}{B_2 - A_2 p_1}$$

Then

$$x_{i,2} = q_2 - p_2 x_{i,3}$$

Thus, $A_2 \leftarrow 0$, $B_2 \leftarrow 1$, $C_2 \leftarrow p_2$, and $D_2 \leftarrow q_2$. Only values for p_2 and q_2 need be stored.

In general, we can define

$$p_j = \frac{C_j}{B_j - A_j p_{j-1}} \tag{15-14}$$

$$q_j = \frac{D_j - A_j q_{j-1}}{B_j - A_j p_{j-1}} \tag{15-15}$$

Then

$$x_{i,j} = q_j - p_j x_{i,j+1} \tag{15-16}$$

with $A_j \leftarrow 0$, $B_j \leftarrow 1$, $C_j \leftarrow p_j$, and $D_j \leftarrow q_j$. Only values of p_j and q_j need be stored. Thus, starting with stage 1, values of p_j and q_j are computed recursively in the order: $p_1, q_1, p_2, q_2, \ldots, p_{N-1}, q_{N-1}, q_N$. For stage N, (15-16) isolates $x_{i,N}$ as

$$x_{i,N} = q_N \tag{15-17}$$

Successive values of x_i are computed recursively by backward substitution from (15-16) in the form

$$x_{i,j-1} = q_{j-1} - p_{j-1} x_{i,j} = r_{j-1} \tag{15-18}$$

Equation (15-18) corresponds to the identity coefficient matrix.

The Thomas algorithm, when applied in this fashion, generally avoids buildup of computer truncation errors because usually none of the steps involves subtraction of nearly equal quantities. Furthermore, computed values of $x_{i,j}$ are almost always positive. The algorithm is highly efficient, requires a minimum of computer storage as noted above, and is superior to alternative matrix-inversion routines. A modified Thomas algorithm for difficult cases is given by Boston and Sullivan.[9] Such cases can occur for columns having large numbers of equilibrium stages and with components whose absorption factors [see (12-48)] are less than unity in one section of stages and greater than unity in another section.

Bubble-Point (BP) Method for Distillation

Frequently, distillation involves species that cover a relatively narrow range of vapor–liquid equilibrium ratios (K-values). A particularly effective solution procedure for this case was suggested by Friday and Smith[7] and developed in detail by Wang and Henke.[1] It is referred to as the bubble-point (BP) method because a new set of stage temperatures is computed during each iteration from

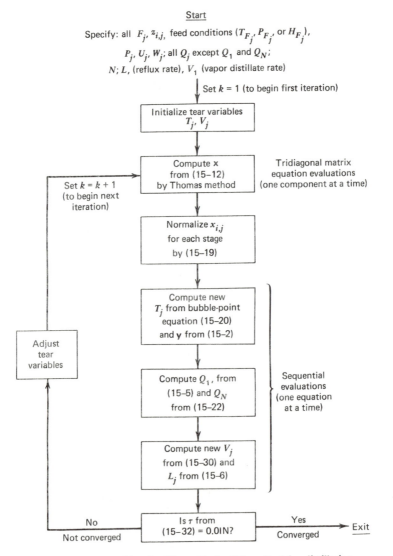

Figure 15.5. Algorithm for Wang–Henke BP method for distillation.

bubble-point equations. In the method, all equations are partitioned and solved sequentially except for the modified M equations, which are solved separately for each component by the tridiagonal matrix technique.

The algorithm for the Wang–Henke BP method is shown in Fig. 15.5. A FORTRAN computer program for the method is available.[10] Problem specifications consist of conditions and stage location of all feeds, pressure at each stage, total flow rates of all sidestreams,* heat transfer rates to or from all stages except stage 1 (condenser) and stage N (reboiler), total number of stages, external bubble-point reflux flow rate, and vapor distillate flow rate. A sample problem specification is shown in Fig. 15.6.

To initiate the calculations, values for the tear variables are assumed. For most problems, it is sufficient to establish an initial set of V_j values based on the assumption of constant molal interstage flows using the specified reflux, distillate, feed, and side-stream flow rates. A generally adequate initial set of T_j values can be provided by computing or assuming both the bubble-point temperature of an estimated bottoms product and the dew-point temperature of an assumed vapor distillate product; or computing or assuming bubble-point temperature if distillate is liquid or a temperature in between the dew-point and bubble-point temperatures if distillate is mixed (both vapor and liquid); and then determining the other stage temperatures by assuming a linear variation of temperatures with stage location.

To solve (15-12) for $\mathbf{x}_i$ by the Thomas method, $K_{i,j}$ values are required. When they are composition dependent, initial assumptions for all $x_{i,j}$ and $y_{i,j}$ values are also needed unless ideal K-values are employed for the first iteration. For each iteration, the computed set of $x_{i,j}$ values for each stage will, in general, not satisfy the summation constraint given by (15-4). Although not mentioned by Wang and Henke, it is advisable to normalize the set of computed $x_{i,j}$ values by the relation

$$(x_{i,j})_{\text{normalized}} = \frac{x_{i,j}}{\displaystyle\sum_{i=1}^{C} x_{i,j}} \tag{15-19}$$

These normalized values are used for all subsequent calculations involving $x_{i,j}$ during the iteration.

A new set of temperatures T_j is computed stage by stage by computing bubble-point temperatures from the normalized $x_{i,j}$ values. Friday and Smith[7] showed that bubble-point calculations for stage temperatures are particularly effective for mixtures having a narrow range of K-values because temperatures are not then sensitive to composition. For example, in the limiting case where all components have identical K-values, the temperature corresponds to the con-

---

* Note that liquid distillate flow rate, if any, is designated as U_1.

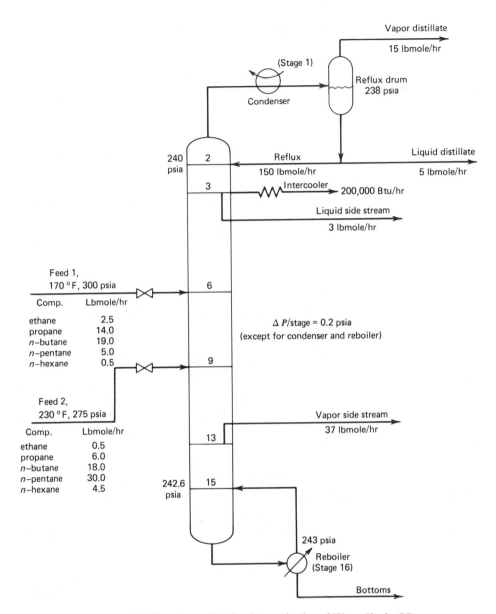

Figure 15.6. Sample specification for application of Wang–Henke BP method to distillation.

ditions of $K_{i,j} = 1$ and is not dependent on $x_{i,j}$ values. At the other extreme, however, bubble-point calculations to establish stage temperatures can be very sensitive to composition. For example, consider a binary mixture containing one component with a high K-value that changes little with temperature. The second component has a low K-value that changes very rapidly with temperature. Such a mixture is methane and n-butane at 400 psia with K-values given in Example 12.8. The effect on the bubble-point temperature of small quantities of methane dissolved in liquid n-butane is very large as indicated by the following results.

Liquid Mole Fraction of Methane	Bubble-Point Temperature, °F
0.000	275
0.018	250
0.054	200
0.093	150

Thus, the BP method is best when components have a relatively narrow range of K-values.

The necessary bubble-point equation is obtained in the manner described in Chapter 7 by combining (15-2) and (15-3) to eliminate $y_{i,j}$ giving

$$\sum_{i=1}^{C} K_{i,j} x_{i,j} - 1.0 = 0 \tag{15-20}$$

which is nonlinear in T_j and must be solved iteratively. An algorithm for this calculation when K-values are dependent on composition is given in Fig. 7.6. Wang and Henke prefer to use Muller's iterative method[11] because it is reliable and does not require the calculation of derivatives. Muller's method requires three initial assumptions of T_j. For each assumption, the value of S_j is computed from

$$S_j = \sum_{i=1}^{C} K_{i,j} x_{i,j} - 1.0 \tag{15-21}$$

The three sets of (T_j, S_j) are fitted to a quadratic equation for S_j in terms of T_j. The quadratic equation is then employed to predict T_j for $S_j = 0$, as required by (15-20). The validity of this value of T_j is checked by using it to compute S_j in (15-21). The quadratic fit and S_j check are repeated with the three best sets of (T_j, S_j) until some convergence tolerance is achieved, say $|T_j^{(n)} - T_j^{(n-1)}|/T_j^{(n)} \le 0.0001$, with T in absolute degrees, where n is the iteration number for the temperature loop in the bubble-point calculation, or one can use $S_j \le 0.0001 \, C$, which is preferred.

Values of $y_{i,j}$ are determined along with the calculation of stage temperatures using the E equations, (15-2). With a consistent set of values for $x_{i,j}$, T_j, and $y_{i,j}$, molal enthalpies are computed for each liquid and vapor stream leaving a stage. Since F_1, V_1, U_1, W_1, and L_1 are specified, V_2 is readily obtained from (15-6), and the condenser duty, a (+) quantity, is obtained from (15-5). Reboiler duty, a (−) quantity, is determined by summing (15-5) for all stages to give

$$Q_N = \sum_{j=1}^{N} (F_j H_{F_j} - U_j H_{L_j} - W_j H_{V_j}) - \sum_{j=1}^{N-1} Q_j - V_1 H_{V_1} - L_N H_{L_N} \qquad (15\text{-}22)$$

A new set of V_j tear variables is computed by applying the following modified energy balance, which is obtained by combining (15-5) and (15-6) twice to eliminate L_{j-1} and L_j. After rearrangement

$$\alpha_j V_j + \beta_j V_{j+1} = \gamma_j \qquad (15\text{-}23)$$

where

$$\alpha_j = H_{L_{j-1}} - H_{V_j} \qquad (15\text{-}24)$$

$$\beta_j = H_{V_{j+1}} - H_{L_j} \qquad (15\text{-}25)$$

$$\gamma_j = \left[\sum_{m=1}^{j-1} (F_m - W_m - U_m) - V_1 \right] \left(H_{L_j} - H_{L_{j-1}} \right) + F_j(H_{L_j} - H_{F_j}) + W_j(H_{V_j} - H_{L_j}) + Q_j \qquad (15\text{-}26)$$

and enthalpies are evaluated at the stage temperatures last computed rather than at those used to initiate the iteration. Written in didiagonal matrix form (15-23) applied over stages 2 to $N - 1$ is:

$$
\begin{bmatrix}
\beta_2 & 0 & 0 & 0 & .. & & 0 \\
\alpha_3 & \beta_3 & 0 & 0 & .. & & 0 \\
0 & \alpha_4 & \beta_4 & 0 & .. & & 0 \\
.. & .. & & & & & \\
.. & .. & & & & & \\
0 & .. & 0 & \alpha_{N-3} & \beta_{N-3} & 0 & 0 \\
0 & .. & 0 & 0 & \alpha_{N-2} & \beta_{N-2} & 0 \\
0 & .. & 0 & 0 & 0 & \alpha_{N-1} & \beta_{N-1}
\end{bmatrix}
\cdot
\begin{bmatrix}
V_3 \\
V_4 \\
V_5 \\
.. \\
.. \\
V_{N-2} \\
V_{N-1} \\
V_N
\end{bmatrix}
=
\begin{bmatrix}
\gamma_2 - \alpha_2 V_2 \\
\gamma_3 \\
\gamma_4 \\
.. \\
.. \\
\gamma_{N-3} \\
\gamma_{N-2} \\
\gamma_{N-1}
\end{bmatrix}
\qquad (15\text{-}27)
$$

Matrix equation (15-27) is readily solved one equation at a time by starting at the top where V_2 is known and working down recursively. Thus

$$V_3 = \frac{\gamma_2 - \alpha_2 V_2}{\beta_2} \qquad (15\text{-}28)$$

$$V_4 = \frac{\gamma_3 - \alpha_3 V_3}{\beta_3} \qquad (15\text{-}29)$$

or, in general

$$V_j = \frac{\gamma_{j-1} - \alpha_{j-1} V_{j-1}}{\beta_{j-1}} \qquad (15\text{-}30)$$

and so on. Corresponding liquid flow rates are obtained from (15-6).

The solution procedure is considered to be converged when sets of $T_j^{(k)}$ and $V_j^{(k)}$ values are within some prescribed tolerance of corresponding sets of $T_j^{(k-1)}$ and $V_j^{(k-1)}$ values, where k is the iteration index. One possible convergence criterion is

$$\sum_{j=1}^{N} \left[\frac{T_j^{(k)} - T_j^{(k-1)}}{T_j^{(k)}} \right]^2 + \sum_{j=1}^{N} \left[\frac{V_j^{(k)} - V_j^{(k-1)}}{V_j^{(k)}} \right]^2 \le \epsilon \qquad (15\text{-}31)$$

where T is the absolute temperature and ϵ is some prescribed tolerance. However, Wang and Henke suggest that the following simpler criterion, which is based on successive sets of T_j values only, is adequate.

$$\tau = \sum_{j=1}^{N} \left[T_j^{(k)} - T_j^{(k-1)} \right]^2 \le 0.01 \, N \qquad (15\text{-}32)$$

Successive substitution is often employed for iterating the tear variables; that is, values of T_j and V_j generated from (15-20) and (15-30), respectively, during an iteration are used directly to initiate the next iteration. However, experience indicates that it is desirable frequently to adjust the values of the generated tear variables prior to beginning the next iteration. For example, upper and lower bounds should be placed on stage temperatures, and any negative values of interstage flow rates should be changed to near-zero positive values. Also, to prevent oscillation of the iterations, damping can be employed to limit changes in the values of V_j and absolute T_j from one iteration to the next to—say, 10%.

Example 15.1.   For the distillation column discussed in Section 7.7 and shown in Fig. 15.7, do one iteration of the BP method up to and including the calculation of a new set of T_j values from (15-20). Use composition-independent K-values from Fig. 7.5.

Solution.   By overall total material balance

$$\text{Liquid distillate} = U_1 = F_3 - L_5 = 100 - 50 = 50 \, \text{lbmole/hr}$$

Then

$$L_1 = (L_1/U_1)U_1 = (2)(50) = 100 \, \text{lbmole/hr}$$

By total material balance around the total condenser

$$V_2 = L_1 + U_1 = 100 + 50 = 150 \, \text{lbmole/hr}$$

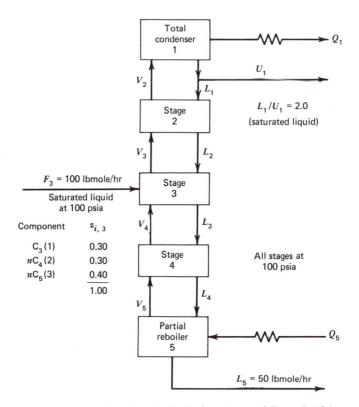

Figure 15.7. Specifications for distillation column of Example 15.1.

Initial guesses of tear variables are

Stage j	V_j, lbmole/hr	T_j, °F
1	(Fixed at 0 by specifications)	65
2	(Fixed at 150 by specifications)	90
3	150	115
4	150	140
5	150	165

From Fig. 7.5 at 100 psia, the K-values at the assumed stage temperatures are

			$K_{i,j}$		
Stage	**1**	**2**	**3**	**4**	**5**
C_3 (1)	1.23	1.63	2.17	2.70	3.33
nC_4 (2)	0.33	0.50	0.71	0.95	1.25
nC_5 (3)	0.103	0.166	0.255	0.36	0.49

The matrix equation (15-12) for the first component C_3 is developed as follows. From (15-8) with $V_1 = 0$, $W = 0$

$$A_j = V_j + \sum_{m=1}^{i-1} (F_m - U_m)$$

Thus, $A_5 = V_5 + F_3 - U_1 = 150 + 100 - 50 = 200$ lbmole/hr. Similarly, $A_4 = 200$, $A_3 = 100$, and $A_2 = 100$ in the same units.

From (15-9) with $V_1 = 0$, $W = 0$

$$B_j = -\left[V_{j+1} + \sum_{m=1}^{j} (F_m - U_m) + U_j + V_j K_{i,j} \right]$$

Thus, $B_5 = -[F_3 - U_1 + V_5 K_{1,5}] = -[100 - 50 + (150)3.33] = -549.5$ lbmole/hr. Similarly, $B_4 = -605$, $B_3 = -525.5$, $B_2 = -344.5$, and $B_1 = -150$ in the same units.

From (15-10), $C_j = V_{j+1} K_{1,j+1}$. Thus, $C_1 = V_2 K_{1,2} = 150(1.63) = 244.5$ lbmole/hr. Similarly, $C_2 = 325.5$, $C_3 = 405$, and $C_4 = 499.5$ in the same units.

From (15-11), $D_j = -F_j z_{1,j}$. Thus, $D_3 = -100(0.30) = -30$ lbmole/hr. Similarly, $D_1 = D_2 = D_4 = D_5 = 0$.

Substitution of the above values in (15-7) gives

$$\begin{bmatrix} -150 & 244.5 & 0 & 0 & 0 \\ 100 & -344.5 & 325.5 & 0 & 0 \\ 0 & 100 & -525.5 & 405 & 0 \\ 0 & 0 & 200 & -605 & 499.5 \\ 0 & 0 & 0 & 200 & -549.5 \end{bmatrix} \begin{bmatrix} x_{1,1} \\ x_{1,2} \\ x_{1,3} \\ x_{1,4} \\ x_{1,5} \end{bmatrix} = \begin{bmatrix} 0 \\ 0 \\ -30 \\ 0 \\ 0 \end{bmatrix}$$

Using (15-14) and (15-15), we apply the forward step of the Thomas algorithm as follows.

$$p_1 = \frac{C_1}{B_1} = 244.5/(-150) = -1.630$$

$$q_1 = \frac{D_1}{B_1} = 0/(-150) = 0$$

$$p_2 = \frac{C_2}{B_2 - A_2 p_1} = \frac{325.5}{-344.5 - 100(-1.630)} = -1.793$$

By similar calculations, the matrix equation after the forward elimination procedure is

$$\begin{bmatrix} 1 & -1.630 & 0 & 0 & 0 \\ 0 & 1 & -1.793 & 0 & 0 \\ 0 & 0 & 1 & -1.170 & 0 \\ 0 & 0 & 0 & 1 & -1.346 \\ 0 & 0 & 0 & 0 & 1 \end{bmatrix} \begin{bmatrix} x_{1,1} \\ x_{1,2} \\ x_{1,3} \\ x_{1,4} \\ x_{1,5} \end{bmatrix} = \begin{bmatrix} 0 \\ 0 \\ 0.0867 \\ 0.0467 \\ 0.0333 \end{bmatrix}$$

Applying the backward steps of (15-17) and (15-18) gives

$$x_{1,5} = q_5 = 0.0333$$

$$x_{1,4} = q_4 - p_4 x_{1,5} = 0.0467 - (-1.346)(0.0333) = 0.0915$$

Similarly

$$x_{1,3} = 0.1938 \qquad x_{1,2} = 0.3475 \qquad x_{1,1} = 0.5664$$

The matrix equations for nC_4 and nC_5 are solved in a similar manner to give

			$x_{i,j}$		
Stage	**1**	**2**	**3**	**4**	**5**
C_3	0.5664	0.3475	0.1938	0.0915	0.0333
nC_4	0.1910	0.3820	0.4483	0.4857	0.4090
nC_5	0.0191	0.1149	0.3253	0.4820	0.7806
$\sum_i x_{i,j}$	0.7765	0.8444	0.9674	1.0592	1.2229

After these compositions are normalized, bubble-point temperatures at 100 psia are computed iteratively from (15-20) and compared to the initially assumed values,

Stage	$T^{(2)}, °F$	$T^{(1)}, °F$
1	66	65
2	94	90
3	131	115
4	154	140
5	184	165

☐

The rate of convergence of the BP method is unpredictable, and, as shown in Example 15.2, it can depend drastically on the assumed initial set of T_j values. In addition, cases with high reflux ratios can be more difficult to converge than cases with low reflux ratios. Orbach and Crowe[12] describe a generalized extrapolation method for accelerating convergence based on periodic adjustment of the tear variables when their values form geometric progressions during at least four successive iterations.

Example 15.2.   Calculate stage temperatures, interstage vapor and liquid flow rates and compositions, reboiler duty, and condenser duty by the BP method for the distillation column specifications given in Example 12.11.

Solution.   The computer program of Johansen and Seader[10] based on the Wang–Henke procedure was used. In this program, no adjustments to the tear variables are made prior to the start of each iteration, and the convergence criterion is (15-32). The

	Assumed Temperatures, °F		Number of Iterations for Convergence	Execution Time on UNIVAC 1108, sec
Case	**Distillate**	**Bottoms**		
1	11.5	164.9	29	6.0
2	0.0	200.0	5	2.1
3	20.0	180.0	12	3.1
4	50.0	150.0	19	3.7

K-values and enthalpies are computed as in Example 12.8. The only initial assumptions required are distillate and bottoms temperatures.

The significant effect of initially assumed distillate and bottoms temperatures on the number of iterations required to satisfy (15-32) is indicated by the results on page 574. Case 1 used the same terminal temperatures assumed to initiate the group method in Example 12.11. These temperatures were within a few degrees of the exact values and were much closer estimates that those of the other three cases. Nevertheless, Case 1 required the largest number of iterations. Figure 15.8 is a plot of τ from (15-32) as a function of the number of iterations for each of the four cases. Case 2 converged rapidly to the criterion of $\tau < 0.13$. Cases 1, 3, and 4 converged rapidly for the first three or four iterations, but then moved only slowly toward the criterion. This was particularly true of Case 1, for which application of a convergence acceleration method would be particularly desirable. In none of the four cases did oscillations of values of the tear variables occur; rather the values approached the converged results in a monotonic fashion.

The overall results of the converged calculations, as taken from Case 2, are shown in Fig. 15.9. Product component flow rates were not quite in material balance with the feed. Therefore, adjusted values that do satisfy overall material balance equations were determined by averaging the calculated values and are included in Fig. 15.9. A smaller value of τ would have improved the overall material balance. Figures 15.10, 15.11, 15.12, and 15.13 are plots of converged values for stage temperatures, interstage flow rates, and mole fraction compositions from the results of Case 2. Results from the other three cases were almost identical to those of Case 2. Included in Fig. 15.10 is the initially assumed linear temperature profile. Except for the bottom stages, it does not deviate significantly from the converged profile. A jog in the profile is seen at the feed stage. This is a common occurrence.

In Fig. 15.11, it is seen that the assumption of constant interstage molal flow rates does not hold in the rectifying section. Both liquid and vapor flow rates decrease in moving down from the top stage toward the feed stage. Because the feed is vapor near the dew point, the liquid rate changes only slightly across the feed stage. Correspondingly, the vapor rate decreases across the feed stage by an amount almost equal to the feed rate. For this problem, the interstage molal flow rates are almost constant in the stripping section. However, the assumed vapor flow rate in this section based on adjusting the rectifying section rate across the feed zone (see Fig. 12.27a) is approximately 33% higher than the average converged vapor rate. A much better initial estimate of the vapor rate in the stripping section can be made as in Example 12.11 by first computing the reboiler duty from the condenser duty based on the specified reflux rate and then determining the corresponding vapor rate leaving the partial reboiler (see Fig. 12.27b).

For this problem, the separation is between C_2 and C_3. Thus, these two components can be designated as the light key (LK) and heavy key (HK), respectively. Thus C_1 is a lighter-than-light key (LLK), and C_4 and C_5 are heavier than the heavy key (HHK). Each of these four designations exhibits a different type of composition profile curve as shown in Figs. 15.12 and 15.13. Except at the feed zone and at each end of the column, both liquid and vapor mole fractions of the light key (C_2) decrease smoothly and continuously from the top of the column to the bottom. The inverse occurs for the heavy key (C_3). Mole fractions of methane (LLK) are almost constant over the rectifying section except near the top. Below the feed zone, methane rapidly disappears from both vapor and liquid streams. The inverse is true for the two HHK components. In Fig. 15.13, it is seen that the feed composition is somewhat different from the composition of either the vapor entering the feed stage from the stage below or the vapor leaving the feed stage.

□

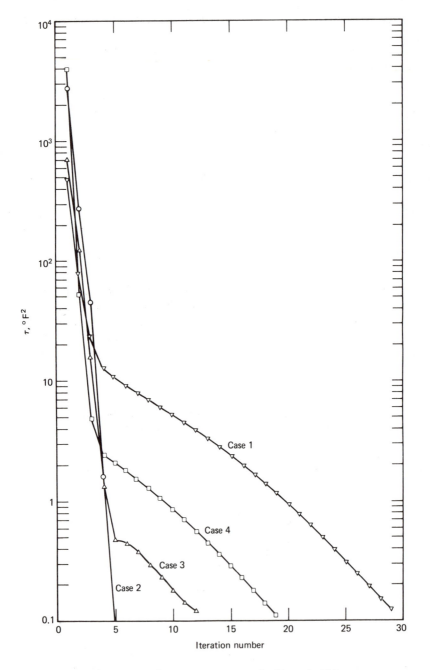

Figure 15.8. Convergence patterns for Example 15.2.

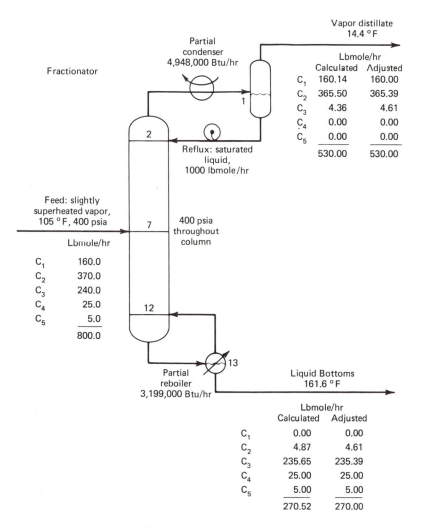

Figure 15.9. Specifications and overall results for Example 15.2.

For problems where a specification is made of the distillate flow rate and the number of theoretical stages, it is difficult to specify the feed-stage location that will give the higest degree of separation. However, once the results of a rigorous calculation are available, a modified McCabe–Thiele plot based on the key components[13] can be constructed to determine whether the feed stage is optimally located or whether it should be moved. For this plot, mole fractions of the light-key component are computed on a nonkey-free basis. The resulting

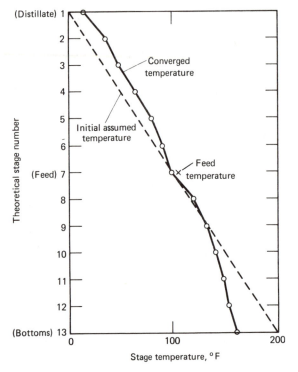

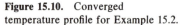

Figure 15.10. Converged temperature profile for Example 15.2.

diagram for Example 15.2 is shown in Fig. 15.14. It is seen that the trend toward a pinched-in region is more noticeable in the rectifying section just above stage 7 than in the stripping section just below stage 7. This suggests that a better separation between the key components might be made by shifting the feed entry to stage 6. The effect of feed-stage location on the percent loss of ethane to the bottoms product is shown in Fig. 15.15. As predicted from Fig. 15.14, the optimum feed stage is stage 6.

Sum-Rates (SR) Method for Absorption and Stripping

The chemical components present in most absorbers and strippers cover a relatively wide range of volatility. Hence, the BP method of solving the MESH equations will fail because calculation of stage temperature by bubble-point determination (15-20) is too sensitive to liquid-phase composition and the stage energy balance (15-5) is much more sensitive to stage temperatures than to interstage flow rates. In this case, Friday and Smith[7] showed that an alternative procedure devised by Sujata[14] could be successfully applied. This procedure, termed the *sum-rates* (SR) *method*, was further developed in conjunction with the tridiagonal matrix formulation for the modified M equations by Burningham and Otto.[15]

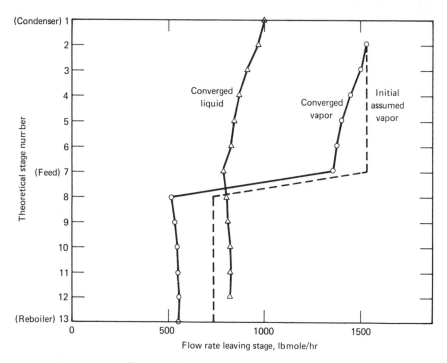

Figure 15.11. Converged interstage flow rate profiles for Example 15.2.

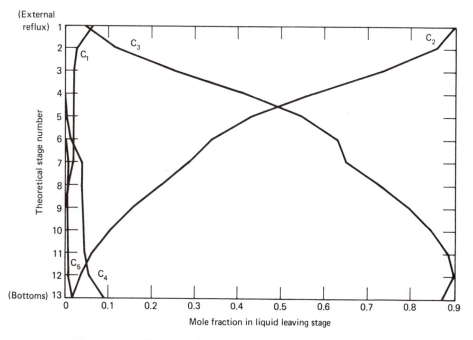

Figure 15.12. Converged liquid composition profiles for Example 15.2.

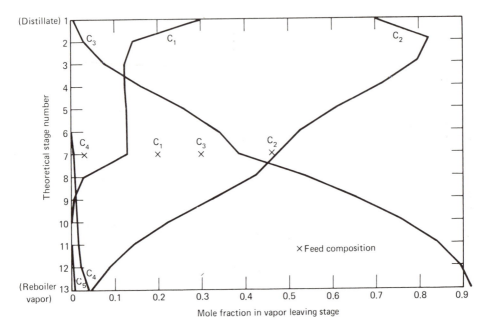

Figure 15.13.   Converged vapor composition profiles for Example 15.2.

Figure 15.16 shows the algorithm for the Burningham–Otto SR method. A FORTRAN computer program for the method is available.[16] Problem specifications consist of conditions and stage locations for all feeds, pressure at each stage, total flow rates of any side streams, heat transfer rates to or from any stages, and total number of stages.

An initial set of tear variables T_j and V_j is assumed to initiate the calculations. For most problems it is sufficient to assume a set of V_j values based on the assumption of constant molal interstage flows, working up from the bottom of the absorber using specified vapor feeds and any vapor side-stream flows. Generally, an adequate initial set of T_j values can be derived from assumed top-stage and bottom-stage values and a linear variation with stages in-between.

Values of $x_{i,j}$ are obtained by solving (15-12) by the Thomas algorithm. However, the values obtained are not normalized at this step but are utilized directly to produce new values of L_j by applying (15-4) in the form referred to as the *sum-rates equation*.

$$L_j^{(k+1)} = L_j^{(k)} \sum_{i=1}^{C} x_{i,j} \qquad (15\text{-}33)$$

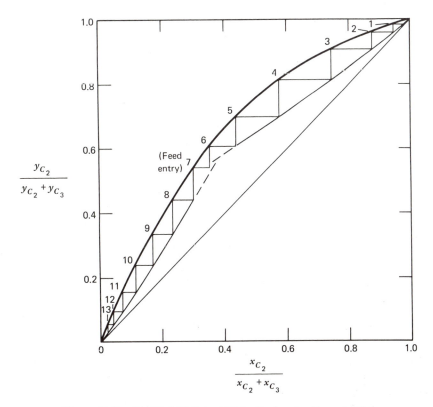

Figure 15.14. Modified McCabe–Thiele diagram for Example 15.2.

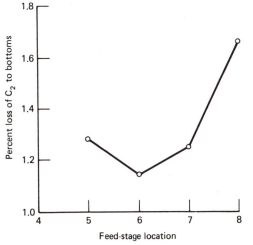

Figure 15.15. Effect of feed–stage location on separation for Example 15.2.

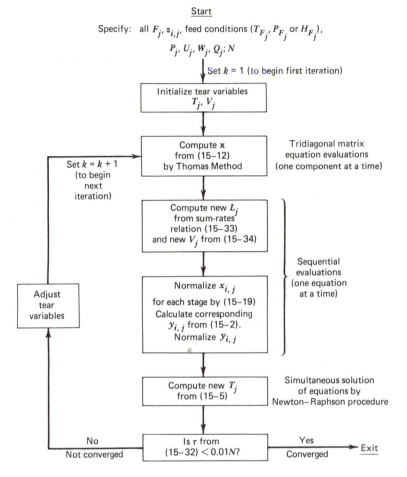

Figure 15.16. Algorithm for Burningham–Otto SR methods for absorption/stripping.

where values of $L_j^{(k)}$ are obtained from values of $V_j^{(k)}$ by (15-6). Corresponding values of $V_j^{(k+1)}$ are obtained from a total material balance, which is derived by summing (15-1) over the C components, combining the result with (15-3) and (15-4) and summing that result over stages j through N to give

$$V_j = L_{j-1} - L_N + \sum_{m=j}^{N} (F_m - W_m - U_m) \qquad (15\text{-}34)$$

Normalized values of $x_{i,j}$ are next calculated from (15-19). Corresponding values of $y_{i,j}$ are computed from (15-2).

A new set of values for stage temperatures T_j is obtained by solving the simultaneous set of energy balance relations for the N stages given by (15-5). The temperatures are embedded in the specific enthalpies corresponding to the unspecified vapor and liquid flow rates. In general, these enthalpies are nonlinear in temperature. Therefore, an iterative solution procedure is required, such as the Newton–Raphson method.[2]

In the Newton–Raphson method, the simultaneous nonlinear equations are written in the form

$$f_i\{x_1, x_2, \ldots, x_n\} = 0 \qquad i = 1, 2, \ldots, n \tag{15-35}$$

Initial guesses, marked by asterisks, are provided for the n variables and each function is expanded about these guesses in a Taylor's series that is terminated after the first derivatives to give

$$0 = f_i\{x_1, x_2, \ldots, x_n\} \tag{15-36}$$

$$\approx f_i\{x_1^*, x_2^*, \ldots, x_n^*\} + \frac{\partial f_i}{\partial x_1}\bigg|^* \Delta x_1 + \frac{\partial f_i}{\partial x_2}\bigg|^* \Delta x_2 + \cdots + \frac{\partial f_i}{\partial x_n}\bigg|^* \Delta x_n \tag{15-37}$$

where $\Delta x_j = x_j - x_j^*$.

Equations (15-36) are linear and can be solved directly for the corrections Δx_i. If the corrections are all found to be zero, the guesses are correct and equations (15-35) have been solved; if not, the corrections are added to the guesses to provide a new set of guesses that are applied to (15-36). The procedure is repeated until all the corrections, and thus the functions, become zero to within some tolerance. In recursion form (15-36) and (15-37) are

$$\sum_{j=1}^{n}\left[\left(\frac{\partial f_i}{\partial x_j}\right)^{(r)} \Delta x_j^{(r)}\right] = -f_i^{(r)} \qquad i = 1, 2, \ldots, n \tag{15-38}$$

$$x_j^{(r+1)} = x_j^{(r)} + \Delta x_j^{(r)} \qquad j = 1, 2, \ldots, n \tag{15-39}$$

Example 15.3. Solve the simultaneous nonlinear equations

$$x_1 \ln x_2 + x_2 \exp(x_1) = \exp(1)$$
$$x_2 \ln x_1 + 2x_1 \exp(x_2) = 2 \exp(1)$$

for x_1 and x_2 to within ± 0.001, by the Newton–Raphson method.

Solution. In the form of (15-35), the two equations are

$$f_1\{x_1, x_2\} = x_1 \ln x_2 + x_2 \exp(x_1) - \exp(1) = 0$$
$$f_2\{x_1, x_2\} = x_2 \ln x_1 + 2x_1 \exp(x_2) - 2 \exp(1) = 0$$

From (15-38), the linearized recursive form of these equations is

$$\left(\frac{\partial f_1}{\partial x_1}\right)^{(r)} \Delta x_1^{(r)} + \left(\frac{\partial f_1}{\partial x_2}\right)^{(r)} \Delta x_2^{(r)} = -f_1^{(r)}$$

$$\left(\frac{\partial f_2}{\partial x_1}\right)^{(r)} \Delta x_1^{(r)} + \left(\frac{\partial f_2}{\partial x_2}\right)^{(r)} \Delta x_2^{(r)} = -f_2^{(r)}$$

The solution of these two equations is readily obtained by the method of determinants to give

$$\Delta x_1^{(r)} = \frac{\left[f_2^{(r)} \left(\frac{\partial f_1}{\partial x_2} \right)^{(r)} - f_1^{(r)} \left(\frac{\partial f_2}{\partial x_2} \right)^{(r)} \right]}{D}$$

and

$$\Delta x_2^{(r)} = \frac{\left[f_1^{(r)} \left(\frac{\partial f_2}{\partial x_1} \right)^{(r)} - f_2^{(r)} \left(\frac{\partial f_1}{\partial x_1} \right)^{(r)} \right]}{D}$$

where

$$D = \left(\frac{\partial f_1}{\partial x_1} \right)^{(r)} \left(\frac{\partial f_2}{\partial x_2} \right)^{(r)} - \left(\frac{\partial f_1}{\partial x_2} \right)^{(r)} \left(\frac{\partial f_2}{\partial x_1} \right)^{(r)}$$

and the derivatives as obtained from the equations are

$$\left(\frac{\partial f_1}{\partial x_1} \right)^{(r)} = \ln(x_2^{(r)}) + x_2^{(r)} \exp(x_1^{(r)}) \qquad \left(\frac{\partial f_2}{\partial x_1} \right)^{(r)} = \frac{x_2^{(r)}}{x_1^{(r)}} + 2 \exp(x_2^{(r)})$$

$$\left(\frac{\partial f_1}{\partial x_2} \right)^{(r)} = \frac{x_1^{(r)}}{x_2^{(r)}} + \exp(x_1^{(r)}) \qquad \left(\frac{\partial f_2}{\partial x_2} \right)^{(r)} = \ln(x_1^{(r)}) + 2x_1^{(r)} \exp(x_2^{(r)})$$

As initial guesses, take $x_1^{(1)} = 2$, $x_2^{(1)} = 2$. Applying the Newton–Raphson procedure, one obtains the following results where at the sixth iteration values of $x_1 = 1.0000$ and $x_2 = 1.0000$ correspond closely to the required values of zero for f_1 and f_2.

r	$x_1^{(r)}$	$x_2^{(r)}$	$f_1^{(r)}$	$f_2^{(r)}$	$(\partial f_1 / \partial x_1)^{(r)}$	$(\partial f_1 / \partial x_2)^{(r)}$	$(\partial f_2 / \partial x_1)^{(r)}$	$(\partial f_2 / \partial x_2)^{(r)}$	$\Delta x_1^{(r)}$	$\Delta x_2^{(r)}$
1	2.0000	2.0000	13.4461	25.5060	15.4731	8.3891	15.7781	30.2494	−0.5743	−0.5436
2	1.4257	1.4564	3.8772	7.3133	6.4354	5.1395	9.6024	12.5880	−0.3544	−0.3106
3	1.0713	1.1457	0.7720	1.3802	3.4806	3.8541	7.3591	6.8067	−0.0138	−0.1878
4	1.0575	0.9579	−0.0059	0.1290	2.7149	3.9830	6.1183	5.5679	−0.0591	0.0417
5	0.9984	0.9996	−0.0057	−0.0122	2.7126	3.7127	6.4358	5.4244	0.00159	0.000368
6	1.0000	1.0000	5.51×10^{-6}	2.86×10^{-6}	2.7183	3.7183	6.4366	5.4366	12.1×10^{-6}	-3.0×10^{-6}
7	1.0000	1.0000	0.0	-2×10^{-9}	2.7183	3.7183	6.4366	5.4366	—	—

☐

As applied to the solution of a new set of T_j values from the energy equation (15-5), the recursion equation for the Newton–Raphson method is

$$\left(\frac{\partial H_j}{\partial T_{j-1}} \right)^{(r)} \Delta T_{j-1}^{(r)} + \left(\frac{\partial H_j}{\partial T_j} \right)^{(r)} \Delta T_j^{(r)} + \left(\frac{\partial H_j}{\partial T_{j+1}} \right)^{(r)} \Delta T_{j+1}^{(r)} = -H_j^{(r)} \qquad (15\text{-}40)$$

where

$$\Delta T_j^{(r)} = T_j^{(r+1)} - T_j^{(r)}| \qquad (15\text{-}41)$$

$$\frac{\partial H_j}{\partial T_{j-1}} = L_{j-1} \frac{\partial H_{L_{j-1}}}{\partial T_{j-1}} \qquad (15\text{-}42)$$

$$\frac{\partial H_j}{\partial T_j} = -(L_j + U_j)\frac{\partial H_{L_j}}{\partial T_j} - (V_j + W_j)\frac{\partial H_{V_j}}{\partial T_j} \tag{15-43}$$

$$\frac{\partial H_j}{\partial T_{j+1}} = V_{j+1}\frac{\partial H_{V_{j+1}}}{\partial T_{j+1}} \tag{15-44}$$

The partial derivatives depend upon the enthalpy correlations that are utilized. For example, if composition-independent polynomial equations in temperature are used then

$$H_{V_j} = \sum_{i=1}^{C} y_{i,j}(A_i + B_i T + C_i T^2) \tag{15-45}$$

$$H_{L_j} = \sum_{i=1}^{C} x_{i,j}(a_i + b_i T + c_i T^2) \tag{15-46}$$

and the partial derivatives are

$$\frac{\partial H_{V_j}}{\partial T_j} = \sum_{i=1}^{C} y_{i,j}(B_i + 2C_i T) \tag{15-47}$$

$$\frac{\partial H_{L_j}}{\partial T_j} = \sum_{i=1}^{C} x_{i,j}(b_i + 2c_i T) \tag{15-48}$$

The N relations given by (15-40) form a tridiagonal matrix equation that is linear in $\Delta T_j^{(r)}$. The form of the matrix equation is identical to (15-12) where, for example, $A_2 = (\partial H_2/\partial T_1)^{(r)}$, $B_2 = (\partial H_2/\partial T_2)^{(r)}$, $C_2 = (\partial H_2/\partial T_3)^{(r)}$, $x_{i,2} \leftarrow \Delta T_2^{(r)}$, and $D_2 = -H_2^{(r)}$. The matrix of partial derivatives is called the *Jacobian correction matrix*. The Thomas algorithm can be employed to solve for the set of corrections $\Delta T_j^{(r)}$. New guesses of T_j are then determined from

$$T_j^{(r+1)} = T_j^{(r)} + t\Delta T_j^{(r)} \tag{15-49}$$

where t is a scalar attenuation factor that is useful when initial guesses are not reasonably close to the true values. Generally, as in (15-39), t is taken as one, but an optimal value can be determined at each iteration to minimize the sum of the squares of the functions

$$\sum_{j=1}^{N} \left[H_j^{(r+1)} \right]^2$$

When all the corrections $\Delta T_j^{(r)}$ have approached zero, the resulting values of T_j are used with criteria like (15-31) or (15-32) to determine whether convergence has been achieved. If not, before beginning a new k iteration, one can adjust values of V_j and T_j as indicated in Fig. 15.16 and previously discussed for the BP method. Rapid convergence is generally observed for the sum-rates method.

Example 15.4. Calculate stage temperatures and interstage vapor and liquid flow rates and compositions by the rigorous SR method for the absorber column specifications given in Example 12.8.

Solution. The digital computer program of Shinohara et al.,[16] based on the Burningham–Otto solution procedure, was used. In this program, K-values and enthalpies are
computed as in Example 12.8. Initial assumptions for the top-stage and bottom-stage
temperatures were 90°F (32.2°C) (entering liquid temperature) and 105°F (40.6°C) (entering
gas temperature), respectively. The corresponding number of iterations to satisfy the
convergence criterion of (15-32) was seven. Values of τ were as follows.

Iteration Number	τ, $(°F)^2$
1	9948
2	2556
3	46.0
4	8.65
5	0.856
6	0.124
7	0.0217

The overall results of the converged calculations are shown in Fig. 15.17. Adjusted
values of product component flow rates that satisfy overall material balance equations are
included. Figures 15.18, 15.19, and 15.20 are plots of converged values for stage temperatures, interstage total flow rates, and interstage component vapor flow rates, respectively. Figure 15.18 shows that the initial assumed linear temperature profile is grossly in
error. Due to the substantial degree of absorption and accompanying high heat of
absorption, stage temperatures are considerably greater than the two entering stream
temperatures. The heat is absorbed by both the vapor and liquid streams. The peak stage
temperature is essentially at the midpoint of the column. Figure 15.19 shows that the bulk
of the overall absorption occurs at the two terminal stages. In Fig. 15.20, it is seen that
absorption of C_1 and C_2 occurs almost exclusively at the top and bottom stages.
Absorption of C_3 occurs throughout the column, but mainly at the two terminal stages.
Absorption of C_4 and C_5 also occurs throughout the column, but mainly at the bottom
where vapor first contacts absorption oil.

□

Isothermal Sum-Rates (ISR) Method for Liquid–Liquid Extraction

Multistage liquid–liquid extraction equipment is operated frequently in an adiabatic manner. When entering streams are at the same temperature and heat of
mixing is negligible, the operation is also isothermal. For this condition, or when
stage temperatures are specified, as indicated by Friday and Smith[7] and shown in
detail by Tsuboka and Katayama,[17] a simplified isothermal version of the
sum-rates method (ISR) can be applied. It is based on the same equilibrium-stage
model presented in Section 15.1. However, with all stage temperatures specified,
values of Q_j can be computed from stage energy balances, which can be
partitioned from the other equations and solved in a separate step following the
calculations discussed here. In the ISR method, particular attention is paid to the
possibility that phase compositions may strongly influence K_{ij} values.

Figure 15.21 shows the algorithm for the Tsuboka–Katayama ISR method.
Liquid-phase and vapor-phase symbols correspond to raffinate and extract,
respectively. Problem specifications consist of flow rates, compositions, and

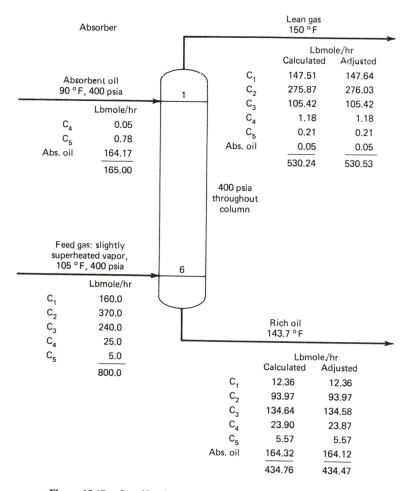

Figure 15.17. Specifications and overall results for Example 15.4.

stage locations for all feeds; stage temperatures (frequently all equal); total flow rates of any side streams; and total number of stages. Stage pressures need not be specified but are understood to be greater than corresponding stage bubble-point pressures to prevent vaporization.

With stage temperatures specified, the only tear variables are V_j values. An initial set is obtained by assuming a perfect separation among the components of the feed and neglecting mass transfer of the solvent to the raffinate phase. This gives approximate values for the flow rates of the exiting raffinate and extract phases. Intermediate values of V_j are obtained by linear interpolation over the N stages. Modifications to this procedure are necessary for side streams or

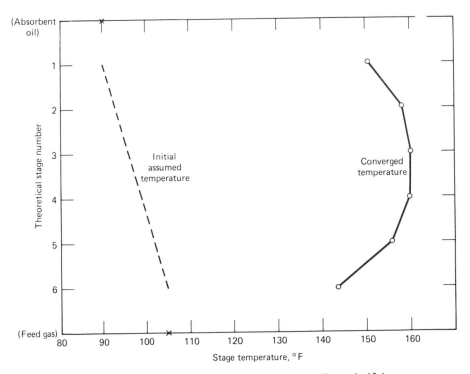

Figure 15.18. Converged temperature profile for Example 15.4.

intermediate feeds. As shown in Fig. 15.21, the tear variables are reset in an outer iterative loop.

The effect of phase compositions is often considerable on K-values (distribution coefficients) for liquid–liquid extraction. Therefore, it is best also to provide initial estimates of $x_{i,j}$ and $y_{i,j}$ from which initial values of $K_{i,j}$ are computed. Initial values of $x_{i,j}$ are obtained by linear interpolation with stage of the compositions of the known entering and assumed exit streams. Corresponding values of $y_{i,j}$ are computed by material balance from (15-1). Values of $\gamma_{iL,j}$ and $\gamma_{iV,j}$ are determined from an appropriate correlation—for example, the van Laar, NRTL, UNIQUAC, or UNIFAC equations discussed in Chapter 5. Corresponding K-values are obtained from the following equation, which is equivalent to (4-31).

$$K_{i,j} = \frac{\gamma_{iL,j}}{\gamma_{iV,j}} \tag{15-50}$$

A new set of $x_{i,j}$ values is obtained by solving (15-12) by the Thomas

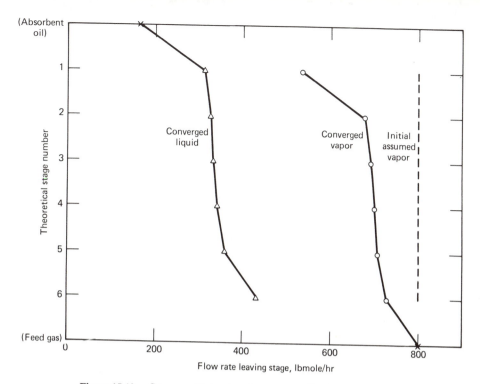

Figure 15.19. Converged interstage flow rate profiles for Example 15.4.

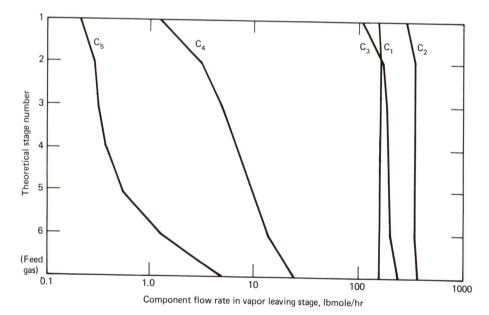

Figure 15.20. Converged component vapor flow rate profiles for Example 15.4.

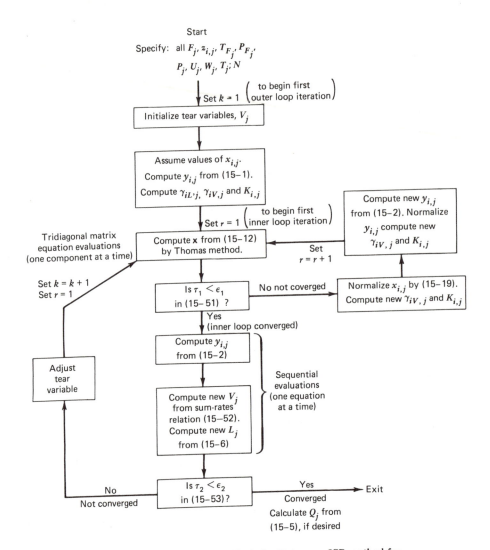

Figure 15.21. Algorithm for Tsuboka–Katayama ISR method for liquid–liquid extraction.

algorithm. These values are compared to the assumed values by computing

$$\tau_1 = \sum_{j=1}^{N} \sum_{i=1}^{C} |x_{i,j}^{(r-1)} - x_{i,j}^{(r)}| \qquad (15\text{-}51)$$

where r is an inner loop index. If $\tau_1 > \epsilon_1$, where, for example, the convergence criterion ϵ_1 might be taken as 0.01 NC, the inner loop is used to improve values of $K_{i,j}$ by using normalized values of $x_{i,j}$ and $y_{i,j}$ to compute new values of $\gamma_{iL,j}$ and $\gamma_{iV,j}$.

When the inner loop is converged, values of $x_{i,j}$ are used to calculate new values of $y_{i,j}$ from (15-2). A new set of tear variables V_j is then computed from the sum-rates relation

$$V_j^{(k+1)} = V_j^{(k)} \sum_{i=1}^{C} y_{i,j} \tag{15-52}$$

where k is an outer loop index. Corresponding values of $L_j^{(k+1)}$ are obtained from (15-6).

The outer loop is converged when

$$\tau_2 = \sum_{j-1}^{N} \left(\frac{V_j^{(k)} - V_j^{(k-1)}}{V_j^{(k)}} \right)^2 \leq \epsilon_2 \tag{15-53}$$

where, for example, the convergence criterion ϵ_2 may be taken as 0.01 N.

Before beginning a new k iteration, we can adjust values of V_j as previously discussed for the BP method. Convergence of the ISR method is generally rapid but is subject to the extent to which $K_{i,j}$ depends upon composition.

Example 15.5. The separation of benzene B from n-heptane H by ordinary distillation is difficult. At atmospheric pressure, the boiling points differ by 18.3°C. However, because of liquid-phase nonideality, the relative volatility decreases to a value less than 1.15 at high benzene concentrations.[18] An alternative method of separation is liquid–liquid extraction with a mixture of dimethylformamide (DMF) and water.[19] The solvent is much more selective for benzene than for n-heptane at 20°C. For two different solvent compositions, calculate interstage flow rates and compositions by the rigorous ISR method for the countercurrent liquid–liquid extraction cascade, which contains five equilibrium stages and is shown schematically in Fig. 15.22.

Solution. Experimental phase equilibrium data for the quaternary system[19] were fitted to the NRTL equation by Cohen and Renon.[20] The resulting binary pair constants in (5-68) and (5-69) are

Binary Pair, ij	τ_{ij}	τ_{ji}	α_{ji}
DMF, H	2.036	1.910	0.25
Water, H	7.038	4.806	0.15
B, H	1.196	−0.355	0.30
Water, DMF	2.506	−2.128	0.253
B, DMF	−0.240	0.676	0.425
B, Water	3.639	5.750	0.203

For Case A, initial estimates of V_j (the extract phase), $x_{i,j}$, and $y_{i,j}$ are as follows, based on a perfect separation and linear interpolation by stage.

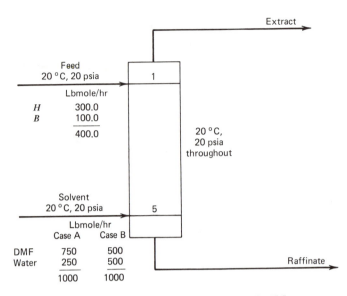

Figure 15.22. Specifications for Example 15.5.

Stage		$y_{i,j}$				$x_{i,j}$			
j	V_j	H	B	DMF	Water	H	B	DMF	Water
1	1100	0.0	0.0909	0.6818	0.2273	0.7895	0.2105	0.0	0.0
2	1080	0.0	0.0741	0.6944	0.2315	0.8333	0.1667	0.0	0.0
3	1060	0.0	0.0566	0.7076	0.2359	0.8824	0.1176	0.0	0.0
4	1040	0.0	0.0385	0.7211	0.2404	0.9375	0.0625	0.0	0.0
5	1020	0.0	0.0196	0.7353	0.2451	1.0000	0.0	0.0	0.0

The converged solution is obtained by the ISR method with the following corresponding stage flow rates and compositions.

Stage		$y_{i,j}$				$x_{i,j}$			
j	V_j	H	B	DMF	Water	H	B	DMF	Water
1	1113.1	0.0263	0.0866	0.6626	0.2245	0.7586	0.1628	0.0777	0.0009
2	1104.7	0.0238	0.0545	0.6952	0.2265	0.8326	0.1035	0.0633	0.0006
3	1065.6	0.0213	0.0309	0.7131	0.2347	0.8858	0.0606	0.0532	0.0004
4	1042.1	0.0198	0.0157	0.7246	0.2399	0.9211	0.0315	0.0471	0.0003
5	1028.2	0.0190	0.0062	0.7316	0.2432	0.9438	0.0125	0.0434	0.0003

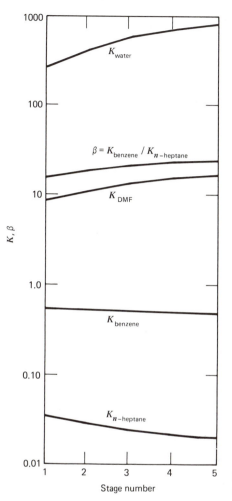

Figure 15.23. Variation of distribution coefficient and relative selectivity for Example 15.5, Case A.

Computed products for the two cases are:

	Extract, lbmole/hr		Raffinate, lbmole/hr	
	Case A	Case B	Case A	Case B
H	29.3	5.6	270.7	294.4
B	96.4	43.0	3.6	57.0
DMF	737.5	485.8	12.5	14.2
Water	249.9	499.7	0.1	0.3
	1113.1	1034.1	286.9	365.9

On a percentage extraction basis, the results are:

	Case A	Case B
Percent of benzene feed extracted	96.4	43.0
Percent of *n*-heptane feed extracted	9.8	1.87
Percent of solvent transferred to raffinate	1.26	1.45

Thus, the solvent with 75% DMF extracts a much larger percentage of the benzene, but the solvent with 50% DMF is more selective between benzene and *n*-heptane.

For Case A, the variations with stage of K-values and the relative selectivity are shown in Fig. 15.23, where the relative selectivity is $\beta_{B,H} = K_B/K_H$. The distribution coefficient for *n*-heptane varies by a factor of almost 1.75 from stage 5 to stage 1, while the coefficient for benzene is almost constant. The relative selectivity varies by a factor of almost two.
□

15.4. Simultaneous Correction (SC) Procedures

The BP and SR methods for vapor–liquid contacting converge only with difficulty or not at all for separations involving very nonideal liquid mixtures (e.g., in extractive distillation) or for cases where the separator is like an absorber or stripper in one section and a fractionator in another section (e.g., a reboiled absorber). Furthermore, BP and SR methods are generally restricted to the very limited specifications stated above. More general procedures capable of solving all types of multicomponent, multistage separation problems are based on the solution of all the MESH equations, or combinations thereof, by simultaneous correction (SC) techniques.

Newton–Raphson Method

In order to develop an SC procedure that uses the Newton–Raphson method, one must select and order the unknown variables and the corresponding functions (MESH equations) that contain them. As discussed by Goldstein and Stanfield,[21] grouping of the functions by type is computationally most efficient for problems involving a large number of components, but few stages. Alternatively, it is most efficient to group the functions according to stage location for problems involving many stages, but relatively few components. The latter grouping is described by Naphtali[22] and was implemented by Naphtali and Sandholm.[23]

The SC procedure of Naphtali and Sandholm is developed in detail because it utilizes many of the mathematical techniques presented in Section 15.3 on tearing methods. A computer program for their method is given by Fredenslund, Gmehling, and Rasmussen.[24]

The equilibrium stage model of Figs. 15.1 and 15.3 is again employed. However, rather than solving the $N(2C+3)$ MESH equations simultaneously, we combine (15-3) and (15-4) with the other MESH equations to eliminate $2N$ variables and thus reduce the problem to the simultaneous solution of $N(2C+1)$ equations. This is done by first multiplying (15-3) and (15-4) by V_j and L_j, respectively, to give

$$V_j = \sum_{i=1}^{C} v_{i,j} \tag{15-54}$$

$$L_j = \sum_{i=1}^{C} l_{ij} \tag{15-55}$$

where we have used the mole fraction definitions

$$y_{i,j} = \frac{v_{i,j}}{V_j} \tag{15-56}$$

$$x_{i,j} = \frac{l_{i,j}}{L_j} \tag{15-57}$$

Equations (15-54), (15-55), (15-56), and (15-57) are now substituted into (15-1), (15-2), and (15-5) to eliminate V_j, L_j, $y_{i,j}$ and $x_{i,j}$ and introduce component flow rates $v_{i,j}$ and $l_{i,j}$. As a result, the following $N(2C+1)$ equations are obtained, where $s_j = U_j/L_j$ and $S_j = W_j/V_j$ are dimensionless side-stream flow rates.

Material Balance

$$M_{i,j} = l_{i,j}(1 + s_j) + v_{i,j}(1 + S_j) - l_{i,j-1} - v_{i,j+1} - f_{i,j} = 0 \tag{15-58}$$

Phase Equilibria

$$E_{i,j} = K_{i,j} l_{i,j} \frac{\sum\limits_{\kappa=1}^{C} v_{\kappa,j}}{\sum\limits_{\kappa=1}^{C} l_{\kappa,j}} - v_{i,j} = 0 \tag{15-59}$$

Energy Balance

$$H_j = H_{L_j}(1 + s_j) \sum_{i=1}^{C} l_{i,j} + H_{V_j}(1 + S_j) \sum_{i=1}^{C} v_{i,j} - H_{L_{j-1}} \sum_{i=1}^{C} l_{i,j-1}$$
$$- H_{V_{j+1}} \sum_{i=1}^{C} v_{i,j+1} - H_{F_j} \sum_{i=1}^{C} f_{i,j} - Q_j = 0 \tag{15-60}$$

where $f_{i,j} = F_j z_{i,j}$

If N and all $f_{i,j}$, T_{F_j}, P_{F_j}, P_j, s_j, S_j, and Q_j are specified, the M, E, and H functions are nonlinear in the $N(2C+1)$ unknown (output) variables $v_{i,j}$, $l_{i,j}$, and T_j for $i = 1$ to C and $j = 1$ to N. Although other sets of specified and unknown variables are possible, we consider these sets first.

Equations (15-58), (15-59), and (15-60) are solved simultaneously by the Newton–Raphson iterative method in which successive sets of the output variables are produced until the values of the M, E, and H functions are driven to within some tolerance of zero. During the iterations, nonzero values of the functions are called *discrepancies* or *errors*. Let the functions and output variables be grouped by stage in order from top to bottom. As will be shown, this is done to produce a block tridiagonal structure for the Jacobian matrix of partial derivatives so that the Thomas algorithm can be applied. Let

$$\mathbf{X} = [\mathbf{X}_1, \mathbf{X}_2, \ldots, \mathbf{X}_j, \ldots, \mathbf{X}_N]^T \tag{15-61}$$

and

$$\mathbf{F} = [\mathbf{F}_1, \mathbf{F}_2, \ldots, \mathbf{F}_j, \ldots, \mathbf{F}_N]^T \tag{15-62}$$

where $\mathbf{X}_j$ is the vector of output variables for stage j arranged in the order:

$$\mathbf{X}_j = [v_{1,j}, v_{2,j}, \ldots, v_{i,j}, \ldots, v_{C,j}, T_j, l_{1,j}, l_{2,j}, \ldots, l_{i,j}, \ldots, l_{C,j}]^T \tag{15-63}$$

and $\mathbf{F}_j$ is the vector of functions for stage j arranged in the order:

$$\mathbf{F}_j = [H_j, M_{1,j}, M_{2,j}, \ldots, M_{i,j}, \ldots, M_{C,j}, E_{1,j}, E_{2,j}, \ldots, E_{i,j}, \ldots, E_{C,j}]^T \tag{15-64}$$

The Newton–Raphson iteration is performed by solving for the corrections $\Delta\mathbf{X}$ to the output variables from (15-38), which in matrix form becomes

$$\Delta\mathbf{X}^{(k)} = -\left[\left(\frac{\partial\mathbf{F}}{\partial\mathbf{X}}\right)^{-1}\right]^{(k)} \mathbf{F}^{(k)} \tag{15-65}$$

These corrections are used to compute the next approximation to the set of output variables from

$$\mathbf{X}^{(k+1)} = \mathbf{X}^{(k)} + t\,\Delta\mathbf{X}^{(k)} \tag{15-66}$$

The quantity $(\overline{\partial\mathbf{F}/\partial\mathbf{X}})$ is the following Jacobian or $(N \times N)$ matrix of blocks of partial derivatives of all the functions with respect to all the output variables.

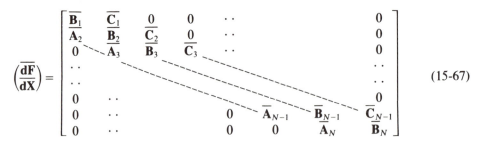

$$\left(\frac{\overline{d\mathbf{F}}}{d\mathbf{X}}\right) = \begin{bmatrix} \overline{\mathbf{B}}_1 & \overline{\mathbf{C}}_1 & 0 & 0 & \cdot\,\cdot & & & 0 \\ \overline{\mathbf{A}}_2 & \overline{\mathbf{B}}_2 & \overline{\mathbf{C}}_2 & 0 & \cdot\,\cdot & & & 0 \\ 0 & \overline{\mathbf{A}}_3 & \overline{\mathbf{B}}_3 & \overline{\mathbf{C}}_3 & \cdot\,\cdot & & & 0 \\ \cdot\,\cdot & & & & & & & \cdot\,\cdot \\ \cdot\,\cdot & & & & & & & \\ 0 & \cdot\,\cdot & & & & & & 0 \\ 0 & \cdot\,\cdot & & & 0 & \overline{\mathbf{A}}_{N-1} & \overline{\mathbf{B}}_{N-1} & \overline{\mathbf{C}}_{N-1} \\ 0 & \cdot\,\cdot & & & 0 & 0 & \overline{\mathbf{A}}_N & \overline{\mathbf{B}}_N \end{bmatrix} \tag{15-67}$$

This Jacobian is of a block tridiagonal form like (15-12) because functions for stage j are only dependent on output variables for stages $j-1$, j, and $j+1$. Each $\overline{\mathbf{A}}$, $\overline{\mathbf{B}}$, or $\overline{\mathbf{C}}$ block in (15-67) represents a $(2C+1)$ by $(2C+1)$ submatrix of partial derivatives, where the arrangements of output variables and functions are given by (15-63) and (15-64), respectively. Blocks $\overline{\mathbf{A}}_j$, $\overline{\mathbf{B}}_j$, and $\overline{\mathbf{C}}_j$ correspond to submatrices of partial derivatives of the functions on stage j with respect to the output variables on stages $j-1$, j, and $j+1$, respectively. Thus, using (15-58), (15-59), and (15-60) and denoting only the nonzero partial derivatives by +, or by row or diagonal strings of $+ \cdots +$, or by the following square or rectangular blocks enclosed by connected strings

$$
\begin{array}{ccc}
+ & \cdots & + \\
\vdots & & \vdots \\
+ & \cdots & +
\end{array}
$$

we find that the blocks have the following form, where + is replaced by a numerical value (-1 or 1) in the event that the partial derivative has only that value.

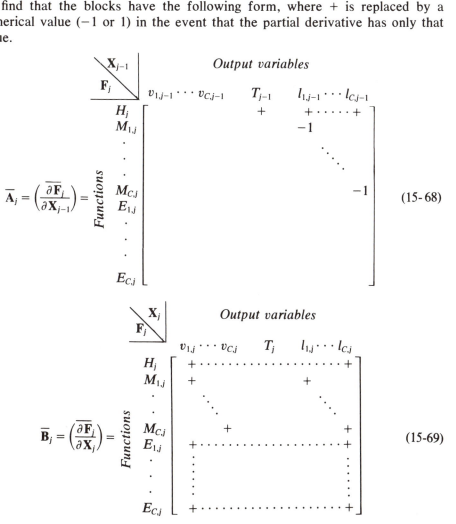

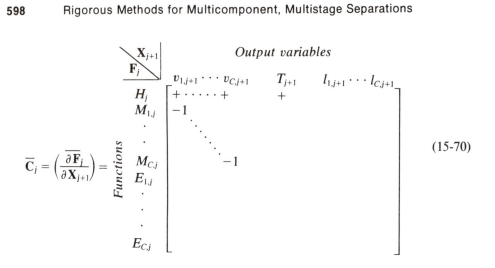

$$\overline{C}_j = \left(\frac{\overline{\partial F_j}}{\partial X_{j+1}}\right) = \text{Functions} \quad (15\text{-}70)$$

Thus, (15-65) consists of a set of $N(2C+1)$ simultaneous linear equations in the $N(2C+1)$ corrections ΔX. For example, the $2C+2$ equation in the set is obtained by expanding function H_2 (15-60) into a Taylor's series like (15-36) around the $N(2C+1)$ output variables. The result is as follows after the usual truncation of terms involving derivatives of order greater than one

$$0(\Delta v_{1,1} + \cdots + \Delta v_{C,1}) - \frac{\partial H_{L_1}}{\partial T_1}\sum_{i=1}^{C} l_{i,1}(\Delta T_1)$$

$$-\left(\frac{\partial H_{L_1}}{\partial l_{1,1}}\sum_{i=1}^{C} l_{i,1} + H_{L_1}\right)\Delta l_{1,1}$$

$$- \cdots - \left(\frac{\partial H_{L_1}}{\partial l_{C,1}}\sum_{i=1}^{C} l_{i,1} + H_{L_1}\right)\Delta l_{C,1} + \left[\left(\frac{\partial H_{V_2}}{\partial v_{1,2}}\right)(1+S_2)\sum_{i=1}^{C} v_{i,2}\right.$$

$$\left. + H_{V_2}(1+S_2)\right]\Delta v_{1,2} + \cdots + \left[\left(\frac{\partial H_{V_2}}{\partial v_{C,2}}\right)(1+S_2)\sum_{i=1}^{C} v_{i,2} + H_{V_2}(1+S_2)\right]\Delta v_{C,2}$$

$$+ \left[\left(\frac{\partial H_{L_2}}{\partial T_2}\right)(1+s_2)\sum_{i=1}^{C} l_{i,2} + \left(\frac{\partial H_{V_2}}{\partial T_2}\right)(1+S_2)\sum_{i=1}^{C} v_{i,2}\right]\Delta T_2$$

$$+ \left[\left(\frac{\partial H_{L_2}}{\partial l_{1,2}}\right)(1+s_2)\sum_{i=1}^{C} l_{i,2} + H_{L_2}(1+s_2)\right]\Delta l_{1,2} + \cdots$$

$$+ \left[\left(\frac{\partial H_{L_2}}{\partial l_{C,2}}\right)(1+s_2)\sum_{i=1}^{C} l_{i,2} + H_{L_2}(1+s_2)\right]\Delta l_{C,2}$$

$$- \left(\frac{\partial H_{V_3}}{\partial v_{1,3}}\sum_{i=1}^{C} v_{i,3} + H_{V_3}\right)\Delta v_{1,3} - \cdots - \left(\frac{\partial H_{V_3}}{\partial v_{C,3}}\sum_{i=1}^{C} v_{i,3} + H_{V_3}\right)\Delta v_{C,3}$$

$$- \frac{\partial H_{V_3}}{\partial T_3}\sum_{i=1}^{C} v_{i,3}\,\Delta T_3 + 0(\Delta l_{1,3} + \cdots + \Delta l_{C,N}) = -H_2 \qquad (15\text{-}71)$$

Although lengthy, equations such as (15-71) are handled readily in computer programs.

As a further example, the entry in the Jacobian matrix for row $(2C + 2)$ and column $(C + 3)$ is obtained from (15-71) as

$$\frac{\partial H_2}{\partial l_{2,1}} = -\frac{\partial H_{L_1}}{\partial l_{2,1}} \sum_{i=1}^{C} l_{i,1} + H_{L_1} \tag{15-72}$$

All partial derivatives are stated by Naphtali and Sandholm.[23]

Partial derivatives of enthalpies and K-values depend upon the particular correlation utilized for these properties and are sometimes simplified by including only the dominant terms. For example, suppose that the Chao–Seader correlation is to be used for K-values. In general,

$$K_{i,j} = K_{i,j} \left\{ P_j, T_j, \frac{l_{i,j}}{\sum\limits_{\kappa=1}^{C} l_{\kappa,j}}, \frac{v_{i,j}}{\sum\limits_{\kappa=1}^{C} v_{\kappa,j}} \right\}$$

In terms of the output variables, the partial derivatives $\partial K_{i,j}/\partial T_j$; $\partial K_{i,j}/\partial l_{i,j}$; and $\partial K_{i,j}/\partial v_{i,j}$ all exist and can be expressed analytically or evaluated numerically if desired. However, for some problems, the terms that include the first and second of these three groups of derivatives may be the dominant terms so that the third group may be taken as zero.

Example 15.6. Derive an expression for $(\partial H_V/\partial T)$ from the Redlich–Kwong equation of state.

Solution. From (4-64)

$$H_V = \sum_{i=1}^{C} (y_i H^\circ_{iV}) + RT \left[Z_V - 1 - \frac{3A^2}{2B} \left(1 + \frac{BP}{Z_V} \right) \right]$$

where H°_{iV}, Z_V, A, and B all depend on T, as determined from (4-60), (4-41), (4-42), (4-43), (4-44), and (4-45). Thus

$$\frac{\partial H_V}{\partial T} = \sum_{i=1}^{C} \left[y_i \left(\frac{\partial H^\circ_{iV}}{\partial T} \right) \right] + R \left[Z_V - 1 - \frac{3A^2}{2B} \left(1 + \frac{BP}{Z_V} \right) \right]$$
$$+ RT \left\{ \left(\frac{\partial Z_V}{\partial T} \right) - \frac{3}{2} \left(\frac{\partial (A^2/B)}{\partial T} \right) \left(1 + \frac{BP}{Z_V} \right) \right.$$
$$\left. - \frac{3A^2}{2B} \left[\frac{P}{Z_V} \left(\frac{\partial B}{\partial T} \right) - \frac{BP}{Z_V^2} \left(\frac{\partial Z_V}{\partial T} \right) \right] \right\}$$

From (4-60) and (4-59)

$$\left(\frac{\partial H^\circ_{iV}}{\partial T} \right) = \sum_{k=1}^{5} (a_k)_i T^{k-1} = (C^\circ_{P_V})_i$$

From (4-43) and (4-45)

$$B = \sum_{i=1}^{C} \left(\frac{y_i \cdot 0.0867 T_{c_i}}{T P_{c_i}} \right)$$

Thus

$$\frac{\partial B}{\partial T} = -\frac{B}{T}$$

From (4-42) to (4-45)

$$\frac{A^2}{B} = \frac{\left[\sum_{i=1}^{C} (y_i \cdot 0.4278^{1/2} \, T_{c_i}^{1.25} / T^{1.25} P_{c_i}) \right]^2}{\sum_{i=1}^{C} (y_i \cdot 0.0867 T_{c_i} / T P_{c_i})}$$

Thus

$$\frac{\partial (A^2/B)}{\partial T} = -\frac{A^2}{BT}$$

From (4-41)

$$Z_V^3 - Z_V^2 + Z_V BP \left(\frac{A^2}{B} - BP - 1 \right) - \frac{A^2}{B} (BP)^2 = 0$$

By implicit differentiation

$$3Z_V^2 \frac{\partial Z_V}{\partial T} - 2Z_V \frac{\partial Z_V}{\partial T} + BP \left(\frac{A^2}{B} - BP - 1 \right) \frac{\partial Z_V}{\partial T}$$
$$+ Z_V P \left(\frac{A^2}{B} - BP - 1 \right) \frac{\partial B}{\partial T} + Z_V BP \left[\frac{\partial (A^2/B)}{\partial T} - P \left(\frac{\partial B}{\partial T} \right) \right]$$
$$- (BP)^2 \frac{\partial (A^2/B)}{\partial T} - A^2 P \left(\frac{\partial B}{\partial T} \right) = 0$$

which when combined with the above expressions for $(\partial B/\partial T)$ and $\partial (A^2/B)/\partial T$ gives

$$\frac{\partial Z_V}{\partial T} = \frac{(Z_V PB/T)[2(A^2/B) - 2BP - 1] - (A^2 BP/T)(P - 1)}{3Z_V^2 - 2Z_V + BP[(A^2/B) - BP - 1]}$$

□

Because the Thomas algorithm can be applied to the block tridiagonal structure (15-67) of (15-70), submatrices of partial derivatives are computed only as needed. The solution of (15-65) follows the scheme in Section 15.3, given by (15-13) to (15-18) and represented in Fig. 15.4, where matrices and vectors $\overline{\mathbf{A}}_j$, $\mathbf{B}_j$, $\overline{\mathbf{C}}_j$, $-\overline{\mathbf{F}}_j$, and $\Delta \mathbf{X}_j$ correspond to variables A_j, B_j, C_j, D_j, and x_j, respectively. However, the simple multiplication and division operations in Section 15.3 are changed to matrix multiplication and inversion, respectively. The steps are as follows

Starting at stage 1, $\overline{\mathbf{C}}_1 \leftarrow (\overline{\mathbf{B}}_1)^{-1} \overline{\mathbf{C}}_1$, $\mathbf{F}_1 \leftarrow (\overline{\mathbf{B}}_1)^{-1} \mathbf{F}_1$, and $\overline{\mathbf{B}}_1 \leftarrow \mathbf{I}$ (the identity submatrix). Only $\overline{\mathbf{C}}_1$ and $\mathbf{F}_1$ are saved. For stages j from 2 to $(N - 1)$, $\overline{\mathbf{C}}_j \leftarrow (\mathbf{B}_j -$

$\overline{A}_j \overline{C}_{j-1})^{-1} \overline{C}_j$, $F_j \leftarrow (\overline{B}_j - \overline{A}_j \overline{C}_{j-1})^{-1} (F_j - \overline{A}_j F_{j-1})$. Then $\overline{A}_j \leftarrow 0$, and $\overline{B}_j \leftarrow I$. Save $\overline{C}_j$ and F_j for each stage. For the last stage, $F_N \leftarrow (\overline{B}_N - A_N \overline{C}_{N-1})^{-1}(F_N - A_N F_{N-1})$, $A_N \leftarrow 0$, $\overline{B}_N \leftarrow I$, and therefore $\Delta X_N = -F_N$. This completes the forward steps. Remaining values of ΔX are obtained by successive backward substitution from $\Delta X_j = -F_j \leftarrow -(F_j - \overline{C}_j F_{j+1})$. This procedure is illustrated by the following example.

Example 15.7. Solve the following matrix equation, which has a block tridiagonal structure, by the Thomas algorithm.

$$
\begin{bmatrix}
1 & 2 & 1 & 2 & 2 & 1 & 0 & 0 & 0 \\
2 & 1 & 1 & 2 & 1 & 0 & 0 & 0 & 0 \\
1 & 2 & 2 & 1 & 2 & 0 & 0 & 0 & 0 \\
0 & 1 & 3 & 1 & 2 & 1 & 1 & 2 & 1 \\
0 & 0 & 1 & 2 & 2 & 0 & 1 & 2 & 0 \\
0 & 0 & 2 & 2 & 1 & 1 & 1 & 1 & 0 \\
0 & 0 & 0 & 0 & 1 & 2 & 2 & 1 & 1 \\
0 & 0 & 0 & 0 & 0 & 2 & 1 & 1 & 1 \\
0 & 0 & 0 & 0 & 0 & 1 & 2 & 1 & 2
\end{bmatrix}
\begin{bmatrix}
\Delta x_1 \\
\Delta x_2 \\
\Delta x_3 \\
\Delta x_4 \\
\Delta x_5 \\
\Delta x_6 \\
\Delta x_7 \\
\Delta x_8 \\
\Delta x_9
\end{bmatrix}
=
\begin{bmatrix}
9 \\
7 \\
8 \\
12 \\
8 \\
8 \\
7 \\
5 \\
6
\end{bmatrix}
$$

Solution. The matrix equation is in the form

$$
\begin{bmatrix}
\overline{B}_1 & \overline{C}_1 & 0 \\
A_2 & \overline{B}_2 & \overline{C}_2 \\
0 & A_3 & \overline{B}_3
\end{bmatrix}
\cdot
\begin{bmatrix}
\Delta X_1 \\
\Delta X_2 \\
\Delta X_3
\end{bmatrix}
= -
\begin{bmatrix}
F_1 \\
F_2 \\
F_3
\end{bmatrix}
$$

Following the above procedure, starting at the first block row,

$$
\overline{B}_1 = \begin{bmatrix} 1 & 2 & 1 \\ 2 & 1 & 1 \\ 1 & 2 & 2 \end{bmatrix}, \quad
\overline{C}_1 = \begin{bmatrix} 2 & 2 & 1 \\ 2 & 1 & 0 \\ 1 & 2 & 0 \end{bmatrix}, \quad
F_1 = \begin{bmatrix} -9 \\ -7 \\ -8 \end{bmatrix}
$$

By standard matrix inversion

$$
(\overline{B}_1)^{-1} = \begin{bmatrix} 0 & 2/3 & -1/3 \\ 1 & -1/3 & -1/3 \\ -1 & 0 & 1 \end{bmatrix}
$$

By standard matrix multiplication

$$
(\overline{B}_1)^{-1}(\overline{C}_1) = \begin{bmatrix} 1 & 0 & 0 \\ 1 & 1 & 1 \\ -1 & 0 & -1 \end{bmatrix}, \quad \text{which replaces } \overline{C}_1
$$

and

$$
(\overline{B}_1)^{-1}(F_1) = \begin{bmatrix} -2 \\ -4 \\ 1 \end{bmatrix}, \text{which replaces } F_1
$$

Also

$$I = \begin{bmatrix} 1_/ & 0 & 0 \\ 0 & 1 & 0 \\ 0 & 0 & 1 \end{bmatrix} \text{replaces } \overline{B}_1$$

For the second block row

$$\overline{A}_2 = \begin{bmatrix} 0 & 1 & 3 \\ 0 & 0 & 1 \\ 1 & 2 & 2 \end{bmatrix}, \overline{B}_2 = \begin{bmatrix} 1 & 2 & 1 \\ 2 & 2 & 0 \\ 2 & 1 & 1 \end{bmatrix}, \overline{C}_2 = \begin{bmatrix} 1 & 2 & 1 \\ 1 & 2 & 0 \\ 1 & 1 & 0 \end{bmatrix}, F_2 = \begin{bmatrix} -12 \\ -8 \\ -8 \end{bmatrix}$$

By matrix multiplication and subtraction

$$(\overline{B}_2 - \overline{A}_2\overline{C}_1) = \begin{bmatrix} 3 & 1 & 3 \\ 3 & 2 & 1 \\ 4 & 1 & 3 \end{bmatrix}$$

which upon inversion becomes

$$(\overline{B}_2 - \overline{A}_2\overline{C}_1)^{-1} = \begin{bmatrix} -1 & 0 & 1 \\ 1 & 3/5 & -6/5 \\ 1 & -1/5 & -3/5 \end{bmatrix}$$

By multiplication

$$(\overline{B}_2 - \overline{A}_2\overline{C}_1)^{-1}\overline{C}_2 = \begin{bmatrix} 0 & -1 & -1 \\ 2/5 & 2 & 1 \\ 1/5 & 1 & 1 \end{bmatrix}, \text{ which replaces } \overline{C}_2$$

In a similar manner, the remaining steps for this and the third block row are carried out to give

$$\begin{bmatrix} \begin{bmatrix} 1 & 0 & 0 \\ 0 & 1 & 0 \\ 0 & 0 & 1 \end{bmatrix} & \begin{bmatrix} 1 & 0 & 0 \\ 1 & 1 & 1 \\ -1 & 0 & -1 \end{bmatrix} & \begin{bmatrix} 0 & 0 & 0 \\ 0 & 0 & 0 \\ 0 & 0 & 0 \end{bmatrix} \\ \begin{bmatrix} 0 & 0 & 0 \\ 0 & 0 & 0 \\ 0 & 0 & 0 \end{bmatrix} & \begin{bmatrix} 1 & 0 & 0 \\ 0 & 1 & 0 \\ 0 & 0 & 1 \end{bmatrix} & \begin{bmatrix} 0 & -1 & -1 \\ 2/5 & 2 & 1 \\ 1/5 & 1 & 1 \end{bmatrix} \\ \begin{bmatrix} 0 & 0 & 0 \\ 0 & 0 & 0 \\ 0 & 0 & 0 \end{bmatrix} & \begin{bmatrix} 0 & 0 & 0 \\ 0 & 0 & 0 \\ 0 & 0 & 0 \end{bmatrix} & \begin{bmatrix} 1 & 0 & 0 \\ 0 & 1 & 0 \\ 0 & 0 & 1 \end{bmatrix} \end{bmatrix} \begin{bmatrix} \Delta X_1 \\ \Delta X_2 \\ \Delta X_3 \\ \Delta X_4 \\ \Delta X_5 \\ \Delta X_6 \\ \Delta X_7 \\ \Delta X_8 \\ \Delta X_9 \end{bmatrix} = - \begin{bmatrix} -2 \\ -4 \\ +1 \\ +1 \\ -22/5 \\ -16/5 \\ -1 \\ -1 \\ -1 \end{bmatrix}$$

Thus, $\Delta X_7 = \Delta X_8 = \Delta X_9 = 1$.

The remaining backward steps begin with the second block row where

$$\overline{C}_2 = \begin{bmatrix} 0 & -1 & -1 \\ 2/5 & 2 & 1 \\ 1/5 & 1 & 1 \end{bmatrix}, \overline{F}_2 = \begin{bmatrix} 1 \\ -22/5 \\ -16/5 \end{bmatrix}$$

$$(F_2 - \overline{C}_2 F_3) = \begin{bmatrix} -1 \\ -1 \\ -1 \end{bmatrix}$$

Thus, $\Delta X_4 = \Delta X_5 = \Delta X_6 = 1$. Similarly, for the first block row, the result is

$$\Delta X_1 = \Delta X_2 = \Delta X_3 = 1$$

☐

Usually, it is desirable to specify certain top- and bottom-stage variables other than the condenser duty and/or reboiler duty.* This is readily accomplished by removing heat balance functions H_1 and/or H_N from the simultaneous equation set and replacing them with discrepancy functions depending upon the desired specification(s). Functions for alternate specifications for a column with a partial condenser are listed in Table 15.1.

If desired, (15-54) can be modified to permit real rather than theoretical stages to be computed. Values of the Murphree vapor-phase plate efficiency must then be specified. These values are related to phase compositions by the definition

$$\eta_j = \frac{y_{i,j} - y_{i,j+1}}{K_{i,j}x_{i,j} - y_{i,j+1}} \tag{15-73}$$

In terms of component flow rates, (15-73) becomes the following discrepancy function, which replaces (15-59).

$$E_{i,j} = \frac{\eta_j K_{i,j} l_{i,j} \sum\limits_{\kappa=1}^{C} v_{\kappa,j}}{\sum\limits_{\kappa=1}^{C} l_{\kappa,j}} - v_{i,j} + \frac{(1-\eta_j)v_{i,j+1}\sum\limits_{\kappa=1}^{C} v_{\kappa,j}}{\sum\limits_{\kappa=1}^{C} v_{k,j+1}} = 0 \tag{15-74}$$

Table 15.1 Alternative functions for H_1 and H_N

Specification	Replacement for H_1	Replacement for H_N
Reflux or reboil (boilup) ratio, (L/D) or (V/B)	$\sum l_{i,1} - (L/D) \sum v_{i,1} = 0$	$\sum v_{i,N} - (V/B) \sum l_{i,N} = 0$
Stage temperature, T_D or T_B	$T_1 - T_D = 0$	$T_N - T_B = 0$
Product flow rate, D or B	$\sum v_{i,1} - D = 0$	$\sum l_{i,N} - B = 0$
Component flow rate in product, d_i or b_i	$v_{i,1} - d_i = 0$	$l_{i,N} - b_i = 0$
Component mole fraction in product, y_{iD} or x_{iB}	$v_{i,1} - \left(\sum v_{i,1} \right) y_{iD} = 0$	$l_{i,N} - \left(\sum l_{i,N} \right) x_{iB} = 0$

* In fact, the condenser and reboiler duties are usually so interdependent that specification of both values is not recommended.

If a total condenser with subcooling is desired, it is necessary to specify the degrees of subcooling, if any, and to replace (15-59) or (15-74) with functions that express identity of reflux and distillate compositions as discussed by Naphtali and Sandholm.[23]

The algorithm for the Naphtali–Sandholm SC method is shown in Fig. 15.24. Problem specifications are quite flexible. Pressure, compositions, flow rates, and stage locations are necessary specifications for all feeds. The thermal condition of each feed can be given in terms of enthalpy, temperature, or molar fraction vaporized. If a feed is found to consist of two phases, the phases can be sent to

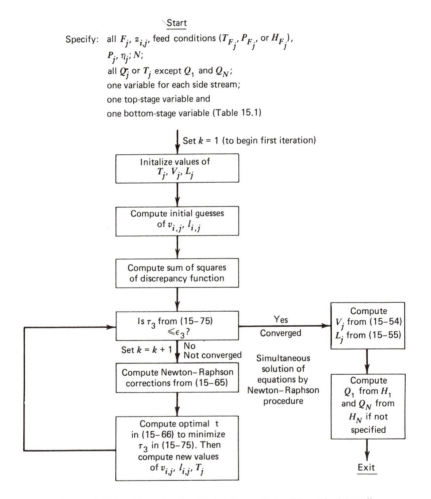

Figure 15.24. Algorithm for Naphtali–Sandholm SC method for all V/L separators.

the same stage or the vapor can be directed to the stage above the designated feed stage. Stage pressures and stage efficiencies can be designated by specifying top- and bottom-stage values. Remaining values are obtained by linear interpolation. By default, intermediate stages are assumed to be adiabatic unless Q_j or T_j values are specified. Vapor and/or liquid side streams can be designated in terms of total flow rate or flow rate of a specified component, or by the ratio of the side-stream flow rate to the flow rate remaining and passing to the next stage. The top- and bottom-stage specifications are selected from Q_1 or Q_N, and/or more generally from the other specifications listed in Table 15.1.

In order to achieve convergence, the Newton–Raphson procedure requires that reasonable guesses be provided for the values of all output variables. Rather than provide all these guesses *a priori*, we can generate them if T, V, and L are guessed for the bottom and top stages and, perhaps, for one or more intermediate stages. Remaining guessed values of T_j, V_j, and L_j are readily obtained by linear interpolation of the given T_j values and computed (V_j/L_j) values. Initial values for $v_{i,j}$ and $l_{i,j}$ are then obtained by either of two techniques. If K-values are composition independent or can be approximated as such, one technique is to compute $x_{i,j}$ values and corresponding $y_{i,j}$ values from (15-12) and (15-2) as in the first iteration of the BP or SR method. A much cruder estimate is obtained by flashing the combined feeds at some average column pressure and a V/L ratio that approximates the ratio of overheads to bottoms products. The resulting mole fraction compositions of the equilibrium vapor and liquid phases are assumed to hold for each stage. The second technique works surprisingly well, but the first technique is preferred for difficult cases. For either technique, the initial component flow rates are computed by using the $x_{i,j}$ and $y_{i,j}$ values to solve (15-56) and (15-57) for $l_{i,j}$ and $v_{i,j}$, respectively.

Based on initial guesses for all output variables, the sum of the squares of the discrepancy functions is computed and compared to a convergence criterion

$$\tau_3 = \sum_{j=1}^{N} \left\{ (H_j)^2 + \sum_{i=1}^{C} [(M_{i,j})^2 + (E_{i,j})^2] \right\} \leq \epsilon_3 \qquad (15\text{-}75)$$

In order that the values of all discrepancies be of the same order of magnitude, it is necessary to divide energy balance functions H_j by a scale factor approximating the latent heat of vaporization (e.g., 1000 Btu/lbmole). If the convergence criterion is computed from

$$\varepsilon_3 = N(2C + 1) \left(\sum_{j=1}^{N} F_j^2 \right) 10^{-10} \qquad (15\text{-}76)$$

resulting converged values of the output variables will generally be accurate, on the average, to four or more significant figures. When employing (15-76), most problems are converged in 10 iterations or less.

Generally, the convergence criterion is far from satisfied during the first iteration when guessed values are assumed for the output variables. For each subsequent iteration, the Newton–Raphson corrections are computed from (15-65). These corrections can be added directly to the present values of the output variables to obtain a new set of values for the output variables. Alternatively, (15-66) can be employed where t is a nonnegative scalar step factor. At each iteration, a single value of t is applied to all output variables. By permitting t to vary from, say, slightly greater than zero up to 2, it can serve to dampen or accelerate convergence, as appropriate. For each iteration, an optimal value of t is sought to minimize the sum of the squares given by (15-75). Generally, optimal values of t proceed from an initial value for the second iteration at between 0 and 1 to a value nearly equal to or slightly greater than 1 when the convergence criterion is almost satisfied. An efficient optimization procedure for finding t at each iteration is the Fibonacci search.[25] If no optimal value of t can be found within the designated range, t can be set to 1, or some smaller value, and the sum of squares can be allowed to increase. Generally, after several iterations, the sum of squares will decrease for every iteration.

If the application of (15-66) results in a negative component flow rate, Naphtali and Sandholm recommend the following mapping equation, which reduces the value of the unknown variable to a near-zero, but nonnegative, quantity.

$$X^{(k+1)} = X^{(k)} \exp \left[\frac{t \Delta X^{(k)}}{X^{(k)}} \right] \qquad (15\text{-}77)$$

In addition, it is advisable to limit temperature corrections at each iteration.

The Naphtali–Sandholm SC method is readily extended to staged separators involving two liquid phases (e.g., extraction) and three coexisting phases (e.g., three-phase distillation), as shown by Block and Hegner,[26] and to interlinked separators as shown by Hofeling and Seader.[27]

Example 15.8. A reboiled absorber is to be designed to separate the hydrocarbon vapor feed of Examples 15.2 and 15.4. Absorbent oil of the same composition as that of Example 15.4 will enter the top stage. Complete specifications are given in Fig. 15.25. The 770 lbmole/hr (349 kgmole/hr) of bottoms product corresponds to the amount of C_3 and heavier in the two feeds. Thus, the column is to be designed as a deethanizer. Calculate stage temperatures, interstage vapor and liquid flow rates and compositions, and reboiler duty by the rigorous SC method. Assume all stage efficiencies are 100%. Compare the degree of separation of the feed to that achieved by ordinary distillation in Example 15.2.

Solution. A digital computer program for the method of Naphtali and Sandholm was used. The K-values and enthalpies were assumed independent of composition and were computed by linear interpolation between tabular values given at 100°F increments from 0°F to 400°F (-17.8 to 204.4°C). The tabular values were computed from the equations given in Example 12.8, except for the following values at 400°F.

	400°F, 400 psia		
Species	**K**	H_V, $\dfrac{\text{Btu}}{\text{lbmole}}$	H_L, $\dfrac{\text{Btu}}{\text{lbmole}}$
C_1	11.0	5631	5388
C_2	5.40	11410	7156
C_3	3.25	15650	9890
C_4	2.00	19790	13790
C_5	1.200	23520	17170
Abs.Oil	0.0019	43590	33410

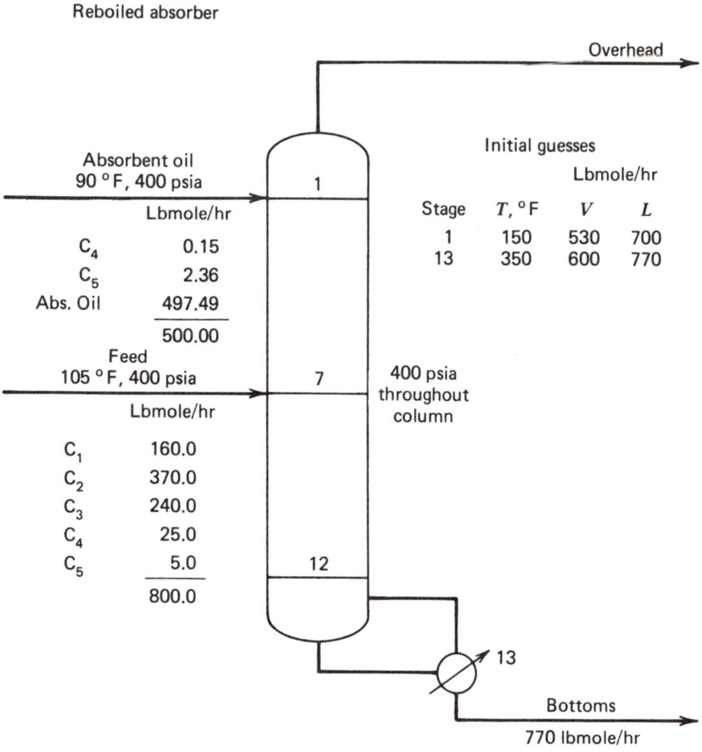

Reboiled absorber

Figure 15.25. Specifications for Example 15.8.

From (15-76), the convergence criterion is

$$\varepsilon_3 = 13[2(6) + 1](500 + 800)^2 10^{-10} = 2.856 \times 10^{-2}$$

Figure 15.26 shows the reduction in the sum of the squares of the 169 discrepancy

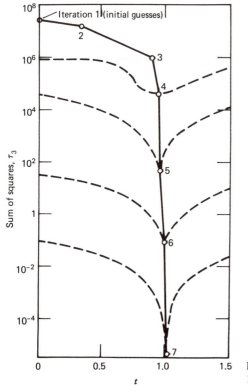

Figure 15.26. Convergence pattern for Example 15.8.

functions from iteration to iteration. Seven iterations were required to satisfy the convergence criterion. The initial iteration was based on values of the unknown variables computed from interpolation of the initial guesses shown in Fig. 15.25 together with a flash of the combined feeds at 400 psia (2.76 MPa) and a V/L ratio of 0.688 (530/770). Thus, for the first iteration, the following mole fraction compositions were computed and were assumed to apply to every stage.

Species	y	x
C_1	0.2603	0.0286
C_2	0.4858	0.1462
C_3	0.2358	0.1494
C_4	0.0153	0.0221
C_5	0.0025	0.0078
Abs.Oil	0.0003	0.6459
	1.0000	1.0000

The corresponding sum of squares of the discrepancy functions τ_3 of 2.865×10^7 was very large. Subsequent iterations employed the Newton–Raphson method. For iteration

2, the optimal value of t was found to be 0.34. However, this caused only a moderate reduction in the sum of squares. The optimal value of t increased to 0.904 for iteration 3, and the sum of squares was reduced by an order of magnitude. For the fourth and subsequent iterations, the effect of t on the sum of squares is included in Fig. 15.26. Following iteration 4, the sum of squares was reduced by at least two orders of magnitude for each iteration. Also, the optimal value of t was rather sharply defined and corresponded closely to a value of 1. The converged solution required 8.2 sec. of execution time on a UNIVAC 1108. An improvement of τ_3 was obtained for every iteration.

In Figs. 15.27 and 15.28, converged temperature and V/L profiles are compared to the initially guessed profiles. In Fig. 15.27, the converged temperatures are far from linear with respect to stage number. Above the feed stage, the temperature profile increases from the top down in a gradual and declining manner. The relatively cold feed causes a small temperature drop from stage 6 to stage 7. Temperature also increases from stage 7 to stage 13. A particularly dramatic increase occurs in moving from the bottom stage in the column to the reboiler, where heat is added. In Fig. 15.28, the V/L profile is also far from linear with respect to stage number. Dramatic changes in this ratio occur at the top, middle, and bottom of the column.

Component flow rate profiles for the two key components (ethane vapor and propane liquid) are shown in Fig. 15.29. The initial guessed values are in very poor agreement with the converged values. The propane liquid profile is quite regular except at the bottom where a large decrease occurs due to vaporization in the reboiler. The ethane vapor profile has large changes at the top, where entering oil absorbs appreciable ethane, and at the feed stage, where substantial ethane vapor is introduced.

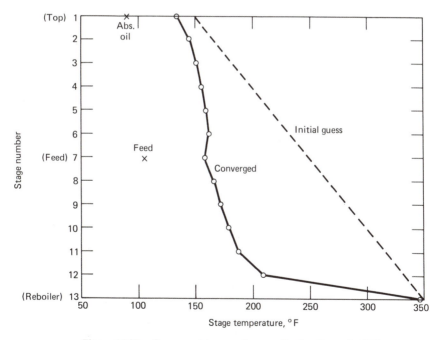

Figure 15.27. Converged temperature profile for Example 15.8.

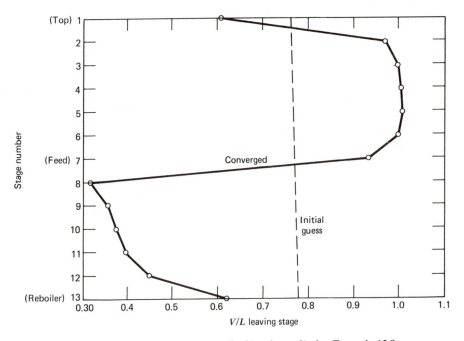

Figure 15.28. Converged vapor–liquid ratio profile for Example 15.8.

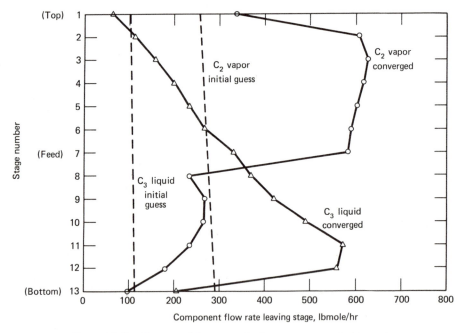

Figure 15.29. Converged flow rates for key components in Example 15.8.

Converged values for the reboiler duty and overhead and bottoms compositions are given in Table 15.2. Also included are converged results for two additional solutions that used the Chao–Seader and Soave–Redlich–Kwong correlations for K-values and enthalpies in place of interpolation of composition-independent tabular properties. With the Soave–Redlich–Kwong correlation, a somewhat sharper separation between the two key components is predicted. In addition, the Soave–Redlich–Kwong correlation predicts a substantially higher bottoms temperature and a much larger reboiler duty. As discussed in Chapter 4, the effect of physical properties on equilibrium stage calculations can be significant.

It is interesting to compare the separation achieved with the reboiled absorber of this example to the separation achieved by ordinary distillation of the same feed in Example 15.2 as shown in Fig. 15.9. The latter separation technique results in a much sharper separation and a much lower bottoms temperature and reboiler duty for the same number of stages. However, refrigeration is necessary for the overhead condenser and the reflux flow rate is twice the absorbent oil flow rate. If the absorbent oil flow rate for the reboiled absorber is made equal to the reflux flow rate, calculations give a separation almost as sharp as for ordinary distillation. However, the bottoms temperature and reboiler duty are increased to almost 600°F (315.6°C) and 60,000,000 Btu/hr (63.3 GJ/hr), respectively.

Table 15.2 Product compositions and reboiler duty for example 15.8

	Composition-Independent Tabular Properties	Chao–Seader Correlation	Soave–Redlich–Kwong Correlation
Overhead component flow rates, lbmole/hr			
C_1	159.99	159.98	159.99
C_2	337.96	333.52	341.57
C_3	31.79	36.08	28.12
C_4	0.04	0.06	0.04
C_5	0.17	0.21	0.18
Abs.oil	0.05	0.15	0.10
	530.00	530.00	530.00
Bottoms component flow rates, lbmole/hr			
C_1	0.01	0.02	0.01
C_2	32.04	36.4	28.43
C_3	208.21	203.92	211.88
C_4	25.11	25.09	25.11
C_5	7.19	7.15	7.18
Abs. oil	497.44	497.34	497.39
	770.00	770.00	770.00
Reboiler duty, Btu/hr	11,350,000	10,980,000	15,640,000
Bottoms temperature, °F	346.4	338.5	380.8

☐

Relaxation Method

Poor initial guesses for the output variables may cause the Newton–Raphson SC procedure to fail to converge within a reasonable number of iterations. A new set of initial guesses for the output variables should then be provided and the procedure restarted. Occasionally, it is impossible to find a set of initial guesses that is successful. An alternative procedure that always converges is the relaxation method of Rose, Sweeny, and Schrodt,[28] as modified by Ball,[29] and further improved by Jelinek, Hlavacek, and Kubicek[30] to accelerate convergence. The relaxation method, which is explained and illustrated in detail by Holland,[5] uses unsteady-state differential equations for energy and component material balances. Starting from any assumed set of initial values, these equations are solved numerically, at each time step, with the phase equilibria equations, to obtain changes in stage temperatures, flow rates, and compositions. However, because the rate of convergence of the relaxation method decreases as the solution is approached, the method is not widely used in practice. For difficult problems, Ketchum[31] has combined the stability of the relaxation method with the speed of the Newton–Raphson SC method to obtain a single algorithm that utilizes an adjustable relaxation factor.

References

1. Wang, J. C., and G. E. Henke, *Hydrocarbon Processing*, **45** (8), 155–163 (1966).

2. Myers, A. L., and W. D. Seider, *Introduction to Chemical Engineering and Computer Calculations*, Prentice-Hall, Inc., Englewood Cliffs, N.J., 1976, 484–507.

3. Lewis, W. K., and G. L. Matheson, *Ind. Eng. Chem*, **24**, 496–498 (1932).

4. Thiele, E. W., and R. L. Geddes, *Ind. Eng. Chem.*, **25**, 290 (1933).

5. Holland, C. D., *Multicomponent Distillation*, Prentice-Hall, Inc., Englewood Cliffs, N.J., 1963.

6. Amundson, N. R., and A. J. Pontinen, *Ind. Eng. Chem.*, **50**, 730–736 (1958).

7. Friday, J. R., and B. D. Smith, *AIChE J.*, **10**, 698–707 (1964).

8. Lapidus, L., *Digital Computation for Chemical Engineers*, McGraw-Hill Book Co., New York, 1962, 254.

9. Boston, J. F., and S. L. Sullivan, Jr., *Can. J. Chem. Eng.*, **50**, 663–669 (1972).

10. Johanson, P. J., and J. D. Seader, *Stagewise Computations—Computer Programs for Chemical Engineering Education*, ed. by J. Christensen, Aztec Publishing Co., Austin, Texas, 1972, 349–389, A-16.

11. Lapidus, L., *Digital Computation for Chemical Engineers*, McGraw-Hill Book Co., New York, 1962, 308–309.

12. Orbach, O., and C. M. Crowe, *Can. J. Chem. Eng.*, **49**, 509–513 (1971).

13. Scheibel, E. G., *Ind. Eng. Chem.*, **38**, 397–399 (1946).

14. Sujata, A. D., *Hydrocarbon Processing*, 40 (12), 137–140 (1961).

15. Burningham, D. W., and F. D. Otto, *Hydrocarbon Processing*, **46** (10), 163–170 (1967).

16. Shinohara, T., P. J. Johansen, and J.

D. Seader, *Stagewise Computations—Computer Programs for Chemical Engineering Education*, ed. by J. Christensen, Aztec Publishing Co., Austin, Texas, 1972, 390–428, A-17.

17. Tsuboka, T., and T. Katayama, *J. Chem. Eng. Japan*, **9**, 40–45 (1976).

18. Hála, E., I. Wichterle, J. Polak, and T. Boublik, *Vapor-Liquid Equilibrium Data at Normal Pressures*, Pergamon Press, Oxford, 1968, 308.

19. Steib, V. H., *J. Prakt. Chem.*, **4**, Reihe, 1965, Bd. 28, 252–280.

20. Cohen, G., and H. Renon, *Can. J. Chem. Eng.*, **48**, 291–296 (1970).

21. Goldstein, R. P., and R. B. Stanfield, *Ind. Eng. Chem.*, *Process Des. Develop.*, **9**, 78–84 (1970).

22. Naphtali, L. M., "The Distillation Column as a Large System," paper presented at the AIChE 56th National Meeting, San Francisco, May 16-19, 1965.

23. Naphtali, L. M., and D. P. Sandholm, *AIChE J.*, **17**, 148–153 (1971).

24. Fredenslund, A., J. Gmehling, and P. Rasmussen, *Vapor-Liquid Equilibria Using UNIFAC, A Group Contribution Method*, Elsevier, Amsterdam, 1977.

25. Beveridge, G. S. G., and R. S. Schechter, *Optimization: Theory and Practice*, McGraw–Hill Book Co., New York, 1970, 180–189.

26. Block, U., and B. Hegner, *AIChE J.*, **22**, 582–589 (1976).

27. Hofeling, B., and J. D. Seader, *AIChE J.*, **24**, 1131–1134 (1978).

28. Rose, A., R. F. Sweeny, and V. N. Schrodt, *Ind. Eng. Chem.*, **50**, 737–740 (1958).

29. Ball, W. E., "Computer Programs for Distillation," paper presented at the 44th National Meeting of AIChE, New Orleans, February 27, 1961.

30. Jelinek, J., V. Hlavacek, and M. Kubicek, *Chem. Eng. Sci.*, **28**, 1825–1832 (1973).

31. Ketchum, R. G., *Chem. Eng. Sci.*, **34**, 387–395 (1979).

Problems

15.1 Revise the MESH equations [(15-1) to (15-6)] to allow for pumparounds of the type shown in Fig. 15.2 and discussed by Bannon and Marple [*Chem. Eng. Prog.*, **74** (7), 41–45 (1978)] and Huber [*Hydrocarbon Processing*, **56** (8), 121–125 (1977)]. Combine the equations to obtain modified M equations similar to (15-7). Can these equations still be partitioned in a series of C tridiagonal matrix equations?

15.2 Use the Thomas algorithm to solve the following matrix equation for x_1, x_2, and x_3.

$$\begin{bmatrix} -160 & 200 & 0 \\ 50 & -350 & 180 \\ 0 & 150 & -230 \end{bmatrix} \cdot \begin{bmatrix} x_1 \\ x_2 \\ x_3 \end{bmatrix} = \begin{bmatrix} 0 \\ -50 \\ 0 \end{bmatrix}$$

15.3 Use the Thomas algorithm to solve the following tridiagonal matrix equation for the **x** vector.

$$\begin{bmatrix} -6 & 3 & 0 & 0 & 0 \\ 3 & -4.5 & 3 & 0 & 0 \\ 0 & 1.5 & -7.5 & 3 & 0 \\ 0 & 0 & 4.5 & -7.5 & 3 \\ 0 & 0 & 0 & 4.5 & -4.5 \end{bmatrix} \cdot \begin{bmatrix} x_1 \\ x_2 \\ x_3 \\ x_4 \\ x_5 \end{bmatrix} = \begin{bmatrix} 0 \\ 0 \\ 100 \\ 0 \\ 0 \end{bmatrix}$$

15.4 On page 162 of their article, Wang and Henke[1] claim that their method of solving the tridiagonal matrix for the liquid-phase mole fractions does not involve subtraction of nearly equal quantities. Prove or disprove their statement.

15.5 One thousand kilogram-moles per hour of a saturated liquid mixture of 60 mole% methanol, 20 mole% ethanol, and 20 mole% n-propanol is fed to the middle stage of a distillation column having three equilibrium stages, a total condenser, a partial reboiler, and an operating pressure of 1 atm. The distillate rate is 600 kgmole/hr, and the external reflux rate is 2000 kgmole/hr of saturated liquid. Assuming that ideal solutions are formed such that K-values can be obtained from the Antoine vapor pressure equation using the constants in Appendix I and assuming constant molal overflow such that the vapor rate leaving the reboiler and each stage is 2600 kgmole/hr, calculate one iteration of the BP method up to and including a new set of T_j values. To initiate the iteration, assume a linear temperature profile based on a distillate temperature equal to the normal boiling point of methanol and a bottoms temperature equal to the arithmetic average of the normal boiling points of the other two alcohols.

15.6 Derive an equation similar to (15-7), but with $v_{i,j} = y_{i,j} V_j$ as the variables instead of the liquid-phase mole fractions. Can the resulting equations still be partitioned into a series of C tridiagonal matrix equations?

15.7 In a computer program for the Wang–Henke bubble-point method, 10100 storage locations are wastefully set aside for the four indexed coefficients of the tridiagonal matrix solution of the component material balances for a 100-stage distillation column.

$$A_j x_{i,j-1} + B_j x_{i,j} + C_j x_{i,j+1} - D_j = 0$$

Determine the minimum number of storage locations required if the calculations are conducted in the most efficient manner.

15.8 Calculate, by the Wang–Henke BP method, product compositions, stage temperatures, interstage vapor and liquid flow rates and compositions, reboiler duty, and condenser duty for the following distillation column specifications.
Feed (bubble-point liquid at 250 psia and 213.9°F):

Component	Lbmole/hr
Ethane	3.0
Propane	20.0
n-Butane	37.0
n-Pentane	35.0
n-Hexane	5.0

Column pressure = 250 psia
Partial condenser and partial reboiler.
Distillate rate = 23.0 lbmole/hr.

Reflux rate = 150.0 lbmole/hr.
Number of equilibrium stages (exclusive of condenser and reboiler) = 15.
Feed is sent to middle stage.

For this system at 250 psia, K-values and enthalpies may be computed over a temperature range of 50°F to 350°F by the polynomial equations of Example 12.8 using the following constants (Amundson and Pontinen[6]).

	K-value Constants				Vapor Enthalpy Constants			Liquid Enthalpy Constants		
Species	α_i	$\beta_i \times 10^4$	$\gamma_i \times 10^6$	$\delta_i \times 10^8$	A_i	B_i	$C_i \times 10^2$	a_i	b_i	$c_i \times 10^2$
C_2	1.665	−1.50	73.5	−3.00	8310	13.2	0.602	6120	9.35	1.5
C_3	0.840	−46.6	49.4	−3.033	11950	12.35	2.21	3620	40.5	−1.32
nC_4	−0.177	49.5	−4.15	2.22	16550	9.01	3.78	6280	25.0	4.65
nC_5	−0.0879	17.7	0.2031	1.310	19200	29.5	0.72	6840	38.5	2.52
nC_6	0.0930	−15.39	10.37	−0.1590	22000	37.0	0	7920	43.5	3.02

15.9 Determine the optimum feed stage location for Problem 15.8.

15.10 Revise Problem 15.8 so as to withdraw a vapor side stream at a rate of 37.0 lbmole/hr from the fourth stage from the bottom.

15.11 Revise Problem 15.8 so as to provide an intercondenser on the fourth stage from the top with a duty of 200,000 Btu/hr and an interreboiler on the fourth stage from the bottom with a duty of 300,000 Btu/hr.

15.12 Using the thermodynamic properties given in Problem 15.8, calculate by the Wang–Henke BP method the product compositions, stage temperatures, interstage vapor and liquid flow rates and compositions, reboiler duty, and condenser duty for the following multiple-feed distillation column, which has 30 equilibrium stages exclusive of a partial condenser and a partial reboiler and operates at 250 psia.

Feeds (both bubble-point liquids at 250 psia):

	Pound-moles per hour	
Component	Feed 1 to stage 15 from the bottom	Feed 2 to stage 6 from the bottom
Ethane	1.5	0.5
Propane	24.0	10.0
n-Butane	16.5	22.0
n-Pentane	7.5	14.5
n-Hexane	0.5	3.0

Distillate rate = 36.0 lbmole/hr.
Reflux rate = 150.0 lbmole/hr.

Determine whether the feed locations are optimal.

15.13   Solve by the Newton–Raphson method the simultaneous nonlinear equations

$$x_1^2 + x_2^2 = 17$$
$$(8x_1)^{1/3} + x_2^{1/2} = 4$$

for x_1 and x_2 to within ± 0.001. As initial guesses, assume

(a) $x_1 = 2$, $x_2 = 5$.
(b) $x_1 = 4$, $x_2 = 5$.
(c) $x_1 = 1$, $x_2 = 1$.
(d) $x_1 = 8$, $x_2 = 1$.

15.14   Solve by the Newton–Raphson method the simultaneous nonlinear equations

$$\sin(\pi x_1 x_2) - \frac{x_2}{2} - x_1 = 0$$

$$\exp(2x_1)\left[1 - \frac{1}{4\pi}\right] + \exp(1)\left[\frac{1}{4\pi} - 1 - 2x_1 + x_2\right] = 0$$

for x_1 and x_2 to within ± 0.001. As initial guesses, assume

(a) $x_1 = 0.4$, $x_2 = 0.9$.
(b) $x_1 = 0.6$, $x_2 = 0.9$.
(c) $x_1 = 1.0$, $x_2 = 1.0$.

15.15   An absorber is to be designed for a pressure of 75 psia to handle 2000 lbmole/hr of gas at 60°F having the following composition.

Component	Mole Fraction
Methane	0.830
Ethane	0.084
Propane	0.048
n-Butane	0.026
n-Pentane	0.012

The absorbent is an oil, which can be treated as a pure component having a molecular weight of 161. Calculate, by the Burningham–Otto SR method, product rates and compositions, stage temperatures, and interstage vapor and liquid flow rate and compositions for the following conditions.

	Number of Equilibrium Stages	Entering Absorbent Flow Rate, lbmole/hr	Entering Absorbent Temperature, °F
(a)	6	500	90
(b)	12	500	90
(c)	6	1000	90
(d)	6	500	60

For this system at 75 psia, K-values may be computed over a temperature range of 50°F to 150°F by the polynomial equations of Example 12.8 using the following constants.

Species	α_i	$\beta_i \times 10^4$	$\gamma_i \times 10^6$	$\delta_i \times 10^8$
C_1	12.25	2500	−625	0
C_2	4.75	−4.27	208	−8.55
C_3	2.35	−131	138	−8.47
nC_4	−0.434	121	−10.2	5.44
nC_5	−0.207	41.6	0.477	3.08
Oil	−0.00009	0.02	0	0

Enthalpies over the same temperature range can be obtained from the polynomial equations of Example 12.8 using the constants of Problem 15.8 for C_2, C_3, nC_4, and nC_5. For C_1 and the oil, the constants are:

Species	Vapor Enthalpy Constants			Liquid Enthalpy Constants		
	A_I	B_I	$C_I \times 10^2$	a_i	b_i	$c_i \times 10^2$
C_1	4800	9	0	3300	11	0
Oil	37400	44	0	12800	73	0

15.16 Use the Burningham–Otto SR method to calculate product rates and compositions, stage temperatures, and interstage vapor and liquid flow rates and compositions for an absorber having four equilibrium stages with the following specifications.

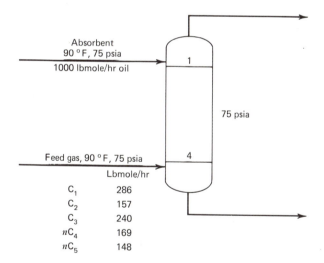

Thermodynamic properties for this system are the same as in Problem 15.15.

15.17 In Example 15.4, temperatures of the gas and oil, as they pass through the absorber, increase substantially. This limits the extent of absorption. Repeat the calculations with a heat exchanger that removes 500,000 Btu/hr from:
(a) Stage 2.

(b) Stage 3.
(c) Stage 4.
(d) Stage 5.

How effective is the intercooler? Which stage is the preferred location for the intercooler? Should the duty of the intercooler be increased or decreased assuming that the minimum stage temperature is 100°F using of cooling water?

15.18 Use the Burningham–Otto SR method to calculate product rates and compositions stage temperatures, and interstage vapor and liquid flow rates and compositions for the absorber shown below. Use the thermodynamic properties of Example 12.8.

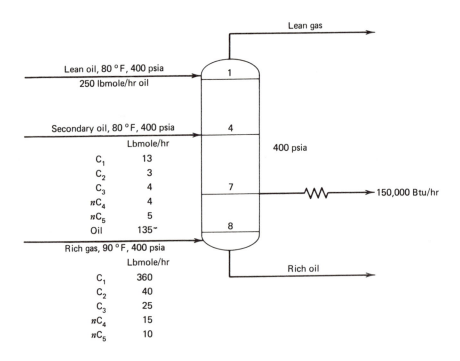

15.19 A mixture of cyclohexane and cyclopentane is to be separated by liquid–liquid extraction at 25°C with methanol. Phase equilibria for this system may be predicted by the van Laar equation with constants given in Example 5.11. Calculate, by the ISR method, product rates and compositions and interstage flow rates and compositions for the conditions below with:

(a) $N = 1$ equilibrium stage.
(b) $N = 2$ equilibrium stages.
(c) $N = 5$ equilibrium stages.
(d) $N = 10$ equilibrium stages.

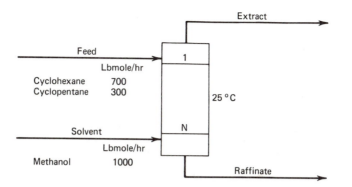

15.20 Solve the following nine simultaneous linear equations, which have a block tridiagonal matrix structure, by the Thomas algorithm.

$$x_2 + 2x_3 + 2x_4 + x_6 = 7$$
$$x_1 + x_3 + x_4 + 3x_5 = 6$$
$$x_1 + x_2 + x_3 + x_5 + x_6 = 6$$
$$x_4 + 2x_5 + x_6 + 2x_7 + 2x_8 + x_9 = 11$$
$$x_4 + x_5 + 2x_6 + 3x_7 + x_9 = 8$$
$$x_5 + x_6 + x_7 + 2x_8 + x_9 = 8$$
$$x_1 + 2x_2 + x_3 + x_4 + x_5 + 2x_6 + 3x_7 + x_8 = 13$$
$$x_2 + 2x_3 + 2x_4 + x_5 + x_6 + x_7 + x_8 + 3x_9 = 14$$
$$x_3 + x_4 + 2x_5 + x_6 + 2x_7 + x_8 + x_9 = 10$$

15.21 Naphtali and Sandholm group the $N(2C + 1)$ equations by stage. Instead, group the equations by type (i.e., enthalpy balances, component balances, and equilibrium relations). Using a three-component, three-stage example, show whether the resulting matrix structure is still block tridiagonal.

15.22 Derivatives of properties are needed in the Naphtali–Sandholm SC method. For the Chao–Seader correlation, determine analytical derivatives for

$$\frac{\partial K_{i,j}}{\partial T_j} \qquad \frac{\partial K_{i,j}}{\partial v_{i,k}} \qquad \frac{\partial K_{i,j}}{\partial l_{i,k}}$$

15.23 A rigorous partial SC method for multicomponent, multistage vapor–liquid separations can be devised that is midway between the complexity of the BP/SR methods on the one hand and the SC methods on the other hand. The first major step in the procedure is to solve the modified M equations for the liquid-phase mole fractions by the usual tridiagonal matrix algorithm. Then, in the second major step, new sets of stage temperatures and total vapor flow rates leaving a stage are computed simultaneously by a Newton–Raphson method. These two major steps are repeated until a sum-of-squares criterion is satisfied. For this partial SC method:

(a) Write the two indexed equations you would use to simultaneously solve for a new set of T_j and V_j.

    (b) Write the truncated Taylor series expansions for the two indexed equations in the T_j and V_j unknowns, and derive complete expressions for all partial derivatives, except that derivatives of physical properties with respect to temperature can be left as such. These derivatives are subject to the choice of physical property correlations.

    (c) Order the resulting linear equations and the new variables ΔT_j and ΔV_j into a Jacobian matrix that will permit a rapid and efficient solution.

15.24 Revise Equations (15-58) to (15-60) to allow two interlinked columns of the type shown below to be solved simultaneously by the SC method. Does the matrix equation that results from the Newton–Raphson procedure still have a block tridiagonal structure?

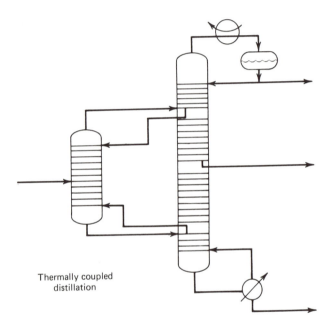

Thermally coupled
distillation

15.25 In Equation (15-63), why is the variable order selected as v, T, l? What would be the consequence of changing the order to l, v, T?

15.26 In Equation (15-64), why is the function order selected as H, M, E? What would be the consequence of changing the order to E, M, H?

15.27 Use the Naphtali–Sandholm SC method with the Chao–Seader correlation for thermodynamic properties to calculate product compositions, stage temperatures, interstage flow rates and compositions, reboiler duty, and condenser duty for the following distillation specifications.

Compare your results with those given in the *Chemical Engineers' Handbook*, Fifth Edition, pp. 13–32 to 13–35. Why do the two solutions differ?

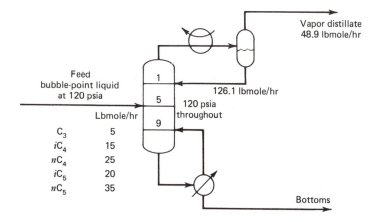

15.28 Calculate by the Naphtali–Sandholm SC method, with the Chao–Seader correlation for thermodynamic properties, the product compositions, stage temperatures, interstage flow rates and compositions, reboiler duty, and condenser duty for the following distillation specifications, which represent an attempt to obtain four nearly pure products from a single distillation operation. Reflux is a saturated liquid.

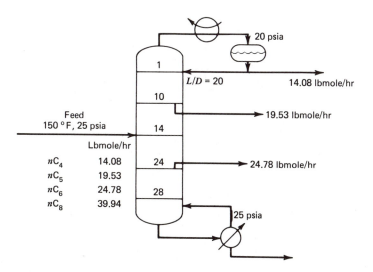

15.29 Calculate by the Naphtali–Sandholm SC method, with the Chao–Seader correlation for thermodynamic properties, the product compositions, stage temperatures, interstage flow rates and compositions, and reboiler duty for the reboiled stripper shown below.

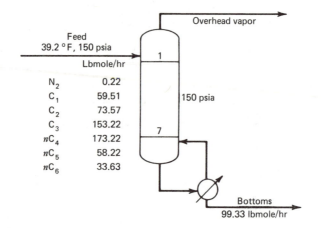

15.30 Toluene and *n*-heptane are to be separated by extractive distillation with phenol. One proposed specification for the operation is shown below. Use the Naphtali–Sandholm SC method, with the Wilson equation for activity coefficients, to calculate the product compositions, stage temperatures, interstage flow rates and compositions, and condenser and reboiler duties. Constants for the Wilson equation can be determined readily for the van Laar constants A_{ij} developed in Example 5.5 by computing from these constants the infinite-dilution activity

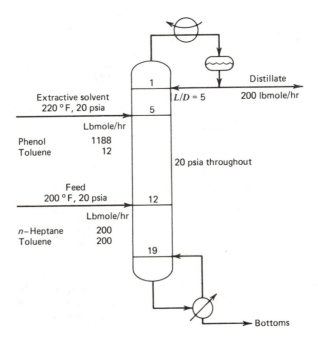

coefficients at some convenient temperature (e.g., 250°F), using the method of Example 5.6 to obtain values of the Wilson constants Λ_{ij} and then obtaining values of $(\lambda_{ij} - \lambda_{ii})$ from Equation (5-39). If the design objective is to obtain a heptane purity of at least 99 mole% and a minimum of 2.0 lbmole/hr of heptane in the toluene product, is the proposed specification adequate? If not, make changes to the feed locations, number of stages, and/or reflux rate and recalculate until the objective is achieved.

15.31 Use the Naphtali–Sandholm SC method, with the Chao–Seader correlation for thermodynamic properties, to determine product compositions, stage temperatures, interstage flow rates and compositions, and reboiler duty for the reboiled absorber shown below. Repeat the calculations without the interreboiler and compare both sets of results. Is the interreboiler worthwhile? Should an intercooler in the top section of the column be considered?

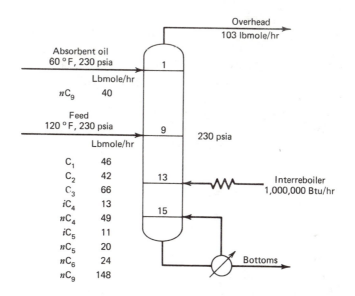

15.32 A saturated liquid feed at 125 psia contains 200 lbmole/hr of 5 mole% iC_4, 20 mole% nC_4, 35 mole% iC_5, and 40 mole% nC_5. This feed is to be distilled at 125 psia with a column equipped with a total condenser and partial reboiler. The distillate is to contain 95% of the nC_4 in the feed, and the bottoms is to contain 95% of the iC_5 in the feed. Use the Naphtali–Sandholm SC method, with the Chao–Seader correlations for thermodynamic properties, to determine a suitable design. Twice the minimum number of equilibrium stages, as estimated by the Fenske equation in Chapter 12, should provide a reasonable number of equilibrium stages.

16

Continuous Differential Contacting Operations: Gas Absorption

---

Heat transfer, dehumidification, absorption, distillation, and extraction operations in general involve two-film resistances. Where there is direct contact between the two fluid streams as in packed or spray columns, the use of an 'overall' value of H.T.U. greatly simplifies calculations...the overall H.T.U. and resulting column height do not vary from one of these systems to another as much as might be supposed.

Allan P. Colburn, 1939

In this chapter, we discuss the design of packed-bed, steady-state, gas-absorption equipment. Unlike the staged columns treated in previous chapters, packed beds are continuous contacting devices that have no· physically distinguishable stages. Despite this, we have a legacy of data and design procedures dealing with the application of the basic stagewise contactor equations to these intrinsically nonstagewise processes. Concepts such as the HETP (Height Equivalent to a Theoretical Plate), HTU (Height of a Transfer Unit), NTP (Number of Theoretical Plates), and the NTU (Number of Transfer Units) are still used to characterize continuous multiphase contacting devices. These descriptors came into being because staged processes were studied before packed beds. It was not illogical, therefore, for early investigators to apply an existing body of knowledge to a new process description, rather than to start afresh and develop uniquely useful data and correlations. This design-by-analogy approach delayed the development of rigorous calculational procedures for continuous differential contactors many years and is an example of why the expedient and economically imperative decision is often, in the long run, the philosophically wrong decision.

A serious reexamination of gas absorption and related operations is now taking place and it is expected that in the next few years more fundamentally oriented design procedures will be available. These methods must be computer-based because absorption is described by differential equations that are usually perversely nonlinear. Those who paved the way to the modern design methods let nature solve the differential equations and were content to describe, with discrete steps (NTP, NTU), the continuum of questions posed by continuous differential contactors.

In this chapter, we present in some detail an elementary treatment of mass transfer, followed by applications to gas absorption that ignore energy balances and are restricted to the absorption of only one component.

Absorption is a process in which material is transferred from a gaseous to a liquid stream. The two streams may travel in axial flow directions, with no axial dispersion, or with varying degrees of axial backmixing—a totally backmixed vessel being a limiting case. In some applications, the two streams may be quiescent.

Absorption is the underlying phenomenon in a multiplicity of industrial processes and equipment.

(a) Evaporation from tanks—a generally undesired process.

(b) Wetted-wall absorbers.

(c) Packed absorbers—countercurrent and cocurrent.

(d) Stagewise absorbers.

(e) Absorbers with simultaneous chemical reaction.

(f) Liquid droplet scrubbers—packed columns in which droplets of liquid fall through a gas stream.

(g) Venturi scrubbers.

Usually, the feed streams are known and the desired product streams are estimated from other process requirements. Equilibrium laws are used to decide whether it is feasible to use an operation like absorption, or whether an alternative method is needed. If the equilibrium calculations establish the feasibility of the operation, then the rates of heat, momentum, and mass transfer determine the design parameters and the extent to which true equilibrium is approached in the equipment.

In continuous contactors, interphase mass transfer occurs by diffusion—a spontaneous, but slow process. The rate of diffusion can be increased by inducing turbulence in the phases. A spoonful of sugar in a glass of quiescent water will sit at the bottom for days before it dissolves, whereas it will dissolve readily upon stirring due to eddy diffusion, which is, roughly speaking, 1000 times faster than molecular diffusion.

Consider Fig. 16.1, which depicts boundary layer flow of a liquid past a

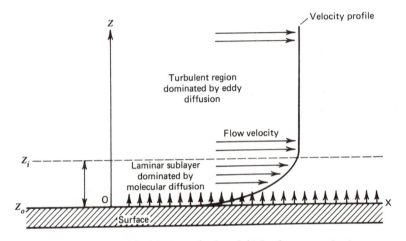

Figure 16.1. Diffusional transfer in a fluid flowing past a slowly dissolving surface.

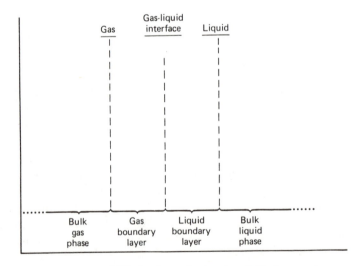

Figure 16.2. Diffusional transfer in a two-fluid phase system.

slowly dissolving solid surface. Suppose a laminar flow pattern is developed in the region between Z_o and Z_i, which constitutes a laminar sublayer. Beyond Z_i, a turbulent flow pattern exists. Molecular diffusion occurs in both the laminar and turbulent regions, but the latter is dominated by eddy diffusion. The limiting case is diffusion into a stagnant fluid, that is, where the velocity is zero everywhere in the fluid, in which case only molecular diffusion is important.

In absorption, instead of a solid surface dissolving into a liquid, we have a gas dissolving into a liquid. The liquid may exist as dispersed droplets or as a flowing film like that shown in Fig. 16.2. Since gas and liquid layers are free to move, two boundary layers can arise at the interface.

16.1 Molecular Diffusion

Fick's law of diffusion,[1] which was developed in the late 1800's, states that molecular diffusion is proportional to the decrease in the concentration gradient; that is, for a gaseous or liquid mixture of two components A and B, the molar flux vector $\mathbf{J}_A$ of A with respect to an observer moving at the stream average velocity is given by

$$\mathbf{J}_A = -D_{AB}\nabla C_A \tag{16-1}$$

The flux $\mathbf{N}_A$ of component A with respect to a stationary observer is

$$\mathbf{N}_A = \mathbf{J}_A + x_A(\mathbf{N}_A + \mathbf{N}_B) = -D_{AB}\nabla C_A + x_A(\mathbf{N}_A + \mathbf{N}_B) \tag{16-2}$$

In these equations C_A is the concentration of component A, D_{AB} is the diffusivity of A in B, and x is the mole fraction. The second term in (16-2) accounts for transport by bulk flow, and the first for the diffusive transport which is superimposed on the bulk flow. Diffusivity is a physical property of the component pair A and B. Values for a number of gases in binary mixtures with air at 25°C and 1 atm (101.3 kPa) pressure can be found in many handbooks. Diffusivities for other gas or liquid mixtures are scarcer and often have to be determined experimentally, as will be illustrated in Example 16.1. In the absence of experimental data, binary diffusivities can be calculated in gases at low pressures by the empirical equation of Fuller, Schettler, and Giddings.[2]

$$D_{BA} = D_{AB} = \frac{0.001\,T^{1.75}(1/M_A + 1/M_B)^{0.5}}{P\left[\left(\sum V_A\right)^{1/3} + \left(\sum V_B\right)^{1/3}\right]^2} \tag{16-3}$$

In (16-3), $(\sum V)$ the diffusion volume for simple molecules is given in Table 16.1. For more complex molecules, the molecular contributions can be obtained by summing the atomic contributions given in Table 16.2.

For dilute nonelectrolyte liquid solutions of A in solvent B, the empirical

Table 16.1 Diffusion volumes for simple molecules, ΣV

H_2	7.07	CO	18.9
D_2	6.70	CO_2	26.9
He	2.88	N_2O	35.9
N_2	17.9	NH_3	14.9
O_2	16.6	H_2O	12.7
Air	20.1	$(CCl_2F_2)^a$	114.8
Ar	16.1	(SF_6)	69.7
Kr	22.8	(Cl_2)	37.7
(Xe)	37.9	(Br_2)	67.2
		(SO_2)	41.1

[a] Parentheses indicate that the value listed is based on only a few data points.

Table 16.2 Atomic diffusion volumes for use in the method of Fuller, Schettler, and Giddings

Atomic and Structural Diffusion Volume Increments, V			
C	16.5	(Cl)	19.5
H	1.98	(S)	17.0
O	5.48	Aromatic ring	−20.2
$(N)^a$	5.69	Heterocyclic ring	−20.2

[a] Parentheses indicate that the value listed is based on only a few data points.

Source. E. N. Fuller, P. D. Schettler, and J. C. Giddings, *Ind. Eng. Chem.*, **58** (5), 18 (1966).

equation of Wilke and Chang[3] applies.

$$D_{AB} = \frac{7.4 \times 10^{-8} T (\Psi_B M_B)^{0.5}}{\mu_B V_A^{0.6}}$$ (16-4)

where

D_{AB} = diffusivity of A in B, cm^2/s.

T = absolute temperature, °K.

μ = viscosity, cp.

V_A = molar volume of solute in $cm^3/gmole$, at its normal boiling point.

Ψ_B = association parameter for solvent B, which ranges in value from 1.0 for unassociated solvents to 2.6 for water, with values of 1.9 for methanol and 1.5 for ethanol.

M_A, M_B = molecular weights of components A and B, respectively.

P = absolute pressure, atm.

Two common physically realizable situations to which (16-2) is applicable are:

1. Equimolar counter diffusion, EMD, in which the number of moles of A diffusing in one direction equals the number of moles of B diffusing in the opposite direction; that is, $\mathbf{N}_A = -\mathbf{N}_B$, and (16-2) reduces to

$$\mathbf{N}_A = -D_{AB}\nabla C_A = -D_{AB}C\nabla x_A \qquad (16\text{-}5)$$

2. Unimolecular diffusion, UMD, of one component A through stagnant component B; that is, $-\mathbf{N}_B = 0$. Then (16-2) reduces to

$$\mathbf{N}_A = -\frac{D_{AB}\nabla C_A}{1 - x_A} = -\frac{D_{AB}C\nabla x_A}{1 - x_A} \qquad (16\text{-}6)$$

which, when applied across a stagnant film extending from Z_1 to Z_2, becomes

$$\int_{Z_1}^{Z_2} \mathbf{N}_A dZ = -D_{AB}C \int_{x_{A_1}}^{x_{A_2}} \frac{dx_A}{(1 - x_A)} \qquad (16\text{-}7)$$

In dilute binary solutions, x_A is small, $(1 - x_A) \approx 1$, and (16-5) and (16-6) are equivalent. In this chapter, (16-5) and (16-6) are applied only to planar geometries so the vector notation will be discarded. We will also assume the diffusion flux to be proportional to the gradient, thus eliminating possible confusion due to the minus sign in (16-5).

16.2 Evaporation into a Stagnant Gas Layer

For the case of evaporation of a pure liquid into a gas or the condensation of a component in a vapor into its own pure liquid, the only resistance to diffusion is in the gas phase, since there are no concentration gradients in the pure liquid. A test tube, beaker, graduated cylinder, tank, or the Great Salt Lake will all eventually empty by evaporation if the liquid contents are not replenished regularly. One knows intuitively that spills evaporate faster if they are spread over a large area, if the temperature is high, or if we blow air on the spill and also that liquid in a bottle takes longer to evaporate the farther the surface of the liquid is from the top of the bottle.

The temperature-dependence of evaporation is a property of liquid A and the gas B into which A is diffusing. An increase in temperature causes an increase in diffusivity. Also, the concentration of component A in the gas at the liquid–gas interface is determined by the vapor pressure of A, which increases rapidly with temperature, thus providing an increased concentration difference that enhances diffusion [see (16-5)]. Blowing air over a liquid surface induces eddy diffusion, causing faster evaporation. Blowing air directly at a liquid surface causes even faster evaporation because of the greater reduction in boundary layer thickness. A barrier, such as stagnant gas in the bottle, decreases the concentration gradient and thus the rate of diffusion. Some of these basic principles are now demonstrated by example.

Example 16.1.   The open beaker of Fig. 16.3 is filled with liquid benzene to 0.5 cm of the top. A gentle 25°C breeze blows across the mouth of the beaker so that the benzene vapor is carried away by convection after it diffuses through the 0.5-cm air layer in the beaker. The total pressure P is 101.3 kPa (1 atm), the vapor pressure, P^s of benzene at 25°C is 13.3 kPa; and its diffusivity in air at 25°C multiplied by the total gas concentration is 3.5×10^{-7} gmole/cm · sec. Calculate the initial rate of evaporation of benzene.

Solution.   This is clearly a case of diffusion of gaseous benzene A through a stagnant layer of air B. The concentration of benzene at the liquid surface $Z_1 = 0$ can be estimated from Dalton's law.

$$y_A = P_A^s/P = 13.3/101.3 = 0.132$$

Applying (16-7) and neglecting any accumulation of benzene in the stagnant air layer in

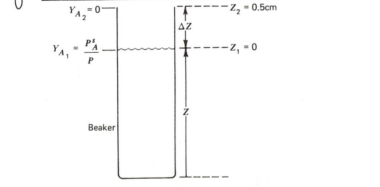

Figure 16.3.   Evaporation of benzene from a beaker into air—open window or fan causing air to blow so that the concentration of benzene A at the top of the beaker is approximately zero.

the beaker so that N_A is constant across ΔZ.

$$N_A \int_0^{0.5} dZ = -(3.5 \times 10^{-7}) \int_{0.132}^{0} \frac{dy_A}{1 - y_A}$$

$$N_A(0.5) = 3.5 \times 10^{-7} \ln\left(\frac{1-0}{1-0.132}\right)$$

Thus

$$N_A = 9.91 \times 10^{-6} \, gmole/sec \cdot cm^2$$

☐

Example 16.2. A beaker filled with an equimolar liquid mixture of ethyl alcohol and ethyl acetate evaporates at 0°C into still air at 101 kPa (1 atm) total pressure. Assuming Raoult's law applies, what will be the composition of the liquid remaining when half the original ethyl alcohol has evaporated?
The following data are available.

	Vapor pressure, kPa at 0°C	Diffusivity in air, m^2/s
Ethyl acetate (AC)	3.23	6.45×10^{-6}
Ethyl alcohol (AL)	1.62	9.29×10^{-6}

Solution. Because of the low vapor pressures, we can consider each component to be evaporating independently through a stagnant layer of air. Thus (16-7) applies, and assuming $(1 - y_{AL}) = (1 - y_{AC}) \approx 1$, we have

$$N_{AL} \int_{Z_1}^{Z_2} dZ = -D_{AL}C \int_{y_{AL_1}}^{y_{AL_2}} dy_{AL} \tag{A}$$

$$N_{AC} \int_{Z_1}^{Z_2} dZ = -D_{AC}C \int_{y_{AC_1}}^{y_{AC_2}} dy_{AC} \tag{B}$$

Integrating (A) and (B) between the liquid surface and some point above the surface where $Z = Z_2$, $y_{AL_2} = 0$, $y_{AC_2} = 0$ and dividing the resulting equations gives

$$\frac{N_{AL}}{N_{AC}} = \left(\frac{D_{AL}}{D_{AC}}\right)\left(\frac{y_{AL_1}}{y_{AC_1}}\right) \tag{C}$$

The fluxes N_{AL} and N_{AC} can be expressed in terms of moles of liquid in the beaker m_{AL} and m_{AC} and time t as

$$N_{AL} = \frac{-dm_{AL}}{A\,dt} \qquad N_{AC} = \frac{-dm_{AC}}{A\,dt} \tag{D}$$

where A is the interfacial surface area. The gas-phase concentrations are related to m_{AL} and m_{AC} by Raoult's and Dalton's laws.

$$y_{AL_1} = \frac{p_{AL}}{P} = \frac{P_{AL}^s}{P}\left(\frac{m_{AL}}{m_{AL} + m_{AC}}\right) \qquad y_{AC_1} = \frac{p_{AC}}{P} = \frac{P_{AC}^s}{P}\left(\frac{m_{AC}}{m_{AC} + m_{AL}}\right) \tag{E}$$

where p_{AL} and p_{AC} are partial pressures.
Substituting (D) and (E) into (C) and integrating, we have

$$\ln\left(\frac{m_{AL_2}}{m_{AL_1}}\right) = \left(\frac{P_{AL}^s}{P_{AC}^s}\right)\left(\frac{D_{AL}}{D_{AC}}\right)\ln\left(\frac{m_{AC_2}}{m_{AC_1}}\right) \tag{F}$$

For 100 moles of initial liquid mixture $m_{AL_1} = 50$ moles, $m_{AL_2} = 25$ moles, and $m_{AC_1} = 50$ moles. Solving (F) for m_{AC_2}, we obtain 19.2. Hence

$$x_{AL_2} = \frac{25}{(25 + 19.2)} = 0.57$$

$$x_{AC_2} = \frac{19.2}{(25 + 19.2)} = 0.43$$

☐

16.3 Wetted-Wall Columns and Mass Transfer Coefficients

Figure 16.4 shows a simplified picture of a wetted-wall column—usually a modified vertical condenser tube with a film of absorbing liquid flowing down the inside wall of the tube—and a stream of gas containing the component to be absorbed flowing up the center and contacting the falling liquid. Related and more complex devices are the climbing-film evaporator—loosely speaking, a distillation column without trays operating at a very high reboiler return ratio—and a rotary evaporator, where the liquid film clings to the inside of a slanted, half-full, rotating, round-bottom flask.

If the gas flow rate is increased in the device in Fig. 16.4, ripples appear in the falling liquid film. As the gas flow rate is increased further, the film breaks and is pushed up and entrained into the gas stream. At low gas rates, if there are no liquid ripples, the interfacial area through which mass transfer occurs is simply the (measurable) liquid surface; and the wetted-wall column can be used for mass transfer or hydrodynamic studies.

In commercial absorption equipment, both the liquid and vapor are usually in turbulent flow and the effective stagnant film thickness ΔZ is not known. The common practice, therefore, is to rewrite (16-2) in terms of an empirical mass transfer (film) coefficient that replaces both the diffusivity and film thickness. Thus, by definition,

$$N_A = k_y(y_A - y_{A_i}) \tag{16-8}$$

for the gas phase, and

$$N_A = k_x(x_{A_i} - x_A) \tag{16-9}$$

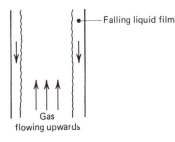

Falling liquid film

Gas
flowing upwards

Figure 16.4. Wetted-wall column.

for the liquid phase, where

y_A = bulk gas concentration.

y_{A_i} = concentration of A in the gas at the gas–liquid interface.

x_A = bulk liquid concentration.

x_{A_i} = concentration of A in the liquid at the liquid–gas interface.

For dilute solutions or equimolar counterdiffusion EMD, (16-5) can be integrated provided that D_{AB} and Z are not functions of x_A to yield

$$N_A = \frac{D_{AB}C}{\Delta Z}(x_{A_i} - x_A) \tag{16-10}$$

Comparing (16-10) and (16-9) we have

$$k_x = \left(\frac{D_{AB}C}{\Delta Z}\right) \tag{16-11}$$

for the liquid phase, and from (16-8) and (16-10)

$$k_y = \left(\frac{D_{AB}C}{\Delta Z}\right) \tag{16-12}$$

for the gas phase.

If, instead of EMD, we have UMD, as in gas absorption of a single component through a stagnant film, then (16-8) and (16-9) must be matched with (16-6) rather than (16-5). Denoting the mass transfer coefficient for UMD by k'_y, we have, from (16-6) and (16-8) for a binary gas mixture of A and B,

$$N_A = k'_y(y_A - y_{A_i}) = \frac{D_{AB}C}{\Delta Z}\int_{y_{A_i}}^{y_A}\frac{dy_A}{(1-y_A)}$$

To convert this equation to a form similar to (16-10), we substitute $dy_A = -dy_B$ and $(1 - y_A) = y_B$ and integrate between y_B and y_{B_i} to obtain

$$N_A = k'_y(y_A - y_{A_i}) = \frac{D_{AB}C}{\Delta Z}\ln\frac{y_{B_i}}{y_B}$$

Noting that

$$\frac{(y_{B_i} - y_B)}{(y_{A_i} - y_A)} = 1$$

and defining y_{BM} the log mean concentration of component B by

$$y_{BM} = \frac{y_{B_i} - y_B}{\ln(y_{B_i}/y_B)}$$

we obtain as a final result

$$N_A = k'_y(y_A - y_{A_i}) = \frac{D_{AB}C}{(\Delta Z)y_{BM}}(y_A - y_{A_i}) \tag{16-13}$$

Comparing (16-12) and (16-13) we see that

$$k'_y = \frac{k_y}{y_{BM}} \tag{16-14}$$

As indicated in the first three rows of Table 16.3, (16-8) and (16-9) can be written in terms of component partial pressure $(p - p_i)$ or concentration $(C_i - C)$ driving forces, the corresponding mass transfer coefficients being designated k_g and k_C, respectively, for the gas, and k_l for the liquid, in terms of a concentration driving force. The relationships between k and k' in the different mass transfer expressions are given at the bottom of Table 16.3. Here P_{BM} is the log mean partial pressure of B, ρ is the gas density, and M is the average molecular

Table 16.3   Defining equations for mass transfer coefficients based on dilute solutions or equimolar diffusion[a]

Because we generally don't know and can't measure interphase compositions, we don't apply these equations.	$N_A = k_g(p_A - p_{A_i})$, (gas) $N_A = k_l(C_{A_i} - C_A)$, (liquid) $N_A = k_C(C_A - C_{A_i})$, (gas) $N_A = k_y(y_A - y_{A_i})$, (gas) $N_A = k_x(x_{A_i} - x_A)$, (liquid)
Instead, with the help of an equilibrium equation (e.g. Henry's law) we use these equations in terms of imaginary but calculable equilibrium compositions.[b]	$N_A = K_G(p_A - p_A^*)$ where $p_A^* = HC_A$ (C_A in bulk liquid) $N_A = K_L(C_A^* - C_A)$ where $C_A^* = \frac{1}{H}p_A$ $N_A = K_y(y_A - y_A^*)$ where $y_A^* = H'x_A$ $N_A = K_x(x_A^* - x_A)$ where $x_A^* = \frac{y_A}{H'}$

Equivalences between UMD and EMD mass transfer coefficients:

$$k_l C = k_l\left(\frac{\rho}{M}\right) = k_x = k'_x x_{BM} = k'_l x_{BM}\frac{\rho}{M} = k'_l x_{BM}C$$

$$k_g P = k_y = \frac{k_C P}{RT} = k_C C = k'_g P_{BM} = k'_y \frac{P_{BM}}{P} = k'_C \frac{P_{BM}}{RT} = k'_y y_{BM}$$

[a] From A. S. Foust and co-workers, *Principals of Unit Operations*, John Wiley & Sons, New York, 1960, 171.
[b] These coefficients are introduced in Section 16.5.

weight. These relationships are valid only for isothermal binary systems, where the equation of state $C = P/RT$ can be used for the gas phase.

It is also possible to use mole (or mass) ratio driving forces rather than mole fractions provided that the mass transfer coefficients are in the corresponding units. The mole ratio-based mass transfer coefficients are approximately related to k_C by

$$k_Y = \frac{k_C P y_{B_1} \bar{y}_B}{RT} \qquad k'_Y = \frac{k'_C P y_{B_1} \bar{y}_B}{RT}$$

$$k_X = \frac{k_C \bar{\rho} x_{B_1}(\bar{x}_B)}{\bar{M}} \qquad k'_X = \frac{k'_C \bar{\rho} x_{B_1} \bar{x}_B}{\bar{M}}$$

Here B_1 denotes outlet concentration and the bars over x, y, ρ and M signify mean values.[4]

The following example will clarify the method for calculating simple (film) mass transfer coefficients.

Example 16.3.   The following data are reported for the absorption of ammonia from an ammonia-air mixture by a strong acid in a wetted-wall column 0.0146 m inside diameter and 0.827 m long. Temperature is 25.6°C and the pressure is 101.3 kPa throughout the column.

Partial pressure of NH_3 in inlet gas = 8.17 kPa
Partial pressure NH_3 in outlet gas = 2.08 kPa
Air rate, kgmole/hr = 0.260

The change in acid strength over the length of the column is unappreciable, and the equilibrium partial pressure of ammonia over the acid is assumed to be negligible. Calculate the mass transfer coefficients, k_y, k'_y, k_g, k'_g, of Table 16.3.

Solution.   The kgmoles of ammonia absorbed per hour by the acid =

$$\left(\frac{0.26 \text{ kgmole air}}{\text{hr}}\right)\left(\frac{8.17}{93.1} - \frac{2.08}{99.2}\right)\frac{\text{kgmole } NH_3}{\text{kgmole air}} = 1.74 \times 10^{-2}\frac{\text{kgmole } NH_3}{\text{hr}}$$

The average mass transfer flux, N_A, for ammonia, using the inside surface area of the column as the area for mass transfer is $(1.74 \times 10^{-2} \text{ kgmole/hr})/[0.827(3.14)(0.0146) \text{ m}^2] = 0.459(\text{kgmole } NH_3/\text{hr} \cdot \text{m}^2)$.

Since we have used an average mass flux, we must, for consistency, use an average driving force $(y_A - y_{A_i})_{avg}$ to compute the mass transfer coefficient. At the top of the column, $(y_A - y_{A_i}) = 2.08/101.3 = 0.0205$; at the bottom, $(y_A - y_{A_i}) = 0.0807$. In the absence of any other information, we take an arithmetic average of the top and bottom ammonia concentration = $(y_A - y_{A_i})_{avg} = 0.0506$.

Thus, from (16-8)

$$k_y = N_A/(y_A - y_{A_i})_{avg} = 0.459/0.0506 = 9.07 \frac{\text{kgmole}}{\text{hr} \cdot \text{mole fraction} \cdot \text{m}^2}$$

When both the equilibrium curve (represented by the y_{A_i} values) and the operating curve (represented by the y_A values) are straight lines along the length of a countercurrent contactor, it can be shown that the correct driving force is actually the log mean value:

$$(y_A - y_{A_i})_{LM} = \frac{(y_A - y_{A_i})_b - (y_A - y_{A_i})_t}{\ln\dfrac{(y_A - y_{A_i})_b}{.(y_A - y_{A_i})_t}}$$

where the subscripts b and t refer to the bottom and top of the column. This equation gives

$$(y_A - y_{A_i})_{avg} = 0.0439 \text{ and } k_y = 10.45 \frac{\text{kgmole}}{\text{hr·mole fraction·m}^2}$$

From Table 16.3, since the pressure is 101.3 kPa = 1 atm, $k_g = k_y$ provided that partial pressures are in atmospheres. If Pascals are used, the conversion factor is required. Since $y_{BM} = (0.9795 - 0.9193)/\ln(0.9795/0.9193) = 0.949$, at 1 atm, from Table 16.3, $k'_y = k_y/y_{BM} = 10.45/0.949 = 11.01$ kgmole/hr·mole fraction·m^2.

Mass transfer coefficients are experimentally determined by methods such as the one used in this example. They must then be correlated with the physical properties of the system, equipment geometry, and flow conditions if they are to be generally useful. Examples of such correlations are given later in the chapter.

☐

In general, the molecular diffusion equations developed and applied in Examples 16.1 and 16.2 are not useful for engineering design because they require that interface concentrations as well as the distance, ΔZ, through which molecular diffusion takes place be known. The empirical (film) coefficients, k_x and k_y, defined by (16-8) and (16-9), do not require ΔZ; but they require the interface concentrations, y_{A_i} and x_{A_i}. In general, they are known only under limiting conditions of dilute solutions or pure fluids. Another difficulty is that the mass flux, N_A, the moles per time per interfacial area, is not easily applied to packed columns because the interfacial area is not known.

Although not directly useful in themselves, the molecular diffusion equations do provide valuable insights into mechanisms by which the mass transfer process occurs and are useful correlating tools.

16.4   Packed Columns

The most common gas absorption system is a packed column operating in a countercurrent mode, as in Fig. 16.5(a), where the streams are quite unmixed and plug flow conditions predominate. The gas containing the component to be absorbed is introduced at the bottom of a vertical pipe filled with inert packing material. The absorbing liquid is introduced at the top of the column and allowed to trickle downwards, under the influence of gravity, over and through the inert packing.

As was pointed out in Chapter 1, countercurrent flow is usually the most desirable flow scheme in a two-phase contactor because it results in the largest

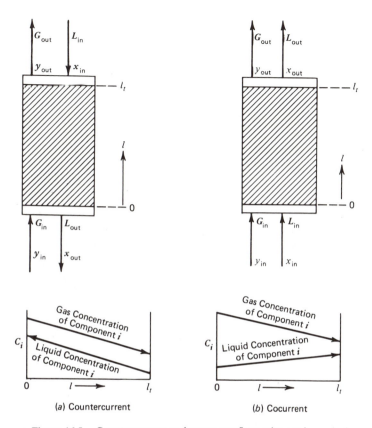

Figure 16.5. Countercurrent and cocurrent flow schemes in packed absorbers. (*a*) Countercurrent. (*b*) Cocurrent.

average driving force for mass transfer. There are systems, however, for which the cocurrent flow absorbers of Fig. 16.5(b) may be more desirable.

The ideal and expected modus operandi of a gas-liquid contactor is one in which the continuous phase is the gas and the dispersed phase is the absorbing liquid. If flow rate and other conditions are such that the liquid phase becomes continuous and the gas phase becomes dispersed, we have the very inefficient and unpredictable condition called flooding.

Flooding is virtually impossible to induce in cocurrent flow, but extremely easy to produce in a countercurrent absorber. Cocurrent absorbers also require a lower pressure drop across the packing for equivalent throughputs, and required rates for the liquid absorbent are usually lower in a cocurrent absorber. Furthermore, if an absorbing liquid is chosen such that a chemical reaction occurs between the liquid and gas phases, conditions may be adjusted so that a constant concentration gradient may be maintained per unit length of absorber in a cocurrent configuration. All these considerations point to a need for careful qualification of the commonly accepted rule of thumb concerning countercurrent scrubbers being more efficient for a given application.

16.5 Diffusion Through Two Films in Series

For the general case of mass transfer between two phases, concentration gradients can exist on each side of the interface. If the two phases are in turbulent flow, the concentration gradients may be significant only in the effective films (laminar sublayer or stagnant regions) on each side of the interface. Thus, the films limit the total mass transfer process. This situation is illustrated in Fig. 16.6.

The boundary between the stagnant film and the turbulent region is actually a transition zone; however, for purposes of application, it is advantageous to think of these fictitious layers as separated by a boundary. In most applications, the distances Z_l and Z_g are not quantities that can be measured; however, as explained above, they are embedded in the mass transfer coefficients defined by (16-8) and

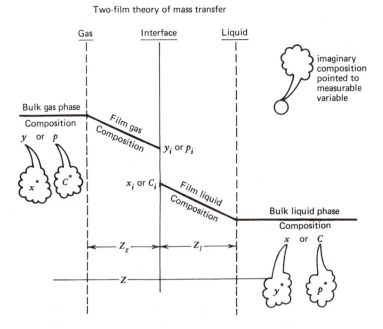

Figure 16.6.   Interface properties in terms of bulk properties.

(16-9). For a given mass transfer rate, the higher the mass transfer coefficient, the thinner the film thickness.

The compositions of the respective phases at the outside boundaries of the fictitious films are essentially those of the bulk conditions. The compositions x_i and y_i at the interface boundary are assumed to be in equilibrium. For example, in terms of Henry's law, $y_i = H'x_i$.

In the absence of chemical reactions and at steady-state conditions, since moles are conserved, the rate of diffusion across the gas-phase film must equal the rate across the liquid phase film. Applying (16-8) and (16-9) to this case, and equating diffusion rates on each side of the interface yields the two-film model of Whitman:

$$N_A = k_y(y_A - y_{A_i}) = k_x(x_{A_i} - x_A) \qquad (16\text{-}16)$$

The difficulty in applying (16-16) to real systems centers about the problem of determining interfacial compositions. In general, the (x_i, y_i) represents the composition of the phases at the interface and, therefore, lies on the equilibrium curve. This invokes the assumption that equilibrium conditions are achieved at the interface, which has been substantiated by experimental evidence.

Application of (16-16) gives for the equation of the slope of the line AB in Fig. 16.7, after dropping the A subscript

$$\frac{k_x}{k_y} = \frac{(y - y_i)}{(x_i - x)} \qquad (16\text{-}17)$$

This slope determines the relative resistances of the two phases to mass transfer. The distance AE is the gas-phase driving force $(y - y_i)$, while AF is the liquid-phase driving force $(x_i - x)$.

If the mass transfer resistance in the gas phase is very low, y_i is approximately equal to y. Then the entire resistance to mass transfer resides in the

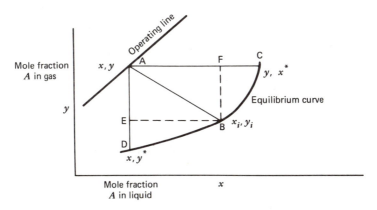

Figure 16.7. Countercurrent absorption tower.

liquid phase. This situation is characteristic of the absorption of a very slightly soluble gas in a liquid phase and is referred to as a *liquid-film resistance controlling process*. If the mass transfer resistance in the liquid phase is very low, x_i is approximately equal to x. This situation is characteristic of processes involving the absorption of very soluble gases in liquids and is referred to as a *gas-film resistance controlled process*. From an engineering point of view it is rather important to know which phase resistance is controlling. If, for example, the liquid phase is offering the major resistance to mass transfer, the capacity of the equipment can be increased appreciably by promoting turbulence in the liquid phase.

If the equilibrium and operating lines are straight, as in Fig. 16.8, then, by geometry, the ratio of gas-phase resistance to total mass transfer resistance is

$$\frac{y - y_i}{y - y^*} = \frac{x^* - x_i}{x^* - x} \tag{16-18}$$

It is convenient to define overall mass transfer coefficients K_y and K_x for the gas and liquid phases, respectively, by

$$N = K_y(y - y^*) = K_x(x^* - x) \tag{16-19}$$

where y^* is the composition of vapor in equilibrium with x, and x^* is the composition of liquid in equilibrium with vapor y.

Equating (16-16), (16-17), and (16-19) gives the following relationship between the overall mass transfer coefficients K_y and the individual mass transfer coefficients k_y and k_x.

$$\frac{1}{K_y} = \frac{(y - y^*)}{k_y(y - y_i)} \tag{16-20}$$

$$= \frac{(y - y_i) + (y_i - y^*)}{k_y(y - y_i)}$$

$$= \frac{1}{k_y} + \frac{1}{k_x}\left(\frac{y^* - y_i}{x - x_i}\right) \tag{16-21}$$

In a similar manner, it is found that

$$\frac{1}{K_x} = \frac{1}{k_x} + \frac{1}{k_y}\left(\frac{x^* - x_i}{y - y_i}\right) \tag{16-22}$$

However, $(y_i - y^*)/(x_i - x)$ is the slope of the equilibrium curve H', which in Fig. 16.8 is constant. Therefore,

$$\frac{1}{K_y} = \frac{1}{k_y} + \frac{H'}{k_x} \tag{16-23}$$

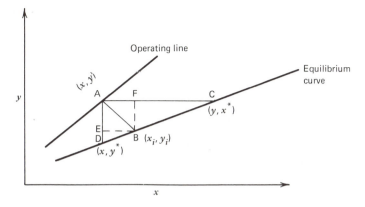

Figure 16.8. Case of linear equilibrium curve and operating line.

The relationships* between K_x, k_x, and k_y can be developed in a similar fashion to give

$$\frac{1}{K_x} = \frac{1}{k_x} + \frac{1}{H'k_y} \tag{16-24}$$

If the equilibrium relationship is not linear, there is no justification for (16-23) and (16-24). Furthermore, it should be noted that the mass transfer coefficients were all defined on the basis of EMD, or UMD under dilute conditions.

Table 16.3 shows some possible forms of mass transfer coefficient definitions and the relationships between the coefficients based on EMD and UMD. In addition to K_y and K_x coefficients, we can also define K_Y and K_X coefficients based on mole ratio $[(Y - Y_i)$ and $(X_i - X)]$ driving forces.

Example 16.4 shows how mass transfer coefficients can be calculated from experimental data.

Example 16.4. At a point in an ammonia absorber using water as the absorbent and operating at 101.3 kPa and 20°C, the bulk gas phase contains 10% NH_3 by volume. At the interface, the partial pressure of NH_3 is 2.26 kPa. The concentration of the ammonia in the body of the liquid is 1% by weight. The rate of ammonia absorption at this point is 0.05 kgmole/hr · m^2.

(a) Given this information and the equilibrium curve of Fig. 16.9, calculate X, Y, Y_i, X_i, X^*, Y^*, K_Y, K_X, k_Y, and k_X.

---

* The identities $1/K_G = 1/k_g + H'/k_l$ and $1/K_L = 1/k_l + 1/H'k_g$ are also valid by definition. In actual practice, surface effects, waves, and other phenomena can complicate these relationships.

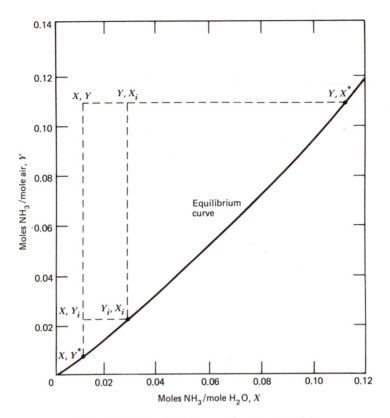

Figure 16.9. Equilibrium curve for Example 16.4.

(b) What percent of the mass transfer resistance is in each phase?

(c) Verify (16-23) for these data.

Solution. Using mole ratios, we find the following.

(a) $X = (1/17)/(99/18) = 0.0107$; $Y = 10/90 = 0.111$; $Y_i = (2.26)/(101 - 2.26) = 0.0229$. X_i, X^*, and Y^* (from Fig. 16.9): $X_i = 0.03$; $X^* = 0.114$; $Y^* = 0.007$. K_Y: $0.05 = K_Y(Y - Y^*) = K_Y(0.111 - 0.007)$; $K_Y = 0.48$ kgmole/hr · m² · mole-ratio driving force. K_X: $0.05 = K_X(X^* - X) = K_X(0.114 - 0.0107)$; $K_X = 0.484$ kgmole/hr · m² · mole ratio driving force. k_Y: $0.05 = k_Y(Y - Y_i) = k_Y(0.111 - 0.0229)$; $k_Y = 0.568$ kgmole/hr · m² · mole ratio driving force. k_X: $0.05 = k_X(X_i - X) = k_X(0.03 - 0.0107)$; $k_X = 2.59$ kgmole/hr · m² · mole-ratio driving force.

(b) From Fig. 16.9, based on the gas-phase concentrations, the percentage of the gas-phase resistance in the total mass transfer resistance from (16-18) is

$$100\frac{[Y - Y_i]}{[Y - Y^*]} = 100\frac{[0.111 - 0.0229]}{[0.111 - 0.007]} = 84.7\%$$

Based on the liquid-phase concentrations, the result is

$$100\frac{[X^* - X_i]}{[X^* - X]} = 100\frac{[0.114 - 0.03]}{[0.114 - 0.0107]} = 81.3\%$$

The difference between 84.7 and 81.3 is due to the slight curvature of the equilibrium line in Fig. 16.9.

(c)   From (16-23), since $H' = (Y_i - Y^*)/(X_i - X) = (0.0229 - 0.007)/(0.03 - 0.0107) = 0.824$,

$$\frac{1}{K_Y} = \frac{1}{k_Y} + \frac{H'}{k_X} = \frac{1}{0.568} + \frac{0.824}{2.59} = 2.079$$

Therefore, $K_Y \approx 0.481$, which checks the directly calculated value.

As written, all transfer coefficients have the units of kilogram-moles per hour per square meter per mole ratio driving force. In the literature, the driving forces are usually in terms of mole fractions or partial pressures. It is possible to convert from one to another via (16-15).

□

16.7   Alternative Models for Mass Transfer in Gas–Liquid Systems

In the preceding discussion, a very simple model of the absorption process has been used: the film model. Other models of increased complexity have been proposed for calculating and correlating mass transfer coefficients for fluid properties and flow regimes. The rationale and assumptions for the more important mass transfer models are now very briefly discussed.

The Film Theory

The *film theory*, as discussed above, assumes the direction of mass transfer to be normal to the interface and the resistance to be due to stagnant films of fluid on both sides of the interface. Accordingly, the model[5,6] predicts k_l or k_g to be proportional to $D^{1.0}$.

The experimental evidence, as summarized by Sherwood, Pigford, and Wilke,[7] indicates that the mass transfer coefficients are more nearly proportional to the molecular diffusivity to the square root power. Nevertheless, the film theory is used in the development of the working equations in this chapter, since the physical picture it depicts is simple and adequate. Actually, it is irrelevant, from a pragmatic point of view, what model is used to develop a working equation based on empirical mass or heat transfer coefficients that must, ultimately, be obtained from experimental data.

The Penetration Theory

The *penetration theory* of Higbie[8] replaces the stagnant fluid by intermittently static and moving eddies that arrive at the interface from the bulk stream, stay for a period of time in the interface (during which molecular diffusion normal to the interface occurs), and leave the interface to mix again with the bulk fluid. During the period that the eddies stay at the interface, the assumption is made that they are static. The penetration theory predicts k_l or k_g to be proportional to $D^{0.5}$.

The Random Surface Renewal Theory

Higbie's penetration theory was modified by Danckwerts,[9] who assumed a residence time distribution for the active eddies at the interface. The same square root dependence of the k_l or k_g value on diffusivity as in the penetration theory is obtained, but the resulting model is much more intellectually satisfying than the original penetration model.

The Film–Penetration Theory

In 1958, Toor and Marcello[10] investigated the effect of removing the short residence time constraint in the film that is implicit in penetration models, so that the film model would become a limiting case of the penetration model. Not unexpectedly, the resulting model predicts a dependence on diffusivity of the k_l or k_g value from $D^{0.5}$ to $D^{1.0}$

The Boundary Layer Theory

The *boundary layer theory*[5] removes the restriction of stationary eddies at the interface and replaces it by the assumption that eddies travel along the interface and that transport occurs by diffusion in the direction normal to the interface and by convection in the direction along the interface. The boundary layer theory for laminar flow predicts k_l and k_g to be proportional to $D^{2/3}$.

16.8   Capacity Coefficients for EMD and Dilute UMD

The overall mass transfer coefficients K_x and K_y have units of (moles)/(time-interfacial area-unit mole fraction driving force). In the case of a wetted-wall column, the interfacial area is known. However, for most types of mass transfer equipment the interfacial area cannot be determined. It is necessary therefore to define a quantity a that is the interfacial area per unit of active equipment volume. Although separate compilations of a can be found in handbooks and vendor literature, this parameter is usually combined with the mass transfer coefficients to define capacity coefficients $(k_x a)$ and $(K_x a)$ for the liquid phase and $(K_y a)$ or $(k_y a)$ for the vapor phase, which then have the dimensions of moles per unit time per unit driving force per unit of active equipment volume. The application of these composite coefficients to the design of packed towers is now demonstrated.

Consider the countercurrent-flow packed tower shown in Fig. 16.10. The

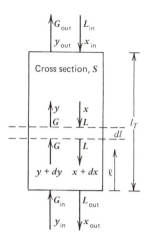

Figure 16.10. Countercurrent-flow packed tower.

compositions and flow rates of the streams are defined by

G = mass or molar flow rate of gas phase.

L = mass or molar flow rate of liquid phase.

y = mass or mole fraction of component A in G.

x = mass or mole fraction of component A in L.

A material balance for component A, which is being absorbed over a differential height of column dl, gives

$$-d(Gy) = d(Lx) = Na\,Sdl \tag{16-25}$$

Introducing (16-19) into (16-25) and forming the mass transfer coefficients $K_x a$ and $K_y a$, we have

$$d(Lx) = (K_x a)(x^* - x)Sdl$$

$$-d(Gy) = (K_y a)(y - y^*)Sdl \tag{16-26}$$

Alternative expressions for (16-26) in terms of interface concentrations and the individual film coefficients k_x and k_y are obtained using (16-16) rather than (16-19) in combination with (16-25). Thus

$$d(Lx) = (k_x a)(x_i - x)Sdl$$

$$-d(Gy) = (k_y a)(y - y_i)Sdl \tag{16-27}$$

Generally speaking, (16-27) is not very useful because of the problem of unknown interface concentrations.

A typical absorption, distillation, or extraction problem consists of determining the total packed height l_T to achieve a separation characterized by mole fractions changing from y_1 to y_2 (or x_1 to x_2). Rearranging (16-26) we obtain

$$l_T = \int_0^{l_T} dl = \int_{x_1}^{x_2} \frac{d(Lx)}{(K_x a)S(x^* - x)} = \frac{L}{(K_x a)S} \int_{x_1}^{x_2} \frac{dx}{(x^* - x)} \qquad (16\text{-}28)$$

$$l_T = \int_0^{l_T} dl = \int_{y_2}^{y_1} \frac{d(Gy)}{(K_y a)S(y - y^*)} = \frac{G}{(K_y a)S} \int_{y_2}^{y_1} \frac{dy}{(y - y^*)} \qquad (16\text{-}29)$$

In (16-28) and (16-29), L and G are constant in the case of equimolar diffusion, EMD, or dilute solution, UMD; hence, they can be taken out of the integral sign for these cases. This permits us to form the groupings $(L/K_x aS)$ and $(G/K_y aS)$, which are indicated in lines 1 and 6, columns 2 and 3 of Table 16.4 and are known, respectively, as H_{OL} and H_{OG} (or HTU), the overall heights of liquid and gas transfer units. They are functions of gas and liquid rates and all hydrodynamic and physical factors pertaining to the ability of a particular device to facilitate mass transfer. The H_{OL} and H_{OG} have the dimension of length (of column): the more efficient the device, the lower the HTU.

The integrals in (16-28) and (16-29) are called, respectively, the *number of overall liquid and gas transfer units* N_{OL} and N_{OG} (or NTU), as listed in Table 16.4, rows 1 and 6, columns 5 and 6. Rows 9 and 4 of Table 16.4 give analogous expressions for HTUs and NTUs derived from (16-27). These are called, respectively, the *height of a liquid (gas) transfer unit* and the *number of liquid (gas) transfer units*. Also included in Table 16.4 are expressions involving concentration, mole ratio, and partial pressure driving forces.

The NTUs, like the number of equilibrium stages in a plate column, are a function of the required separation and the departures from thermodynamic equilibrium, as evidenced by the numerator and denominator under the integral. The NTUs represent the amount of contacting required to accomplish a phase enrichment, divided by the driving force in that phase. They are closely related to the number of theoretical stages required to achieve a stated separation. In actual practice, the two quantities are numerically close. For the special case when the operating and equilibrium lines are straight and parallel, they are equal.

This conceptually simple and elegant approach to the design of continuous differential contacting mass transfer equipment was first developed by Chilton and Colburn in the early 1930s. It enables us to establish the height of a packed column by multiplying a number related to the mass transfer capability of the equipment, the HTU, by a dimensionless number, the NTU, which is closely related to the number of theoretical stages. The HTU or Ka values must be obtained from literature tabulations, experimental data, or correlations. Particular care must be exercised to assure that the dimensionless NTU used is compatible with the particular HTU. For example, in Perry's *Handbook*, fifth

Table 16.4 Alternative mass transfer coefficient groupings for gas absorption

Driving Force	Height of a Transfer Unit, HTU			Number of Transfer Units, NTU		
	Symbol	EM Diffusion or Dilute UM Diffusion	UM Diffusion	Symbol	EM Diffusion[a] or Dilute UM Diffusion	UM Diffusion
1. $(y - y^*)$	H_{OG}	$\dfrac{G}{K_y aS}$	$\dfrac{G}{K'_y a(1-y)_{LM}S}$	N_{OG}	$\displaystyle\int \dfrac{dy}{(y-y^*)}$	$\displaystyle\int \dfrac{(1-y)_{LM}dy}{(1-y)(y-y^*)}$
2. $(p - p^*)$	H_{OG}	$\dfrac{G}{K_G aPS}$	$\dfrac{G}{K'_G a(1-y)_{LM}PS}$	N_{OG}	$\displaystyle\int \dfrac{dp}{(p-p^*)}$	$\displaystyle\int \dfrac{(1-p)_{LM}dp}{(1-p)(p-p^*)}$
3. $(Y - Y^*)$	H_{OG}	$\dfrac{G'}{K_Y aS}$	$\dfrac{G'}{K_Y aS}$	N_{OG}	$\displaystyle\int \dfrac{dY}{(Y-Y^*)}$	$\displaystyle\int \dfrac{dY}{(Y-Y^*)}$
4. $(y - y_i)$	H_G	$\dfrac{G}{k_y aS}$	$\dfrac{G}{k'_y a(1-y)_{LM}S}$	N_G	$\displaystyle\int \dfrac{dy}{(y-y_i)}$	$\displaystyle\int \dfrac{(1-y)_{LM}dy}{(1-y)(y-y_i)}$
5. $(p - p_i)$	H_G	$\dfrac{G}{k_g aS}$	$\dfrac{G}{k'_g a(1-p)_{LM}PS}$	N_G	$\displaystyle\int \dfrac{dp}{(p-p_i)}$	$\displaystyle\int \dfrac{(1-p)_{LM}dp}{(1-p)(p-p_i)}$
6. $(x^* - x)$	H_{OL}	$\dfrac{L}{K_x aS}$	$\dfrac{L}{K'_x a(1-x)_{LM}S}$	N_{OL}	$\displaystyle\int \dfrac{dx}{(x^*-x)}$	$\displaystyle\int \dfrac{(1-x)_{LM}dx}{(1-x)(x^*-x)}$
7. $(C^* - C)$	H_{OL}	$\dfrac{L}{K_L aS}$	$\dfrac{L}{K'_L a(1-C)_{LM}PS/M}$	N_{OL}	$\displaystyle\int \dfrac{dC}{(C^*-C)}$	$\displaystyle\int \dfrac{(1-C)_{LM}dC}{(1-C)(C^*-C)}$
8. $(X^* - X)$	H_{OL}	$\dfrac{L'}{K_x aS}$	$\dfrac{L'}{K_x aS}$	N_{OL}	$\displaystyle\int \dfrac{dX}{(X^*-X)}$	$\displaystyle\int \dfrac{dX}{(X^*-X)}$
9. $(x_i - x)$	H_L	$\dfrac{L}{k_x aS}$	$\dfrac{L}{k'_x a(1-x)_{LM}S}$	N_L	$\displaystyle\int \dfrac{dx}{(x_i-x)}$	$\displaystyle\int \dfrac{(1-x)_{LM}dx}{(1-x)(x_i-x)}$
10. $(C_i - C)$	H_L	$\dfrac{L}{k_L aS}$	$\dfrac{L}{k'_L a(1-C)_{LM}PS/M}$	N_L	$\displaystyle\int \dfrac{dC}{(C_i-C)}$	$\displaystyle\int \dfrac{(1-C)_{LM}dC}{(1-C)(C_i-C)}$

[a] The substitution $K_y = K'_y y_{BLM}$ or its equivalent can be made.

647

edition, on p. 18–43, overall absorption coefficients for absorption from air of SO_2 in water are given as $K_L a$ lbmole/hr · ft³ · lbmole/ft³. Thus, the corresponding HTU and NTU are obtained from row 7 of Table 16.4. On page 18–45 of the same reference, $K_G a$ for the CO_2–NaOH–H_2O system is plotted with dimensions of pound-moles per hour per cubic foot per atmosphere, so that the HTU and NTU expressions should be taken from row 2 of Table 16.4. In general, it is always safest to refer back to the original data and equations from which values of Ka or HTU were obtained; otherwise, order-of-magnitude errors may result: *caveat emptor.*

Before equations for the general case of UMD are developed, we might briefly mention which HTU and NTU expressions are used for what design situation.

(a) Gas absorption—Because we usually work here with the gas composition in equilibrium with the liquid (hence concentration differences in the gas phase are large and concentration differences in the liquid are small) N_{OG} and H_{OG} are natural choices. However, $K_L a$ correlations are frequently available for sparingly soluble gases.

(b) Stripping—When removing an absorbed gas from a liquid, the concentration gradients $(x - x^*)$ in the liquid phase are logical variables to work with, so the H_{OL} and N_{OL} concepts are most commonly used.

(c) Humidification—Resistance to mass transfer is gas-phase controlled, and N_G and H_G expressions can be used because y_i can be obtained from vapor pressure data.

(d) Extraction—One is constrained to use the N_{OL} and H_{OL} concept.

(e) Distillation—Most commonly, transfer units are calculated on the basis of the vapor phase.

The following example demonstrates how HTU and NTU values can be calculated from experimental data.

Example 16.5. Air containing 1.6% sulfur dioxide by volume is being scrubbed with pure water in a packed column 1.5 m² in cross-sectional area and 3.5 m in height at a pressure of 1 atm. Total gas flow rate is 0.062 kgmole/sec, the liquid flow rate is 2.2 kgmole/sec, and the outlet gas concentration is $y_T = 0.004$. At the column temperature, the equilibrium relationship is given by $y^* = 40x$.

(a) What is L/L_{min}?

(b) Calculate N_{OG} for this system, and compare your answer to the number of theoretical stages required.

(c)   Obtain H_{OG} and the HETP.

(d)   Calculate $K_g a$.

   Solution.   At these low gas concentrations $(1-y) \approx 1$ and $P_{BM} = 1$ atm so the EMD (dilute UMD) equations apply.

(a)   Neglecting transport between the water and air phases, the total kilogram-moles of SO_2 absorbed per second is:

$$(0.016)(0.062) - (0.004)(0.062)(0.984/0.996) = 0.000747 \text{ kgmole/sec.}$$

   In the outlet water, the SO_2 mole fraction is $x_B = 0.000747/2.2 = 0.00034$.

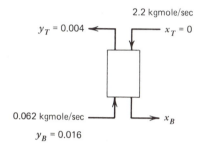

   The minimum water rate L corresponds th the outlet water leaving in equilibrium with the inlet gas and is achieved with an infinite packed height. Thus, at the bottom of an infinite tower, $x_B^* = 0.016/40 = 0.0004$ and $L_{min} = 0.000747/0.0004 = 1.87$ kgmole/sec. Thus, $L/L_{min} = 2.2/1.87 = 1.18$. A common rule-of-thumb is to select the liquid rate so that $L/L_{min} = 1.25$.

(b)   To obtain the number of transfer units, (16-29) can be integrated graphically, as is done in a later example. For this problem, however, because the system is dilute in SO_2, both the equilibrium and the operating line relationships are linear, and (16-29) can be integrated analytically as follows.

$$N_{OG} = \int_{y_T}^{y_B} \frac{dy}{y - y^*} \tag{A}$$

From Fig. 16.11, $(y - y^*)$ is given by the linear relationship

$$(y - y^*) = (y - y_T)\left[\frac{(y - y^*)_B - (y - y^*)_T}{y_B - y_T}\right] + (y - y^*)_T \tag{B}$$

Substitution of (B) into (A) and integration gives

$$N_{OG} = \frac{y_B - y_T}{(y - y^*)_B - (y - y^*)_T}\left[\ln\frac{(y - y^*)_B}{(y - y^*)_T}\right] \tag{C}$$

or

$$N_{OG} = \frac{y_B - y_T}{(y - y^*)_{LM}} \tag{D}$$

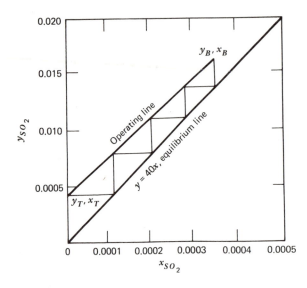

Figure 16.11. Solution for Example 16.5.

Thus, using (C)

$$N_{OG} = \frac{(0.016 - 0.004)}{[0.016 - (40)(0.000338)] - (0.004 - 0)} \ln \frac{[0.016 - (40)(0.000338)]}{(0.004 - 0)} = 3.8$$

The number of theoretical stages, as obtained by the McCabe–Thiele graph of Fig. 16.11, is 3.8. In this example, the slopes of the operating lines and equilibrium lines are sufficiently close (35.3 vs. 40) that N_{OG} and the number of theoretical stages are essentially identical.

(c)   HETP = 3.5 m/3.8 stages = 0.92 m

$$H_{OG} = \frac{l_T}{N_{OG}} = \frac{3.5}{3.8} = 0.92 \text{ m}$$

We see that HETP and H_{OG} are also essentially identical, for the same reason as N_{OG} and the number of theoretical stages are equal.

(d)   From Table 16.4

$$H_{OG} = \frac{G}{K_y a S} \qquad K_y a = \frac{0.062}{0.92(1.5)} = 0.0449 \frac{\text{kgmole}}{\text{m}^3 s(\Delta y)}$$

This is also equal to $K_g a$ since $P = 1$ atm (if the pressure is in Pascals we must divide $K_y a$ by 101.3 kPa/atm).

□

16.9   UM Diffusion in Nondilute Systems

To obtain the general UMD equations in columns 4 and 7 of Table 16.4 we let $L' = L(1 - x)$ and $G' = G(1 - y)$, where L' and G' are the constant flow rates of the inert (solvent) liquid or (carrier) gas; then

$$d(Gy) = G'd\left(\frac{y}{1-y}\right) = G'\frac{dy}{(1-y)^2} = G\frac{dy}{(1-y)} \tag{16-30}$$

$$d(Lx) = L'd\left(\frac{x}{1-x}\right) = L'\frac{dx}{(1-x)^2} = L\frac{dx}{(1-x)} \tag{16-31}$$

Equations (16-28) and (16-29) now become

$$l_T = \int_{x_1}^{x_2}\left(\frac{L}{K'_x aS}\right)\frac{dx}{(1-x)(x^*-x)} = \frac{L}{K'_x aS}\int_{x_1}^{x_2}\frac{dx}{(1-x)(x^*-x)} \tag{16-32}$$

$$l_T = \int_{y_2}^{y_1}\left(\frac{G}{K'_y aS}\right)\frac{dy}{(1-y)(y-y^*)} = \frac{G}{K'_y aS}\int_{y_2}^{y_1}\frac{dy}{(1-y)(y-y^*)} \tag{16-33}$$

where the mass transfer coefficients have been primed to signify UM diffusion.

If the numerators and denominators of (16-33) and (16-32) are multiplied by $(1-y)_{LM}$ and $(1-x)_{LM}$, respectively, where $(1-y)_{LM}$ is the log mean of $(1-y)$ and $(1-y^*)$ and $(1-x)_{LM}$ is the log mean of $(1-x)$ and $(1-x^*)$, we obtain the expressions in rows 6 and 1 of columns 4 and 7, Table 16.4.

$$l_T = \int_{x_1}^{x_2}\left[\frac{L}{K'_y a(1-x)_{LM}S}\right]\frac{(1-x)_{LM}dx}{(1-x)(x^*-x)} = \frac{L}{K'_x a(1-x)_{LM}S}\int_{x_1}^{x_2}\frac{(1-x)_{LM}dx}{(1-x)(x^*-x)} \tag{16-34}$$

$$l_T = \int_{y_2}^{y_1}\left[\frac{G}{K'_y a(1-y)_{LM}S}\right]\frac{(1-y)_{LM}dy}{(1-y)(y-y^*)} = \frac{G}{K'_y a(1-y)_{LM}S}\int_{y_2}^{y_1}\frac{(1-y)_{LM}dy}{(1-y)(y-y^*)} \tag{16-35}$$

In these equations, $K'_y(1-y)_{LM}$ is equal to the concentration-independent K_y, and $K'_x(1-x)_{LM}$ is equal to the concentration-independent K_x according to the identities in Table 16.3. If there is appreciable absorption, G will decrease from the bottom to top of the absorber. However, the values of Ka are also a function of flow rate (for many types of equipment $Ka \sim G^{0.8}$) so the ratio G/Ka is approximately constant and the HTU groupings $(L/K'_x a(1-x)_{LM}S)$ and $(G/K'_y a(1-y)_{LM}S)$ can often be taken out of the integral sign without incurring errors larger than those inherent in the experimental measurement of Ka. Usually, average values of G, L, and $(1-y)_{LM}$ are used.

Another approach is to leave all of the terms in (16-34) or (16-35) under the integral sign and evaluate l_T by a stepwise or graphical integration. In either case, to obtain the terms $(y-y^*)$ or (x^*-x), the equilibrium and operating lines must be established. The equilibrium curve is established from appropriate thermodynamic data or correlations. To establish the operating line, which will not be straight if the solutions are concentrated, the appropriate material balance equations must be developed. With reference to Fig. 16.10, an overall balance around the upper part of the absorber gives

$$G + L_{in} = G_{out} + L \qquad (16\text{-}36)$$

Similarly a balance around the upper part of the absorber for the component being absorbed, assuming a pure liquid absorbent, gives

$$Gy = G_{out}y_{out} + Lx \qquad (16\text{-}37)$$

An absorbent balance around the upper part of the absorber is

$$L_{in} = L(1 - x) \qquad (16\text{-}38)$$

Combining (16-36), (16-37), and (16-38) to eliminate G and L gives

$$y = \frac{G_{out}y_{out} + [L_{in}x/(1 - x)]}{G_{out} + [L_{in}x/(1 - x)]} \qquad (16\text{-}39)$$

Equation (16-39) allows the operating line y versus x to be calculated from a knowledge of terminal conditions only.

A simpler approach to the problem of concentrated gas or liquid mixture is to linearize the operating line by expressing all concentrations in mole ratios and the gas and liquid flows in terms of inerts; that is, $G' = (1 - y)G$, $L' = (1 - x)L$. Then, we have in place of (16-34) and (16-35)

$$l_T = \int_{X_1}^{X_2} \left(\frac{L'}{K_X aS}\right)\left(\frac{dX}{X^* - X}\right) = \frac{L'}{K_X aS} \int_{X_1}^{X_2} \frac{dX}{(X^* - X)} \qquad (16\text{-}40)$$

$$l_T = \int_{Y_2}^{Y_1} \left(\frac{G'}{K_Y aS}\right)\frac{dY}{(Y - Y^*)} = \frac{G'}{K_Y aS} \int_{Y_2}^{Y_1} \frac{dY}{(Y - Y^*)} \qquad (16\text{-}41)$$

This set of equations is compatible with the mass transfer coefficients calculated in Example 16.4 and is listed in rows 3 and 8 of Table 16.4.

Example 16.6.   To remove 95% of the ammonia from an air stream containing 40% ammonia by volume, 488 lbmole/hr of a certain solution per 100 lbmole/hr of entering gas are to be used, which is greater than the minimum requirement. Equilibrium data are given in Fig. 16.12. Pressure is 1 atm and temperature is 298°K. Calculate the number of transfer units by:

(a)   (16-35) using a curved operating line determined from (16-39).

(b)   (16-41) using mole ratios.

Solution

(a) Take as a basis $L_{in} = 488$. Then $G = 100 - (40)(0.95) = 62$, and $y_{OUT} = (0.05)(40)/62 = 0.0323$. From (16-39), it is possible to construct the curved operating line of Fig. 16.12. For

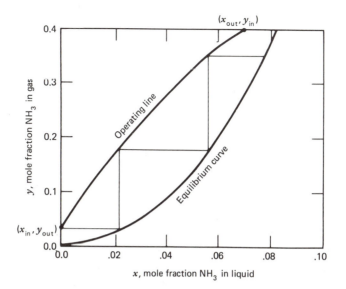

Figure 16.12. Equilibrium curve, ammonia solution–air system; $P = 1$ atm, $T = 298°K$.

example, if $x = 0.04$

$$y = \frac{(62)(0.0323) + [(488)(0.04)/(1 - 0.04)]}{62 + [(488)(0.04)/(1 - 0.04)]} = 0.27$$

It is now possible to calculate values of y, y^*, $(1 - y)_{LM} = [(1 - y) - (1 - y^*)]/\ln[(1 - y)/(1 - y^*)]$, and $(1 - y)_{LM}/(1 - y)(y - y^*)$ for use in (16-35).

y	y^*	$(y - y^*)$	$(1 - y)$	$(1 - y)_{LM}$	$\dfrac{(1 - y)_{LM}}{(1 - y)(y - y^*)}$
0.03	0.002	0.028	0.97	0.99	36.47
0.05	0.005	0.045	0.95	0.97	22.68
0.10	0.01	0.09	0.90	0.94	11.60
0.15	0.025	0.125	0.85	0.91	8.56
0.20	0.04	0.16	0.80	0.89	6.95
0.25	0.08	0.17	0.75	0.85	6.66
0.30	0.12	0.18	0.70	0.82	6.51
0.35	0.17	0.18	0.65	0.73	6.24
0.40	0.26	0.14	0.60	0.67	7.97

Note that $(1 - y) \approx (1 - y)_{LM}$, these two terms frequently being canceled out of the NTU equations, particularly when y is small.

Figure 16.13 is a plot of y versus $(1 - y)_{LM}/[(1 - y)(y - y^*)]$ to determine N_{OG}. The

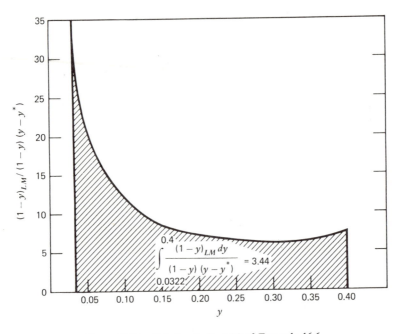

Figure 16.13.   Solution to Part (a) of Example 16.6.

integral on the right-hand side of (16-35), between $y = 0.4$ and $y = 0.0322$, is $3.44 = N_{OG}$. This is approximately 1 more than the number of theoretical stages of 2.8 as seen in the steps of Fig. 16.12.

(b) It is a simple matter to obtain values for $Y = y/(1-y)$, $Y^* = y^*/(1-y^*)$, $(Y-Y)^*$ and $(Y-Y^*)^{-1}$.

y	Y	y^*	Y^*	$(Y-Y^*)^{-1}$
0.03	0.031	0.002	0.002	34.48
0.05	0.053	0.005	0.005	20.83
0.1	0.111	0.01	0.010	9.9
0.15	0.176	0.025	0.026	6.66
0.20	0.250	0.04	0.042	4.8
0.25	0.333	0.08	0.087	4.06
0.30	0.43	0.12	0.136	3.40
0.35	0.54	0.17	0.205	2.98
0.40	0.67	0.26	0.310	2.78

Graphical integration of the right-hand-side integral of (16-41) is carried out by determining the area under the curve of $(Y-Y^*)^{-1}$ versus Y between $Y = 0.67$ and $Y = 0.033$. The result is $N_{OG} = 3.46$.

Parts (a) and (b) must predict the same column height

$$l_T = \left(\frac{G}{K_y'a(1-y)_{BM}S}\right)(3.44) = \frac{G'}{K_YaS}(3.46)$$

Since $K_y'(1-y)_{BM} = K_y$, and K_y is related to K_Y by (16-15)

$$K_Y = K_C\bar{y}_B\frac{P}{RT} = K_y\bar{y}_B$$

we have

$$\frac{G(3.44)}{K_yaS} = \frac{G'(3.46)}{K_y\bar{y}_BaS}$$

Using arithmetic average values for G and $\bar{y}_B$, we have

$$\frac{((100+62)/2)(3.44)}{K_yaS} = \frac{60(3.46)}{K_yaS(0.784)}$$

or

$$\frac{279}{K_yaS} = \frac{265}{K_yaS}$$

Considering the accuracy of the graphical methods and the use of arithmetic mean values for G and $\bar{y}_B$, this is reasonable agreement.
□

16.10 Choice of Packing and Calculation of Column Diameter

The examples in the last few sections involved situations where a specified amount of gas was to be absorbed. A liquid rate, based on some multiple of an exit saturation corresponding to the minimum liquid rate at infinite transfer units, was then specified. Thus, the liquid and gas flow were fixed and the NTU could be determined. A column height could be obtained if values of HTU or k were available from experimental measurements or correlations. The HTU values, however, depend upon type of packing and column diameter. Thus we must now specify the packing to be used and calculate the diameter of the column required to handle the liquid and gas flows. It can never be taken for granted that the L and G obtained from material balance and equilibrium relations can be translated into a piece of hardware. Packed columns can only accommodate a restricted range of L/G values. With a gas flow of 500 lb/ft$^2 \cdot$ hr in a 30-in.-diameter bed, for example, a liquid rate above 50,000 lb/hr may result in flooding, while a liquid rate below 15,000 lb/hr may result in much of the column's packing running dry, with a corresponding loss in efficiency. There are also ancillary limitations with respect to pressure drop.

All of the flow parameters and the column diameter are a function of the

Table 16.5   Comparison of mesh, ring, and saddle packings

Mesh packings
> Lowest HETP, therefore most efficient.
> Excellent initial liquid flow distribution is imperative.
> Poor redistribution of liquid.
> High cost.

Ring Packings
> Generally metal or plastic.
> Most common in distillation duty.
> Good turndown ratio.
> Easy to machine from sheet metal by pressing.
> Raschig Rings tend to induce channeling.
> Low cost.

Saddle Packings
> Generally ceramic or plastic.
> Most common in absorption duty.
> Good liquid distribution.
> Good corrosion resistance.
> Most common with aqueous corrosive fluids.
> Saddles are best for redistributing liquids.
> Low cost.

size and type of packing. There are three common types of packing material: mesh, rings, and saddles. The most common rings are Raschig Rings, Pall Rings, and Hy-Pack Rings, and the most common saddles are Berl Saddles and Intalox Saddles. Table 16.5 compares mesh, ring, and saddle packings.

Once the type of packing is selected, it is necessary to specify a nominal size, which should not be more than one-eighth of the column diameter in order to minimize channeling of the liquid toward the column wall. Corresponding to the nominal size is a packing factor F, which characterizes the flow capacity. Values of F derived from experiment data are given for several different types of packing in Table 16.6 from Eckert.[11] As shown, the larger the packing, the smaller the F factor. In general, allowable gas rates are inversely proportional to the square root of the packing factors

$$\frac{G_2}{G_1} = \left(\frac{F_1}{F_2}\right)^{0.5}$$

This equation can be used to approximate the effect of the packing on the diameter of the column, or to predict new flow rates when packings are changed.

In Fig. 16.14, a generalized pressure drop correlation for packed beds is given based on the early work of Sherwood et al.[12] as modified by Eckert.[11] The pressure drop, which appears as a parameter in Fig. 16.14, is a key variable in

Table 16.6 Packing factors

Type of Packing	Material	Nominal Packing Size, In.										
		$\frac{1}{4}$	$\frac{3}{8}$	$\frac{1}{2}$	$\frac{5}{8}$	$\frac{3}{4}$	1	$1\frac{1}{4}$	$1\frac{1}{2}$	2	3	$3\frac{1}{2}$
Super Intalox	Ceramic	—	—	—	—	—	60	—	—	30	—	—
Super Intalox	Plastic	—	—	—	—	—	33	—	—	21	16	—
Intalox Saddles	Ceramic	725	330	200	—	145	98	—	52	40	22	—
Hy-Pak Rings	Metal	—	—	—	—	—	42	—	—	18	15	—
Pall Rings	Plastic	—	—	—	97	—	52	—	40	25	—	16
Pall Rings	Metal	—	—	—	70	—	48	—	28	20	—	16
Berl Saddles	Ceramic	900^i	—	240^i	—	170^h	110^h	—	65^h	45^i	—	—
Raschig Rings	Ceramic	1,600^{b,i}	1,000^{b,i}	580^c	380^c	255^c	155^d	125^{c,i}	95^e	65^f	37^{g,i}	—
Raschig Rings, $\frac{1}{32}$-in. wall	Metal	700^i	390^i	300^i	170	155	115^i	—	—		—	—
Raschig Rings, $\frac{1}{16}$-in. wall	Metal	—	—	410	290	220	137	110^i	83	57	32^i	—
Tellerettes	Plastic	—	—	—	—	—	40	—	—	20	—	—
Maspak	Plastic	—	—	—	—	—	—	—	—	32	20	—
Lessing exp.	Metal	—	—	—	—	—	—	—	30	—	—	—
Cross partition	Ceramic	—	—	—	—	—	—	—	—	—	70	—

^{b} $\frac{1}{16}$ wall. ^{c} $\frac{3}{32}$ wall. ^{d} $\frac{1}{8}$ wall. ^{e} $\frac{3}{32}$ wall. ^{f} $\frac{1}{4}$ wall. ^{g} $\frac{3}{8}$ wall. ^{h} Packing factors obtained in 16- and 30-in. I.D. towers. ^{i} Extrapolated.

Source. J. S. Eckert, *Chem. Eng.*, **82** (8), 70–76 (1975).

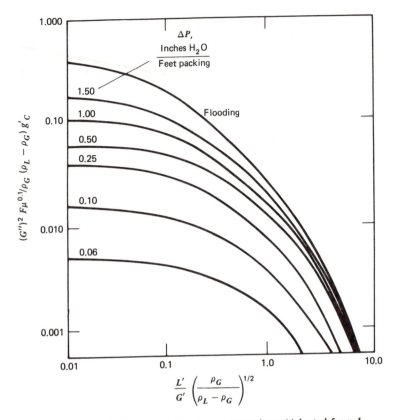

Figure 16.14. Eckert correlation for pressure drop. (Adapted from J. S. Eckert, *Chem. Eng.*, **82** (8), 70–76 (1975).]

packed column design. If ΔP is greater than approximately 1.5 in. of water pressure per foot of packing, flooding is approached. On the other hand, an unduly low (less than 0.25 in./ft pressure drop) signifies poor gas–liquid contacting, and possibly dry spots in the column.

The foregoing suggests that column diameters and packing selection can be based on pressure drop and, indeed, this is commonly done. Figure 16.15 shows the computational steps. We start with the L/G obtained from the material balance, choose a preliminary packing size, and obtain F from Table 16.6. Next we decide on the ΔP. A reasonable value for absorbers and strippers is 0.4 in./ft, which, as shown in Fig. 16.16, corresponds to about 60 to 70% of capacity (flooding).

Next, using Fig. 16.14 with L/G, F, fluid properties, and ΔP fixed, we read

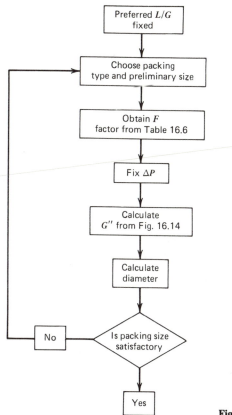

Figure 16.15.   Algorithm for tower sizing.

the ordinate, which is

$$\frac{(G'')^2 F \mu^{0.1}}{\rho_G (\rho_L - \rho_G) g_c}$$

and calculate the column cross-sectional area from G'' where

$G'' =$ gas rate, lb/hr $\cdot$ ft^2.

$F =$ packing factor.

$\rho_G =$ gas density, lb/ft^3.

$\rho_L =$ liquid density, lb/ft^3.

$\mu =$ viscosity, liquid, cp.

$g_c =$ gravitational constant, 4.17×10^8 ft/hr^2.

From the column cross-sectional area the final design parameter, the column diameter, is determined, since G'' can be translated into a cross-sectional area by dividing it into the G obtained by material balance. This presumes that the L/G calculated by material balance can be found on the abscissa of Fig. 16.14. If

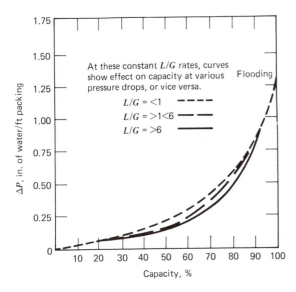

Figure 16.16. Pressure drop-capacity relationship–permissible operating rate with packed bed. [Adapted from J. S. Eckert, *Chem. Eng.*, 82 (8), 70–76 (1975).]

it cannot, either the process specification or the absorbent should be changed if the equipment of choice is to remain a packed column. If the column diameter is unreasonably large or small or inappropriate to the size packing assumed (a 1-*m* packing in a 3-*m*-diameter column, for example) the design calculations must be repeated.

Example 16.7.   Choose a packing and calculate the diameter for the column of Example 16.6. Assume the absorbing solution has the properties of water.

Solution.   It is preferable to make the calculations at the bottom of the tower where the flow rates are highest.

$$L = (488 \text{ lbmole } H_2O)(18) + (38 \text{ moles } NH_3)(17) = 9430 \text{ lb/hr}$$

$$G = (60 \text{ lbmole air})(29) + (40 \text{ lbmole } NH_3)(17) = 2420 \text{ lb/hr}$$

$$\rho_G = 0.077 \text{ lb/ft}^3 \qquad \rho_L = 63 \text{ lb/ft}^3 \qquad \mu = 2 \text{ cp}$$

Choosing 1.5-in. ceramic Intalox Saddles, we see that F from Table 16.6 is 52. For $\Delta P = 0.4$ in. H_2O/ft and $L/G[\rho_G/(\rho_L - \rho_G)]^{1/2} = 0.14$, from Fig. 16.14, $(G'')^2 F \mu^{0.1}/\rho_G(\rho_L - \rho_G)g_c = 0.037$

$$G'' = \left[\frac{(0.037)(0.077)(63 - 0.077)(4.17 \times 10^8)}{(52)(2)^{0.1}} \right]^{1/2} = 1158 \frac{\text{lb}}{\text{hr} \cdot \text{ft}^2}$$

The diameter is

$$\left[\frac{(2420/1158)4}{3.14} \right]^{1/2} = 1.63 \text{ ft}$$

The ratio of column diameter to packing diameter is $(1.63)(12)/1.5 = 13$, which is satisfactory.

☐

16.11   Prediction of Mass Transfer Capacity Coefficients

It was shown earlier in this chapter how overall mass transfer coefficients and HTU values can be experimentally determined. This is the preferred procedure because, as shown in Fig. 16.17 from the data of Eckert[11] for the system air with

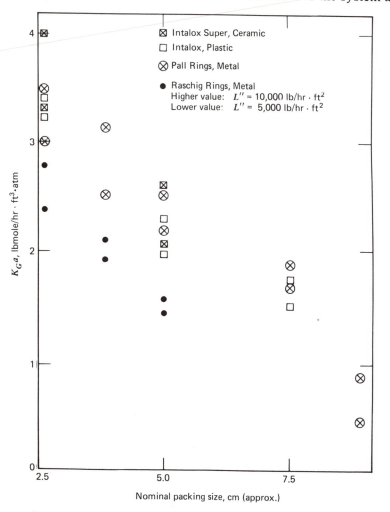

Figure 16.17. Mass transfer coefficients, $K_G a$, selected packings. System: CO_2—1%, NaOH—4%, 25% conversion to CO_2—75°F. [Data from J. S. Eckert, *Chem. Eng.*, **82** (8), 70–76 (1975).]

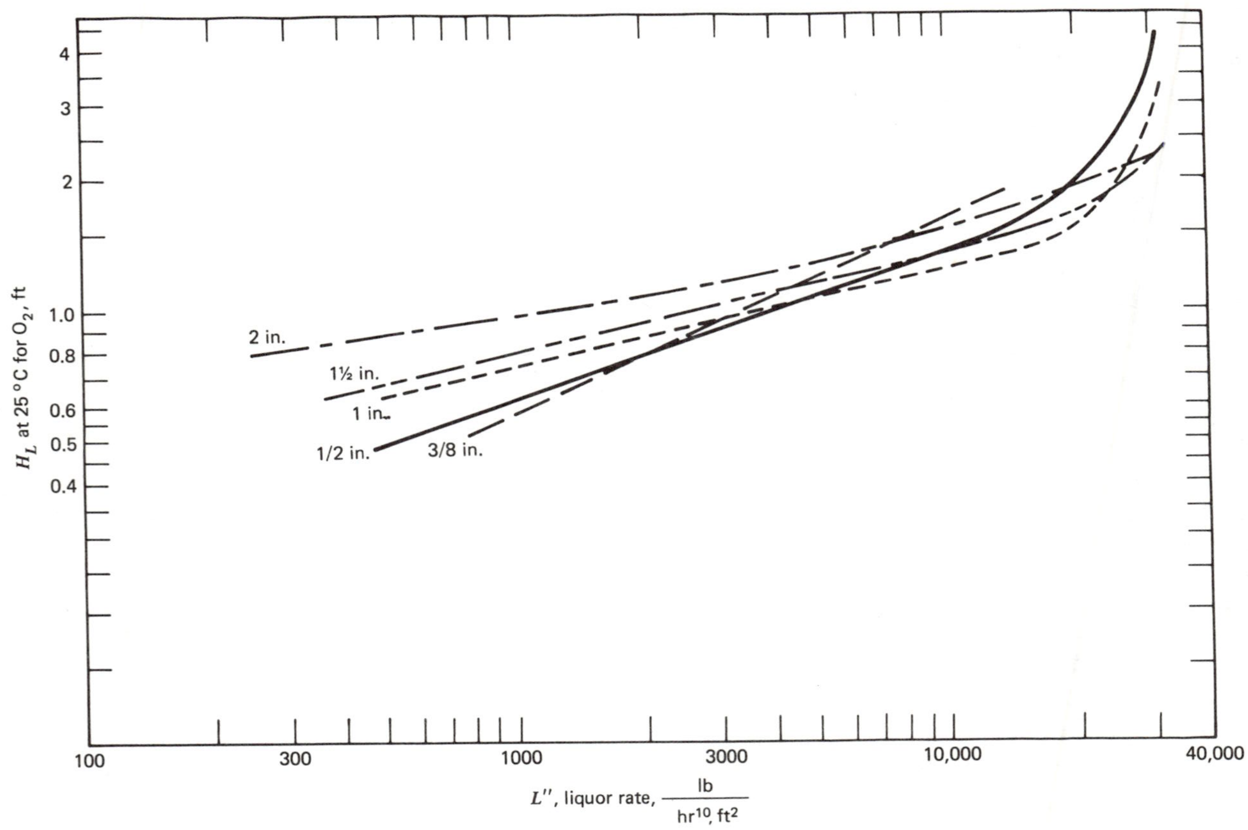
4
3
2
1.0
0.8
0.6
0.5
0.4
H_L at 25 °C for O_2, ft
2 in.
1½ in.
1 in.
1/2 in.
3/8 in.
100
300
1000
3000
10,000
40,000
L'', liquor rate, $\frac{\text{lb}}{\text{hr}^{10}\text{, ft}^2}$

Figure 16.18. Values of H_L for air–O_2–H_2O system and Raschig Rings. [Adapted from T. K. Sherwood and F. A. L. Halloway, *Trans AIChE*, **36**, 39–70 (1940).]

1% CO_2–H_2O with 4% NaOH and 25% conversion to carbonate, for a fixed gas rate of 500 lb/hr · ft² in a 30-in.-diameter column, values of K_Ga depend strongly on packing type and size and on liquid flow rate. Although most of the fourfold variation shown is probably due to changes in a, the square feet of interface/cubic foot, it is not easy to separate the two variables in K_Ga. Once a value of K_Ga has been determined experimentally for a certain system and packing, Fig. 16.17 can be used as a first approximation to obtain values of K_Ga for a different packing type or size by a simple ratio procedure.

In the absence of any experimental data at all, overall mass transfer coefficients for nonchemically reacting systems can be predicted from values of individual film mass transfer coefficients coupled with (16-23) or its HTU analog

$$H_{OG} = H_G + \left(\frac{H'G}{L}\right)H_L \qquad (16\text{-}42)$$

which is obtained by substituting the definitions in Table 16.4 into (16-23).

The individual film capacity coefficients k_ya and k_xa are traditionally obtained from experiments in which one or the other of the film resistances is negligible. In the absorption or desorption of very insoluble gases, gas-phase resistance is negligible and measurements lead directly to values of k_x. Gas-phase resistances are obtained by similar measurements for very soluble gases. Classical data of this type include that of Sherwood and Holloway,[13] shown in Fig. 16.18 for H_L for the desorption of CO_2, O_2, and H_2 from water by air in a 20-in.-diameter tower packed with ceramic Raschig Rings or Berl Saddles, and that of Fellinger,[14] as shown for one packing in Fig. 16.19, for H_G (and H_{OG}) for the absorption of NH_3 from air with water in an 18-in.-diameter tower packed with ceramic rings or Berl Saddles. These data, together with those of many

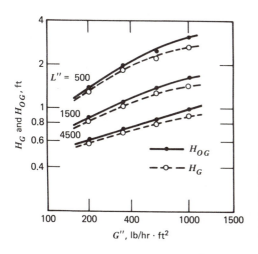

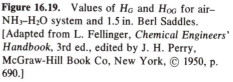

Figure 16.19. Values of H_G and H_{OG} for air–NH_3–H_2O system and 1.5 in. Berl Saddles. [Adapted from L. Fellinger, *Chemical Engineers' Handbook*, 3rd ed., edited by J. H. Perry, McGraw-Hill Book Co, New York, © 1950, p. 690.]

other investigations, were empirically correlated by Cornell, Knapp, and Fair[15] to absolute average deviations of 9.7% and 23.2% for H_L and H_G, respectively. These correlations, based largely on data for air–solute–water systems, were subsequently verified by Cornell, Knapp, Close, and Fair[16] for application to the distillation of several binary organic systems in columns ranging in diameter from 8 to 24 in. and packed with Raschig Rings, Berl Saddles, and Intalox Saddles. In the correlations of Cornell and co-workers, values of H_L in feet are predicted from

$$H_L = \phi C_F (\text{Sc}_L)^{0.5} \tag{16-43}$$

where ϕ depends on the liquid rate as shown in Figs. 16.20 and 16.21 for Raschig Rings and Berl Saddles, respectively, and C_F is a correction factor that equals 1 for gas rates below 40% of flooding and is shown in Fig. 16.22 for other conditions, where the percent flooding can be determined from Fig. 16.14 by

$$\text{Percent Flooding} = \left(\frac{100 G''}{G''_f}\right)_{\text{const. } L''} \tag{16-44}$$

The term Sc in (16-43) is the Schmidt number for the solute

$$\text{Sc} = \left(\frac{\mu}{\rho D}\right) \tag{16-45}$$

Thus, H_L is inversely proportional to the square root of the molecular diffusivity. Values of H_G in feet are predicted from

$$H_G = \frac{\Psi(\text{Sc}_G)^{0.5} D_T^{1}\left(\frac{l_T}{10}\right)^{1/3}}{(L'' f_1 f_2 f_3)^{m_1}} \tag{16-46}$$

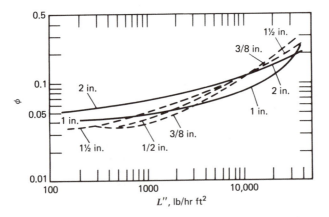

Figure 16.20. H_L correlation for various sizes of Raschig Rings. [Cornell et al., *Chem. Eng. Progr.*, **56** (7), 68 (1960).]

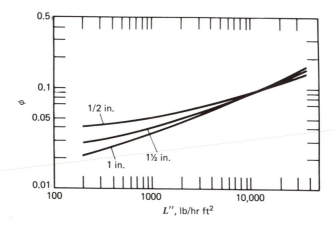

Figure 16.21. H_L correlation for various sizes of Berl Saddles. [Cornell et al., *Chem. Eng. Progr.*, **56** (7), 68 (1960).]

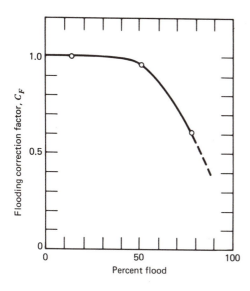

Figure 16.22. Liquid-film correlation factor for operation at high percent of flood. [Cornell et al., *Chem. Eng. Progr.*, **56** (7), 68 (1960).]

where Ψ depends on the percent of flooding as shown in Figs. 16.23 and 16.24 for Raschig Rings and Berl Saddles, respectively; D_T is column diameter in feet; l_T is the column height in feet; L'' is the liquid mass velocity in lb/hr · ft². The f-factors are property corrections referred to pure water at 20°C and given by

$$f_1 = \left(\frac{\mu_L}{2.42}\right)^{0.16} \tag{16-47}$$

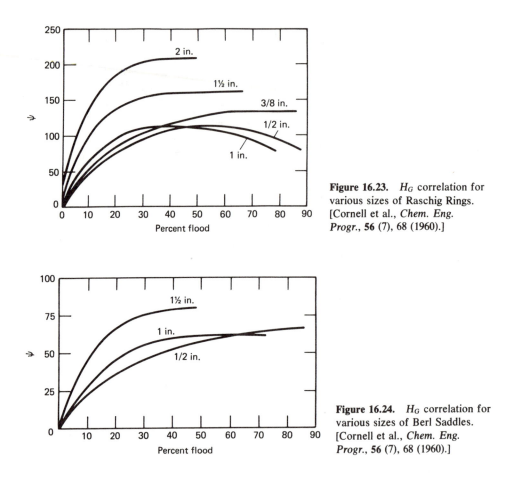

Figure 16.23. H_G correlation for various sizes of Raschig Rings. [Cornell et al., *Chem. Eng. Progr.*, **56** (7), 68 (1960).]

Figure 16.24. H_G correlation for various sizes of Berl Saddles. [Cornell et al., *Chem. Eng. Progr.*, **56** (7), 68 (1960).]

where μ_L is liquid-phase viscosity in pound-moles per foot per hour

$$f_2 = \left(\frac{62.4}{\rho_L}\right)^{1.25} \tag{16-48}$$

where ρ_L is liquid-phase density in pound-moles per cubic foot

$$f_3 = \left(\frac{72.8}{\sigma}\right)^{0.8} \tag{16-49}$$

where σ is surface tension in dynes per centimeter. Also

$n_1 = 1.24$ for Raschig Rings and 1.11 for Berl Saddles

$m_1 = 0.6$ for Raschig Rings and 0.5 for Berl Saddles

Example 16.8. Predict H_{OG} and calculate the packed height for the column in Examples 16.6 and 16.7. Assume the absorbing solution is water, and do the calculation both by the correlation of Fellinger and by the correlation of Cornell.

Solution.

Method 1.   Use experimental data of Fellinger[14] in Fig. 16.19. From Example 16.7

$$L''_{avg} = \frac{488(18) + 9430}{2[3.14(1.63)^2/4]} = 4360 \text{ lb/hr} \cdot \text{ft}^2$$

$$G''_{avg} = \frac{[(60)(29) + (2)(17)] + 2420}{2[3.14(1.63)^2/4]} = 1000 \text{ lb/hr} \cdot \text{ft}^2$$

From Fig. 16.19 for the same system, but with 1.5-in. Berl Saddles, $H_{OG} = 1.0$ ft. From Fig. 16.17, noting that $K_G a$ is inversely proportional to H_{OG}, we can apply a correction to predict the value of H_{OG} for 1.5-in. Intalox Saddles.

$$H_{OG} = 1.0\left(\frac{1.8}{2.3}\right) = 0.78 \text{ ft}$$

From Example 16.6, $N_{OG} = 3.46$. Thus

$$\text{Packed height} = H_{OG} N_{OG} = 0.78(3.46) = 2.70 \text{ ft}$$

Method 2.   Using correlations of Cornell et al. as given by (16-43) to (16-49) and Figs. 16.20 to 16.24, we will first make calculations for Berl Saddles and then use Fig. 16.17 to correct for type of packing. The average mole fraction of NH_3 in the liquid is

$$\frac{0 + [(0.95)(40)]/[488 + (0.95)(40)]}{2} = 0.036$$

Therefore, assume the liquid solution is sufficiently dilute so that it has the properties of water at 25°C

$$\mu = 0.008904 \text{ gm/cm} \cdot \text{sec}$$

$$\rho = 1.0 \text{ gm/cm}^3$$

From Perry's *Handbook*, fifth edition, p. 3-244

$$D \text{ for } NH_3 \text{ in water} = 1.7 \times 10^{-5} \text{ cm}^2/\text{sec}$$

From (16-45)

$$\text{Sc}_L = \frac{0.008904}{(1.0)(1.7 \times 10^{-5})} = 524$$

From Fig. 16.21, for 1.5-in. packing, and $L''_{avg} = 4360 \text{ lb/hr} \cdot \text{ft}^2$

$$\phi = 0.063$$

From Fig. 16.16, for $\Delta P = 0.4$ in./ft and $(L''/G'')_{avg} = 4360/100 = 4.36$

$$\text{Percent flooding} = 70\%$$

From Fig. 16.22

$$C_F = 0.75$$

From (16-43)

$$H_L = 0.063(0.75)(524)^{0.5} = 1.08 \text{ ft}$$

Now calculate H_G from (16-46). From Fig. 16.24

$$\Psi = 80$$

From Perry's *Handbook*, third edition, p. 539, Sc for NH_3 in air at 25°C and 1 atm is 0.66. From Example 16.7

$$D_T = 1.63 \text{ ft}$$

From (16-47)

$$f_1 = \left[\frac{(0.8904)(2.42)}{2.42} \right]^{0.16} = 0.982$$

From (16-48)

$$f_2 = \left(\frac{62.4}{62.4} \right)^{1.25} = 1.0$$

From (16-49)

$$f_3 = \left(\frac{72.8}{72} \right)^{0.8} = 1.01$$

For Berl Saddles, $n_1 = 1.11$, $m_1 = 0.5$. Take $l_T = 2.70$ ft from the result of Method 1. From (16-46)

$$H_G = \frac{80(0.66)^{0.5}(1.63)^{1.11}(2.73/10)^{1/3}}{[4360(0.982)(1.0)(1.01)]^{0.5}} = 1.10 \text{ ft}$$

Now use (16-42) to calculate H_{OG}. From Table 16.3, the Henry's law constant is defined as

$$H' = \frac{y^*}{x}$$

for a straight equilibrium line. From Fig. 16.12, we see that H' is not a constant because the equilibrium line is curved. Take H' as the limiting slope in the dilute region. Thus

$$H' \approx 1.0$$

From (16-42), using values of L and G at the dilute (top) end of the column so that

$$L = 488 \text{ lbmole/hr}$$

$$G = 100 - (0.95)(40) = 62 \text{ lbmole/hr}$$

we have

$$H_{OG} = 1.10 + \left[\frac{(1.0)(62)}{488} \right](1.08) = 1.24 \text{ ft (for Berl Saddles)}$$

Correcting, as in Method 1, for type of packing, we have

$$H_{OG} = 1.24 \left(\frac{1.8}{2.3} \right) = 0.97 \text{ ft}$$

$$\text{Packed height} = 3.46(0.97) = 3.35 \text{ ft}$$

which is somewhat higher than the value computed by Method 1.

□

16.12 Rigorous Calculation Methods for Packed Gas Absorbers

The preceding method for design of packed gas absorbers is primitive in the sense that it:

1. Ignores or makes seriously unrealistic assumptions about the energy balance around the absorber, thereby assuming or implying either:
   (a) Isothermal behavior.
   or
   (b) An equilibrium curve that is not a function of temperature.

2. Makes assumptions to allow the formulation and integration of mass balance and mass transfer integrals.

3. Applies the HTU and NTU discrete concepts to many cases where G, k, or other variables are not constant with axial flow length.

4. Makes very difficult, if not impossible, the development of calculation methods for more than one absorbing liquid component, one stripping gas component, and/or one absorbed component solute.

Of particular importance is the assumption of isothermal behavior. Because most absorbers operate adiabatically with the liberation of heat as absorption occurs, the liquid-phase temperature to a great extent and the gas-phase temperature to some extent can experience temperature rises which can reduce gas solubility and make absorption more difficult.

Stockar and Wilke[17] cite an example of the absorption of acetone from air, entering at 15°C, by water. Assuming isothermal conditions, the required packed height was found to be 6.4 ft. By rigorous calculations, the liquid phase and vapor phase exit the absorber at 24°C and 16.5°C, respectively. These increased temperatures cause the required packed height to almost double to a value of 11.9 ft.

The development of more rigorous models may proceed along three different paths:

1. Application of the sum-rates method developed in Chapter 15 for the equilibrium and flow rate calculation to estimate the number of theoretical equilibrium stages, coupled with some form of HETP calculation based on experience or available correlations.

2. Application of differential contactor equations to a differential section of the gas absorber and integration of the resulting set of coupled differential equations.

3. Application of statistical design techniques to develop correlations over a range of specific systems, equipment characteristics, and operating conditions. A computer is essential to apply the correlation equations to practical design cases.

References

1. Fick, A., *Ann. Phys.* (Leipzig) **94**, 59 (1855).

2. Fuller, E. N., P. D. Schettler, and J. C. Giddings, *Ind. Eng. Chem.*, **58** (5), 18 (1966).

3. Wilke, C. R., and P. Chang, *AIChE J.*, **1**, 264 (1955).

4. Foust, A. S., et al., *Principles of Unit Operations*, J. Wiley & Sons, Inc., New York, 1960, 171.

5. Schlichting, H., *Boundary Layer Theory*, Pergamon Press, London, 1955, 130.

6. Bird, R. B., W. E. Stewart, and E. N. Lightfoot, *Transport Phenomena*, John Wiley & Sons, New York, 1960.

7. Sherwood, T. K., R. L. Pigford, and C. R. Wilke, *Mass Transfer*, McGraw–Hill Book Co., New York, 1975.

8. Higbie, R., *Trans. AIChE*, **31**, 365 (1935).

9. Danckwerts, P. V., *Ind. Eng. Chem.*, **43**, 1460 (1951).

10. Toor, H. L., and J. M. Marchello, *AIChE J.*, **4**, 97 (1958).

11. Eckert, J. S., *Chem. Eng.*, **82** (8), 70–76 (1975).

12. Sherwood, T. K., G. H. Shipley, and F. A. L. Holloway, *Ind. Eng. Chem.*, **30**, 765–769 (1938).

13. Sherwood, T. K., and F. A. L. Holloway, *Trans. AIChE*, **36**, 39–70 (1940).

14. Fellinger, L., in *Chemical Engineers' Handbook*, 3rd ed., ed. by J. H. Perry, McGraw–Hill Book Co., New York, 1950, 690.

15. Cornell, D., W. G. Knapp, and J. R. Fair, *Chem. Eng. Progr.*, **56** (7), 68–74 (1960).

16. Cornell, D., W. G. Knapp, H. J. Close, and J. R. Fair, *Chem. Eng. Progr.*, **56** (8), 48–53 (1960).

17. Stockar, U. V., and C. R. Wilke, *Ind. Eng. Chem., Fundam.*, **16** (1), 88–103 (1977).

18. Whitman, W. G., *Chem. and Met. Eng.*, **29** (4), 71–79 (1923).

Problems

16.1 Water in an open dish exposed to dry air at 25°C is found to vaporize at a constant rate of 0.04 gm/hr · cm². Assuming the water surface to be at the wet-bulb temperature of 11.0°C, calculate the effective gas-film thickness (i.e., the thickness of a stagnant air film that would offer the same resistance to vapor diffusion as is actually encountered at the water surface). The diffusivity for water–air at the mean film temperature is 0.24 cm²/sec. The vapor pressure of water at 11.0°C is 9.84 torr.

16.2 An open tank, 10 ft in diameter and containing benzene at 25°C, is exposed to air in such a manner that the surface of the liquid is covered with a stagnant air film estimated to be 0.2 in. thick. If the total pressure is 1 atm and the air temperature is 25°C, what loss of material in pounds per day occurs from this tank?

The specific gravity of benzene at 60°F is 0.877. The concentration of benzene at the outside of the film is so low that it may be neglected. For benzene, the vapor pressure at 25°C is 100 torr, and the diffusivity in air is 0.08 cm²/sec.

16.3 An insulated glass tube and condenser are mounted on a reboiler containing benzene and toluene. The condenser returns liquid reflux in such a manner that it runs down the wall of the tube. At one point in the tube the temperature is 170°F, the vapor contains 30 mole% toluene, and the liquid reflux contains 40 mole% toluene. The effective thickness of the stagnant vapor film is calculated to be 0.1 in. The molal latent heats of benzene and toluene are equal. Calculate the rate at which toluene and benzene are being interchanged at this point in the tube in pound moles per hour per square foot.

Diffusivity of toluene in benzene = 0.2 ft²/hr.
$P = 1$ atm total pressure (in the tube).
Vapor pressure of toluene at 170°F = 400 torr.

16.4 Air at 25°C with a dew-point temperature of 0°C flows past the open end of a vertical tube filled with liquid water maintained at 25°C. The tube has an inside diameter of 0.83 in., and the liquid level was originally 0.5 in. below the top of the tube. The diffusivity of water in air at 25°C is 0.256 cm²/sec.
(a) How long will it take for the liquid level in the tube to drop 3 in.?
(b) Make a plot of the liquid level in the tube as a function of time for this period.

16.5 The diffusivity of toluene in air was determined experimentally by allowing liquid toluene to vaporize isothermally into air from a partially filled vertical tube 3 mm in diameter. At a temperature of 39.4°C, it took 96×10^4 sec for the level of the toluene to drop from 1.9 cm below the top of the open tube to a level of 7.9 cm below the top. The density of toluene is 0.852 gm/cm³, and the vapor pressure is 57.3 torr at 39.4°C. The barometer reading was 1 atm. Calculate the diffusivity and compare it with the value predicted from (16-3). Neglect the counterdiffusion of air.

16.6 An open tube, 1 mm in diameter and 6 in. long, has pure hydrogen blowing across one end and pure nitrogen blowing across the other. The temperature is 75°C.
(a) For equimolar counterdiffusion, what will be the rate of transfer of hydrogen into the nitrogen stream (gmole/sec)?
(b) Assuming the flow is uniform at all points in any cross section of the tube, calculate the net flow of gas (gmole/sec) if the number of moles of hydrogen passing into the nitrogen is maintained at 10 times the number of moles of nitrogen passing into the hydrogen. What is the direction of the molar flow? Of the mass flow?
(c) For both Parts (a) and (b), plot the mole fraction of hydrogen against distance from the end of the tube past which nitrogen is blown.

16.7 Some HCl gas diffuses across a film of air 0.1 in. thick at 20°C. The partial pressure of HCl on one side of the film is 0.08 atm and zero on the other. Estimate the rate of diffusion, as gram-moles of HCl per second per square centimeter if the

total pressure is:
(a) 10 atm.
(b) 1 atm.
(c) 0.1 atm.
The diffusivity of HCl in air at 20°C and 1 atm is 0.145 cm²/sec.

16.8 In a test on the vaporization of H_2O into air in a wetted-wall column, the following data were obtained.

Tube diameter, 1.46 cm.
Wetted-tube length, 82.7 cm.
Air rate to tube at 24°C and 1 atm, 720 cm³/sec.
Temperature of inlet air, 57°C.
Temperature of outlet air, 33°C.
Temperature of inlet water, 25.15°C.
Temperature of outlet water, 25.35°C.
Partial pressure of water in inlet air, 6.27 torr, and in outlet air, 20.1 torr.

The value for the diffusivity of water vapor in air is 0.22 cm²/sec at 0°C and 1 atm. The mass velocity of air is taken relative to the pipe wall.

Calculate K_g for the wetted-wall column.

16.9 The following data were obtained by Chamber and Sherwood [*Ind. Eng. Chem.*, **29**, 1415 (1937)] on the absorption of ammonia from an ammonia–air system by acid in a wetted-wall column 0.575 in. in diameter and 32.5 in. long.

Inlet acid ($2N$ H_2SO_4) temperature, °F	76
Outlet acid temperature, °F	81
Inlet air temperature, °F	77
Outlet air temperature, °F	84
Total pressure, atm	1.00
Partial pressure NH_3 in inlet gas, atm	0.0807
Partial pressure NH_3 in outlet gas, atm	0.0205
Air rate, lbmole/hr	0.260

The operation was countercurrent with the gas entering at the bottom of the vertical tower and the acid passing down in a thin film on the inner wall. The change in acid strength was inappreciable, and the vapor pressure of ammonia over the liquid may be assumed to have been negligible. Calculate the absorption coefficient k_g from the data.

	Bottom	Top
Water temperature, °F	120	126
Water vapor pressure, psia	1.69	1.995
Mole fraction H_2O in air	0.001609	0.0882
Total pressure, psia	14.1	14.3
Air rate, lbmole/hr	0.401	0.401
Column area, ft²	0.5	0.5
Water rate, lbmole/hr (approximate)	20	20

16.10 A new type of cooling-tower packing is being tested in a laboratory column. At two points in the column, 0.7 ft apart, the data at the bottom of page 672 have been taken. Using the data given, calculate mass transfer coefficient $K_y a$ and H_{OG} that can be used to design a large, packed-bed cooling tower.

16.11 A mixture of benzene and dichloroethane is used to test the efficiency of a packed column that contains 10 ft of packing and operates adiabatically at atmospheric pressure. The liquid is charged to the reboiler, and the column is operated at total reflux until equilibrium is established. At equilibrium, liquid samples from the distillate D and reboiler B, as analyzed by refractive index, give the following compositions for benzene: $x_D = 0.653$, $x_B = 0.298$.

Calculate the value of HETP in inches for this packing. What are the limitations on using this calculated value for design?

Data for x-y at 1 atm (in benzene mole fractions)

x	0.1	0.2	0.3	0.4	0.5	0.6	0.7	0.8	0.9
y	0.11	0.22	0.325	0.426	0.526	0.625	0.720	0.815	0.91

16.12 An SO_2–air mixture is being scrubbed with water in a countercurrent packed tower operating at 20°C, the spent gas leaving at atmospheric pressure. Solute-free water enters the top of the tower at a constant rate of 1000 lb/hr and is well distributed over the packing. The liquor leaving contains 0.6 lb SO_2/100 lb of solute-free water. The partial pressure of SO_2 in the spent gas leaving the top of the tower is 23 torr. The mole ratio of water to air is 25. The necessary equilibrium data can be found in Perry's *Chemical Engineers' Handbook*.
(a) What percent of the SO_2 originally in the entering gases is absorbed in the tower?
(b) In operating the tower it was found that the rate coefficients k_g, k_l, and K_L remained substantially constant throughout the tower, having the following values.

$$k_l = 1.3 \text{ ft/hr}$$

$$k_g = 0.195 \text{ lbmole/hr} \cdot \text{ft}^2 \cdot \text{atm}$$

$$K_L = 0.6 \text{ ft/hr}$$

At that point in the tower where the liquid concentration is 0.001 lbmole SO_2 per pound-mole of water, what is the liquid concentration at the gas–liquid interface in pound-moles per cubic foot? Assume that the solution has the same density as H_2O.

16.13 Exit gas from a chlorinator consists of a mixture of 20 mole% chlorine in air. This concentration is to be reduced to 1% chlorine by water absorption in a packed column to operate isothermally at 20°C and atmospheric pressure. Calculate for 100 kgmole/hr of feed gas:

Data for x-y at 20°C (in chlorine mole fractions)

x	0.0001	0.00015	0.0002	0.00025	0.0003
y	0.006	0.012	0.024	0.04	0.06

(a) Minimum water rate, kg/hr.

(b) N_{OG} for twice the minimum water rate.

16.14 One thousand cubic feet per hour of a 10 mole% NH_3 in air mixture is required to produce nitrogen oxides. This mixture is to be obtained by desorbing an aqueous 20 wt% NH_3 solution with air at 20°C. The spent solution should not contain more than 1 wt% NH_3.

Calculate the volume of packing required for the desorption column. Vapor-liquid equilibrium data from Example 16.4 can be used and $K_Ga = 4\ \text{lbmole/hr} \cdot \text{ft}^3 \cdot \text{atm}$ (partial pressure).

16.15 You are asked to design a packed column to continuously recover acetone A from air by absorption with water at 60°F. The air contains 3 mole% acetone and a 97% recovery is desired. The gas flow rate is 50 ft³/min at 60°F, 1 atm. The maximum allowed gas superficial velocity in the column is 2.4 ft/sec.

It may be assumed that in the range of operation $Y^* = 1.75X$, where Y and X are mole ratios (acetone to pure carrier).

Calculate:

(a) The minimum water-to-air molar flow rate ratio.

(b) The maximum acetone concentration possible in the aqueous solution.

(c) The number of theoretical stages for a flow rate ratio of 1.4 times the minimum.

(d) The corresponding number of overall gas transfer units.

(e) The height of packing, assuming $K_Ya = 12.0\ \text{lbmole/hr} \cdot \text{ft}^3 \cdot$ molar ratio difference.

(f) The height of packing as a function of the molar flow rate ratio (you may maintain G and HTU constant).

16.16 Ammonia, present at a partial pressure of 12 torr in an air stream saturated with water vapor at 68°F and 1 atm, must be removed to the extent of 99.6% by water absorption at the same temperature and pressure.

Two thousand pounds of dry air per hour are to be handled.

(a) Calculate the minimum amount of water necessary (see Fig. 16.9 for equilibrium data).

(b) Assuming you operate at 10 times the minimum water flow and at one half the flooding gas velocity, compute the dimensions of a column, packed with 1-in. ceramic Raschig Rings, that will do the job.

16.17 From a 10 mole% NH_3-in-air mixture, 95% of the ammonia is to be removed by countercurrent scrubbing with $0.1N\ H_2SO_4$ at 1 atm, 68°F. The entering gas rate is 735 lb/hr $\cdot$ ft². Calculate:

(a) The minimum acid rate.

(b) The number of theoretical stages at 1.2 times the minimum acid rate.

(c) The tower diameter in feet for a reasonable pressure drop.

(d) The number of overall gas transfer units N_{OG}.

(e) The H_{OG} based on a K_Ga of 16 lbmole/hr $\cdot$ ft³ $\cdot$ atm.

(f) The column height.

(g) The HETP.

16.18 Consider a distillation column separating ethanol from water at 1 atm. The following specifications are set.

Feed—10 mole% ethanol (bubble-point liquid).
Bottoms—1 mole% ethanol.
Distillate—80 mole% ethanol (saturated liquid).
Reflux ratio—1.5 times the minimum.
Constant molal overflow may be assumed and vapor–liquid equilibrium data are given in Problem 8.32.

(a) How many theoretical plates would be required above and below the feed if a plate column were used?

(b) How many transfer units would be required above and below the feed if a packed column were used?

(c) Assuming the plate efficiency is approximately 80% and the plate spacing is 12 in., find how high the plate column would be.

(d) Using HTU values of 1.5 ft, find how high the packed column would have to be.

(e) Assuming that you had HTU data only on the benzene–toluene system available, how would you go about applying the data to obtain the HTU for the ethanol–water system?

16.19 A 2 mole% NH_3-in-air mixture at 68°F and 1 atm is to be scrubbed with water in a tower packed with 1-in. ceramic Intalox Saddles. The inlet water rate will be 2401 lb/hr · ft², and the inlet gas rate 240 lb/hr · ft². Assume the tower temperature remains constant at 68°F, at which the gas-solubility relationship follows Henry's law, $p = Hx$, where p is the partial pressure of ammonia over the solution, x is the mole fraction of ammonia in the liquid, and H is the Henry's law constant equal to 2.7 atm/mole fraction.

(a) Calculate the required height for absorption of 90% of the NH_3.

(b) Calculate the minimum water rate for absorbing 98% of the NH_3.

(c) The use of 1/2-in. rather than 1-in. Raschig Rings has been suggested. What would you expect in the way of changes in: K_Ga, pressure drop, maximum liquid rate, K_L, column height, column diameter, H_{OG}, and N_{OG}?

16.20 You are to design a packed column to absorb CO_2 from air in fresh dilute–caustic solution. The entering air contains 3 mole% CO_2 and a 97% recovery of CO_2 is desired. The gas flow rate is 5000 ft³/min at 60°F, 1 atm. It may be assumed that at the range of operation $Y^* = 1.75X^*$ where Y and X are mole ratios of CO_2 to carrier. A column diameter of 30 in. with 2-in. Intalox packing can be assumed for the initial design estimates. Assume the caustic solution has the properties of water. Calculate:

(a) The minimum caustic solution-to-air molal flow rate ratio.

(b) The maximum possible concentration of CO_2 in caustic.

(c) The number of theoretical plates at $(L/G) = 1.4$ times minimum.

(d) The caustic rate.

(e) The pressure drop per foot of column height. (What does this result suggest?)

(f) The overall number of gas transfer units N_{OG} by integration.

(g) The height of packing, using the K_Ga from Fig. 16.17.

16.21 Determine the size of a countercurrently operated packed tower required to recover 99% of the ammonia from a gas mixture that contains 6 mole% NH_3 in air. The tower, packed with 1-in. metal Pall Rings, must handle 2000 ft³/min of gas as measured at 68°F and 1 atm. The entering water absorbent rate will be twice the theoretical minimum, and the gas velocity will be such that it is 50% of the flooding velocity. Assume isothermal operation at 68°F and 1 atm. The height of the tower is to be calculated using the "transfer unit" concept on the assumption that there are no mass transfer data available in the literature for this particular system. Equilibrium data are given in Fig. 16.9.

16.22 A tower, packed with 1-in. Raschig Rings, is to be designed to absorb SO_2 from air by scrubbing with water. The design specifications are as shown in the sketch below assuming that neither air nor water will be transferred between phases. Equilibrium data for SO_2 solubility in water at 30°C and 2 atm (Perry's *Chemical Engineers' Handbook*, fourth edition, Table 14.31, p. 14-6) have been fitted by the least-squares method to the following equation.

$$y = 12.697x + 3148.0x^2 - 4.724 \times 10^5 x^3 + 3.001$$
$$\times 10^7 x^4 - 6.524 \times 10^8 x^5$$

(a) Derive the following molar material balance operating line for SO_2 mole fractions.

$$x = 0.0189\left[\frac{y}{1-y}\right] - 0.0010$$

(b) Write a computer program for calculating the number of required transfer units based on the overall gas-phase resistance and solve for the value. Use Simpson's rule to integrate numerically any integrals that cannot be handled analytically.

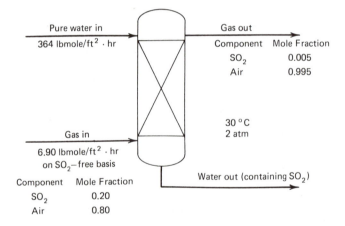

Pure water in	Gas out	
364 lbmole/ft² · hr	Component	Mole Fraction
	SO_2	0.005
	Air	0.995

30 °C
2 atm

Gas in
6.90 lbmole/ft² · hr
on SO_2–free basis

Component	Mole Fraction
SO_2	0.20
Air	0.80

Water out (containing SO_2)

17

Energy Conservation and Thermodynamic Efficiency

The increase in availability, ΔB, is entirely determined by the initial and final states of the materials, together with the temperature and pressure of the medium (infinite surroundings). However the engineer still retains the liberty to make σ (created entropy), and therefore w_t (total equivalent work requirement of the process), as small as possible. This requires a suitable choice of the path between the given initial and final states, i.e. the choice of the details of the operation.

Kenneth G. Denbigh, 1956

Except where an expendable mass separating agent is available and can be allowed to contaminate the products, all separation processes require energy addition in the form of heat and/or work. Historically, energy costs for separating mixtures have always been important compared to depreciation costs of equipment. However, in recent years, energy costs have become even more relatively significant. It is of interest therefore to determine the theoretical minimum energy requirement for conducting a separation and to seek a practical process that approaches this limit or minimizes the use of expensive forms of energy. To accomplish the former objective, we employ thermodynamic analysis. The achievement of the latter goal represents a considerable challenge, but a number of interesting schemes have been devised. The practical application of such analysis and search has long been important in cryogenic separation processes such as the separation of air, and recently considerable interest has been shown in extending thermodynamic analysis to other processes operating at low temperatures, as well as processes operating at temperatures above ambient.

677

17.1 Minimum Work of Separation

Consider the continuous steady-state flow system, involving m streams, shown in Fig. 17.1, in which single-phase streams j flowing in are separated, without chemical reaction, into two or more single-phase streams k flowing out that have compositions that differ from each other and from the inlet stream(s). For the streams, the molal flow rates are n, the component mole fractions are z_i, the molar enthalpies are H, and the molar entropies are S. Total heat rate Q_t and total work rate W_t can flow in or out of the system. Following the usual thermodynamic convention, if heat is transferred to the system from the surroundings, it is positive; if work is done by the system on the surroundings, it is positive. If kinetic, potential, surface, and other energy changes due to the process are neglected, application of the first law of thermodynamics gives the energy balance

$$\sum_{in} n_j H_j + Q_t = \sum_{out} n_k H_k + W_t \tag{17-1}$$

An ideal process that is useful for comparison is one that is isothermal and reversible. This implies that heat transfer takes place between the system and the surroundings, where both are at the same temperature T_o. Therefore, by the second law of thermodynamics

$$Q_t = T_o \left[\sum_{out} n_k S_k - \sum_{in} n_j S_j \right] \tag{17-2}$$

If (17-2) is substituted into (17-1), we obtain an expression for $(-W_{min})$, the

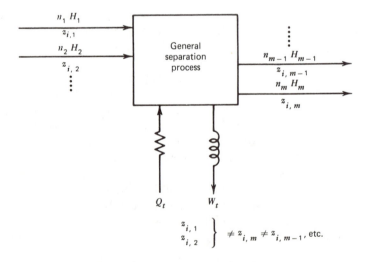

Figure 17.1. General separation process.

minimum rate of work that must be done by the surroundings on the system,

$$-W_{min} = \sum_{out} n_k H_k - \sum_{in} n_j H_j - T_o \left[\sum_{out} n_k S_k - \sum_{in} n_j S_j \right] \tag{17-3}$$

Thus, the minimum rate of work depends only on the feed and product conditions.

If use is made of the definition of the Gibbs free energy from classical thermodynamics, when $T = T_o$

$$G = H - T_o S \tag{17-4}$$

the energy balance becomes

$$-W_{min} = \sum_{out} n_k G_k - \sum_{in} n_j G_j \tag{17-5}$$

Thus, the minimum rate of work equals the change in Gibbs free energy between the feed(s) and products. The Gibbs free energy of a mixture is the mole fraction summation of the partial molal free energies of the components.

$$G = \sum_i z_i \bar{G}_i \tag{17-6}$$

The partial molal free energies are, in turn, related at temperature T_o to the component fugacities by

$$\bar{G}_i = \bar{G}_i^o + RT_o[\ln f_i - \ln f_i^o] \tag{17-7}$$

If we assume that entering and exiting streams are all at the same pressure P, the standard state quantities $\bar{G}_i^o$ and f_i^o are unique for each component. Then, by combining (17-5), (17-6), and (17-7), the minimum rate of work in terms of feed and product component fugacities becomes

$$-W_{min} = RT_o \left[\sum_{out} n_k \left(\sum_i z_{i,k} \ln f_{i,k} \right) - \sum_{in} n_j \left(\sum_i z_{i,j} \ln f_{i,j} \right) \right] \tag{17-8}$$

Ideal Gas Mixtures

For gas mixtures that form ideal solutions and follow the ideal gas law, $z_i = y_i$ and $f_i = y_i P$, and (17-8) reduces to

$$-W_{min} = RT_o \left\{ \sum_{out} n_k \left[\sum_i y_{i,k} \ln(y_{i,k}) \right] - \sum_{in} n_j \left[\sum_i y_{i,j} \ln(y_{i,j}) \right] \right\} \tag{17-9}$$

which indicates that the minimum rate of work does not depend upon pressure or upon relative volatilities of the components being separated. Furthermore, the minimum rate of work is still finite for a perfect separation. For example, for the perfect separation of a binary gas mixture of components A and B into gas products at the temperature and pressure of the feed, (17-9) reduces to the following dimensionless minimum work group.

$$\frac{-W_{min}}{n_F R T_o} = -[(y_A)_F \ln (y_A)_F + (y_B)_F \ln (y_B)_F] \tag{17-10}$$

where subscript F refers to the feed condition. For an equimolar feed, a maximum value of 0.6931 is obtained from (17-10) for the dimensionless minimum work function.

Example 17.1. In Fig. 17.2, a continuous process separates a 60 mole% mixture of propylene in propane at ambient conditions into two products containing 99 mole% propylene and 95 mole% propane, respectively. The products are also at ambient temperature and pressure. Determine the minimum rate of work required.

Solution. Both product temperatures and pressures are equal to those of the feed. Because the components are similar in molecular structure and the pressure is 1 atm, all streams entering and leaving the process are ideal gas solutions that follow the ideal gas law. Therefore, (17-9) can be used to compute the minimum rate of work. Note that the temperature of the surroundings T_o is also the ambient temperature, which is taken as 530°R.

$$-W_{min} = (1.987)(530)\{(351)[0.99 \ln(0.99) + 0.01 \ln(0.01)]$$
$$+ (249)[0.05 \ln(0.05) + 0.95 \ln(0.95)] - (600)[0.60 \ln(0.60)$$
$$+ 0.40 \ln(0.40)]\} = 352,500 \text{ Btu/hr or } 5875 \text{ Btu/lbmole of feed}$$

done by the surroundings on the system.

□

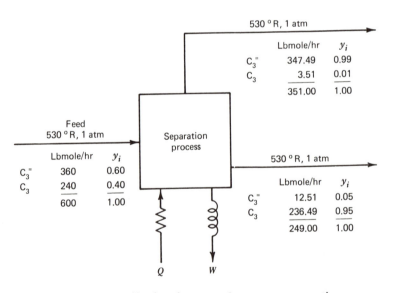

Figure 17.2. Specifications for a propylene–propane separation process.

Liquid Mixtures at Low Pressures

Another useful limiting case of (17-8) is that of liquid mixtures at pressures near ambient or below such that $z_i = x_i$ and $f_i = \gamma_i x_i P_i^s$, where γ_i is the liquid-phase activity coefficient and P_i^s is the vapor pressure. In this case, (17-8) reduces to

$$-W_{min} = RT_o \left\{ \sum_{out} n_k \left[\sum_i x_{i,k} \ln(\gamma_{i,k} x_{i,k}) \right] \right.$$
$$\left. - \sum_{in} n_j \left[\sum_i x_{i,j} \ln(\gamma_{i,j} x_{i,j}) \right] \right\} \qquad (17\text{-}11)$$

Again, $-W_{min}$ is not influenced by pressure or relative volatilities, except for their influence on the activity coefficient. For the perfect separation of a binary liquid mixture into liquid products, (17-11) reduces to

$$\frac{-W_{min}}{n_F RT_o} = -[x_{A,F} \ln(\gamma_{A,F} x_{A,F}) + x_{B,F} \ln(\gamma_{B,F} x_{B,F})] \qquad (17\text{-}12)$$

Again, $-W_{min}$ is finite in value. As might be expected, mixtures with $\gamma_i > 1$ require less work than mixtures with $\gamma_i < 1$, where unlike pairs of unlike molecules tend to be attracted more than pairs of like molecules. From (17-12), when $\gamma_{A,F} x_{A,F} = 1$ and $\gamma_{B,F} x_{B,F} = 1$, $(-W_{min}) = 0$ because the components of the feed are immiscible and a perfect separation is already achieved.*

Example 17.2.   The specifications for a methanol M–water W separation are shown in Fig. 17.3. Determine the minimum rate of work required for the process.

Solution.   Conditions are such that all streams are liquid. At 530°R, liquid-phase activity coefficients may be computed from the van Laar equations

$$\log \gamma_M = \frac{0.25}{[1 + 1.25 x_M/x_W]^2}$$

$$\log \gamma_W = \frac{0.20}{[1 + 0.80 x_W/x_M]^2}$$

For the compositions in Fig. 17.3, the activity coefficients, as computed from these equations, are

		γ	
Component	Feed	Methanol-Rich Product	Waste Product
Methanol	1.08	1.00	1.75
Water	1.20	1.57	1.00

* From Section 5.8, for liquid–liquid equilibrium between phases I and II, $\gamma_i^I x_i^I = \gamma_i^{II} x_i^{II}$. If phase I is pure i, then $\gamma_i^I = 1$, and $\gamma_i^{II} x_i^{II} = 1$.

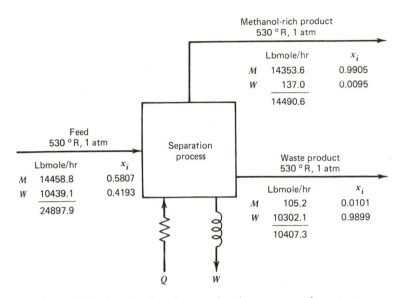

Figure 17.3. Specifications for a methanol–water separation process.

To calculate the minimum rate of work, it is of interest to divide (17-11) into an ideal contribution and an excess contribution due to departures from an ideal solution. The resulting equation is

$$
\begin{aligned}
-W_{min} = RT_o &\left\{ \sum_{out} n_k \left[\sum_i x_{i,k} \ln x_{i,k} \right] - \sum_{in} n_j \left[\sum x_{i,j} \ln x_{i,j} \right] \right\} \\
+ RT_o &\left\{ \sum_{out} n_k \left[\sum x_{i,k} \ln \gamma_{i,k} \right] - \sum_{in} n_j \left[\sum x_{i,j} \ln \gamma_{i,j} \right] \right\}
\end{aligned} \tag{17-13}
$$

$$
\begin{aligned}
&= (1.987)(530)\{(14490.6)[0.9905 \ln 0.9905 + 0.0095 \ln 0.0095] \\
&\quad + (10407.3)[0.0101 \ln 0.0101 + 0.9899 \ln 0.9899] \\
&\quad - (24897.9)[0.5807 \ln 0.5807 + 0.4193 \ln 0.4193]\} \\
&\quad + (1.987)(530)\{(14490.6)[0.9905 \ln 1.00 + 0.0095 \ln 1.57] \\
&\quad + (10407.3)[0.0101 \ln 1.75 + 0.9899 \ln 1.00] \\
&\quad - (24897.9)[0.5807 \ln 1.08 + 0.4193 \ln 1.20]\} \\
&= 16,393,000 - 2,887,000 = 13,506,000 \text{ Btu/hr}
\end{aligned}
$$

The positive deviation from an ideal solution causes a reduction of 17.6% in the minimum rate of work from that for an ideal solution.

☐

The separation processes in Figs. 17.1, 17.2, and 17.3 indicate heat transfer as an additional mode of energy transfer between the process system and the surroundings. Once the minimum rate of work is determined, the corresponding rate of heat transfer can be calculated from the energy balance given by (17-1).

For a process involving only gas streams that enter and leave the process at the surroundings temperature and at identical pressures and are ideal solutions that follow the ideal gas law, then no change in enthalpy occurs (because heat of mixing is zero), and from (17-1) the rate of heat transfer from the process to the surroundings is equal to the minimum rate of work done by the surroundings on the system. Thus, in Example 17-1, heat must be rejected from the process to the surroundings at the rate of 352,500 Btu/hr.

For the case of liquid streams that form ideal solutions and enter and leave the process at the temperature of the surroundings and at a near-ambient pressure, the rate of heat transfer from the process will again equal the minimum rate of work on the process. When the liquid streams form nonideal solutions, the rate of heat transfer will be different from the minimum rate of work because the sum of the exit stream enthalpies will not equal the sum of the inlet stream enthalpies. In that case, (17-1) becomes

$$Q = -(-W_{min}) + \sum_{out} n_k H_k^E - \sum_{in} n_j H_j^E \tag{17-14}$$

where H^E is the excess enthalpy. For nonideal solutions that follow the van Laar equation, the excess enthalpy is obtained from (5-37)

$$H^E = RT_o \sum_i x_i \ln \gamma_i \tag{17-15}$$

Thus,

$$\sum_{out} n_k H_k^E - \sum_{in} n_j H_j^E = RT_o \left\{ \sum_{out} n_k \left[\sum_i x_{i,k} \ln \gamma_{i,k} \right] - \sum_{in} n_j \left[\sum_i x_{i,j} \ln \gamma_{i,j} \right] \right\} \tag{17-16}$$

But the right-hand term in (17-16) is identical to the second term on the right-hand side of (17-13). Thus, the change in excess enthalpy is equal to the excess contribution to the minimum rate of work. Consequently, the rate of heat transfer is equal to only the ideal contribution to the minimum rate of work.

For solutions that exhibit positive deviations from ideality, the excess enthalpy is positive; that is, the heat of mixing is endothermic. For separation processes involving such solutions, the net heat of mixing reduces the minimum rate of work from that for ideal solutions. The rate of heat rejection is equal to the sum of the net exothermic heat of separation and the minimum rate of work. In Example 17.2, the net exothermic heat of separation is 2,887,000 Btu/hr and the rate of heat transfer from the process to the surroundings is 16,393,000 Btu/hr.

17.2 Net Work Consumption and Thermodynamic Efficiency

It is useful to compare the minimum rate of work for a separation process to the actual rate of work. However, the comparison is complicated because many separation processes use heat, rather than work, as the energy separating agent. The difficulty is overcome by determining a net work consumption for the actual process, as discussed by Robinson and Gilliland.[2] This involves the conversion of heat into work by a reversible heat engine (e.g., a Carnot cycle engine) that rejects heat, or absorbs heat from the surroundings at T_o.

Consider, for example, the bottom section of the distillation column shown in Fig. 17.4a. Instead of work, the energy separating agent is heat transferred to the reboiler. If the heating medium (e.g., steam) in the reboiler is at a constant temperature T_s, the work equivalent to the reboiler duty is obtained from the reversible heat engine, shown in Fig. 17.4b. For such an engine, classical thermodynamics gives the relationship

$$W_{eq} = Q_{in}\left(\frac{T_s - T_o}{T_s}\right) \tag{17-17}$$

Thus, if $T_s > T_o$ the Q_{in} required is greater than W_{eq}. When the heating medium temperature T changes during heat transfer, a differential form of (17-17) must be employed

$$dW_{eq,out} = \left(\frac{T - T_o}{T}\right)dQ_{in} \tag{17-18}$$

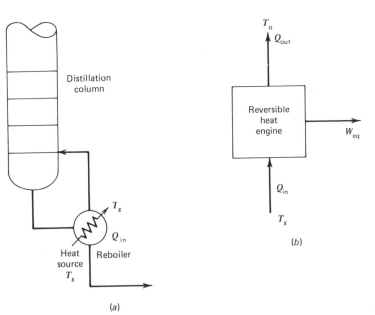

Figure 17.4.   Work equivalent of heat addition to a distillation column. (a) Heat transfer to reboiler of a distillation column. (b) Reversible heat engine.

A rigorous expression for the relationship between the net work consumption and the minimum (or maximum) rate of work required for a given process was derived by Denbigh[3] for the general case of streams entering and leaving the process at different temperatures and pressures and with any number of heat reservoirs at different temperatures, including the infinite medium at T_o. From the first law of thermodynamics for a continuous steady-state process:

$$\Delta H = Q_o + \sum_i Q_i + \sum (- W_s) \tag{17-19}$$

where ΔH refers to changes to the process streams; that is, $\Delta H = \sum_{out} n_k H_k - \sum_{in} n_j H_j$; W_s is shaft or other external work (but not flow work) done by the process on the surroundings (or vice versa if negative); Q_o is the heat transferred to (or from if negative) the process from the infinite medium at T_o; and Q_i are heat transfer rates to (or from if negative) the process from reservoirs i at temperatures T_i. The second law of thermodynamics for all changes caused by the process can be stated in terms of the following equality.

$$\Delta S + \Delta S_o + \sum_i \Delta S_i = \Delta S_{irr} \tag{17-20}$$

where ΔS, ΔS_o, ΔS_i, and ΔS_{irr} refer to changes to the process, the infinite medium at T_o, the various heat reservoirs at temperatures T_i, and the created entropy due to irreversibility. But the entropy changes of the infinite medium and the other reservoirs are simply $\Delta S_o = - Q_o/T_o$ and $\Delta S_i = - Q_i/T_i$. Substitution of these expressions into (17-20) followed by combination of the resulting equation with (17-19) to eliminate Q_o gives a combined statement of the first and second laws of thermodynamics that is equivalent to but different from that of Denbigh[3].

$$\Delta B = \Delta H - T_o \, \Delta S = \sum_i (1 - T_o/T_i)Q_i + \sum (- W_s) - T_o \, \Delta S_{irr} \tag{17-21}$$

B is generally referred to as the *availability function*, and the $T_o \Delta S_{irr}$ term is generally referred to as the *lost work LW*.[4] Thus

$$\Delta B = \sum (1 - T_o/T_i)Q_i + \sum (- W_s) - LW \tag{17-22}$$

The term involving Q_i is identical to (17-17) and corresponds to the equivalent work that could be produced by supplying Q_i from a heat reservoir at T_i to a reversible heat engine that exhausts to the infinite medium at T_o. It is convenient to combine this equivalent work with the shaft work to obtain a net work consumption

$$(- W_{net}) = \sum (1 - T_o/T_i)Q_i + \sum (- W_s)$$

$$= \Delta B + LW \tag{17-23}$$

For a reversible separation process, $LW = 0$ and $(-W_{net}) = \Delta B$. For an actual separation process, $LW > 0$, $(-W_{net}) > \Delta B$, and a thermodynamic efficiency can be defined as the ratio of the change in availability function to the net work consumption for the actual process, provided that ΔB is positive.

$$\eta = \frac{\Delta B}{(-W_{net})} \qquad \text{if } \Delta B = (+) \qquad (17\text{-}24)$$

If all streams enter and leave the process as gases at the same temperature and pressure and form ideal gas solutions and follow the ideal gas law, $(-W_{min})$ in (17-9) is equal to ΔB. If all streams enter and leave the process as liquids at the same temperature and the same low pressure, $(-W_{min})$ in (17-11) is equal to ΔB. Otherwise ΔB must be calculated from entering and leaving stream enthalpies and entropies.

Processes that involve operations other than separation steps may result in a ΔB that is negative, thus indicating that it would be possible to obtain useful work from a reversible process. In that case, for the actual process it will still be true that $LW > 0$ and $(-W_{net}) > \Delta B$. However, the thermodynamic efficiency should now be defined as a ratio of the net work consumption for the actual process to the change in availability function

$$\eta = \frac{(-W_{net})}{\Delta B} \qquad \text{if } \Delta B = (-) \qquad (17\text{-}25)$$

The application of (17-25) can lead to negative efficiencies when $LW > |\Delta B|$.

Example 17.3. An actual, but not optimal, process for the propylene–propane separation specified in Fig. 17.2 for Example 17.1 is shown in Fig. 17.5. The process is a modification of one discussed by Tyreus and Luyben.[5] Conventional distillation is used at a bottom pressure of 300 psia, so cooling water can be employed to condense overhead vapor for reflux. Because the relative volatility is low, varying from 1.08 to 1.14 from the top stage to the bottom stage, a reflux ratio of 15.9 is required at operation near the minimum reflux. Because of high product purities as well as the low relative volatility, 200 stages are required at 100% tray efficiency. With 24-in. tray spacing, this necessitates a division of the separator into two columns in series. Therefore, an intercolumn pump is required in addition to the reflux pump. Total pressure drop for the two columns is 20 psia. Feed at ambient conditions is compressed to column feed pressure in two isentropic stages with an intercooler inbetween. Vapor from the second-stage compressor is cooled, condensed, and sent to a surge tank, from which it is pumped to the bottom section of the distillation operation. Vapor products at specified ambient conditions are obtained by using a partial condenser and a total reboiler, with adiabatic valves to drop the pressure to ambient. Temperatures downstream of the valves are below ambient, so cooling-water heaters are used to bring the products to ambient temperature.

Two heat reservoirs, external to the processing system, are required. One is cooling water for the intercooler, aftercooler, partial condenser, propane heater, and propylene heater. Although the temperature of the cooling water in passing through these heat exchangers increases, we will select a single average temperature of 70°F (530°R), equal

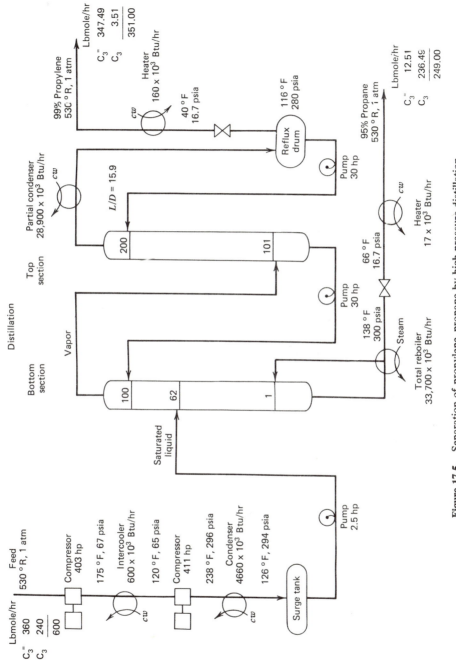

Figure 17.5. Separation of propylene–propane by high-pressure distillation.

Table 17.1 Net work consumption for example 17.3

Item	Rate of Work Done by Surroundings, hp	Rate of Heat Transfer to Process, Btu/hr	Percent of Heat Transfer Available for Work	Equivalent Rate of Work Done on the Process, Btu/hr.
Compressor 1	403	—	—	1,030,000
Compressor 2	411	—	—	1,050,000
Distillation feed pump	2.5	—	—	6,400
Intercolumn pump	30	—	—	76,000
Reflux pump	30	—	—	76,000
Total reboiler	—	33,700,000	22.1	7,433,800
Partial condenser	—	−28,900,000	0.0	0
Intercooler	—	−600,000	0.0	0
Aftercooler-condenser	—	−4,660,000	0.0	0
Propane heater	—	17,000	0.0	0
Propylene heater	—	160,000	0.0	0
Valves	0	—	—	0

NET WORK CONSUMPTION ≅ 9,672,200 Btu/hr

to T_o of the infinite medium. The other reservoir is saturated steam at 220°F (680°R) for the total reboiler.

Calculate the net work consumption, the thermodynamic efficiency, and the lost work. Discuss possible means of improving the thermodynamic efficiency.

Solution. Because the feed and product conditions are those of Example 17.1, $\Delta B = (- W_{min})$ in that example. The net work consumption in British thermal units per hour is computed from (17-23) by summing the shaft work and the equivalent work. For example, the shaft work for the first-stage compressor is

$$(- W_s) = (403 \text{ hp})\left(2545 \frac{\text{Btu}}{\text{hp} \cdot \text{hr}}\right) = 1,030,000 \text{ Btu/hr}$$

Results for other contributions to the total shaft work are given in Table 17.1.

The equivalent work for the total reboiler is computed from (17-17) using T_s equal to 680°R, the saturated steam temperature. Thus

$$(- W_{eq}) = 33,700,000 \text{ Btu/hr}\left(1 - \frac{530}{680}\right) = 7,433,800 \text{ Btu/hr}$$

That is, 22.1% of the reboiler duty could be theoretically converted to work by a reversible heat engine.

Work equivalents for all other heat exchanges are zero because T_s in (17-17) (or T_i in 17-23) is equal to T_o.

From Table 17.1, the net work consumption is 9,672,200 Btu/hr. From (17-24)

$$\eta = (352,500)/(9,672,200) \times 100\% = 3.64\%$$

The lost work from (17-23) is

$$LW = 9,672,200 - 352,500 = 9,319,700 \text{ Btu/hr}$$

or

$$\frac{9,319,700}{2545} = 3,662 \text{ hp}$$

Possible methods for increasing the thermodynamic efficiency involve:

1. Reduction of column operating pressure so as to reduce compressor horsepower.

2. Elimination of aftercooler-condenser following the second-stage compressor so as to reduce reboiler duty.

3. Replacement of valves with expansion engines to produce work.

□

17.3   Reduction of Energy Requirements in Distillation

Usually, the largest energy costs in separation processes are associated with compressors, reboilers, and condensers cooled with refrigerant. Elaborate schemes for effecting energy economies for these items of equipment have received considerable attention and are discussed by Robinson and Gilliland[2] and King.[6] Unfortunately, these schemes often lead to additional equipment capital costs that more than offset savings in utility costs. However, because utility costs have recently been increasing at a faster rate than equipment costs, several schemes that are discussed in this section have received attention in recent literature and appear to be economically feasible for some large-scale processes.

Multieffect Distillation

A general system for multieffect distillation is shown in Fig. 17.6. The feed is split more or less equally among the N columns, which operate in parallel but at different pressures. By reducing column operating pressures successively from left to right, overhead vapor from a higher-pressure column can be condensed in the reboiler of a lower-pressure column. If condenser and reboiler duties of adjacent columns are balanced, utilities are required only in the reboiler of the highest-pressure column and in the condenser of the lowest-pressure column. The number of effects and the column pressures must provide reasonable temperature driving forces in the reboiler of the second to Nth column; to avoid approaching the critical temperature or decomposition temperature; and to permit, if possible, the use of steam in the reboiler of the first effect and cooling water in the condenser of the Nth effect. When the relative volatility remains essentially constant over the pressure range, the utility requirements for an N-effect system are reduced by a factor of $1/N$ compared to those for a single-effect system.

Tyreus and Luyben[5] investigated the application of double-effect distillation to the separation of methanol from water and to the propane separation problem

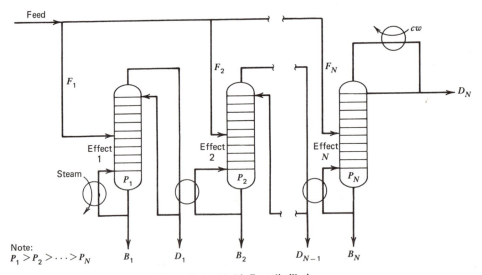

Note:
$P_1 > P_2 > \cdots > P_N$

Figure 17.6. Multieffect distillation.

Table 17.2 Net work consumption for double-effect process of Fig. 17.7

Item	Rate of Work Done by Surroundings, hp	Rate of Heat Transfer to Process, Btu/hr	Percentage of Heat Transfer Available for Work	Equivalent Rate of Work Done on Process, Btu/hr
Compressor 1	403	—	—	1,030,000
Compressor 2	411	—	—	1,050,000
Compressor 3	43	—	—	110,000
Effect 1 feed pump	1.3	—	—	3,300
Effect 2 feed pump	1.2	—	—	3,100
Effect 1 intercolumn pump	22	—	—	56,000
Effect 1 reflux pump	22	—	—	56,000
Effect 2 intercolumn pump	16	—	—	41,000
Effect 2 reflux pump	16	—	—	41,000
Total reboiler	—	18,200,000	22.1	4,014,700
Partial condenser	—	−15,590,000	0	0
Intercooler 1	—	−600,000	0	0
Condenser 1	—	−2,400,000	0	0
Intercooler 2	—	−771,000	0	0
Condenser 2	—	−1,285,000	0	0
Propane heater	—	17,000	0	0
Propylene heater	—	160,000	0	0
Valves	0	—		0

NET WORK CONSUMPTION ≅ 6,405,100 Btu/hr

of Fig. 17.2. A process for the latter case is shown in Fig. 17.7,* which should be compared to the single-effect case shown in Fig. 17.5. Calculations for the net work consumption of the double-effect distillation process, using the same heat reservoirs as in Example 17.3, are summarized in Table 17.2. The resulting thermodynamic efficiency for the double-effect process is 5.50% compared to 3.64% computed in Example 17.3 for the single-effect process. The lost work is reduced by 35% from 3,662 hp to 2,378 hp.

It is perhaps of greater practical interest to compare the daily utility costs of the double-effect and single-effect processes for propylene–propane separation. We assume the following costs and use electric motors for all compressors and pumps.

Utility	Cost	Equivalent Cost, $/10^6$ Btu
Steam (17.2 psia saturated condensed at same pressure)	$1.60/1000 lb	1.66
Cooling Water (20°F rise)	$0.04/1000 gal	0.24
Electricity	$0.04/kWh	11.72

Based on a 24-hr operating day, the resulting utility costs in Table 17.3 show a reduction of 30% for the double-effect scheme in Fig. 17.7 compared to that of the single-effect scheme in Fig. 17.5. The comparison is dominated by the steam requirement. The potential utility savings for the double-effect scheme for a 350-operating-day year is $228,600/yr, which for a reasonable rate of return on investment may offset the additional investment required for the double-effect scheme.

Heat Pumps for Low-Temperature Distillation

Rather than setting the pressure of a distillation column at a level sufficiently high to permit the use of cooling water in the overhead condenser, one may specify a lower pressure and use a refrigerant in the condenser. For example, the separation of propylene–propane, as specified in Fig. 17.2, can be conducted by low-temperature distillation at a 100-psia column overhead pressure, as shown in Fig. 17.8, if a feed system such as shown in Fig. 17.9 is provided. There, the feed mixture is compressed in two stages with an intercooler. A refrigerant-cooled condenser, which follows a water-cooled aftercooler, prepares a saturated liquid feed for the distillation operation. In Fig. 17.8, refrigerant must be supplied to the partial condenser to condense the overhead vapor to obtain reflux at 43°F. At

---

* Each tower shown in Fig. 17.7 should be divided into two sections in the manner shown in Fig. 17.5. Also, necessary liquid pumps, surge drums, and reflux drums are not shown in Fig. 17.7.

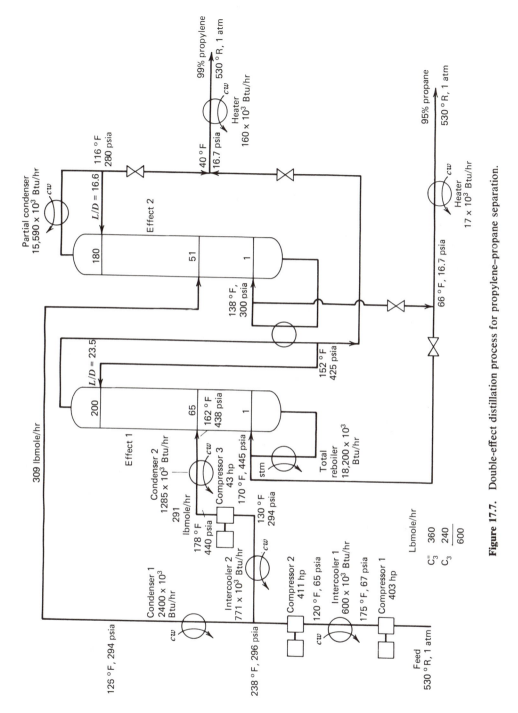

Figure 17.7. Double-effect distillation process for propylene–propane separation.

Table 17-3 Daily utility costs for propylene–propane separation

	Costs, $/day	
Utility	Single-Effect Scheme (Fig. 17.5)	Double-Effect Scheme (Fig. 17.7)
Steam	1341	724
Cooling water	198	120
Electricity	628	670
TOTAL	2167	1514

100 psia, the average relative volatility between propylene and propane is 1.20, which is 8.1% greater than the value of 1.11 at the nominal 300-psia pressure of the distillation operation in Fig. 17.5. The higher relative volatility reduces the number of trays from 200 to 115 and the reflux ratio from 15.9 to 8.76, with corresponding decreases in condenser and reboiler duties.

As discussed by Freshwater[7] and Null,[8] energy requirements of low-temperature distillation operations can often be reduced by retaining a single effect and using a heat pump to "pump" heat from the condenser to the reboiler. Three basic schemes given by Null are shown in Fig. 17.10. In all three an expansion valve and a compressor are used to alter condensing and/or boiling temperatures so that the heat rejected in the condenser can be used to provide the heat needed in the reboiler. Although they are not shown in Fig. 17.10, auxiliary condensers or reboilers may be necessary when condenser and reboiler duties are not matched.

In the simplest scheme, an external refrigerant is used in a closed cycle involving the overhead condenser and bottoms reboiler. The former is the evaporator for the refrigerant and the latter is the refrigerant condenser. The application of the scheme to the propylene–propane separation problem of Fig. 17.2 is shown in Fig. 17.11. Propane is used as the external refrigerant. By evaporating it at 35°F and condensing it at 110°F, the required duties of the partial condenser and total reboiler are exactly matched. Thus, in place of cooling water and steam, 1572 hp are required (assuming isentropic compression of the refrigerant).

When the distillate is a good refrigerant, the scheme in Fig. 17.10b can be employed. The column overhead vapor is compressed so that its condensation temperature is greater than the boiling temperature of the column bottoms product. The heat given off by condensation of the column overhead vapor can then be used in the reboiler. Condensate leaving the reboiler is flashed across an expansion valve to column top pressure to provide reflux and distillate product. Excess vapor is recycled to the compressor. This scheme is often referred to as *vapor recompression*; its application to the propylene–propane separation problem is shown in Fig. 17.12, where the overhead vapor, which is compressed to

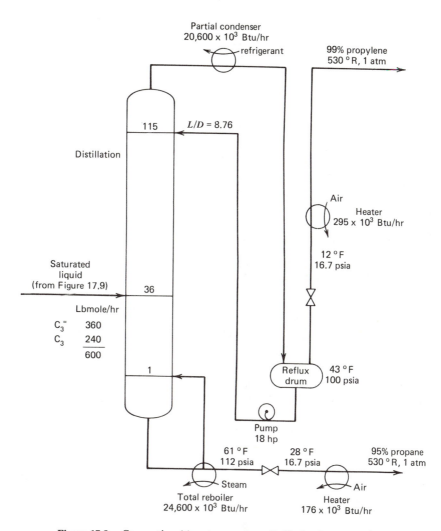

Figure 17.8. Conventional low-temperature distillation for separation of propylene–propane system.

215 psia, falls short of providing all the heat necessary in the total reboiler. Accordingly, an auxiliary steam-heated reboiler is shown.

When the bottoms product is a good refrigerant, the scheme in Fig. 17.10c is a possible candidate for reducing energy consumption. Bottoms liquid is flashed across the expansion valve to a pressure corresponding to a saturation temperature of the distillate. The overhead condenser then doubles as the reboiler.

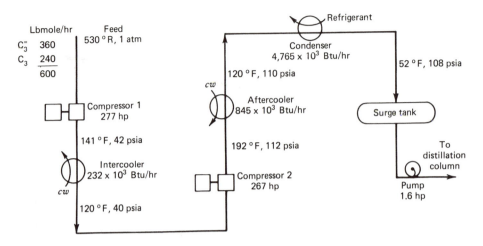

Figure 17.9. Feed system for separation of propylene–propane system by distillation at 100 psia.

Vapor produced in the condenser is compressed back to column bottom pressure before entering the column. Figure 17.13 shows the application of this scheme to the propylene–propane separation problem. The bottoms liquid is flashed to 72 psia to remove the required heat in the condenser. Additional heat added during isentropic compression is insufficient to make up the difference between reboiler and condenser duties. Therefore, the auxiliary steam-heated reboiler is needed.

A comparison of the thermodynamic efficiency and the daily utility costs for the propylene–propane separation schemes shown in Figs. 17.8, 17.9, 17.10, 17.11, 17.12, and 17.13 is given in Table 17.4 based on the use of 220°F steam where required,* the same utility costs given previously for the calculations of Table 17.3, negligible utility cost for air used in heating, and the external propane refrigerant cycle shown in Fig. 17.14, where the refrigerant evaporates at 35°F. This cycle is used in the condenser of Fig. 17.9 and the partial condenser of Fig. 17.8. The best thermodynamic efficiency is achieved for the heat pump scheme using reboiler liquid flashing. That efficiency of 8.10%, while low, is considerably greater than the 3.64% efficiency for the high-pressure distillation process of Fig. 17.5. The other two heat pump schemes have thermodynamic efficiencies that are significantly lower than the best efficiency. Daily utility costs for all three heat pump schemes are significantly lower than for conventional low-temperature distillation. The heat pump arrangement using reboiler liquid flashing

---

* This temperature may cause film boiling of 61°F propane in the reboilers, in which case the steam can be used indirectly to heat water, an oil, or other heating medium for use in the reboiler.

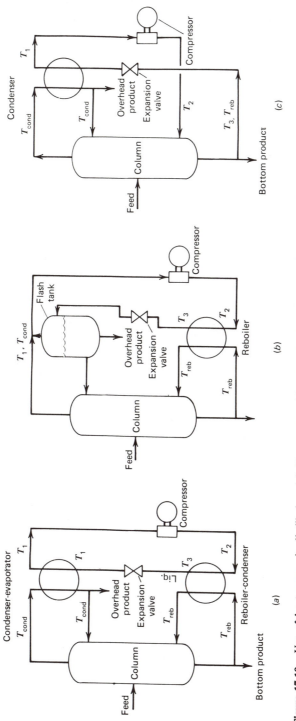

Figure 17.10. Use of heat pumps in distillation. (*a*) Heat pump with external refrigerant. (*b*) Heat pump with compression of overhead vapor. (*c*) Heat pump with reboiler liquid flashing. [H. R. Null, *Chem. Eng. Progr.,* **72** (7), 58–64 (1976).]

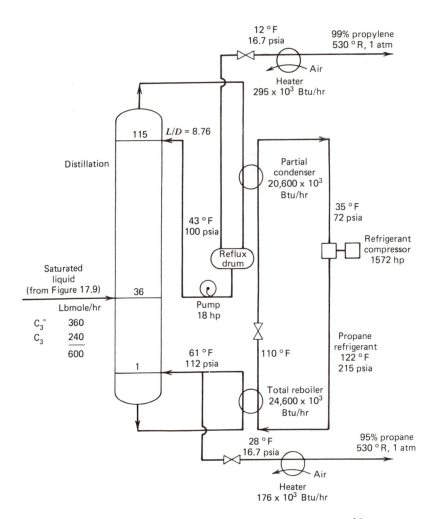

Figure 17.11. Low-temperature distillation using heat pump with external propane refrigerant for separation of propylene–propane system.

also gives the lowest daily utility cost, which is also significantly lower than the cost for the double-effect scheme given in Table 17.3.

Distillation with Secondary Reflux and Boilup

In conventional distillation, heat is added only to the reboiler at the bottom of the column where the temperature is highest, and heat is removed only from the condenser at the top of the column where the temperature is lowest. Because of

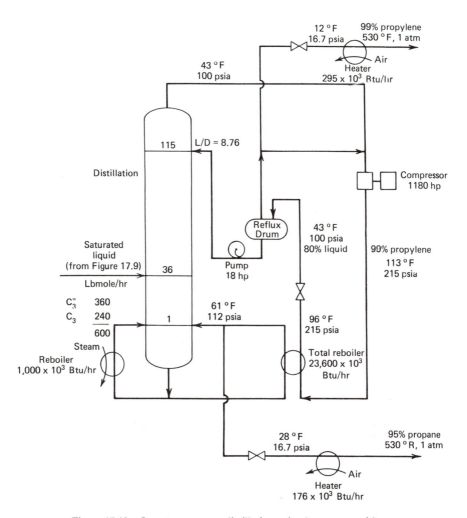

Figure 17.12. Low-temperature distillation using heat pump with compression of overhead vapor for separation of propylene–propane system.

the low thermodynamic efficiency of conventional distillation and because heating costs increase with increasing temperature and cooling costs increase with decreasing temperature, consideration is often given to the use of inter-condensers and interreboilers operating at intermediate temperature levels, as shown in Fig. 17.15a, particularly where a large temperature difference exists between the two ends of the column. For this arrangement, as shown by

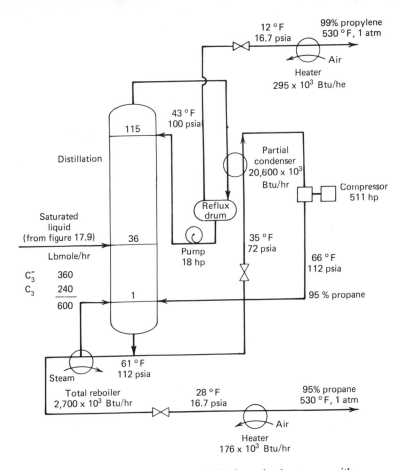

Figure 17.13. Low temperature distillation using heat pump with reboiler liquid flashing for separation of propylene–propane system.

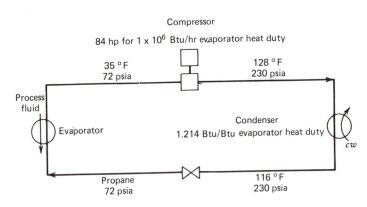

Figure 17.14. External refrigerant cycle.

Table 17.4 Thermodynamic efficiency and daily utility costs for propylene-propane separation at low temperature

	Conventional,[a] (Fig. 17.8)	Heat Pump with External Refrigerant,[a] (Fig. 17.11)	Heat Pump with Compression of Overhead Vapor,[a] (Fig. 17.12)	Heat Pump with Reboiler Liquid Flashing,[a] (Fig. 17.13)
Thermodynamic efficiency, percent	2.87	5.46	6.21	8.10
Utilities, $/day				
Steam	980	0	40	108
Cooling water	184	40	40	40
Electricity	1928	1815	1535	1056
TOTAL	3092	1855	1615	1204

[a] Includes feed preparation in Fig. 17.9 and refrigerant cycle in Fig. 17.14.

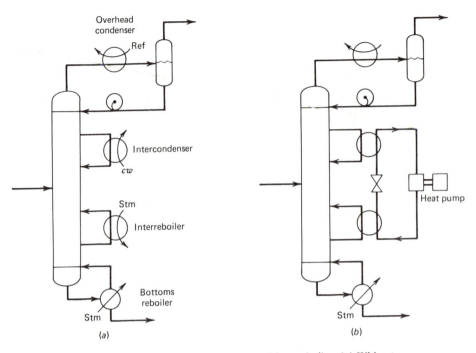

Figure 17.15. Use of intercondenser and interreboiler. (*a*) Without heat pump. (*b*) With heat pump.

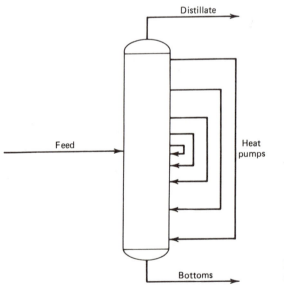

Figure 17.16. Multiple heat-pump scheme.

Petterson and Wells,[9] savings in heating and cooling utility costs tend to be offset somewhat by additional capital costs.

A further improvement is possible if a heat pump is added between the intercondenser and interreboiler as discussed by Freshwater[10] and shown in Fig. 17.15b. Flower and Jackson[11] point out that the thermodynamic efficiency of such a heat pumping arrangement can be further improved by using a large number of heat pumps, placed, for example, as shown in Fig. 17.16, so as to crowd the heat supply and removal from the column as closely as possible in a region near the feed point. Thus, most of the heat is pumped through much smaller temperature differences; and reflux and boilup are generated in the central region of the column. An alternative method for generating secondary reflux and boilup, called *SRV distillation*, has been developed and evaluated by

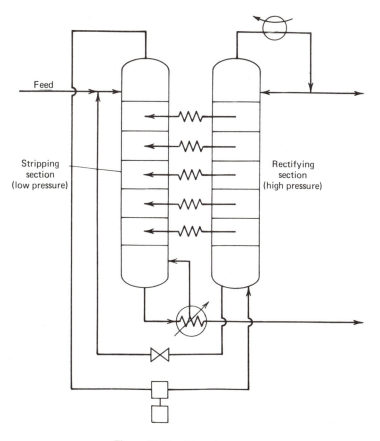

Figure 17.17. SRV distillation.

Mah, Nicholas, and Wodnik[12], and appears in a patent by Haselden.[13] In this scheme, shown in Fig. 17.17, the rectifying section of the column is operated at a pressure greater than that of the stripping section. The pressure difference is set so that resulting temperature will permit heat to be transferred between pairs of stages in the two sections, as desired. As a result of distributing heat sources and sinks throughout the column, top condenser and bottom reboiler duties are significantly reduced and replaced by intermediate-stage heat exchange such that the liquid reflux rate steadily increases as it proceeds down the rectifying section; and the vapor rate steadily increases up the stripping section. Based on an overall reduction in utility requirements, SRV distillation appears particularly attractive for cryogenic separation of close-boiling mixtures into nearly pure products.

Heat Exchange Integration

When the feed to a distillation column is a subcooled liquid and/or a large difference in temperature exists between the top of the column and the bottom of the column, the products are often used to preheat the feed and thereby reduce the reboiler duty. An example is shown in Fig. 17.18 for a sequence of

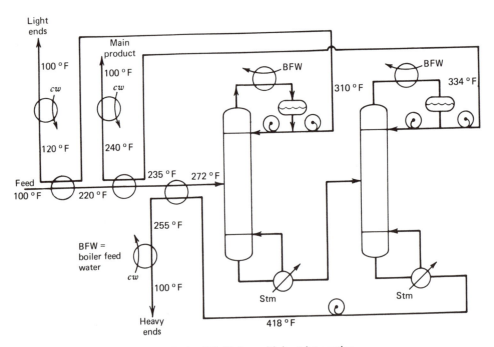

Figure 17.18. Distillation with heat integration.

two distillation columns, where the three products are used to preheat the feed to the first column.

Such sequences of distillation columns can provide additional opportunities for conserving energy by exchanging heat between the condenser of one column and the reboiler of another column. Examples of such schemes and methods for determining optimal energy-integrated sequences of two-product distillation columns are given by Rathore, Van Wormer, and Powers[14] and Umeda, Niida, and Shiroko.[15] Sequences containing other than conventional two-product distillation columns are evaluated by Tedder and Rudd.[16]

References

1. Dodge, B. F., and C. Housum, *Trans. AIChE*, **19**, 117–151 (1927).

2. Robinson, C. S., and E. R. Gilliland, *Elements of Fractional Distillation*, 4th ed, McGraw–Hill Book Co., New York, 1950, 162–174.

3. Denbigh, K. G., *Chem. Eng. Sci.*, **6**, 1–9 (1956).

4. de Nevers, N., and J. D. Seader, "Mechanical Lost Work, Thermodynamic Lost Work, and Thermodynamic Efficiencies of Processes," paper presented at the AIChE 86th National Meeting, Houston, Texas, April 1–5, 1979.

5. Tyreus, B. D., and W. L. Luyben, *Hydrocarbon Processing*, **54** (7), 93–96 (1975).

6. King, C. J., *Separation Processes*, McGraw–Hill Book Co., New York, 1971, Chapter 13.

7. Freshwater, D. C., *Trans. Instn. Chem. Eng.*, **29**, 149–160 (1951).

8. Null, H. R., *Chem. Eng. Prog.*, **72** (7), 58–64 (1976).

9. Petterson, W. C., and T. A. Wells, *Chem. Eng.*, **84** (20), 78–86 (1977).

10. Freshwater, D. C., *Brit. Chem. Eng.*, **6**, 388–391 (1961).

11. Flower, J. R., and R. Jackson, *Trans. Instn. Chem. Eng.*, **42**, T249–T258 (1964).

12. Mah, R. S. H., J. J. Nicholas, and R. B. Wodnik, *AIChE J.*, **23**, 651–658 (1977).

13. Haselden, G. G., U.S. Patent 4,025,398; May 24, 1977.

14. Rathore, R. N. S., K. A. Van Wormer, and G. J. Powers, *AIChE J.*, **20**, 940–950 (1974).

15. Umeda, T., K. Niida, and K. Shiroko, "A Thermodynamic Approach to Heat Integration in Distillation Systems," paper presented at the 85th National Meeting of AIChE, June 4–8, 1978, Philadelphia, Pa.

16. Tedder, D. W., and D. R. Rudd, *AIChE J.*, **24**, 303–315 (1978).

Problems

17.1 Consider a binary ideal-gas mixture at T_o. Calculate and plot the dimensionless minimum work function as a function of feed composition for:
(a) A perfect separation.

(b) A separation giving 98 mole% pure products.
(c) A separation giving 90 mole% pure products.
How sensitive is the minimum work of separation to the product compositions?

17.2 Prove by calculus that the maximum value of the dimensionless minimum work function for a binary ideal-gas mixture at T_o occurs for an equimolal feed.

17.3 Calculate the minimum rate of work in watts for the gaseous separation at ambient conditions indicated in the following diagram.

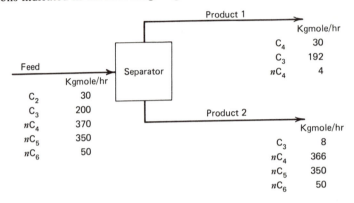

17.4 Calculate the minimum rate of work in watts for the gaseous separation at ambient conditions of the feed indicated below into the three products shown.

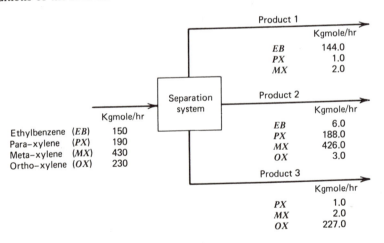

17.5 For the liquid-phase separation at ambient conditions of a 35 mole% mixture of acetone (1) in water (2) into 99 mole% acetone and 98 mole% water products, calculate the minimum rate of work in watts/kgmole of feed. Liquid-phase activity coefficients at ambient conditions are correlated reasonably well by the van Laar equations (5-26) with $A_{12} = 2.0$ and $A_{21} = 1.7$. What would the minimum rate of work be if acetone and water formed an ideal liquid solution?

17.6 For the ambient (25°C, 1 atm) separation indicated below, calculate the minimum rate of work in watts. Liquid-phase activity coefficients for the acetone–water system are given in Problem 17.5.

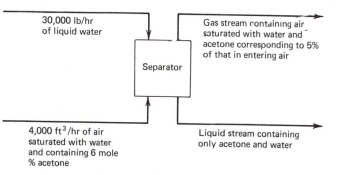

30,000 lb/hr of liquid water

Gas stream containing air saturated with water and acetone corresponding to 5% of that in entering air

Separator

4,000 ft³/hr of air saturated with water and containing 6 mole % acetone

Liquid stream containing only acetone and water

17.7 For the adiabatic flash operation shown below, calculate:
(a) Change in availability function ($T_o = 100°F$).
(b) Net work consumption.
(c) Lost work.
(d) Thermodynamic efficiency.

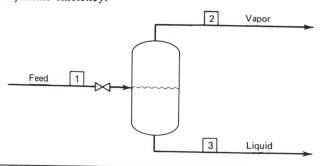

2 Vapor

Feed 1

3 Liquid

	Flow rate, lb mole/hr		
	Stream 1	Stream 2	Stream 3
H_2	0.98	0.95	0.03
N_2	0.22	0.21	0.01
Benzene	0.08	0	0.08
Cyclohexane	91.92	0.69	91.23

	Stream 1	Stream 2	Stream 3
Temperature, °F	120	119.9	119.9
Pressure, psia	300	15	15
Enthalpy, 1000 Btu/hr	−3642.05	−14.27	−3627.78
Entropy, 1000 Btu/hr · °R	4.920	0.094	4.860

17.8 A partial condenser operates as shown below. Assuming $T_o = 70°F$, calculate:
(a) Condenser duty.
(b) Change in availability function.
(c) Net work consumption.
(d) Lost work.
(e) Thermodynamic efficiency.

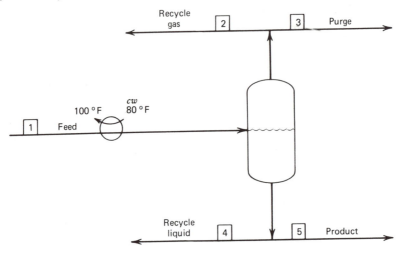

	Flow rate, lbmole/hr				
	Stream 1	**Stream 2**	**Stream 3**	**Stream 4**	**Stream 5**
H_2	72.53	65.15	5.80	0.60	0.98
N_2	7.98	7.01	0.62	0.13	0.22
Benzene	0.13	0	0	0.05	0.08
Cyclohexane	150.00	1.61	0.14	56.33	91.92
	Stream 1	**Stream 2**	**Stream 3**	**Stream 4**	**Stream 5**
Temperature, °F	392	120	120	120	120
Pressure, psia	315	300	300	300	300
Enthalpy,1000 Btu/hr	−2303.29	241.76	21.61	−2231.84	−3642.05
Entropy, 1000 Btu/hr · °R	14.68	2.13	0.19	3.02	4.92

17.9 For the propylene–propane separation shown below, calculate:
(a) Reboiler duty (condenser duty is given).
(b) Change in availability function.
(c) Net work consumption.
(d) Lost work.
(e) Thermodynamic efficiency.

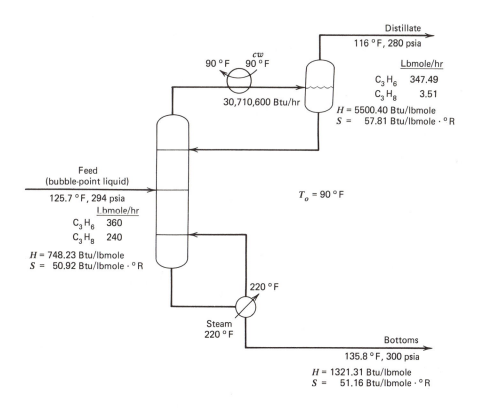

17.10 For the adjusted material balance of Example 15.2, as shown in Fig. 15.9 (but using the Chao–Seader correlation for thermodynamic properties), stream temperatures, enthalpies, and entropies are as follows.

	Feed	Vapor Distillate	Liquid Bottoms
Temperature, °F	150.0	17.03	171.73
Enthalpy, 1000 Btu/hr	4054.36	1849.78	563.06
Entropy, 1000 Btu/hr · °R	41.61	23.42	14.54

If the condenser has a duty of 5,000,000 Btu/hr and is cooled by refrigerant at a constant temperature of 0°F, the column heat loss is 100,000 Btu/hr, the reboiler is heated by steam at a constant temperature of 250°F, and the temperature of the infinite medium is 80°F, calculate:
(a) Reboiler duty.
(b) Change in availability function.
(c) Net work consumption.
(d) Lost work.
(e) Thermodynamic efficiency.

17.11 A ternary hydrocarbon mixture of nC_6, nC_8, and nC_{10} is separated by thermally

coupled distillation. For the design calculations shown below, calculate:
(a) Reboiler duty.
(b) Change in availability function.
(c) Net work consumption.
(d) Lost work.
(e) Thermodynamic efficiency.

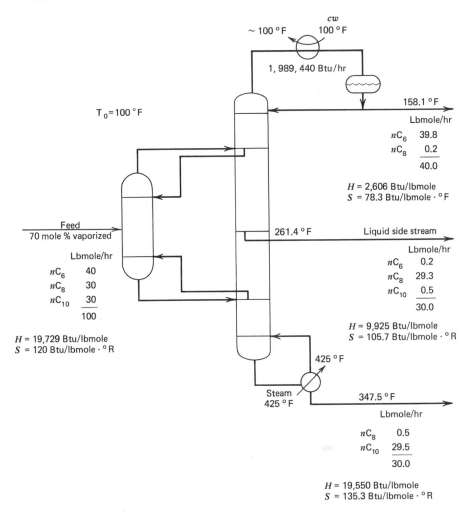

17.12 As indicated in the sketch below, a mixture of propane and n-pentane is to be separated by distillation at 100 psia. Two designs are to be made by the Ponchon–Savarit method. In the first design, a total overhead condenser and a partial bottoms reboiler are to be used with saturated reflux at a flow rate of twice the minimum value. In the second design, a total overhead condenser, an intercondenser, an interreboiler, and a partial bottom reboiler are to be used in the configuration of Fig. 17.15a such that the combined condenser duties of the second design are equal to the overhead condenser duty of the first design. For both designs, assume coolant temperatures that give a 10°F closest approach for each condenser and heating medium temperatures that give an 80°F closest approach for each reboiler. Locations of the intercondenser and the interreboiler should be carefully considered. For each design, calculate:

(a) Change in availability function ($T_o = 100°F$).
(b) Net work consumption.
(c) Lost work.
(d) Thermodynamic efficiency.

Compare the two designs and comment.

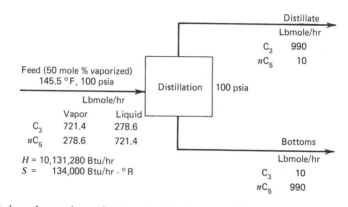

Equilibrium thermodynamic data at 100 psia are as follows.

			Enthalpy, 1000 Btu/lbmole		Entropy, 1000 Btu/lbmole · °R	
x_{C_3}	y_{C_3}	T, °F	Vapor	Liquid	Vapor	Liquid
0.9900	0.9990	59.2	5.652	−1.140	0.060	0.047
0.9500	0.9946	62.1	5.722	−1.074	0.060	0.048
0.9000	0.9885	65.9	5.815	−0.983	0.060	0.049
0.7000	0.9542	83.6	6.280	−0.515	0.062	0.054
0.5000	0.8890	107.2	7.005	0.213	0.064	0.059
0.3000	0.7456	141.0	8.337	1.414	0.069	0.064
0.1000	0.3775	192.2	11.362	3.515	0.079	0.070
0.0500	0.2116	208.6	12.691	4.258	0.083	0.072
0.0100	0.0464	223.0	14.021	4.933	0.086	0.073

17.13 Mah, Nicholas, and Wodnik [*AIChE J.*, **23**, 651–658 (1977)] consider the separation of *trans*-2-butene from *cis*-2-butene by both conventional and SRV distillation for the specifications indicated in the sketch below. Their latter distillation scheme is like that of Fig. 17.17 except that a cooler is placed directly after the valve to bring the two-phase mixture, which results from the valve pressure drop, back to its bubble point. Results of their calculations based on the use of a modified Wang–Henke method are as follows.

	Conventional Distillation	SRV Distillation
Rectifier pressure, kPa	121.6	293.8
Distillate temperature, °K	279.9	306.0
Condenser duty, W	29.392×10^6	16.905×10^6
Stripper pressure, kPa	121.6	121.6
Bottoms temperature, °K	281.5	281.5
Reboiler duty, W	30.023×10^6	19.484×10^6
Compressor duty, W	—	2.864×10^6
Valve cooler duty, W	—	4.684×10^6
Valve outlet temperature, °K	—	280.7

Assume that a heating medium is available at 331.5°K and that cooling media are available to give minimum approach temperatures of 5.6°K.

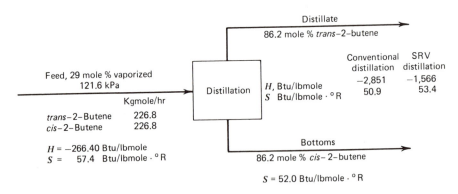

Calculate for each scheme:
(a) Change in availability function ($T_o = 299.8$°K).
(b) Net work consumption.
(c) Lost work.
(d) Thermodynamic efficiency.

17.14 Consider the hypothetical perfect separation of a mixture of ethylene and ethane into pure products by distillation as shown below. Two schemes are to be considered: conventional distillation and distillation using a heat pump with reboiler liquid flashing. In both cases the column will operate at a pressure of 200 psia, at which the average relative volatility is 1.55. A reflux ratio of 1.10 times minimum, as computed from the Underwood equation, is to be used. Other conditions for the scheme using reboiler liquid flashing are shown below. Calculate for each scheme:
(a) Change in availability function ($T_o = 100$°F).
(b) Net work consumption.

(c) Lost work.
(d) Thermodynamic efficiency.

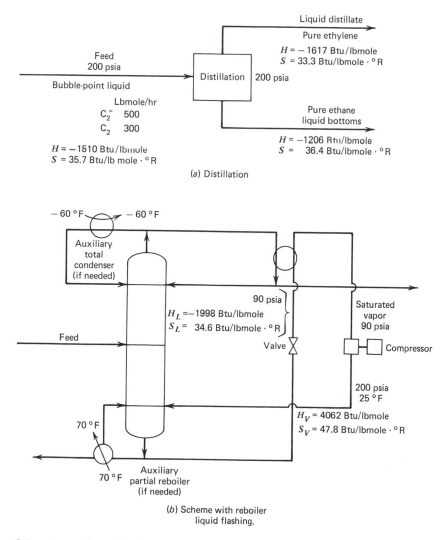

(a) Distillation

(b) Scheme with reboiler
liquid flashing.

Other thermodynamic data are:

	Latent heat of vaporization, Btu/lbmole
Ethylene at 200 psia	4348
Ethane at 200 psia	4751
Ethane at 90 psia	5473

I. Physical Property Constants and Coefficients

This appendix is a physical property data bank for 176 chemicals that are divided into inorganic and organic groups. Within each group, the succession of elements is alphabetical in the empirical formula. However, organic chemicals begin with carbon and hydrogen according to the convention in the *Handbook of Chemistry and Physics*.

In Part A, the following physical property constants are listed:

M = molecular weight.

T_b = normal boiling point, °R.

T_c = critical temperature, °R.

P_c = critical pressure, psia.

Z_c = critical compressibility factor.

ω = Pitzer's acentric factor.

δ = Hildebrand's solubility parameter at 25°C, $(\text{cal/cm}^3)^{1/2}$.

v_L = liquid molal volume at superscripted temperature (°C), cm^3/gmole; values in parentheses are for hypothetical liquid.

In Part B, physical property coefficients are listed for:

1. Ideal gas heat capacity.

$$C_{P_V}^o = a_1 + a_2 T + a_3 T^2 + a_4 T^3 + a_5 T^4, \text{ Btu/lbmole} \cdot °F,$$

where $T = °F$.

2. Antoine vapor pressure.

$$\ln \frac{P_i^s}{P_c} = A_1 - \frac{A_2}{T' + A_3}$$

where P_i^s = vapor pressure and $T' = °F$.

A value listed for example, as 0.8334 E-3 means 0.8334×10^{-3} or 0.0008334. Most of the values in Parts A and B of Appendix I were taken, with permission by the Monsanto Company, from the data records of the FLOWTRAN simulation program.*

---

* Seader, J. D., W. D. Seider, and A. C. Pauls, *FLOWTRAN Simulation—An Introduction*, 2nd ed., CACHE, 77 Massachusetts Avenue, Cambridge, MA 02139, 1977.

Part A

Number	Empirical Formula	Name	M	T_b	T_c	P_c	Z_c	ω	δ	v_L
Inorganic Chemicals										
1	Ar	Argon	39.948	157.1	271.2	705.4	0.293	−0.0034	5.330	
2	Br_2	Bromine	159.808	597.5	1051.5	1499.0	0.298	0.1242	11.442	51.2^{20}
3	CCl_4	Carbon tetrachloride	153.823	629.5	1001.5	661.3	0.277	0.1938	9.338	97.1^{25}
4	CO	Carbon monoxide	28.010	147.0	239.3	507.4	0.289	0.048	3.1300	$(35.2)^{25}$
5	$COCl_2$	Phosgene	98.916	505.3	819.3	823.0	0.279	0.203	8.4170	71.6^{20}
6	CO_2	Carbon dioxide	44.011	350.4	547.6	1070.5	0.273	0.177	7.1200	$(44.0)^{25}$
7	CS_2	Carbon disulfide	76.131	574.9	993.6	1146.3	0.285	0.115	9.8640	58.9^{0}
8	C_2OCl_4	Trichloroacetyl chloride	181.833	704.1	1061.4	594.6	0.275	0.359	12.0540	112.2^{20}
9	ClH	Hydrogen chloride	36.461	338.6	584.2	1198.5	0.267	0.133	7.0110	30.6^{-85}
10	Cl_2	Chlorine	70.906	430.4	750.9	1118.4	0.278	0.0743	8.708	45.4^{-34}
11	HI	Hydrogen iodide	127.912	428.0	761.7	1190.4	0.305	0.0290	8.270	45.6^{-36}
12	H_2	Hydrogen	2.016	36.7	59.7	190.8	0.321	0.0	0.0	$(31.0)^{25}$
13	H_2O	Water	18.015	671.7	1165.1	3206.7	0.232	0.3477	18.0	18.1^{20}
14	H_2S	Hydrogen sulfide	34.080	383.1	672.4	1306.5	0.283	0.0868	8.8	34.3^{60}
15	H_3N	Ammonia	17.031	431.5	730.2	1653.7	0.248	0.2582	12.408	26.7^{70}
16	Ne	Neon	20.183	49.1	80.1	395.3	0.306	−0.0299	0.0	
17	NO	Nitric oxide	30.006	218.5	324.0	940.5	0.267	0.5877	0.0	
18	NO_2	Nitrogen dioxide	46.006	530.1	775.8	1469.6	0.247	0.8499	16.208	31.8^{20}
19	N_2	Nitrogen	28.013	139.3	227.3	492.9	0.289	0.0206	4.440	$(53.0)^{25}$
20	N_2O	Nitrous oxide	44.013	330.7	557.5	1053.7	0.277	0.1601	5.474	35.9^{-89}
21	O_2	Oxygen	31.999	162.3	278.6	736.9	0.291	0.0250	4.0	$(28.4)^{25}$
22	O_2S	Sulfur dioxide	64.063	473.7	775.2	1144.8	0.267	0.2402	6.0	44.0^{-10}
23	O_3S	Sulfur trioxide	80.058	572.2	883.6	1196.8	0.252	0.4384	15.329	45.0^{45}
Organic Chemicals										
24	$CHCl_3$	Chloroform	119.378	602.8	965.8	793.6	0.277	0.2117	9.236	80.2^{20}
25	CHN	Hydrogen cyanide	27.026	538.0	822.0	718.6	0.172	0.3752	12.192	39.3^{20}
26	CH_2O	Formaldehyde	30.026	457.1	747.3	984.6	0.222	0.2298	10.604	36.8^{-20}
27	CH_3Cl	Methyl chloride	50.488	448.1	749.3	968.5	0.270	0.1530	8.585	55.2^{20}
28	CH_3I	Methyl iodide	141.939	568.2	950.7	1061.1	0.283	0.1925	9.863	62.3^{20}
29	CH_4	Methane	16.043	201.0	343.9	673.1	0.289	0.0	5.680	$(52.0)^{25}$
30	CH_4O	Methanol	32.042	607.8	923.7	1153.6	0.228	0.5556	14.510	40.5^{20}
31	CH_5N	Methylamine	31.058	480.1	774.1	1081.6	0.260	0.2852	10.479	44.2^{-13}
32	C_2HCl_3	Trichloroethylene	131.389	648.1	979.5	727.5	0.278	0.4281	9.263	89.9^{20}

Part A (cont'd)

Organic Chemicals

Number	Empirical Formula	Name	M	T_b	T_c	P_c	Z_c	ω	δ	V_L
33	C_2HCl_3O	Dichloroacetyl chloride	147.388	685.2	1039.1	668.5	0.271	0.3645	12.679	96.2[16]
34	C_2H_2	Acetylene	26.038	339.1	555.0	890.3	0.267	0.1917	5.329	42.3[-84]
35	$C_2H_2Cl_2O$	Chloroacetyl chloride	112.943	687.9	1054.8	740.7	0.255	0.3194	13.856	79.5[20]
36	C_2H_3Cl	Vinyl chloride	62.499	467.0	776.8	774.5	0.266	0.0929	7.717	64.5[-14]
37	C_2H_3ClO	Acetyl chloride	78.498	583.3	914.3	832.7	0.270	0.3238	12.485	71.0[20]
38	$C_2H_3Cl_3$	1,1,2-Trichloroethane	113.405	969.5	1101.9	701.2	0.267	0.2273	9.628	92.6[20]
39	C_2H_3N	Acetonitrile	41.053	638.6	986.2	701.0	0.194	0.3234	12.049	52.5[20]
40	C_2H_4	Ethylene	28.054	305.0	509.5	742.2	0.284	0.0872	5.801	(61.0)[25]
41	$C_2H_4Cl_2$	1,1-Dichloroethane	98.960	594.8	941.7	734.8	0.274	0.2450	8.913	84.7[25]
42	$C_2H_4Cl_2$	1,2-Dichloroethane	98.960	641.9	1010.8	778.9	0.267	0.3064	9.828	79.2[16]
43	C_2H_4O	Acetaldehyde	44.054	528.4	830.1	805.3	0.238	0.2882	9.844	56.6[20]
44	C_2H_4O	Ethylene oxide	44.054	510.6	842.7	1043.4	0.260	0.2121	10.271	49.0[0]
45	$C_2H_4O_2$	Acetic acid	60.052	705.0	1070.6	839.1	0.220	0.4536	10.051	57.2[20]
46	$C_2H_4O_2$	Methyl formate	60.052	548.9	876.9	870.7	0.259	0.2562	10.018	61.7[20]
47	C_2H_5Cl	Ethyl chloride	64.515	513.8	828.7	764.2	0.268	0.1918	8.471	75.1[20]
48	C_2H_6	Ethane	30.070	332.2	550.0	709.8	0.282	0.1064	6.050	(68.0)[25]
49	C_2H_6O	Dimethyl ether	46.069	447.0	720.1	764.2	0.271	0.1960	7.608	69.1[20]
50	C_2H_6O	Ethanol	46.069	632.7	929.3	925.3	0.250	0.6341	12.915	58.4[20]
51	$C_2H_6O_2$	Ethylene glycol	62.069	847.1	1161.4	1091.9	0.242	0.1177	16.604	55.7[20]
52	C_2H_6S	Dimethyl sulfide	62.130	558.9	905.5	802.4	0.268	0.1951	9.045	73.3[20]
53	C_2H_6S	Ethyl mercaptan	62.130	554.7	898.5	796.5	0.271	0.1856	8.933	74.1[20]
54	C_2H_7N	Ethylamine	45.085	521.5	821.1	815.6	0.264	0.2861	9.427	66.0[20]
55	C_3H_3N	Acrylonitrile	53.064	630.8	934.5	512.9	0.186	0.3853	11.029	65.8[20]
56	C_3H_4	Methylacetylene	40.065	449.9	724.3	816.2	0.271	0.2150	8.010	56.7[-50]
57	C_3H_4	Propadiene	40.065	429.6	721.7	747.2	0.284	0.0631	6.854	61.6[25]
58	C_3H_6	Propylene	42.081	405.8	657.2	667.0	0.279	0.1421	6.208	79.0[25]
59	C_3H_6O	Acetone	58.080	592.7	917.0	693.7	0.247	0.3035	9.566	73.5[20]
60	$C_3H_6O_2$	Ethyl formate	74.080	589.5	915.3	680.4	0.258	0.2784	9.311	79.9[16]
61	$C_3H_6O_2$	Methyl acetate	74.080	594.7	912.4	680.9	0.256	0.3269	9.014	79.3[20]
62	$C_3H_6O_2$	Propionic acid	74.080	745.5	1102.8	778.9	0.249	0.5322	12.385	74.6[20]
63	C_3H_7NO	Dimethylformamide	73.095	767.1	1074.4	683.1	0.236	0.7458	11.775	77.0[20]
64	C_3H_8	Propane	44.097	416.0	665.9	617.4	0.278	0.1538	6.400	84.0[25]
65	C_3H_8O	Isopropanol	60.096	639.8	915.0	691.0	0.249	0.6614	11.572	76.5[20]
66	C_3H_8O	n-Propanol	60.096	666.7	966.4	737.1	0.250	0.6111	12.050	74.7[20]

No.	Formula	Name							
67	C$_3$H$_9$N	Trimethylamine	59.112	496.9	779.9	590.8	0.272	0.2008	7.070	93.4[20]
68	C$_4$H$_4$	Vinylacetylene	52.076	501.7	821.3	704.8	0.264	0.0970	10.229	73.3[0]
69	C$_4$H$_4$S	Thiophene	84.136	643.3	1062.3	705.4	0.246	0.0670	9.654	78.6[16]
70	C$_4$H$_5$N	Methacrylonitrile	67.091	654.2	998.2	563.3	0.228	0.2823	8.576	83.3[20]
71	C$_4$H$_6$	Dimethylacetylene	54.092	540.4	879.7	737.4	0.270	0.1359	7.937	78.3[20]
72	C$_4$H$_6$	Ethylacetylene	54.092	506.3	834.7	683.2	0.269	0.0610	7.937	83.2[16]
73	C$_4$H$_6$	1,2-Butadiene	54.092	511.2	834.7	578.1	0.276	0.0987	7.950	83.7[25]
74	C$_4$H$_6$	1,3-Butadiene	54.092	483.8	765.0	628.0	0.272	0.2028	6.940	88.0[25]
75	C$_4$H$_8$	1-Butene	56.108	480.2	755.3	583.0	0.274	0.2085	6.766	95.6[25]
76	C$_4$H$_8$	cis-2-Butene	56.108	498.3	779.7	610.0	0.272	0.2575	6.760	91.2[25]
77	C$_4$H$_8$	Isobutene	56.108	479.3	752.2	580.0	0.274	0.1975	6.760	95.4[25]
78	C$_4$H$_8$	trans-2-Butene	56.108	493.3	770.7	595.0	0.273	0.2230	6.760	93.8[25]
79	C$_4$H$_8$O	Isobutyraldehyde	72.107	606.9	909.7	609.0	0.261	0.3800	9.199	91.4[20]
80	C$_4$H$_8$O	Methyl ethyl ketone	72.107	635.0	964.2	603.0	0.251	0.3188	9.199	89.6[20]
81	C$_4$H$_8$O$_2$	n-Butyric acid	88.107	785.6	1130.7	648.1	0.249	0.6030	11.861	92.0[20]
82	C$_4$H$_8$O$_2$	Ethyl acetate	88.107	630.5	941.9	556.0	0.254	0.3718	8.974	97.8[20]
83	C$_4$H$_8$O$_2$	Methyl propionate	88.107	635.6	955.1	578.0	0.256	0.3500	9.046	96.3[20]
84	C$_4$H$_8$O$_2$	Propyl formate	88.107	638.1	968.6	589.3	0.260	0.3154	9.024	96.7[16]
85	C$_4$H$_9$NO	Dimethyl acetamide	87.120	792.3	1182.0	583.7	0.230	0.3762	10.788	93.0[25]
86	C$_4$H$_{10}$	Isobutane	58.124	470.6	734.7	529.1	0.276	0.1825	6.730	105.5[25]
87	C$_4$H$_{10}$	n-Butane	58.124	490.8	765.3	550.7	0.274	0.1954	6.634	101.4[25]
88	C$_4$H$_{10}$O	Isobutanol	74.123	686.9	985.9	623.0	0.256	0.5917	10.949	92.4[20]
89	C$_4$H$_{10}$O	n-Butanol	74.123	703.6	1013.2	640.5	0.256	0.5903	11.440	91.5[20]
90	C$_4$H$_{10}$O	t-Butyl alcohol	74.123	640.0	912.0	576.1	0.255	0.6071	10.316	94.2[20]
91	C$_4$H$_{10}$O	Diethyl ether	74.123	553.9	840.2	523.2	0.264	0.2800	7.544	104.0[20]
92	C$_4$H$_{10}$O$_3$	Diethylene glycol	106.122	933.8	1225.9	668.0	0.244	1.1747	13.551	95.1[20]
93	C$_5$H$_4$O$_2$	Furfural	96.085	782.8	1182.8	714.2	0.243	0.4239	11.986	82.8[20]
94	C$_5$H$_{10}$	2-Methyl-1-butene	70.135	547.8	850.0	514.4	0.274	0.2000	7.055	108.7[25]
95	C$_5$H$_{10}$	2-Methyl-2-butene	70.135	561.1	870.0	527.6	0.273	0.2120	7.055	106.7[25]
96	C$_5$H$_{10}$	3-Methyl-1-butene	70.135	527.8	831.0	507.0	0.278	0.1490	7.055	112.8[25]
97	C$_5$H$_{10}$	Cyclopentane	70.135	580.4	921.2	655.0	0.274	0.1966	8.010	94.7[25]
98	C$_5$H$_{10}$	1-Pentene	70.135	545.6	853.0	586.0	0.273	0.2198	7.055	110.4[25]
99	C$_5$H$_{10}$	cis-2-Pentene	70.135	558.2	860.6	512.0	0.272	0.2060	7.055	107.8[25]
100	C$_5$H$_{10}$	trans-2-Pentene	70.135	557.1	857.0	508.4	0.272	0.2090	7.055	109.0[25]
101	C$_5$H$_{10}$O	Diethyl ketone	86.134	674.9	1009.7	542.3	0.256	0.3407	8.898	105.8[20]
102	C$_5$H$_{10}$O$_2$	n-Propyl acetate	102.134	674.5	988.9	483.6	0.253	0.3936	8.729	115.1[20]
103	C$_5$H$_{12}$	Isopentane	72.151	541.8	829.8	483.0	0.270	0.2104	7.020	117.4[25]
104	C$_5$H$_{12}$	n-Pentane	72.151	556.6	845.6	489.5	0.269	0.2387	7.020	116.1[25]
105	C$_5$H$_{12}$	Neopentane	72.151	508.8	780.8	464.0	0.276	0.1950	7.020	123.3[25]
106	C$_6$H$_3$Cl$_3$	1,2,4-Trichlorobenzene	181.449	876.0	1322.9	578.2	0.262	0.3358	9.956	124.8[20]

Number	Empirical Formula	Name	M	T_b	T_c	P_c	Z_c	ω	δ	V_L
Organic Chemicals										
107	$C_6H_4Cl_2$	m-Dichlorobenzene	147.004	803.1	1231.1	562.9	0.252	0.3073	9.554	114.1[20]
108	$C_6H_4Cl_2$	o-Dichlorobenzene	147.004	813.9	1255.1	595.3	0.256	0.2720	9.815	112.6[20]
109	$C_6H_4Cl_2$	p-Dichlorobenzene	147.004	804.7	1232.6	566.5	0.253	0.2822	9.645	117.8[55]
110	C_6H_5Br	Bromobenzene	157.010	772.9	1206.3	655.4	0.245	0.2508	9.753	105.0[20]
111	C_6H_5Cl	Chlorobenzene	112.559	729.7	1138.1	656.0	0.266	0.2545	9.623	101.8[20]
112	C_6H_5I	Iodobenzene	204.011	830.7	1298.1	655.8	0.266	0.2470	9.782	110.0[4]
113	C_6H_6	Benzene	78.114	635.9	1012.7	714.2	0.272	0.2116	9.158	89.4[25]
114	C_6H_6O	Phenol	94.113	819.0	1251.1	889.1	0.279	0.4201	12.106	88.9[40]
115	C_6H_7N	Aniline	93.129	823.1	1257.8	768.6	0.261	0.3830	11.461	91.1[20]
116	C_6H_{12}	Cyclohexane	84.162	637.0	995.3	591.5	0.272	0.2149	8.193	108.7[25]
117	C_6H_{12}	Methylcyclopentane	84.162	621.0	959.0	549.0	0.271	0.2316	7.847	113.1[25]
118	C_6H_{12}	1-Hexene	84.162	606.0	920.0	471.7	0.269	0.2463	7.400	125.8[25]
119	C_6H_{14}	2,2-Dimethylbutane	86.178	581.2	880.9	450.5	0.275	0.2312	6.712	122.7[25]
120	C_6H_{14}	2,3-Dimethylbutane	86.178	596.1	900.5	455.4	0.272	0.2447	6.967	131.2[25]
121	C_6H_{14}	n-Hexane	86.178	615.4	914.2	440.0	0.266	0.2972	7.266	131.6[25]
122	C_6H_{14}	2-Methylpentane	86.178	600.2	896.5	440.1	0.269	0.2771	7.018	132.0[25]
123	C_6H_{14}	3-Methylpentane	86.178	605.6	907.8	453.1	0.270	0.2745	7.132	129.8[25]
124	$C_6H_{14}O_4$	Triethylene glycol	150.176	1008.7	1282.2	481.0	0.243	1.2715	12.677	133.2[15]
125	C_7H_8	Toluene	92.141	690.8	1069.1	587.8	0.263	0.2415	8.914	106.8[25]
126	C_7H_8O	o-Cresol	108.140	835.5	1255.6	726.0	0.272	0.4299	11.139	105.2[40]
127	C_7H_{14}	Methylcyclohexane	98.189	673.4	1030.2	504.4	0.271	0.2362	7.825	128.3[25]
128	C_7H_{14}	Ethylcyclopentane	98.189	677.9	1025.0	492.8	0.268	0.2712	7.739	128.8[25]
129	C_7H_{14}	1-Heptene	98.189	660.3	963.9	412.2	0.262	0.3467	7.168	140.9[20]
130	C_7H_{16}	n-Heptane	100.205	668.9	972.3	396.9	0.261	0.3403	7.430	147.5[25]
131	C_8H_8	Styrene	104.152	752.9	1146.4	559.0	0.261	0.2885	9.211	115.0[20]
132	C_8H_{10}	Ethylbenzene	106.168	736.8	1115.5	540.0	0.265	0.2981	8.783	123.1[25]
133	C_8H_{10}	m-Xylene	106.168	742.1	1114.6	510.0	0.264	0.3086	8.818	123.5[25]
134	C_8H_{10}	o-Xylene	106.168	751.6	1138.0	530.0	0.266	0.2904	8.987	121.2[25]
135	C_8H_{10}	p-Xylene	106.168	740.7	1112.8	500.0	0.265	0.3304	8.769	124.0[25]
136	C_8H_{16}	Ethylcyclohexane	112.216	728.9	1084.7	453.9	0.265	0.3041	7.739	143.1[25]
137	C_8H_{16}	n-Propylcyclopentane	112.216	727.4	1062.5	406.5	0.253	0.3386	7.894	143.7[16]
138	C_8H_{18}	n-Octane	114.232	717.9	1024.9	362.1	0.258	0.3992	7.551	163.5[25]
139	$C_8H_{18}O_5$	Tetraethylene glycol	194.229	1065.8	1432.4	304.4	0.205	0.8162	12.113	172.1[15]
140	C_9H_8	Indene	116.163	819.2	1245.5	553.6	0.250	0.3064	9.647	116.6[20]

No.	Formula	Name								
141	C_9H_{10}	Indan	118.179	810.3	1225.9	526.6	0.251	0.2912	9.334	122.6[20]
142	C_9H_{10}	Methylstyrene	118.179	797.7	1192.6	500.0	0.255	0.3191	9.002	129.7[20]
143	C_9H_{12}	1-Ethyl-2-methylbenzene	120.195	788.7	1172.0	441.0	0.247	0.2970	8.839	136.4[20]
144	C_9H_{12}	n-Propylbenzene	120.195	778.3	1149.0	464.1	0.261	0.3446	8.661	139.4[20]
145	C_9H_{18}	n-Propylcyclohexane	126.243	773.8	1114.5	369.2	0.248	0.3617	7.886	159.2[20]
146	C_9H_{20}	n-Nonane	128.259	763.1	1071.0	331.0	0.254	0.4439	7.649	179.6[25]
147	$C_{10}H_8$	Naphthalene	128.174	884.0	1347.0	576.1	0.258	0.2934	9.738	132.0[90]
148	$C_{10}H_{10}$	1-Methylindene	130.190	851.7	1266.2	483.0	0.247	0.3291	9.323	144.1[20]
149	$C_{10}H_{10}$	2-Methylindene	130.190	866.1	1286.2	486.5	0.246	0.3367	9.485	142.1[35]
150	$C_{10}H_{12}$	Dicyclopentadiene	132.206	797.7	1188.7	773.9	0.254	0.2767	8.398	156.1[20]
151	$C_{10}H_{14}$	n-Butylbenzene	134.222	821.6	1188.8	418.7	0.258	0.3929	8.425	175.5[20]
152	$C_{10}H_{14}$	1,2-Dimethyl-3-ethylbenzene	134.222	840.8	1224.1	453.6	0.262	0.3968	8.916	
153	$C_{10}H_{20}$	n-Butylcyclohexane	140.27	817.4	1162.4	353.9	0.252	0.4035	7.90	
154	$C_{10}H_{22}$	n-Decane	142.286	805.1	1114.0	306.0	0.251	0.4869	7.722	196.0[25]
155	$C_{11}H_{10}$	1-Methylnaphthalene	142.201	932.0	1384.5	517.6	0.254	0.3607	9.770	139.4[20]
156	$C_{11}H_{10}$	2-Methylnaphthalene	142.201	925.6	1371.4	508.1	0.256	0.3647	9.660	
157	$C_{11}H_{24}$	n-Undecane	156.313	844.3	1152.0	282.0	0.248	0.521	7.790	212.2[25]
158	$C_{12}H_8$	Acenaphthalene	152.196	977.7	1434.5	467.2	0.237	0.3733	10.018	169.3[16]
159	$C_{12}H_{10}$	Diphenyl	154.212	951.1	1420.0	557.0	0.276	0.3638	9.891	155.8[74]
160	$C_{12}H_{12}$	2,7-Dimethylnaphthalene	156.228	965.1	1400.7	467.4	0.257	0.4232	9.760	
161	$C_{12}H_{14}$	1,2,3-Trimethylindene	158.244	909.3	1296.4	384.1	0.242	0.4271	8.955	
162	$C_{12}H_{26}$	n-Dodecane	170.328	881.0	1188.3	261.6	0.245	0.561	7.840	228.6[25]
163	$C_{13}H_{10}$	Fluorene	166.223	1027.9	1480.1	434.2	0.234	0.4512	10.136	
164	$C_{13}H_{14}$	1-Methylethylnaphthalene	170.255	986.7	1393.4	408.6	0.233	0.5044	10.03	
165	$C_{13}H_{14}$	2,3,5-Trimethylnaphthalene	170.255	1004.7	1418.9	408.5	0.232	0.5044	10.121	
166	$C_{13}H_{28}$	n-Tridecane	184.367	915.5	1219.0	250.0	0.242	0.6002	7.890	244.9[25]
167	$C_{14}H_{10}$	Phenanthrene	178.234	1103.0	1581.8	420.4	0.228	0.4396	10.524	
168	$C_{14}H_{30}$	n-Tetradecane	198.394	948.1	1251.0	230.0	0.240	0.640	7.920	261.3[25]
169	$C_{15}H_{12}$	1-Phenylindene	192.261	1071.3	1518.6	391.0	0.230	0.4644	9.933	
170	$C_{15}H_{14}$	2-Ethylfluorene	194.277	1047.9	1459.9	357.5	0.230	0.5175	9.636	
171	$C_{15}H_{32}$	n-Pentadecane	212.421	978.8	1278.0	220.0	0.237	0.6743	7.960	277.8[25]
172	$C_{16}H_{10}$	Fluoranthene	202.256	1199.1	1685.9	378.1	0.221	0.4930	10.426	
173	$C_{16}H_{10}$	Pyrene	202.256	1143.3	1605.7	378.1	0.224	0.4988	10.222	
174	$C_{16}H_{12}$	1-Phenylnaphthalene	204.272	1076.7	1512.2	381.8	0.228	0.5034	10.283	
175	$C_{16}H_{34}$	n-Hexadecane	226.448	1007.9	1305.0	206.0	0.236	0.7078	7.990	294.1[25]
176	$C_{18}H_{12}$	Chrysene	228.294	1298.1	1788.5	346.4	0.213	0.5676	10.691	

Number	Empirical Formula	Name	a_1	a_2	a_3	a_4	a_5	A_1	A_2	A_3
Inorganic Chemicals										
1	Ar	Argon	4.9647	0.0	0.0	0.0	0.0	5.42578	1499.0	463.9195
2	Br_2	Bromine	8.552	0.8334 E-3	-0.5053 E-06	0.1098 E-09	0.0	5.53786	5563.552	409.318
3	CCl_4	Carbon tetrachloride	18.36077	0.1932811 E-01	-0.256049 E-04	0.173446 E-07	-0.4549234 E-11	6.199663	6110.034	441.2806
4	CO	Carbon monoxide	6.956012	0.591124 E-04	0.5075809 E-06	0.7641183 E-09	-0.6540363 E-12	5.712089	1385.883	462.6165
5	$COCl_2$	Phosgene	13.76767	0.1093685 E-01	-0.1226482 E-04	0.7900403 E-08	-0.2079814 E-11	5.870425	4545.777	412.5742
6	CO_2	Carbon dioxide	8.398605	-0.6475766 E-02	-0.3555025 E-05	0.1194595 E-08	-0.1851702 E-12	6.470232	3521.259	455.869
7	CS_2	Carbon disulfide	11.63044	0.123864 E-01	0.4778188 E-05	-0.8568489 E-08	0.2966357 E-11	6.263774	6389.82	486.9933
8	C_2OCl_4	Trichloroacetyl chloride	23.70705	0.2731643 E-01	-0.3266946 E-04	0.2107581 E-07	-0.5450088 E-11	5.752994	5751.971	363.6816
9	ClH	Hydrogen chloride	6.969	0.2236 E-03	0.7333 E-06	-0.1776 E-09	0.0	5.688540	3244.028	442.2680
10	Cl_2	Chlorine	7.973	0.18901 E-02	-0.12101 E-05	0.26526 E-09	0.0	5.186832	3699.272	417.709
11	HI	Hydrogen iodide	6.948176	0.145827 E-05	0.1104365 E-05	-0.5203962 E-09	0.7254255 E-13	5.362089	3840.331	424.3549
12	H_2	Hydrogen	6.647816	0.2472647 E-02	-0.4557635 E-05	0.3117701 E-08	-0.6643678 E-12	5.602657	418.1773	474.214
13	H_2O	Water	7.985742	0.4631191 E-03	0.1402841 E-05	-0.6578387 E-09	0.9795288 E-13	6.53247	7173.79	389.4747
14	H_2S	Hydrogen sulfide	8.031194	0.9868632 E-03	0.2388543 E-05	-0.159311 E-08	0.320329 E-12	5.445487	3535.867	432.235
15	H_3N	Ammonia	8.2765	0.39006 E-02	0.35245 E-06	-0.27402 E-09	0.0	6.152480	4253.826	418.9528
16	Ne	Neon	4.9647	0.0	0.0	0.0	0.0	5.560555	476.5174	464.7596
17	NO	Nitric oxide	7.255110	0.4551648 E-03	0.2640019 E-06	-0.1225439 E-09	0.1332261 E-13	8.142411	2374.39	434.2132
18	NO_2	Nitrogen dioxide	8.495494	0.4921745 E-02	-0.1239651 E-05	-0.2704296 E-09	0.1240708 E-12	12.76017	11907.65	616.0504
19	N_2	Nitrogen	6.947158	0.6609477 E-04	0.5693395 E-06	0.3226862 E-10	-0.9683259 E-13	5.316656	1184.797	454.5328
20	N_2O	Nitrous oxide	8.767171	0.6369103 E-02	-0.3621806 E-05	0.1572094 E-08	-0.2566121 E-12	5.353828	2285.606	416.6338
21	O_2	Oxygen	6.986501	0.5581101 E-03	0.1399925 E-05	-0.1093827 E-08	0.2299662 E-12	5.15038	1404.466	452.1536
22	O_2S	Sulfur dioxide	9.134	0.532 E-02	-0.2323 E-05	0.3527 E-09	0.0	5.966623	4293.005	401.685
23	O_3S	Sulfur trioxide	10.964	0.1251 E-01	-0.6523 E-05	0.1328 E-08	0.0	4.828507	3199.785	234.315
Organic Chemicals										
24	$CHCl_3$	Chloroform	14.84	0.1245 E-01	-0.6495 E-05	0.1259 E-08	0.0	5.894057	5289.385	393.075
25	CHN	Hydrogen cyanide	8.194594	0.526853 E-02	-0.3123949 E-05	0.1256972 E-08	-0.2109817 E-12	6.97745	5672.753	443.761
26	CH_2O	Formaldehyde	8.209533	0.1838523 E-02	0.1697531 E-05	-0.1107293 E-08	0.2015218 E-12	5.525307	3890.399	402.114
27	CH_3Cl	Methyl chloride	8.964398	0.1006818 E-01	-0.8251344 E-07	-0.2858358 E-08	0.1064404 E-11	5.504941	3906.036	414.1839
28	CH_3I	Methyl iodide	9.71643	0.1105462 E-01	-0.3302857 E-05	-0.173936 E-09	0.3161217 E-12	5.235173	4840.0	399.9808
29	CH_4	Methane	8.245223	0.3806333 E-02	0.8864745 E-05	-0.7461153 E-08	0.1822959 E-11	5.14135	1742.638	452.974
30	CH_4O	Methanol	9.801084	0.8430642 E-02	0.6669185 E-05	-0.8208981 E-08	0.2500638 E-11	7.513334	6468.101	396.2652
31	CH_5N	Methylamine	11.192	0.1571 E-01	-0.43782 E-05	0.50592 E-09	0.0	5.973955	4216.777	389.9955
32	C_2HCl_3	Trichloroethylene	17.8198	0.1891162 E-01	-0.1588157 E-04	0.7817208 E-08	-0.1645299 E-11	5.814331	5636.178	391.6901
33	C_2HCl_3O	Dichloroacetyl chloride	21.08973	0.2396316 E-01	-0.2415185 E-04	0.1440388 E-07	-0.3583915 E-11	6.437229	6108.109	369.1736
34	C_2H_2	Acetylene	9.89	0.8273 E-02	-0.3783 E-05	0.7457 E-09	0.0	6.109766	3305.991	444.4562
35	$C_2H_3Cl_2O$	Chloroacetyl chloride	18.47242	0.2060988 E-02	-0.1563425 E-04	0.7731953 E-08	-0.1717742 E-11	5.875686	5522.555	341.0137
36	C_2H_3Cl	Vinyl chloride	11.52336	0.1776248 E-01	-0.9084331 E-05	0.2459001 E-08	-0.1792789 E-12	5.033325	3650.037	399.3080
37	C_2H_3ClO	Acetyl chloride	15.8551	0.1725661 E-01	-0.7116638 E-05	0.1060025 E-08	0.1484303 E-12	5.090718	4413.716	360.1358

No.	Formula	Name	A	B	C	D	E		
38	$C_2H_3Cl_3$	1,1,1-Trichloroethane	19.37069	0.2598217 E-01	-0.1790523 E-04	0.7161791 E-08	-0.1193551 E-11	6.877873	411.7562
39	C_2H_3N	Acetonitrile	11.5361	0.1232096 E-01	-0.1200452 E-05	-0.2031388 E-08	0.7743102 E-12	6.393303	411.6977
40	C_2H_4	Ethylene	9.326018	0.1393934 E-01	0.1010831 E-05	-0.7516552 E-08	0.3615367 E-11	5.27791	433.9156
41	$C_2H_4Cl_2$	1,1-Dichloroethane	16.548	0.22237 E-01	-0.92549 E-05	0.15584 E-08	0.0	5.34928	371.5705
42	$C_2H_4Cl_2$	1,2-Dichloroethane	17.38636	0.1985922 E-01	-0.8175161 E-05	0.1405784 E-08	0.9333614 E-13	5.768529	383.4458
43	C_2H_4O	Acetaldehyde	11.90924	0.1481123 E-01	0.9439146 E-06	-0.501431 E-08	0.172932 E-11	6.49856	419.9494
44	C_2H_4O	Ethylene oxide	10.14899	0.1762188 E-01	0.4347253 E-05	-0.1164914 E-07	0.4326094 E-11	5.630604	399.9853
45	$C_2H_4O_2$	Acetic acid	14.63924	0.229877 E-01	-0.1021997 E-04	0.2589452 E-08	-0.2804407 E-12	7.203594	410.1814
46	$C_2H_4O_2$	Methyl formate	14.31964	0.2105721 E-01	-0.3231842 E-05	-0.5463847 E-08	0.2608819 E-11	4.906897	396.03
47	C_2H_5Cl	Ethyl chloride	13.436	0.21267 E-01	-0.64893 E-05	0.6832 E-09	0.0	5.696013	404.411
48	C_2H_6	Ethane	11.51606	0.140309 E-01	0.854034 E-05	-0.1106078 E-07	0.3162199 E-11	5.383894	434.898
49	C_2H_6O	Dimethyl ether	15.91995	0.1599677 E-01	0.7899362 E-05	-0.1293051 E-07	0.439304 E-11	5.77524	415.2581
50	C_2H_6O	Ethanol	14.04853	0.2153149 E-01	-0.2153442 E-05	-0.4607259 E-38	0.1893692 E-11	7.43437	359.3826
51	$C_2H_6O_2$	Ethylene glycol	18.11978	0.2404298 E-01	0.1265575 E-05	-0.1072173 E-17	0.4248091 E-11	7.258288	311.8854
52	C_2H_6S	Dimethyl sulfide	16.23989	0.1919445 E-01	-0.3534516 E-06	-0.5604401 E-08	0.2328362 E-11	5.610757	396.909
53	C_2H_6S	Ethyl mercaptan	15.73943	0.2175969 E-01	-0.4178643 E-05	-0.2550827 E-08	0.1279568 E-11	5.631324	394.2954
54	C_2H_7N	Ethylamine	14.61884	0.2301337 E-01	0.5042536 E-05	-0.1298503 E-07	0.470142 E-11	5.841118	377.3129
55	C_3H_3N	Acrylonitrile	13.77061	0.1984967 E-01	-0.8937427 E-05	0.193298 E-08	-0.2797383 E-13	6.038654	396.7914
56	C_3H_4	Methylacetylene	13.17061	0.1775777 E-01	-0.7296423 E-05	0.1921976 E-08	-0.242782 E-12	5.179851	387.1001
57	C_3H_4	Propadiene	12.6505	0.1928835 E-01	-0.6452827 E-05	-0.1674653 E-08	0.1813724 E-11	2.443058	317.5695
58	C_3H_6	Propylene	16.13621	0.2106998 E-01	0.249845 E-05	-0.1146863 E-07	0.5247386 E-11	5.44467	418.4319
59	C_3H_6O	Acetone	20.94868	0.2340064 E-01	-0.1479392 E-05	-0.4143552 E-08	0.1323724 E-11	6.244412	397.5290
60	$C_3H_6O_2$	Ethyl formate	21.04651	0.273312 E-01	-0.146055 E-05	-0.8906131 E-08	0.3692011 E-11	5.94826	377.2637
61	$C_3H_6O_2$	Methyl acetate	20.17881	0.2395065 E-01	0.154732 E-05	-0.9770737 E-08	0.3676818 E-11	6.253272	385.1996
62	$C_3H_6O_2$	Propionic acid	20.8472	0.2781613 E-01	-0.2700043 E-05	-0.7554358 E-09	0.323127 E-11	6.838962	340.8683
63	C_3H_7NO	Dimethylformamide	15.58683	0.3068996 E-01	-0.4238415 E-05	-0.6471348 E-03	0.287579 E-11	5.298043	313.254
64	C_3H_8	Propane	18.703	0.2504953 E-01	0.1404258 E-04	-0.3526261 E-07	0.1864467 E-10	5.353418	414.488
65	C_3H_8O	Isopropanol	18.71145	0.3374798 E-01	-0.8097677 E-05	-0.4524869 E-08	0.2586334 E-11	7.180215	327.2873
66	C_3H_8O	n-Propanol	19.09535	0.2788756 E-01	0.8585266 E-06	-0.9785116 E-08	0.3679178 E-11	6.683944	303.9864
67	C_3H_9N	Trimethylamine		0.3856445 E-01	-0.136958 E-04	0.2035707 E-08	-0.1282133 E-13	5.552747	398.7676
68	C_4H_4	Vinylacetylene	5.297392	0.171349 E-01	-0.11425 E-05	-0.3775036 E-08	0.1440128 E-11	5.297392	381.4085
69	C_4H_4S	Thiophene	19.23704	0.3754804 E-01	-0.2222312 E-04	0.6967862 E-09	-0.709689 E-12	5.520119	366.4536
70	C_4H_5N	Methacrylonitrile	16.76823	0.2314625 E-01	-0.2793527 E-05	-0.45453 E-08	0.1908719 E-11	5.824528	366.5642
71	C_4H_6	Dimethylacetylene	17.38834	0.2517102 E-01	-0.8454522 E-05	0.2326427 E-08	-0.4824367 E-12	5.637212	388.6972
72	C_4H_6	Ethylacetylene	17.15982	0.2778978 E-01	-0.116434 E-04	0.3040097 E-08	-0.3742009 E-12	5.606554	387.8234
73	C_4H_6	1,2-Butadiene	16.47292	0.26213 E-01	-0.5937819 E-05	-0.2361299 E-08	0.1011447 E-11	6.184923	419.7079
74	C_4H_6	1,3-Butadiene	17.96141	0.3392027 E-01	-0.1392511 E-04	-0.4786315 E-08	0.5611304 E-11	5.69864	409.9979
75	C_4H_8	1-Butene	16.54537	0.3297022 E-01	-0.605339 E-05	-0.569809 E-08	0.2826942 E-11	5.58272	404.741
76	C_4H_8	cis-2-Butene	18.93086	0.2966393 E-01	0.5471621 E-05	-0.6036954 E-08	0.8437562 E-11	5.456285	394.4956
77	C_4H_8	Isobutene	18.44267	0.3101008 E-01	-0.4930015 E-05	-0.1337556 E-07	0.3200886 E-11	5.616762	405.9166
78	C_4H_8	trans-2-Butene	20.75125	0.2761259 E-01	0.3097447 E-05	-0.1104319 E-07	0.5631517 E-11	5.487073	400.0703
79	C_4H_8O	Isobutyraldehyde	22.30644	0.3284207 E-01	-0.3152369 E-06		0.4330567 E-11	6.508438	379.294
80	C_4H_8O	Methyl ethyl ketone		0.2959044 E-01	-0.3506391 E-06	-0.8248579 E-08	0.3030054 E-11	5.885353	356.3804
81	$C_4H_8O_2$	n-Butyric acid	23.98619	0.3727046 E-01	-0.3625636 E-05	-0.1019473 E-07	0.4365542 E-11	7.444973	330.4845
82	$C_4H_8O_2$	Ethyl acetate	24.90819	0.3329732 E-01	0.7316711 E-06	-0.1247032 E-07	0.4824152 E-11	6.3307	373.48

Number	Empirical Formula	Name	a_1	a_2	a_3	a_4	a_5	A_1	A_2	A_3
Organic Chemicals										
83	$C_4H_8O_2$	Methyl propionate	24.90819	0.3329732 E-01	0.7316711 E-06	-0.1247032 E-07	0.4824152 E-11	6.280916	5462.483	372.7137
84	$C_4H_8O_2$	Propyl formate	24.90819	0.3329732 E-01	0.7316711 E-06	-0.1247032 E-07	0.4824152 E-11	5.956599	5225.881	363.3314
85	C_4H_9NO	Dimethyl acetamide	24.63017	0.4015238 E-01	-0.5139106 E-05	-0.913829 E-08	0.4017948 E-11	4.812879	5194.59	279.0869
86	C_4H_{10}	Isobutane	20.41853	0.3462286 E-01	0.1415619 E-04	-0.4246126 E-07	0.2296993 E-10	5.611805	3870.419	409.949
87	C_4H_{10}	n-Butane	20.79783	0.3143287 E-01	0.1928511 E-04	-0.4588652 E-07	0.2380972 E-10	5.741624	4126.385	409.5179
88	$C_4H_{10}O$	Isobutanol	17.908	0.3086043 E-01	0.2615677 E-04	-0.3727992 E-07	0.1247926 E-10	7.134107	5843.713	310.811
89	$C_4H_{10}O$	n-Butanol	22.86768	0.4421951 E-01	-0.1487031 E-04	-0.4480266 E-09	0.139999 E-11	6.303186	5225.324	274.4291
90	$C_4H_{10}O$	t-Butyl alcohol	23.80689	0.4378233 E-01	-0.1467142 E-04	-0.8051186 E-09	0.1541653 E-11	6.180797	4482.792	274.7841
91	$C_4H_{10}O$	Diethyl ether	23.43495	0.3456899 E-01	0.6985382 E-05	-0.1902477 E-07	0.6906888 E-11	5.976844	4614.723	388.8028
92	$C_4H_{10}O_3$	Diethylene glycol	26.63047	0.4356575 E-01	0.6483778 E-05	-0.2359434 E-07	0.8904581 E-11	9.238843	10861.34	363.888
93	$C_5H_4O_2$	Furfural	20.38649	0.3085299 E-01	-0.4999507 E-05	-0.7442546 E-08	0.3466989 E-11	8.278279	9658.592	469.7835
94	C_5H_{10}	2-Methyl-1-butene	23.51432	0.416842 E-01	-0.6797959 E-05	-0.1216479 E-07	0.7590583 E-11	5.655016	4383.237	387.8014
95	C_5H_{10}	2-Methyl-2-butene	22.08508	0.3993158 E-01	-0.3039504 E-05	-0.1105363 E-07	0.511218 E-11	5.701977	4531.679	386.7617
96	C_5H_{10}	3-Methyl-1-butene	24.98621	0.4546002 E-01	-0.2709002 E-04	0.1719799 E-07	-0.6539064 E-11	5.554918	4210.159	394.7537
97	C_5H_{10}	Cyclopentane	16.21714	0.4643893 E-01	0.5947453 E-05	-0.2901838 E-07	0.1338546 E-10	5.429031	4662.062	384.6612
98	C_5H_{10}	1-Pentene	23.09041	0.4065747 E-01	-0.7117898 E-05	-0.5289083 E-08	0.1708107 E-11	5.452355	4336.393	388.5999
99	C_5H_{10}	cis-2-Pentene	21.12406	0.4169782 E-01	-0.3053253 E-05	-0.1195115 E-07	0.4474386 E-11	5.625241	4412.304	382.3599
100	C_5H_{10}	trans-2-Pentene	22.95406	0.3853433 E-01	-0.1306019 E-05	-0.1368152 E-07	0.6325655 E-11	5.722722	4491.719	387.2968
101	$C_5H_{10}O$	Diethyl ketone	24.50798	0.4231144 E-01	-0.117334 E-05	-0.1375883 E-07	0.5488549 E-11	6.159896	5631.04	361.7281
102	$C_5H_{10}O_2$	n-Propyl acetate	28.73052	0.4269 E-01	-0.9989801 E-07	-0.1517592 E-07	0.5975479 E-11	6.302585	5596.313	355.8455
103	C_5H_{12}	Isopentane	24.94637	0.4446726 E-01	0.7054883 E-05	-0.3344167 E-07	0.1774503 E-10	5.49978	4221.154	387.287
104	C_5H_{12}	n-Pentane	25.64627	0.389176 E-01	0.2397294 E-04	-0.5842615 E-07	0.3079918 E-10	5.853654	4598.287	394.4148
105	C_5H_{12}	Neopentane	25.46761	0.461177 E-01	0.1147232 E-04	-0.4605143 E-07	0.2506343 E-10	5.692011	4148.025	404.42
106	$C_6H_3Cl_3$	1,2,4-Trichlorobenzene	25.75694	0.4831912 E-01	-0.3089766 E-05	0.1057239 E-07	-0.1390436 E-11	6.592514	8014.486	364.6567
107	$C_6H_4Cl_2$	m-Dichlorobenzene	22.87009	0.4456102 E-01	-0.2128264 E-04	0.2990787 E-08	0.7361333 E-12	11.37776	15451.49	682.6164
108	$C_6H_4Cl_2$	o-Dichlorobenzene	22.87009	0.4456102 E-01	-0.2128264 E-04	0.2990787 E-08	0.7361333 E-12	5.981208	6798.548	345.1776
109	$C_6H_4Cl_2$	p-Dichlorobenzene	22.87009	0.4456102 E-01	-0.2128264 E-04	0.2990787 E-08	0.7361333 E-12	6.813621	7773.248	396.0557
110	C_6H_5Br	Bromobenzene	20.36329	0.4011145 E-01	-0.1108768 E-04	-0.4813553 E-08	0.2980795 E-11	5.881747	6571.377	365.2812
111	C_6H_5Cl	Chlorobenzene	19.98323	0.4080291 E-01	-0.1166762 E-04	-0.4590812 E-08	0.2862703 E-11	5.88808	6222.905	372.2756
112	C_6H_5I	Iodobenzene	20.57765	0.3968462 E-01	-0.1067708 E-04	-0.50094 E-08	0.2927215 E-11	5.72827	6854.36	348.1382
113	C_6H_6	Benzene	16.39282	0.4020369 E-01	0.6925399 E-05	-0.4114202 E-07	0.2398098 E-10	5.658375	5307.813	379.456
114	C_6H_6O	Phenol	19.91816	0.4992518 E-01	-0.2451622 E-04	0.460471 E-08	0.4113815 E-12	6.555719	7250.359	321.6074
115	C_6H_7N	Aniline	20.11747	0.4538924 E-01	-0.5743054 E-05	-0.1216273 E-07	0.5366204 E-11	6.44519	7386.331	346.6331
116	C_6H_{12}	Cyclohexane	21.00016	0.5627391 E-01	0.1129438 E-04	-0.3606168 E-07	0.1482606 E-10	5.473055	5030.253	371.1755

No.	Formula	Name								
117	C_6H_{12}	Methylcyclopentane	375.9433	4936.44	5.567563	0.164543 E-10	-0.3442294 E-07	0.5935187 E-05	0.5465972 E-01	22.02735
118	C_6H_{12}	1-Hexene	374.7552	4783.217	5.711574	0.3952236 E-11	-0.9298869 E-08	-0.7512748 E-05	0.4926029 E-01	27.87277
119	C_6H_{14}	2,2-Dimethylbutane	381.1012	4486.167	5.50245	0.9570332 E-11	-0.2478134 E-07	0.1806208 E-05	0.550671 E-01	29.64918
120	C_6H_{14}	2,3-Dimethylbutane	380.1172	4672.79	5.61351	0.1323425 E-11	-0.2494276 E-07	-0.2028283 E-05	0.5595826 E-01	29.27018
121	C_6H_{14}	n-Hexane	382.794	5085.758	6.039243	0.1346731 E-10	-0.2763996 E-07	0.3048799 E-05	0.5199263 E-01	30.17847
122	C_6H_{14}	2-Methylpentane	375.5984	4700.639	5.7088	0.2132741 E-10	-0.3870868 E-07	0.5716877 E-05	0.3351181 E-01	30.30218
123	C_6H_{14}	3-Methylpentane	376.5611	4700.526	5.7023	0.1391842 E-10	-0.2860809 E-07	0.366338 E-05	0.5189874 E-01	30.17174
124	$C_6H_{14}O_4$	Triethylene glycol	356.928	11860.26	9.707385	0.1149462 E-10	-0.2983416 E-07	0.5449502 E-05	0.6252653 E-01	38.88318
125	C_7H_8	Toluene	374.745	5836.287	5.944251	0.2585787 E-10	-0.4628264 E-07	0.9961368 E-05	0.4639546 E-01	21.17722
126	C_7H_8O	o-Cresol	286.9147	6394.925	5.749559	0.5990957 E-11	-0.1323952 E-07	-0.7496517 E-05	0.5181666 E-01	24.15791
127	C_7H_{14}	Methylcyclohexane	370.0705	5338.374	5.608872	0.1837708 E-10	-0.3885885 E-07	0.5750553 E-05	0.6729289 E-01	27.06952
128	C_7H_{14}	Ethylcyclopentane	364.7743	5369.766	5.698096	-0.8292246 E-10	0.1674966 E-06	-0.1351091 E-03	0.9883192 E-01	24.63008
129	C_7H_{14}	1-Heptene	362.5801	5212.626	5.922457	0.6897952 E-11	-0.1448889 E-07	-0.724699 E-05	0.5769426 E-01	32.68419
130	C_7H_{16}	n-Heptane	359.5259	5278.902	5.98627	0.1454746 E-10	-0.293693 E-07	0.1213345 E-04	0.608752 E-01	34.96845
131	C_8H_8	Styrene	358.5947	6329.575	6.071326	0.8297016 E-12	0.3432486 E-08	-0.25693 E-04	0.5843 E-01	24.82866
132	C_8H_{10}	Ethylbenzene	349.8527	5852.905	5.747492	0.3331962 E-10	-0.5839197 E-07	0.1239678 E-04	0.5526271 E-01	26.37827
133	C_8H_{10}	m-Xylene	354.6467	6049.457	5.949452	0.2628339 E-10	-0.4900955 E-07	0.122123 E-04	0.5188146 E-01	26.42788
134	C_8H_{10}	o-Xylene	354.0417	6141.641	5.922098	0.1949676 E-10	-0.3659655 E-07	0.5908631 E-05	0.5103585 E-01	27.89247
135	C_8H_{10}	p-Xylene	355.99	6033.046	5.94371	0.276508 E-10	-0.5289838 E-07	0.1658367 E-04	0.4982215 E-01	26.39862
136	C_8H_{16}	Ethylcyclohexane	355.93	5751.059	5.769319	0.2469959 E-10	-0.4481198 E-07	0.7343807 E-05	0.7610546 E-01	32.07366
137	C_8H_{16}	n-Propylcyclopentane	352.481	5754.086	5.957854	-0.10736 E-11	0.9086445 E-08	-0.3576344 E-04	0.8393195 E-01	30.7991
138	C_8H_{18}	n-Octane	360.26	5947.491	6.4141	0.1749419 E-10	-0.345095 E-07	0.1479927 E-05	0.6930903 E-01	39.77987
139	$C_8H_{18}O_5$	Tetraethylene glycol	439.2803	14787.51	10.89268	0.141849 E-10	-0.3636893 E-07		0.8159435 E-01	50.93845
140	C_9H_8	Indene	371.053	7190.945	6.176081	0.4361004 E-11	-0.7372481 E-08	-0.162819 E-04	0.6007207 E-01	26.11577
141	C_9H_{10}	Indan	357.368	6821.739	6.049353	0.5886033 E-11	-0.1126863 E-07	-0.1535148 E-04	0.6712433 E-01	27.27218
142	C_9H_{10}	Methylstyrene	352.369	6638.097	6.087191	0.9784106 E-11	-0.204508 E-07	0.1423367 E-04	0.5092828 E-01	40.40845
143	C_9H_{12}	1-Ethyl-2-methylbenzene	351.7732	6541.084	6.202647	-0.3666534 E-12	0.5325566 E-08	-0.2692206 E-04	0.6787234 E-01	32.7457
144	C_9H_{12}	n-Propylbenzene	340.82	6180.323	5.919976	-0.8488248 E-12	0.8236172 E-08	-0.333376 E-04	0.7397426 E-01	30.85809
145	C_9H_{18}	n-Propylcyclohexane	343.4018	6079.604	6.022487	-0.3686077 E-11	0.6920047 E-08	-0.3752096 E-04	0.9596314 E-01	36.93022
146	C_9H_{20}	n-Nonane	330.96	6662.655	6.22189	0.2114216 E-11	-0.4134716 E-08	0.2963375 E-04	0.7738344 E-01	44.6198
147	$C_{10}H_8$	Naphthalene	305.1725	7090.598	5.464939	0.1536979 E-11	0.2599732 E-08	-0.303242 E-04	0.7107626 E-01	26.38315
148	$C_{10}H_{10}$	1-Methylindene	343.832	7223.491	6.139705	0.5436545 E-11	-0.9807097 E-08	-0.1746832 E-04	0.696464 E-01	29.92083
149	$C_{10}H_{10}$	2-Methylindene	340.573	6479.445	6.165742	0.5540522 E-11	-0.990292 E-08	-0.181527 E-04	0.7131178 E-01	29.5297
150	$C_{10}H_{12}$	Dicyclopentadiene	353.37	6534.868	5.97896	0.3171173 E-11	-0.1037675 E-08	-0.3064237 E-04	0.8077651 E-01	35.24185
151	$C_{10}H_{14}$	n-Butylbenzene	330.3357	6534.868	6.089988	-0.951656 E-12	0.919402 E-08	-0.3733832 E-04	0.8354658 E-01	35.57803
152	$C_{10}H_{14}$	1,2-Dimethyl-3-Ethylbenzene	330.4982	6838.277	6.180938	0.9362406 E-11	-0.2192313 E-07	-0.7317236 E-05	0.7623887 E-01	32.18688
153	$C_{10}H_{20}$	n-Butylcyclohexane	329.0528	6369.972	6.094088	-0.214681 E-12	0.6648178 E-08	-0.3964926 E-04	0.1046007 E-01	41.74714
154	$C_{10}H_{22}$	n-Decane	317.6512	6213.998	6.33557	0.2256651 E-10	-0.4415409 E-07	0.2049705 E-05	0.8602711 E-01	49.42138
155	$C_{11}H_{10}$	1-Methylnaphthalene	323.8246	7676.997	6.080512	0.7089941 E-11	-0.1428024 E-07	-0.1560299 E-04	0.7475587 E-01	29.3571

Part B (cont'd)

Number	Empirical Formula	Name	a_1	a_2	a_3	a_4	a_5	A_1	A_2	A_3
Organic Chemicals										
156	$C_{11}H_{10}$	2-Methylnaphthalene	29.3571	0.7475587 E-01	-0.1560299 E-04	-0.1428024 E-07	0.7089941 E-11	6.101162	7630.143	325.2701
157	$C_{11}H_{24}$	n-Undecane	54.25211	0.9427374 E-01	0.2727295 E-05	-0.4957262 E-07	0.2544979 E-10	7.21247	7475.258	350.7821
158	$C_{12}H_8$	Acenaphthlene	30.48077	0.7758508 E-01	-0.2664829 E-04	-0.5063562 E-08	0.4429521 E-11	6.216864	8047.648	313.565
159	$C_{12}H_{10}$	Diphenyl	32.02568	0.9116691 E-01	-0.4408667 E-04	0.7505404 E-08	0.9059193 E-12	6.194778	7947.647	317.1246
160	$C_{12}H_{12}$	2,7-Dimethylnaphthalene	33.16177	0.8467461 E-01	-0.1714886 E-04	-0.1655051 E-07	0.8138765 E-11	6.707037	8521.498	331.2805
161	$C_{12}H_{14}$	1,2,3-Trimethylindene	37.49874	0.8872509 E-01	-0.1961391 E-04	-0.1486368 E-07	0.7642588 E-11	6.452743	7497.763	321.886
162	$C_{12}H_{26}$	n-Dodecane	59.0528	0.1029143	0.2243202 E-05	-0.5378563 E-07	0.279613 E-10	6.561135	6739.22	292.574
163	$C_{13}H_{10}$	Fluorene	34.75668	0.811078 E-01	-0.2232686 E-04	-0.1036036 E-07	0.6152866 E-11	8.197664	11632.66	435.9648
164	$C_{13}H_{14}$	1-Methylethylnaphthalene	36.92984	0.9421013 E-01	-0.181649 E-04	-0.1913785 E-07	0.9260544 E-11	6.678618	8234.079	295.939
165	$C_{13}H_{14}$	2,3,5-Trimethylnaphthalene	36.96391	0.9458171 E-01	-0.1866141 E-04	-0.188505 E-07	0.9196502 E-11	6.667748	8366.407	292.104
166	$C_{13}H_{28}$	n-Tridecane	63.85792	0.111595	0.1116408 E-05	-0.5633991 E-07	0.2630431 E-10	6.65466	6995.694	281.4988
167	$C_{14}H_{10}$	Phenanthrene	34.43893	0.9086123 E-01	-0.2523548 E-04	-0.1184692 E-07	0.7013306 E-11	5.543959	7914.989	244.9414
168	$C_{14}H_{30}$	n-Tetradecane	68.69802	0.1196599	0.269646 E-05	-0.6341006 E-07	0.3319885 E-10	6.75784	7203.471	269.269
169	$C_{15}H_{12}$	1-Phenylindene	39.22844	0.101567	-0.2937448 E-04	-0.1146747 E-07	0.7190597 E-11	6.502465	8771.184	284.878
170	$C_{15}H_{14}$	2-Ethylfluorene	40.82861	0.1076498	-0.286231 E-04	-0.1479421 E-07	0.8512086 E-11	6.695303	8621.0	283.702
171	$C_{15}H_{32}$	n-Pentadecane	73.51018	0.1281686	0.2315556 E-05	-0.6703845 E-07	0.3505542 E-10	6.82225	7400.305	257.5534
172	$C_{16}H_{10}$	Fluoranthene	40.95336	0.1036185	-0.4006217 E-04	-0.2144235 E-08	0.4437308 E-11	6.571647	9789.787	257.606
173	$C_{16}H_{10}$	Pyrene	38.96862	0.1053988	-0.3875499 E-04	-0.4808038 E-08	0.5453756 E-11	6.603641	9365.547	267.09
174	$C_{16}H_{12}$	1-Phenylnaphthalene	37.71548	0.1059667	-0.2424246 E-04	-0.189117 E-07	0.9723825 E-11	7.078733	9631.87	312.9423
175	$C_{16}H_{34}$	n-Hexadecane	78.32123	0.1367191	0.1674373 E-05	-0.7030701 E-07	0.3676587 E-10	6.92955	7569.57	245.2032
176	$C_{18}H_{12}$	Chrysene	42.55082	0.1204639	-0.3831014 E-04	-0.1145987 E-07	0.8077914 E-11	6.810698	10647.46	229.481

II. Sources of Computer Programs

This appendix presents, in three parts, references to sources of computer programs that are useful in making calculations of equilibrium-stage separation operations. The first part refers to programs that deal directly with subject matter in this book and for which computer program listings are either printed in the references or are available from the author of the program. In Part B we reference some widely used industrial-type computer-aided design programs that include subroutines for many of the calculation procedures described in this book. Finally, in Part C, references are given that contain extensive descriptions of computer programs for chemical engineers.

Part A. Computer Programs for Which Listings Are Available (Usually in FORTRAN)

These programs deal with the subject matter in this book.

Subject	Reference Number
Thermodynamic properties (Chapters 4, 5)	
Soave–Redlich–Kwong	2
Chao–Seader/Grayson–Streed	1,2
van Laar	7
Wilson	3,7
UNIQUAC	3
UNIFAC	3
Equilibrium flash (Chapter 7)	
Bubble and dew point	1, 2, 5, 6, 7
Flash	1, 2, 5, 6, 7
MSEQ	1
Approximate multicomponent methods (chapter 12)	
Fenske–Underwood–Gilliland	1
Group method	1
Rigorous multicomponent methods (Chapter 15)	
Bubble-point method (Wang–Henke)	4
Sum-rates	9
Simultaneous correction (Naphtali–Sandholm)	3
Continuous differential contacting operations (Chapter 16)	
Packed-column absorption and stripping	8

1. "Chemical Engineering Simulation System (CHESS)," Professor R. L. Motard, Department of Chemical Engineering, Washington University, St. Louis, Missouri 63130 (1971).

2. "GPA K and H Computer Program, Calculating Phase Equilibria for Hydrocarbon and Selected Non-Hydrocarbon Systems," (by J. H. Erbar) Gas Processors Association, 1812 First Place, 15 East Fifth Street, Tulsa, Oklahoma 74103 (1974).

3. Fredenslund, A., J. Gmehling, and P. Rasmussen, *Vapor-Liquid Equilibria Using UNIFAC: A Group Contribution Method*, Elsevier Scientific Publishing Co., Amsterdam, 1977.

4. Johansen, P. J., and J. D. Seader, "Multicomponent Distillation by the Wang-Henke Method," in *Computer Programs for Chemical Engineering, Volume VII, Stagewise Computations*, ed. by J. Christensen, Aztec Publishing Co., P. O. Box 5574, Austin, Texas 78763 (1972).

5. Kalb, C. E., and J. D. Seader, "Equilibrium Flash Vaporization by the Newton–Raphson Method," in *Computer Programs for Chemical Engineering, Volume V, Thermodynamics*, R. Jelinek, Ed., Aztec Publishing Co., P. O. Box 5574, Austin, Texas 78763 (1972).

6. Myers, A. L., and W. D. Seider, *Introduction to Chemical Engineering and Computer Calculations*, Prentice–Hall, Inc., Englewood Cliffs, N. J., 1976.

7. Prausnitz, J. M., C. A. Eckert, R. V. Orye, and J. P. O'Connell, *Computer Calculations for Multicomponent Vapor-Liquid Equilibria*, Prentice–Hall, Inc., Englewood Cliffs, N. J., 1967.

8. Sherwood, T. K., R. L. Pigford, and C. R. Wilke, *Mass Transfer*, McGraw–Hill Book Co., New York, 1975, 525–542.

9. Shinohara, T., P. J. Johansen, and J. D. Seader, "Multicomponent Stripping and Absorption," in *Computer Programs for Chemical Engineering, Volume VII, Stagewise Computations*, ed. by J. Christensen, Aztec Publishing Co., P. O. Box 5574, Austin, Texas 78763 (1972).

Part B. Industrial-Type Computer-Aided Design Programs

These programs, which contain subroutines for many of the types of calculation procedures discussed in this book, can be accessed from computer networks and/or can be licensed and installed on the user's computer.

1. CONCEPT: The Concept Group, Chemical Engineering Department, Uni-

versity of Texas, Austin, Texas 78712; and AAA Technology and Specialties Co., Inc., P. O. Box 37189, Houston, Texas 77036.

2. DESIGN/2000: ChemShare Corp., P. O. Box 6706, Houston, Texas 77005.

3. FLOWTRAN: Monsanto Company, Engineering Technology—F4EE, 800 N. Lindbergh Blvd., St. Louis, Missouri 63166.

4. GPS-II: McDonnell-Douglas Automation Company, Box 516, St. Louis, Missouri 63166 (program developed by Phillips Petroleum Company).

5. MADCAP: Systems Analysis Control and Design Activity, Faculty of Engineering Science, The University of Western Ontario, London N6A 5B9 Ontario, Canada.

6. PDS: School of Chemical Engineering, Oklahoma State University, Stillwater, Oklahoma 74074.

7. PROCESS: Simulation Sciences, Inc., 1440 North Harbor Blvd., Fullerton, California 92635.

Part C. Extensive Annotated Bibliographies of Computer Programs

These bibliographies are contained in the following articles.

1. Peterson, J. N., C. C. Chen, and L. B. Evans, *Chem. Eng.*, **85** (13), 145–152, 154 (June 5, 1978); **85** (15), 69–82 (July 3, 1978); **85** (17), 79–86 (July 31, 1978); **85** (19), 107–115 (August 28, 1978); **86** (11), 167–173 (May 21, 1979).

2. Hughson, R. V., and E. G. Steymann, *Chem. Eng.*, **78** (16), 66–86 (July 12, 1971); **78** (29), 63–72 (December 27, 1971); **80** (19), 121–132 (August 20, 1973); **80** (21), 127–138, 140 (September 17, 1973).

AUTHOR INDEX

Abrams, D. S., 231, 232
Ackerman, G. H., 523
Akell, R. B., 83
Amundson, N. R., 612
Asselineau, L., 232

Bachelor, J. B., 490
Badger, W. L., 119
Bailes, P. J., 85
Ball, W. E., 613
Banchero, J. T., 119
Barnes, F. J., 490
Beattie, J. A., 177
Benedict, M., 177, 178
Berg, L., 231, 233
Berndt, R. J., 39
Beveridge, G. S. C., 613
Beyer, C. H., 136
Bieber, H., 115
Billingsley, D. S., 491
Bird, R. B., 670
Bliss, H., 43
Block, B., 369
Block, U., 613
Boatright, R. G., 547
Bogart, M. J. P., 369
Bolles, W. L., 490
Boston, J. F., 612
Boublik, T., 613
Boyes, A. P., 56
Bradford, J. P., 523
Brian, P. L. T., 233
Bridgeman, O. C., 177

Brinkley, W. K., 491
Brinkman, N. D., 231
Brown, G. G., 231, 490, 491, 501
Brown, G. M., 177
Bruin, S., 232
Buell, C. K., 547
Burningham, D. W., 612
Butler, J. A. V., 231

Cantiveri, L., 71
Carle, T. C., 38
Carlson, H. C., 231
Cavett, R. H., 178
Chaiyavech, P., 178
Chang, P., 670
Chao, K. C., 177, 230
Chen, C. C., 727
Chertow, B., 121
Chilton, C. H., 38, 39, 126, 176, 230, 297
Chueh, P. L., 177
Close, H. J., 670
Cohen, G., 232, 613
Colburn, A. P., 231, 624
Considine, D. M., 38
Cooper, C. M., 490
Cornell, D., 670
Cox, R. P., 136
Craig, L. C., 41
Crowe, C. M., 612
Cukor, P. M., 231
Curl, R. F., Jr., 177

Dadyburjer, D. B., 277

Dallin, D. E., 297
Dankwerts, P. V., 670
Deal, C. H., 232
Denbigh, K. G., 677
de Nevers, N., 705
De Priester, C. L., 277
Derr, E. L., 232
Dodge, B. F., 705
Donnell, J. W., 490
Drickamer, H. G., 231, 523
Duffin, J. H., 262
Dunlop, J. G., 404
Dunn, C. L., 231
Duran, J. L., 232

Eagleton, L. C., 43
Eckert, C. A., 64, 231, 232, 726
Eckert, J. S., 81, 84, 177, 657, 660, 661, 670
Edmister, W. C., 131, 177, 178, 230, 231, 491
Ellerbe, R. W., 369
Ellis, S. R. M., 56, 84
Erbar, J. H., 177, 178, 230, 726
Evans, H. D., 297
Evans, L. B., 727
Ewell, R. H., 231

Fair, J., 85, 490, 522, 670
Farrington, P. S., 178
Fear, J. V. D., 38
Fellinger, L., 663
Feneske, M. R., 427, 490
Fick, A., 670
Findlay, A., 131
Flower, J. R., 705
Foote, E. H., 81
Foust, A. S., 634, 670
Francis, A. W., 131
Frank, J. C., 178
Franklin, W. B., 491
Frazier, J. P., 231
Fredenslund, A., 232, 613, 726
Freshwater, D. C., 705
Friday, J. R., 612
Fuller, E. N., 628, 670

Gautreaux, M. F., 523
Gayler, R., 523
Getes, R. L., 612
Geyer, G. R., 178
Gibbs, J. W., 89

Gibson, C. H., 490
Giddings, J. C., 628, 670
Gilissen, F. A. H., 523
Gilliland, E. R., 121, 262, 490, 372, 705
Glanville, J. W., 230, 231
Gmehling, J., 231, 613, 726
Goff, G. H., 178
Goldstein, R. P., 613
Grayson, H. G., 176, 230
Guerreri, G., 490
Guffey, C. G., 233
Gully, A. J., 176

Hála, E., 231, 613
Hale, E., 405
Han, M. S., 177
Hanson, D. N., 262, 490
Harrison, J. M., 231
Haselden, G. G., 705
Hayden, J. G., 177
Headington, C. E., 33, 39
Heaven, D. L., 35, 39, 548
Hegner, B., 613
Heidemann, R. A., 233
Hendry, J. E., 32, 39, 547
Henke, G. E., 612
Henley, E. J., 15, 16, 22, 91, 94, 115, 272, 262
Hermsen, R. W., 230
Higbie, R., 670
Hildebrand, J. H., 230, 233
Hiraizumi, Y., 548
Hiranuma, M., 232
Hirata, M., 231
Hlavacek, V., 613
Hofeling, B., 613
Hoffman, W. H., 121
Holland, C. D., 369, 491, 612
Holloway, F. A. L., 523, 662, 670
Holmes, M. J., 231
Holub, R., 232
Horton, G., 491
Horvath, P. J., 340, 359
Hougen, O. A., 88, 99, 176, 177
Houlihan, R., 523
Housum, C., 705
Hudson, J. W., 231
Huggins, C. M., 177
Hughes, R. R., 32, 39, 297, 547
Hughson, R. V., 727
Hunter, T. G., 410, 421

Innes, E. D., 38

Jackson, R., 705
Janecke, E., 131
Jelinek, J., 613
Johansen, P. J., 612, 726
Johnson, C. A., 178
Johnson, G. D., 178, 233
Jones, H. K. D., 232
Jones, J. B., 73
Jones, W. D., 73
Joy, D. S., 233

Kalb, C. E., 726
Kaliaguine, S., 232
Karr, A. E., 523
Katayama, T., 232, 233, 613
Kato, M., 233
Kehde, H., 178
Ketchum, R. G., 613
Keyes, D. B., 131
King, C. J., 490, 527, 547, 548, 705
Kinney, G. F., 421
Kirkbride, C. G., 491
Kirkpatrick, S. D., 126
Kiser, R. W., 233
Kister, A. T., 177
Knapp, W. G., 670
Kobayashi, R., 178
Kobe, K. A., 490, 523
Kochar, N. K., 231
Koppany, C. R., 230
Korablina, T. P., 490
Kremser, A., 490
Kubicek, M., 613
Kwauk, M., 239, 262
Kwong, J. N. S., 177
Kyle, B. G., 233

Lacey, W. N., 177
Landau, L., 523
Lapidus, L., 612
Larson, C. D., 232
Lau, A. D., 237
Lee, B., 177
Leggett, C. W., 523
Lenoir, J. M., 230
Lewis, G. M., 142
Lewis, W. K., 121, 612
Li, N. N., 22

Lightfoot, E. N., 670
Lippman, R. F., 177, 230
List, H. L., 124
Lo, C., 233
Lo, T. C., 523
Lockhart, F. J., 491, 523, 548
Logsdail, D. H., 523
Long, R. B., 22
Lu, B. C. Y., 232
Luyben, W. L., 705
Lynch, C. C., 39
Lyster, W. N., 491

McCabe, W. L., 306, 340
McKetta, J. J., 20, 490, 523
McLaren, D. B., 47
Maclennan, W. H., 233
McNeil, L. J., 297
Maddox, R. N., 176
Mah, R. S. H., 705
Mandhane, J. M., 233
Marchello, J. M., 670
Marek, J., 548
Marina, J. M., 232
Mark, H. F., 20
Martin, H. Z., 490
Matheson, G. L., 612
Maude, A. H., 38
Maxwell, J. B., 93, 131
Mazurina, N. I., 490
Mehra, V. S., 177
Melpolder, F. W., 33, 39
Mertl, I., 232
Meyers, A. L., 612, 726
Michelsen, M. L., 232
Millar, R. W., 231
Molokanov, Y. K., 490
Motard, R. L., 297, 726
Muller, D. E., 297
Murphree, E. V., 335, 340

Nagata, I., 231, 232
Naphtali, L. M., 613
Nash, A. W., 410, 421
Neumaites, R. R., 81
Nichols, J. H., 523, 705
Niida, K., 705
Nikiforov, G. A., 490
Nilsson, N. J., 547
Nishimura, H., 548

Nitta, T., 232
Null, H. R., 231, 232, 705

O'Connell, H. E., 501, 523
O'Connell, J. P., 177, 231
Ohe, S., 231
Okamoto, K. K., 131
Oliver, E. D., 44, 84, 490, 523
Olney, R. B., 523
Onken, U., 231
Orbach, O., 612
Orye, R. V., 177, 231, 726
Othmer, D. F., 20
Otto, F. D., 612

Pauls, A. C., 714
Peng, D. Y., 178
Perry, R. H., 38, 39, 126, 230, 231, 297
Persyn, C. L., 230
Peterson, H. C., 136
Peterson, J. N., 727
Petterson, W. C., 705
Pick, J., 232
Pierotti, G. J., 231
Pigford, R. L., 670, 726
Pitzer, K. S., 230
Plowright, J. D., 230
Polák, J., 231, 613
Ponchon, M., 400
Pontinen, A. J., 612
Post, O., 41
Powers, G. J., 46, 705
Pratt, H. R. C., 523
Prausnitz, J. M., 39, 176, 177, 187, 183, 230,
  231, 232, 233, 237, 726

Rachford, H. H., Jr., 273
Ragatz, R. A., 99, 176
Ramalho, R. S., 232
Rao, R. J., 422
Raschig, F., 57
Rasmussen, P., 232, 613, 726
Rathore, R. N. S., 705
Rayleigh, J. W. S., 361, 369
Redlich, O., 140, 177, 178
Reed, C. E., 262
Reid, R. C., 39, 177
Reman, G. H., 523
Renon, H., 232, 613
Rice, J. D., 273

Roberts, N. W., 523
Robinson, C. S., 373, 490, 705
Robinson, D. B., 178
Robinson, R. L., 177, 230, 232
Rod, V., 548
Rodrigo, B. F. R., 547
Rollinson, L. R., 81
Rose, A., 369, 613
Rosen, E. M., 91, 94, 272
Rubin, L. C., 177, 178
Rudd, D. F., 46, 705

Sage, B. H., 177, 178, 230
Sandholm, D. P., 613
Savarit, R., 400
Sawistowski, H., 422
Schechter, R. S., 613
Scheibel, E. G., 612
Schettler, P. D., 628, 670
Schlichting, H., 670
Schmidt, A. Y., 124
Schreiber, L. B., 232
Schreve, N., 131
Schrodt, V. N., 613
Schubert, R. F., 340, 359
Scott, R. L., 233
Seader, J. D., 230, 297, 547, 548, 612, 613,
  705, 714, 726
Sherwood, T. K., 39, 177, 523, 662, 670, 726
Shetlar, M. D., 233
Shinohara, T., 612, 726
Shipley, G. H., 523
Shiras, R. N., 231, 490
Shiroko, K., 705
Sieder, W. D., 612, 714, 726
Siirola, J. J., 46
Sinor, J. E., 231
Skelland, A. H. P., 422
Smith, B. D., 68, 262, 270, 400, 491, 556, 612
Smith, V. C., 231
Smith, W., 84, 422
Smoker, E. H., 340, 369
Soave, G., 177
Soczek, C. A., 231
Somerville, G. F., 262
Souders, M., Jr., 1, 231, 490, 501, 522, 547
Staffin, H. K., 15, 16, 262
Stanfield, R. B., 613
Starling, K. E., 177, 180
Steib, V. H., 613

Stephanopoulos, G., 548
Sternling, C. V., 297
Stewart, D. M., 38
Stewart, W. E., 670
Steymann, E. G., 727
Stigger, E. K., 523
Stockar, U. V., 670
Stocking, M., 176
Strand, C. P., 523
Strangio, V. A., 490
Streed, C. W., 177, 230
Strubl, K., 232
Stupin, W. J., 491
Sujata, A. D., 612
Sullivan, S. L., 491, 612
Svoboda, V., 232
Sweeny, R. F., 613

Tai, T. B., 232
Takeuchi, S., 232
Tanigawa, S., 70, 84
Tao, L. C., 231
Tassios, D. P., 231, 232
Taylor, D. L., 131
Tavana, M., 490
Tedder, D. W., 548, 705
Thiele, E. W., 306, 340, 612
Thodos, G., 177
Thompson, R. W., 527, 547, 548
Thomson, D. W., 231
Thornton, J. D., 523
Timmermans, J., 39
Todd, W. G., 47, 490
Toor, H. L., 670
Treybal, R. E., 85, 490
Tsuboka, T., 232, 613
Tyreus, B. D., 705

Umeda, T., 705

Underwood, A. J. V., 490
Upchurch, J. C., 47

van Laar, J. J., 231
Van Ness, H. C., 231
Van Winkle, M., 47, 178, 231, 490, 523
Van Wormer, R. B., 705
Venkataranam, A., 422
Verburg, H., 523

Wang, J. C., 612
Waterman, W. W., 231
Watson, K. M., 88, 99, 177, 178
Weast, R. C., 39
Webb, G. B., 177
Weber, J. H., 231
Wehe, A. H., 233
Wehner, J. F., 422
Wells, T. A., 705
West, E. W., 178
Westerberg, A. W., 548
White, R. N., 231
Whitman, W. G., 670
Wichterle, I., 178, 231, 613
Wilke, C. R., 670, 726
Williams, G. C., 523
Wilson, G. M., 178, 231, 232
Winn, F. W., 490
Winward, A., 85
Wodnick, R. B., 705
Wohl, K., 231

Yamada, T., 232
Yarborough, L., 177, 178, 231
Yasuda, M., 233
Yerazunis, S., 230
Younger, A. H., 523

Zuiderweg, F. J., 523

SUBJECT INDEX

Absorption, 9, 12
  column design, 644-670
  equilibrium, 123-127
  factor, 49, 445, 459, 461, 463
  group method, 459-470
  McCabe-Thiele method, 318-321
  reboiled, 8, 12, 606-611
  sum-rates (SR) method, 578-586
Acentric factor, 159, 714-724
Active area, 502
Activity, 143
Activity coefficient, 145
  correlations, 183-230
  infinite-dilution, 200
  liquid phase, 184
Adiabatic:
  flash, 286
  mixing process, 373
Adiabatic equilibrium stage, degrees of freedom, 240-242
Adsorption, 11, 14
  calculation procedures, 120-123, 463
  commercial separations, 15
Algorithmic synthesis, 544-546
Alkylation effluent, 430
Amagat's law, 147
Amundson-Pontinen method, 562
Anion membrane, 24
Antoine, equation, 159, 163
  constants, 714-724
Approximate methods for design, 427-493
ASTM distillation curve, 103
Athabasca tar sands, 2

Athermal solution, 184
Auxiliary operations, 3
Availability function, 685
Azeotropes, distillation, 9
  heterogeneous, 101
  homogeneous, 101
  maximum-boiling, 99, 208-209
  minimum-boiling, 98, 100, 207-208
  thermodynamic activity coefficients, 208
Azeotropic distillation, 9, 13

Baffle, 71
  antijump, 71
  column for extraction, 77
  picket fence splash, 71
Batch distillation, 361-369
  constant reflux, 364
  variable reflux, 367
Beattie-Bridgman equation, 148
Benedict-Webb-Rubin equation, 148
Binary interaction parameter, 172
Block flow diagrams, 3
Block tridiagonal matrix, 596
Blowing, 55
Boiler, degrees of freedom, 244-245
Boilup, 12
Boundary-layer theory, 644
BP method, 362
Bubble-cap, 69-70
  design, 70
  riser, 70
  skirt, 70

trays, 69-72
Bubble point, 90
  calculation, 281-284
  method, 362
Burningham-Otto method, 578-586
  algorithm, 582
  computer program, 725

CACHE Corporation, 714
Capacity, coefficients, 55, 644
  of contacting devices, 503-506, 516-521
  factor, 503
Cascade, countercurrent, 561
  equations, 483
  tray, 67
Centrifugal contactor, 82
Channeling, 17, 53
Chao-Seader correlation, 172, 183-189
  computer program, 725
Cheapest first heuristic, 545
Chemical potential, 141
CHESS, 726
Chromatographic separations, 20, 22
Cloud-point titration, 106
Cocurrent flow, 27, 28
  in absorption, 637
Column: capacity, 503
  diameter, 502
Complex units, degrees of freedom, 251
Compressibility factor, 148, 151, 714-724
Computer programs, 725-727
CONCEPT, 726
Condensation, partial, 7, 270-297
Condenser: degrees of freedom, 244-245
  partial, 327
  total, 327
  type, selection, 432-434
Constant molar overflow, 313
Contact, equilibrium, 24
Contactors: liquid-liquid, 516-522
  vapor-liquid, 502-515
Continuous contacting, 37, 624-640
Convergence, acceleration, 278
  criteria, 571, 591
  partial, 576
  pressure, 169
Corresponding states, principle, 147-149
Cost, effect of reflux on, 451
  factors, 340
  of separators, 534

utilities, 701
Countercurrent flow, 27, 31, 307
  in absorption, 637
Crosscurrent flow, 27, 28
  on stages, 502
Crystallization, 10, 14

Dalton's law, 92, 147
Debutanizer, 431
Degrees of freedom, 89, 239
  for operation elements and units, 246-247
Dehumidification, 13
Demister pads, 72
De Priester K charts, 277, 278
Derived thermodynamic properties, 153-161
Description rule, 250-251
DESIGN, computer program, 727
Design variables, 239-262
  independence, 239
Desublimation, 11, 14
Dew point, 90
  calculation, 281-284
Dialysis, 20, 21
Didiagonal matrix, 570
Difference point, net flow, 379
Differential distillation, 362-364
Diffusion, equimolar (EMD), 629, 647
  gaseous, 20, 21
  gas-film controlling, 640
  liquid-film controlling, 640
  molecular, 627-629
  pressure, 20, 21
  thermal, 20, 22
  two-film, 638-643
  unimolecular (UMD), 629, 647, 650
  volumes, 628
Diffusivity, 627-628
Discrepancies, 596, 603
Disk-baffle column, 77
Dispensers, liquid, 61
Distillation, 8, 12
  azeotropic, 9, 13
  batch, 361-368
  binary, 322-340, 375-394
  BP method, 566-578
  curve, 103
  energy requirements, 677
  extractive, 8, 12, 102
  group method, 484
  low-temperature, 692

multicomponent, 428-459, 484-490, 556-602
multieffect, 690
SC method, 594-611
sequence, 529
SR method, 578-586
SRV, 703
steam, 9
Distribution: coefficient, 26
diagrams, 108, 111
feed, 58
Divider, degrees of freedom, 245, 248
Dorrco, hydro-softener, 17
Downcomer, 68
design, 502
flooding, 503
Drawoff sumps and seals, 71
Drums, reflux and flash, 514
Drying, 10, 13
Dumping, 54
Du Pont Co., 74, 125

Edmister equations for absorption and stripping
factors, 463
group method, 480-490
Efficiency: stage, 55, 394
of trays, 335, 394, 509-514, 521-522
Electrodialysis, 20, 23
Electrolysis, 20, 23
EMD diffusion, 629
Empirical method for distillation, 428-459
Energy: conservation, 677-705
requirement for separation, 677
separation agent (ESA), 7, 529
Enricher, 481
Enriching section, 12
Enthalpy, 154-157
concentration diagrams, 128, 129, 373, 377
from C-S correlation, 190
excess, 184
from NRTL equation, 214
from UNIQUAC equation, 216
from van Laar equation, 200
from Wilson equation, 211
Entrainment, 54
flooding, 503
Entropy, excess, 184
Enumeration algorithm, 248-250
Environmental factors, 35
Equations of state, 140
Beattie-Bridgman, 148

Benedict-Webb-Rubin (B-W-R), 148
generalized, 148
ideal gas law, 148
Redlich-Kwong (R-W), 148
Soave-Redlich-Kwong (S-R-K), 169-176
Van der Waal, 148
virial, 148
Equation-tearing methods, 562-594
Equilateral triangular diagram, in extraction,
106, 415
Equilibrium, 89
diagrams, 88-121
heterogeneous, 89
homogeneous, 89
liquid-liquid, 146-147, 223-230
ratio, 145
stage, degrees of freedom, 242-244
stage models, 557
vapor-liquid, 145-146
Equilibrium stage, 25
concept, 24
contact, 24
theoretical model, 557-560
Equipment: capacity, 50
cost, 51
operability, 51
pressure drop, 51
ESA, energy separation agent, 7
Ethyl alcohol, Keyes process, 101-102, 114-
115
Ethylene glycol: dehydration, 27
hydration, 5
Evaporation, 11, 13
through stagnant gas, 629-632
EVF, 103
Evolutionary synthesis, 540
Excess free energy, 184
Exhauster, 481
Extensive variables, 239
Extract, 103
reflux, 415
Extraction column: capacity, 521
efficiency, 521
Extraction, liquid-liquid, 8, 10, 13
equilibrium diagrams, 103
equipment, 76-84
extract reflux, 415
group method, 475-480
infinite stages, 415
McCabe-Thiele method, 315-318

minimum stages, 410-415
Ponchon method, 394
raffinate reflux, 415
when preferred to distillation, 79-80
solvents, 78-79
sum-rates method (ISR), 586-594
triangular diagrams, 410-421
Extractive distillation, 8, 12
Evolutionary synthesis, 540-544

F factors, 63
False-position method, 284-285
Fair correlation for entrainment flooding,
505-506
Feeds, multiple, 337
Feed stage location: in McCabe-Thiele method,
328-329
in multicomponent distillation, 455-457,
577-578
in Ponchon method, 384
optimum, 329
Fenske equation, 436
Fenske-Underwood-Gilliland method, 428-459
algorithm, 429
computer program, 725
Fibonacci search, 606
Fick's law, 627
Film: diffusion, 629, 638
penetration theory, 644
Flash, 270
adiabatic, 271, 286-287
drums, 514
equations, 272
equilibrium, 7, 270-297
graphical methods, 272-273
isothermal, 271, 273-278
multistage, 289-293
specifications, other, 288-289
vaporization, 7, 8
Flooding, 53, 503
correlation, 506
in extraction, 519
in packed columns, 658
Flory-Huggins correction, 185
Flow modes for columns, 502
FLOWTRAN, 714
Foam fractionation, 20, 21
Foaming, 54
Forbidden splits, 529-539
Fractionation Research Inc., 49, 505

Free energy, 141, 184, 679
excess, 184
partial molal, 679
Friday and Smith study, 562, 578, 586
Fugacity, 141
coefficient, 143
mixture coefficient, 143, 160
pure species coefficient, 143, 157

Gas absorption, 624-670
Gas-liquid systems, 123
Gaussian elimination, 564
GCOS process, 2
Gibbs, free energy, 141, 679
phase rule, 89
Gilliland correlation, 450-455
Glitsch and Sons, 48, 63, 66, 69
GPA, computer programs, 726
GPS II, 727
Graphical solutions, multistage systems, 306-340
McCabe-Thiele methods, 306
Ponchon-Savarit method, 372
Group contribution methods, 218-220
Group methods, 459-490
Edmister method, 218
equations, 483
for extraction, 475

Heat capacity, 714-724
Heat exchange integration, 704-705
Heat pumps, 692-698
Heavy key, 429
Heavy phase, 49
Height of transfer unit (HTU), 55, 647
in extraction, 521
Henry's law, 123, 639
HETP, 55
HETS, in extraction, 521
Heuristics, in synthesis, 540
Hold-down plates, 60
Holdup, in extraction, 519
Horton-Franklin method, 463
HTU, 55
Humidification, 13
Hydraulic gradient, 66
Hydrochloric acid process, 3
Hydrogen bonding, 193
Hydronyl Co., 58, 60

Ideal gas law, 147

for mixtures, 674
Ideal K-value, 164
Ideal solution(s), 143, 161-165
Ion exchange, 16-19, 321-322
Intalox saddles, 57
Intensive variables, 239-240
Intercondenser, 702
Intercooler, 702
Interphase mass transfer, 6
Interreboiler, 702
Intraphase mass transfer, 19, 20
Investment, fixed, 34
Ion exchange, 16
    calculation procedure, 120-123, 321

Jacobian matrix, 596
Janecke diagram, 108, 111, 397

K-value, 26
    hydrocarbons at 80 psia, 438
    light hydrocarbons, 277, 278
Key component, 428
Key operations, 2
Keyes process, 101
Kirkbride equation, 456
Kirk-Othmer encyclopedia, 10, 20
Koch cascade tower, 81
Koch Co., 58, 62, 66
Koch-Sulzer packing, 62
Kremser method, 459-463
    plot, 462

Leaching, 11, 14
    diagrams, 116
Lever-arm rule, 96
Lewis-Matheson method, 560
Lewis-Randall solutions, 143
Light hydrocarbons from casinghead gas, process, 4
Light key, 428
Light phase, 49
Linde, Division of Union Carbide, 74
    double column, 359
Liquid distributors, 58, 59
    Rosette type, 59
Liquid-liquid equilibrium, 146-147, 223-230
Liquid-liquid extraction, 8, 10, 13. *See also* Extraction
    capacity of columns, 516
    centrifugal devices, 83

equilibrium, 103-116, 147, 223
equipment, 76-84
holdup, 519
mechanically assisted gravity devices, 82-83
sum-rates method, 586
two-solvent, 10
Liquid maldistribution, 55
Liquid-solid equilibrium, 116-120
Local composition concept, 203-206
Lost work, 685
Low-temperature distillation, 692-698

McCabe-Thiele diagrams: for adsorption, 320
    for binary distillation, 326, 329, 330, 332, 334
    for countercurrent multistage separator, 309
    for extraction, 317
    for ion exchange, 322
    for rectifying column, 314
McCabe-Thiele method for distillation, 310-315, 322-340
    feed conditions, 326-327
    feed-stage location, 328-329
    minimum reflux ratio, 323, 328-331
    minimum stages, 323, 328-330
    overall column, 322-340
    partial condenser, 327
    q-line, 325-327
    rectifying section, 310-315
    side streams, multiple feeds, 336-339
    stage efficiency, 335-336
MADCAP, 727
Manholes, in towers, 72
Margules equation, 198
Mass separating agent (MSA), 7, 529
Mass transfer: coefficients, 632, 638-641, 647
    controlling process, 640
    defining equations, 634
    devices, 50
    film, 634
    models, 643-644
    overall, 640
    prediction, 661-668
Mechanical equilibrium, 26
Membrane processes, 20
MESH equations, 559-560
Minimum: absorbent rate, 465
    stripping rate, 472
Minimum equilibrium stages, 435
Minimum reflux, 323, 441
    by McCabe-Thiele method, 323, 328-331

by Ponchon method, 386
by Underwood equations, 441-450
Minimum work of separation, 678-683
Mitsui Shipbuilding and Engineering Co., 71
Mixer, degrees of freedom, 245
Mixer-settler, 80
  degrees of freedom, 245
Mixing rules, 147, 152, 171-172
Molar gas volumes, 714-724
Molecular diffusion, 628
Molecular weight, 714-724
Monsanto Co., 714
Montz tray, 65
MSA, *see* Mass separation agent
MSEQ algorithm, 293
Muller's method, 284, 569
Multicomponent distillation, 556-602
  bubble-point (BP) method, 566
  equation tearing, 562
  general strategy, 560
  simultaneous-correction (SC) method, 594
  sum-rates (SR) method, 578
  theoretical models, 557
Multicomponent extraction, 586
Multicomponent separation methods:
  approximate, 427-491
  rigorous, 556-602
Multieffect distillation, 690-692
Multipass trays, 512
Multiphase contacting, equipment, 47-86
Multiple feeds, 392
Multiple miscibility gaps, 228
Multistage separations, 30
Murphree plate efficiency, 335, 394

Naphtali-Sandholm method, 594-611
  algorithm, 604
  computer program, 725
Net-flow point, 379
Net work consumption, 683-689
Newton method for convergence, 274-275
Newton-Raphson method, 562, 583-584,
  594
NGAA K-charts, 726
Nonideal liquid mixtures, 192-197
Nonkey components, distribution at total
  reflux, 439-441
  at actual reflux, 457-459
Norton Co., 57, 59, 62
NRTL equation, 198, 212-215

for liquid extraction, 227
NTP, number of theoretical plates, 624
NTU, number of transfer units, 624, 647

O'Connell correlation for tray efficiency,
  510
Oldshue-Rushton column, 82
Operability, 34
Operating expenses, 34
Operating line, 308
Ordered branch searches, 545
Osmosis, reverse, 20, 21
Overflow, 117, 411

Packed columns, 38, 636-638
  comparison with plate columns, 72-73
  diameter, 658-660
  for extraction, 81
  flooding, 637
  internals, 58-63
Packing, 57
  choice of, 655
  comparisons, 656
  dumped, 50
  factor, 655, 657
  grid, 50
  mesh, 50
Pall rings, 57
Partial condensation, 7, 8, 271, 273-278
Partial condenser, 387
PDS, 727
Penetration theory, 643
Permeation, 20, 21
Phase equilibrium diagrams, 90, 91, 94,
  95, 98, 99, 100, 104, 105, 106, 115, 121,
  123, 124, 127, 129
  from equations of state, 140-176
Phase rule, Gibbs, 89
Physical properties: constants, 714-724
  criteria for separator selection, 32
  light hydrocarbons, 33
  paraffin hydrocarbons, 36
  to predict phase equilibrium, 91
Pinch: point, 418, 441
  zone, 418, 442
Plait point, 104, 107
Plate, *see* Stages; Trays
Plate columns, 65
  comparison with packed columns, 72-73
  for extraction, 81

Polar species, effect on liquid mixtures, 193-194
Ponchon-Savarit method, 372-402
  enriching section, 379-380
  feed-stage location, 384-385
  for extraction, 394-398
  minimum reflux ratio, 386
  minimum stages, 385-386
  overall column, 381-383
  partial condenser, 387
  side streams, multiple feeds, 392-394
  stage efficiency, 394
  stripping section, 380-381
Poynting correction, 159
Pressure: column operating, 432-434
  convergence, 169
  critical, 714-724
Pressure drop, 17, 51
Pressure, optimum, in distillation, 422
PROCESS, 727
Process design data sheet, 48
Process flow diagrams, 3
Pumparounds, 559

q-line equation, 326

Rachford-Rice flash algorithm, 273-274
Raffinate, 103
  reflux, 415
Raoult's law, 92
  K-values, 161
Raschig rings, 57
Rayleigh equation, 363
RDC column, 82, 516
Reboiled absorption, 8, 12
Reboiled stripper, 9, 13, 464
Reboilers: partial, 327
  total, 327
Reciprocating plate contactor, 516-522
Recovery fraction, 461
Rectifying section, 293
Redlich-Kwong equation, 148, 599
Reflux, 12
  drum, 514
  total, 323, 435
Refluxed stripper, 9, 13
Reflux ratio: cost, 451
  external, 310
  internal, 310
  minimum, 323, 443

  secondary, 698
Regular solution, 183-185
Relative selectivity, 26
Relative volatility, 26, 33, 97
Relaxation method, 611-612
Retainer plates, 61
Right-triangular diagram, 108, 410-415
Ripple trays, 65
Rotating disk contactor, 516-522
RPC columns, 516

Saddle packing, 57
Safety, in design, 34
Savannah River plant, 23, 73
Scatchard-Hamer equation, 198
Scheibel column, 82
Schmidt number, 664
Secondary reflux and boilup, 698-704
Selection of separation operations, 32, 534
  factors in selection, 34-35
Selectivity, relative, 145
Separation operations, 8-11, 20
Separation process, 1-47
  synthesis, 527-547
Separator sequences, 530
Sequence of separators, 527-528
  direct, 529
  indirect, 529
Shell Chemical Co., 513
Shortcut design methods, 427-493
Shower trays, 77
Side streams, 392
Sieve trays, 68-69
Simultaneous correction (SC) methods, 594-612
Smoker method, 331
Soave-Redlich-Kwong (S-R-K): computer program, 725
  equation, 169
Social factors, 35
Solubility parameter, 185
Solutropy, 115
Solvents, 78-79
Souders and Brown capacity factor, 503
Splash baffles, 68
Splash panels, 71
Splits, unique, 531
Splitter, degrees of freedom, 248
Spray column, 75-76
SRV distillation, 703

Stage(s), capacity, 501-523
    efficiency, 25, 55, 335, 501-523
    equilibrium, 25
    minimum number, 323, 328-330, 385-386,
        435-437
    theoretical, 25
Starling-Han B-W-R correlation, 172
Steam distillation, 9
Stevens Institute of Technology, 71
Stream variables, 240
Stripping, 9, 12
    group method, 470-475
    inert gas, 464
    reboiled, 9, 13, 464
    refluxed, 9, 13
    section, 12, 293
    stripping factor, 461, 463
    sum-rates (SR) method, 578-586
    vapor, 12
Sublimation, 14
Sujata method for absorption, 578
Sum-rates (SR) method: computer pro-
        gram, 725
    equation, 580
Surface-renewal theory, 645
Synthesis of separation sequences, 35,
        527-547
    algorithmic, 544
    by evolutionary techniques, 540
    by heuristics, 540

TBP distillation curve, 103
Temperature: boiling point, 714-724
    critical, 714-724
Thermal equilibrium, 26
Thermodynamic: efficiency, 683-686
    first law, 678
    quantities for phase equilibria, 144
    second law, 685
Theta method of Holland, 562
Thiele-Geddes method, 560
Thomas algorithm, 563-565
Tie lines, 107
Total reflux 323, 436
    in McCable-Thiele method, 323, 328-380
    in Ponchon method, 385, 386
Tray, 50
    capacity, 503-509
    efficiency, 509-514
    multipass, 512

spacing, 507
technology, 73
Triangular diagrams, 104
Tridiagonal matrix algorithm, 563-565
Tsuboka-Katayama ISR method, 586-594
    algorithm, 590
Turbogrid, tray, 65
Turndown Ratio, 65, 507

UMD diffusion, 629
Underflow, 117, 411
Underwood euqations, 443, 445
UNIFAC, computer program, 725
    extraction applications, 228
    method, 218-220
Uniflux tray, 65
UNIQUAC: computer program, 725
    equation, 198, 215-217
    extraction applications, 228

Valve trays, 65-67
    flooding, 506
Van der Waal equation, 148
van Laar equation, 197-200
    computer program, 725
    modification by Null, 199
Vaporization, flash, 7
Vapor-liquid equilibrium, 90-103, 145-146
Vapor pressure, from equations of state,
        158-159
Variable specifications, 256
    design case, 261
    for multicomponent distillation, 428
    for separation operations, 239, 256-260
Variance, 239
Venturi scrubber, 625
Venturi tray, 65
Virial equation of state, 148
Volatility, relative, 145
Volume: critical, 714-724
    fraction, local, 206
    molal, of liquid, 162
VST cap, 70

Wang-Henke method, 566-578
    algorithm, 566
    computer program, 725
Water softening, 17
Weeping, 55
Wegstein method, 278

Weir, 58
Wetted-wall column, 632-636
Wilke-Chang equation, 628
Wilson equation, 198, 203-
        211
    computer program, 725

Winn equation, 437

x-y diagram, 91

Zone: melting, 20, 22
    refining, 24